Emissions Control Catalysis

Emissions Control Catalysis

Special Issue Editors

Ioannis V. Yentekakis
Philippe Vernoux

MDPI • Basel • Beijing • Wuhan • Barcelona • Belgrade • Manchester • Tokyo • Cluj • Tianjin

Special Issue Editors
Ioannis V. Yentekakis
Technical University of Crete
Greece

Philippe Vernoux
Université de Lyon
France

Editorial Office
MDPI
St. Alban-Anlage 66
4052 Basel, Switzerland

This is a reprint of articles from the Special Issue published online in the open access journal *Catalysts* (ISSN 2073-4344) (available at: https://www.mdpi.com/journal/catalysts/special_issues/emissions_catalysis).

For citation purposes, cite each article independently as indicated on the article page online and as indicated below:

LastName, A.A.; LastName, B.B.; LastName, C.C. Article Title. *Journal Name* **Year**, *Article Number*, Page Range.

ISBN 978-3-03936-036-9 (Pbk)
ISBN 978-3-03936-037-6 (PDF)

Contents

About the Special Issue Editors

Ioannis V. Yentekakis is currently a Full Professor of Physical Chemistry and Catalysis, School of Environmental Engineering, Technical University of Crete (TUC), Greece. He received a Chemical Engineering Diploma (1983) and a Ph.D. (1988) from the University of Patras (UP). His academic career is associated with Princeton University (Post-Doc), ICE-HT/FORTH (Senior Researcher), the Department of Chemical Engineering UP (Assistant Professor), and Cambridge University, U.K. (Visiting Professor). His current interest concerns development of novel nano-structured catalysts and eco-friendly processes with applications in added-value chemicals production, environmental protection, renewable energy generation, fuel cells, hydrogen energy, CO2 hydrogenation, natural gas (NG), and biogas valorization. He has authorized more than 110 journal publications (>4500 citations, h-index = 38), 3 international patents, 10 books, and has Guest Edited 5 Special Issues. He has served as department chairman, member of the University Council and the Senate, and chairman or organizer of many scientific conferences. He has coordinated more than 30 funded research projects.

Philippe Vernoux is a senior researcher at CNRS (Ph.D. in Electrochemistry in 1998 and habilitation in catalysis in 2006). Dr. Vernoux has published more than 130 refereed papers, 18 patents in domains of solid state electrochemistry, environmental catalysis, and electrochemical promotion of catalysis (H-index = 32, >3400 citations). Dr. Vernoux has supervised 22 Ph.D. students and 20 post-docs. He has Guest Edited 5 Special Issues and gave 39 invited lectures and 180 oral communications. He has strong experience in the coordination of national and international projects (EFEPOC Marie Curie Project, ADEME PIREP, ANR PIREP2, EPOX, DYCAT, etc.). His current interest concerns development of catalysts and electrocatalysts for vehicle exhaust treatment, air cleaning, wastes valorization, and H2 production. He is currently the head of the CARE group (Catalytic and Atmospheric Reaction for the Environment) of the Institute of Researches for Catalysis and Environment of Lyon in France.

Editorial

Emissions Control Catalysis

Ioannis V. Yentekakis [1,*] and Philippe Vernoux [2]

1 Laboratory of Physical Chemistry & Chemical Processes, School of Environmental Engineering, Technical University of Crete, 73100 Chania, Greece
2 Université de Lyon, Institut de Recherches sur la Catalyse et l'Environnement de Lyon, UMR 5256, CNRS, Université Claude Bernard Lyon 1, 2 Avenue A. Einstein, 69626 Villeurbanne, France; philippe.vernoux@ircelyon.univ-lyon1.fr
* Correspondence: yyentek@isc.tuc.gr; Tel.: +30-282-103-7752

Received: 1 October 2019; Accepted: 21 October 2019; Published: 31 October 2019

---

1. Overview

Important advances have been achieved over the past years in agriculture, industrial technology, energy, and health, which have contributed to human well-being. However, some of these improvements in our lives were accompanied by a great threat to the environment and public health, with photochemical smog, stratospheric ozone depletion, acid rain, global warming, and finally climate change being the most well-known major issues, as a result of a variety of pollutants emitted through these human activities. The indications are that we are already at a tipping point that might lead to non-linear, sudden environmental change on a global scale.

Aiming to ensure that "we live well, within the planet's ecological limits" the United Nation's Sustainable Development Goals [1] and the EU's Environmental Action Plan [2] include calls to action and priority objectives to protect life on land and water and tackling climate change. In a full harmony, scientists around the world are developing tools and techniques to understand, monitor, protect, and improve the environment, and to ensure that our achievements must not adversely affect our environment and at the same time we must mitigate any damage that has already occurred. In turn, new tools, technology, and advanced materials are continuously added to our quiver that enable us to effectively control emissions, either of mobile (e.g., cars) or stationary (industrial) sources, and to improve the quality of outdoor and indoor air, with catalysis to play a major role on these efforts, i.e., as a leading technology for improving quality of life, health, and environment.

Emissions Control Catalysis in the area of Environmental Catalysis [3,4] is continuously growing up, providing novel multifunctional, nano-structured materials with active metal species at sizes of the level of nanoparticles, nanoclusters or even single atoms [5–8]. These materials are also often promoted by several ways (i.e., by surface- or support-induced promotion [9–13] and by electrochemical promotion [14–16], among others) in order to be very active and selective for the abatement of a variety of pollutants and greenhouse gases. Regarding the latter, representative cases concern the abatement of CO, NO_x, N_2O, NH_3, CH_4, higher hydrocarbons, volatile organic compounds (VOCs), particulate matter (PM), and specific pollutants emitted by industry (e.g., SO_x, H_2S, dioxins, and aromatic hydrocarbons) or landfill and wastewater treatment plants (biogas).

The concept of Cyclic Economy is currently of growing attention in emission control catalysis strategies for the production of useful chemicals and fuels from the controlled pollutants. The CO or CO_2 hydrogenation for the production of CH_4 or liquid fuels, biogas or methane reforming for the production of syngas, and/or H_2, oxidative coupling of methane for the production of ethylene are a few examples of such efforts.

2. Special Issue Contributions and Highlights

The present "Emissions Control Catalysis" Special Issue was successful in collecting 21 high-quality contributions by several research groups around the world, which covers recent research progress in the field of the catalytic control of air pollutants emitted by mobile or stationary sources, not limited only to the abatement of pollutants but also including research dedicated to cyclic economy strategies. These papers among others are ranging from the synthesis and physicochemical-textural-structural characterization of the materials, activity–selectivity–durability evaluation under environmentally important reactions, fundamental understanding of structure–activity relationships or other metal–metal and metal–support interactions on the multifunctional materials involved. In particular, three comprehensive review articles covering several major topics and directions in emissions control catalysis subject were published in this special issue. The review of the guest editors (Yentekakis I., and Vernoux P.) and co-workers [17] concerns mainly the of CO, CHs, and NO_x emissions abatement from stoichiometric, lean burn, and diesel engines exhausts, addressing the literature that concerns the electropositive promotion by alkalis or alkaline earths of platinum group metals (PGMs) that have found to be extremely effective for the related three-way and lean-burn reactions and catalytic chemistry. The review of Smirniotis P. and co-workers [18] concerns the selective catalytic reduction of NO_x with NH_3 (NH_3-SCR of NO_x) focusing to low temperature applications that is a highly desirable perspective, and finally the review of Bogaerts A. and co-workers [19] covering a hot recent trend in emissions control implicating cyclic economy strategies, that is the conversion and utilization of CO_2 for the production of value-added chemicals. On the other hand, a major part of contributions (9/21) concerns original research on nitrogen oxides reduction processes [20–28], reflecting the fact that this topic still remains hot among the targets of environmental catalysis. Five out of 21 studies concern CO and hydrocarbons oxidation processes [29–33] while the remained 4/21 concern CO_2 capture/recycling processes under the view of cyclic economy [34–37].

Review Contributions

Alkali metals as surface-promoters of platinum group metals (PGMs) have shown remarkable promotional effects on the vast majority of reactions related to the emissions control catalysis either they are applied by traditional routes (e.g., impregnation) or electrochemically, via the so called, concept of the electrochemical promotion of catalysis (EPOC) or non-faradaic electrochemical modification of catalytic activity (NEMCA effect) [14,15]. In the comprehensive review of Yentekakis et al. [17] published in this Special Issue, more than 120 papers on the subject were collected and discussed comparatively after classification in groups, on the basis of the specific reaction promoted, e.g., (CO, HCs, or H_2)-SCR of NO_x, and CO or hydrocarbons oxidation. The authors also present and analyze, by means of indirect (kinetics) and direct (spectroscopic) evidences, a mechanistic model for the mode of action of electropositive promoters, which consistently interprets all the observed promoting phenomena. Concluding, they claim that this very pronounced (in some cases extraordinarily) alkali-induced promotion in emissions control catalysis prompts to the development of novel catalyst formulations for a more efficient and cost-effective control of the emissions of automotives and stationary combustion processes.

Since the emission standards for NO_x are becoming more stringent to keep our atmosphere clean and severe pollution regulations being imposed around the world, the challenge of a cost-effective, i.e., low-temperature, selective catalytic reduction (LT-SCR) of NO_x by NH_3 with a combination of high NO_x activity and N_2 selectivity in a wide operation temperature window and good resistance to SO_2/H_2O have attracted paramount attention. Supported and mixed transition metal oxides have been widely investigated for LT-SCR technology. However, these catalytic materials have some drawbacks, especially in terms of catalyst poisoning by H_2O or/and SO_2. Hence, the development of catalysts for the LT-SCR process is still under active investigation throughout seeking better performance. Extensive research efforts have been made to develop new advanced materials for this technology. The comprehensive review of Smirniotis and co-workers [18] collected and comparatively analyses more than 140 publications in the topic, covering the description of the influence of operating conditions, materials, and promoters on the LT-SCR performance, as well as active sites, reaction intermediates, and mechanistic implications evidenced by using isotopic labeling and in situ FT-IR studies.

Global climate change as a result of the greenhouse effects of the increasing emissions of the so-called greenhouse gases, with CO_2 to be the most representative one, is of major concern currently. CO_2 capture, conversion, and utilization is one of the possible solutions to reduce CO_2 concentration in the atmosphere. Among other methods, this can be accomplished by direct catalytic hydrogenation of CO_2, producing value-added products. In their comprehensive review, Bogaerts A. and co-workers [19], focused mainly in the last 5-years progress on the topic summarized the literature research (~125 papers) and the current priorities on CO_2 hydrogenation to CO, CH_4, CH_3OH, DME, olefins, and higher hydrocarbons value-added chemicals by heterogeneous catalysis and plasma catalysis. Although, its energy efficiency is still deterring for commercialization, plasma catalysis has currently attracted sufficient attention, due to its simple operating conditions (ambient temperature and atmospheric pressure) and unique advantages in activating inert molecules; its potential advantages and current limitations on CO_2 hydrogenation process have been accounted in the review.

3. Original Contributions

3.1. NO_x *Abatement Related Results*

A major part of the contributions (9/21) involves research on the selective catalytic reduction (SCR) of NO and/or its direct catalytic decomposition. In particular:

Shen D. and co-workers [20] investigated the deNO_x activity on a series of bimetallic Cu–Mn molecular sieve catalysts (Cu–Mn/SAPO-34) with different loadings of Cu and Mn components during the selective catalytic reduction (SCR) of NO with NH_3 at low temperatures (ca. 120–330 °C), including the effects of H_2O vapor and/or SO_2. Among the catalysts tested, the performance of 2 wt% Cu-6 wt% Mn/SAPO-34 one found to be superior, achieving 72% NO conversion at 120 °C and even better (90%) at 180–330 °C. The reversible negative effect of H_2O on NO conversion was attributed to the competitive adsorption of H_2O and NH_3 on Lewis acid sites; this poisoning was diminished upon increasing the reaction temperature to 300 °C. A permanent poisoning effect of SO_2 on deNO_x activity found is strongly dependent on the reaction temperature, becomes more pronounced at lower ones and is further enhanced by H_2O co-feed; this is assigned to the formation of $(NH_4)_2SO_4$, which results in the plug of active sites and a decrease of surface area.

Gao Y. and co-workers [21] synthesized Mn-Co/TiO_2 and Mn-Fe/TiO_2 nanocatalysts by a hydrothermal method which were characterized by a variety of methods including Brunner–Emmet–Teller (BET)/Barrett–Joyner–Halenda (BJH) analysis of N_2 adsorption/desorption isotherms at −196 °C, transmission electron microscope (TEM), X-ray diffraction (XRD), H_2-temperature-programmed reduction (TPR), NH_3-temperature-programmed desorption (TPD), and X-ray photoelectron spectroscopy (XPS) that enabled the authors for a comprehensive comparison of the catalysts nanostructure characteristics and their de-NOx catalytic performance, gained insight into the structure-activity relationships. The Mn–Co/TiO_2 catalyst offered superior structure characteristics than Mn–Fe/TiO_2: higher surface area and active components distribution, diminished crystallinity, reduced nanoparticle size and also higher Mn^{4+}/Mn^{n+} ratio, confirming its better oxidation ability and larger amount of Lewis and Brønsted acid sites on the Mn–Co/TiO_2 surface. As a result, Mn–Co/TiO_2 nanocatalyst displayed superior SCR of NO with NH_3 on both activity and selectivity in the temperature range of 75–250 °C. Kinetics data revealed that both Eley–Rideal (E–R) and Langmuir–Hinshelwood (L–H) mechanisms were implicated in NH_3-SCR process over Mn–Fe/TiO_2 and Mn–Co/TiO_2 catalysts.

Han J. and co-workers [22] investigated the enhanced deNO_x performance and stability of sulfated sintered ore catalysts (SSOC) during the selective catalytic reduction of NO_x with NH_3. The maximum deNOx efficiency found was about 92% at 300 °C, NH_3/NO = 1 and 5000 h^{-1} gas hourly space velocity (GHSV). A systematic characterization of the materials by means of X-ray fluorescence spectrometry (XRF), Brunauer–Emmett–Teller (BET) analyzer, X-ray diffraction (XRD), X-ray photoelectron spectroscopy (XPS), and diffuse reflectance infrared spectroscopy (DRIFTS) was conducted to provide an in depth understanding of the NH_3-SCR reaction mechanism and to explain the denitration performance and stability of SSOC. The existence of more Brønsted acid sites at the surface of SSOC found to be responsible for the improved adsorption capacity of NH_3 and NO over the SSOC surface that accomplished the formation of amide species ($-NH_2$), NH_4^+ species, NO_2 molecules in a gaseous or weakly adsorbed state, and nitrates. The reaction between $-NH_2$, NH_4^+, and NO (E–R mechanism) and the reaction of the coordinated ammonia with the adsorbed NO_2 (L–H mechanism) were attributed to NO_x reduction.

Song C. and co-workers [23] studied the promotional effect of Ce and/or Zr incorporation by ion exchange on Cu/ZMS-5 catalysts for the NH_3-SCR of NO. The cerium and zirconium addition promotes the activity of catalysts; the cerium-rich catalysts exhibiting superior SCR activities compared to the zirconium-rich ones. The improved low temperature activity of the $CuCe_xZr_{1-x}O_y$/ZSM-5 catalysts in comparison to the unpromoted Cu/ZSM-5 (the former achieving >95% NO conversion at 175–468 °C, the later at 209–405 °C) was attributed to an increase of the reactive lattice oxygen content and reducibility of the catalysts via the Ce^{3+}/Ce^{4+} redox couple and its interaction with cooper species. Moreover, the presence of zirconium in the catalysts promotes surface copper enrichment, prevents copper crystallization and causes suppression of N_2O formation, increasing N_2-selectivity of the system.

He H. and co-workers [24] synthesized $CeZr_{0.5}Ti_aO_x$ (with a = 0, 1, 2, 5, and 10) catalysts by a stepwise precipitation approach, which were studied on the NH_3-SCR of NO_x. In all contents, Ti addition was beneficial to the deNO_x catalytic performance. Superior behavior was obtained by the $CeZr_{0.5}Ti_2O_x$ particular catalyst composition. Controlling pH and precipitation time during the two steps involved in the preparation method enabled the authors to achieve a catalyst with enhanced acidity (favorable for NH_3 adsorption in NH_3-SCR processes) and high dispersion of CeO_2 onto the surface of ZrO_2-TiO_2 synthesized first. The as-prepared $CeZr_{0.5}Ti_2O_x$ catalyst was characterized by superior redox properties, enhanced adsorption and activation of NO_x and NH_3 and enhanced surface adsorbed oxygen such as O_2^{2-} and O^- belonging to defect-oxide or a hydroxyl-like group; all these factors positively affecting its SCR of NO_x with NH_3 performance.

Olson L. and co-workers [25] investigated the poisoning effect of a phosphorous containing atmosphere on the NO_x storage capacity of a Pt/Ba/Al_2O_3 structured (i.e., washcoated on a ceramic monolith) catalyst. A significant loss of the NO_x storage capacity was caused by phosphorous exposure characterized by a progressively decreasing axial distribution of phosphorous concentration from the inlet to the outlet of the monolith. The values of the specific surface area and pore volume of phosphorous-poisoned monolithic catalysts followed an inverse order: lower at the inlet, higher at the outlet of the monolith. Additional features of the axial phosphorous accumulation detected were: a higher surface accumulation at the inlet of the monolith mostly appeared in the form of P_4O_{10}, the presence of more metaphosphate (PO_3^-) in the middle section of the monolith, and a less surface accumulation of phosphorous at the outlet of the monolith due to its extended diffusion into the washcoat. In respect to the poisoning effect of phosphorous on the SCR of NO_x it was revealed that the formations of N_2 and N_2O were decreased in favor of NH_3 production; the reaction is more influenced by the phosphorous poisoning than the ammonia formation from the stored nitrates.

The direct NO decomposition activity on PdO or PtO supported on Co_3O_4 spinel was studied by Reddy et al. [26] in an attempt to discover means of enhancing the activity of Co_3O_4 spinel, one of the most active single-element oxide catalysts for NO decomposition at high temperatures (typically > 650 °C). In fact, the authors demonstrated a four-fold higher promotion on the NO decomposition activity of PdO- rather than of PtO-modified Co_3O_4 at 650 °C, accompanied by superior selectivity towards N_2 as well. Structural and surface analysis measurements using a variety of methods (e.g., XRD, XPS, H_2-TPR, and in situ FT-IR) showed an enhanced reducibility of PdO/Co_3O_4 with an increased thermal stability of surface adsorbed NO_x species, both considered to contribute on the promotion observed. In opposite, PtO enters into the Co_3O_4 structure, without notable influences on the redox and NO adsorption properties of Co_3O_4, resulting in marginal promotion compared to PdO. The PdO promotion followed volcano behavior with an optimal PdO loading of 3 wt%.

Zhang et al. [27] investigated the NO_x storage capacity of a series of Pd/BEA catalysts with various Pd loadings for cold-start applications. In situ FTIR measurements using CO and NH_3 enable the authors to identify two isolated Pd^{2+} species, Z^--Pd^{2+}-Z^- and Z^--$Pd(OH)^+$, on exchanged sites of zeolites, as the main active sites for NO trapping. Among these active sites a superior NO_x storage capacity of Z^--Pd^{2+}-Z^- was demonstrated, which is caused by the different resistance to H_2O. Atom utilization of Pd can be improved by using lower Pd loading, with an optimum at 0.5 wt%, since this leads to a sharp decline of Z^--$Pd(OH)^+$ attributed to the 'exchange preference' for Z^--Pd^{2+}-Z^- in BEA.

Finally, Ingel et al. [28] in a different approach for controlling N_2O emissions in ammonia oxidation process at high temperature, proposed the design of experiments and response surface methodology to study this process. The reactor's load, the temperature of reaction and the number of catalytic gauzes were selected as independent variables, whereas ammonia oxidation efficiency and N_2O concentration in nitrous gases were assumed as dependent variables (response). Statistically significant mathematical models were developed from the achieved results, which describe the effect of independent variables on the analyzed responses. The ammonia oxidation efficiency value depends on the reactor's load and the number of catalytic gauzes but not on the temperature in the studied range (870–910 °C). The concentration of N_2O in nitrous gases depends on all three parameters. The developed models were used for the multi-criteria optimization with the application of desirability function. Sets of parameters were achieved for which optimization assumptions were met: maximization of ammonia oxidation efficiency and minimization of the N_2O amount being formed in the reaction. As authors claim, the presented methodology can be used to minimize the primary N_2O emission at high ammonia oxidation efficiency. It can be applied for optimization of operating parameters of ammonia oxidation reactor with two types of catalysts: catalytic gauzes and catalyst for high temperature of N_2O decomposition. As a result, it is possible to obtain the set of independent variables ensuring low N_2O emission and to meet the binding environmental regulations.

3.2. CO, CH_4, and Other Hydrocarbons Oxidation Reactions

Avgouropoulos and co-workers [29] synthesized a series of atomically dispersed copper-ceria nanocatalysts via appropriate tuning of a novel hydrothermal method and investigated their activity on CO oxidation, which was found to be strongly dependent on the nanostructured morphology, oxygen vacancy concentration, and nature of atomically dispersed Cu^{2+} clusters. A number of techniques including electron paramagnetic resonance (EPR) spectroscopy, X-ray diffraction (XRD), N_2 adsorption, scanning electron microscopy (SEM), Raman spectroscopy, and ultraviolet-visible diffuse reflectance spectroscopy (UV-Vis DRS) were employed in the characterization of the synthesized materials. The aim was to find the key factors that govern the physicochemical properties of the synthesized materials during preparation and then to provide convincing structure-activity correlations. Elevated temperatures and low concentrations of NaOH (≤0.1 M) during preparation have led to more active ceria-based catalysts in CO oxidation. This was explained with the obtained morphology, the nature of oxygen vacancies and dispersed copper species, and to a lesser extent, with the specific surface area of the materials and the concentration of defects.

Briois and co-workers [30] in an attempt to provide means of replacing expensive platinum group metals with cost-effective perovskite type materials for catalytic oxidation reactions, prepared thin catalytic coatings of Sr and Ag doped lanthanum perovskites, $La_{1-x-y}Sr_xAg_yCoO_{3-\alpha}$ (x = 0.13–0.28, and y = 0.14–0.48), on alumina substrates by using the cathodic co-sputtering magnetron method in reactive condition. Such thin porous catalytic film arrangements can optimize the surface/bulk ratio by combining a large gas exposure surface area with an extremely low loading, thus saving raw materials. The sputtering method was optimized to generate crystallized and thin perovskite films. The authors found that high Ag contents has a strong impact on the morphology of the coatings, favoring the growth of covering films with a porous wire-like morphology that showed a good catalytic activity for CO oxidation. Interestingly, the optimal composition ($La_{0.40}Sr_{0.1}Ag_{0.48}Co_{0.93}O_3$) displayed similar catalytic performance than this of a Pt film, and was also efficient for CO and NO abatement in a simulated Diesel exhaust gas mixture, demonstrating the promising catalytic properties of such nanostructured thin sputtered perovskite films.

Methane, a substantially more potent greenhouse gas than CO_2, is the main compound of natural gas, which is lately used with a continuous increased rate in various industrial processes and as an alternative fuel for heavy-duty transportation not excluding light-duty tracks. As a consequence, control of CH_4 emissions via catalytic deep oxidation has attracted considerable renewed attention, given that methane appears the lowest reactivity among alkanes. Under this view, Iliopoulou and co-workers [31] synthesized a series of novel Co–Ce mixed oxide catalysts in an effort to enhance synergistic effects that could improve their redox and oxygen storage properties and, thus, their activity in methane deep oxidation. The effect of the synthesis method (hydrothermal or precipitation) and Co loading (0, 2, 5, and 15 wt%) on the catalytic efficiency and stability was investigated. Hydrothermally synthesized Co_3O_4/CeO_2 catalysts appeared superior performance due to their improved physicochemical properties (smaller crystallite size, larger surface area, and enhanced reducibility). In respect to Co loading, the optimum performance was observed over a 15 wt% Co/CeO_2 catalyst, which also presented sufficient tolerance to water presence.

Based on the fact that Pd is one of the most active catalysts for complete methane oxidation Baranova and coworkers [32] used the concept of electrochemical promotion of catalysis (EPOC) to further promote the reaction over palladium nano-structured catalysts deposited on yttria-stabilized zirconia (YSZ) solid electrolyte. Anodic polarization (O^{2-} supply to the catalyst) resulted to a rate enhancement up to ~3 at 450 °C with an apparent Faradaic efficiency as high as 3000 (for a current application as low as 1 μA). Electrochemical promotion on this catalytic system showed persistent behavior the catalyst remained under promotion for a long period of time after interruption of the external bias induced EPOC. Increasing polarization time resulted in a longer-lasting persistent promotion (p-EPOC); more time was required for the reaction rate to reach its initial un-promoted value. The phenomenon was attributed to the continuing promotion by the stored oxygen in palladium oxide formed during the anodic polarization.

Taking into account that sulfur poisoning is one of the most important factors deteriorating the efficiency of diesel exhaust after-treatment systems and that bare TiO_2 appears high sulfur resistivity, Zhang et al. [33] prepared TiO_2–CeO_2 composites by co-precipitation and studied their sulfur resistance and catalytic activity in the oxidation of diesel soluble organic fraction (SOF). They found that TiO_2-modification of CeO_2 significantly improves the catalytic SOF purification efficiency of CeO_2 besides the fact that this ceria doping does not downgrade the excellent sulfur resistance of bare TiO_2; the prepared TiO_2–CeO_2 exhibited superior sulfur resistance than e CeO_2 and commercial CeO_2–ZrO_2–Al_2O_3. TiO_2–CeO_2 characterization by X-ray diffraction (XRD) and Raman spectroscopy indicate that cerium ions can enter into the TiO_2 lattice, without forming complete CeO_2 crystals. Moreover, as confirmed by XPS and H_2-TPR, the synthesized TiO_2–CeO_2 composites appeared enhanced oxygen storage capacities (OSC) that considered responsible for their better SOF oxidation activity.

3.3. CO_2 Capture/Recycling: Combining Emissions Control with Added-Value Chemical Production (Cyclic Economy)

De Lucas-Consuegra and co-workers [34] developed a low-temperature (below 90 °C) proton exchange membrane (Sterion) electrochemical cell for the electrocatalytic conversion of gaseous CO_2 to liquid fuels. This novel system achieved gas-phase electrocatalytic reduction of CO_2 over a Cu-based cathode by using water electrolysis-derived protons generated in-situ on an IrO_2 anode. Three Cu-activated carbon cathodes with varying Cu loading (10, 20, and 50 wt% Cu–AC), and thus particle size, were tested. Products distribution was a function of the Cu loading and particle size of the Cu–AC cathode; methyl formate, acetaldehyde, and methanol were being the main reaction products, respectively, over 50, 20, and 10 wt% Cu–AC. The membrane electrode assembly (MEA) containing the cathode with the largest Cu loading and particle size (50 wt% Cu–AC, 40 nm) showed the highest CO_2 electrocatalytic activity per mole of Cu (and the lowest energy consumption values for the conversion of CO_2, 119 kW·h·mol^{-1}), which was attributed to the lower Cu–CO bonding strength over large Cu particles.

Wang H., Lu J.-X. and co-workers [35] fabricated an electrocatalytic cell consisting of a 0.5 M $KHCO_3$ aqueous solution as electrolyte saturated with CO_2 by bubbling, a CuO/TiO_2-Nafion as working electrode, and Pt as counter and reference electrodes in order to study the electroreduction of CO_2 to added-value multi-carbon oxygenate products (ethanol, acetone, and n-propanol). The non-noble metal electrocatalyst CuO/TiO_2 was in situ reduced to Cu/TiO_2, which efficiently catalyzed CO_2 reduction, offering a maximum overall faradaic efficiency of 47.4% at a potential of −0.85 V vs. reversible hydrogen electrode (RHE). The catalytic activity for CO_2 electroreduction was strongly dependent on the CuO contents of the catalysts as-prepared, resulting in different electroactive surface areas. The significantly improved CO_2 reduction activity of CuO/TiO_2 was attributed to the high CO_2 adsorption ability of TiO_2 component of the working electrode.

Since hydrogen is currently considered as an efficient and environmentally benign energy carrier, among others, considerable attention is played by scientists worldwide for its sufficient production from hydrocarbon feedstocks (natural gas (NG), liquefied petroleum gas (LPG), etc.), biogas and bio-alcohols. To this end, Tang D. and co-workers [36] applied chemical looping reforming (CLR) as a prospective alternative for hydrogen production via ethanol steam reforming, which is characterized by energy efficiency and inherent CO_2 capture. Taking into account that oxygen carriers (OCs) with sufficient oxygen mobility and sintering resistance still remain the main challenges for the development of high-performance materials in the CLR process, the authors explore the performance of Ni/CeO_2 nanorod (NR) synthesized by a hydrothermal method as an OC in the CLR of ethanol. Using a variety of characterization techniques, they showed that the as-prepared Ni/CeO_2-NR possesses the desired properties for CLR, i.e., high Ni dispersion, abundant oxygen vacancies, and strong metal-support interaction, all factors improving catalytic activity. Testing the material in CLR process successfully offered a H_2 selectivity of 80% in 10-cycle stability test. The authors concluded that the small particle size and abundant oxygen vacancies contributed to improve water gas shift reaction, the high oxygen mobility of CeO_2–NR effectively eliminated surface coke on the Ni particle, and the covered interfacial Ni atoms closely anchored on the underlying surface oxygen vacancies on the (111) facets of CeO_2–NR enhance the anti-sintering capability.

Ioannidou et al. [37] contributed with a detailed and comparative catalytic-kinetic study of the performance of modified X-Ni/GDC electrodes (where X = Au, Mo, and Fe), in the form of half-electrolyte supported cells, in the reverse water gas shift reaction (RWGS). The importance of the RWGS reaction ($H_2 + CO_2 \rightarrow H_2O + CO$) is well known, since it takes place in most of the hydrocarbon processing reactions (e.g., hydrocarbons reforming processes) as well as in co-electrolysis of H_2O and CO_2 (CO_2 utilization) in solid oxide electrolysis cells (SOECs) yielding synthesis gas ($CO + H_2$), as considered in the present study. Solid oxide electrolysis is a contemporary process for CO_2 capture/recycling, which is proven as an attractive method to provide CO_2 neutral synthetic hydrocarbon fuels. The X-Ni/GDC catalysts were tested at open circuit conditions in order to elucidate their catalytic activity towards the production of CO; one of the products of the H_2O/CO_2 co-electrolysis reaction. The CO production rate increases by increasing the operating temperature and the partial pressure of H_2 in the reaction mixture. Fe and Mo modification enhances CO production, and 2 wt% Fe-Ni/GDC and 3 wt% Mo-Ni/GDC electrodes were superior compared to the other samples, in the whole studied temperature range (800–900 °C) reaching thermodynamic equilibrium. No carbon formation was detected.

Conflicts of Interest: The authors declare no conflict of interest.

References

1. EU 7th Environment Action Programme to 2020. Published 2014-02-21. Available online: http://ec.europa.eu/environment/action-programme/ (accessed on 23 October 2019). [CrossRef]
2. United Nations: Sustainable Development Goals. Available online: https://www.un.org/sustainabledevelopment/sustainable-development-goals/ (accessed on 23 October 2019).
3. Centi, G.; Ciambelli, P.; Perathoner, S.; Russo, P. Environmental Catalysis: Trends and outlook. *Catal. Today* **2002**, *75*, 3–15. [CrossRef]
4. Environmental Catalysis. A Section of Catalysts MDPI (ISSN 2073-4344). Available online: https://www.mdpi.com/journal/catalysts/sections/environmental_catalysis (accessed on 1 October 2019).
5. Yang, X.-F.; Wang, A.; Qiao, B.; Li, J.; Liu, J.; Zhang, T. Single-atom catalysts: A new frontier in heterogeneous catalysis. *Acc. Chem. Res.* **2013**, *46*, 1740–1748. [CrossRef] [PubMed]
6. Flytzani-Stephanopoulos, M.; Gates, B.C. Atomically dispersed supported metal catalysts. *Ann. Rev. Chem. Biomol. Eng.* **2012**, *3*, 545–574. [CrossRef] [PubMed]
7. Goula, G.; Botzolaki, G.; Osatiashtiani, A.; Parlett, C.M.A.; Kyriakou, G.; Lambert, R.M.; Yentekakis, I.V. Oxidative thermal sintering and redispersion of Rh nanoparticles on supports with high oxygen ion lability. *Catalysts* **2019**, *9*, 541. [CrossRef]
8. Liu, L.; Corma, A. Metal catalysts for heterogeneous catalysis: Form single atoms to nanoclusters and nanoparticles. *Chem. Rev.* **2018**, *118*, 4981–5079. [CrossRef] [PubMed]
9. Matsouka, V.; Konsolakis, M.; Lambert, R.M.; Yentekakis, I.V. In situ DRIFTS study of the effect of structure (CeO_2-La_2O_3) and surface (Na) modifiers on the catalytic and surface behaviour of Pt/γ-Al_2O_3 catalyst under simulated exhaust conditions. *Appl. Catal. B Environ.* **2008**, *84*, 715–722. [CrossRef]
10. Vernoux, P.; Leinekugel-Le-Cocq, A.-Y.; Gaillard, F. Effect of the addition of Na to Pt/Al_2O_3 catalysts for the reduction of NO by C_3H_8 and C_3H_6 under lean-burn conditions. *J. Catal.* **2003**, *219*, 247–257. [CrossRef]
11. Pliangos, A.; Yentekakis, I.V.; Papadakis, V.G.; Vayenas, C.G.; Verykios, X.E. Support-induced promotional effects on the activity of automotive exhaust catalysts: 1. The case of oxidation of light hydrocarbons (C_2H_4). *Appl. Catal. B Environ.* **1997**, *14*, 161–173. [CrossRef]
12. Papadakis, V.G.; Pliangos, C.A.; Yentekakis, I.V.; Verykios, X.E.; Vayenas, C.G. Development of high performance, Pd-based, three way catalysts. *Catal. Today* **1996**, *29*, 71–75. [CrossRef]
13. Konsolakis, M.; Drosou, C.; Yentekakis, I.V. Support mediated promotional effects of rare earth oxides (CeO_2 and La_2O_3) on N_2O decomposition and N_2O reduction by CO and C_3H_6 over Pt/Al_2O_3 structured catalysts. *Appl. Catal. B* **2012**, *123–124*, 405–413. [CrossRef]
14. Vayenas, C.G.; Bebelis, S.; Yentekakis, I.V.; Lintz, H.-G. Non-faradaic electrochemical modification of catalytic activity: A status report. *Catal. Today* **1992**, *11*, 303–438. [CrossRef]
15. Vayenas, C.G.; Ladas, S.; Bebelis, S.; Yentekakis, I.V.; Neophytides, S.; Yi, J.; Karavasilis, C.; Pliangos, C. Electrochemical promotion in catalysis: Non-faradaic electrochemical modification of catalytic activity. *Electrochim. Acta* **1994**, *39*, 1849–1855. [CrossRef]
16. Vernoux, P.; Lizarraga, L.; Tsampas, M.N.; Sapountzi, F.M.; de Lucas-Consuegra, A.; Valverde, J.-L.; Souentie, S.; Vayenas, C.G.; Tsiplakides, D.; Balomenou, S.; et al. Ionically Conducting Ceramics as Active Catalyst Supports. *Chem. Rev.* **2013**, *113*, 8192–8260. [CrossRef] [PubMed]
17. Yentekakis, I.; Vernoux, P.; Goula, G.; Caravaca, A. Electropositive promotion by alkalis or alkaline earths of Pt-group metals in emissions control catalysis: A status report. *Catalysts* **2019**, *9*, 157. [CrossRef]
18. Liu, M.; Yi, Y.; Wang, L.; Guo, H.; Bogaerts, A. Hydrogenation of carbon dioxide to value-added chemicals by heterogeneous catalysis and plasma catalysis. *Catalysts* **2019**, *9*, 275. [CrossRef]
19. Damma, D.; Ettireddy, P.; Reddy, B.; Smirniotis, P. A review of low temperature NH_3-SCR for removal of NO_x. *Catalysts* **2019**, *9*, 349. [CrossRef]
20. Liu, G.; Zhang, W.; He, P.; Guan, S.; Yuan, B.; Li, R.; Sun, Y.; Shen, D. H_2O and/or SO_2 tolerance of Cu-Mn/SAPO-34 catalyst for NO reduction with NH_3 at low temperature. *Catalysts* **2019**, *9*, 289. [CrossRef]
21. Gao, Y.; Luan, T.; Zhang, S.; Jiang, W.; Feng, W.; Jiang, H. Comprehensive comparison between nanocatalysts of Mn−Co/TiO_2 and Mn−Fe/TiO_2 for NO catalytic conversion: An insight from nanostructure, performance, kinetics, and thermodynamics. *Catalysts* **2019**, *9*, 175. [CrossRef]
22. Chen, W.; Hu, F.; Qin, L.; Han, J.; Zhao, B.; Tu, Y.; Yu, F. Mechanism and performance of the SCR of NO with NH_3 over sulfated sintered ore catalyst. *Catalysts* **2019**, *9*, 90. [CrossRef]

23. Liu, Y.; Song, C.; Lv, G.; Fan, C.; Li, X. Promotional effect of cerium and/or zirconium doping on Cu/ZSM-5 Ccatalysts for selective catalytic reduction of NO by NH_3. *Catalysts* **2018**, *8*, 306. [CrossRef]
24. Shan, W.; Geng, Y.; Zhang, Y.; Lian, Z.; He, H. A CeO_2/ZrO_2-TiO_2 catalyst for the selective catalytic reduction of NO_x with NH_3. *Catalysts* **2018**, *8*, 592. [CrossRef]
25. Jonsson, R.; Mihai, O.; Woo, J.; Skoglundh, M.; Olsson, E.; Berggrund, M.; Olsson, L. Gas-Phase Phosphorous Poisoning of a Pt/Ba/Al2O3 NOx Storage Catalyst. *Catalysts* **2018**, *8*, 155. [CrossRef]
26. Reddy, G.K.; Peck, T.C.; Roberts, C.A. "PdO vs. PtO"—The influence of PGM oxide promotion of Co_3O_4 spinel on direct NO decomposition activity. *Catalysts* **2019**, *9*, 62. [CrossRef]
27. Zhang, B.; Shen, M.; Wang, J.; Wang, J.; Wang, J. Investigation of various Pd species in Pd/BEA for cold start application. *Catalysts* **2019**, *9*, 247. [CrossRef]
28. Inger, M.; Dobrzyńska-Inger, A.; Rajewski, J.; Wilk, M. Optimization of ammonia oxidation using response surface methodology. *Catalysts* **2019**, *9*, 249. [CrossRef]
29. Kappis, K.; Papadopoulos, C.; Papavasiliou, J.; Vakros, J.; Georgiou, Y.; Deligiannakis, Y.; Avgouropoulos, G. Tuning the catalytic properties of copper-promoted nanoceria via a hydrothermal method. *Catalysts* **2019**, *9*, 138. [CrossRef]
30. Arab Pour Yazdi, M.; Lizarraga, L.; Vernoux, P.; Billard, A.; Briois, P. Catalytic properties of double substituted lanthanum cobaltite nanostructured coatings prepared by reactive magnetron sputtering. *Catalysts* **2019**, *9*, 381. [CrossRef]
31. Darda, S.; Pachatouridou, E.; Lappas, A.; Iliopoulou, E. Effect of preparation method of Co-Ce catalysts on CH_4 combustion. *Catalysts* **2019**, *9*, 219. [CrossRef]
32. Hajar, Y.; Venkatesh, B.; Baranova, E. Electrochemical promotion of nanostructured palladium catalyst for complete methane oxidation. *Catalysts* **2019**, *9*, 48. [CrossRef]
33. Zhang, N.; Yang, Z.; Chen, Z.; Li, Y.; Liao, Y.; Li, Y.; Gong, M.; Chen, Y. Synthesis of sulfur-resistant TiO_2-CeO_2 composite and its catalytic performance in the oxidation of a soluble organic fraction from diesel exhaust. *Catalysts* **2018**, *8*, 246. [CrossRef]
34. De Lucas-Consuegra, A.; Serrano-Ruiz, J.; Gutiérrez-Guerra, N.; Valverde, J. Low-temperature electrocatalytic conversion of CO_2 to liquid fuels: Effect of the Cu particle size. *Catalysts* **2018**, *8*, 340. [CrossRef]
35. Yuan, J.; Zhang, J.; Yang, M.; Meng, W.; Wang, H.; Lu, J. CuO Nanoparticles Supported on TiO_2 with High Efficiency for CO_2 Electrochemical Reduction to Ethanol. *Catalysts* **2018**, *8*, 171. [CrossRef]
36. Li, L.; Jiang, B.; Tang, D.; Zheng, Z.; Zhao, C. Hydrogen Production from Chemical Looping Reforming of Ethanol Using Ni/CeO_2 Nanorod Oxygen Carrier. *Catalysts* **2018**, *8*, 257. [CrossRef]
37. Ioannidou, E.; Neophytides, S.; Niakolas, D. Experimental clarification of the RWGS reaction effect in H_2O/CO_2 SOEC co-electrolysis conditions. *Catalysts* **2019**, *9*, 151. [CrossRef]

Review

Electropositive Promotion by Alkalis or Alkaline Earths of Pt-Group Metals in Emissions Control Catalysis: A Status Report

Ioannis V. Yentekakis [1,*], Philippe Vernoux [2], Grammatiki Goula [1] and Angel Caravaca [2]

1 Laboratory of Physical Chemistry & Chemical Processes, School of Environmental Engineering, Technical University of Crete (TUC), 73100 Chania, Crete, Greece; mgoula@science.tuc.gr

2 Univ. Lyon, Université Claude Bernard Lyon 1, CNRS—IRCELYON—UMR 5256, 2 Avenue A. Einstein, 69626 Villeurbanne, France; philippe.vernoux@ircelyon.univ-lyon1.fr (P.V.); angel.caravaca@ircelyon.univ-lyon1.fr (A.C.)

* Correspondence: yyentek@isc.tuc.gr

Received: 23 December 2018; Accepted: 29 January 2019; Published: 5 February 2019

Abstract: Recent studies have shown that the catalytic performance (activity and/or selectivity) of Pt-group metal (PGM) catalysts for the CO and hydrocarbons oxidation as well as for the (CO, HCs or H_2)-SCR of NO_x or N_2O can be remarkably affected through surface-induced promotion by successful application of electropositive promoters, such as alkalis or alkaline earths. Two promotion methodologies were implemented for these studies: the Electrochemical Promotion of Catalysis (EPOC) and the Conventional Catalysts Promotion (CCP). Both methodologies were in general found to achieve similar results. Turnover rate enhancements by up to two orders of magnitude were typically achievable for the reduction of NO_x by hydrocarbons or CO, in the presence or absence of oxygen. Subsequent improvements (ca. 30–60 additional percentage units) in selectivity towards N_2 were also observed. Electropositively promoted PGMs were also found to be significantly more active for CO and hydrocarbons oxidations, either when these reactions occur simultaneously with deNO_x reactions or not. The aforementioned direct (via surface) promotion was also found to act synergistically with support-mediated promotion (structural promotion); the latter is typically implemented in TWCs through the complex (Ce–La–Zr)-modified γ-Al_2O_3 washcoats used. These attractive findings prompt to the development of novel catalyst formulations for a more efficient and cost-effective control of the emissions of automotives and stationary combustion processes. In this report the literature findings in the relevant area are summarized, classified and discussed. The mechanism and the mode of action of the electropositive promoters are consistently interpreted with all the observed promoting phenomena, by means of indirect (kinetics) and direct (spectroscopic) evidences.

Keywords: platinum; palladium; Rhodium; iridium; NO; N_2O; propene; CO; methane; alkali; alkaline earth; platinum group metals; deNO_x chemistry; lean burn conditions; TWC; catalyst promotion; EPOC

List of Contents

1. Introduction

The development of new catalyst formulations for a more efficient and cost-effective control of unburned hydrocarbons, CO and NO_x pollutants emitted by mobile and stationary combustion processes is currently an urgent need [1–3]. Commercial three-way catalytic converters (TWCs) have been highly successful in controlling NO_x, CO and hydrocarbon emissions from conventional stoichiometric gasoline engines. However, some problems, that emerged during their four decades of implementation, concerning the economy of their production process, their composition, their operational/life time behaviour and their recycling still need to be resolved. The most significant issues of TWC technology are the following:

1. Commercial TWCs use formulations based variously on two or three noble metals (Pt, Pd and Rh) where Rh is essential for an efficient control of NO_x emissions (Rh is highly effective and therefore the key component in TWCs for the NO dissociation/reduction; Pt and Pd, although very active for CO and hydrocarbons oxidations, are almost ineffective for NO_x reduction). However, Rh is rare, therefore its successful reduction or even replacement by another De-NO_x efficient catalyst in TWC formulations is highly desirable.
2. A small but significant portion of NO is still converted to the undesirable by-product N_2O in commercial TWCs; Nitrous oxide emission control, due to N_2O harmful impact on stratospheric ozone depletion and its outstanding global warming potential (the latter is about 310 and 21 times greater than that of CO_2 and CH_4, respectively [4,5]), is a current challenge in environmental catalysis technology.
3. TWCs recycling for the noble metals recovery is nowadays marginally profitable, mainly due to the costly final separation of the recovered noble metals; simpler TWCs formulations (i.e., consisting of only one noble metal) are expected to substantially reduce both TWCs' production and recycling costs.
4. Commercial TWCs are no longer efficient in controlling NO_x emissions from the advanced lean-burn and diesel engines that operate at net-oxidising conditions.

In general, catalytic systems with suitable efficiency under oxygen rich conditions are a great challenge in emissions control technology. Besides lean-burn and diesel engines' effluent gases, such conditions are also typical in several stationary processes of significant environmental footprint, such as in soil fuels combustion processes for heat production in industry, in solid wastes combustion processes and certain chemical processes such as ammonia oxidation processes (e.g., nitric acid plants), adipic acid and other specific industries.

Prompted by these issues, numerous efforts are currently reported in open literature, by several research groups, on discovering means for enhancing the catalytic performance of Pt and Pd or even Rh, for reactions related to CO, NO_x and hydrocarbons abatement under stoichiometric and/or oxygen rich conditions. A notable part of these studies is focussed on the reduction or replacement of the Rh usage due to its scarcity and consequently higher cost compared to Pt and Pd.

It has been recently shown that the catalytic performance of precious metals can be strongly promoted by alkalis and alkaline earths for reactions related to emissions control catalysis. Two promotion methods were implemented towards this aim: (i) the Electrochemical Promotion of Catalysis (EPOC) concept or NEMCA effect (Non-Faradaic Electrochemical Modification of Catalytic Activity), proposed three decades ago by Vayenas and co-workers (e.g., [6–13]) and (ii) the Conventional Catalysts' Promotion (CCP), applicable by means of highly dispersed (supported) catalysts, as typically used in industrial applications (e.g., [14,15]). On these bases, rational/more efficient catalyst formulations and catalytic systems have been designed with significantly widened operational and/or implementation windows.

The present review summarises and comparatively analyses these literature results, paying attention on both fundamental and practical issues emerged. Mechanistic considerations (involving the mode of action of electropositive promoters in emissions control catalytic chemistry) are described in confrontation with recent spectroscopic and surface analysis studies evidencing their rightness.

2. Promotion Methodologies and Catalysts Formulations/Designs

Materials formulations and designs, experimental setup and other specific characteristics and/or rules regarding the two promotion methods, EPOC and CCP, are briefly described below.

2.1. Electrochemical Catalysts Formulations. The Electrochemical Promotion of Catalysis (EPOC) Concept

It has been shown by Vayenas and co-workers (e.g., [6,7]) that the chemisorptive and catalytic properties of metal catalysts, in the form of thin polycrystalline porous metal films, interfaced with solid electrolytes (X^z-ionic conductors: X is the type of ion transferred, z is its positive or negative charge) are subjected to in situ controlled electrochemical promotion via external bias applications, that is, currents or voltages imposed between the catalyst metal film and a separate counter electrode deposited on the same solid electrolyte specimen, which constitute a solid state electrochemical galvanic cell schematically shown in Figure 1.

In this way, applying small electric currents (typically some tens of μA units) or potentials (typically −2 to +2 V) between the working (catalyst film) and the counter electrodes via a Galvanostat/Potentiostat (Figure 1), ions are supplied from (or to) the solid electrolyte to (or from) the catalyst-electrode surface. Ions flux direction is determined by the direction of the applied current in combination with their charge, while their flux rate by the Faraday's law, $r_z = I/zF$ (I is the applied current in A; z is the ion's charge; F is the Faraday's constant = 96,485 Cb/mol). As we shall see in detail in Section 3.3, there is compelling evidence that these ions (together with their compensating charge in the metal thus forming surface dipoles) spill over (migrate) onto the gas-exposed catalyst surface, establishing an effective electrochemical double layer on the catalytically active surface altering its electron availability (work function). This strongly affects the binding strength of chemisorbed reactants and intermediates resulting to a significant modification on catalytic reaction rate (and or selectivity), typically greatly larger than the rate of ions flux that created these modifications (i.e., non-Faradaic behaviour), the so-called NEMCA (Non-Faradaic Electrochemical Modification of Catalytic Activity) phenomenon or EPOC (Electrochemical romotion of Catalysis).

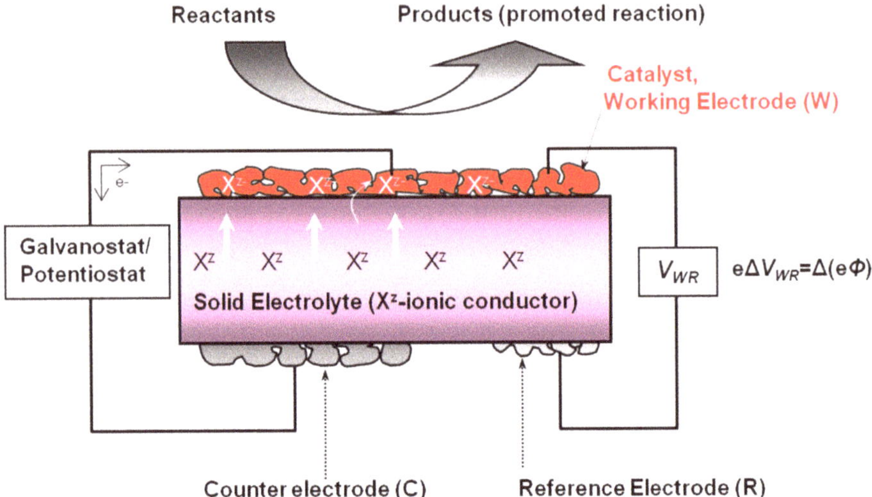

Figure 1. Schematic representation of a solid state electrochemical cell, through which the EPOC concept can be applied. X: type of ion transferred from the solid electrolyte to the catalyst surface and z: ion charge (e.g., O^{2-}, Na^+, K^+, H^+).

The EPOC was then applied to a large number of catalytic reactions [6,7,13], including reactions related to emissions control catalysis (e.g., [16–22]). Numerous EPOC studies regarding electrochemically imposed electropositive promotion by alkalis of Pt-group metals' catalysed emissions control reactions have also been reported, which are subject of the present review.

Alkali-conducting solid electrolytes such as Na^+ or K^+ substituted β- and β''-aluminas, (Na or K)$_{1+X}Al_{11}O_{17+X/2}$ and (Na or K)$_{1+X}M_XAl_{11-X}O_{17}$, respectively [6], are used for the construction of electrochemical galvanic cells with the following configuration (Figure 1):

PGM-film electrode / alkali-conducting solid electrolyte / inert counter electrode (e.g., Au) (1)

These cells are exposed to the reacting gas mixture by means of a continuous flow well-mixed reactor, as that shown in Figure 2, described as "single-pellet" (reactor) configuration and behaved similarly to a classical Continuous-Stirred-Tank-Reactor (CSTR) [12]. Controlled amounts of promoting species (alkali cations) can therefore be in situ supplied through the solid electrolyte onto the catalyst surface, causing significant alterations on the catalytic properties of the latter.

For the sake of simplicity, inert (in respect to the studied reaction) counter and reference electrodes (e.g., Au) are typically applied in EPOC studies conducted in reactor configurations similar to this shown in Figure 2; avoiding their participation in the catalytic reaction vectors (activity, selectivity) simplifies the data analysis, interpretation and understanding. It is also worth noting that besides their role as an in situ and reversible promoting species sources, solid electrolytes involved in the configuration of EPOC galvanic cells do not actually play any additional role on the promotional phenomena obtained.

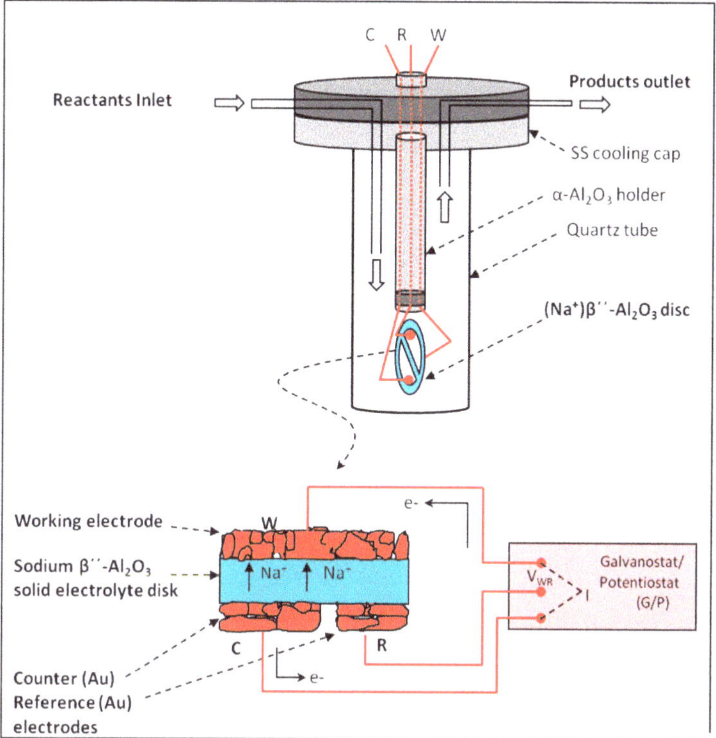

Figure 2. The single pellet reactor configuration for electrochemical promotion studies [12].

2.1.1. Operation Modes of the Electrochemical Promotion of Catalysis (EPOC)

(i) Galvanostatic transient operation:

One can use the galvanostatic transient mode of EPOC operation, described in detail in ref. [16], in order to be able to calculate the coverage, θ, of the promoting species (electrochemically supplied) on the catalyst surface as a function of time. Briefly, transient galvanostatic operation concerns the application of a constant current, I, between the catalyst and counter electrodes via the galvanostat. In cells of the type described by the formula (1), imposition of a constant negative current corresponds to a constant flux of alkali cations to the catalyst surface at a rate equal I/F (in moles of alkali/s).

The obtained alkali coverage, θ_{Alk}, can therefore be calculated from Faraday's law [6,16],

$$\theta_{Alk}\ (\%) = 100\cdot(-It/FN_o) \tag{2}$$

where, t is the time of current (I) application, F is the Faraday's constant and N_o is the number of catalyst active sites independently measured via surface titration. Recording the metal-solid electrolyte polarization transient during the galvanostatic experiment that is,, the time variation of catalyst potential by means of a reference electrode, V_{WR} (Figure 1), this permits establishment of the relationship $\theta_{Alk}(V_{WR})$ between alkali coverage and catalyst potential V_{WR} (Figure 3).

This relationship is in general dependent on gaseous composition but this dependence is relatively weak because the catalyst work function $e\Phi$ varies according to [6,7,11]:

$$e\Delta V_{WR} = \Delta(e\Phi) \tag{3}$$

and is determined primarily by the coverage of alkali, due to the large dipole moment of alkalis on the Pt-group metal surfaces.

Galvanostatic operation concerns a continuous supply and accumulation of alkali on the catalyst surface (Equation (2)). Current interruption at a desired alkali coverage can then provide an optimally alkali-promoted catalyst surface. Alkali species are not typically consumed during the catalytic reaction, which allows working under stable reaction conditions. Hence, the promotional effect induced by the alkali coverage remains practically constant, even after the interruption of the applied current.

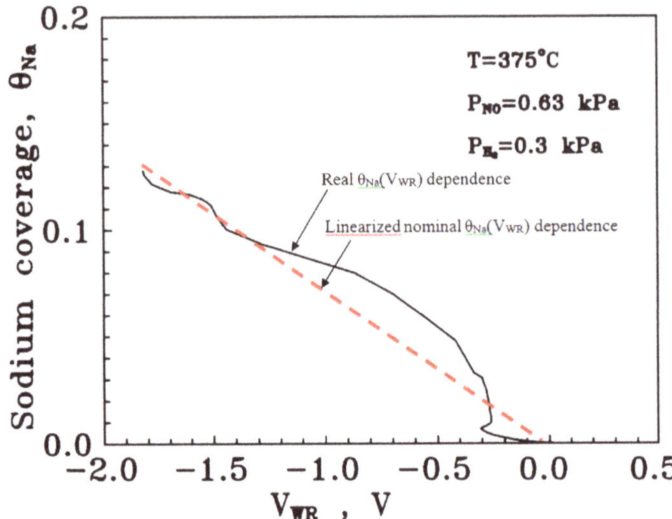

Figure 3. Dependence of alkali coverage on catalyst potential V_{WR} under certain reaction conditions (Reproduced with permission from Ref. [19]. Copyright 1997, Elsevier).

It is worth noting that galvanostatic EPOC operation provides a rapid method for assessing the response of the reaction rate to promoter coverage (Figure 4); it consequently offers a very quick and easy way to find, at any set of conditions, the optimal promoter loading. This information could then be used for the direct design of optimally promoted conventional catalysts formulations for practical applications (Figure 4) [21].

(ii) Potentiostatic operation (a steady-state mode of operation):

The potentiostatic mode of EPOC operation corresponds to catalyst operation under the imposition of a constant catalyst polarization (V_{WR}) via the potentiostat. This is a steady-state mode of operation: since the desired catalyst potential V_{WR} is imposed via the potentiostat, a very rapid transfer of the appropriate amount of alkali cations to the catalyst surface takes place, which supplies the required amount of alkali in order to develop and stabilize the externally imposed polarization value V_{WR}; then the current is sharply vanished, asymptotically approaching the zero value.

Since the $\theta_{Alk}(V_{WR})$ relationship between alkali coverage and catalyst potential can be obtained through a galvanostatic transient conducted at the desired reaction conditions (as described in (*i*); e.g., Figure 3), the potentiostatic mode of operation is more suitable and commonly used in EPOC studies. It allows to work under steady-state conditions at several constant catalyst potentials, that is, at several constant coverages of the promoter. The steady-state reaction rate response on promoted catalyst surfaces, upon varying of the other reaction parameters (e.g., temperature, reactants' composition, etc.) can then be studied (e.g., [6,7]).

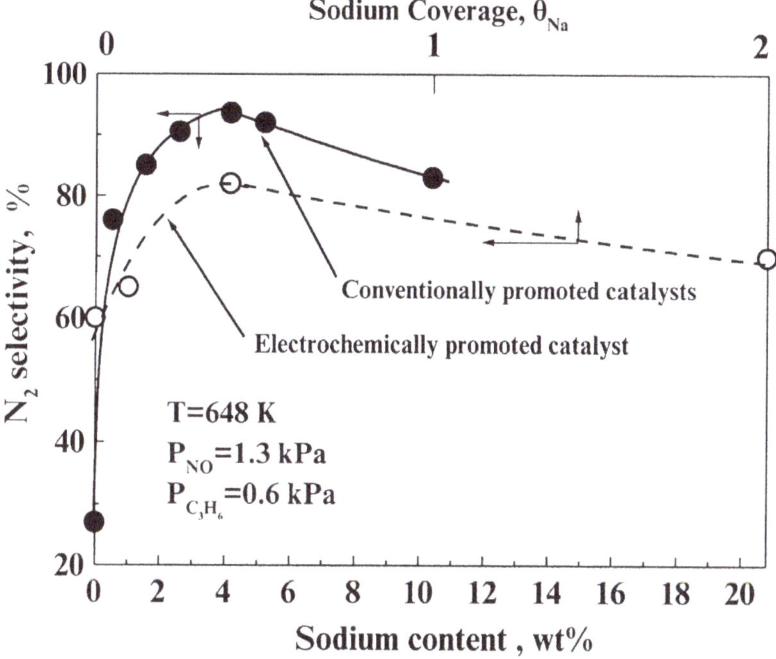

Figure 4. Comparison of electrochemically promoted (EPOC) and conventionally promoted (CCP) Pt with respect to N_2-selectivity under the NO + C_3H_6 reaction at T = 375 °C and [NO] = 1.3%, [C_3H_6] = 0.6% feed. Similar well-correlated dependences of the variation of N_2-selectivity on nominal sodium coverage (θ_{Na}, upper abscissa) were obtained by the two methods of promotion (EPOC and CCP) applied. EPOC was performed in a Pt/(Na)$\beta''Al_2O_3$/Au galvanic cell (Reprinted with permission from Ref. [22]. Copyright 2000, Elsevier).

2.1.2. Certain Characteristics of EPOC Concept:

(i) Reversibility of the electrochemically imposed promotional phenomena:

By reversing the direction of the applied current in the galvanic cell represented in Figure 1 or even by setting the potential value that corresponds to an alkali-free catalyst surface (Figure 3), the catalyst restores its initial un-promoted intrinsic properties. In other words, the electrochemically imposed promotional phenomena are totally reversible; EPOC provides an in situ, controlled and reversible way of catalyst promotion [6,7].

(ii) EPOC can be used as a fast probe for the optimal promoter loading determination:

Inter alia, EPOC can be successfully used as an effective research tool for evaluating the influence of a candidate promoter to a catalytic system and at the same time for determining the optimal promoter loading (coverage) at any reaction conditions of our interest. After the understanding that electrochemical and conventional promotions in catalysis, EPOC and CCP, are subjected to the same physicochemical rules (this has been proved both experimentally [21,22] and theoretically [23,24]), the findings and knowledge obtained by the application of the electrochemical promotion can be successfully applied for the design of effective conventional catalyst formulations for practical applications [21,22,25]. This strategy enabled us to design novel catalyst formulations extremely active and selective in emission control catalysis, for example, optimally alkali-promoted noble metal catalysts [21].

2.2. Conventional Catalyst Formulations. The Conventional Catalysts Promotion (CCP) Method

The wet impregnation method was used in most cases for the production of conventional type (i.e., highly dispersed on large surface area carriers), alkali-promoted noble metal catalysts. Typical carriers used were γ-Al_2O_3, SnO_2, yttria stabilized zirconia (YSZ) and rare earth oxides (REOs: La_2O_3, CeO_2) modified γ-Al_2O_3. The preparation of the conventional promoted catalysts was performed either by two subsequent impregnation steps (where the noble metal was first deposited, followed by that of the promoter) or by simultaneous, one-step impregnation, in a solution containing both the noble metal and the promoter precursors, typically nitrate but also chloride salts). When using only nitrate precursors, pre-treatment procedures typically involve a high temperature (ca. 500–600 °C) treatment step in air for the decomposition of the nitrates (active phase and promoter) and a subsequent treatment of the resulted catalysts under reaction conditions (i.e., under the studied reaction) for several hours or days in order to ensure stable operation. On the other hand, when a chloride precursor is used, pre-treatment in H_2 flow at temperatures typically 400–450 °C for several hours (>2 h) is applied as a first step for the precursor decomposition and the removal of the residual chlorine (e.g., [14,15]); the inhibitory behaviour of the latter on the reactions related to emissions control catalysis is well known. Then, for performance stabilization purposes, operation at certain reaction conditions is applied, as well [14,15].

2.2.1. Estimation of the Promoter Coverage

The coverage of the promoter species on the active catalyst surface is a significant factor not only for ranking but also for better understanding its promoting effects. Its knowledge is also useful for comparison purposes between electrochemically and conventionally promoted catalysts. For the former case (EPOC) we have already shown (Equation (2)) that the coverage of the alkali promoter can be directly estimated via the Faraday law applied to the data of a galvanostatic transient. In the latter case (CCP), a "nominal" percentage surface coverage of the alkali promoter (θ_{Alk}) could be calculated from the alkali wt% content of the catalyst (e.g., [14,21]) by the assumption that all the promoter is present at the surface and distributed uniformly over the entire available area (noble metal + support), without any incorporation into the bulk and that an one-to-one correlation between alkali adatoms and active metal sites exists. Hence, the following equation can be used:

$$\theta_{\text{Alk}}\,(\%) = (\text{wt\% of Alkali on the catalyst})\cdot(N_{AV}/M_{\text{Alk}}\cdot A\cdot d) \tag{4}$$

where N_{AV} is the Avogadro number, M_{Alk} is the molecular weight of alkali used, A is the BET catalyst surface area (m^2/g) and d is the surface density (atoms/m^2) of the active metal (PGM) (e.g., $d_{Pt} = 1.53 \times 10^{19}$, $d_{Pd} = 1.27 \times 10^{19}$, $d_{Rh} = 1.33 \times 10^{19}$ for {111} crystallites). Considering d in Equation (4) as the surface density of the alkali adatom used (i.e., 8.84×10^{19}, 3.314×10^{19}, 1.8×10^{19}, 1.115×10^{19} or 1.45×10^{19} for Li, Na, K, Cs or Rb, respectively; values calculated on the basis of the alkali ionic radius), one can probably be led to a better estimation of the nominal alkali coverage [22,25]. In any case and since the issues about the alkali distribution (preference between active metal or support surface, formation of 2D and/or 3D aggregates) are not clear yet, it is apparent that Equation (4) leads to a quite rough estimation of alkali coverage, which however, in the absence of more rigorous estimations or experimental measurements, can be of some interest and applicable as a first approximation upon comparing the EPOC and CCP methods of promotion on a catalytic system (e.g., [22,25]).

The fixed bed, single pass, quartz (or stainless steel) tube reactor configuration is typically used for testing the conventional type catalysts (in the form of powder, small particle clusters or even structured honeycomb monolithic configurations) at the desired reactions. On-line gas chromatography, on-line mass spectroscopy and continuous NO_x chemiluminescence's analysis were used for the analysis of the influent and effluent reactor streams in most studies.

2.2.2. EPOC and CCP Comparison Issues

In order to have a solid basis for comparison between electrochemically promoted or conventional promoted catalysts, the rate enhancement ratio (ρ), is introduced in both cases, defined as:

$$\rho = r_x/r_o \quad (x = \text{EPOC or CCP}) \tag{5}$$

r_o is the unpromoted rate and r_x is the promoted by alkali rate, where the alkali species are either supplied electrochemically (x = EPOC) or conventionally (x = CCP). ρ values as high as 420 (42,000% rate increase) have been reported for some important emissions control catalytic systems (e.g., $NO + C_3H_6/Pt$) by means of alkali promotion. It must be emphasized here that in literature studies (either EPOC or CCP) involving ρ values estimation, particular attention was devoted in order the catalytic activity data to be acquired in the intrinsic kinetic regime, not influenced by mass transfer limitations (i.e., under low conversions, no limiting reactant, differential reactor operation, small catalyst particles, etc.)

In the case of EPOC an additional parameter, the Faradaic efficiency (Λ) is also introduced [6], which can describe the magnitude of promotion; however, this has no meaning in CCP studies:

$$\Lambda = (r_{EPOC} - r_o)/(I/zF) = \Delta r/(I/zF) \tag{6}$$

where F is the Faradaic constant and I is the current. The term I/zF (z is the promoting ion charge; $z = +1$ in the case of alkali promotion) corresponds to the rate of promoting ions supplied electrochemically to the catalyst surface according to the Faraday's law.

Another useful parameter to quantify of the magnitude of the EPOC is the promotion index Pi [16,26]:

$$Pi = (r_{EPOC} - r_o)/\Delta\theta_i \tag{7}$$

where $\Delta\theta_i$ is the coverage of the promoting species such as Na^+.

3. Results and Discussion

3.1. *Promotion of Simple, "Model" Reactions*

Literature studies concerning "model" (i.e., simple) reactions, catalysed by Pt-group metals under electropositive promotion by alkalis or alkaline earths, conducted either by EPOC or CCP, are listed in Table 1 and discussed below.

3.1.1. CO Oxidation

Several publications concern the promotion of the PGMs-catalysed CO oxidation by alkalis. Significant promotional effects have been achieved either on thin metal film electrocatalysts interfaced with an alkali-contacting solid electrolyte (EPOC), playing the role of in situ source of alkali cations or on conventional type highly dispersed catalysts dosed with several amounts of alkali promoter (CCP). However, the promoting phenomena were found to be strongly depended on the CO/O_2 gas phase composition and on the alkali loading of the catalyst.

In particular, Yentekakis et al. [16] using the continuous flow, well-mixed, single-pellet reactor described in Figure 2 have shown that the rate of CO oxidation can be markedly enhanced via EPOC by up to 600% (at T = 350 °C) under CO-rich conditions (i.e., at conditions where CO oxidation kinetics obey negative order in CO, positive order in O_2) over a Pt film electrocatalyst interfaced with a Na^+-conducting $\beta''Al_2O_3$ solid electrolyte (Table 1). Those rate enhancements were achieved at sodium coverages on Pt surface of about 2–4%, while higher sodium coverages were found to depress the catalyst activity, thus leading to a "volcano" type behaviour upon increasing the coverage of the promoting species (Figure 5, curve a). However, at O_2-rich conditions (i.e., when the rate is of positive order in CO) only poisoning phenomena occurred upon increasing Na coverage on the catalyst surface (Figure 5 curve b). According to the rules of catalyst promotion, Na promoters modify the chemisorptive properties of CO and O_2 reactants on Pt active sites. At CO-rich conditions and for Na coverages up to ~4%, the promotional effect was attributed to a Na-induced enhancement in oxygen chemisorption as a result of a strengthening of the Pt–O bond. The rate poisoning behaviour at higher Na coverages was attributed to active sites blocking phenomena due to the formation of a CO–Na–Pt surface complex.

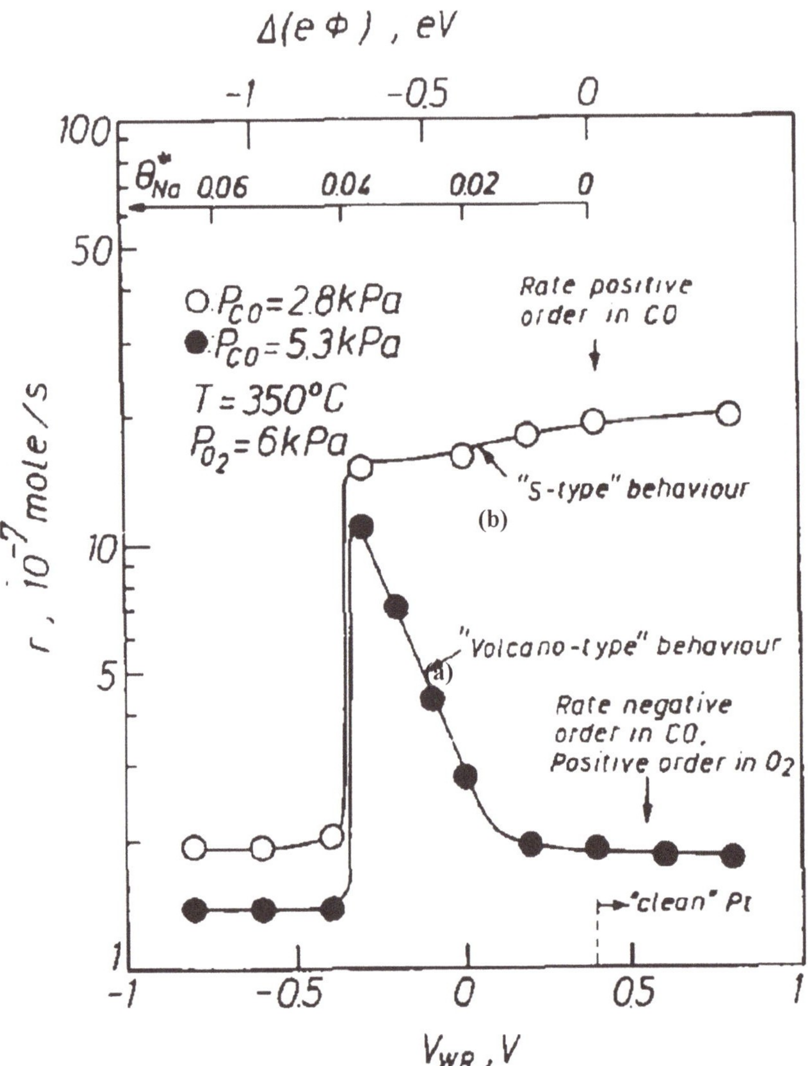

Figure 5. Effect of Na coverage θ_{Na} on the rate of CO oxidation at oxygen lean (a: T = 350 °C, $[O_2]$ = 6%, [CO] = 5.3%; filled symbols) and at oxygen rich ((**b**): T = 350 °C, $[O_2]$ = 6%, [CO] = 2.8%; open symbols) conditions. The volcano type behaviour upon increased Na coverage is obvious in case (**a**). Data was acquired in a Pt/(Na)$\beta''$$Al_2O_3$/Au galvanic cell (Reprinted with permission from Ref. [16]; Copyright 1994, Elsevier).

Another EPOC study was reported by De Lucas-Consuegra et al. [27] on the effect of electrochemically imposed potassium (K) on the Pt-catalysed CO oxidation at 200–350 °C. Rate enhancement ratios of about $\rho = 11$ (1100%) at T = 278 °C and light-off temperature decreases by ~40 °C were achieved at equimolar reactant composition, [CO] = $[O_2]$ = 5000 ppm (Figure 6 and Table 1).

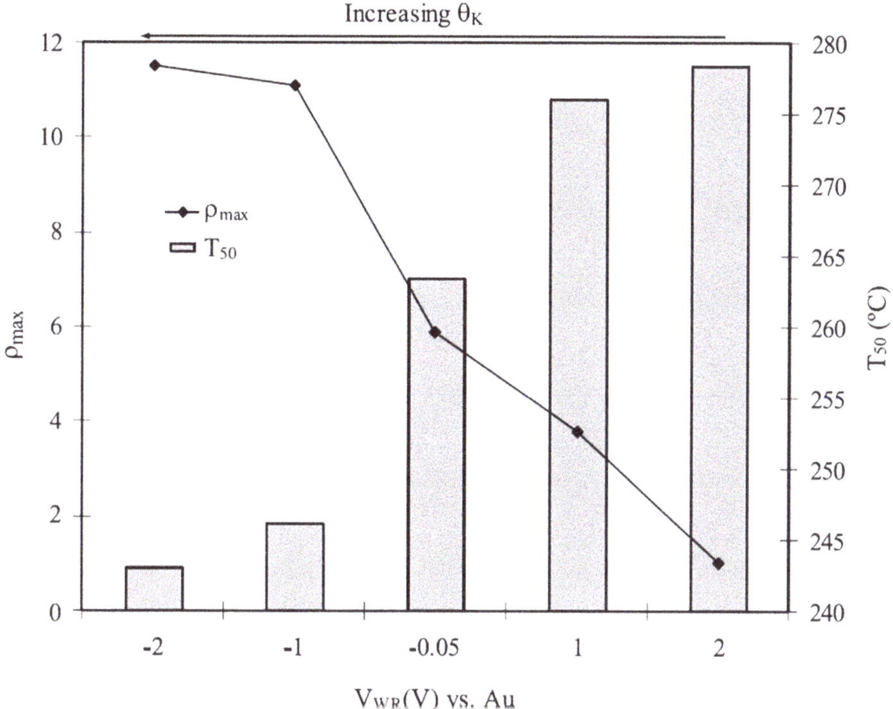

Figure 6. Effect of catalyst overpotential (VWR) on the maximum rate enhancement ratio (ρ_{max}) and on T_{50} of CO oxidation on Pt. Conditions: [CO] = $[O_2]$ = 0.5% He balance, F_t = 108 cm^3/min. Data was acquired in a Pt/(K)$\beta''Al_2O_3$/Au galvanic cell. (Reprinted with permission from Ref. [27]; Copyright 2008, Elsevier).

The maximum promotional effect was found under the application of −2 V; the appropriate parameters necessary for the estimation of the K coverage corresponding to this catalyst polarization are not available. The reversible promotional phenomena were attributed to the formation and decomposition of potassium compounds on the Pt surface (probably potassium carbonates and/or oxides) that enhanced the adsorption of oxygen at the expense of CO on a surface predominately covered by CO under unpromoted conditions in the reaction temperature range studied.

Regarding CCP studies, CO oxidation activity enhancements were also reported by Mirkelamoglu and Karakas [28,29] over conventional type 0.1wt% Na-promoted 1wt% Pd/SnO_2 catalyst tested at two different [CO]/[O] ratios, 1.25 and 0.5. The authors found that in the case of CO-rich gas composition ([CO]/[O] = 1.25), Pd can be up to 2.6-fold times (260%) more active on CO oxidation at 150 °C (Table 1), while non or moderate promotional effects were observed at O_2-rich conditions ([CO]/[O] = 0.5). A surface segregation of Pd atoms increasing the reactive sites for CO oxidation, together with the formation of super-oxide species observed over Na-PdO/SnO_2 were considered to be the origin of the observed promotional effects on the Na-dosed catalyst [29].

The effect of alkali oxide additives (Na_2O, K_2O) on the alumina support of Pt/Al_2O_3 catalysts on CO oxidation was investigated by Lee and Chen [30]. They concluded that the addition of alkalis on Al_2O_3 influences the basicity of the catalyst, which in turn results to higher oxidation activities. Higher promotional effect was achieved by potassium, which resulted in a decrease of CO light-off temperature by 90 °C under CO/O_2 stoichiometric conditions. The study, however, was performed under simulated two-stroke motorcycle emissions (involving C_3H_6, H_2, CO_2 and H_2O in the feed, besides CO and O_2) at the stoichiometric point and oxygen-deficient environments. Therefore, this work will be further analysed in the proper Section 3.2.1.

The influence of alkaline earths on CO oxidation activity was also investigated over Pd catalysts supported on Ba-modified γ-Al_2O_3 or CeO_2-ZrO_2 carriers by Tanikawa and Egawa [31]. They found that the activity of Pd/γ-Al_2O_3 catalyst can be significantly improved by increasing the Ba promoter loading; an amount of 10–15wt% Ba decreases the light-off temperature by ~45 °C compared to the un-modified catalyst (Table 1). However, an adverse effect upon Ba addition was observed for Pd/CeO_2-ZrO_2 catalysts; an amount of 10 wt% Ba leads to an increase of light-off temperature by ~100 °C. The pronounced effect of Ba over Pd/γ-Al_2O_3 catalysts was ascribed to the weakening of CO adsorption strength, whereas its inhibiting role over Pd/CZ catalysts was attributed to the suppression of Pd interaction with ceria-zirconia support.

In addition, the pronounced effect of electropositive promoters on CO preferential oxidation (PROX) under H_2-rich conditions has been recently demonstrated [32–41]. Minemura et al. [32,33] showed that the addition of alkalis (Li, Na, K, Rb and Cs) in a 2wt% Pt/γ-Al_2O_3 catalyst results in very significant promotional effects on the selective oxidation of CO under H_2-rich environments (Table 1). The promotional effects followed "volcano" type behaviour upon increasing alkali loading, thus providing an optimal loading for each alkali; a 10–15 alkali/Pt loading ratio was found to optimize the promotional effects of Na and K, while the optimal alkali/Pt ratios were 3 and 5 for Cs and Rb, respectively. For a constant alkali/Pt loading equal to 3, the order on the promoting effect on CO conversion (and TOFs) was found to be Cs > Rb > K > Na > Li ~alkali-free Pt, while the order of the selectivity to CO_2 was K > Rb, Na > Cs > Li. At higher alkali loadings (alkali/Pt = 10), the order of the turnover frequency rates of CO oxidation were K > Na > Rb > Cs~Li > alkali-free Pt, with a similar order for the selectivity towards CO_2. The changes in the order of the activity of the catalysts were due to over-promotion, that is, for some alkalis (e.g., Rb and Cs) their optimal loading was exceeded under the conditions used. In these studies, turnover rate enhancement ratio values up to $\rho = 10$ for the optimal K promotion at T = 100 °C [32] and even better (up to $\rho = 20$) at 80 °C [33] were achieved. Further studies by the research group, focused on the optimal promoter (K) when applying different supports (Al_2O_3, SiO_2, ZrO_2, Nb_2O_5 and TiO_2) for the dispersion of Pt, have shown that the most remarkable promoting effects are obtained on Pt/γ-Al_2O_3 [34] (Table 1). Corroborative characterization studies by means of a variety of techniques, e.g., Transmission Electron Microscopy (TEM), Extended X-ray Absorption Fine Structure (EXAFS), X-ray Absorption Near Edge Structure (XANES), Fourier-Tranform Infrared Spectroscopy (FTIR), enable the authors to consistently interpret the promotion features [35,36]. FTIR spectroscopic evidences for a K-induced weakening in the strength of CO adsorption on Pt and drastic changes on the adsorption site of CO (bridge and three-fold hollow CO species) found on the Pt (10wt% K)/γ-Al_2O_3 catalyst were considered responsible for the observed promotional effects attributed to the change in the electronic state of K-modified platinum [35,36]. OH co-adsorbed species (originating form H_2 and O_2 interaction under PROX conditions) that promote the CO oxidation were found by in situ FTIR on the highly active K-modified Pt/γ-Al_2O_3 catalyst [36]. The inhibiting effects on PROX reaction under over-promotion conditions (beyond the optimal promoter loading) were attributed to aggregation of Pt metal particles caused by the larger amount of alkali metal used [36].

CO preferential oxidation was also found to be very pronounced over Pt clusters dispersed on alkali (Na, Rb or Cs)-modified SiO_2 by Pedrero et al. [37]. Optimum promotional effects were obtained with Cs at a surface concentration of 1.6 atoms/nm^2; CO oxidation turnover rates (TOFs) more than one order of magnitude higher than those obtained on the un-promoted Pt/SiO_2 catalysts were recorded, which were accompanied by CO oxidation selectivity >90% (Table 1). In accordance to the authors, the beneficial effect of alkalis can be interpreted based both on the electronic interactions between alkali atoms and metals and on the inhibition of spill over-mediated H_2 oxidation pathways induced by alkalis.

The effect of Na promoter on CO PROX reaction was also investigated over bimetallic PtCo/Al_2O_3 catalysts by Kwak et al. [38]. They found moderate promotion of PROX activity on monometallic Pt catalyst while substantial rate promotion on the bimetallic PtCo counterpart catalyst up to a sodium loading of 2.0wt%; the CO-PROX selectivity was not influenced in both cases (Table 1). They concluded that the formation of Na-O-Al bonds by incorporation of Na ions into the alumina lattice, that happens at low sodium content, suppresses the formation of the surface spinel cobalt species and promotes the formation of bimetallic Pt-Co species; at higher Na content (ca. 3wt%) bulk sodium particles on the catalyst surface interact with the active species inhibiting PROX activity.

On the other hand, a notable increase on both CO conversion and CO_2 selectivity under PROX conditions was attained with Mg-promoter over Pt/Al_2O_3 catalysts by Cho et al. [39] (Table 1). The maximum values of CO conversion and CO_2 selectivity for Pt-Mg/Al_2O_3 catalysts were 93.1 and 62%, respectively, at 170 °C, compared to 70.2 and 46.8% at 200 °C over the un-promoted Pt/Al_2O_3 catalysts. The effect of CO_2 and H_2O in the feed was also studied in this work under PROX conditions, as well as the CO oxidation reaction in the absence of excess H_2 under Mg-promotion. The superior performance of Mg-modified catalysts was attributed to an increase in Pt electron density, as well as to an increase of hydroxyl groups on catalyst surface.

Finally, de Lucas-Consuegra and co-workers [40] demonstrated the pronounced effect of electropositive modifiers (K^+ ions) on the preferential oxidation of CO via EPOC by the use of a Pt/K^+-conducting-βAl_2O_3/Au galvanic cell. At a 195 °C and a feed composition of $CO/O_2/H_2$ = 0.4%/0.2%/16%, the authors found up to ρ = 1.3 rate enhancements, achievable at a potassium coverage of θ_K~2–4%. A concomitant increase in CO_2 selectivity of ~10% was recorded as well (Table 1). The effect of the alkali on the chemisorptive bonds of CO, H and O on Pt surface was considered to be the origin of the observed promotional phenomena.

For more information, a short review has been recently published regarding the promotional effect of alkali promoters (CCP) on PGMs (Pt, Ru and Ir)-catalysed CO PROX reaction [41].

A comparative overview of the literature findings included in Section 3.1.1 and in Table 1 leads to the following general remarks: Among PGMs only Pt and Pd were investigated so far for CO oxidation under electropositive promotion (by alkalis or alkaline earths). No studies exist that were performed at similar conditions using both EPOC and CCP methods, preventing us from making a direct one-to-one comparison of the outputs of the two methods. The general view is that alkalines efficiently promote the Pt and Pd catalysed CO oxidation only at conditions where the reaction rate is negative order in CO, positive order in O_2 (the so-called CO-rich conditions at which CO coverage predominates on the catalyst surface; not necessary at $CO/O_2 > 1$), while non, moderate or only poisoning was found under O_2-rich conditions. Even in the former case, the promotion follows volcano type behaviour upon increasing promoter loading, reversed to poisoning for high promoter loadings; optimal promoter loadings were typically low. The maximum rate enhancement ratios were reported for Pt/CO+O_2 catalytic system under EPOC: ρ = 6 with Na at 350 °C [16] and ρ = 11 with K at 278 °C [27].

Volcano-type promotion was the main feature of the electropositive promotion of PGMs under the preferential CO oxidation (PROX), as well. Moreover, for similar reaction and promoter loading conditions larger alkalis (Cs, Rb, K) offered superior promotion. For this reaction two studies exist that concern usage of the same promoter (K) and active metal (Pt), one performed using CCP the other EPOC. Although ρ values obtained by CCP were significantly larger (ρ = 10 [32–36]) than that obtained by EPOC (ρ = 1.3 [40]), the applied conditions were quite different, preventing us from making a creative comparison.

The alkaline-induced pronounced enhancement on oxygen adsorption compared to that of CO was a key argument in most of the aforementioned studies upon explaining their findings, whereas the formation of large surface alkali complexes was considered responsible for the rate poisoning at high alkali loadings.

Table 1. Electropositive promotion of PGMs-catalysed CO oxidation ($CO+O_2$) and Preferential CO oxidation (PROX: $CO+O_2$+excess H_2).

Reactants	Catalyst, (promotion method applied)	Promoter	Reaction conditions	Promotion highlights and optimal achievements	Ref.
CO, O_2	Pt-film over (Na)$\beta''Al_2O_3$ solid electrolyte (EPOC)	Na	T = 300–450 °C [CO] = 0–4% [O_2] = 0–6%	✓ Volcano behaviour of promotion upon increasing θ_{Na}, under CO-rich conditions. ✓ Promotion occurs at θ_{Na} = 2–4%, under CO-rich conditions; moderate inhibition at low θ_{Na} = 2–4%, under O_2-rich conditions; strong inhibition for θ_{Na} > 4% in both cases. ➢ ρ = 6 at θ_{Na} = 4%, T = 350 °C, CO-rich conditions ([CO] = 5.3%, [O_2] = 6%)	[16]
CO, O_2	Pt-film over (K)$\beta''Al_2O_3$ (EPOC)	K	T = 200–350 °C [CO] = [O_2] = 500 ppm	✓ Significant K-induced promotion, providing a system proper for LT CO oxidation. ➢ ρ = 11 at T = 278 °C and ΔT_{50} = −40 °C; [CO] = [O_2] = 500 ppm	[27]
CO, O_2	1 wt% PdO/SnO_2 (CCP)	Na	T = 150 °C [CO]/[O] = 1.25 [CO]/[O] = 0.5 ([CO] = 2400 ppm) Na-loading: 0.1 wt%	✓ Significant promotion at CO-rich conditions; more effective on pre-reduced catalysts ➢ Achievements on pre-oxidized catalysts: ρ = 1.5, ΔT_{50} = −12 °C (at [CO]/[O] = 1.25); ρ = 0.66, ΔT_{50} = +4 °C (at [CO]/[O] = 0.5) ➢ Achievements on pre-reduced catalysts: ρ = 2.6, ΔT_{50} = −25 °C (at [CO]/[O] = 1.25); ρ = 1.3, ΔT_{50} = −8 °C (at [CO]/[O] = 0.5)	[28] [29]
CO, O_2	1 wt% Pd/γ-Al_2O_3 and 1 wt% Pd/CZ (CCP)	Ba	T = 50–250 °C [CO] = 2.2%, [O_2] = 1.1% Ba-loadings: 0, 1, 5.5, 10, 15 wt%	✓ Promotion on Pd(Ba)/γ-Al_2O_3; poisoning on Pd(Ba)/CZ ➢ Promotion: ΔT_{50} = −45 °C, with 10–15wt% Ba loading on Pd(Ba)/γ-Al_2O_3 ➢ Poisoning: ΔT_{50} = +100 °C, with 10wt% Ba loading on Pd(Ba)/CZ	[31]

Table 1. *Cont.*

Reactants	Catalyst, (promotion method applied)	Promoter	Reaction conditions	Promotion highlights and optimal achievements	Ref.
CO, O_2, H_2 excess	2 wt% Pt/(γ-Al_2O_3, SiO_2, ZrO_2, Nb_2O_5 or TiO_2) (CCP)	Li, Na, K, Rb, Cs	T = 100–160 °C [CO] = [O_2] = 0.2% Alkali loadings: A/Pt molar ratio = 0–20	✓ Volcano-type promotion upon increasing alkali loading. K was the superior promoter. ✓ For A/Pt < 3 the promoting effect of alkalis followed the order: Cs > Rb > K > Na > Li ✓ For A/Pt > 5 the promoting effect of alkalis followed the order: K > Na > Rb > Cs > Li ✓ An A/Pt=10–15, 3 and 5 optimizes Na or K, Cs and Rb promotion, respectively. ➢ $\rho = 10$ with K, with a loading of K/Pt = 10 molar ratio	[32] [33] [34] [35] [36]
CO, O_2, H_2 excess	1.6 wt% Pt/SiO_2 (CCP)	Na, Rb, Cs	T = 110 °C [CO] = [O_2] = 1%, [H_2] = 70% Cs surface density = 0–6 atoms/nm^2	✓ Volcano-type promotion upon increasing alkali loading. Cs was the superior promoter. ✓ Promotion is maximized at a Cs loading = 1.6 Cs/nm^2 of catalyst surface. ➢ $\rho = 10$ and ΔS_{CO} = +30% (from 60%→90%) with an 1.6 Cs/nm^2 alkali surface density.	[37]
CO, O_2, H_2 excess	1 wt% Pt/γ-Al_2O_3 1 wt% Pt-1.8wt%Co/γ-Al_2O_3 (CCP)	Na	T = 25–300 °C [CO] = 0.1%, [O_2] = 0.1%, [H_2] = 1% Na loading: 0.5–3 wt% Na	✓ Moderate promotion of Na on Pt/Al_2O_3 catalyst. ✓ Significant volcano-type promotion (maximized at 2 wt% Na) on PtCo/Al_2O_3 catalyst. ➢ ΔX_{CO} = 75% (from 25%→100%) on 2 wt% Na promoted PtCo/Al_2O_3.	[38]

Table 1. *Cont.*

Reactants	Catalyst, (promotion method applied)	Promoter	Reaction conditions	Promotion highlights and optimal achievements	Ref.
CO, O_2, H_2 excess	2wt%Pt/γ-Al_2O_3 (CCP)	Mg	T = 100–250 °C [CO] = 1%, [O_2] = 0.75%, [H_2] = 65% [CO_2] = 20%, [H_2O] = 2% Mg loading = 3 wt% Mg	✓ Both CO oxidation and PROX reactions were significantly promoted by 3wt% Mg. ✓ For PROX reaction both CO conversion and selectivity were enhanced in 150-230 °C. ➢ $\Delta T_{50} = -20$ °C for CO + O_2 reaction at moist conditions (2% H_2O on stream). ➢ $\Delta X_{CO} > 35\%$ and $\Delta S_{CO} > 15\%$ for PROX reaction (max $X_{CO} = 93\%$ and max $S_{CO} = 62\%$ at 170 °C) on 3wt% Mg-promoted catalyst.	[39]
CO, O_2, H_2 excess	Pt film on (K)βAl_2O_3 solid electrolyte (EPOC)	K	T = 195 °C [CO] = 0.4%, [O_2] = 0.2%, [H_2] = 16% $\theta_K = 0$–4%	✓ Both PROX activity and selectivity were significantly enhanced by K addition. ✓ K effect allows activating the Pt at lower temperatures for PROX. ✓ The inhibiting effect of H_2 on PROX was attenuated. The inhibiting effect of temperature on CO oxidation selectivity was attenuated. ➢ $\rho = 1.3$ and ~10% increase in CO oxidation selectivity, within $\theta_K = 2$–4%.	[40]

3.1.2. Light Hydrocarbons Oxidation

(i) Alkenes oxidation

The oxidation of ethylene over polycrystalline Pt films has been extensively studied under electropositive promotion (EPOC) by Na via a Na^+-conducting $\beta''Al_2O_3$ solid electrolyte, by Vayenas and co-workers [18,42] (Table 2). Both promoting and poisoning effects induced by alkali have been observed: low Na coverages (typically <8%) were found to cause increases in the reaction rate (up to $\rho = 2$), while larger Na coverages led to a significant decrease of rate, which eventually falls well below the initial value corresponding to Na-free Pt surface (Figure 7). The data was interpreted in terms of (i) Na-enhanced oxygen chemisorption and (ii) poisoning of the surface by accumulation of Na compounds.

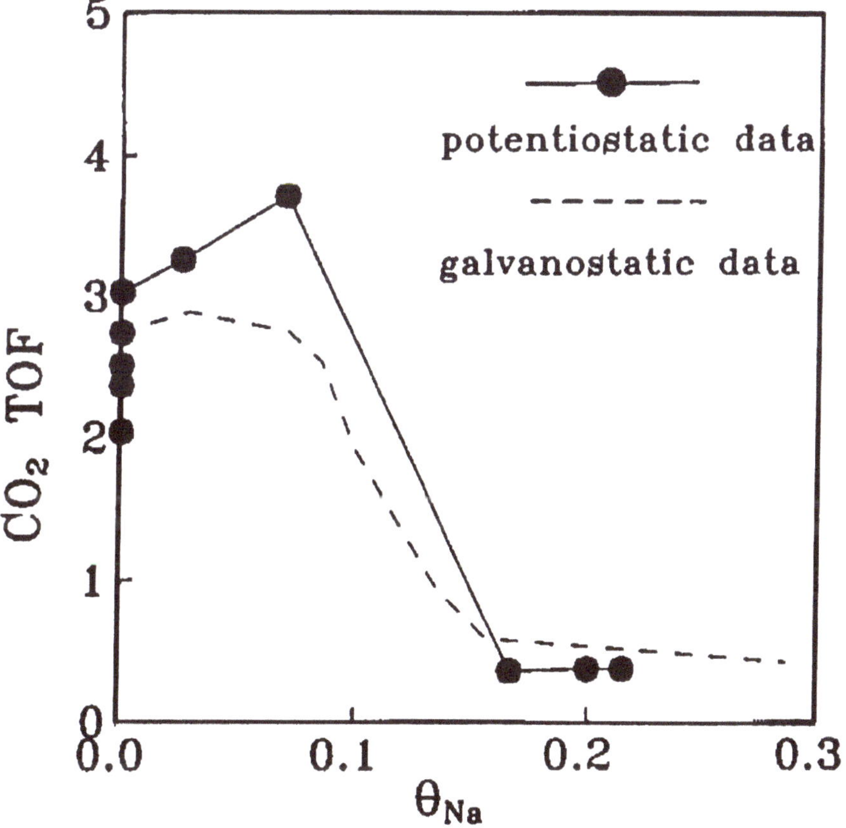

Figure 7. Rate as a function of Na coverage for both potentiostatic and galvanostatic experiments for $[O_2] = 8.0\%$ $[C_2H_4] = 4.2\%$ (balance He) mixture at 352 °C. Data was acquired in a $Pt/(Na)\beta''Al_2O_3/Au$ galvanic cell (Reprinted with permission from Ref. [18]; Copyright 1996, Elsevier).

According to the authors, low Na coverage gives rise to an enhanced chemisorption of O_2 at the expense of ethylene, resulting in an increased rate. At high sodium coverages both (i) and (ii) factors operate synergistically to poison the system: the increased strength of the Pt–O bond and the coverage of the catalytic surface by compounds of Na strongly suppress the rate. Kinetic results obtained over a Pt(111)–Na model catalyst, dosed with several amounts of Na by vapour deposition (CCP), were found to mirror the behaviour of the $Pt/\beta''Al_2O_3$ catalyst promoted by the EPOC concept [18]. This strongly suggested that Na pumped from the solid electrolyte via electrical polarization is indeed the key species for the promoting or poisoning phenomena. In addition, these results demonstrated that electrochemical promotion of catalysis (EPOC) and surface promotion by conventional means (CCP) obey the same physiochemical rules in the case of alkaline promoting cations. This is well documented nowadays but it was unclear at those early steps of electrochemical promotion studies.

The Pt-catalysed propene oxidation has been also extensively studied by Filkin et al. [20] under in situ electrochemical promotion (EPOC) by sodium, using polycrystalline Pt films deposited on Na^+-conducting $\beta''Al_2O_3$. Depending on propene and oxygen reactants partial pressures, fully reversible promotion and poisoning effects were observed as a function of sodium coverage as follows. As $[O_2]/[C_3H_6]$ decreases, the effects of Na promotion rapidly increase: for $[O_2]/[C_3H_6] = 5.7$, where the activity of Na-free Pt is highest, the maximum enhancement ratio $\rho_{max} = r_{max}$(on Na-promoted Pt)/r_o(on Na-free Pt) is equal to 1.1, which then increases to $\rho_{max} = 2.3$ at $[O_2]/[C_3H_6] = 2.0$, where Na-free Pt activity is lower. These maxima enhancement ratios where obtained at moderate Na coverages (not specified) while higher Na coverages cause poisoning of the propene oxidation rate. The larger poisoning was obtained in the case of $[O_2]/[C_3H_6] = 5.7$ (Table 2). A Na-modified chemisorption of the reactants considered (Na enhances oxygen chemisorption and inhibits propene chemisorption) on a Langmuir-Hinselwood type reaction accounted all the experimental findings. Using X-ray Photoelectron Spectroscopy (XPS), Auger Electron Spectroscopy (AES) and postreaction Na K edge XANES (X-ray Absorption Near Edge Structure) the authors have demonstrated that under reaction conditions the promoter phase consists of small amounts of 3D sodium carbonate crystallites, while thick layers of this sodium surface compound were responsible for rate inhibition in the poisoning regime. Notably, these promoter and poisoning phases, although stable at reaction temperatures investigated, were readily destroyed via electrochemical Na pumping away from the catalyst by the application of electrical polarizations of opposite sign [20].

Other studies involving the propene oxidation over Pt films using the EPOC concept as well, were reported by Vernoux and co-workers [43,44]. In these studies Pt was interfaced with NASICON ($Na_3Zr_2Si_2PO_{12}$), an alternative Na^+-conductor, acting therefore as a Na source for its electrochemical pumping to the Pt catalyst surface. The authors used two distinguished propene/oxygen gas phase compositions, near stoichiometric (C_3H_6/O_2 = 0.04%/0.2%) and excess oxygen (C_3H_6/O_2 = 0.04%/8.3%) conditions, to investigate the effect of Na on the catalytic system under consideration at T ~ 300 °C (Table 2). They found that under oxygen excess Na coverages larger than 3% cause a slight poisoning of the propene oxidation rate ($\rho \sim 0.8$). However, under near-stoichiometric conditions, sodium had strong beneficial effect ($\rho = 3.5$) on the Pt-catalysed propene oxidation rate, which is maximized at a $\theta_{Na} \sim 3.6\%$; the rate was then slightly decreased when higher (up to 6%) sodium coverages were supplied. Cyclic voltammetry studies indicated that sodium promoter exists in the form of Na_2CO_3 and $NaHCO_3$ surface phases during reaction [43], in agreement with earlier studies by Filkin et al. [20].

In this line, de Lucas-Consuegra et al. [45,46] have more recently studied the propene oxidation under EPOC of a Pt film interfaced with a K^+-conducting βAl_2O_3 solid electrolyte, that is, using potassium as promoter species. They found that the reaction can be strongly promoted at the temperature interval of ~200–300 °C, under both near-stoichiometric and oxygen-rich conditions. Rate enhancement ratios of the order of $\rho = 7$ were achieved for a catalyst polarization of about −2 V (Table 2); parameters' values necessary for the estimation of the corresponding alkali coverage are not available in the paper. Cyclic voltammetry, FTIR and SEM-EDX studies enable the authors to identify the formation of stable potassium oxide and superoxide phases upon electrical polarization. These species were responsible for the promotional phenomena. In addition, due to the stability of the potassium surface compounds formed, a permanent promotional effect was observed.

The effect of electropositive modifiers on hydrocarbons oxidation over conventional highly dispersed catalysts (CCP) was also investigated by Yentekakis and co-workers [47]. In particular, they studied the propene oxidation over Na-promoted Pt(Na)/γ-Al_2O_3 catalysts under oxygen excess

(1000 ppm propene + 5% O_2). Significant activity enhancements were observed for a Na-loading of ~1.5–2.6 wt % (or 0.68–1.13 × 10^{-3} mmol effective Na/g_{cat}) [47]. Optimal promotion was achieved for 2.6 wt% Na loading, corresponding to a nominal sodium coverage of θ_{Na}~23% (Equation (4) by using d_{Na} = 3.314 × 10^{19} atoms/m^2), causing a significant shift of the propene light-off temperature to about 60 °C lower values (Figure 8, Table 1). Rate enhancement ratios as high as ρ~10 can be calculated from these results. It is worth noting that the aforementioned promotional effects were recorded under oxygen excess conditions, typical of lean-burn combustion engines. The Na-induced enhancement of oxygen adsorption on the Pt surface, which is predominately covered by propene even at these excess oxygen gas phase conditions, was considered to be responsible for the observed promotional effect.

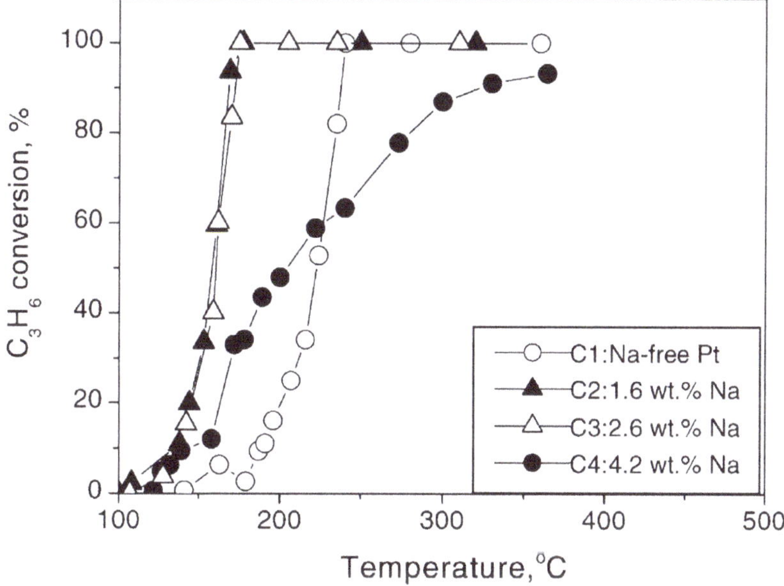

Figure 8. The effect of temperature on the conversion of C_3H_6 Na-dosed Pt(Na)/γ-Al_2O_3 catalysts under C_3H_6 + O_2 reaction at oxygen rich conditions: 1000 ppm C_3H_6 + 5% O_2; catalyst weight 70 mg; total flow rate was varying as F_t = 210, 126, 62.5 and 52 cm^3/min for C1, C2, C3 and C4 catalysts, respectively, in order to keep the effective contact time of C_3H_6 constant at 4 s. (The effective contact time is defined as: surface Pt atoms/reactant molecules/s) (Reprinted with permission from Ref. [47]; Copyright 2005, Elsevier).

(ii) Alkanes oxidation

An EPOC study regarding the propane oxidation on Pt-based catalysts under Na-promotion supplied via a (Na)β"Al_2O_3 solid electrolyte has been reported by Kotsionopoulos and Bebelis [48]. The activity was dramatically inhibited by Na-addition even at low Na coverages, over a wide range of propane/oxygen ratios (Table 2). Under stoichiometric conditions ([O_2]/[C_3H_8] = 1%/0.2%) and T = 320 °C, a Na coverage as small as θ_{Na}~2.5% leads to a significant decrease of the reaction rate of 625 times. The poisoning effect was more pronounced at higher temperatures (e.g., 360, 400 and 440 °C). This poisoning effect of electropositive promoters on Pt for alkanes oxidation can be attributed to the weak adsorption of alkanes compared to oxygen. A latter study published by the same authors dealt with the electrochemical characterization of the aforementioned catalytic system [49]. In this work they observed, in good agreement with some of the studies mentioned above, that the sodium promoter is on the metal surface mainly forming carbonate, bicarbonate and oxide phases.

Table 2. Electropositive promotion of PGMs-catalysed alkenes and alkanes oxidation.

Reactants	Catalyst and (Promotion method applied)	Promoter	Reaction conditions	Promotion highlights and optimal achievements	Ref.
Alkenes oxidation (Alkene + O_2)					
C_2H_4, O_2	Pt-film over (Na)$\beta''Al_2O_3$ (EPOC)	Na	T = 291 °C; $[C_2H_4]$ = 0.021%, $[O_2]$ = 5%; θ_{Na} = 0–3%	✓ Only rate inhibition was found at such extreme fuel lean conditions. ➢ A Na coverage of θ_{Na} > ~1.5% suffices to cause a 70% rate inhibition (poisoning).	[42]
C_2H_4, O_2	Pt film over (Na)$\beta''Al_2O_3$ solid electrolyte (EPOC)	Na	T = 300–470 °C; $[C_2H_4]$ = 4.2%, $[O_2]$ = 8 and 16.2%; θ_{Na} = 0–35%	✓ Na induces both promotion and poisoning. Moderate θ_{Na}, typically<8%, causes promotion, larger θ_{Na} causes rate inhibition. Results obtained on a Pt(111)-Na model catalyst mirror the behaviour of the EPOC system. ✓ No promotional and/or poisoning effects occur at high temperatures (T > ~460 °C). ➢ ρ~2 with θ_{Na} = 8% at 280 °C and $[C_2H_4]$ = 4.2%, $[O_2]$ =1 6.9%; ρ~2 with θ_{Na} = 8% at 352 °C, $[C_2H_4]$ = 4.2%, $[O_2]$ = 8%	[18]
C_3H_6, O_2	Pt film over (Na)$\beta''Al_2O_3$ solid electrolyte (EPOC)	Na	T = 340 °C $[C_3H_6]$ = 0.62%, $[O_2]$ = 0–7%	✓ Volcano type promotion optimized at moderate Na coverages. Promotion effects increase with decreasing $[O_2]/[C_2H_4]$ (promotion is superior under fuel-rich conditions). High Na coverages deactivate catalyst due to excessive accumulation of 3D Na_2CO_3 species; these species can be readily removed electrochemically, accomplishing catalyst re-activation. ➢ P = 1.1 at $[O_2]/[C_2H_4]$ = 5.7, T = 340 °C; ρ = 2.2 at $[O_2]/[C_2H_4]$ = 4.0, T = 340 °C; ρ = 2.3 at $[O_2]/[C_2H_4]$ = 2.0, T = 340 °C	[20]

Table 2. *Cont.*

Reactants	Catalyst and (Promotion method applied)	Promoter	Reaction conditions	Promotion highlights and optimal achievements	Ref.
C_3H_6, O_2	Pt film over NaSICon ($Na_3Zr_2Si_2PO_{12}$) solid electrolyte (EPOC)	Na	T = 300 °C, θ_{Na} = 0–6% - C_3H_6/O_2 = 0.04%/0.2% (stoich.) - C_3H_6/O_2 = 0.04%/8.3% (O_2-rich)	✓ Strong promotion by Na under near-stoichiometric C_3H_6/O_2 conditions; no promotion or moderate poisoning under oxygen excess (lean-burn) conditions. Under reaction conditions promoter phase exists as Na_2CO_3 and $NaHCO_3$. ➢ ρ = 3.5 with θ_{Na} = 3.6% at $[C_3H_6]/[O_2]$ = 0.04/0.2, T = 300 °C; ρ = 0.8 with θ_{Na} > ~2% at $[C_3H_6]/[O_2]$ = 0.04/8.3, T = 300 °C	[43] [44]
C_3H_6, O_2	Pt film over (K)βAl_2O_3 solid electrolyte (EPOC)	K	T=190–310 °C $[C_3H_6]$ = 2000 ppm, $[O_2]$ = 1–7%	✓ Strong promotion by K (better than that of Na) under both near-stoichiometric and oxygen-rich conditions. ✓ Formation of stable and more effective promoter phases (potassium oxides and superoxides). ➢ ρ = 7, propene conversion increased form 10% to 70%, at 200 °C and $[C_3H_6]/[O_2]$ = 2000ppm/7% (O_2–rich) ➢ ρ = 3 at 220 °C and $[C_3H_6]/[O_2]$ = 2000ppm/1% (near-stoichiometric).	[45] [46]
C_3H_6, O_2	0.5 wt% Pt/γ-Al_2O_3 (CCP)	Na	T = 100–500 °C $[C_3H_6]$ = 1000 ppm, $[O_2]$ = 5% Na-loadings: 0, 1.6, 2.6, 4.2 wt%	✓ Volcano type promotion under excess oxygen conditions (optimal Na-loading at ~2.6wt% Na). ✓ C_3H_6 oxidation is inhibited for higher Na loadings; poisoning becomes even worse in case of NO co-feed. ➢ ΔT_{50} = −60 °C at $[C_3H_6]/[O_2]$ = 0.1%/5% (O_2-rich conditions) with 2.6wt% Na (θ_{Na}~23%); ρ~10 at T = 120–200 °C.	[47]

Table 2. *Cont.*

Reactants	Catalyst and (Promotion method applied)	Promoter	Reaction conditions	Promotion highlights and optimal achievements	Ref.
Alkanes oxidation (alkane + O_2)					
C_3H_8, O_2	Pt film over (Na)$\beta''$$Al_2O_3$ solid electrolyte (EPOC)	Na	T = 320–440 °C $[C_3H_8]$ = 0.2%, $[O_2]$ = 1% θ_{Na} = 0–3%	✓ Na addition to the catalyst causes only poisoning of the rate. Low θ_{Na} = 1–3% are enough to cause up to 3 orders of magnitude rate poisoning (ρ~0.002). Under reaction conditions Na forms carbonate, bicarbonate and oxide phases. ➢ Strong poisoning: ρ~0.002 for θ_{Na}>~2.5% at T = 440 °C.	[48] [49]
CH_4, O_2	5 (or 2.5) wt% Pd/γ-Al_2O_3 (CCP)	Na, K, Ba, Ca	T = 250–600 °C; $[CH_4]$ = 500 ppm, $[O_2]$ = 8%, $[H_2O]$ = 5%, $[CO_2]$ = 5% Loadings: Ca: 0.38, 0.76, 1.5 wt%; Ba = 5 wt%; K = 1.48 wt%; Na = 0.87 wt%	✓ Both beneficial and detrimental effects on CH_4 oxidation were found depending on the promoter, catalyst composition and pre-treatment. Promotion was achieved with Ca-doping and after cooling in inert atmosphere. ✓ No promotion (even inhibition) was observed by Na and K addition at the specific loadings used. ➢ Highest activities for 2.5wt%Pd/γ-Al_2O_3 and 5wt%Pd/γ-Al_2O_3 were obtained with 0.38 and 0.76 wt% Ca addition, respectively, that is, at a weight ratio Ca:Pd equal to 0.15 in both cases.	[50]

Olsson and co-workers [50] studied the effect of the addition of alkalis (Na, K) and alkaline earths (Ba, Ca) on Pd/Al_2O_3 catalyst for methane combustion (CCP). Both beneficial and detrimental effects on methane oxidation were observed. High loadings of alkaline modifiers caused reaction inhibition. No promotion was induced by Na and K modifiers, while Ba- and Ca- doping caused reaction inhibition. However, when using a specific catalyst pre-treatment, which leads to the removal of surface hydroxyl species, promotional effects of Ca were recorded at Ca loadings where they previously observed reaction inhibition. The Ca/Pd ratio was found to be critical for the aforementioned promotion. A Ca/Pd ratio of 0.15 was favourable. The modification of the redox properties of palladium and the formation of PdO were considered the origin of this promotion, which overcomes the alkaline-induced inhibition of methane adsorption (activation) on increased electronic density catalyst particles surfaces which occurs in parallel.

Comparatively overviewing the literature findings of Section 3.1.2 summarized in Table 2, the following general features and conclusions can emerge.

Alkali addition on PGMs-based catalyst is detrimental for alkanes oxidation reactions that was attributed to the low adsorption propensity of alkanes on PGMs surfaces which becomes worse due to the alkali induced promotion of oxygen adsorption and the concomitant O-poisoning of the surfaces.

On the opposite, alkenes oxidation on PGMs is substantially promoted by electropositive promoters, attributed to the opposite trend of their adsorption compared to alkanes: The alkali-induced enhancement on oxygen adsorption counterbalances its coverage on a PGM surface, predominantly covered by the alkene molecules on the unpromoted surface, resulting to an enhancement on reaction probability (promotion).

The alkenes oxidation followed volcano type promotion upon increasing promoter loading, resulting from the competitive adsorption of the reactants (Langmuir-Hinshelwood type reaction) and the formation of 3D alkali carbonates and their excessive accumulation on the surface at high promoter loadings.

The main promotion characteristics were qualitatively similar between EPOC and CCP studies, although the intensity of the promotion was larger in the latter case: maximum rate enhancement ratios (ρ) obtained were 3.5 and 10 by using EPOC and CCP, respectively. An ultimate contact between nanodispersed active phase particles and promoter species, which is favoured in CCP, is most probably responsible for this superiority.

3.1.3. NO Reduction by CO

The first study of the effect of electropositive promoters on the Pt-catalysed NO reduction by CO was reported by Palermo et al. [17]. The EPOC was applied by means of a Pt/Na^+-conducting $\beta''Al_2O_3/Au$ cell. The promotional effect of Na on the Pt/NO + CO system was strongly dependent on the reaction temperature, gas composition and Na coverage. For instance, at high CO/NO ratios the effect of Na was marginal and according to the authors this was attenuated by CO island formation and limited availability of chemisorbed NO. At low CO/NO ratios the effect of Na was also relatively small, because, as authors conclude, the CO + O reaction is limited by the low CO coverage. However, at the intermediate CO/NO regime, where the above constrains do not apply, Na strongly accelerates the reaction: rate gains of up to ρ = 13 for N_2 production compared to the unpromoted Na-free Pt rate were achieved (Table 3). Significant increases to N_2 selectivity of the system (up to 30%) were also obtained (Figure 9). Significant progressive changes in the apparent activation energy of the NO + CO reaction upon varying the promoter coverage, similar to that obtained with the Pt/O_2 + CO catalytic system [16], were also recorded in this study. The observed promotional phenomena, on both rate and selectivity, were attributed to the enhanced NO versus CO chemisorption and the extended dissociation of the chemisorbed NO molecules, both induced by Na-promoter.

Table 3. Electropositive promotion of PGMs-catalysed NO reduction by CO (NO + CO).

Reactants	Catalyst and (Promotion method applied)	Promoter	Reaction conditions	Promotion highlights and maximum achievements	Ref.
NO, CO	Pt film over (Na)$\beta''Al_2O_3$ solid electrolyte (EPOC)	Na	T = 300–400 °C [CO] = 0–1.5%, [NO] = 0–1.5% [NO]/[CO] = 0.75%/0.75% θ_{Na} = 3–20%	✓ The promotional effects of Na are maximized at intermediate CO/NO values, while at high and low CO/NO values promotion was attenuated due to the formation of CO islands and the low CO coverage, respectively. ✓ Significant Na-induced changes on the reaction apparent activation energy recorded. ➢ $\rho_{N_2} \sim 13$, $\rho_{CO_2} \sim 4.8$, $\Delta S_{N2} \sim +45\%$ at T = 348 °C for θ_{Na} = 3–20%	[17]
NO, CO	Rh film over (Na)$\beta''Al_2O_3$ solid electrolyte (EPOC)	Na	T = 307 °C [CO] = [NO] = 1%	✓ Sodium coverage (θ_{Na}), catalyst work function ($\Delta\Phi$) and catalyst potential (V_{WR}) are directly correlated; θ_{Na} and $\Delta\Phi$ varying linearly with V_{WR}. Low sodium coverages ca. 1–2% are able to create significant promotion of the Rh-catalysed NO+CO (but less than that obtained on Pt catalysts). ➢ $\rho_{N_2} \sim 3.1$, $\rho_{CO_2} \sim 1.4$, $\rho_{N_2O} = 0.3$, $\Delta S_{N_2} \sim +50\%$ (24%→80%) at T = 307 °C, for θ_{Na}~1–2%	[51] [52]
NO, CO	0.5 wt% Pt/γ-Al_2O_3 (CCP)	K, Rb, Cs	T = 150–500 °C [CO] = 0–3%, [NO] = 0–3% Promoter loadings used: wt% Rb: 1.9, 9.7, 15.5 wt% Cs: 9.0, 15.0, 24.0 wt% K: 2.7, 4.4, 8.3	✓ All K, Rb, Cs alkalis exert a pronounced promotion; however, Rb promotion is superior to K and Cs. ✓ In comparison with Rb, larger molar loadings of K were needed to produce similar reductions in light-off temperature. In both cases promotional effects increased monotonically with increasing alkali loading. ➢ $\rho_{N_2} \sim 110$, $\rho_{CO_2} \sim 45$, $\rho_{N_2O} = 7$, $\Delta S_{N_2} = +58\%$ (24%→82%) at T = 350 °C, [CO] = 0.5%, [NO] = 1%; 15.5 wt% Rb ➢ $\rho_{N_2} \sim 45$, $\rho_{CO_2} \sim 15$, $\rho_{N_2O} = 4$, $\Delta S_{N_2} = +51\%$ (24%→75%) at T = 350 °C, [CO] = 0.5%, [NO] = 1%; 8.8 wt% K ➢ $\Delta T_{50} = -150$ °C with Rb loading >9.7 wt%; $\Delta T_{50} = -115$ °C with K loading > 4.4 wt%; $\Delta T_{50} = -115$ °C with Cs loading > 15 wt%, at [NO] = [CO] = 1000 ppm.	[53]

Table 3. *Cont.*

Reactants	Catalyst and (Promotion method applied)	Promoter	Reaction conditions	Promotion highlights and maximum achievements	Ref.
NO, CO	0.5wt%Pd/YSZ (CCP)	Na	T = 352 °C [CO] = 0–4%, [NO] = 0–4% Na loadings: 0.017, 0.034, 0.068 and 0.102 wt% (or 6/1, 3/1, 1.5/1 and 1/1 Pd/Alkali)	✓ Volcano type promotion was obtained with an optimal Na-loading equal to 0.034wt% Na (i.e., Pd/alkali molar ratio = 3/1), taking into account both activity and N_2- selectivity. ➢ $\rho_{N_2} \sim 2$, $\rho_{CO_2} \sim 1.5$, $\rho_{N_2O} = 1$, $\Delta S_{N_2} = +20\%$ (from ~50% to ~70%) at T = 352 °C, [CO] = [NO] = 1% for 0.034 wt% Na.	[54]
NO, CO, O_2	1 wt% Pd/(γ-Al_2O_3 or CZ) (CCP)	Ba	T = 50–300 °C [CO] = 2.2%, [NO] = 0.2% [O_2] = 1% Ba-loadings: 0, 1, 5 and 15 wt% Ba	✓ Ba-induced promotional effects on Pd/γ-Al_2O_3; progressively increased activity with increasing Ba-loading. ✓ Ba-induced poisoning effects on Pd/CZ; progressively decreased activity with increasing Ba-loading. ➢ $\Delta T_{50(CO)} = -40$ °C and $\Delta T_{50(NO)} = -33$ °C with 15wt% Ba on Pd/γ-Al_2O_3 (promotion) ➢ $\Delta T_{50(CO)} = +97$ °C and $\Delta T_{50(NO)} = +99$ °C with 10wt% Ba on Pd/CZ (strong poisoning)	[55]

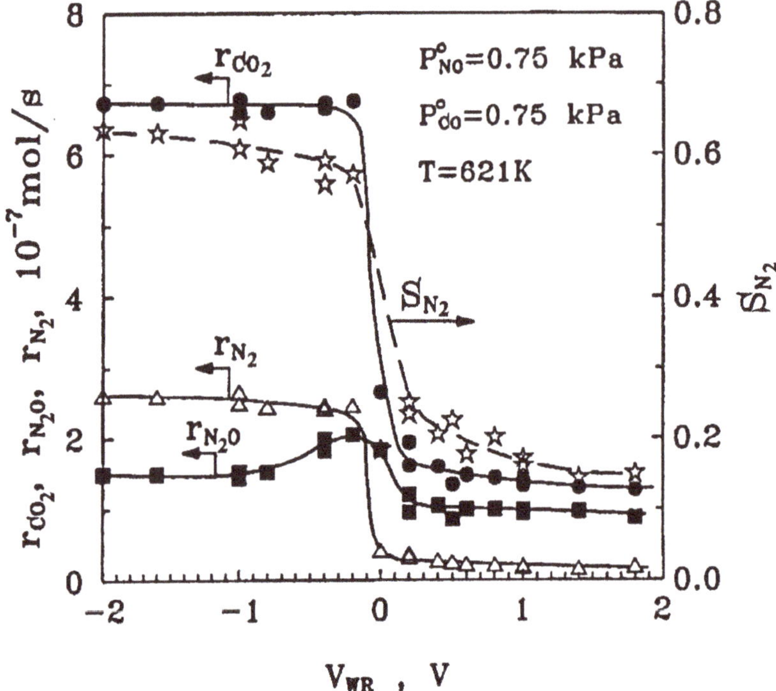

Figure 9. Effect of catalyst potential on the CO_2, N_2 and N_2O formation rates and the selectivity towards N_2 during the EPOC by Na of the Pt-catalysed NO + CO reaction. Conditions: [NO] = [CO] = 0.75%, T = 348 °C. Data was acquired in a Pt/(Na)β″Al_2O_3/Au galvanic cell. (Reprinted with permission from Ref. [17]; Copyright 2005, Elsevier).

Another EPOC (by Na) study using a Rh/(Na)$\beta''$$Al_2O_3$/Au galvanic cell was performed by Williams et al. [51,52]. They showed that the NO reduction by CO can also be promoted over Rh catalyst. Rate enhancement ratios of $\rho_{N_2} = 3.1$, $\rho_{CO_2} = 1.4$ and $\rho_{N_2O} = 0.3$ were observed at T = 307 °C and stoichiometric CO/NO gas reaction conditions ([NO] = [CO] = 1%) for a sodium coverage as low as 2% (Table 3), together with a large increase in N_2-selectivity: from 24% on the electrochemically cleaned Na-free Rh to 80% on Rh electrochemically dosed with θ_{Na} = 2%. These promotional effects were attributed to the Na-induced NO dissociation, which was considered as the rate limiting step for the NO reduction by CO process. Using XPS and work function ($\Delta\Phi$) measurements the authors also demonstrated that the Na coverage on the catalyst surface can be controlled through the externally imposed catalyst potential; a linear relationship of catalyst work function and θ_{Na} with catalyst potential was experimentally demonstrated in this study [51,52].

Konsolakis et al. [53] studied the NO + CO reaction on (K, Rb and Cs)-promoted, conventional (CCP), highly dispersed Pt/γ-Al_2O_3 catalysts, over a wide range of temperature (ca. 150–500 °C), partial pressures of reactants and promoter loadings (Table 3). All these alkalis were found to be very active promoters for the reaction, however, optimal promotion was achieved by rubidium: rate increases by factors as high as $\rho_{N_2} = 110$ and $\rho_{CO_2} = 45$ for the production of N_2 and CO_2, respectively, relative to alkali-free Pt were obtained, accompanied by substantial increases in N_2-selectivity by up to ~60% (Figure 10) [53]. The effects of K and Cs-promotion mirror those of Rb-promotion but they were less pronounced. Light-off measurements in the temperature range of 150–500 °C and under stoichiometric conditions (1000 ppm NO + 1000 ppm CO) showed that the NO conversion performance of alkali-free Pt is very poor (never exceeding 60%), whereas the 0.5wt%Pt(Rb)/γ-Al_2O_3 catalyst promoted with 9.7wt% Rb (optimally promoted catalyst), exhibited 100% NO and CO conversions with 100% N_2-selectivity at ~350 °C. In addition, 115 °C decrease on the NO light-off temperature (T_{50}) was obtained for K and Cs, while a larger decrease in T_{50} by ~150 °C was obtained by Rb-promotion (Figure 11).

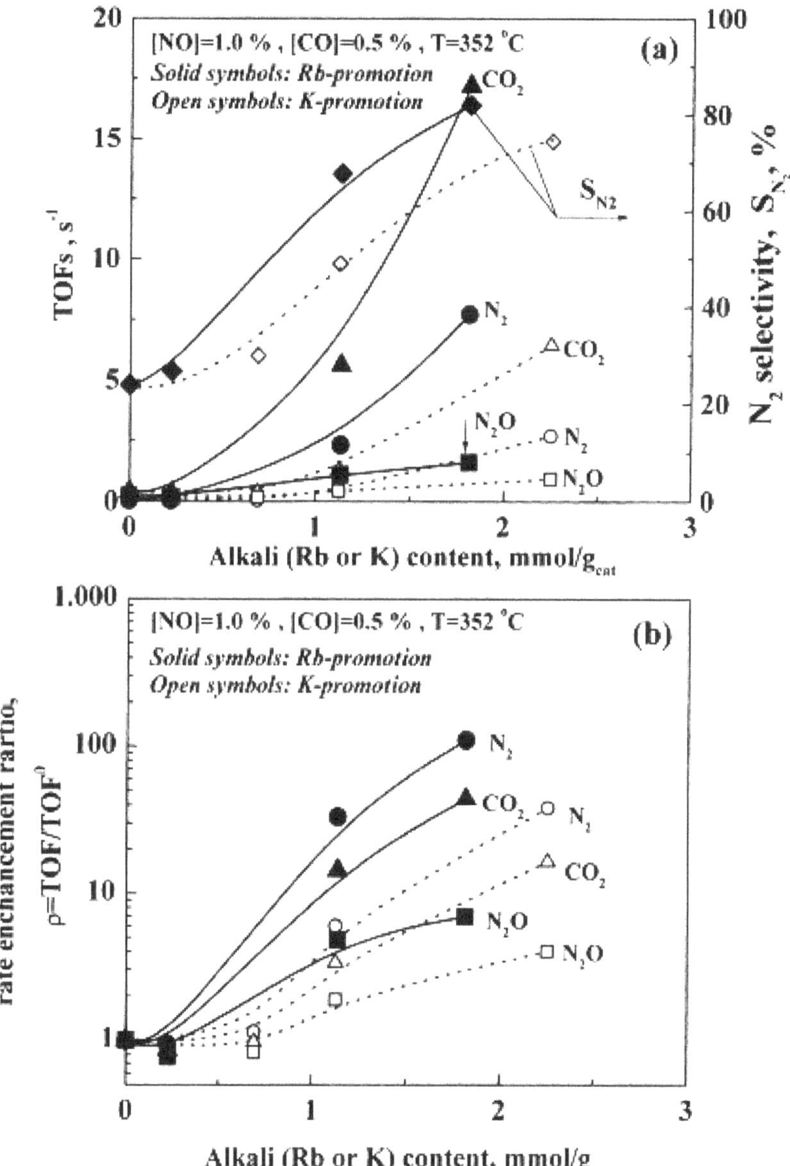

Figure 10. (**a**) Effect of Rb or K content of Pt/γ-Al_2O_3 catalyst on the turnover (TOF) formation rates of N_2, CO_2 and N_2O and on N_2-selectivity at T = 352 °C, [NO] = 1%, [CO] = 0.5%. (**b**) Corresponding enhancement ratio $\rho = TOF/TOF^0$ values. Solid lines and symbols refer to Rb-promotion, dashed lines and open symbols refer to K-promotion. (Reprinted with permission from Ref. [53]; Copyright 2001, Elsevier).

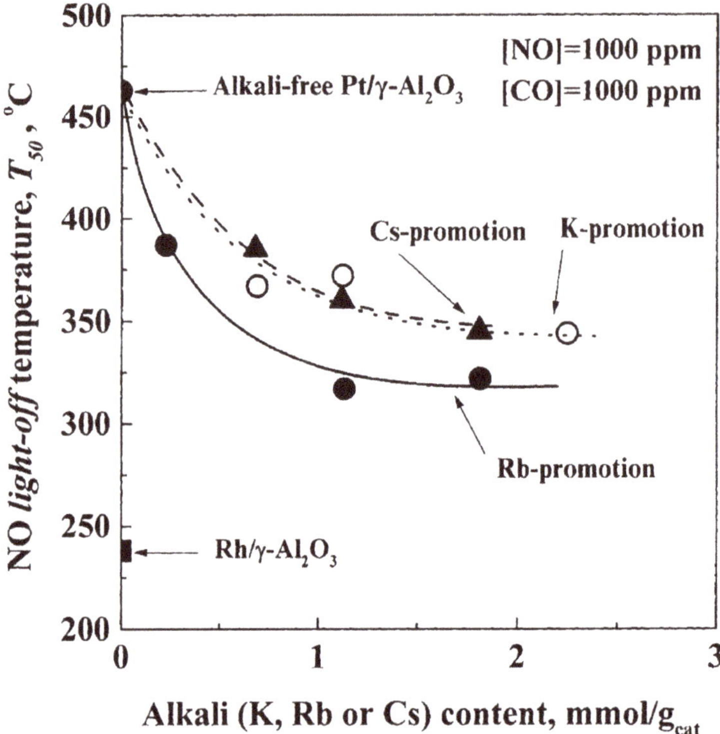

Figure 11. Effect of K-, Rb- or Cs-promotion on the NO light-off temperature (T_{50}) for Pt/NO + CO system. Conditions: [NO] = 1000 ppm, [CO] = 1000 ppm, total gas flow rate F_t = 80 cm^3STP/min; catalyst weight 8(±0.2) mg. ($w/F_t = 6 \times 10^{-3}$ g s/cm^3). (Reprinted with permission from Ref. [53]; Copyright 2001, Elsevier).

The observed promotional effects were ascribed to alkali-induced changes in the chemisorption bond strengths of CO, NO and NO dissociation products. The superior promotion of Rb compared to that of K was attributed to its larger size and thus to the greater electric field experienced by co-adsorbed species located at an adjacent site. On the other hand, the unexpected lower efficiency of Cs promotion compared to that of Rb was attributed to a change in the adsorption site of NO due to the very large Cs ion [53].

More recently Konsolakis and Yentekakis [54] reported on the promotional effect of Na on the Pd-catalysed NO + CO reaction (CCP). A series of Na-promoted Pd catalysts were prepared on yttria-stabilized zirconia (YSZ; 10mol% Y_2O_3/ZrO_2) carrier (i.e., Pd(Na)/YSZ), with different loadings of alkali-promoter. Both promoting and poisoning effects, regarding the Na-loading, were found in a wide range of CO/NO ratios. For example at equimolar CO/NO composition ([CO] = [NO] = 1%) and T = 352 °C, the optimal promoter loading was found to be between 0.02–0.03wt% of Na (corresponding to a nominal sodium coverage of θ_{Na}~3%; Equation (4) with $d_{Na} = 3.314 \times 10^{19}$ atoms/m^2) which leads to N_2 and CO_2 formation rate enhancements by up to ~200%, accompanied by a significant improvement in N_2-selectivity of about 20% in comparison to the un-promoted Pd/YSZ catalyst (Table 3). The promotional effects were interpreted in a similar manner to the aforementioned studies [17,51–53].

Tanikawa and Egawa [55] studied the effect of Ba-promoter on NO + CO reaction in the presence of O_2 over Pd dispersed on γ-Al_2O_3 and ceria-zirconia supports (CCP). They found that the activity of Pd/γ-Al_2O_3 was gradually enhanced by increasing the Ba loading up to 15wt%; the light-off temperatures (T_{50}) for CO oxidation and NO reduction were decreased by 40 and 33 °C, respectively, over the optimally modified Pd(15%Ba)/γ-Al_2O_3. However, Ba addition on Pd/CZ catalyst decreased its activity for both NO reduction and CO oxidation reactions (Table 3).

Finally, Lepage et al. [56] studied the effect of electropositive promoters (monovalent H^+, Na^+, Rb^+ and Cs^+ or divalent Mg^{2+}, Ca^{2+}, Sr^{2+} and Ba^{2+} cations) over Rh nanoparticles supported on a series of Y zeolites for the NO reduction by CO by means of infrared spectroscopy of NO adsorption to monitor the electronic modifications of Rh nanoparticles. The observed decrease in NO reduction ignition temperature (T_{10}) with increasing ionic radius-to-charge ratio was related to the degree of electron transfer from Rh particles to the adsorbed NO molecules. The shift of the IR band of linearly adsorbed NO to lower wavenumbers with increasing the radius/charge ratio of promoter elements substantiated this argument.

A comparative overview of the literature involved in this section and in Table 3 gives prominence to the following general features: (i) Promotion either induced by EPOC or by CCP is substantial and has qualitatively similar characteristics, although CCP leads to much larger rate enhancements; ρ values in the region of 45–110 and around 13 were obtained by using CCP and EPOC, respectively. The ultimate contact and thus the higher interaction of the well dispersed active phase particles with the promoter species may be the cause of the mentioned differences on promotion intensity. (ii) Heavier alkalis were typically more effective than lighter alkalis, with Rb being the best. (iii) Among Pt, Pd and Rh precious metals, alkali promotion was significantly more effective on Pt than on Rh and Pd. (iv) Alkali-induced enhancement of the dissociative chemisorption of NO was considered as a key factor for the interpretation of the aforementioned pronounced promotion.

3.1.4. NO Reduction by Hydrocarbons or H_2

(i) NO reduction by alkenes

NO reduction by hydrocarbons under in situ electrochemical promotion (EPOC) by alkalis was firstly demonstrated by Harkness and Lambert [57] and Yentekakis et al. [58]. Using ethylene as reductant, Harkness and Lambert [57] showed that the rate of the NO + ethylene reaction over a Pt film interfaced with Na^+-conducting $\beta''Al_2O_3$, linearly increased with θ_{Na} al low and intermediate sodium coverages (typically $\theta_{Na} < 0.3$), achieving a plateau for higher values. Rate enhancement values ρ up to 15 were obtained (Table 4).

On the other hand Yentekakis et al. [58] showed very pronounced NO reduction rate enhancements (using propene as the reductant) for the Pt-catalysed NO+propene reaction upon Na promotion over similar Pt/(Na)$\beta''Al_2O_3$ electrocatalyst. This effect was studied at 375 °C in a wide range of P_{NO} (0–7 kPa) and $P_{C_3H_6}$ (0–0.4 kPa) partial pressures. The promotional effect of Na was found to be significant in the whole range of conditions studied. The coverage of the promoting species was determined via galvanostatic transients. These transients showed that the promotional effects maximized at a Na coverage of θ_{Na}~0.4, for which value the maximum enhancements in CO_2 and N_2 rates were ~7 times larger than the rate of the unpromoted Pt, while the N_2-selectivity increased from 60% on the clean Pt surface to 80% on the optimal Na-promoted Pt surface (Table 4). Higher sodium coverages ($\theta_{Na} > 1$) caused a gradual decrease of the promoting phenomena. This suggests that nominal sodium coverages greater than one exhibit activity much higher than clean Pt because a substantial amount of Na is present as 3D crystallites, as demonstrated by XPS. A Na-induced strengthening of the NO chemisorpting bond relative to propene was indicated by the kinetic data. The accompanied weakening of the N–O bond and the concomitant extended NO dissociation were considered for the interpretation of the observed promotional effects. XPS data confirmed that under reaction conditions the promoter phase consists of a mixture of $NaNO_2$ and $NaNO_3$ 3D crystallites. At high Na loadings these surface promoter compounds inhibit reaction due to active sites blocking phenomena.

EPOC was also used by Williams et al. [59] for the study of NO reduction by propene over Rh film catalyst interfaced with (Na)$\beta''Al_2O_3$. They found that Na pumping onto Rh surface at a coverage as low as ~ 2% causes a very large increase in the N_2-selectivity from 45% on the Na-free Rh surface to 82% on the Na-promoted surface together with beneficial effects on the reaction rate: rate enhancement ratio values of ρ = 2.4, 1.7 and 0.4 were recorded for the N_2, CO_2 and N_2O productions, respectively (Table 4). In a subsequent publication [60] the authors examined the impact of O_2 co-feed. It was found that upon increasing [O_2], the promotional effects of Na were progressively attenuated and finally reversed to poisoning for [O_2] = 2%. Similar considerations as in Reference [57] and [58] were used for the interpretation of the promotional effects, as well.

Using CCP, the group of Yentekakis [14,54] showed that the NO + propene reaction over Pd dispersed on 8mol% Y_2O_3 stabilized ZrO_2 support (YSZ) can be substantially promoted by dosing the catalyst with alkalis. In Reference [14], the rate increased by an order of magnitude accompanied by selectivity increases from ~75% over the Na-free Pd/YSZ catalyst to >95% over the optimally Na-promoted Pd/YSZ catalyst in the temperature interval of 250–450 °C. The promotional effects were found to be optimized at a sodium content of 0.06wt% (corresponding to nominal sodium coverage of ~7%; Equation (4)), while over-promotion attenuated the rate (volcano type behaviour). These promotional effects caused a decrease in the NO light-off temperature (T_{50}) of about 110 °C between the optimally promoted and unpromoted Pd catalyst (Table 4). A strengthening of the chemisorption bond of NO relative to propene that facilitates weakening of the N–O bond of the adsorbed NO molecules and thus their dissociative chemisorption was considered as the origin of the promotional effects. In ref. [54] the effects of Li, Na, K and Cs alkalis on Pd-catalysed NO + C_3H_6 reaction were comparatively studied: Four catalysts dosed with the same alkali/Pd = 1/1.5 atom/atom loading were prepared and their behaviour was compared in a wide range of [NO] at constant [C_3H_6] = 0.8% and T=380 °C. All these catalyst were found significantly better in terms of activity and selectivity in comparison to the unpromoted Pd/YSZ, in the whole range of conditions investigated; Na provided the most significant promotional effects (Table 4).

Systematically expanding their studies on conventional promoted catalysts (CCP), Yentekakis and co-workers showed that Pt nanoparticles dispersed on γ-Al_2O_3 are subjected to extraordinarily effective promotion by alkalis (Li, Na, K, Rb and Cs) [15,61] and alkaline earths (Ba) [62,63] in both activity and N_2-selectivity during the NO+propene reaction in the temperature interval of TWCs interest (ca. 200–550 °C). In References [15,61] rate increases by two orders of magnitude were achieved, while the selectivity towards N_2 was improved from ~20% over the alkali-free unpromoted Pt catalyst, to >95% over the optimally alkali-promoted catalysts. Best promotional effects were achieved by Rb-promotion, for which rate increases as high as 420-, 280- and 25-fold were obtained for the formation rates of N_2, CO_2 and N_2O, respectively, in comparison to the performance of the unpromoted Pt/γ-Al_2O_3 catalyst (Table 4). Volcano type behaviour was found to be followed upon increases of promoter loadings for all cases. Thus, promotion was found to be maximized for a Li content of 1.25wt%, a K content of 7.1wt%, a Rb content of9.7 wt% a Cs content of 15wt% (Figure 12) [61] and a Na content of 4.18wt% [15].

This very spectacular promotion in light-off diagrams (see for example Figure 13) showed that the marginal (~10%) NO conversion efficiency of un-promoted Pt/γ-Al_2O_3 can be readily proved by alkalis to values up to 100% in the temperature window of TWCs interest, offering at the same time selectivities towards N_2 >95% [15,61]. However, besides temperature, NO_x removal efficiency of automotive catalytic converters is greatly affected by the presence of oxygen in the exhausts gases, in particular under excess O_2 conditions. Therefore, the NO + C_3H_6 + excess O_2 reaction is of higher practical significance than the NO + C_3H_6 one. Hence, it was decided to analyse this reaction and the effect of O_2 in a distinguishable section of the article (Section 3.2).

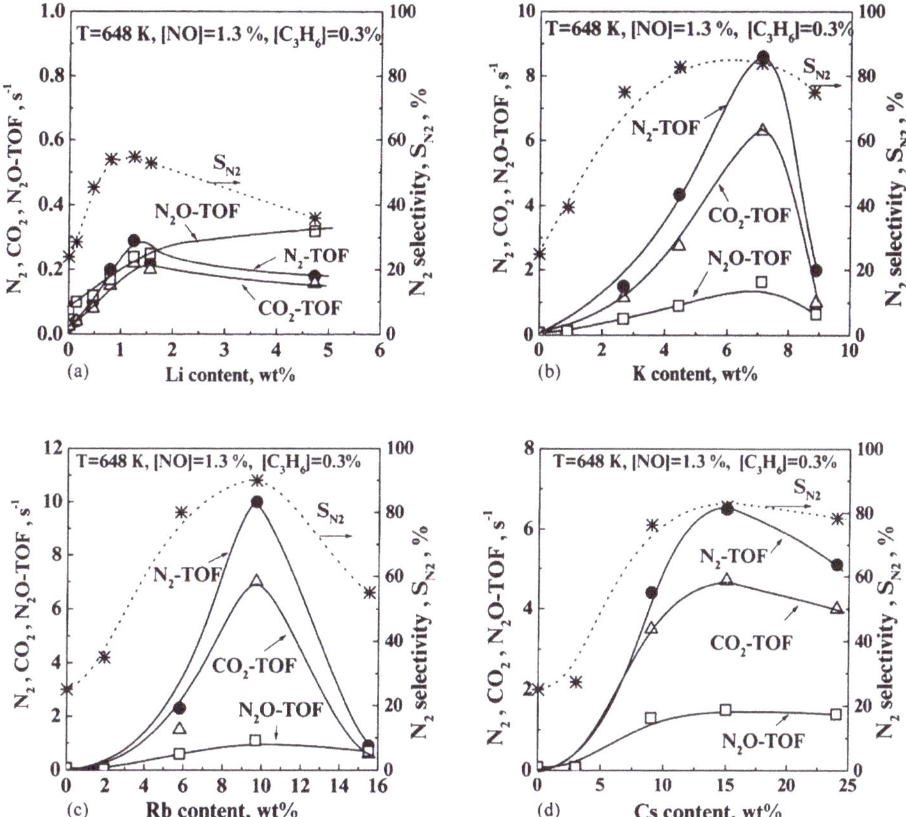

Figure 12. The effect of Li (**a**), K (**b**), Rb (**c**) and Cs (**d**) content of 0.5wt%Pt/γ-Al_2O_3 catalyst on the turnover (TOF) formation rates of N_2, CO_2 and N_2O and on N_2-selectivity. Conditions: T = 375 °C, [NO] = 1.3%, [C_3H_6] = 0.3% (Reprinted with permission from Ref. [61]; Copyright 2001, Elsevier).

In the case of Ba-promotion the optimal Ba-loading was found to be 15.2wt% Ba (Table 4) [62]. Using a 1000 ppm NO + 1000 ppm C_3H_6 feed composition (these concentrations are close to those encountered in practical applications and commonly used in the literature as referred conditions) the authors showed that the very poor NO light-off behaviour of Ba-free Pt/γ-Al_2O_3 catalysts, which gave a lower than 10% NO conversion in the whole 200–550 °C temperature interval, can be dramatically improved (in a similar manner to alkali-promotion [15,61]), achieving 100% NO conversion at relatively low temperatures accompanied by N_2-selectivities near to 100%. Further studying the Ba-promoted Pt/γ-Al_2O_3 system, the authors focussed on the effect of residual chloride, which concerns catalysts prepared from chlorine-containing precursors [63]; the detrimental effect of residual chloride on deNOx catalysis is well known. Very strikingly, it was found that when Ba is used to promote catalysts prepared from chloride-containing precursors, all the original chloride is retained, yet the catalysts exhibit very strong promotion and in every aspect their behaviours are identical to that of the chloride-free Ba-promoted catalyst [63]. This significant from the practical point of view attenuation of the inhibiting effect of chloride was rationalized in terms of formation of a stable 2D $BaCl_2$ phase on the catalysts surface, where the promoting effect of Ba overwhelms the poisoning effect of Cl and a large net promotion is resulted.

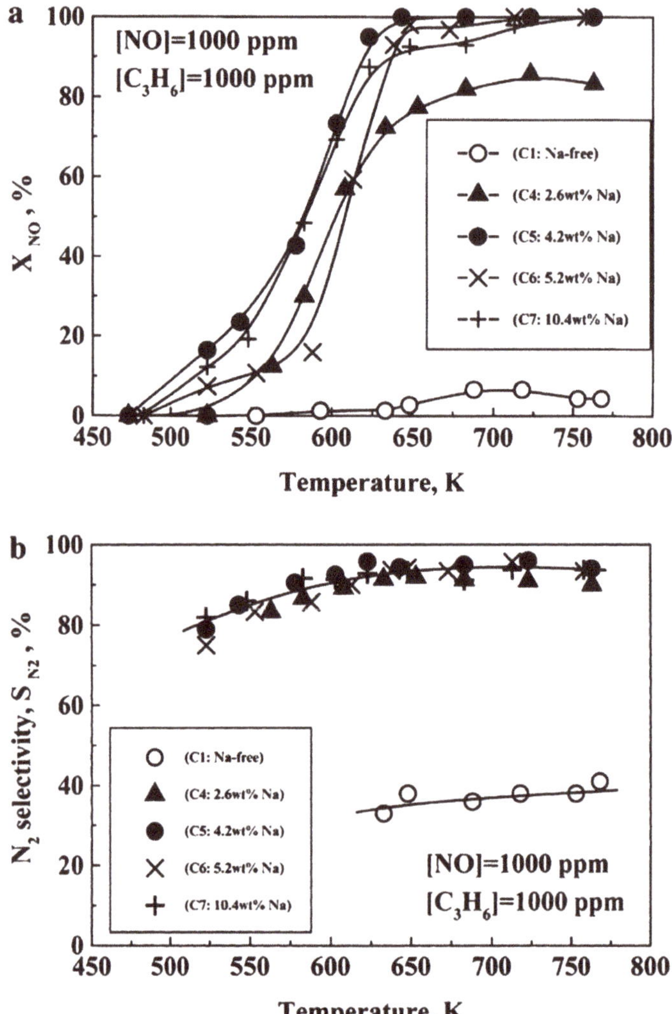

Figure 13. The conversion of NO (**a**) and the corresponding selectivity towards N_2 (**b**) for Na-promoted 0.5wt%Pt/γ-Al_2O_3 catalysts, as a function of temperature at constant reactor inlet conditions: [NO] = 1000 ppm, [C_3H_6] = 1000 ppm; total flow rate F_t = 80 cm^3STP/min; catalyst weight 8(±0.2) mg. (Reprinted with permission from Ref. [15]; Copyright 1999, Elsevier).

Macleod et al. [64] studied the NO + C_3H_6 reaction over dispersed Rh/γ-Al_2O_3 catalysts (CCP). Rh is one of the noble metals used in TWCs and its use is related to the effective NO reduction in the three-way catalytic chemistry. They found that NO reduction chemistry on Rh can also be improved by alkalis. For instance, at 375 °C, a 7.3wt% Na content in the catalyst was found to be optimal promoter loading, leading to a ~3-fold increase in Rh activity (Table 4). Significant increases in N_2-selectivity, up to 90% were also obtained, when the unpromoted Rh/γ-Al_2O_3 appeared to have only 53%. In addition, Na was found to suppress de formation of CO and HCN. Higher Na loadings were poisoning the catalyst bellow to its initial Na-free performance. However, it is worth noticing the substantially lower degree of promotional effect of alkalis on the NO reduction chemistry of Rh compared to that observed for Pt and Pd.

(ii) NO reduction by alkanes

Yentekakis and co-workers [65,66] studied the reduction of NO by methane over Na-dosed Pd catalyst (CCP) dispersed on yttria stabilized zirconia (YSZ) at temperatures between 350–500 °C and a wide range of reactants concentrations ([NO] = 0–1.5%, [CH_4] = 0–25%). It was observed that the reaction clearly exhibits Langmuir-Hinshelwood type kinetics over Pd with characteristic rate maxima reflecting competitive adsorption of NO and methane; NO adsorption however was found to be much more pronounced than that of CH_4 within the temperature range investigated. Sodium was found to cause strong poisoning of the reaction at any loading due to a Na-induced oxygen poisoning of the catalyst (derived by NO decomposition) (Table 4). The very different response of alkali promotion for the NO reduction by alkanes (strong poisoning effects) and by alkenes (strong promotional effects) is attributed to the different trends of the adsorption of the two kinds of hydrocarbons (alkanes and alkenes) on the Pt-group metal surfaces [67,68].

It is worth noting the similarities on the effect of electropositive promotion of Pt-group metals (PGM) between the alkanes or alkenes oxidation by dioxygen (Section 3.1.2) and by NO (Section 3.1.4): In both cases when the oxidation process involves alkenes, electropositive promoters induce strong promotion, whilst when the reductant agent is an alkane, electropositive promoters lead to poisoning. This is understandable in view of the different trend of alkenes and alkanes adsorption on the PGM surfaces (strong and weak, respectively [67,68]) and the influence of the promotion on it [15,20,47,61,65,66] and the fact that in both cases the oxidant species are mainly based on adsorbed atomic oxygen derived by the dissociative chemisorption of molecular oxygen or NO [47].

(iii) NO reduction by H_2 or $CO+H_2$

Besides hydrocarbons, hydrogen is also one of the gases present in automotive exhaust stream. It could also be externally supplied, by the use of a hydrocarbon reformer or another source of safe in situ hydrogen generator, for the reduction of NO_x from stationary power sources and chemical plants. Notably, the NH_3-Selective Catalytic Reduction process (NH_3-SCR) which is currently on of the most performing technologies for deNO_x processes in stationary power applications and chemical plants, experiences drawbacks related to the storage and slip of ammonia, although it may be pointed out that urea (as an NH_3 carrier) is now commercially available for use on diesel truck and some lighter vehicles. These issues can explain the growing interest and efforts for the use of H_2 as a reducing agent of NO_x in the last years (e.g., [69–72]). Most of these studies have used Pt-group metals as catalysts.

The first EPOC work in this field was performed by Marina et al. [19], who studied the effect of Na on the Pt-catalysed NO reduction by H_2 with (Na)$\beta''Al_2O_3$ as a source of Na promoter species. Rate enhancement ratio values up to $\rho_{N_2} = 30$ were achieved for the formation of N_2, accompanied by significant increases in N_2-selectivity. At 375 °C, Na coverage of only 6% increased the N_2-selectivity from ~36% on the clean Na-free Pt surface to ~75% on the Na-promoted Pt surface (Table 4). The promotional effects of Na were applicable in a wide range of H_2/NO gas phase compositions (ca. 0.1–3.0) and temperature (ca. 300–450 °C).

Burch and Coleman [73] studied the effect of sodium for the selective catalytic reduction of NO by hydrogen in the presence of excess oxygen (lean-burn conditions) over Na_2O-modified conventional, highly dispersed Pt/Al_2O_3 and Pt/SiO_2 catalysts (CCP), at temperatures representative of automotive *"cold-start"* conditions (<200 °C). They found that small sodium loadings significantly increased the NO conversion rate while larger sodium loadings led to severe poisoning. Under oxygen excess and low temperatures (<~140 °C) the selectivity to N_2 was practically unaffected by the addition of sodium, while an adverse effect was found at higher temperatures (Table 4).

Machida et al. [74] investigated the effect of alkalis (Na, K, Cs) and alkaline earths (Mg, Ca, Ba) on Pt/ZSM5 catalysts (CCP) for the selective reduction of NO by H_2 under O_2 excess (0.08% NO, 0.08–0.56% H_2, 10% O_2). The optimum promotional effects were obtained with Na promoter at loadings of 10–15wt%, which exhibited NO_x conversion higher than 90% and N_2 selectivity of ~50%. Based on in situ Diffuse Reflectance Infrared Fourier Transform Spectroscopy (DRIFTS) studies, it was suggested that Na promotes the adsorption of NO as NO_2-type species, which would be an intermediate towards N_2 formation.

The H_2-SCR process over Pt for automotive emissions control, however, exhibits an important disadvantage: CO, which also coexists in exhaust stream, is strongly adsorbed on Pt surface causing inhibition of NO_x reduction. Pioneer CCP work by Lambert and co-workers [75–78] showed that Pd overcomes this poisoning effect of CO leading to good NO_x conversions at mixed H_2 + CO feeds. In addition, they found that Pd dispersed on 10wt% TiO_2-modified Al_2O_3 offers even more improved deNO_x efficiency in comparison to the Pd/Al_2O_3 catalyst. The effective temperature window of these new catalyst formulations was however relatively narrow (ca. 150–200 °C). In a similar manner, Yentekakis and co-workers [79] studied the lean NO_x reduction by H_2 + CO over potassium-modified Pd/Al_2O_3 and Pd/Al_2O_3-TiO_2 catalysts. They found that potassium is a very active promoter of the Pd/Al_2O_3 catalyst for H_2-SCR process leading to significant enhancements of both NO_x conversion activity and N_2-selectivity of Pd. A potassium loading of ~0.25 wt% was found to optimize the catalyst performance in both activity and selectivity, leading to a catalyst formulation that offered up to ~90% NO_x conversion with a N_2-selectivity of the order of >85% for the temperature range of 180–300 °C (Table 4).

(iv) Direct NO decomposition

Limited studies can be found on this topic, probably due to the necessity of high temperatures for the direct NO decomposition. Wang et al. [80] studied the direct decomposition of NO into nitrogen and oxygen over Na-modified Pd/Al_2O_3 catalysts (CCP) in a flow reactor. They found that the NO reduction activity of Pd was significantly promoted by pre-coating the Al_2O_3 support with NaOH before the impregnation with Pd. A loading of 13wt% NaOH on a 4.8wt% Pd/Al_2O_3 catalyst enhanced the nitrogen yield from 15% up to 60% at T = 627 °C (Table 4). The observed NO decomposition rate enhancement was attributed to the strengthening of the NO adsorption bond on Pd sites.

Overviewing the literature included in Section 3.1.4 and Table 4 the following general remarks can emerge:

- Alkanes oxidation, either using dioxygen (Section 3.1.2) or NO (this section) behaves similarly under electropositive promotion by alkalis: addition of alkali on PGM surfaces causes only poisoning. Interpretations given are similar in both cases, based on the low adsorption propensity of alkanes on PGM surfaces and on the O-poisoning of the surfaces resulted by the alkali-induced enhancement of NO dissociative chemisorption.
- On the opposite, electropositive promotion of the NO + alkenes reactions is very pronounced; apparently it is the greater observed, for emissions control catalysis model reactions reviewed in the present work, offering ρ values as high as ρ_{N_2} = 420 and ρ_{CO_2} = 280.
- Some studies performing one-by-one comparison of EPOC and CCP by means of Na promoted Pt/NO + C_3H_6 system have demonstrated the very close similarities on the promotion output characteristics of the two methods and have convincingly argued for the similar origin of promotion in both cases.
- Once again (see Section 3.1.3) the intensity of CCP was superior to that of EPOC, indicating that an intimate contact between active phase and promoter particles succeeded by nanostructured catalysts seems to play a significant role on the promotion intensity.
- Heavier alkalis were again (see Section 3.1.3) more effective than the lighter ones; Rb was the best promoter.
- Among Pt, Pd and Rh, alkali promotion was more effective on Pt, lower on Pd and even lower on Rh. As we shall see (Section 3.3) this is related to the initial, unpromoted propensity of each metal on the dissociative adsorption of NO.
- Promotion by alkalis was also found to be substantial on the NO reduction by H_2 as well as on the direct NO decomposition reaction, although only few studies appeared in the open literature so far.
- Catalytic activity promotion was always accompanied by substantial increases on the system selectivity towards N_2.

Table 4. Electropositive promotion of PGMs-catalysed NO reduction by hydrocarbons (alkenes or alkanes), H_2 or H_2 + CO.

Reactants	Catalyst and (Promotion method applied)	Promoter	Reaction conditions	Promotion highlights and maximum achievements	Ref.
NO reduction by alkenes (NO + Alkene)					
NO, C_2H_4	Pt film over (Na)$\beta''Al_2O_3$ solid electrolyte (EPOC)	Na	T = 275–450 °C [NO] = 0.5–1%, [C_2H_4] = 1–3% θ_{Na} = 0–40%	✓ Significant promotion in a wide range of gas compositions (C_2H_4/NO = 1–6); in all compositions promotion was monotonically increased by increasing θ_{Na} up to 40%. Significant changes in apparent activation energy by varying θ_{Na}. ➢ ρ_{CO_2} = 15 for θ_{Na} = ~30% and [C_2H_4]=1.5%, [NO] = 0.65% at T = 402 °C	[57]
NO, C_3H_6	Pt film over (Na)$\beta''Al_2O_3$ solid electrolyte (EPOC)	Na	T = 330–400 °C [NO] = 0–6%, [C_3H_6] = 0–0.4% θ_{Na} = 0–100%	✓ Significant volcano type promotion observed with rate increases by an order of magnitude; optimal promotion achieved at θ_{Na} ~30%. Significant changes in apparent activation energy upon varying θ_{Na} were found. Under reaction conditions, the promoter phase consists of a mixture of $NaNO_2$/$NaNO_3$, which at high Na- loadings cause poisoning. ➢ ρ_{N_2} ~ ρ_{CO_2} = 7, ρ_{N_2O} = 3, ΔS_{N_2} = +20% (60%→80%), for θ_{Na} = ~30%, [C_3H_6] = 0.6%, [NO] = 1.3%, T = 375 °C	[58]
NO, C_3H_6	Rh film over (Na)$\beta''Al_2O_3$ solid electrolyte (EPOC)	Na	T = 350 °C [NO] = 1–9%, [C_3H_6] = 1%, [O_2] = 0–2%	✓ Na promotes the Rh-catalysed NO+propene reaction, especially its selectivity towards N_2. The impact of [NO]/[C_3H_6] ratio was studied: promotion was superior for [NO]/[C_3H_6] = 1%/1%. The impact of O_2 co-feed was also tested: upon increasing [O_2] the promotional effects of Na were progressively decreased, reversing to poisoning at [O_2] = 2%. Under promoted conditions the promoter phase consists of carbonate, nitrate or both. ➢ ρ_{N_2} = 2.4, ρ_{CO_2} = 1.7, ρ_{N_2O} = 0.4, ΔS_{N_2} = +37% (45%→82%), for θ_{Na} = ~20%, [C_3H_6] = [NO] = 1%, T = 350 °C.	[59] [60]

Table 4. *Cont.*

Reactants	Catalyst and (Promotion method applied)	Promoter	Reaction conditions	Promotion highlights and maximum achievements	Ref.
NO, C_3H_6	0.5 wt% Pd/YSZ (CCP)	Na, Li, K, Cs	T = 250–450 °C [NO] = 0–8%, $[C_3H_6]$ = 0–8% Na loadings: 0–0.102 wt% (nominal θ_{Na} = 0–30%)	✓ Rate increases by an order of magnitude were achieved. Promotion followed volcano type behaviour (optimal promotion at ~20% θ_{Na}). In respect to Li, Na, K, Cs alkalis studied Na was superior promoter. Significant variation of the apparent activation energy with Na loading; the promoted induced isokinetic effect was demonstrated. ➢ $\rho_{N_2} \sim \rho_{CO_2} \sim 7$, $\rho_{N_2O} \sim 1.5$, ΔS_{N_2} = +20% (75%→95%), for θ_{Na} = ~20%, [NO] = 1%, $[C_3H_6]$ = 0.8%, T = 380 °C ➢ ΔT_{50} = −110 °C at $[C_3H_6]$ = [NO] = 0.8% with the optimally promoted catalyst, θ_{Na} = 20% (0.068 wt% Na).	[14] [54]
NO, C_3H_6	0.5 wt% Pt/γ-Al_2O_3 (CCP)	Na, Li, K, Rb, Cs	T = 200–500 °C [NO] = 0–7%, $[C_3H_6]$ = 0.3, 0.6% Alkali loadings: Li: 0–4.7 wt%; Na: 0–10.4 wt%; K: 0–8.8 wt%; Rb: 0–15.5 wt%; Cs: 0–24 wt%.	✓ Extraordinarily effective promotion, by up to two orders of magnitude, with all alkalis (the highest found so far in emissions control catalysis: 420-fold rate increases). Volcano type behaviour; the optimal alkali loading depends on its entity. Among Li, Na, K, Rb, Cs alkalis studied superior promotion was offered by Rb. ➢ ρ_{N_2} = 420, ρ_{CO_2} = 280, ρ_{N_2O} = 25, ΔS_{N_2} > +75% (20%→> 95%), with 9.7wt%Rb at [NO] = 1.3%, $[C_3H_6]$ = 0.3%, T = 375 °C ➢ Optimal alkali loadings: Li: 1.25 wt%, Na: 4.16 wt%, K: 7.1 wt%, Rb: 9.7 wt%, Cs: 15 wt%.	[15] [61] [21] [22]

Table 4. *Cont.*

Reactants	Catalyst and (Promotion method applied)	Promoter	Reaction conditions	Promotion highlights and maximum achievements	Ref.
NO, C_3H_6	0.5 wt% Pt/γ-Al_2O_3 (CCP)	Ba	T = 200–500 °C [NO] = 0–5%, [C_3H_6] = 0–2% Ba loadings: 0, 3.5, 9.7, 15.2, 22.3wt%	✓ Significant, volcano type promotion with rate increase by up to two orders of magnitude; optimal promoter loading at 9.7wt% Ba. ✓ Significant changes in apparent activation energy upon varying Ba loading. ➢ ρ_{N_2} = 165, ρ_{CO_2} = 110, ρ_{N_2O} = 9, ΔS_{N_2} = +49% (45%→94%) with 9.7wt% Ba at [NO] = 1.3%, [C_3H_6] = 0.3%, T = 450 °C. ΔT_{50} = −124 °C for NO conversion	[62] [63]
NO, C_3H_6	0.5 wt% Rh/γ-Al_2O_3 (CCP)	Na	T = 200–550 °C [NO] = 0–1%, [C_3H_6] = 0.1% Na loadings: 0, 1.8, 3.0, 7.3, 11.3 wt%	✓ Significant, volcano type promotion -but not such impressive as in the case of Pt [15,61–63]; optimal promoter loading: 7.3wt% Na. ✓ The formation of CO and HCN by-products was also suppressed by Na-promotion. ➢ ρ_{N_2} = 3.3, ρ_{CO_2} = 2.7, ρ_{N_2O} = 2, ΔS_{N_2} = +29% (43%→72%) with 7.3wt% Na at [NO] = [C_3H_6] = 0.3%, T = 375 °C	[64]
NO reduction by alkanes (NO + Alkane)					
NO, CH_4	0.5 wt% Pd/YSZ (CCP)	Na	T = 350–450 °C [NO] = 0–1.5%, [CH_4] = 0–20% Na loadings: 0, 0.017, 0.034, 0.068, 0.102 wt%	✓ Pd by itself is an effective catalyst for NO reduction by CH_4. Na induced only poisoning in the whole region of [NO] and [CH_4] concentrations studied; the weak adsorption of CH_4 compared to that of O_2 and the O-poisoning of Pd surface due to the strengthening of the Pd-O bond induced by Na, readily explains this behaviour. ➢ Only Na-induced poisoning on the system was obtained.	[65] [66]

Table 4. *Cont.*

Reactants	Catalyst and (Promotion method applied)	Promoter	Reaction conditions	Promotion highlights and maximum achievements	Ref.
NO reduction by H_2 and/or CO (NO + H_2, NO + H_2 + CO) in the absence or presence of O_2					
NO, H_2	Pt film on (Na)$\beta''Al_2O_3$ solid electrolyte (EPOC)	Na	T = 300–450 °C [NO] = 0–1.6%, [H_2] = 0–1% θ_{Na} = 0–12%	✓ Significant, volcano type, promotion effects of Na to NO+H_2/Pt catalytic system; optimal promoter coverage θ_{Na}~6% ✓ Significant variations in the apparent activation energy of the reaction. ➢ ρ_{N_2} = 30, ρ_{N_2O} = 6, ΔS_{N_2} = +40% (36→75%) with θ_{Na} ~6% at [NO] = 0.63% [H_2] = 0.3%, T = 375 °C	[19]
NO, H_2, O_2	1 wt% Pt/γ-Al_2O_3 1 wt% Pt/SiO_2 (CCP)	Na (Mo)	T = 50–200 °C [NO] = 0.1%, [H_2] = 0.4%, [O_2] = 6% Na_2O-laodings: 0, 0.27, 1, 5 and 10 wt%	✓ The study conducted at lean conditions and at temperatures representative of automotive "cold-start" conditions. ✓ Volcano type promotion, optimized at a Na_2O loading = 0.27 wt%. Addition of MoO_3 as co-promoter enhances Pt performance for all the loadings tested (2.5, 10, 23, 46 wt% MoO_3) ➢ ΔX_{NO} = 20% (75%→90%) with 0.27 wt% Na_2O-loading at T = 125 °C. ➢ Best catalyst: 1 wt% Pt/10 wt% MoO_3/0.27 wt% Na_2O/Al_2O_3 offered X_{NO} = 95% at 100 °C and a ΔT_{50} = −50 °C.	[73]
NO, H_2, O_2	1 wt% Pt/ZSM-5 (CCP)	Na, K, Cs, Mg, Ca, Ba	T = 40–110 °C [NO] = 0.08%, [H_2] = 0.08–0.56%, [O_2] = 10% O_2 Na loadings: 0, 3, 5, 10, 15 and 20 wt%	✓ Optimum promotion achieved by Na. Optimal promoter loading: 10–15 wt% Na. ➢ NO_x conversion > 90% with a selectivity towards N_2~50%.	[74]

Table 4. *Cont.*

Reactants	Catalyst and (Promotion method applied)	Promoter	Reaction conditions	Promotion highlights and maximum achievements	Ref.
NO, H_2, CO, O_2	0.5 wt% Pd/γ-Al_2O_3 0.5 wt% Pd/γ-Al_2O_3-TiO_2 (CCP)	K	T = 50–400 °C [NO] = 0.1%, [CO] = 0.25%, [H_2] = 0.75%, [O_2] = 6% O_2 K loadings: 0, 0.1, 0.25, 0.5, 1.0 and 3.0 wt% K	✓ Synergy between support-mediated (modification of Al_2O_3 with TiO_2) and surface-induced (K -addition) promotions providing very pronounced *de*NOx performance. ✓ K-promotion over the doubly-promoted catalyst appeared volcano type behaviour with optimal K-loading=0.25wt%. ➢ 90% NOx conversion with >85% N_2-selectivity in a wide range of temperatures (180–300 °C) were achieved on the optimally doubly-promoted catalyst: 0.5 wt% Pd(0.25 wt% K)/γ-Al_2O_3-10 wt% TiO_2. Also, ΔT_{50} = −40 °C	[79]
Direct NO decomposition					
NO	(0.6–10 wt%)Pd/γ-Al_2O_3 (CCP)	Na	T = 550–900 °C [NO] = 4% in He Na loadings (as NaOH): 0, 1.3, 4.3, 6.5 and 13 wt% NaOH	✓ Pre-coadding of γ-Al_2O_3 with NaOH causes significant promotion of the direct NO decomposition activity of Pd/γ-Al_2O_3 catalyst. ✓ Promotion was progressively increased with increasing promoter loading. ➢ ΔX_{NO} = 45% (15%→60%) at T = 625 °C over the 13 wt% NaOH -promoted 4.8 wt% Pd/γ-Al_2O_3 catalyst.	[80]

3.1.5. N_2O Decomposition and/or Reduction

It is well known that nitrous oxide (N_2O) is absent in automotive engines exhausts gases, although this can be a by-product (in low concentrations) of the NO reduction reactions taking place in the after-treatment devices of cars (catalytic converters), primarily formed during the "cold start" and "intermediate temperature" periods. However, N_2O is a powerful greenhouse gas (global warming potential ~310) and also the dominant stratospheric ozone-depleting gas nowadays. These account for the currently devoted high attention on the N_2O emissions control. Besides mobile sources, other anthropogenic sources of N_2O emissions involve stationary fuels combustion processes, agricultural activities and industrial plants (e.g., adipic and nitric acid plants) where it is produced in fairly high concentrations; the latter rationalizes why a major portion of deN_2O studies concerns gas mixtures with relatively high N_2O concentrations (ca. 0.1–1 vol.%; Table 5), not actually mirror cars emissions.

(i) Direct N_2O decomposition

No studies of electrochemical catalyst promotion (EPOC) were found in literature regarding the direct N_2O decomposition. In the next paragraphs we summarize some examples of the CCP studies for this reaction.

The pronounced effect of electropositive promoters on the direct N_2O decomposition performance (de-N_2O) of noble metals supported catalysts was demonstrated by several groups [81–85] (Table 5). In particular, Haber et al. [81] investigated the effect of Li, Na, K and Cs on N_2O decomposition over Rh catalysts supported on alkali-modified Al_2O_3 carrier. They found that the deposition of alkali metals on Al_2O_3 support results in a notable increase of Rh dispersion, which improved de-N_2O efficiency. In almost all cases the optimum promotional effects were obtained at ~0.078 mol% alkali (Table 5).

Parres-Esclaped et al. [82] studied the N_2O decomposition over Rh catalysts supported either on bare γ-Al_2O_3 or on alkaline earth modified alumina (5 wt% Mg- and Sr-doped γ-Al_2O_3). Enhanced deN_2O performance was obtained with Rh/(Sr)Al_2O_3 catalysts, which achieved total N_2O conversion at 300 and 350 °C, in the absence and presence of O_2, respectively, compared to 375 and 450 °C over un-promoted catalysts (Table 5). However, Mg addition was not beneficial to deN_2O Rh/Al_2O_3 activity. The latter was attributed to the formation of magnesium aluminate, while the former was attributed to the enhanced Rh dispersion (determined by TEM) on account of the Sr-modifier.

Combined structural and surface promotion was applied by Konsolakis et al. [83] to facilitate the deN_2O performance of Pt/γ-Al_2O_3 catalyst. To this end, rare earth oxides (CeO_2, La_2O_3) were used as structural promoters of the γ-Al_2O_3 support and potassium (K) as surface promoter of Pt, providing a doubly-promoted Pt(K)/Al_2O_3-(CeO_2-La_2O_3) catalyst composite. The study was conducted on macro-structured, cordierite honeycomb monolith washcoated catalytic arrangements, suitable for practical implementations. The effect of K-promoter was studied in the range of 0–2wt% K loadings, while the impact of O_2, CO and H_2O in the feed stream was also investigated (Table 5). It was found that incorporation of 20wt% CeO_2-La_2O_3 in the γ-Al_2O_3 support (ACZ) results in dramatic enhancement of the deN_2O catalyst activity which is further improved by potassium addition. The promotional effects of K were found to monotonically increase upon increasing its loading on the structurally promoted Pt/ACZ catalyst in the range of 0-2 wt% studied. As a result, the final doubly-promoted catalyst with 2wt% K loading achieved 100% N_2O conversion at a temperature of only 440 °C; notably the bare Pt/γ-Al_2O_3 did not exceed ~35% N_2O conversion even at 600 °C (Figure 14a). In respect to the impact of CO, O_2, H_2O co-feed, the authors found the following deN_2O efficiency sequence in all promoted or non-promoted catalysts: N_2O + CO >> N_2O > N_2O + O_2 > N_2O + O_2 + CO (Figure 14b). Water co-feed was detrimental to deN_2O performance but in a partially reversible manner after its removal from the feed stream; the catalytic efficiency was totally restored by adding H_2 into N_2O + H_2O feed.

Papista et al. [84] studied the promoting impact of potassium addition to Ir/γ-Al_2O_3 catalyst on N_2O decomposition under both deficient and O_2 excess conditions. Varying K loading in the range of 0–1wt%, the authors found poisoning effect on the deN_2O performance of iridium in the absence of oxygen in the feed. The opposite was true in the case of oxygen excess conditions; a pronounced promotion was recorded showing that K prohibits the oxygen-poisoning of the catalysts; the latter self-poisoning is a typical problem in deN_2O catalytic chemistry, since the removal of the N_2O decomposition-derived adsorbed oxygen on the catalyst surface is a key reaction step of the process. The alkali promotion found in the case of oxygen excess conditions follows volcano behaviour providing therefore an optimal promoter loading of 0.5 wt% K, that resulted in a significant decrease of T_{50}, equal to $\Delta T_{50} = -80$ °C (Table 5). The effect of alkali on the adsorption properties of Ir involving strengthening of the Ir–O bond of adsorbed O species with concomitant changes on the oxidation state of the metal was considered upon interpretation of the promotional and/or poisoning effects observed.

Finally, Goncalves and Figueiredo [85] studied the effect of potassium on the catalytic activity of Pt supported on activated carbon (ROX 0.8) for the simultaneous reduction of NO + N_2O mixtures. They found a synergistic effect between K and Pt which led to high activity at relatively low temperatures. For instance, at 350 °C, a very stable and close to 100% conversion of both reactants was recorded over a 5wt% K/0.1wt% Pt catalyst, in opposite to the monometallic 0.5wt% K or 0.1wt% Pt counterparts (Table 5). The latter showed significantly lower conversion activity of about 25% for NO and 90% and 0% for N_2O, respectively. The improved redox properties of the bimetallic K/Pt formulation was responsible for this synergy on enhancing both deNO and deN_2O reactions [85].

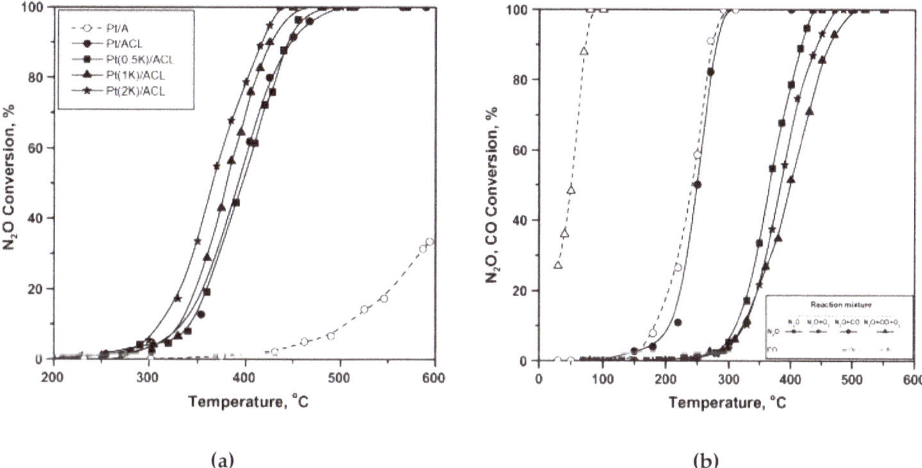

Figure 14. (**a**): N_2O conversion profiles as a function of temperature (light-off curves) for bare Pt/γ-Al_2O_3, structurally promoted Pt/ACL and doubly-promoted Pt(K)/ACL catalysts under direct decomposition of N_2O; Feed: 0.1% N_2O, balance He; GHSV = 10,000 h^{-1}. (**b**): N_2O (and CO) conversion profiles obtained over the optimally doubly-promoted Pt(2wt%K)/ACL catalyst under direct N_2O decomposition or N_2O + CO reactions in the absence or presence of O_2, that is, under the feeds: 0.1% N_2O, balance He; 0.1% N_2O + 2% O_2, balance He; 0.1% N_2O + 0.1% CO, balance He; 0.1% N_2O + 0.1% CO + 2% O_2, balance He. GHSV = 10,000 h^{-1}. (Reprinted with permission from Ref. [83]; Copyright 2013, Elsevier).

(ii) N_2O reduction by hydrocarbons

Very recently, the effect of alkali modifiers on another very important reaction related both to the TWC chemistry and N_2O abatement, that is, the N_2O reduction by hydrocarbons, was extensively studied by Pekridis et al. [25,86]. In particular, the role of potassium (K) promoter on the surface and catalytic properties of Pd/Al_2O_3 (CCP) catalysts for N_2O reduction by alkanes (CH_4, C_3H_8) or alkenes (C_3H_6) was investigated [86]. Potassium strongly enhanced the N_2O reduction by propane or propene, resulting in notably lower N_2O light off temperatures (~100 °C) compared to the un-promoted catalyst (Table 5). However, a slight inhibition upon K-promotion was obtained when CH_4 was used as a reducing agent, implying that the extent of K-promotion is strongly related with the nature of the reducing agent. A comprehensive surface characterization study, involving XPS, in situ DRIFT spectroscopy of CO adsorption and FTIR-pyridine adsorption, was undertaken to correlate the de-N_2O performance with the catalyst surface characteristics. The results revealed that the electronic, acidic and structural properties of Pd/Al_2O_3 catalyst can be substantially modified by potassium, which in turn affects the reactants chemisorption bonds and as a consequence the overall surface activity. Specifically, addition of K on Pd/Al_2O_3 decreases the adsorption strength of electron-donor adsorbates, such as C_3H_8 and C_3H_6, enhancing the N_2O adsorption through an electron transfer from metal sites to N_2O antibonding orbitals. Both factors act synergistically towards increasing active sites for N_2O adsorption/decomposition. However, in the case of CH_4, K enhances the adsorption strength of N_2O and its dissociation products (O_{ads}), at the expense of weakly bonded CH_4, leading to poisoning by strongly bonded oxygen atoms.

In correlation with the above study, the EPOC concept (through a Pd-film electrocatalyst interfaced with (K)$\beta''Al_2O_3$ solid electrolyte) was also used as a probe technique for exploring in detail the response of Pd metal to alkali (K) promotion in the same reaction [25]. The obtained results matched very well the promotion characteristics of the previous CCP study that involved conventional highly dispersed catalysts, demonstrating experimentally again that electrochemical promotion is an efficient tool for a rapid investigation of the promoter effect on a catalytic system. The obtained information could be further used to design and optimize efficient and "smart" catalysts formulations.

The N_2O reduction by propene was also studied by de Lucas-Consuegra et al. [87,88], over Pt-based electro-catalysts, electrochemically promoted by potassium by the use of a (K)$\beta''Al_2O_3$ solid electrolyte. The authors found significant rate enhancement ratios (ρ_{max}~10) at temperatures between 400–450 °C and a reaction mixture composition of $N_2O/C_3H_6/O_2$ = 1000 ppm/2000 ppm/2000 ppm (Table 5). Potassium coverages (estimated by appropriate galvanostatic transients) higher than θ_K~70% were necessary in order to achieve such large rate enhancements. They also found that these pronounced promotional effects are similar for a wide range of O_2 concentration in the reaction mixture (2000–7000 ppm), while at larger O_2 concentrations (>9000 ppm) the promoting phenomena were practically vanished. A very significant observation of this work was also the demonstration that potassium addition strongly decreased the inhibiting effect of water vapor on the Pt-catalysed $N_2O/C_3H_6/O_2$ catalytic system.

Overviewing the literature findings included in Section 3.1.5 and Table 5 the following general conclusions can emerge:

- Alkali addition on PGM catalysts typically promotes the direct N_2O decomposition reaction (in the absence or presence of O_2 in the feed stream) or its reduction by reducing agents.
- Among Pt, Rh and Ir catalysts, alkali-promotion of the N_2O decomposition reaction was found to be very effective and superior on Pt, less effective on Rh and detrimental (poisoning) on Ir; in the case of Ir, alkali-induced promotion was obtained only under excess oxygen conditions, which was attributed to changes on the oxidation state of the metal.
- Comparative EPOC and CCP studies of alkali promotion on the N_2O reduction by the use of several reducing agents (CH_4, C_3H_8 and C_3H_6) demonstrated the great similarities on the output promotional characteristics of the two methods.

Table 5. Electropositive promotion of PGMs-catalysed N_2O decomposition and/or reduction reactions.

Reactants	Catalyst and (Promotion method applied)	Promcter	Reaction conditions	Promotion highlights and maximum achievements	Ref.
N_2O	0.1 wt %Rh/γ-Al_2O_3 (CCP)	Li, Na K, Cs	T = 250–450 °C $[N_2O]$ = 1% Alkali loadings (as alkali oxides): 0.033, 0.066, 0.078, 0.099 wt%	✓ Alkali addition enhanced Rh dispersion and thus deN_2O catalytic activity. Best promoter was Cs. ✓ Volcano type promotion for all alkalis used; optimal promoter loading ~0.078wt% of alkali oxide. ➢ ΔT_{50} = −60 °C for Li and Na promotion; ΔT_{50} = −70 °C for K promotion; ΔT_{50} = −90 °C for Cs promotion	[81]
N_2O, presence or absence of O_2	0.5 wt% Rh/γ-Al_2O_3 (CCP)	Mg, Sr	T = 250–450 °C $[N_2O]$ = 0.1%; $[N_2O]$ = 0.1% + $[O_2]$ = 5%. Mg- or Sr-loading: 5 wt%	✓ Significant promotion by Sr either in the absence or in the presence of O_2. ✓ Small inhibition and marginal promotion by Mg in absence and presence of O_2, respectively. ➢ ΔT_{50} = −45 and −50 °C in the absence and presence of O_2, respectively, with a Sr-loading = 5wt%. ➢ ΔT_{50} = +10 and −15 °C in the absence and presence of O_2, respectively, with a Mg-loading = 5wt%.	[82]
N_2O, presence or absence of O_2, CO, H_2O	0.5 wt% Pt/Al_2O_3-(CeO_2-La_2O_3) (CCP)	K	T = 200–600 °C $[N_2O]$ = 0.1%; $[N_2O]$ = 0.1% + $[O_2]$ = 2%; $[N_2O]$ = 0.1% + [CO] = 1%; $[N_2O]$ = 0.1% + [CO] = 1% + $[O_2]$ = 2%. K-loadings: 0, 0.5, 1.0 and 2.0 wt%	✓ K was found to be an effective promoter of the Pt-catalysed N_2O decomposition under both reducing and oxidizing conditions. H_2O in the feed caused detrimental but to a great extent reversible (upon removal) influence on deN_2O efficiency over both K-free and K-promoted catalysts. ✓ Promotion was monotonically enhanced with increasing K-loading. ➢ ΔT_{100}= −60 °C under N_2O feed; ΔT_{100} = −50 °C under N_2O + O_2 feed; ΔT_{100} = −45 °C under N_2O + CO + O_2 feed	[83]
N_2O presence or absence of O_2	0.5 wt% Ir/γ-Al_2O_3 (CCP)	K	T = 300–600 °C $[N_2O]$ = 0.1%; $[N_2O]$ = 0.1% + $[O_2]$ = 2%. K-loadings: 0, 0.25, 0.5 and 1.0wt%	✓ K induces poisoning effects on the Ir-catalysed N_2O decomposition at O_2-deficient conditions. ✓ In opposite, K is an effective promoter under O_2 excess conditions; volcano type promotion with an optimal K-loading equal to 0.5 wt% K. ➢ ΔT_{50} = −80 °C under $[N_2O]$ = 0.1% + $[O_2]$ = 2% and 0.5 wt% K loading.	[84]

Table 5. *Cont.*

Reactants	Catalyst and (Promotion method applied)	Promoter	Reaction conditions	Promotion highlights and maximum achievements	Ref.
N_2O, NO	(0.1 and 0.5 wt%)Pt/(ROX 0.8) (CCP)	K	T = 50–500 °C [NO] = 0.1%; $[N_2O]$ = 0.05%; and $[N_2O]$ = 0.05% + [NO] = 0.1% K-loading: 3 and 5 wt%	✓ K addition simultaneously promotes the conversions of NO and N_2O at relatively low temperatures. ➢ 100% conversion of both N_2O and NO over 5 wt% K/0.1 wt% Pt/Al_2O_3 under $[N_2O]$ = 0.05% + [NO] = 1% at T = 350 °C.	[85]
$N_2O + CH_4$, $N_2O + C_3H_8$, $N_2O + C_3H_6$, presence or absence of O_2	2 wt% Pd/γ-Al_2O_3,and Pd film on (K)β″Al_2O_3 solid electrolyte (CCP vs EPOC)	K	T=100–500 °C $[N_2O]$ = 0.15% $[CH_4]$ = 0.6% $[C_3H_8]$ = 0.2% $[C_2H_6]$ = 0.2% ($[O_2]$ = 3%) K-loadings: 0, 2.5, 4.5, 7.0, 9.0 wt%	✓ Both structural and electronic properties of Pd particles were modified by K addition. ✓ K addition strongly promotes N_2O reduction by both C_3H_8 and C_3H_6. ✓ K addition slightly inhibits N_2O reduction by CH_4. ✓ CCP and EPOC induced K-promotion behaviours were found to match well each other. ➢ ΔT_{50} = +25 °C under $[N_2O]$ = 0.15% + $[CH_4]$ = 0.6% and 9 wt% K loadings ➢ ΔT_{50}= −100 °C under $[N_2O]$ = 0.15% + $[C_3H_8]$ = 0.2% and 9 wt% K loadings ➢ ΔT_{50} = −135 °C under $[N_2O]$ = 0.15% + $[C_2H_6]$ = 0.2% and 7–9 wt% K-loadings.	[25] [86]
N_2O, C_3H_6, O_2	Pt film on (K)β″Al_2O_3 solid electrolyte (EPOC)	K	T = 180–580 °C $[N_2O]$ = 0.1%, $[C_2H_6]$ = 0.2%, $[O_2]$ = 0.2–1.0% θ_K = 0–90%	✓ Strong K-induced promotion was observed. This was progressively attenuated and finally reversed to poisoning upon increasing O_2 in the feed stream. K-promotion prevented deN_2O activity of Pt from the poisoning effect of H_2O. ➢ ρ_{N_2O} = 38.5, $\rho_{C_3H_6}$ = 1.2 under $[N_2O]$ = 0.1% + $[C_2H_6]$ = 0.2% + $[O_2]$ = 0.2%, T = 450 °C, θ_K~40% (V_{WR} = −1 V)	[87] [88]

3.2. Electropositive Promotion of PGMs Operated Under Simulated Practical Conditions

3.2.1. Simulated TWC Conditions

Given that alkalis and alkaline earths were found to strongly promote reactions related with the three-way catalytic (TWC) chemistry, testing of these new catalyst formulations at conditions that simulate automotive exhaust conditions was a reasonable consequence. To this end, Konsolakis et al. [89] studied the promotional effects of Na on $Pt/\gamma\text{-}Al_2O_3$ catalysts operated under simulated exhaust conditions at the stoichiometric point (SP), that is, at the value equal to 1 of the stoichiometric number (S), calculated as:

$$S = (2[O_2] + [NO])/([CO] + 9[C_3H_6]) \tag{8}$$

using a feed mixture consisting (v/v) 1000 ppm NO + 1067 ppm C_3H_6 + 7000 ppm CO + 7800 ppm O_2.

A series of Na-promoted $Pt/\gamma\text{-}Al_2O_3$ catalysts were tested under these conditions at temperatures from 200–500 °C typical of TWCs. Typical results of the promoter effect on the overall NO, CO and C_3H_6 conversions and selectivity obtained at 400 °C, are shown in Figure 15. The conversions of all three pollutants, NO, CO and C_3H_6, reach 100% at sodium contents > 4.2wt%. In addition, Na promoter causes a significant decrease in the light off temperature of ~100 °C, together with significant improvements (~75%) in N_2-selectivity (Table 6). The promotional effect was attributed to the modified relative adsorption strengths of reactants and reaction intermediates, as explained above in the previous sections [89].

In a similar motive, Macleod et al. [90,91] studied the Na promotion of $Pd/\gamma\text{-}Al_2O_3$ and $Rh/\gamma\text{-}Al_2O_3$ catalysts operated under simulated TWC conditions over a range of stoichiometries from fuel rich to fuel lean conditions (0.90 < S < 1.1). They observed significantly different responses of Na addition in the two cases (Table 6). Na addition to $Pd/\gamma\text{-}Al_2O_3$ significantly improves its overall performance: although the recorded promotional effects were not as high as those observed in Reference [89] for Na-promoted $Pt/\gamma\text{-}Al_2O_3$, results were again very spectacular.

A 7wt% Na loading (optimal promoter loading for the conditions used) offered a ~50 °C lower light off temperature than the unpromoted Pd when tested under stoichiometric feed conditions (1000 ppm NO + 1067 ppm C_3H_6 + 7000 ppm CO + 7800 ppm O_2). Beneficial effects of sodium addition were also recorded under both fuel rich and fuel lean conditions (Table 6). In the former case, Na was found to prevent catalyst poisoning by carbon deposition therefore maintaining high NO conversions, while in the latter case Na reduced the formation of N_2O therefore leading to enhanced selectivities toward N_2. However, the authors found that Na addition to Rh has a detrimental effect causing severe activity poisoning and N_2-selectivity reduction over the greater part of the temperature interval of 200–400 °C investigated [92]. It was attributed to the enhanced oxygen adsorption on the Rh active sites at the expense of the hydrocarbon, whilst in the case of Pd the beneficial effects were considered as a consequence of Na-promoted NO dissociation, together with inhibition of self-poisoning due to excessive propene adsorption [91]. It is worth noting that NO dissociative chemisorption takes place spontaneously on Rh but not on Pd.

Tanaka et al. [92] have also reported the beneficial effects of the addition of a small amount of Na_2O to $(1.67wt\%Pt)(12.3wt\%MoO_3)/SiO_2$ catalysts investigated under simulated exhaust from automobile engine and three-way behaviour around the stoichiometric point (SP). They found that the active window of SP for the Na-promoted $PtMoNa/SiO_2$ catalyst is wider than that of the Na-free counterpart, that is, $PtMo/SiO_2$ and Pt/SiO_2 catalysts (Table 6). The observed improvements were attributed to the wider redox ratio window and to the depressed (by Mo and Na) oxidation of Pt, even under oxygen excess.

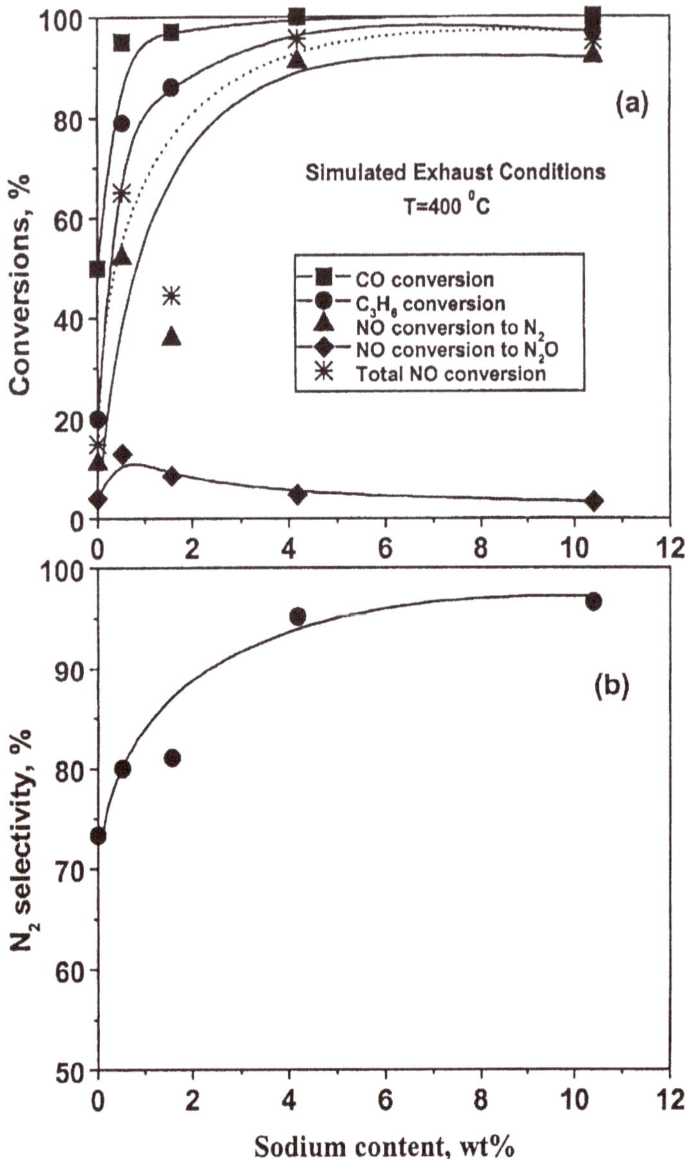

Figure 15. Testing Na-promoted Pt/γ-Al_2O_3 catalysts at simulated exhaust conditions of conventional stoichiometric gasoline engines (T=400 °C, 1000 ppm NO, 1067 ppm CO and 7800 ppm O_2; F_t = 80 cm^3STP/min; m_{cat} = 8 mg). Effect of Na loading on reactants conversions (**a**) and selectivity towards N_2 (**b**). (Reprinted with permission from Ref. [89]; Copyright 2000, Elsevier).

Shinjon et al. [93] have investigated the effect of an alkaline earth (Ba) modifier on the catalytic activity of Pt/γ-Al_2O_3 and Rh/γ-Al_2O_3 catalysts under simulated automotive exhaust conditions at the stoichiometric point (1% CO, 0.3% H_2, 0.1% C_3H_6, 0.1% NO, 0.75% O_2, 3% H_2O, 12% CO_2). They found that the TWC-performance of Pt/γ-Al_2O_3 catalysts was improved by Ba, offering ~30 °C lower light off temperatures compared to those for the un-promoted catalysts (Table 6). However, the overall performance of Rh catalysts was deteriorated with Ba addition. It was attributed to the fact that Ba addition to Pt catalysts suppressed the strong hydrocarbon chemisorption, enhancing simultaneously the oxygen adsorption. On the other hand, Ba addition to Rh catalysts results in strong adsorption of oxygen species, suppressing hydrocarbon chemisorption and thus the reaction between them.

Kobayashi et al. [94] investigated the influence of Ba and Sr alkaline earth metals on a commercial Pd-only TWC catalyst (Pd/Al_2O_3-based TWC catalyst of N.E. ChemCat Corp.). Improved three-way catalytic performance was obtained regarding CO and NO_x conversions (Table 6). The enhanced basicity and the electron-donation ability of these basic elements were considered to play significant role in the improvement of TWC performance. A higher electron density around $Pd^{(II)}$ was revealed by XPS measurements in the Ba-promoted Pd/Al_2O_3 catalyst in comparison with the Ba-free counterpart. The stabilization of PdO that may suppress palladium species sintering and the electron enrichment palladium particles that endows Pd to behave like Rh were concluded as the main promoting factors in this study [94].

A NO-free gas mixture of CO, C_3H_6, H_2, O_2, CO_2 and H_2O, simulating two-stroke motorcycle emissions (Table 6), was considered by Lee and Chen [30] for the investigation of the impact of Na_2O and K_2O addition on a 0.4wt% Pt/γ-Al_2O_3 catalyst operated at 150–450 °C under stoichiometric and oxygen-deficient conditions (by varying the oxygen to reductants stoichiometric number; S = 1, 0.31 and 0.17). Significant enhancements on the CO and C_3H_6 conversions were achieved by the addition of both alkalis under operation at the stoichiometric point (S = SP = 1); superior promotion was the one induced by potassium. Under O_2-deficient conditions alkalis addition enhanced CO conversion but not that of propene, which was reduced. The presence of water in the gas mixture was found to have a strong positive impact on catalyst performance, in particular on K-promoted catalyst and for CO conversion (Table 6).

Yentekakis and co-workers [95,96] developed monolithic type catalytic converters with a washcoat containing only one precious metal (Pt-only TWC), which was optimally promoted by Na. Catalytic performance tests of this novel TWC under simulated exhaust conditions in a wide temperature range (150–500 °C) exhibited similar performance and even better N_2 selectivity (~100% in the whole temperature range) compared to that of a commercial bimetallic (Pt/Rh)-TWC catalyst (Figure 16A, Table 6); notably, the aforementioned TWC had a 4.5-fold lower noble metal loading in comparison to the commercial one. These novel material formulation and design for TWCs is subjected to much lower production cost (use of only one noble metal at much lower loading, without the necessity of scarce Rh in their constitution) and cost-effective recycling.

Most recent tests of these novel TWCs at more severe temperature conditions (>800 °C) have also shown a significant resistance to deactivation (Figure 16B) [97,98], implying that the promoter phase (Na) does not practically escape from the catalyst composite even at very high temperatures, due to its stabilization via the formation of new β' and β'' sodium-alumina phases as verified by TEM and XRD studies [97,98]. These phases, in direct interaction with the active phase (Pt), then act as a spontaneous (thermal diffusion driven) Na promoter source during TWC operation causing permanent promotion [97,98].

Overviewing the literature findings presented in this section and in Table 6, one can readily conclude that electropositive promotion of PGM-catalysts under simulated automotive exhaust conditions follows, in general terms, the promotional characteristics corresponding to the simple (model) reactions of the complex reactions scheme taking place into the converter. However, other possible synergistic or competitive interactions between the reactants and reactions' intermediates on a densely populated surface with all these competitive species can be at work affecting the overall performance.

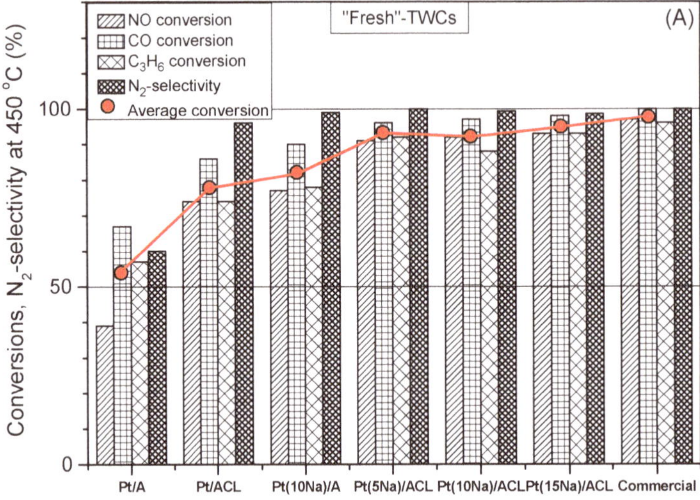

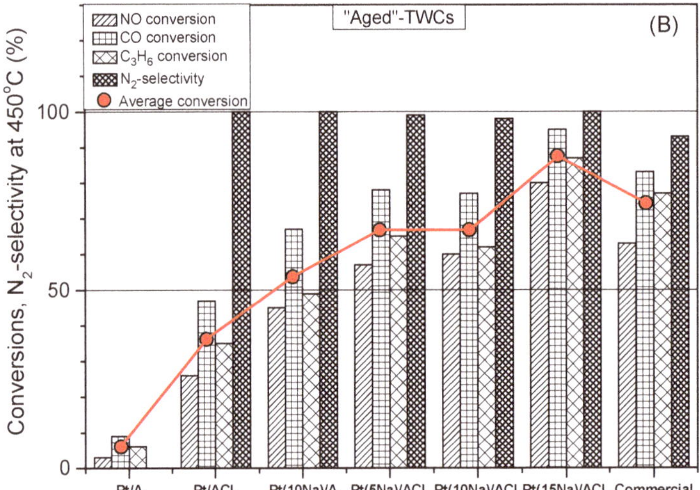

Figure 16. Three-way catalytic performance at 450 °C of "fresh" (**A**) and "aged" (**B**) TWC samples at simulated stoichiometric gasoline engines exhaust conditions (0.1% NO + 0.7% CO + 0.1067% C_3H_6 + 0.78% O_2, balanced with He at 1 bar; F_t = 3200 cm^3 (STP)/min). (Reproduced with permission from Ref. [98]; Copyright 2011, Elsevier).

Table 6. Electropositive promotion of PGM-catalysts operated under simulated automotive exhaust conditions.

Reactants	Catalyst and (promotion method applied)	promoter	Conditions	Promotion highlights and achievements	Ref.
NO, C_3H_6, CO, O_2	0.5wt%Pt/γ-Al_2O_3 (CCP)	Na	T = 200–500 °C (at simulated TWC conditions) [NO] = 1000 ppm, [CO] = 7000 ppm, [C_3H_6] = 1067 ppm, [O_2] = 7800 ppm; (at S = SP = 1) w/F = 6 × 10^{-3} g·s/cm^3 Na loadings: 0, 0.52, 1.57, 4.18 and 10.4 wt% Na	✓ Na promotes simultaneously the conversions of all three pollutants, CO, C_3H_6 and CO and enhances the N_2-selectivity at all promoter loadings used. ➢ CO, C_3H_6 and NO conversions reached 100%, 100% and > 91% at T = 400 °C and for Na-loadings >4.18wt%; the corresponding values on the unpromoted Pt/γ-Al_2O_3 were only 50, 20 and 11%. ➢ $\Delta T_{50\ (CO)} = -79$ °C, $\Delta T_{50\ (C3H6)} = -77$ °C, $\Delta T_{50\ (NO)} = -93$ °C and ΔS_{N_2} = +81%.	[89]
NO, C_3H_6, CO, O_2	0.5 wt% Pd/γ-Al_2O_3 and 0.5 wt% Rh/γ-Al_2O_3 (CCP)	Na	T = 100–340 °C (at simulated TWC conditions) [NO] = 1000 ppm, [CO] = 7000 ppm, [C_3H_6] = 1067 ppm, 6970 ≤ [O_2] ≤ 8630 ppm (around SP: 0.9≤ S ≤1.1) w/F = 15 × 10^{-3} g·s/cm^3 Na loadings for Pd/Al_2O_3: 0, 1.8, 3.5, 7.0, 12.0 wt% Na loadings for Rh/Al_2O_3: 0, 3.0 and 7.3 wt%.	✓ The studies involve the Na impact on Pd and Rh operated at TWC conditions over a range of the stoichiometric number (S) values from fuel-rich to lean. ✓ Na promotion was beneficial under both oxidizing and reducing conditions for Pd catalysts. Optimal loadings >3.5wt% Na: ➢ $\Delta T_{50\ (CO)} = -48$ °C, $\Delta T_{50\ (C3H6)} = -35$ °C, $\Delta T_{50\ (NO)} = -38$ °C (at SP = 1) and ΔS_{N_2} = +35% (at S > 1.04: oxidizing conditions). ✓ In opposite to Pd, Na addition on Rh was detrimental.	[90] [91]
NO, C_3H_6, CO, O_2, H_2, CO_2, H_2O	1.67 wt% Pt/12.3wt% MoO_3-SiO_2 (CCP)	Na_2O	T = 100–600 °C w/F = 9.91 × 10^{-3} g·s/cm^3 Na_2O loadings: 0.05, 0.1, 1.0 wt%. Simulated oxidizing feed, S = 9.91:4.3% O_2, 0.12% NO, 800ppm C_3H_6, 0.12% CO, 400ppm H_2, 12.3% CO_2, 3% H_2O. Simulated exhaust feed, S-scan: 0.40–1.21%O_2, 0.12% NO, 490–62 ppm C_3H_6, 0.45–1.50% CO, 0.15–0.50% H_2, 10.0% CO_2, 3.0% H_2O.	✓ Under nearly stoichiometric conditions, a wider redox ratio window of NO_x conversion was obtained with PtMoNa/SiO_2 catalyst compared to Pt/SiO_2. ✓ Under oxidizing conditions PtMoNa/SiO_2 catalyst resulted to a shift of all CO, C_3H_6 and NO_x conversion profiles to higher temperatures. However, a wider temperature window with higher NOx conversion was obtained.	[92]

Table 6. *Cont.*

Reactants	Catalyst and (promotion method applied)	promoter	Conditions	Promotion highlights and achievements	Ref.
NO, C_3H_6, CO, O_2, H_2, CO_2, H_2O	1.67 wt% Pt/γ-Al_2O_3 0.9 wt% Rh/γ-Al_2O_3 (CCP)	Ba	T = 100–400 °C (at simulated TWC conditions) [NO] = 1000 ppm, [C_3H_6] = 334 ppm, [CO] = 1%, [H_2] = 0.3%, [CO_2] = 12%, [H_2O] = 3%, [O_2] = 0.75% GHSV = 200000 h^{-1} Ba-loading: 5.8, 11.5, 23 and 46 wt% Ba	✓ Ba addition significantly promotes Pt/γ-Al_2O_3 activity under simulated exhaust conditions. The opposite is true for Rh/γ-Al_2O_3. ➢ $\Delta T_{50(NOx)}$ = −30 °C over 46wt% Ba-promoted Pt/γ-Al_2O_3 ➢ $\Delta T_{50(NOx)}$ = +15 °C over 46wt% Ba-promoted Rh/γ-Al_2O_3	[93]
Exhaust gas from an automotive engine	5wt% Pd-commercial TWC catalyst supplied by N.E. ChemCat Corp. (CCP)	Ba, Sr, La	T=300–600 °C Real stoichiometric automotive engine exhaust gas GHSV = 68000 h^{-1} Ba, Sr od La-loading: 5wt%	✓ The TWC performance of the Pd-based commercial catalyst was significantly improved by the addition of Ba, Sr or La. ✓ The performance improvements followed the order: Ba > Sr > La. ✓ Alkaline earths promotion was related with the stabilization of PdO phase and its electron enrichment, which endows Pd to behave like Rh.	[94]
C_3H_6, CO, H_2, O_2, CO_2, H_2O (NO-free)	0.4 wt% Pt/γ-Al_2O_3 (CCP)	Na, K	T = 150–450 °C (simulated two-stroke motorcycle emissions). [CO] = 1–4.14%, [C_3H_6] = 0.08–0.7%, [O_2] = 0.96–0.9%, [H_2] = 0.2%, [CO_2] = 10%, [H_2O] = 0 or 10%. Stoichiometric numbers studied S = 1, 0.31, 0.17 Na-loading: 4.5 wt%; K-loading: 7.6 wt%	✓ Significant promotion by both Na_2O and K_2O; superior that of K_2O. ✓ Under oxygen-deficient conditions (S = 0.17 and 0.31), H_2O addition enhances both CO and C_3H_6 conversions (in particular X_{CO} and over K-promoted catalyst) ➢ $\Delta T_{50(CO)}$ = −90 °C $\Delta T_{50(C3H6)}$ = −110 °C at S = SP = 1 and absence of H_2O. ➢ ΔX_{CO} = 51 and 48 at S = 0.31, without and with H_2O, respectively. ➢ ΔX_{CO} = 22 and 48 at S = 0.17, without and with H_2O, respectively.	[30]

Table 6. *Cont.*

Reactants	Catalyst and (promotion method applied)	promoter	Conditions	Promotion highlights and achievements	Ref.
NO, C_3H_6, CO, O_2, CO_2, H_2O	0.1 wt% Pt/(γ-Al_2O_3-CeO_2-La_2O_3-cordierite monolith) versus a Commercial 0.37Pt/0.08Rh-TWC(CCP)	Na	T = 150–500 °C (simulated TWC conditions) [NO] = 0.1%, [CO] = 0.7%, [C_3H_6] = 0.1067%, [O_2] = 0.78%, [H_2O] = 10%, [CO_2] = 10% GHSV = 50500 h^{-1} Na-loading: 1 and 2 wt% Na (i.e., 5 and 10 wt% in the washcoat)	✓ The Na-promoted monometallic (Pt-only) TWC, with 4.5-fold less PGMs loading compared to bimetallic commercial one, exhibits similar performance with the commercial and better N_2 selectivity (~100% in the whole temperature range 150–500 °C). ✓ Marginal impact of H_2O and CO_2 on the performance of Na/Pt-TWC. ✓ A synergy between surface promoter (Na) and structural promoters (CeO_2, La_2O_3) was resulted, leading to this attractive promotion. ✓ Significant resistant to deactivation, even upon severe aging (>800 °C). ✓ Na promotion remains at work even after aging due to the formation of β', β'' sodium alumina phases, which act as permanent spontaneous sources of Na promoter species during TWC operation.	[95] [96] [97] [98]

3.2.2. Oxygen Excess Conditions (Simulated Lean-Burn and Diesel Exhausts Gases)

Some other EPOC studies regarding the electrochemical activation of Pt-based electrochemical catalysts for NO_x reduction by propene under oxygen excess were performed by the groups of Vernoux and de-Lucas Consuegra. In this sense, Vernoux et al. observed a significant promotional effect of Na promoters electrochemically supplied from a NASICON solid electrolyte support [99]. The study was carried out under lean-burn conditions (oxygen excess, typical of Diesel engines). In such a system, electrochemical promotion is shown to strongly enhance both the catalytic activity and the selectivity to N_2 (from 41% to 61%) using very low overpotentials (−100 mV) at 300 °C (Table 7).

In the same line, the group of de Lucas Consuegra performed several EPOC studies by using Na and K promoters by means of $(Na^+)\beta'Al_2O_3$ and $(K^+)\beta Al_2O_3$ solid electrolytes, respectively. First, they studied the effect of reaction temperature and O_2 concentration in the feed stream on Pt catalysts promoted with Na species. Under lean burn conditions, at low reaction temperatures (220 °C), rate enhancement ratios up to 1.4 were observed for the NO reduction rate (Table 7) [100,101]. Nevertheless, as the reaction temperature increased, the promotional effect decreased, even resulting in a transition to a regime where the alkali metal induced poisoning. This progressive suppression of the promotional effect was due to an increase of the oxygen coverage on the Pt surface with the temperature, which led to a C_3H_6 adsorption inhibition. However, at all explored reaction temperatures, the presence of sodium ions induced a large increase of the N_2 selectivity (up to 90%), minimizing the N_2O formation.

Regarding the effect of the O_2 concentration, a study was performed at low temperatures (240 °C) under O_2 concentrations ranging from 0.5% to 5%. The promotional effect of sodium on the overall catalytic activity for NO removal was progressively lowered with increasing oxygen concentrations, as a result of a strong inhibition of propene adsorption and a relative increase of the oxygen coverage. In all cases, the presence of sodium ions induced an increase in the nitrogen selectivity by promoting the NO adsorption and subsequent dissociation.

In a new investigation, de Lucas Consuegra et al. [102] studied for the first time the promotional effect of K (EPOC) under the presence of steam in the feed stream under lean burn conditions, as a new approach to the development of efficient catalysts under real working conditions of combustion engines. In this study, they demonstrated that the presence of K promoters could be practically used to decrease the inhibitory effect of water for the reduction of NO by propene. In addition, as previously demonstrated in other studies, they verified that the K species form stable nitrate species on the Pt surface. This discovery led to a new series of studies performed by the same research team, leading with the NO_x-storage and Reduction technology (NSR) by means of alkali-based Pt electrochemical catalysts.

Briefly, the NSR process was developed in the early 1990s by Toyota. It is based on the ability of alkali and alkali earth elements to store Nox in form of nitrites/nitrates under lean burn conditions (oxygen excess). Then, by the injection of extra fuel, the stored NO_x are released and reduced with hydrocarbons, CO or H_2, to produce N_2, CO_2 and H_2O. The alkali promoters function in two ways in these "smart" materials: first, they act as electronic promoters to enhance the NO oxidation to NO_2 (which has been identified as the rate determining step of the overall process) and to store the NO_2 produced in form of nitrites/nitrates (as previously demonstrated).

In these studies, de-Lucas Consuegra et al. demonstrated, for the very first time, the possibility to perform the NSR process via EPOC, by the use of Pt catalysts supported on K^+-βAl_2O_3 solid electrolytes (to supply the K promoter species) [103,104]. The studies were performed under lean burn conditions and even under the presence of steam. The 2 main conclusions from these studies were:

- Potassium ions electrochemically transferred to the Pt catalyst play a double role in the NSR process, as a promoter for the NO oxidation reaction and as storage sites through the formation of potassium nitrates.
- The maximum yield of $Pt/(K)\beta Al_2O_3$ to effectively reduce NO_x to N_2 was obtained at T = 300 °C (Figure 17). This temperature is suitable for both, gasoline and diesel engines working under lean burn conditions.

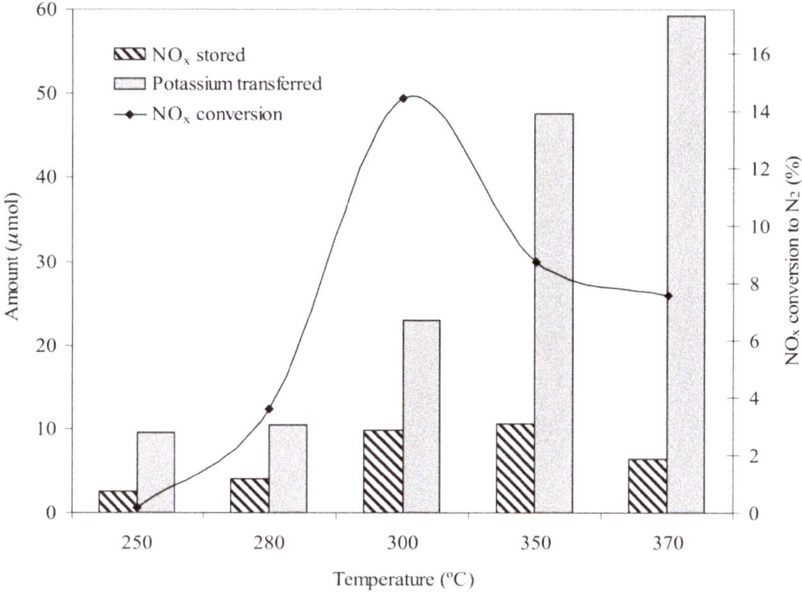

Figure 17. Influence of the reaction temperature on the amount of NO_x stored, potassium transferred and on the NO_x conversion to N_2 during NO_x storage/reduction experiments. Lean phase ($NO/C_3H_6/O_2$:1000 ppm/1000 ppm/5% O_2), 6 min of duration, $V_{cell} = -1.5$ V. Rich phase ($NO/C_3H_6/O_2$:1000 ppm/1000 ppm/0.5% O_2), 5 min of duration, $V_{cell} = 3$V. Data was acquired in a Pt/(K)βAl_2O_3/Au galvanic cell (Reprinted with permission from Ref. [103]; Copyright 2011, Elsevier).

In respect to conventional highly dispersed catalysts and using CCP, early studies by Burch and Watling that involved the influence of a number of promoters, including alkalis (namely K and Cs) and alkaline earths (namely La, Mg, Ba), on the Pt-catalyzed propene SCR of NO_x under lean burn conditions have showed that addition of one of the following: 2.42wt% K_2O, 7.25wt% Cs_2O, 7.93wt% BaO, 2.07wt% MgO and 8.35wt% La_2O_3 in 1wt%Pt/γ-Al_2O_3 catalyst results in suppression of the catalytic activity of Pt (NO_x conversion efficiency) and to marginal effects on N_2/N_2O selectivity (Table 7) [105]. However, later studies of the de-Nox activity of supported Pt catalysts under lean burn conditions have shown very pronounced promotional effects by alkalis or alkaline earths: Vernoux et al. [106] and Yentekakis et al. [47], systematically investigated the resulted promotional effects of Na addition on supported Pt/γ-Al_2O_3 catalysts (CCP) during NO + propene + excess O_2 as a function of the Na-loading, have shown significant beneficial effects of the alkali on both de-Nox activity and N_2/N_2O selectivity at low and intermediate Na-loadings (Table 7). In both studies, the effects (promoting or poisoning, depending on promoter loading) were understandable in terms of the influence of the alkali promoter on the relative adsorption strengths of reactants and intermediates on a highly populated surface with electrophilic and electrophonic adsorbates. In particular, Yentekakis et al. [47] demonstrated that Na acts beneficially to de-Nox process in a narrow window of Na loading (ca. 0–2.6wt% Na), while overpromotion to the optimal value of ~2.6wt% Na causes significant inhibition of the catalyst performance. The recorded promotional effects can be summarized as follows: Na widened the temperature window of the C_3H_6 + NO + O_2 reaction towards lower temperatures by ~50 °C, accompanied by an enhancement in N_2-selectivity by ~40 additional percentage points; the promotion is optimized at a sodium loading of 2.6wt% Na on the 0.5wt%Pt/γ-Al_2O_3 catalyst (Figure 18). Obviously, the promoter loadings chosen by Burch and Watling were out from this optimal window, explaining the poisoning effects observed by the authors [105].

Table 7. Electropositive promotion of PGMs-catalysed lean-NO_x reduction by C_3H_6 (the so-called simulated lean-burn or diesel exhaust conditions).

reactants	Catalyst and (promotion method applied)	Promoter	Conditions	Promotion highlights and achievements	Ref.
NO, C_3H_6, O_2 excess	Pt film on NASICON solid electrolyte (EPOC)	Na	T = 200–400 °C [NO] = 0.2%, $[C_3H_6]$ = 0.2%, $[O_2]$ = 5% Na coverage: non estimated (catalyst potential V_{WR} is given)	✓ Na promotion induced strong enhancements on both activity and N_2-selectivity under lean-burn conditions at quite low T. ➢ $\rho_{N_2} \sim \rho_{CO_2} \sim 1.7$, $\rho_{N_2O} \sim 1.1$, ΔS_{N_2} = +20% (41%→61%) at T = 295 °C, ΔV_{WR} = −200 mV.	[99]
NO, C_3H_6, O_2 excess	Pt impregnated on (Na)$\beta''Al_2O_3$ solid electrolyte (EPOC)	Na	T = 220–300 °C [NO] = 0.2%, $[C_3H_6]$ = 0.2%, $[O_2]$ = 0.5, 1 and 5%. θ_{Na} = 0–8%	✓ Na significantly promotes the low temperature (T = 220 °C) lean-burn activity and N_2-selectivity of Pt impregnated catalyst on Na^+ conducting solid electrolyte. At higher T (>240 °C) Na supply to the catalyst can lead even to poisoning of the activity but not of the selectivity which is increased (reaching ~90%). ➢ Promotion : $\rho_{N_2} \sim \rho_{CO_2} \sim 1.4$, ΔS_{N_2} = +20% (60→75%) at T = 220 °C, $[O_2]$ = 5%, θ_{Na} = 7–8% ➢ Poisoning: $\rho_{N_2} \sim 0.3$, $\rho_{CO_2} \sim 0.02$, ΔS_{N_2} = +50% (40→90%) at T = 300 °C and θ_{Na} = 7–8%	[100] [101]
NO, C_3H_6, O_2 excess, H_2O	Pt impregnated on (K)βAl_2O_3 solid electrolyte (EPOC)	K	T = 180–400 °C [NO] = 0.2%, $[C_3H_6]$ = 0.2%, $[O_2]$ = 5%, $[H_2O]$ = 5% θ_K: varying, non-estimated	✓ K promotion depresses the inhibitory effect of H_2O on the lean NO_x reduction by C_3H_6. ✓ K promoter species form stable nitrates on Pt surface, thus can operate as a NO_x-Storage Reduction (NSR) species.	[102]
NO, C_3H_6, O_2 excess	Pt film on (K)βAl_2O_3 solid electrolyte (EPOC)	K	T = 250–370 °C [NO] = 0.1%, $[C_3H_6]$ = 0.1%, $[O_2]$ = 5% θ_K: varying, non-estimated	✓ The electrochemically assisted NOx-Storage Reduction (NSR) process by a Pt/(K)βAl_2O_3/Pt electrochemical cell is demonstrated. ✓ Electrochemically transferred K play a double role: promoter of the lean NO_x reduction and NO_x-storage/release species. ➢ Maximum deNO_x to N_2 yield was obtained at T = 300 °C.	[103] [104]

Table 7. *Cont.*

reactants	Catalyst and (promotion method applied)	Promoter	Conditions	Promotion highlights and achievements	Ref.
NO, C_3H_6, O_2 excess	1 wt% Pt/γ-Al_2O_3 (CCP)	K, Cs, La, Mg, Ba	T = 200–500 °C [NO] = 0.1%, [C_3H_6] = 0.1%, [O_2] = 5% Alk-loadings: A/Pt = 10/1 molar ratio	✓ Performance inhibition effects of alkaline materials addition were found at the specific loading used for each alkaline (2.42, 7.25, 7.93, 2.07 and 8.35wt% of K_2O, Cs_2O, BaO, MgO and La_2O_3, respectively).	[105]
NO, C_3H_6 or C_3H_8, O_2 excess	0.9 wt% Pt/γ-Al_2O_3 (CCP)	Na	T = 200–650 °C [NO] = 0.2%, [HC] = 0.2% (HC: C_3H_6 or C_3H_8), [O_2] = 5% GHSV = 20,000h^{-1} Na-loadings: 0.12, 1 and 5 wt%	✓ Under C_3H_6/NO/O_2 reaction strong Na-induced promotion was obtained (optimal Na-loading 5wt%). In opposite, under C_3H_8/NO/O_2 reaction severe inhibition was induced by Na addition. ➢ Promotion : $\Delta T_{50(C3H6)} = -39$ °C under C_3H_6/NO/O_2 reaction on the 5 wt% Na – modified catalyst; $\Delta S_{N_2} \sim +20\%$ for T > 280 °C. ➢ Poisoning: $\Delta T_{50(C3H8)} = 191$ °C under C_3H_8/NO/O_2 reaction on the 5wt%Na-modified catalyst.	[106]
NO, C_3H_6, O_2 excess	0.5 wt% Pt/γ-Al_2O_3 (CCP)	Na	T = 200–450 °C [NO] = 0.1%, [C_3H_6] = 0.1%, [O_2] = 5% Na-loadings: 0, 1.6, 2.6 and 4.2 wt%	✓ Strong promotion by Na with an optimal loading of 2.6wt% Na. Higher Na-loadings caused strong inhibition (volcano behaviour; over-promotion). NO_x conversion window is widened and moved to lower temperatures. ➢ $\Delta T_{50(C3H6)} \sim \Delta T_{50(NOx)} = -65$ °C; $\Delta S_{N2} = +40\%$ at T > 320 °C for 2.6wt% Na loading	[47]
NO, C_3H_6, CO, H_2O, CO_2, O_2 excess	Ir-black (CCP)	Na	T = 150–450 °C [NO] = 300 ppm, [C_3H_6] = 1800 ppm, [CO] = 450 ppm, [O_2] = 8%, [H_2O] = 10%, [CO_2] = 10.7% Na-loadings: 0–10wt%	✓ Na-loadings > ~1 wt% enhanced the selectivity towards N_2 in the whole temperature range (150–450 °C) but not affected the activity. ✓ Na-loadings > 3 wt% caused rate inhibition for all CO, NO_x, C_3H_6 conversions.	[107]
NO, C_3H_6, (O_2)	Ir film on (K)β″Al_2O_3 solid electrolyte (EPOC)	K	T = 250–400 °C [NO] = 0.2%, [C_3H_6] = 0.2%, [O_2] = 0–5% θ_K = 0–100%	✓ Marginal effects of K addition on the Ir surface at reducing (absence of O_2) conditions. Strong K-induced poisoning at oxidizing conditions for both propene oxidation and NO_x reduction with significant decreases in N_2-selectivity, as well.	[108]

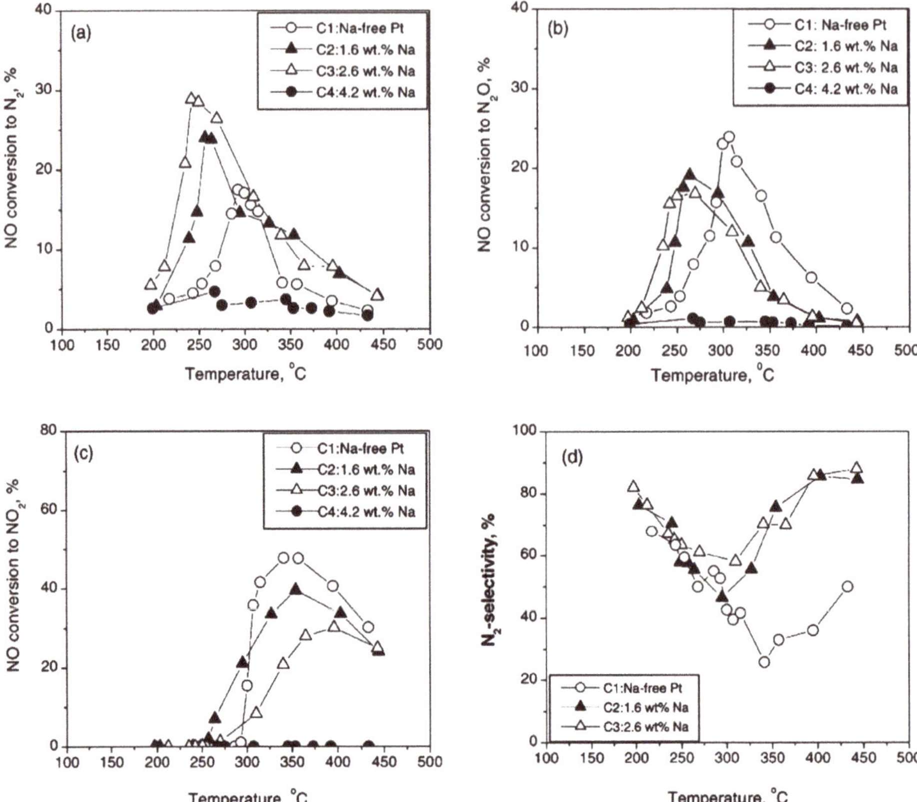

Figure 18. C_3H_6 + NO + O_2 reaction on Na-promoted Pt/γ-Al_2O_3 catalysts. The effect of temperature on the conversion of NO to various N-containing products and the corresponding N_2-selectivity: (**a**) NO conversion to N_2; (**b**) NO conversion to N_2O; (**c**) NO conversion to NO_2; (**d**) N_2-selectivity. Conditions: 1000 ppm NO, 1000 ppm C_3H_6, 5% O_2; reactant contact time 4 s. (Reprinted with permission from Ref. [47]; Copyright 2005, Elsevier).

The performance of another noble metal, iridium, the three-way catalytic chemistry of which has several promising characteristics, have been studied under Na-promotion by Wogerbauer et al. [107]. A simulated exhaust gas mixture consisted of 300 ppm NO, 0.18% propene, 450 ppm CO, 8% O_2, 10% H_2O, 10.7% CO_2, balance N_2 was used, while the temperature and Na promoter loading were varied between 150–450 °C and 0–10wt%, respectively. They found that under the conditions used and for Na loadings higher than ~3wt%, Na caused an inhibition on the Ir-black catalytic activity for CO, propene and NO_x conversions (Table 7). However, the selectivity towards N_2 was significantly improved up to 100% for samples with Na contend > ~1wt% in the whole temperature range.

More recently, Goula et al. studied the effect of electrochemically supplied potassium on a porous Ir-film catalyst under the NO + C_3H_6 + O_2 reaction at a variety of oxygen concentrations [108]. In this EPOC study the Ir-film was interfaced with a (K)$\beta''Al_2O_3$ solid electrolyte on which the catalyst was supported by spattering. It was found that K addition on the Ir surface is detrimental on catalyst performance (propene oxidation, NO reduction efficiency and N_2-selectivity) for all oxygen concentrations investigated. Only at zero O_2 concentration, that is,, at net reducing conditions of the reactive mixture, a slight promoting effect of potassium on the N_2-selectivity but not on the rates was observed (Table 7). This very different effect of alkali-promotion, compared to that observed on Pt and Pd noble metals under similar conditions (on which strong promotional effects were recorded), was understandable in terms of the electronic influence of co-adsorbed potassium on the adsorption strengths of the neighbour reactants on the Ir surface: the excessive enhancement of oxygen adsorption on Ir sites in the expense of the hydrocarbon adsorption caused O-poisoning of the surface.

Overviewing Section 3.2.2 and the involved literature listed in Table 7, it is apparent that as we have seen in Section 3.1.4, alkali-promotion of the reduction of NO by propene remains very effective even under excess O_2 conditions, at concentrations > ~5% which simulate lean-burn and diesel engines exhaust gases. However, at excess oxygen conditions promotional effects on both activity and selectivity are significantly attenuated and also limited in quite narrows temperature and/or promoter loading windows. Volcano type behaviour of promotion is once again recorded upon increasing alkali loading and EPOC or CCP promotion characteristics were qualitatively similar.

3.3. *Mechanistic Implications: The mode of Action of Electropositive Promoters*

3.3.1. Main Promotion Characteristics and Mechanistic Implications

The above described results show that electropositive promoters (alkalis and alkaline earths) can markedly affect the catalytic properties of Pt-group metals during CO or hydrocarbons oxidation and NO reduction by CO or hydrocarbons in the absence or presence of oxygen, as well as of other environmentally important reactions (see Tables 1–7). The promotional benefits on the catalytic performance (activity, selectivity) achieved were similar, independently of the promotion method used, that is,, the electrochemical promotion of catalysis (EPOC) or the conventional catalyst promotion one (CCP). The intensity of promotion, the optimal promoter loading and other characteristics of the induced promotional effects are subjected to the specific reaction/catalyst system and the individual reaction conditions applied (Tables 1 and 2).

This electropositive promotion (by alkalis and alkaline earths) of PGMs for emissions control catalysis reaction was found to be very pronounced not only when applied to model reaction systems (Tables 1–5) but also under complex reaction systems that mirror practical applications, that is, stoichiometric gasoline, lean-burn and diesel engines exhaust conditions (Tables 6 and 7). The latter boosts the practical importance of the subject and direct implementations on practical systems appear attractive and promising.

Model reactions studies and their individual characteristics (Section 3.1) enabled us to better understand the mechanism of the promoters' action. EPOC has also a significant role on this issue, since it gave us the opportunity of in situ adjustment of the promoter loading, recording at the same time its influence on the catalytic performance. Recall that some studies were explicitly focussed on experimentally establishing that EPOC and CCP are similar in origin and are subjected to the same physicochemical rules; the only practical different is the procedure used for the ptomoter supply [21,22,25]. To this end, Yentekakis and co-workers [21,22] have first shown that the promotional effects of sodium, electrochemically induced (via EPOC) on the NO + C_3H_6/Pt catalytic system, mirror those obtained on a highly dispersed Pt/γ-Al_2O_3 catalyst, promoted by means of CCP method at similar conditions. Therefore, EPOC and CCP will be hereafter considered as identical phenomena; the mechanistic model described below that interprets the mode of action of electropositive promoters, may be considered identical valid for both promotion methods. The vast majority of the publications reviewed here have been based on this model to consistently interpret individual promotional phenomena appeared at the specific reaction systems investigated. The main body of the model is described below and several direct spectroscopic evidences supporting its consistency are presented.

Using EPOC (via a galvanic cell as it was described by the formula (1) and Figure 1) one can electrochemically control (Faraday law) the migration of promoting species (e.g., Na^+, Figure 19) from the solid electrolyte onto the catalyst/gas interface, where an effective and overall neutral double layer is formed through this ions back-spill over imposed electrochemically (Figure 19).

Actually, the electrocatalytic reaction taking place at the three-phase-boundary metal-solid electrolyte-gas that leads to the formation of the effective double layer is [6,7,16,109]:

$$Na^{+}(\text{from } \beta''Al_2O_3) + e^{-}(\text{from metal}) \rightarrow [Na^{\delta+}\text{-}\delta^{-}](\text{on metal}) \quad (9)$$

where $[Na^{\delta+}–\delta^{-}]$ denotes the overall neutral Na species present at the metal-gas interface (effective double layer): the Na adatom on the metal surface carries a charge δ^{+} and δ^{-} is the image charge in the metal; the charge δ is about 0.8–0.5, decreasing with increasing Na coverage [7].

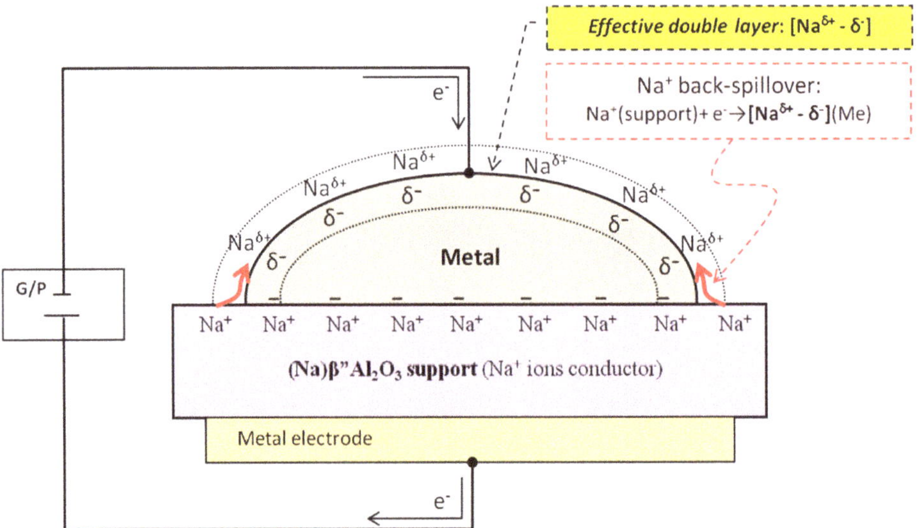

Figure 19. Schematic of the effective double layer approach of catalyst promotion [6,7,110]. An electron enriched (lower work function) catalyst surface is developed due to the uniformly dispersed $Na^{\delta+}$ adatoms.

It is worth noting that an identical configuration is expected on the metal catalyst surface when Na is introduced to the catalyst by conventional methods (e.g., impregnation, vapor deposition, etc.), as recently demonstrated by Lambert and co-workers [110,111] who showed using XPS that the electrochemically-supplied sodium is identical with gas-supplied (by evaporation) of $Na^{\delta+}$ cations on the catalyst surface.

Since alkali ad-atoms are imposed on the catalyst surface, they are spread over the entire metal surface due to strong repulsive dipole-dipole interactions, establishing a homogeneous, neutral, effective double layer. This double layer corresponds to an electrically modified metal surface, practically an electron enriched (lower work function) metal surface due to the δ^{-} charge located at the metal side interface half-layer. This electrically modified surface interacts with the co-adsorbed reactants, reaction intermediates and products, changing their binding energies (in respect to those on a promoter-free metal surface), producing pronounced alterations in catalytic performance. Therefore, the main feature of the electropositive promotion of PGM by alkalis and alkaline earths is the resulted electron enriched, lower work function, PGM surface.

Therefore, the concomitant alterations on the adsorption characteristics of the co-adsorbed reactants on such a modified metal surface are of course subjected to the electronic properties of the reactants: electrophilic (electron acceptor) or electrophobic (electron donor): both theory [112] and experiment (e.g., [106,113]) have shown that electropositively promoted (by alkalis) Pt-group metal (PGM) surfaces appear strengthening in the bonds of the metal—electron acceptor (electrophilic) adsorbates, for example, PGM–NO, PGM–O_2, PGM–N, PGM–O, PGM–CO and weakening in the bonds of the metal-electron donor (electrophobic) adsorbates, for example, PGM-hydrocarbons and their fragments, PGM-CO. Donation and/or backdonation issues of electron charge between adsorbates and metal surfaces during the formation of a chemisorptive bond and their impact on the bond strength have been thoroughly explained by Vayenas and Brosda in Ref. [114] and in brief summarized in the following statements: "*Electron backdonation to bonding singly occupied orbitals of an electron acceptor reactant lying below the metal Fermi level* (E_F) *results in strengthening of the metal-adsorbate bond, while electron backdonation to antibonding orbitals, leads to weakening of the chemisorptive bond and destabilization of the adsorbate. Also electron donation to a metal from a singly occupied orbital on an electron donor adsorbate lying above Fermi level of the metal leads in general to strengthening of the chemisorptive bond. There are cases, such as the CO chemisorption on transition metals, where both donation of electrons (from the adsorbate to the metal) and backdonation of electronic charge (from the metal to the adsorbate) play an important role in the chemisorptive bond formation (e.g., Blyholder model for CO chemisorption [115])*"; (The latter explains why someone can find CO to be considered as an electron-acceptor or even as an electron-donor adsorbate). These modifications on the reactants chemisorptive bonds are accompanied by alterations in the activation energies of the catalytic reactions and in some cases on their mechanism (e.g., on the rate determination step), resulted to dramatic changes on their intrinsic activity and/or selectivity. The specific features of the promotion are subjected to the specific catalytic systems under consideration and to the reaction conditions imposed.

Taking into account that the vast majority of the reactions involved in emission control catalysis processes obey Langmuir-Hinshelwood mechanism (competitive adsorption of the reactants), the following example could be of usefulness for a better understanding of the mechanism of action of the promoters under the concept of the effective double layer approach. Let's suggest a starting rate at point 1 on the rate versus [A] curve I (Figure 20), as determined by the applied reaction conditions on the unpromoted reaction A + B, that are considered to follow a typical Langmuir-Hinshelwood (LH) mechanism with the well-known characteristic rate maximum (point 2) due to the competitive adsorption of the reactants. The reaction rate is proportional to the product of the reactants' coverages, $\theta_A \cdot \theta_B$ (Equation (10)), maximized at equilibrated coverages θ_A and θ_B on the catalyst surface:

$$r = k_o \exp(-E_a/RT)\theta_A \cdot \theta_B \quad (10)$$

where $k_o\exp(-E_a/RT) = k$ is the temperature dependence of the rate constant k with a pre-exponential factor k_o and an apparent activation energy E_a.

Applying (chemically or electrochemically) a promoter, the created effective double layer (Figure 19) causes alterations on the adsorption strengths of the reactants, modifying their coverages θ_A and θ_B thus giving the possibility of their adjustment at the equilibrated values $\theta_A{}^*$ and $\theta_B{}^*$ that minimizes reaction probability and consequently rate (point 2):

$$r^* = k_o \exp(-E_a/RT)\theta_A{}^* \cdot \theta_B{}^* \quad (11)$$

However, this is the minimum effect that can be offered by this course and actually Equation (11) and consequently path 1→2 are not real: Due to the induced changes on the chemisorptive bonds of the reactant, alteration on the apparent activation energy of the reaction is also expected from the unpromoted value E_a to a modified value $E_a{}^*$, leading the reaction rate to follow a new substantially enhanced behaviour (curve II, Figure 20, for example, point 3), following Equation (12):

$$r^{**} = k_o \exp(-E_a{}^*/RT)\theta_A{}^* \cdot \theta_B{}^* \quad (12)$$

Notably, in the cases where reaction mechanism modifications are induced by the promoter as well (e.g., change of the rate determination step), promotion can receive unprecedented enhancements (promoted rate r^{***} on curve III, Figure 20, e.g., point 4).

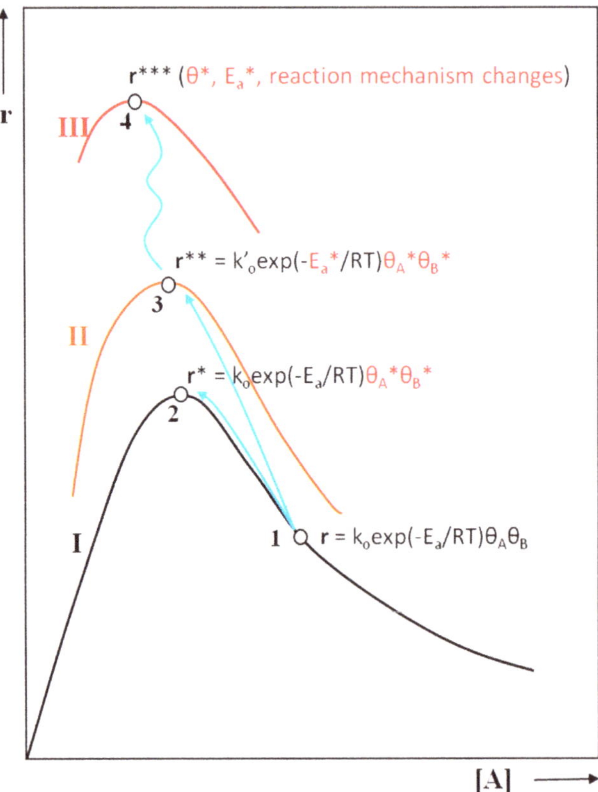

Figure 20. Schematic representation of the mechanism of promotion on a Langmuir-Hinshelwood type reaction.

An example which mirrors all the aforementioned issues is the alkali-promoted NO + C_3H_6 reaction on Pt [15,58,61]; reaction rate enhancements up to 40,000% were achieved accompanied with dramatic chances on the selectivity towards N_2, as well (Table 4). The promotion mechanism on this system is described as follow (Figure 21):

First, it is well established that NO + C_3H_6 reaction over Pt obeys Langmuir-Hinshelwood type kinetics with characteristic rate maxima reflecting competitive adsorption of the two NO and C_3H_6 reactants on Pt active sites; these rate maxima occurred at very low C_3H_6/NO ratios (<0.08), that is, very high NO concentrations relative to C_3H_6 are necessary in order to achieve a comparable coverage of both reactants on the Pt surface, thus maximizing the rate [58,62], reflecting a weaker adsorption of NO on Pt sites relative to C_3H_6 [58,62,67,68] mainly due to the presence of π-electrons in the alkene structure. As a consequence, in a wide range of reactants partial pressures and temperatures conditions the un-promoted (alkali-free) Pt surface is predominantly covered by propene and propene-derived fragments (Figure 21a), therefore the un-promoted reaction rate, r_o, is quite low.

Addition of alkali, which is accompanied by the formation of the [Alkali$^{\delta+}$–δ^-] effective double layer, provides an electron enriched (lower work function) metal surface that in effect favours the adsorption of electrophilic co-adsorbates (NO) and inhibit that of electrophobic co-adsorbates (C_3H_6 and its fragments). According to the following Reaction (13)–(18) scheme,

$$NO(g) \rightarrow NO_{ads} \quad \text{(enhanced by alkali addition)} \tag{13}$$

$$NO_{ads} \rightarrow N_{ads} + O_{ads} \quad \text{(enhanced by alkali addition)} \tag{14}$$

$$C_3H_6(g) \rightarrow C_3H_{6,ads} \quad \text{(inhibited by alkali addition)} \tag{15}$$

$$N_{ads} + N_{ads} \rightarrow N_2(g) \tag{16}$$

$$N_{ads} + NO_{ads} \rightarrow N_2O(g) \tag{17}$$

$$O_{ads} + \text{hydrocarbonaceous species (or CO)} \rightarrow CO_2(g) + H_2O(g) \tag{18}$$

the resulted effects are (Figure 21b): (i) strengthening of the Pt-NO bond with a concomitant enhancement population of NO molecules on the alkali-modified surface (Reaction (13)) –the surface populations of NO and propene are therefore equilibrated; (ii) strong enhancement of the dissociative adsorption of NO (Reaction (14)) as a result of the Pt–NO bonds strengthening and the concomitant weakening of the N–O bond in the adsorbed NO molecules—the surface is enriched with very active atomic oxygen species; (iii) weakening of the strength of the Pt–C_3H_6 bond—an equilibrated population of more active (due to their weaker adsorption) hydrocarbonaceous species is resulted (Reaction (15), Figure 21b).

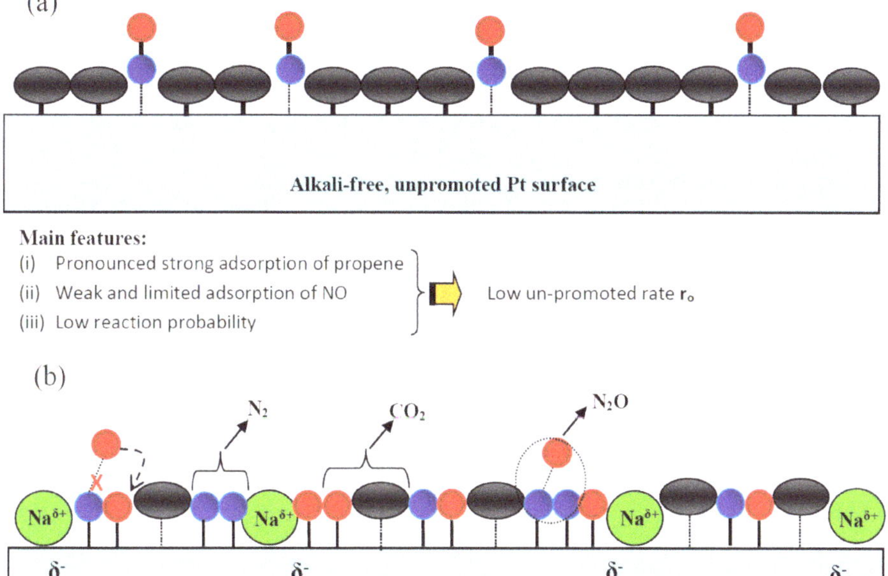

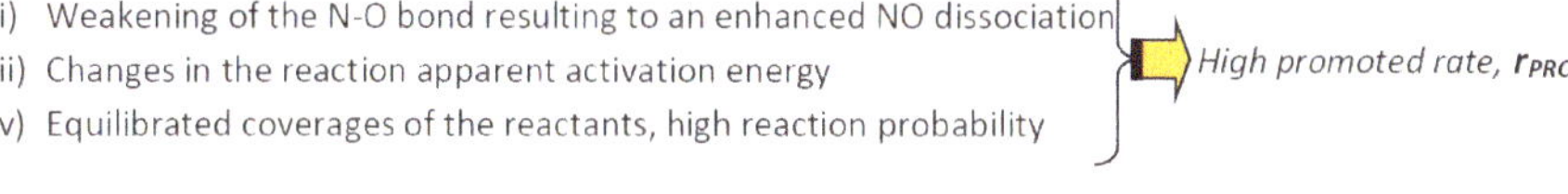

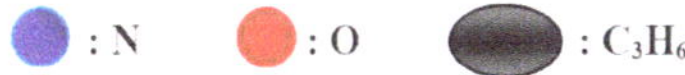

Figure 21. Schematic of the mechanism of action of alkalis as promoters on the Pt-catalysed NO + C_3H_6 reaction. Expected distribution of reactants and/or intermediates over the unpromoted surface (**a**) and the Na-promoted surface (**b**).

All these factors operate together in increasing the reaction probability of NO and propene molecules, leading to an enhanced promoted rate, r_p, over the alkali-modified surface. These changes on adsorbed species distribution and/or population (coverage), their bond-strengths and activation energies and on reaction rate-determination-step (rds) were successfully used to convincingly interpret the rate enhancement values, $\rho = r_{CCP}/r_o$, as high as 420 and 280 for N_2 and CO_2 productions have been reported for this catalytic system under electropositive promotion by alkalis or alkaline earths [15,61,62]. Notably, these gains in catalytic activity did not reflect the maxima of the L-H rate curves, *that is*, were not restricted by the optimal balance of the competitive reactants' coverages. These were extraordinarily larger, due to the synergy of the aforementioned factors.

The observed increase in N_2-selectivity upon electropositive promotion can be also readily understood in terms of the above considerations. According to the reaction network (13)–(18), the production rates of N_2 and N_2O depend on the extent of NO dissociation (reaction (14)), which is followed by the elementary Reactions (16) and (17). The observed increase in the N_2-selectivity upon promoter addition is a consequence of increased NO dissociation, that is, less molecular NO and more atomic N on the surface. According to Lang et al. [112], the electrostatic field produced by Na^+ can shift the $\pi*$-orbital energy of NO adsorbed in the vicinity of an alkali ion below the Pt Fermi level. Then, valence electrons from the metal can populate the NO $\pi*$-orbital energy, resulting in weakening of the N–O bond and a strengthening of the Pt–N bond. Both factors favour Reaction (16) over Reaction (17), leading to higher N_2 selectivity.

In the case of CO + O_2 reaction the strong promotion observed under CO-rich reaction conditions (e.g., [16,27]), was understandable by means of the pronounced strengthening of the O adsorption bond, compared to that of CO: at such conditions where CO coverage predominates and the formation of CO islands suppresses reaction probability, the alkali-induced pronounced adsorption of O species causes destruction of CO islands thus increasing reaction probability [16].

In the case of alkanes oxidation, it is well known that alkanes, in particular CH_4, have low propensity for adsorption (activation) on PGM surfaces. The adsorption of these electrophobic adsorbates is being even worse on alkali-modified electron enriched surfaces. This makes the generally observed alkali-induced poisoning of alkanes oxidation reactions readily understandable, no matter of what oxidant agent (oxygen or NO) used [48,49,65,66].

The opposite was true for alkenes oxidation (by O_2 or NO), due to their strong propensity to adsorb on PGMs. Thus, weakening their adsorption and at the same time strengthening the adsorption of the electrophilic limiting reactant (O) on the alkali modified PGM surfaces are factors that operate synergistically creating ideal promotion conditions (e.g., [18,20,43–47]).

Comparing the alkali promotion of the NO reduction by alkenes, the Ir~Rh < Pd < Pt increasing order can be observed: marginal promoting effects have been obtained during Rh- or Ir-catalysed NO reduction by propene [59,60,64,107,108], in opposite to the very large on Pd [14] and to the extraordinarily large on Pt (e.g., [15,58,61,62,106]) This is related to the different propensity of NO adsorption on these metals: on Ir and Rh, NO has high propensity of adsorption, which is also dissociative; on Pd less and partly dissociative, while on Pt it is even less and non-dissociative. Therefore, enhancing NO adsorption by alkali-promotion on Ir and Rh surfaces has not further practical value (NO adsorption/dissociation is close to ideal on the unpromoted Ir and Rh), while on Pt the non-dissociative and quite limited NO adsorption can become much stronger and fully dissociative, enabling Pt to behave like Rh or Ir (e.g., [15,47,61,62,89]).

Comparing the amount of an alkali necessary to optimize promotion for a certain reaction on different Pt group metals, the general trend is that only few surface concentrations are needed in the case of Rh and Ir (ca. θ_{Alkali}~1–3%), quite more for Pd (θ_{Alkali}~5–10%) and much more for Pt (θ_{Alkali}~15–40%) (e.g., [14,15,18,22,51,57,58,108]). This is readily understandable in similar terms used above: Rh and Ir electron availability (work function, WF) is by its own close to the optimal value for the reactions under consideration and only few amounts of alkalis are necessary for the optimal promoter loading adjustment. The opposite is true for Pd and in particular for Pt; these two metals showing a higher WF than Rh and Ir.

For a certain catalytic system and operating conditions, the amount of alkali that is needed to optimize promotion depends on the chemical identity of the alkali: heavier alkalis appear to be more effective that the lighter ones (e.g., [61]). This is fully consistent to the effective double layer approach considered and to the theoretical predictions of Lang et al. [112], which have shown that the larger the alkali cation the greater the effect its electric field has on an electron acceptor adsorbate (e.g., NO).

In the cases of more complex reactions systems, as for instance those simulating automotive exhaust gas mixtures, the explanation of the promotional effects is typically more complicated and may involve all the possible effects, synergistic or competitive, induced through the modification of the metal work function on all electrophilic and/or electrophobic reactant species and intermediates on such a densely populated surface with all these competitive species (e.g., [47,79,89,106]).

It is also of worth noting that: (i) Alkalis and alkaline earths are typically permanent promoters as a result of the fact that they do not participate in the reactions networks. This is of high technological importance: since the optimal alkali amount is incorporated on the catalyst and independently of the method used for its supply (EPOC or CCP), its amount remains practically constant for a very long time offering promotion. This explains why Faradaic Efficiency values as large as $\Lambda > 10^5$ have been measured in EPOC by alkalis studies (e.g., [16–18,42]). (ii) Volcano type behaviour of the promotional effects as a function of the promoter loading was an additional common feature of the titled promotion, that is, outgoing of the optimal amount of the promoter (the so-called over-promotion) the reaction is gradually inhibited rather than promoted (e.g., [15–17,47,58]). Two main factors were considered to be responsible for this behaviour: (a) the extended strengthening of the adsorption bond of the electrophilic adsorbate (electronic effect) and (b) the formation of extended surface complexes of the promoter with the reactant species and/or reaction intermediates that can block active sites (geometric effect). These factors can operate together suppressing the rate in case of over-promoted catalysts. However, it must be noticed that the formation of such 2D or 3D surface alkali complex compounds (alkali nitrites, nitrates, carbonates, oxides or even superoxides, depending on the reaction atmosphere), that can be valid even at low promoter loadings, do not actually cancel the role of alkali as a promoter (e.g., [20,43–46,51,52,57–60]). On the opposite, stabilization on the catalysts surfaces, even at elevated temperatures of catalytic interest, of normally volatile alkali metals is achieved by the formation of such stable surface alkali complex compounds.

3.3.2. Direct Spectroscopic and Other Analytical Technique Evidences

Besides of the numerous kinetic studies demonstrating the usefulness of the electropositive promotion of PGM in emissions control catalysis, particular emphasis to direct evidences for the origin and mechanism of this promotion have been devoted by several research groups using a variety of surface science or other analytical techniques, including: in situ diffuse reflectance infrared Fourier transform spectroscopy (DRIFTS) [95–98,113], x-ray photoelectron spectroscopy (XPS) (e.g., [51,52,111]), temperature programmed desorption (TPD) [106,116–120], scanning tunnelling microscopy (STM) [121–123] and work function [51,52] measurements.

To this end, Koukiou et al. [113], studying the interaction of NO with Na-modified Pt surfaces over a conventional, highly dispersed, Pt catalyst, namely $Pt(Na)/\gamma\text{-}Al_2O_3$, by means of in situ DRIFTS, have shown that increasing sodium loading causes a pronounced and progressive red shift of the N–O stretching frequency (Figure 19) associated with molecular NO adsorbed on the Pt component of the supported catalyst.

At the highest sodium loading (10wt% Na; corresponding to an upper limit of about $\theta_{Na} = 0.5$ nominal sodium fractional coverage on Pt particles [102]) a 1680 cm^{-1} species is observed that corresponds to an activated $NO^{\delta-}$ species with bond order 2, that is, a negatively charged adsorbate with increased electron density in the π^* antibonding orbital. The rigorous model developed by Lang et al. [112] was considered for understanding the effect of alkali on Pt surface. They have shown that the inhomogeneous electric field associated with adsorbed alkali ions acts to depress the energy of the antibonding π^* orbital of electron-accepting (electrophilic) co-adsopbates (e.g., CO, NO) bellow the metal Fermi level. Therefore, the red shifts showed in Figure 22, as authors argue [113], are due to the resulting metal→adsorbate charge transfer that acts to weaken the N–O bond, red-shifting its vibration frequency and promoting its dissociation, rather than to alkali-induced changes in the relative populations of various forms of adsorbed NO.

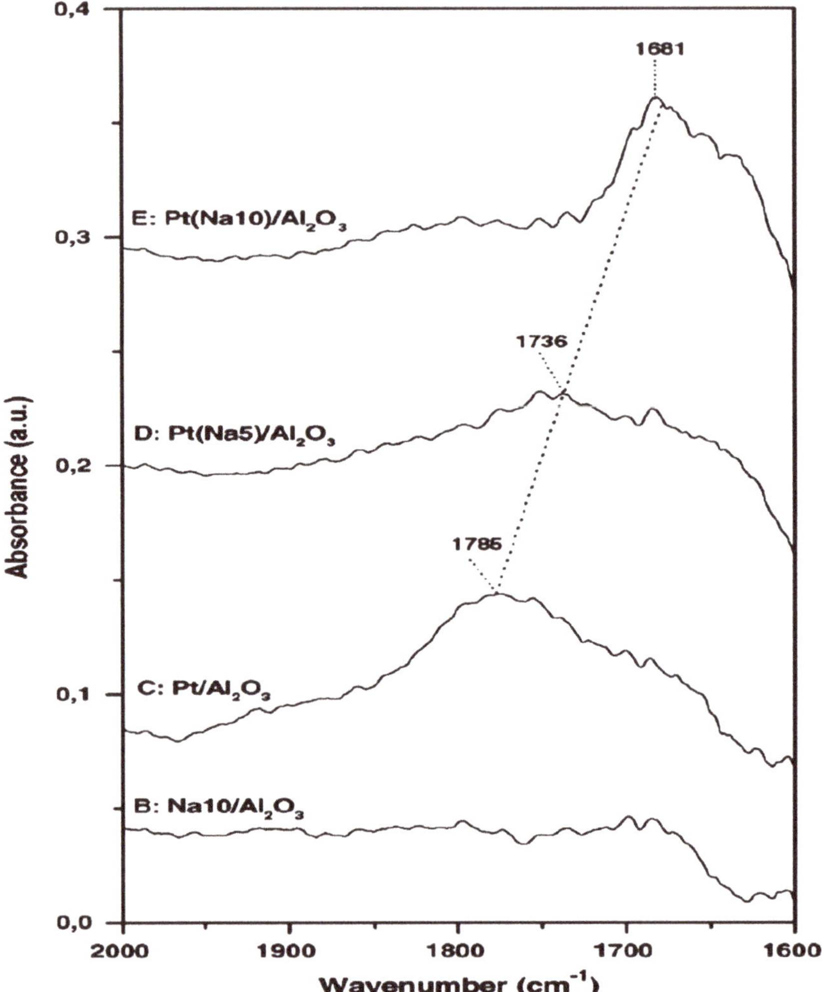

Figure 22. DRIFT spectra (expanded scale for the region 2000–1600 cm^{-1} and subtracting spectrum of pure Al_2O_3) obtained with 10wt%Na/γ-Al_2O_3 (**B**), Pt/γ-Al_2O_3 (**C**), Pt(Na5)/γ-Al_2O_3 (**D**) and Pt(Na10)/γ-Al_2O_3 (**E**) samples after NO adsorption for 1 min at 27 °C. (Reprinted with permission from Ref. [113]; Copyright 2007, Elsevier).

The chemisorption characteristics of NO on Na-dosed Pt{111} surface has been also studied by Harkness and Lambert [116] by means of temperature-programmed desorption (TPD). It was clearly demonstrated that the presence of Na on the Pt{111} single crystal causes an increase in the adsorption energy of NO (i.e., strengthening of the metal-NO bond) accompanied by a monotonically increased

NO dissociation (i.e., N_2 desorption) as a function of Na coverage. In particular, NO-TPD spectra on Na-dosed Pt{111} (from 0 to 1.2ML of Na coverage) have showed two characteristic peaks, indicated as features (a) and (b); sodium caused an increase in the desorption peak (a) temperature from 360 K on the clean Pt{111} to 380 K at 0.25 ML Na-dosed Pt{111}. Even at very low Na coverages (>0.1 ML), the second NO desorption feature (b) from sodium affected sites is appeared at ~520K which is grew at the expense of the feature (a) and progressively shifted to higher desorption temperature, up to 610 K, upon increasing Na coverage to 1.2 ML. At the same time, the feature (a) is progressively attenuated to zero (practically vanished at~>1ML of Na); A progressively increased N_2 peak is concomitant to the aforementioned TPD experiment, with the N_2 desorption peak to remain constant at 540K up to 1ML of Na, after which it is shifted to 540 K for 1.2 ML (i.e., 100% dissociation of adsorbed NO can be obtained at sufficiently high, ca 1 ML, sodium coverage) [116].

Similar TPD studies by Garfungel et al. [117], involving the interaction of NO with K-dosed Pt{111}, yielded the same evidences: an alkali-induced increase in the Pt–NO bond strength accompanied by alkali-promoted NO dissociation was demonstrated.

Temperature-programmed desorption (TPD) studies after adsorption of oxygen on Na-dosed Pt/Al_2O_3 catalysts where performed by Vernoux et al. [106] in order to elucidate possible influences of the Na addition on the Pt–O bond strength. They demonstrated that the adsorption strength of the atomic oxygen (an electron acceptor adsorbate) on the Pt component of the catalyst is enhanced by Na addition; their data clearly showed that the temperature of the oxygen desorption peak correspond to Pt–O interaction increases with increasing Na loading of the catalyst, directly evidencing a strengthening of the Pt–O bond [106].

A number of early and recent studies concern the chemisorption of CO on alkali-modified PGM surfaces [95–97,117–120]. To this end, Bertolini et al. [118] studied the chemisorption of CO on K-modified Pt(100) surfaces by means TPD and Auger Electron Spectroscopy (AES). The progressive increase on the CO desorption peak temperature (i.e., strengthening of the Pt–CO) as a function of K coverage, observed at low K coverages θ_K (ca. 0–0.4), was attributed to long range uniform electronic modifications, that is, work function changes, due to charge transfer; $K^{\delta+}$ ($\delta \approx 1$) adspecies appear on the K-dosed surface, uniformly distributed due to repulsion between these charged species. Such an electron enriched (lower work function) Pt surface strengthens the adsorption of CO via substantial charge donation form the K-modified Pt surface into the $2\pi^*$ orbitals of CO [118]. At high K coverages (ca. 0.45–1) short range interactions between the CO and K adspecies, that is, an even more direct bonding of CO with K (formation of CO-K-Pt surface complexes), were considered.

Pitchon et al. studied the effect of (Li, Na, K and Cs)-dosed Pd/SiO_2 on CO adsorption characteristics [119]. They demonstrated drastic changes in the infrared spectrum of CO adsorbed on Pd sites; a significant weakening on the strength of the C–O bond of adsorbed CO molecules caused the appearance of a new νCO band at low infrared frequency—the extent of the interaction was depended by the nature of the alkali. These influences were assigned to a localized interaction "Pd–CO–alkali" rather than to long range electronic modifications.

Using FTIR spectroscopy, Liotta et al. studied the effect of sodium on the adsorption of CO on Pd-based catalysts supported on two different supports (SiO_2 and model or natural pumices) [120]. Depending on the support used, the appearance of both electronic and geometric effects—that attributed to the different localization of sodium ions in the catalysts- were evidenced; the geometric effects were predominant in the Pd/SiO_2 catalysts, whereas electronic effects were most important than the geometric ones in the Pd/pumise catalysts. It was again considered in this study that an electronic density transfer to Pd is the common effect of the presence of alkali either it is subsequently added to the catalysts or is present as a structural component of the support, although, the authors observed important geometric effects in the former case. The electronic effect, that produces a red shift (towards lower frequencies) of the CO bands of the IR spectra of chemisorbed CO, owing to an enhanced transfer of electron density from the metal to the π^* molecular orbitals of CO, was again invoked [120].

Electron enriched Pt surfaces resulted by combined application of alkali-induced surface promotion and support-mediated promotion by Ce-based mixed oxides were also recently demonstrated via in situ DRIFTS studies by Yentekakis and co-workers [95–97]. Over doubly promoted $Pt(Na)/Al_2O_3$-CeO_2-La_2O_3 catalysts operated under TWC conditions, the authors demonstrated a substantial population of adsorbed species at 2060 cm^{-1} attributed to CO on reduced Pt^0 sites when CeO_2-La_2O_3 was incorporated into the Al_2O_3 support. This feature was substantially exacerbated when Na promoter was also incorporated in the catalyst formulation, while at the same time the formation of active Pt–NCO intermediates at about 2180 cm^{-1}, which resulted from an enhanced NO decomposition, was facilitated by alkali-promotion on the $Pt(Na)/Al_2O_3$-CeLa catalyst (Figure 21). On the un-modified Pt/Al_2O_3 catalyst counterpart, CO species adsorbed on positively charged Pt sites ($Pt^{\delta+}$–CO) were assigned (Figure 23). These features, fully compatible with the way of action of alkali promotion analysed so far, were considered as a convincing explanation of the substantially enhanced TWC performance of this complex reactions/catalyst system.

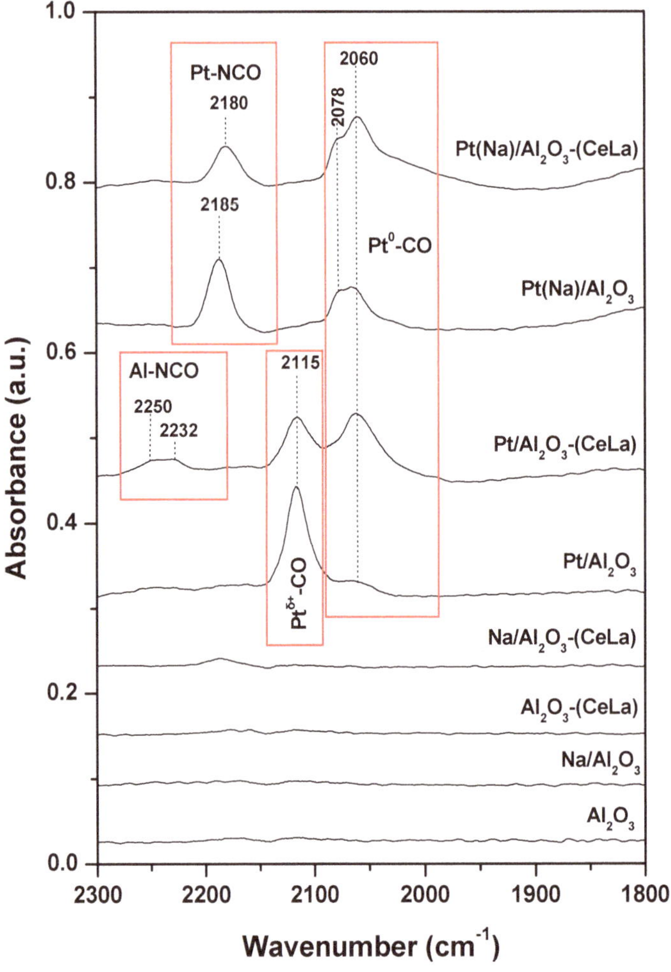

Figure 23. In-situ DRIFT spectra during simulated TWC reaction conditions (1000 ppm NO + 1067 ppm C_3H_6 + 7000 ppm CO + 7800 ppm O_2; F_t = 80 cm^3/min; T = 200 °C). (Reproduced with permission from Ref. [96]. Copyright 2008, Elsevier).

Finally, a number of studies involving x-ray photoelectron spectroscopy (XPS) published by the groups of Lambert and Yentekakis (e.g., [20,51,52,57–60]) have doubtlessly shown that in EPOC studies that concern PGMs films interfaced with an alkali conducting solid electrolyte, the promoting species are indeed alkaline ions reversibly supplied by the external bias onto the catalyst surface.

The authors have also shown that under reaction conditions the electrochemically supplied alkali ions form stable surface compounds (e.g., alkali nitrites, nitrates and carbonates) that can act as promoters even under such formulations and which, in excess (over-promotion), induce geometric, active sites blocking, phenomena and thus poisoning effects on PGMs activity [20,51,52,57–60]. Combining XPS and ultraviolet photoelectron spectroscopy (UPS), the authors demonstrated a linear relationship of the catalyst work function ($\Delta\Phi$) and the alkali coverage (θ_{Na}) with the catalyst potential overpotential (ΔV_{WR}) (Figure 24) [51,52] and confirmed that electrochemically supplied alkali ions on the catalyst surface are identical in behaviour (and chemical state) with the alkali supplied by vacuum deposition [21,22]. This makes the close similarities found between EPOC and CCP methodologies of catalyst promotion readily understandable [15,21,22,25,58] and prompts for the use of EPOC as an effective and rapid method for exploring the effects of a range of promoters and for assessing the response of the reaction rate/selectivity to promoter coverage (finding out its optimal loadings at any set of conditions) before applying them to the design of efficient conventional catalyst formulations.

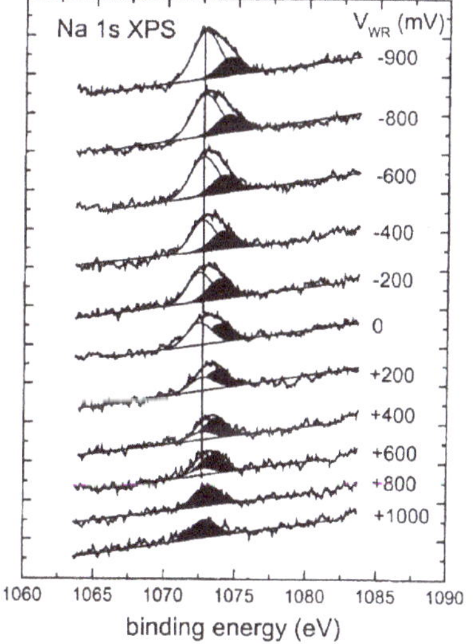

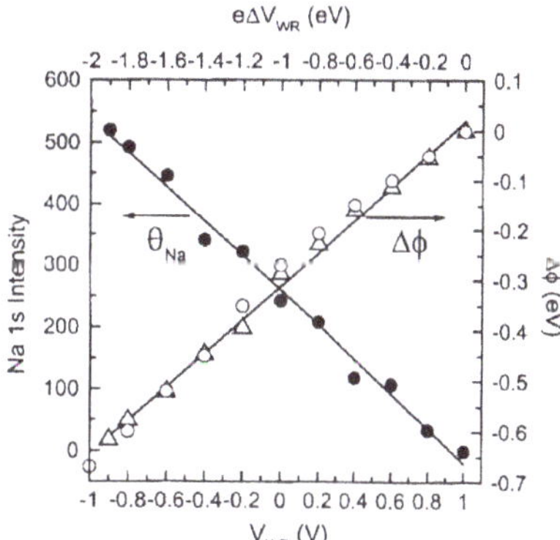

Figure 24. Na 1s XPS spectra versus catalyst potential (V_{WR}) for Rh film interfaced with (Na)$\beta''Al_2O_3$ Na^+-ions conductor, at 307 °C and UHV conditions. The inset shows the integrated Na1s XPS intensity due to sodium on the Rh surface and associated work function change of the Rh film as a function of V_{WR}. Data was acquired using a Rh/(Na)$\beta''Al_2O_3$/Au galvanic cell (Reproduced with permission from Ref. [51]. Copyright 2001, American Chemical Society).

A recent review of the literature using surface analysis techniques to shed light on the electropositive promotion via the EPOC concept has recently provided by Gonzalez-Cobos and de Lucas-Consuegra [124].

All the aforementioned in this chapter studies, unambiguously lead to the main conclusion that alkalis adatoms on PGM surfaces provide electron enriched metal sites (lower work function surfaces), which favour the adsorption of electron donor (electrophilic) co-adsorbates and inhibit the adsorption of electron acceptor (electrophobic) co-adsorbates. The formation of alkali-reactants-metal surface complexes and therefore geometric type influences, besides the above electronic ones, can be at work in cases where the conditions assist such a course. These issues were considered in the vast majority of electropositive promotion studies by alkalis or alkaline earths to explain promotion phenomena observed on PGM-catalysed emissions control catalysis reactions.

4. Conclusions and Perspectives

The electropositive promotion, by alkalis or alkaline earths, has been found to be an effective tool for promoting the catalytic performance of Platinum Group Metals for the vast majority of important reactions involved in emissions control catalysis systems, such as CO and alkenes oxidations, NO reduction by CO, alkenes or H_2, N_2O decomposition and/or reduction; the only exceptions were the alkanes oxidation by dioxygen or NO.

Up to two orders of magnitude rate increases were achievable, in particular with electropositive-promoted Pt, while in de-Nox reactions the selectivity towards N_2 was simultaneously improved towards values approaching 100% over the "optimally promoted" catalysts.

Promotional achievements were large not only for model reactions but also for gas mixtures that simulate practical systems, such as stoichiometric gasoline, lean-burn or diesel engines exhaust streams.

The intensity of promotion was generally followed the order Pt > Pd > Rh~Ir, thus providing Pt or Pd-based catalysts to operate like Rh or even better in emissions control catalytic reactions. These achievements were lead to the synthesis of novel simple in synthesis and cost-effective monometallic (Pt or Pd-only) with low precious metal loading, extremely active, stable and readily recycling, three-way catalyst formulations that were compared well with commercial bi- or three-metallic (Rh, Pt, Pd)-TWCs for the control of stoichiometric gasoline engines exhaust emissions. The absence of rare and expensive Rh in the formulations of three-way converters has therefore been introduced with the concomitant environmental and economic gains.

Electropositive promotion of PGM was also found to be effective at excess oxygen conditions (e.g., lean-burn of diesel engines exhaust gases), without however in that case to obtain solutions fully satisfying the requirements.

Latterly, doubly promoted catalyst formulations, combing electropositive promotion by alkalis and support-mediated promotion by mixed oxide supports (including combinations of Al_2O_3 with CeO_2, ZrO_2, La_2O_3 or TiO_2) have been found to provide even better promotion and stability for emissions control implementations. Such synergistic promotion catalysts' designs could lead to control systems readily adapted on a number of specific emissions of stationary sources, that have high environmental footprint and their control is an urgent issue.

From the theoretical point of view, electropositive promotion of PGM was successfully understood and interpreted in terms of the effective double layer approach, that is, a homogeneous, neutral, effective double layer established by the alkali (or alkaline earths) adatoms that are spread over the entire PGM surface providing an electron enriched (lower work function) metal surface due to the δ^- charge located at the metal side interface half-layer, which interacts with the co-adsorbed reactants, reaction intermediates and products, changing their binding energies producing pronounced alterations in catalytic performance. To this end, strengthening in the bonds of the metal—electron acceptor adsorbates and weakening in the bonds of the metal—electron donor adsorbates are resulted, which are accompanied by alterations in the activation energies of the reactions between them, leading to dramatic changes of the intrinsic catalytic performance (activity and/or selectivity). In particular, in emissions control catalysis processes involving NO_x reduction, the promotion of the dissociative chemisorption of the NO molecules, which is a key reaction initiating step in $deNO_x$ processes, facilitated by the alkali-induced strengthening of the PGM–NO and the accompanied weakening of the N–O bond of the adsorbed NO molecules, convincingly interpret the enormous effective promotion of both $deNO_x$ activity and N_2-selectivity offered via electropositive promotion of these catalytic systems.

Author Contributions: I.V.Y. was the primary author, writing this review and contributing to the conception, design and comparative interpretation of the literature. P.V., A.C. and G.G. provided assistance in writing, discussing and interpreting results and updating the article. All authors have a significant research contribution in the field, publishing a number of papers that are included and analysed in the present review. All authors read and approved the final version of the manuscript.

Acknowledgments: I.V.Y. and G.G. acknowledge support of the European Union and Greek national funds through the Operational Program "Competitiveness, Entrepreneurship and Innovation," under the call "RESEARCH-CREATE-INNOVATE" (project code: T1EΔK-00782). The authors thank all co-workers, colleagues and researchers contributed in the field. This work is dedicated to Professor Costas G. Vayenas, to whom the authors express their gratitude for discovering and introducing to science the EPOC phenomenon (described also as "the NEMCA effect"), which obviously is of high fundamental and practical importance and has opened new horizons in the frontiers of Catalysis and Electrochemistry.

Conflicts of Interest: The authors declare no conflict of interest.

References

1. Kaspar, J.; Fornasiero, P.; Hickey, N. Automotive catalytic converters: Current status and some perspectives. *Catal. Today* **2003**, *77*, 419–449. [CrossRef]
2. Farrauto, R.J.; Heck, R.M. Catalytic converters: State of the art and perspectives. *Catal. Today* **1999**, *51*, 351–360. [CrossRef]
3. Yentekakis, I.V.; Konsolakis, M. Three-way Catalysis (Book Chapter). In *Perovskites and Related Mixed Oxides: Concepts and Applications*; Wiley-VCH, Vergal GmbH & Co. KGaA: Weinheim, Germany, 2016; pp. 559–586. Available online: https://onlinelibrary.wiley.com/doi/pdf/10.1002/9783527686605.ch25 (accessed on 12 November 2015).
4. Wuebbles, D.J. Nitrous Oxide: No Laughing Matter. *Science* **2009**, *326*, 56–57. [CrossRef] [PubMed]
5. Ravishankara, A.R.; Daniel, J.S.; Portmann, R.W. Nitrous Oxide (N_2O): The Dominant Ozone-Depleting Substance Emitted in the 21st Century. *Science* **2009**, *326*, 123–125. [CrossRef] [PubMed]
6. Vayenas, C.G.; Bebelis, S.; Yentekakis, I.V.; Lintz, H.-G. Non-faradaic electrochemical modification of catalytic activity: A status report. *Catal. Today* **1992**, *11*, 303–442. [CrossRef]
7. Vayenas, C.G.; Bebelis, S.; Pliangos, C.; Brosda, S.; Tsiplakides, D. *Electrochemical Activation of Catalysis*; Kluwer Academic/Plenum: New York, NY, USA, 2001.
8. Stoukides, M.; Vayenas, C.G. The effect of electrochemical oxygen pumping on the rate and selectivity of ethylene oxidation on polycrystalline silver. *J. Catal.* **1981**, *70*, 137–146. [CrossRef]
9. Yentekakis, I.V.; Vayenas, C.G. The effect of electrochemical oxygen pumping on the steady-state and oscillatory behaviour of CO oxidation on polycrystalline Pt. *J. Catal.* **1988**, *111*, 170–188. [CrossRef]
10. Vayenas, C.G.; Bebelis, S.; Neophytides, S. Non-faradaic electrochemical modification of catalytic activity. *J. Phys. Chem.* **1988**, *92*, 5083–5085. [CrossRef]
11. Vayenas, C.G.; Bebelis, S.; Ladas, S. Dependence of catalytic rates on catalyst work function. *Nature* **1990**, *343*, 625–627. Available online: https://www.nature.com/articles/343625a0 (accessed on 15 February 1990). [CrossRef]
12. Yentekakis, I.V.; Bebelis, S. Study of the NEMCA effect in a single-pellet catalytic reactor. *J. Catal.* **1992**, *137*, 278–283. [CrossRef]
13. Vernoux, P.; Lizarraga, L.; Tsampas, M.N.; Sapountzi, F.M.; De Lucas-Consuegra, A.; Valverde, J.-L.; Souentie, S.; Vayenas, C.G.; Tsiplakides, D.; Balomenou, S.; et al. Ionically Conducting Ceramics as Active Catalyst Supports. *Chem. Rev.* **2013**, *113*, 8192–8260. [CrossRef] [PubMed]
14. Yentekakis, I.V.; Lambert, R.M.; Tikhov, M.; Konsolakis, M.; Kiousis, V. Promotion by sodium in emission control catalysis: A kinetic and spectroscopic study of the Pd-catalysed reduction of NO by propene. *J. Catal.* **1998**, *176*, 82–92. [CrossRef]
15. Yentekakis, I.V.; Konsolakis, M.; Lambert, R.M.; Macleod, N.; Nalbantian, L. Extraordinarily effective promotion by sodium in emission control catalysis: NO reduction by propene over Na-promoted Pt/γ-Al_2O_3. *Appl. Catal. B* **1999**, *22*, 123–133. [CrossRef]
16. Yentekakis, I.V.; Moggridge, G.; Vayenas, C.G.; Lambert, R.M. *In situ* controlled promotion of catalyst surfaces via NEMCA: The effect of Na on the Pt-catalysed CO oxidation. *J. Catal.* **1994**, *146*, 292–305. [CrossRef]
17. Palermo, A.; Lambert, R.M.; Harkness, I.R.; Yentekakis, I.V.; Marina, O.A.; Vayenas, C.G. Electrochemical promotion by Na of the platinum-catalysed reaction between CO and NO. *J. Catal.* **1996**, *161*, 471–479. [CrossRef]

18. Harkness, I.R.; Hardacre, C.; Lambert, R.M.; Yentekakis, I.V.; Vayenas, C.G. Ethylene oxidation over platinum: *In situ* electrochemically controlled promotion using Na–β'' alumina and studies with a Pt(111)/Na model catalyst. *J. Catal.* **1996**, *160*, 19–26. [CrossRef]
19. Marina, O.A.; Yentekakis, I.V.; Vayenas, C.G.; Palermo, A.; Lambert, R.M. *In situ* controlled promotion of catalyst surfaces via NEMCA: The effect of Na on the Pt-catalysed NO reduction by H_2. *J. Catal.* **1997**, *166*, 218–228. [CrossRef]
20. Filkin, N.C.; Tikhov, M.S.; Palermo, A.; Lambert, R.M. A kinetic and spectroscopic study of the *in situ* electrochemical promotion by sodium of the platinum-catalysed combustion of propene. *J. Phys. Chem. A* **1999**, *103*, 2680–2687. [CrossRef]
21. Konsolakis, M.; Palermo, A.; Tikhov, M.S.; Lambert, R.M.; Yentekakis, I.V. Electrochemical vs. Conventional promotion: A new tool to design effective, highly dispersed conventional catalysts. *Ionics* **1998**, *4*, 148–156. [CrossRef]
22. Yentekakis, I.V.; Konsolakis, M.; Lambert, R.M.; Palermo, A.; Tikhov, M. Successful application of electrochemical promotion to the design of effective conventional catalyst formulations. *Solid State Ionics* **2000**, *136/137*, 783–790. [CrossRef]
23. Nicole, J.; Tsiplakides, D.; Pliangos, C.; Verykios, X.E.; Comninelis, C.; Vayenas, C.G. Electrochemical promotion and metal–support interactions. *J. Catal.* **2001**, *204*, 23–34. [CrossRef]
24. Brosda, S.; Vayenas, C.G.; Wei, J. Rules of chemical promotion. *Appl. Catal. B* **2006**, *68*, 109–124. [CrossRef]
25. Pekridis, G.; Kaklidis, N.; Konsolakis, M.; Athanasiou, C.; Yentekakis, I.V.; Marnellos, G.E. A comparison between electrochemical and conventional catalyst promotion: The case of N_2O reduction by alkanes or alkenes over K-modified Pd catalysts. *Solid State Ionics* **2011**, *192*, 653–658. [CrossRef]
26. Pliangos, C.; Yentekakis, I.V.; Verykios, X.E.; Vayenas, C.G. Non-Faradaic electrochemical modification of catalytic activity: VIII. Rh-catalysed C_2H_4 oxidation. *J. Catal.* **1995**, *154*, 124–136. [CrossRef]
27. De Lucas-Consuegra, A.; Dorado, F.; Valverde, J.L.; Karoum, R.; Vernoux, P. Electrochemical activation of Pt catalyst by potassium for low temperature CO deep oxidation. *Catal. Commun.* **2008**, *9*, 17–20. [CrossRef]
28. Mirkelamoglu, B.; Karakas, G. CO oxidation over palladium- and sodium-promoted tin dioxide: Catalyst characterization and temperature-programmed studies. *Appl. Catal. A* **2005**, *281*, 275–284. [CrossRef]
29. Mirkelamoglu, B.; Karakas, G. The role of alkali-metal promotion on CO oxidation over PdO/SnO_2 catalysts. *Appl. Catal. A* **2006**, *299*, 84–94. [CrossRef]
30. Lee, C.-H.; Chen, Y.-W. Effect of basic additives on Pt/Al_2O_3 for CO and propylene oxidation under oxygen-deficient conditions. *Ind. Eng. Chem. Res.* **1997**, *36*, 1498–1506. [CrossRef]
31. Tanikawa, K.; Egawa, C. Effect of barium addition on CO oxidation activity of palladium catalysts. *Appl. Catal A* **2011**, *403*, 12–17. [CrossRef]
32. Minemura, Y.; Kuriyama, M.; Ito, S.; Tomishige, K.; Kunimori, K. Additive effect of alkali metal ions on preferential CO oxidation over Pt/Al_2O_3. *Catal. Commun.* **2006**, *7*, 623–626. [CrossRef]
33. Minemura, Y.; Ito, S.; Miyao, T.; Naito, S.; Tomishige, K.; Kunimori, K. Preferential CO oxidation promoted by the presence of H_2 over $K–Pt/Al_2O_3$. *Chem. Commun.* **2005**, *11*, 1429–1431. [CrossRef] [PubMed]
34. Tanaka, H.; Kuriyama, M.; Ishida, Y.; Ito, S.; Tomishige, K.; Kunimori, K. Preferential CO oxidation in hydrogen-rich stream over Pt catalysts modified with alkali metals: Part I. Catalytic performance. *Appl. Catal. A* **2008**, *343*, 117–124. [CrossRef]
35. Kuriyama, M.; Tanaka, H.; Ito, S.; Kubota, T.; Miyao, T.; Naito, S.; Timoshige, K.; Kunimori, K. Promoting mechanism of potassium in preferential CO oxidation on Pt/Al_2O_3. *J. Catal.* **2007**, *252*, 39–48. [CrossRef]
36. Tanaka, H.; Kuriyama, M.; Ishida, Y.; Ito, S.; Kubota, T.; Miyao, T.; Naito, S.; Tomishige, K.; Kunimori, K. Preferential CO oxidation in hydrogen-rich stream over Pt catalysts modified with alkali metals: Part II. Catalyst characterization and role of alkali metals. *Appl. Catal. A* **2008**, *343*, 125–133. [CrossRef]
37. Pedrero, C.; Waku, T.; Iglesia, E. Oxidation of CO in H_2–CO mixtures catalysed by platinum: Alkali effects on rates and selectivity. *J. Catal.* **2005**, *233*, 242–255. [CrossRef]
38. Kwak, C.; Park, T.-J.; Suh, D.J. Effects of sodium addition on the performance of $PtCo/Al_2O_3$ catalysts for preferential oxidation of carbon monoxide from hydrogen-rich fuels. *Appl. Catal. A* **2005**, *278*, 181–186. [CrossRef]
39. Cho, S.-H.; Park, J.-S.; Choi, S.-H.; Kim, S.-H. Effect of magnesium on preferential oxidation of carbon monoxide on platinum catalyst in hydrogen-rich stream. *J. Power Sources* **2006**, *156*, 260–266. [CrossRef]

40. De Lucas-Consuegra, A.; Princivalle, A.; Caravaca, A.; Dorado, F.; Guizard, C.; Valverde, J.L.; Vernoux, P. Preferential CO oxidation in hydrogen-rich stream over an electrochemically promoted Pt catalyst. *Appl. Catal. B* **2010**, *94*, 281–287. [CrossRef]
41. Liu, K.; Wang, A.; Zhang, T. Recent advances in preferential oxidation of CO reaction over platinum group metal catalysts. *ACS Catal.* **2012**, *2*, 1165–1178. [CrossRef]
42. Vayenas, C.G.; Bebelis, S.; Despotopoulou, M. Non-faradaic electrochemical modification of catalytic activity 4. The use of β''-Al_2O_3 as the solid electrolyte. *J. Catal.* **1991**, *128*, 415–435. [CrossRef]
43. Vernoux, P.; Gaillard, F.; Lopez, C.; Siebert, E. In-situ electrochemical control of the catalytic activity of platinum for the propene oxidation. *Solid State Ionics* **2004**, *175*, 609–613. [CrossRef]
44. Billard, A.; Vernoux, P. Electrochemical catalysts for hydrocarbon combustion. *Top. Catal.* **2007**, *44*, 369–377. [CrossRef]
45. De Lucas-Consuegra, A.; Dorado, F.; Valverde, J.L.; Karoum, R.; Vernoux, P. Low-temperature propene combustion over Pt/K-βAl_2O_3 electrochemical catalyst: Characterization, catalytic activity measurements and investigation of the NEMCA effect. *J. Catal.* **2007**, *251*, 474–484. [CrossRef]
46. De Lucas-Consuegra, A.; Dorado, F.; Jimenez-Borja, C.; Caravaca, A.; Vernoux, P.; Valverde, J.L. Use of potassium conductors in the electrochemical promotion of environmental catalysis. *Catal. Today* **2009**, *146*, 293–298. [CrossRef]
47. Yentekakis, I.V.; Tellou, V.; Bontzolaki, G.; Rapakousios, I.A. A comparative study of the C_3H_6 + NO + O_2, C_3H_6 + O_2 and NO + O_2 reactions in excess oxygen over Na-modified Pt/γ-Al_2O_3 catalysts. *Appl. Catal. B* **2005**, *56*, 229–239. [CrossRef]
48. Kotsionopoulos, N.; Bebelis, S. *In situ* electrochemical modification of catalytic activity for propane combustion of Pt/β''-Al_2O_3 catalyst-electrodes. *Top. Catal.* **2007**, *44*, 379–389. [CrossRef]
49. Kotsionopoulos, N.; Bebelis, S. Electrochemical characterization of the Pt/β''–Al_2O_3 system under conditions of in situ electrochemical modification of catalytic activity for propane combustion. *J. Appl. Electrochem.* **2010**, *40*, 1883–1891. [CrossRef]
50. Auvray, X.; Lindholm, A.; Milh, M.; Olsson, L. The addition of alkali and alkaline earths metals to Pd/Al_2O_3 to promote methane compustion. Effect of Pd and Ca loading. *Catal. Today* **2018**, *299*, 212–218. [CrossRef]
51. Williams, F.J.; Palermo, A.; Tikhov, M.S.; Lambert, R.M. Electrochemical promotion by sodium of the rhodium-catalysed NO + CO reaction. *J. Phys. Chem. B* **2000**, *104*, 11883–11890. [CrossRef]
52. Williams, F.J.; Palermo, A.; Tikhov, M.S.; Lambert, R.M. Mechanism of alkali promotion in heterogeneous catalysis under realistic conditions: Application of electron spectroscopy and electrochemical promotion to the reduction of NO by CO and by propene over rhodium. *Surf. Sci.* **2001**, *482–485*, 177–182. [CrossRef]
53. Konsolakis, M.; Yentekakis, I.V.; Palermo, A.; Lambert, R.M. Optimal promotion by rubidium of the CO + NO reaction over Pt/γ-Al_2O_3 catalysts. *Appl. Catal. B* **2001**, *33*, 293–302. [CrossRef]
54. Konsolakis, M.; Yentekakis, I.V. NO reduction by propene or CO over alkali-promoted Pd/YSZ catalysts. *J. Hazard. Mater.* **2007**, *149*, 619–624. [CrossRef] [PubMed]
55. Tanikawa, K.; Egawa, C. Effect of barium addition over palladium catalyst for CO–NO–O_2 reaction. *J. Mol. Catal. A* **2011**, *349*, 94–99. [CrossRef]
56. Lepage, M.; Visser, T.; Soulimani, F.; Inglesias-Juez, A.; Weckhuysen, B.M. Promotion Effects in the Reduction of NO by CO over Zeolite-Supported Rh Catalysts. *J. Phys. Chem. C* **2010**, *114*, 2282–2292. [CrossRef]
57. Harkness, I.R.; Lambert, R.M. Electrochemical Promotion of the NO + Ethylene Reaction over Platinum. *J. Catal.* **1995**, *152*, 211–214. [CrossRef]
58. Yentekakis, I.V.; Palermo, A.; Filkin, N.; Tikhov, M.S.; Lambert, R.M. *In situ* electrochemical promotion by sodium of the platinum-catalysed reduction of NO by propene. *J. Phys. Chem. B* **1997**, *101*, 3759–3768. [CrossRef]
59. Williams, F.J.; Palermo, A.; Tikhov, M.S.; Lambert, R.M. Electrochemical promotion by sodium of the rhodium-catalysed reduction of NO by propene: Kinetics and spectroscopy. *J. Phys. Chem. B* **2001**, *105*, 1381–1388. [CrossRef]
60. Williams, F.J.; Tikhov, M.S.; Palermo, A.; Macleod, N.; Lambert, R.M. Electrochemical promotion of rhodium-catalysed NO reduction by CO and by propene in the presence of oxygen. *J. Phys. Chem. B* **2001**, *105*, 2800–2808. [CrossRef]
61. Konsolakis, M.; Yentekakis, I.V. Strong promotional effects of Li, K, Rb and Cs on the Pt-catalysed reduction of NO by propene. *Appl. Catal. B* **2001**, *29*, 103–113. [CrossRef]

62. Konsolakis, M.; Yentekakis, I.V. The reduction of NO by propene over Ba-promoted Pt/γ-Al_2O_3 catalysts. *J. Catal.* **2001**, *198*, 142–150. [CrossRef]
63. Yentekakis, I.V.; Lambert, R.M.; Konsolakis, M.; Kallithrakas-Kontos, N. On the Effects of Residual Chloride and of Barium Promotion on Pt/γ-Al_2O_3 Catalysts in the Reduction of NO by Propene. *Catal. Lett.* **2002**, *81*, 181–185. [CrossRef]
64. Macleod, N.; Isaac, J.; Lambert, R.M. Sodium promotion of the NO+C_3H_6 reaction over Rh/γ-Al_2O_3 catalysts. *J. Catal.* **2000**, *193*, 115–122. [CrossRef]
65. Yentekakis, I.V.; Lambert, R.M.; Konsolakis, M.; Kiousis, V. The effect of sodium on the Pd-catalysed reduction of NO by methane. *Appl. Catal. B* **1998**, *18*, 293–305. [CrossRef]
66. Yentekakis, I.V.; Konsolakis, M.; Kiousis, V.; Lambert, R.M.; Tikhov, M. Promotion by sodium in emission control catalysis: The difference between alkanes and alkenes in the Pd-catalysed reduction of NO by hydrocarbons. *G-NEST Int. J.* **1999**, *1*, 121–130.
67. Burch, R.; Watling, T.C. The difference between alkanes and alkenes in the reduction of NO by hydrocarbons over Pt catalysts under lean-burn conditions. *Catal. Lett.* **1997**, *43*, 19–23. [CrossRef]
68. Burch, R.; Sullivan, J.A.; Watling, T.C. Mechanistic considerations for the reduction of NOx over Pt/Al_2O_3 and Al_2O_3 catalysts under lean-burn conditions. *Catal. Today* **1998**, *42*, 13–23. [CrossRef]
69. Goula, M.A.; Charisiou, N.D.; Papageridis, K.N.; Delimitis, A.; Papista, E.; Pachatouridou, E.; Iliopoulou, E.F.; Marnellos, G.; Konsolakis, M.; Yentekakis, I.V. A comparative study of the H_2-assisted selective catalytic reduction of nitric oxide by propene over noble metal (Pt, Pd, Ir)/γ-Al_2O_3 catalysts. *J. Environ. Chem. Eng.* **2016**, *4*, 1629–1641. [CrossRef]
70. Machida, M.; Ikeda, S.; Kurogi, D.; Kijima, T. Low temperature catalytic NO_x-H_2 reactions over Pt/TiO_2-ZrO_2 in an excess oxygen. *Appl. Catal. B* **2001**, *35*, 107–116. [CrossRef]
71. Costa, C.N.; Savva, P.G.; Andronikou, C.; Lambrou, P.S.; Polychronopoulou, K.; Belessi, V.C.; Stathopoulos, V.N.; Pomonis, P.J.; Efstathiou, A.M. An Investigation of the NO/H_2/O_2 (lean De-NOx) reaction on a highly active and selective Pt/$La_{0.7}Sr_{0.2}Ce_{0.1}FeO_3$ catalyst at low temperatures. *J. Catal.* **2002**, *209*, 456–471. [CrossRef]
72. Polychronopoulou, K.; Efstathiou, A.M. NO_x control via H_2-selective catalytic reduction (H_2-SCR) technology for stationary and mobile applications. *Recent Pat. Mater. Sci.* **2012**, *5*, 87–104. [CrossRef]
73. Burch, R.; Coleman, M.D. An Investigation of Promoter Effects in the Reduction of NO by H_2 under Lean-Burn Conditions. *J. Catal.* **2002**, *208*, 435–447. [CrossRef]
74. Machida, M.; Watanabe, T. Effect of Na-addition on catalytic activity of Pt-ZSM-5 for low-temperature NO–H_2–O_2 reactions. *Appl. Catal. B* **2004**, *52*, 281–286. [CrossRef]
75. Macleod, N.; Lambert, R.M. Low-temperature NO_x reduction with H_2+CO under oxygen-rich conditions over a Pd/TiO_2/Al_2O_3 catalyst. *Catal. Commun.* **2002**, *3*, 61–65. [CrossRef]
76. Macleod, N.; Lambert, R.M. Lean NO_x reduction with CO + H_2 mixtures over Pt/Al_2O_3 and Pd/Al_2O_3 catalysts. *Appl. Catal. B* **2002**, *35*, 269–279. [CrossRef]
77. Macleod, N.; Lambert, R.M. In situ ammonia generation as a strategy for catalytic NOx reduction under oxygen rich conditions. *Chem. Commun.* **2003**, *11*, 1300–1301. [CrossRef]
78. Macleod, N.; Lambert, R.M. An in situ DRIFTS study of efficient lean NO_x reduction with H_2 + CO over Pd/Al_2O_3: The key role of transient NCO formation in the subsequent generation of ammonia. *Appl. Catal. B* **2003**, *46*, 483–495. [CrossRef]
79. Konsolakis, M.; Vrontaki, M.; Avgouropoulos, G.; Ioannides, T.; Yentekakis, I.V. Novel doubly-promoted catalysts for the lean NO_x reduction by H_2 + CO: Pd(K)/Al_2O_3–(TiO_2). *Appl. Catal. B* **2006**, *68*, 59–67. [CrossRef]
80. Wang, C.-B.; Chang, J.-G.; Wu, R.-C.; Yeh, C.-T. Promotion effect of coating alumina supported palladium with sodium hydroxide on the catalytic conversion of nitric oxide. *Appl. Catal. B* **1998**, *17*, 51–62. [CrossRef]
81. Haber, J.; Nattich, M.; Machej, T. Alkali-metal promoted rhodium-on-alumina catalysts for nitrous oxide decomposition. *Appl. Catal. B* **2008**, *77*, 278–283. [CrossRef]
82. Parres-Esclaped, S.; Lopez-Suerez, F.E.; Bueno-Lopez, A.; Illan-Gomez, M.J.; Ura, B.; Trawcczynski, J. Rh–Sr/Al_2O_3 catalyst for N_2O decomposition in the presence of O_2. *Top. Catal.* **2009**, *52*, 1832–1836. [CrossRef]
83. Konsolakis, M.; Aligizou, F.; Goula, G.; Yentekakis, I.V. N_2O decomposition over doubly-promoted Pt(K)/Al_2O_3-(CeO_2-La_2O_3) structured catalysts: On the compined effects of promotion and feed composition. *Chem. Eng. J.* **2013**, *230*, 286–295. [CrossRef]

84. Papista, E.; Pachatouridou, E.; Goula, M.A.; Marnellos, G.E.; Iliopoulou, E.; Konsolakis, M.; Yentekakis, I.V. Effect of alkali promoters (K) on nitrous oxide abatement over Ir/Al2O3 catalysts. *Top. Catal.* **2016**, *59*, 1020–1027. [CrossRef]
85. Goncalves, F.; Figueiredo, J.L. Sinergistic effect between Pt and K in the catalytic reduction of NO and N_2O. *Appl. Catal. B* **2006**, *62*, 181–192. [CrossRef]
86. Pekridis, G.; Kaklidis, N.; Konsolakis, M.; Iliopoulou, E.F.; Yentekakis, I.V.; Marnellos, G. Correlation of surface characteristics with catalytic performance of potassium promoted Pd/Al_2O_3 catalysts: The case of N_2O reduction by alkanes or alkenes. *Top. Catal.* **2011**, *54*, 1135–1142. [CrossRef]
87. De Lucas-Consuegra, A.; Dorado, F.; Jimenez-Borja, C.; Valerde, J.L. Influence of the reaction conditions on the electrochemical promotion by potassium for the selective catalytic reduction of N_2O by C_3H_6 on platinum. *Appl. Catal. B* **2008**, *78*, 222–231. [CrossRef]
88. De Lucas Consuegra, A.; Dorado, F.; Jimenez-Borja, C.; Valverde, J.L. Electrochemical promotion of Pt impregnated catalyst for the treatment of automotive exhaust emissions. *J. Appl. Electrochem.* **2008**, *38*, 1151–1157. [CrossRef]
89. Konsolakis, M.; Macleod, N.; Isaac, J.; Yentekakis, I.V.; Lambert, R.M. Strong promotion by Na of Pt/γ-Al_2O_3 catalysts operated under simulated exhaust conditions. *J. Catal.* **2000**, *193*, 330–337. [CrossRef]
90. Macleod, N.; Isaac, J.; Lambert, R.M. Sodium promotion of Pd/γ-Al_2O_3 catalysts operated under simulated "three-way" conditions. *J. Catal.* **2001**, *198*, 128–135. [CrossRef]
91. Macleod, N.; Isaac, J.; Lambert, R.M. A comparison of sodium-modified Rh/γ-Al_2O_3 and Pd/γ-Al_2O_3 catalysts operated under simulated TWC conditions. *Appl. Catal. B* **2001**, *33*, 335–343. [CrossRef]
92. Tanaka, T.; Yokota, K.; Isomura, N.; Doi, H.; Sugiura, M. Effect of the addition of Mo and Na to Pt catalysts on the selective reduction of NO. *Appl. Catal. B* **1998**, *16*, 199–208. [CrossRef]
93. Shinjoh, H.; Tanabe, T.; Sobukawa, H.; Sugiura, M. Effect of Ba addition on catalytic activity of Pt and Rh catalysts loaded on γ-alumina. *Top. Catal.* **2001**, *16/17*, 95–99. [CrossRef]
94. Kobayashi, T.; Yamada, T.; Kayano, K. Effect of basic metal additives on NO_x reduction property of Pd-based three-way catalyst. *Appl. Catal. B* **2001**, *30*, 287–292. [CrossRef]
95. Yentekakis, I.V.; Konsolakis, M.; Rapakousios, I.A.; Matsouka, V. Novel electropositively promoted monometallic (Pt-only) catalytic converters for automotive pollution control. *Top. Catal.* **2007**, *42–43*, 393–397. [CrossRef]
96. Matsouka, V.; Konsolakis, M.; Lambert, R.M.; Yentekakis, I.V. In situ DRIFTS study of the effect of structure (CeO_2-La_2O_3) and surface (Na) modifiers on the catalytic and surface behaviour of Pt/γ-Al_2O_3 catalyst under simulated exhaust conditions. *Appl. Catal. B* **2008**, *84*, 715–722. [CrossRef]
97. Matsouka, V.; Konsolakis, M.; Yentekakis, I.V.; Papavasiliou, A.; Tsetsekou, A.; Boukos, N. Thermal aging behavior of Pt-only TWC converters under simulated exhaust conditions: Effect of rare earths (CeO_2, La_2O_3) and alkali (Na) modifiers. *Top. Catal.* **2011**, *54*, 1124–1134. [CrossRef]
98. Papavasiliou, A.; Tsetsekou, A.; Matsouka, V.; Konsolakis, M.; Yentekakis, I.V.; Boukos, N. Synergistic structural and surface promotion of monometallic (Pt) TWCs: Effectiveness and thermal aging tolerance. *Appl. Catal. B* **2011**, *106*, 228–241. [CrossRef]
99. Vernoux, P.; Gaillard, F.; Lopez, C.; Siebert, E. Coupling catalysis to electrochemistry: A solution to selective reduction of nitrogen oxides in lean-burn engine exhausts? *J. Catal.* **2003**, *217*, 203–208. [CrossRef]
100. Dorado, F.; de Lucas-Consuegra, A.; Jimenez, C.; Valverde, J.L. Influence of the reaction temperature on the elecrochemical promoted catalytic behaviour of platinum impregnated catalysts for the reduction of nitrogen oxides under lean burn conditions. *Appl. Catal. A* **2007**, *321*, 86–92. [CrossRef]
101. Dorado, F.; de Lucas-Consuegra, A.; Vernoux, P.; Valverde, J.L. Electrochemical promotion of platinum impregnated catalyst for the selective catalytic reduction of NO by propene in presence of oxygen. *Appl. Catal. B* **2007**, *73*, 42–50. [CrossRef]
102. De Lucas Consuegra, A.; Caravaca, A.; Dorado, F.; Valverde, J.L. Pt/K–βAl_2O_3 solid electrolyte cell as a "smart electrochemical catalyst" for the effective removal of NO_x under wet reaction conditions. *Catal. Today* **2009**, *146*, 330–335. [CrossRef]
103. De Lucas Consuegra, A.; Caravaca, A.; Sanchez, P.; Dorado, F.; Valverde, J.L. A new improvement of catalysis by solid-state electrochemistry: An electrochemically assisted NO_x storage/reduction catalyst. *J. Catal.* **2008**, *259*, 54–65. [CrossRef]

104. De Lucas Consuegra, A.; Caravaca, A.; Martin de Vidales, M.J.; Dorado, F.; Balomenou, S.; Tsiplakides, D.; Vernoux, P.; Valverde, J.L. An electrochemically assisted NO_x storage/reduction catalyst operating under fixed lean burn conditions. *Catal. Commun.* **2009**, *11*, 247–251. [CrossRef]
105. Burch, R.; Watling, T.C. The effect of promoters on Pt/Al_2O_3 catalysts for the reduction of NO by C_3H_6 under lean-burn conditions. *Appl. Catal. B* **1997**, *11*, 207–216. [CrossRef]
106. Vernoux, P.; Leinekugel-Le-Cocq, A.-Y.; Gaillard, F. Effect of the addition of Na on Pt/Al_2O_3 catalysts for the reduction of NO by C_3H_8 and C_3H_6 under lean-burn conditions. *J. Catal.* **2003**, *219*, 247–257. [CrossRef]
107. Wogerbauer, C.; Maciejewski, M.; Schubert, M.M.; Baiker, A. Effect of sodium on the catalytic properties of iridium black in the selective reduction of NOx by propene under lean-burn conditions. *Catal. Lett.* **2001**, *74*, 1–7. [CrossRef]
108. Goula, G.; Katzourakis, P.; Vakakis, N.; Papadam, T.; Konsolakis, M.; Tikhov, M.; Yentekakis, I.V. The effect of potassium on the Ir/C_3H_6+NO+O_2 catalytic system. *Catal. Today* **2007**, *127*, 199–206. [CrossRef]
109. Vayenas, C.G.; Koutsodontis, C.G. Non-Faradaic electrochemical activation of catalysis. *J. Chem. Phys.* **2008**, *128*, 182506–182513. [CrossRef]
110. Lambert, R.M. *Catalysis and Electrocatalysis at Nanoparticles*; Wieckowski, A., Savinova, E., Vayenas, C.G., Eds.; Dekker: New York, NY, USA, 2003; p. 583.
111. Lambert, R.M.; Palermo, A.; Williams, F.J.; Tikhov, M.S. Electrochemical promotion of catalytic reactions using alkali ion conductors. *Solid State Ionics* **2000**, *136–137*, 677–685. [CrossRef]
112. Lang, N.D.; Holloway, S.; Norskov, J.K. Electrostatic adsorbate-adsorbate interactions: The poisoning and promotion of the molecular adsorption reaction. *Surf. Sci.* **1985**, *150*, 24–38. [CrossRef]
113. Koukiou, S.; Konsolakis, M.; Lambert, R.M.; Yentekakis, I.V. Spectroscopic evidence for the mode of action of alkali promoters in Pt-catalyzed de-NOx chemistry. *Appl. Catal. B* **2007**, *76*, 101–106. [CrossRef]
114. Vayenas, C.G.; Brosda, S. Electron donation–backdonation and the rules of catalytic promotion. *Top. Catal.* **2014**, *57*, 1287–1301. [CrossRef]
115. Blyholder, G. Molecular orbital view of chemisorbed carbon monoxide. *J. Phys. Chem.* **1964**, *68*, 2772–2778. [CrossRef]
116. Harkness, I.R.; Lambert, R.M. Chemisorption and reactivity of nitric oxide on Na-dosed platinum{111}. *J. Chem. Soc. Faraday Trans.* **1997**, *93*, 1425–1429. [CrossRef]
117. Garfunkel, E.L.; Maj, J.J.; Frost, J.C.; Farias, M.H.; Somorjai, G.A. Interaction of potassium with .pi.-electron orbital containing molecules on platinum(111). *J. Phys. Chem.* **1983**, *87*, 3629–3635. [CrossRef]
118. Bertolini, J.C.; Derichele, P.; Massardier, J. The influence of potassium and the role of coadsorbed oxygen on the chemisorptive properties of Pt(100). *Surf. Sci.* **1985**, *160*, 531–541. [CrossRef]
119. Pitchon, V.; Primet, M.; Praliaud, H. Alkali addition to silica-supported palladium: Infrared investigation of the carbon monoxide chemisorptions. *Appl. Catal.* **1990**, *62*, 317–334. [CrossRef]
120. Liotta, F.L.; Martin, G.A.; Deganello, G. The influence of alkali metal ions in the chemisorption of CO and CO_2 on supported palladium catalysts: A fourier transform infrared spectroscopic study. *J. Catal.* **1996**, *164*, 322–333. [CrossRef]
121. Vayenas, C.G.; Archonta, D.; Tsiplakides, D. Scanning tunneling microscopy observation of the origin of electrochemical promotion and metal–support interactions. *J. Electroanal. Chem.* **2003**, *554–555*, 301–306. [CrossRef]
122. Makri, M.; Vayenas, C.G.; Bebelis, S.; Besocke, K.H.; Gavalca, C. Atomic resolution STM imaging of electrochemically controlled reversible promoter dosing of catalysts. *Surf. Sci.* **1996**, *369*, 351–359. [CrossRef]
123. Archonta, D.; Frantzis, A.; Tsiplakides, D.; Vayenas, C.G. STM observation of the origin of electrochemical promotion on metal catalyst-electrodes interfaced with YSZ and β''-Al_2O_3. *Solid State Ionics* **2006**, *26–32*, 2221–2225. [CrossRef]
124. Gonzalez-Cobos, J.; de Lucas-Consuegra, A. A Review of surface analysis techniques for the investigation of the phenomenon of electrochemical promotion of catalysis with alkaline ionic conductors. *Catalysts* **2016**, *6*, 15. [CrossRef]

Review

A Review of Low Temperature NH_3-SCR for Removal of NO_x

Devaiah Damma [1], Padmanabha R. Ettireddy [2], Benjaram M. Reddy [3] and Panagiotis G. Smirniotis [1,*]

1 Chemical Engineering, College of Engineering and Applied Science, University of Cincinnati, Cincinnati, OH 45221-0012, USA; dammadh@ucmail.uc.edu
2 Cummins Inc., Columbus, IN 47201, USA; ettireddypr@gmail.com
3 Catalysis and Fine Chemicals Department, CSIR-Indian Institute of Chemical Technology (IICT), Uppal Road, Hyderabad, Telangana 500007, India; bmreddy@iict.res.in
* Correspondence: smirnipp@ucmail.uc.edu; Tel.: +1-513-556-1474; Fax: +1-513-556-3473

Received: 6 March 2019; Accepted: 4 April 2019; Published: 10 April 2019

Abstract: The importance of the low-temperature selective catalytic reduction (LT-SCR) of NO_x by NH_3 is increasing due to the recent severe pollution regulations being imposed around the world. Supported and mixed transition metal oxides have been widely investigated for LT-SCR technology. However, these catalytic materials have some drawbacks, especially in terms of catalyst poisoning by H_2O or/and SO_2. Hence, the development of catalysts for the LT-SCR process is still under active investigation throughout seeking better performance. Extensive research efforts have been made to develop new advanced materials for this technology. This article critically reviews the recent research progress on supported transition and mixed transition metal oxide catalysts for the LT-SCR reaction. The review covered the description of the influence of operating conditions and promoters on the LT-SCR performance. The reaction mechanism, reaction intermediates, and active sites are also discussed in detail using isotopic labelling and in situ FT-IR studies.

Keywords: low-temperature selective catalytic reduction; NH_3-SCR; de-NO_x catalysis; SO_2/H_2O tolerance; transition metal-based catalysts

---

1. Introduction

The non-renewable fossil fuels are continuing to remain the dominant energy source in power plants and automobiles to satisfy the ever-growing energy demands. However, the combustion of fossil fuels mainly generates nitrogen oxide (NO_x) pollutants (NO, NO_2, and N_2O and their derivatives) which can cause acid rain, photochemical smog, ozone depletion, and eutrophication problems [1–4]. Due to the negative impacts of NO_x, the mitigation of NO_x emissions is of paramount importance for environmental protection. Several technologies are available to reduce NO_x emissions by using catalytic materials and among them, selective catalytic reduction of NO_x with NH_3 (NH_3-SCR) has been widely applied due to its high NO_x removal efficiency [5–7]. Usually, the flue gas temperature of the industrial process is as low as 300 °C and, thus, the SCR catalyst must be active in the low-temperature regime (100–300 °C). V_2O_5–$WO_3(MoO_3)/TiO_2$ is the typical and efficient catalyst and has been commercialized for NH_3-SCR technology for medium temperature process [8,9]. However, this catalyst has some intrinsic drawbacks such as narrow and high working temperature window (350–400 °C), and low N_2 selectivity in the high-temperature range [3,10,11]. Therefore, many researchers continue to develop highly active catalysts for low-temperature NH_3-SCR in a wide temperature window.

With this perspective, several transition metal oxide-based catalysts have been extensively investigated for low-temperature NH_3-SCR reaction due to their excellent redox properties, low price, and high thermodynamic stability. Especially, the easy gain and loss of electrons in the *d* shell of the transition metal ions could be responsible for the facile redox properties [12–14]. For example, the Cr/TiO_2 [15], Cr-MnO_x [8], Fe-MnO_x [16], Mn/TiO_2 [17,18], Fe_xTiO_y [19], MnO_x/CeO_2 [20], and Cu/TiO_2 [21] catalysts were shown to exhibit good SCR activity in the low temperature range. In our earlier work, we investigated the low-temperature NH_3-SCR in the presence of excess O_2 on the TiO_2 supported V, Cr, Mn, Fe, Co, Ni, and Cu oxides and found the catalytic performance decreased in the following order of Mn > Cu $\geq$ Cr $\gg$ Co > Fe $\gg$ V $\gg$ Ni [16]. Particularly, manganese-containing catalysts have attracted much attention due to its variable valence states and excellent redox ability [2,5]. In the recent past, we published a series of papers on Mn-based SCR catalysts that showed a highly promising deNO_x potential in the low-temperature region [5,22–26]. However, these catalysts are very sensitive to the presence of SO_2 in the feed and exhibit lower N_2 selectivity [8,27–29]. Hence, the development of catalysts that show both good low-temperature activity and high SO_2/H_2O durability is of great importance for the NH_3-SCR reaction. In general, there are two plausible strategies available to enhance low-temperature NH_3-SCR performance. One strategy is to modify the transition metal oxide with one or multiple metal oxides, which could enhance the active sites for the reaction by inducing the synergistic effect [30–33]. The other approach is to synthesize the supported materials to disperse the transition metal-based oxides which can increase the activity by metal-support interactions [26,34–37]. Recently, many supported and mixed transition metal catalyst formulations have been studied to improve the low-temperature SCR performance, as well as resistance to SO_2/H_2O.

In this study, we systematically reviewed the recent advancements in developing the transition metal-based catalysts for low-temperature NH_3-SCR reaction. This review also demonstrated the action of different promoters and supports on the catalytic performance and SO_2/H_2O tolerance of the transition metal-based catalysts in NH_3-SCR of NO_x. The reported catalysts were divided into four categories, such as binary, ternary/multi, supported single, and supported binary/multi-transition metal-based catalysts.

2. Binary Transition Metal-Based Catalysts

Various transition-metal oxides have been proved to be active for the NH_3-SCR at low-temperature. However, the catalytic performance on single transition metal oxide is far from satisfactory due to their low specific surface area and thermal instability [38–41]. The addition of dopants is a common method to improve the drawbacks associated with pure transition metal oxide. Hence, much progress has done to improve the low-temperature SCR activity of transition metal oxides by mixing or doping with other metal oxides. In recent years, Mn, Fe, Co, Ni, and Cu-based binary oxide catalysts have been extensively studied for low-temperature NH_3-SCR reaction due to their attractive catalytic performance [19,33,41–49]. Particularly, Mn-based binary oxides are popular and proven to be effective catalysts for low-temperature NH_3-SCR reaction [42,50,51]. Recently, Xin et al. [52] designed bifunctional V_a-MnO_x (where a represents the molar ratios of V / (V + Mn)) catalysts composed of Mn_2O_3 and $Mn_2V_2O_7$ phases that significantly improved both NO_x conversion and N_2 selectivity in comparison with Mn_2O_3 at low-temperature (Figure 1). Although $Mn_2V_2O_7$ showed an excellent N_2 selectivity, the NO_x conversion is much lower on it. Especially, above 90% NO_x conversion and 80% N_2 selectivity was observed in the temperature region of 120–240 °C over the $V_{0.05}$-MnO_x catalyst. The $V_{0.05}$-MnO_x catalyst also found to be exhibit higher NO_x conversion to N_2 as compared to the mechanically mixed Mn_2O_3 + $Mn_2V_2O_7$ sample which has the same component content to $V_{0.05}$-MnO_x (Figure 1). This finding indicated that the synergism between Mn_2O_3 and $Mn_2V_2O_7$ exists in the chemically prepared $V_{0.05}$-MnO_x rather than the mechanically mixed Mn_2O_3 + $Mn_2V_2O_7$ sample. Moreover, the mechanically mixed Mn_2O_3 + $Mn_2V_2O_7$ sample showed higher activity in comparison to mechanically mixed MoO_3 + $Mn_2V_2O_7$ sample, suggesting that the presence of Mn_2O_3 phase in the catalyst is necessary for NH_3-SCR reaction. In conjunction with in situ IR characterization and DFT (density functional theory) calculation results, the authors concluded that the Mn_2O_3 phase of the catalyst could activate NH_3 into NH_2 intermediate, which then transferred to the $Mn_2V_2O_7$ phase of the catalyst and reacted with gaseous NO into NH_2NO. Finally, the generated NH_2NO intermediate on the $Mn_2V_2O_7$ phase exclusively decomposed to the N_2 rather than the undesired byproduct, N_2O, which is formed due to the deep oxidation of adsorbed NH_3 on Mn_2O_3.

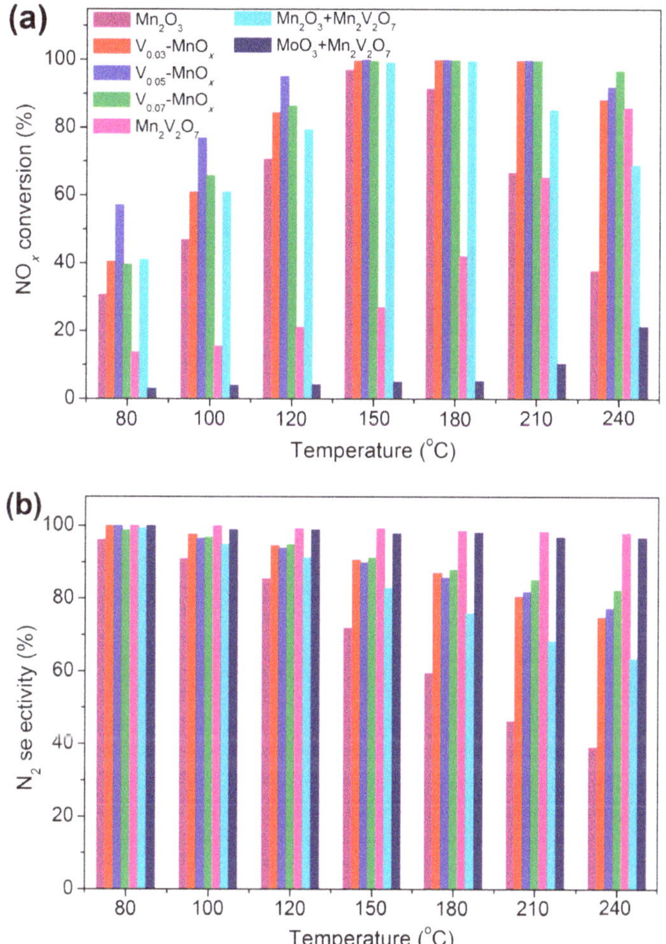

Figure 1. (**a**) NO_x conversion and (**b**) N_2 selectivity for V_a-MnO_x, Mn_2O_3, $Mn_2V_2O_7$, and reference samples. Reprinted from Reference [52]. Copyright 2018, with Permission from American Chemical Society.

Han and co-workers [53] fabricated triple-shelled $NiMn_2O_4$ hollow spheres (Figure 2a,b) by using a solvothermal method and tested their ability for low-temperature NH_3-SCR reaction. As shown in Figure 2c, the prepared $NiMn_2O_4$ hollow spheres ($NiMn_2O_4$-S) showed the best catalytic activity with NO_x conversion of above 90% over a wide temperature range from 100 °C to 225 °C as compared to the $NiMn_2O_4$ nanoparticles ($NiMn_2O_4$-P). The triple-layer shell structure of the $NiMn_2O_4$-S catalyst generates a larger surface area (165.3 m^2 g^{-1}) that exposes more active sites (such as surface Mn^{4+} and surface adsorbed oxygen species), which are responsible for its superior activity. Additionally, the $NiMn_2O_4$-S catalyst displayed outstanding stability and good tolerance to H_2O and SO_2 (Figure 2d).

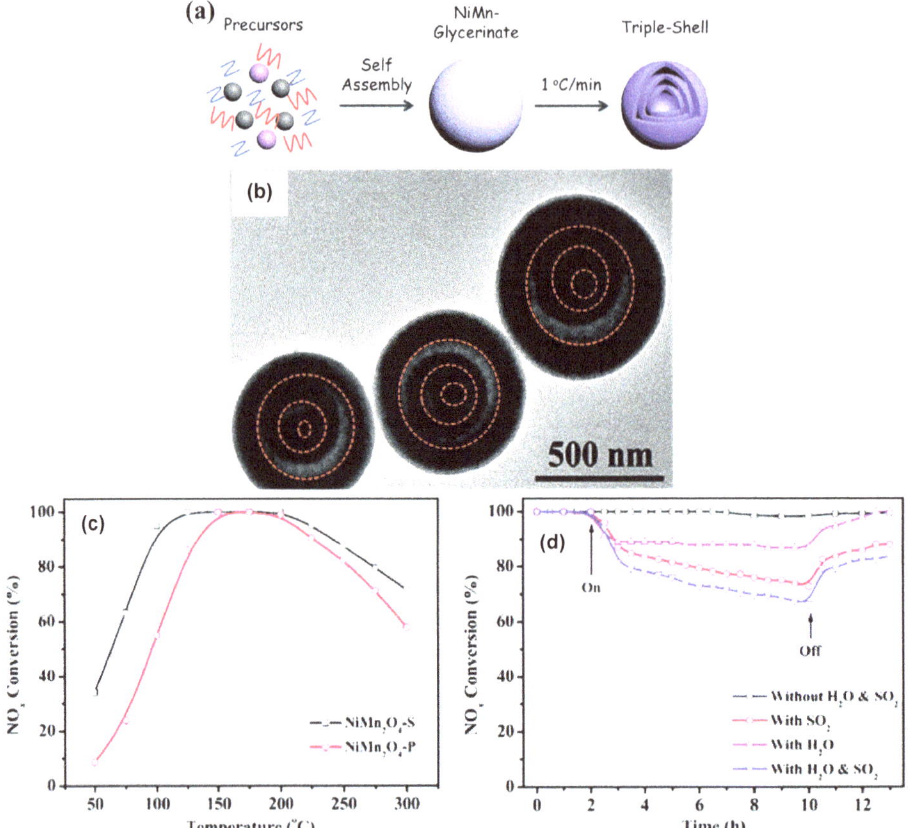

Figure 2. (**a**) Scheme of the preparation of triple shelled $NiMn_2O_4$ hollow spheres and (**b**) TEM image of $NiMn_2O_4$-S; (**c**) NO_x conversion over the $NiMn_2O_4$ catalysts and (**d**) durability tests of the $NiMn_2O_4$-S catalyst at 150 °C. Reaction conditions: [NO] = [NH_3] = 500 ppm, [O_2] = 5 vol%, [SO_2] = 100 ppm (when used), [H_2O] = 5 vol% (when used), balanced with Ar, GHSV = 68,000 h^{-1}. Adapted from Reference [53]. Copyright 2018, with Permission from Royal Society of Chemistry.

Gao et al. [54] investigated the low-temperature NH_3-SCR reaction over the hydroxyl-containing Me-Mn binary oxides (Me = Co, Ni) prepared by a combined complexation–esterification method. It was found that the NO_x conversion decreased in the order of Mn_3O_4-Co_3O_4-OH (Co-MnO_x binary oxide) > Mn_2O_3-$NiMnO_3$-OH (Ni-MnO_x binary oxide) > Mn_2O_3-OH, while the N_2 selectivity increased in the sequence of Mn_3O_4-Co_3O_4-OH < Mn_2O_3-OH < Mn_2O_3-$NiMnO_3$-OH. Although the Co and Ni elements in the catalysts delay the poisoning of SO_2 as compared to MnO_x sample, the Co-MnO_x and Ni-MnO_x binary oxides are deactivated by SO_2 over the postponement due to the formation of metal sulfate and ammonia hydrogensulfite species. In another study, Sun and co-workers [55] prepared $Mn_{0.66}M_{0.33}O_x$ catalysts (M = Fe, Zn, Cu) and a series of $Fe_\alpha Mn_{1-\alpha}O_x$ (α = 1, 0.25, 0.33, 0.50, 0 mol%) catalysts and examined for NH_3-SCR at low-temperatures. The results demonstrated that the $Fe_{0.33}Mn_{0.66}O_x$ catalyst displayed the superior NH_3-SCR activity (NO_x removal efficiency > 90%) in a wide temperature range (75–225 °C) among the $Cu_{0.33}Mn_{0.66}O_x$, $Zn_{0.33}Mn_{0.66}O_x$, and $Fe_\alpha Mn_{1-\alpha}O_x$ (α = 1, 0.25, 0.50, 0 mol%) catalysts. The authors proposed that the distortion of the catalyst structure by Fe doping could play a key role in improving the NH_3-SCR performance over the $Fe_{0.33}Mn_{0.66}O_x$ catalyst.

Rare-earth metal oxides have been frequently adopted to modify the MnO_x as an efficient low-temperature NH_3-SCR catalyst due to their incomplete 4f and empty 5d orbitals [50,51,56]. Fan et al. [57] synthesized Gd-modified MnO_x catalysts with Gd/Mn molar ratio of 0.05, 0.1, and 0.3 to improve the catalytic performance and sulfur resistance in the NH_3-SCR reaction at low-temperature. The MnGdO-2 catalyst (the mole ratio of Gd/Mn = 0.1) found to show the optimal NO conversion and N_2 selectivity among the investigated catalysts. The addition of a proper amount of Gd into MnO_x could enhance the concentrations of surface Mn^{4+} and chemisorbed oxygen species, and increase the amount and the strength of surface acid sites, which lead to better low-temperature catalytic performance than the others. Furthermore, the MnGdO-2 catalyst had an excellent tolerance to SO_2/H_2O as compared to pure MnO_x sample (Figure 3). Their results demonstrate that the doping of Gd could restrains the transformation of MnO_2 to Mn_2O_3 and the generation of $MnSO_4$, obstructs the decrease in Lewis acid sites and the increase in Brønsted acid sites, and eases the competitive adsorption between the NO and SO_2 and, thus, improves the resistance to SO_2.

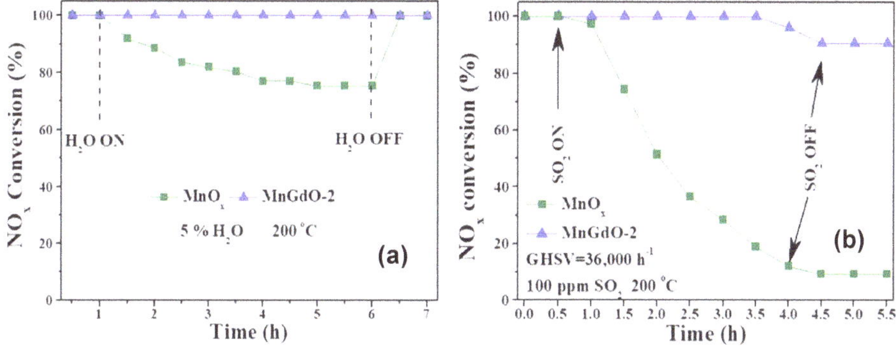

Figure 3. The (**a**) resistance to water vapor poisoning test and (**b**) resistance to sulfur poisoning test. (Reaction condition: [NO] = [NH_3] = 500 ppm, [O_2] = 5 vol%, balanced with N_2, [H_2O] = 5 vol%, [SO_2] = 100 ppm, and GHSV = 36,000 h^{-1}). Reprinted from Reference [57]. Copyright 2018, with Permission from Elsevier.

Li et al. [58] developed hollow MnO_x-CeO_2 binary nanotubes as efficient low-temperature NH_3-SCR catalysts via an interfacial oxidation-reduction process using $KMnO_4$ aqueous solution and $Ce(OH)CO_3$ nanorod as both template and reducing agent without any other intermediate. They reported that the MnO_x-CeO_2 hollow nanotube catalyst with 3.75 g of $Ce(OH)CO_3$ template (denoted it as MnO_x-CeO_2-B) exhibited outstanding performance with more than 96% NO_x conversion in the temperature range of 100–180 °C. The best activity of the MnO_x-CeO_2-B catalyst was due to its ample number of surface Mn^{4+} and O species, and hollow and porous structures that provide abundant Lewis acid sites and large surface area. Additionally, MnO_x-CeO_2-B catalyst showed an excellent resistance to H_2O and SO_2 (Figure 4) and especially, the great SO_2 tolerance was ascribed to the hierarchically porous and hollow structure that inhibits the deposition of ammonium sulfate species, and the doping of ceria that acts as an SO_2 trap to limit sulfation of the main active phase.

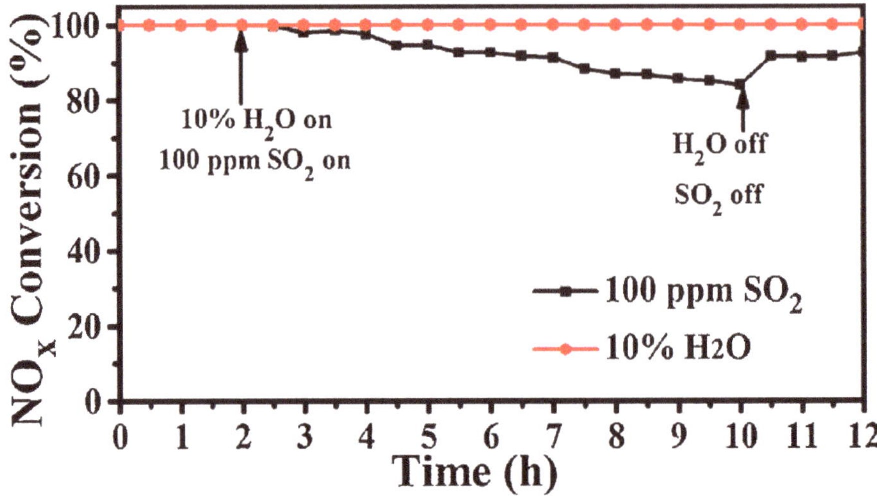

Figure 4. H_2O tolerance and SO_2 tolerance of the MnO_x-CeO_2-B hollow nanotube. Reaction conditions: $[NO_x] = [NH_3] = 1000$ ppm, $[O_2] = 5\%$, N_2 as balance gas, and GHSV = 30,000 h^{-1}. Reprinted from Reference [58]. Copyright 2018, with Permission from Elsevier.

Fe-based binary catalysts have also been studied as NH_3-SCR catalysts due to their high activity, excellent resistance to H_2O and SO_2, outstanding environmentally friendly performance, lower cost, and higher abundance [59–62]. Mu et al. [63] synthesized a series of vanadium-doped Fe_2O_3 catalysts and evaluated the effect of V on the low-temperature NH_3-SCR activity of hematite. The NH_3-SCR activity and N_2 selectivity are greatly enhanced after the incorporation of vanadium into Fe_2O_3 and the $Fe_{0.75}V_{0.25}O_\delta$ catalyst with a Fe/V mole ratio of 3/1 showed the best catalytic performance over a wide temperature window and strong resistance to H_2O and SO_2. They found that the charge transfer from Fe to V due to the electron inductive effect between Fe and V which could enhance the redox ability and surface acidity thereby superior NH_3-SCR activity at low-temperature. The in situ DRIFTS and kinetic studies suggested that the SCR reaction followed the Langmuir–Hinshelwood mechanism below 200 °C, while an Eley−Rideal mechanism dominated at and above 200 °C. Li and co-workers [64] reported novel iron titanium (CT-FeTi) catalyst, prepared by a CTAB-assisted process, showing good deNO_x efficiency and H_2O resistance at low-temperature as compared to the FeTi catalyst that prepared without adding CTAB. The addition of CTAB during the CT-FeTi catalyst preparation not only promotes to form the uniform mesoporous structure to avoid being excessively enlarged in the presence of H_2O, but also enhances the adsorption of bridging nitrate and NH_3 species on Lewis acid sites. Thus, the authors concluded that the CTAB acted as a "structural" and "chemical" promoter in improving the NH_3-SCR activity and H_2O resistance at low-temperature.

Recently, Co-based spinel catalysts have shown to exhibit a remarkable low-temperature NH_3-SCR activity, N_2 selectivity, and tolerance to SO_2/H_2O [30,65–67]. Meng et al. [48] synthesized a highly efficient $Co_aMn_bO_x$ (where a/b is the molar ratio of Co/Mn) mixed oxide catalysts and investigated the effects of the Co/Mn molar ratio on the low-temperature NH_3-SCR reaction. The $Co_aMn_bO_x$ mixed oxides showed higher NH_3-SCR activity than either MnO_x or CoO_x alone due to their improved redox properties and surface acid sites by the synergistic effects between the Co and Mn species. Particularly, the catalyst with Co/Mn molar ratio of 7:3 ($Co_7Mn_3O_x$) exhibited the greatest activity (>80% NO_x conversion) in a temperature window of 116–285 °C as compared to the catalysts with Co/Mn molar ratio of 5:5 ($Co_5Mn_5O_x$) and 3:7 ($Co_3Mn_7O_x$). They considered that the high NO + O_2 adsorption ability and enhanced redox properties of the $Co_7Mn_3O_x$ catalyst, emerging from its $MnCo_2O_{4.5}$ spinel phase and higher surface area, were beneficial to augment the NH_3-SCR performance by forming nitrate species on the catalyst surface. Although $Co_7Mn_3O_x$ catalyst had better resistance to H_2O/SO_2 than the $Co_3Mn_7O_x$ and MnO_x, the tolerance to SO_2 poisoning still need to be improved for practical use. Nevertheless, it was found that the deactivated $Co_7Mn_3O_x$, $Co_3Mn_7O_x$, and MnO_x catalysts in SO_2 stream can be regenerated simply by washing with water. Based on their results, the authors also proposed the NH_3-SCR reaction mechanism over the $Co_7Mn_3O_x$ catalyst, which is shown in Scheme 1. The reaction was initiated by adsorption and activation of gaseous oxygen on oxygen vacancies (symbol □), which was then transformed into lattice oxygen O^{2-} (Step 1); This lattice oxygen was diffused to the catalyst surface and then it had become surface active oxygen (O*) (Step 2); Gaseous NO was adsorbed and subsequently reacted with O* to form NO_2/NO_3^- intermediates (Step 3); Meanwhile, NH_3 was activated to $-NH_2$ and NH_4^+ species by Mn^{4+} (Step 4); Finally, NO_2/NO_3^- intermediates reacted with the NH species to produce reaction products, N_2, and H_2O (Step 5); By the electron transfer from Mn^{3+} to Co^{3+} (Step 6); the catalyst was recovered to its original state (Step 7); Thus, the synergistic effect between the Co and Mn plays a key role in improving the NH_3-SCR activity over $Co_7Mn_3O_x$ catalyst.

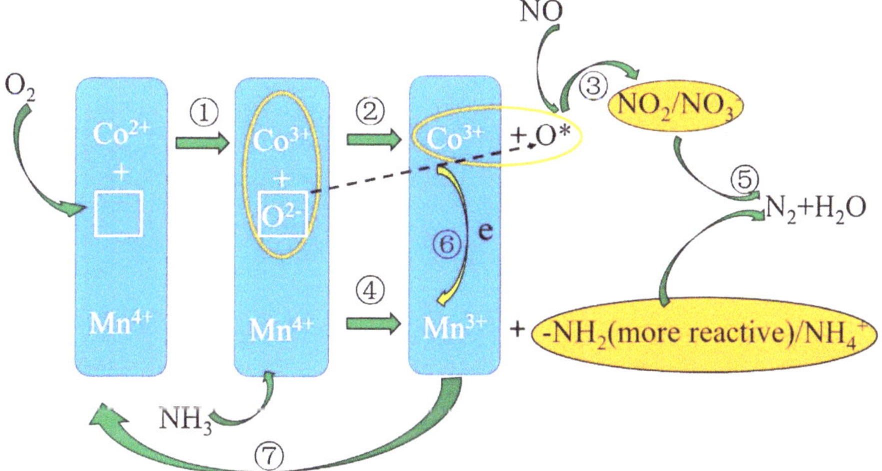

Scheme 1. The proposed mechanism of the NH_3-SCR reaction over the $Co_7Mn_3O_x$ catalyst and the synergetic catalytic effect between Mn and Co cations. Reprinted from Reference [48]. Copyright 2018, with Permission from Elsevier.

Mesoporous materials have been proved as promising catalysts for NH_3-SCR reaction since they can facilitate to promote effective diffusion of reactants towards the active sites [30,65,66,68]. With this perspective, Hu et al. [47] developed mesoporous 3D nanosphere-like Mn-Co-O catalysts through a template-free approach and evaluated for low-temperature NH_3-SCR reaction. It was found that the synthesized Mn-Co-O samples showed excellent NH_3-SCR activity in a broad working temperature window of 75 to 325 °C (NO_x conversion above 80%). They ascribed this outstanding performance to the strong and abundant acid sites, the strong adsorption of NO_x, robust redox properties, the formation of more oxygen vacancies and metal-metal interactions between the cobalt and manganese species.

Besides Mn, Fe, and Co oxides, CuO_x has also been considered in the bimetallic catalyst formulations for low-temperature NH_3-SCR reaction [69,70]. For instance, Ali et al. [71] reported the Cu_x-$Nb_{1.1-x}$ (x = 0.45, 0.35, 0.25, 0.15) bimetal oxides catalysts and found that the Cu/Nb ratio was crucial in enhancing the NH_3-SCR activity. As shown in Figure 5a,b, all binary Cu_x-$Nb_{1.1-x}$ oxides exhibited significantly higher activity than the CuO_x and Nb_2O_5, and among the Cu_x-$Nb_{1.1-x}$ samples, $Cu_{0.25}$-$Nb_{0.85}$ catalyst displayed the best performance in a wide temperature window of 180–330 °C (>90% NO conversion). Even at a high GHSV of 105,000 h^{-1}, the optimal $Cu_{0.25}$-$Nb_{0.85}$ catalyst showed a good NO removal efficiency (above 90% NO conversion) from 210 °C to 360 °C (Figure 5c). Although the SO_2/H_2O streams in the feed gas have some adverse impact on $Cu_{0.25}$-$Nb_{0.85}$, still the catalyst showed excellent resistance to SO_2/H_2O with reversible deactivation (Figure 5d). The superior NH_3-SCR performance and SO_2/H_2O tolerance of $Cu_{0.25}$-$Nb_{0.85}$ catalyst were attributed to its high acid amount and NO adsorption capacity.

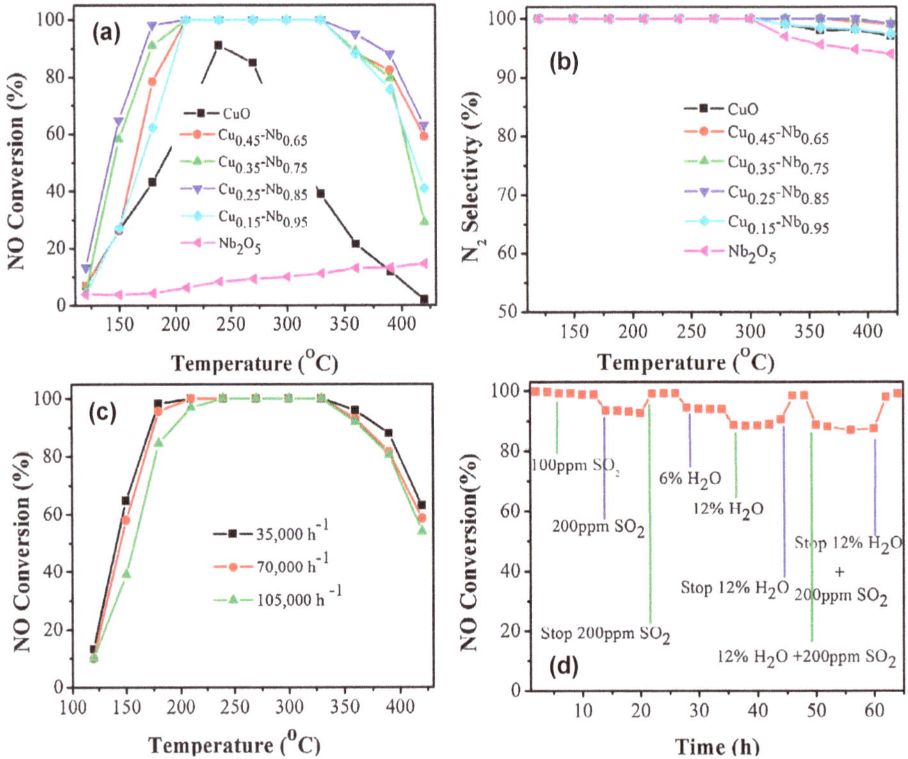

Figure 5. (**a**) NO conversion; (**b**) N_2 selectivity over Cu_x-$Nb_{1.1\text{-}x}$ (x = 0.45, 0.35, 0.25, 0.15) as a function of temperature under a GHSV of 35,000 h^{-1}; (**c**) effect of GHSV on NO conversion over $Cu_{0.25}$-$Nb_{0.85}$, and (**d**) effect of SO_2, H_2O, and SO_2 + H_2O on NO conversion over $Cu_{0.25}$-$Nb_{0.85}$ at 200 °C under a GHSV of 35,000 h^{-1}. (Reaction conditions: [NO] = [NH_3] = 1000 ppm, [O_2] = 3% and N_2 balance), and Effect of SO_2, H_2O, and SO_2 + H_2O on NO conversion over $Cu_{0.25}$-$Nb_{0.85}$ at 200 °C under a GHSV of 35,000 h^{-1}. Reprinted from Reference [71]. Copyright 2018, with Permission from Elsevier.

3. Ternary and Multi-Transition Metal-Based Catalysts

The catalytic performance of single transition metal oxides can also be improved by mixing with two or multi other metal oxides. Thus, transition metal oxides are widely reported to fabricate ternary- or multi-metal-based low-temperature NH_3-SCR catalysts, which could improve the catalytic activity by the enlarged synergetic interactions [14,72–87]. Fang et al. [88] investigated the low-temperature NH_3-SCR reaction over the $Fe_{0.3}Mn_{0.5}Zr_{0.2}$ catalyst and found that it showed an excellent deNO_x activity with 100% NO conversion in the temperature range of 200–360 °C as compared to the $Fe_{0.5}Zr_{0.5}$ and $Mn_{0.5}Zr_{0.5}$ samples. Moreover, the $Fe_{0.3}Mn_{0.5}Zr_{0.2}$ catalyst had outstanding stability and good tolerance to SO_2 (Figure 6), which they attributed to the strong interactions among Fe, Mn, and Zr species. However, the durability of the catalyst in the presence of both SO_2 and H_2O need to be tested to investigate its feasibility in practical use.

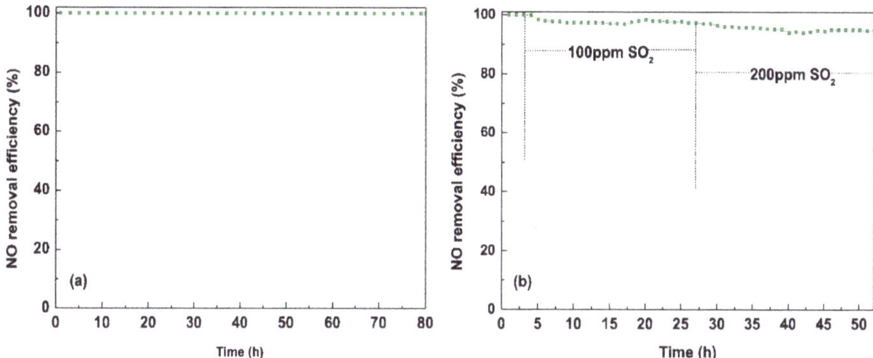

Figure 6. (**a**) NO removal efficiency of $Fe_{0.3}Mn_{0.5}Zr_{0.2}$ as a function of time at 200 °C and (**b**) effect of SO_2 on the NO removal over $Fe_{0.3}Mn_{0.5}Zr_{0.2}$ at 200 °C. Reprinted from Reference [88]. Copyright 2017, with Permission from Elsevier.

Guo and co-workers [89] studied the effect of Sb doping on the activity of $MnTiO_x$ catalyst for NH_3-SCR reaction. The results showed that Sb modification has greatly improved the NH_3-SCR performance of $MnSbTiO_x$ catalysts in comparison to the $MnTiO_x$ and $SbTiO_x$ samples. Particularly, the $MnSbTiO_x$-0.2 (Sb/Mn molar ratio = 0.2) catalyst exhibited the best activity with above 90% NO_x conversion in the temperature range of 138–367 °C as it had good adsorption and activation properties for NH_3 and NO_x reactants in SCR. It can be seen from Figure 7 that the addition of Sb dramatically improved the SO_2 and H_2O resistance of $MnTiO_x$ catalyst. Although the NO_x conversion over the $MnSbTiO_x$-0.2 catalyst slightly decreased in the presence of SO_2 and H_2O, it recovered to almost the original level after stopping the SO_2/H_2O supply.

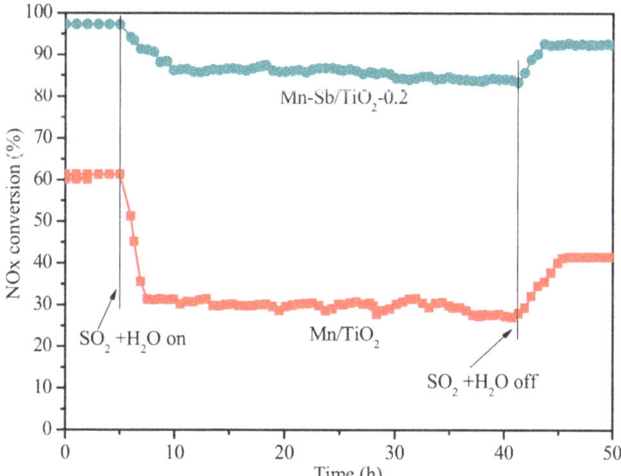

Figure 7. SO_2 resistance of $MnTiO_x$ and $MnSbTiO_x$-0.2 catalysts at 150 °C. Reaction conditions: 600 ppm NO, 600 ppm NH_3, 5% O_2, 5% H_2O, 100 ppm SO_2, balance Ar, GHSV = 108,000 h^{-1}. Reprinted from Reference [89]. Copyright 2018, with Permission from Elsevier.

Shi et al. [90] synthesized a series of $Ni_yCo_{1-y}Mn_2O_x$ microspheres (MSs) (y = 0.1, 0.3, 0.5, 0.7, 0.9) for NH_3-SCR using a hydrothermal method. It was observed that the activity of all ternary MSs was greater than binary $CoMn_2O_x$ and $NiMn_2O_x$, and $Ni_{0.7}Co_{0.3}Mn_2O_x$ showed the best NH_3-SCR performance among the $Ni_yCo_{1-y}Mn_2O_x$ catalysts. Although the $Ni_{0.7}Co_{0.3}Mn_2O_x$ catalyst exhibited good resistance to H_2O, it had poor SO_2 tolerance which needs to be improved. Wu and co-workers [91] compared the DeNO$_x$ performance of MnO_2/CoAl-LDO and CoMnAl-LDO mixed metal oxides prepared from CoAl-MnO_2-LDH and CoMnAl-LDH templates by ion-exchange/redox reaction and hexamethylenetetramine (HMT) hydrolysis methods, respectively. The CoAl-MnO_2-LDH showed higher NH_3-SCR activity in a broad temperature window of 90–300 °C (Figure 8a) as well as better stability and SO_2/H_2O resistance (Figure 8b) than the CoMnAl-LDO, which was attributed to its larger specific surface area, stronger redox ability, more quantitative acid sites, and abundant active components.

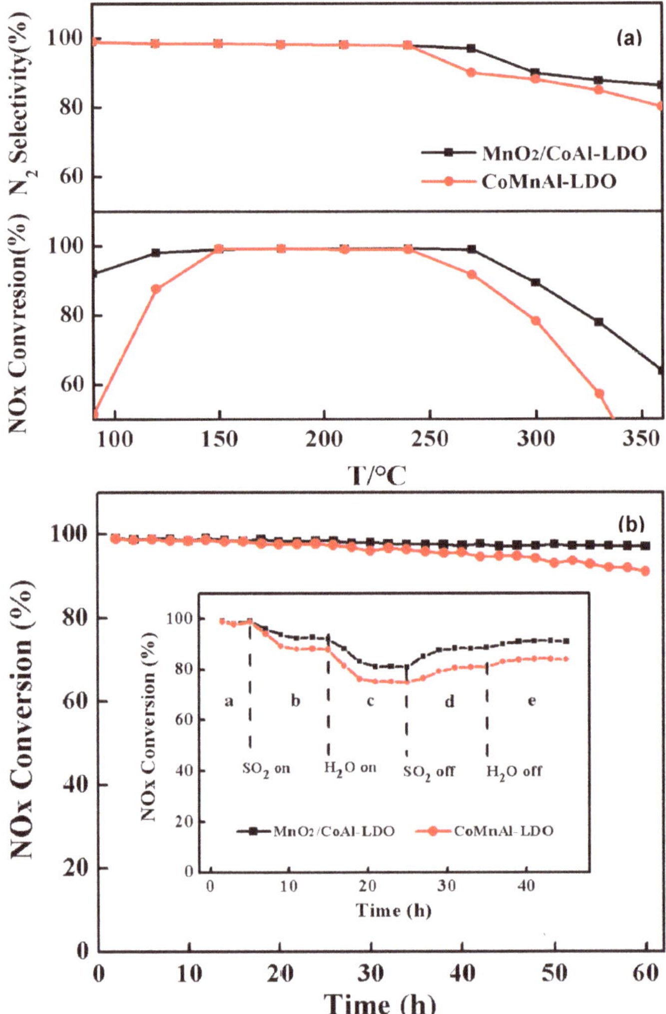

Figure 8. (**a**) NO$_x$ conversion and N_2 selectivity over MnO_2/CoAl-LDO and CoMnAl-LDO catalysts prepared by calcination at 500 °C; and (**b**) the stability and SO_2/H_2O resistance test (inset) of MnO_2/CoAl-LDO and CoMnAl-LDO catalysts at 240 °C. Reaction conditions: [NH_3] = 600 ppm, [NO] = 600 ppm, [O_2] = 5 vol%, 100 ppm SO_2 (when used), 10 vol% H_2O. (when used) balanced by N_2. Reprinted from Reference [91]. Copyright 2019, with Permission from Elsevier.

Leng and collaborators [92] synthesized $Mn_{0.2}TiO_x$, $Ce_{0.3}TiO_x$ and series of $Mn_aCe_{0.3}TiO_x$ (a = 0.1, 0.2, 0.3) catalysts and investigated their applicability for low-temperature NH_3-SCR reaction. They demonstrated that the low-temperature NH_3-SCR activity of $Mn_aCe_{0.3}TiO_x$ was greatly improved after incorporation of Mn, and the $Mn_{0.1}Ce_{0.3}TiO_x$ catalyst displayed the best performance (with 100% NO conversion and above 90% N_2 selectivity) in the temperature range of 175–400 °C even at high GHSV of 80,000 h^{-1}. The outstanding performance of $Mn_{0.1}Ce_{0.3}TiO_x$ catalyst in NH_3-SCR resulted from its enhanced acidity and chemisorbed oxygen, and suitable redox property derived from Ce^{3+} + $Mn^{4+} \leftrightarrow Ce^{4+} + Mn^{3+}$ reaction. Furthermore, the NO conversion over the $Mn_{0.1}Ce_{0.3}TiO_x$ decreased and stabilized at 82% after the introduction of 100 ppm SO_2 and 6% H_2O and restored to almost 100% NO conversion after stopping the supply of SO_2 and H_2O (Figure 9), suggesting that the catalyst had excellent resistance to SO_2/H_2O and the effects were reversible.

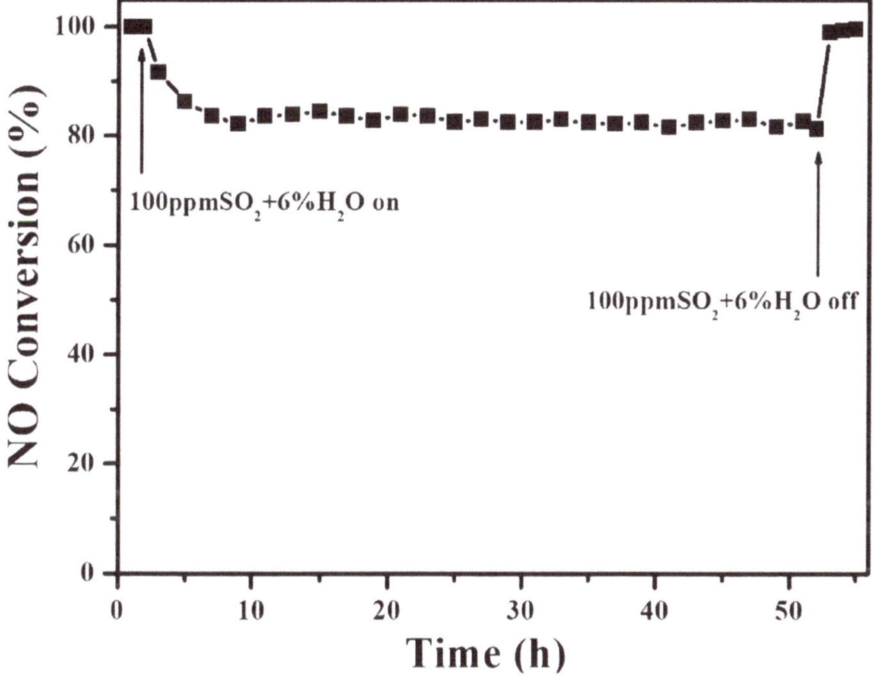

Figure 9. Effect of H_2O and SO_2 on NO conversion over the $Mn_{0.1}Ce_{0.3}TiO_x$ catalyst at 200 °C (1000 ppm NO, 1000 ppm NH_3, 3% O_2, balance N_2, GHSV = 40,000 h^{-1}). Reprinted from Reference [92]. Copyright 2018, with Permission from Elsevier.

Ali et al. [93] developed a series of Nb-promoted Fe_x-$Nb_{0.5\text{-}x}$-$Ce_{0.5}$ (x = 0.45, 0.4, 0.35) oxides for NH_3-SCR. The best activity (>90% NO conversion and near 100% N_2 selectivity) in the broad temperature window of 180–400 °C as well as excellent SO_2/H_2O resistance (Figure 10) was observed for $Fe_{0.4}$-$Nb_{0.1}$-$Ce_{0.5}$ catalyst. The authors considered the strong interaction among Nb, Fe, and Ce oxides leading to the enhancement of BET surface area, redox ability, acid amount, and NO adsorption capacity, which could be responsible for the outstanding performance of the catalyst.

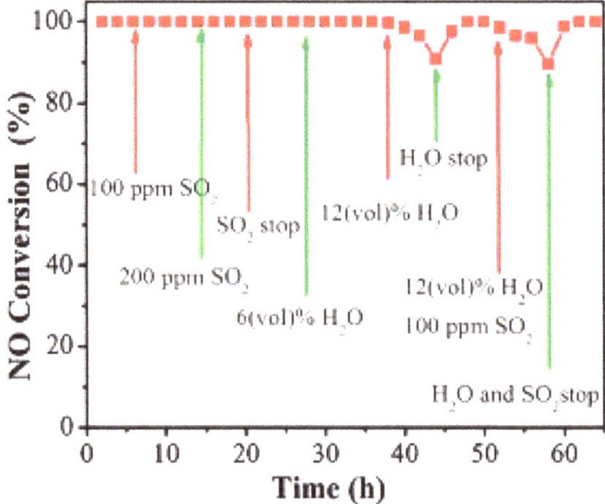

Figure 10. Effect of SO_2, H_2O, and SO_2 + H_2O on NO conversion over the $Fe_{0.4}$-$Nb_{0.1}$-$Ce_{0.5}$ catalyst at 220 °C. Reprinted from Reference [93]. Copyright 2018, with Permission from Elsevier.

Sun and co-workers [94] reported the multimetallic Sm- and/or Zr-doped MnO_x-TiO_2 catalysts for NH_3-SCR reaction. As shown in Figure 11, the Sm and Zr co-doped MnO_x-TiO_2 ($MSZTO_x$) catalyst had better activity (≈100% NO conversion and >95% N_2 selectivity) in a wide temperature range (125–275 °C) with an excellent H_2O/SO_2 tolerance than the $MSTO_x$ (MnO_x-SmO_x-TiO_2), $MZTO_x$ (MnO_x-ZrO_x-TiO_2) and MTO_x (MnO_x-TiO_2) catalysts. The authors claimed that the enhanced redox properties and acidic sites play a crucial role in improving the NH_3-SCR performance of $MSZTO_x$ catalyst. Yan and co-workers [95] fabricated $Cu_{0.5}Mg_{1.5}Mn_{0.5}Al_{0.5}O_x$ catalyst from layered double hydroxides and found that it showed better activity in a wide temperature range together with superior SO_2 and H_2O tolerance than conventional Mn/γ-Al_2O_3. The improved performance of $Cu_{0.5}Mg_{1.5}Mn_{0.5}Al_{0.5}O_x$ was attributed to the high specific surface area, high reducibility of MnO_2 and CuO species, an abundance of acid sites, and the good dispersion of MnO_2 and CuO species.

Chen et al. [96] investigated the NH_3-SCR reaction over a series of $Co_{0.2}Ce_xMn_{0.8-x}Ti_{10}$ (x = 0, 0.05, 0.15, 0.25, 0.35, and 0.40) oxides catalysts and observed that the $Co_{0.2}Ce_{0.35}Mn_{0.45}Ti_{10}$ catalyst exhibited the best catalytic performance with 100% NO conversion and over 91% N_2 selectivity in a broad temperature window of 180–390 °C. Although NO_x conversion decreased to some extent after introducing SO_2 and H_2O, the $Co_{0.2}Ce_{0.35}Mn_{0.45}Ti_{10}$ catalyst showed excellent resistance to SO_2/H_2O with reversible inhibition effect (Figure 12). It was concluded that the interactions among Ce, Co, Mn, and Ti oxides led to more surface Brønsted acid and Lewis acid sites, NO_x adsorption sites and modest redox ability which could play a crucial role to improve the NH_3-SCR activity of $Co_{0.2}Ce_{0.35}Mn_{0.45}Ti_{10}$.

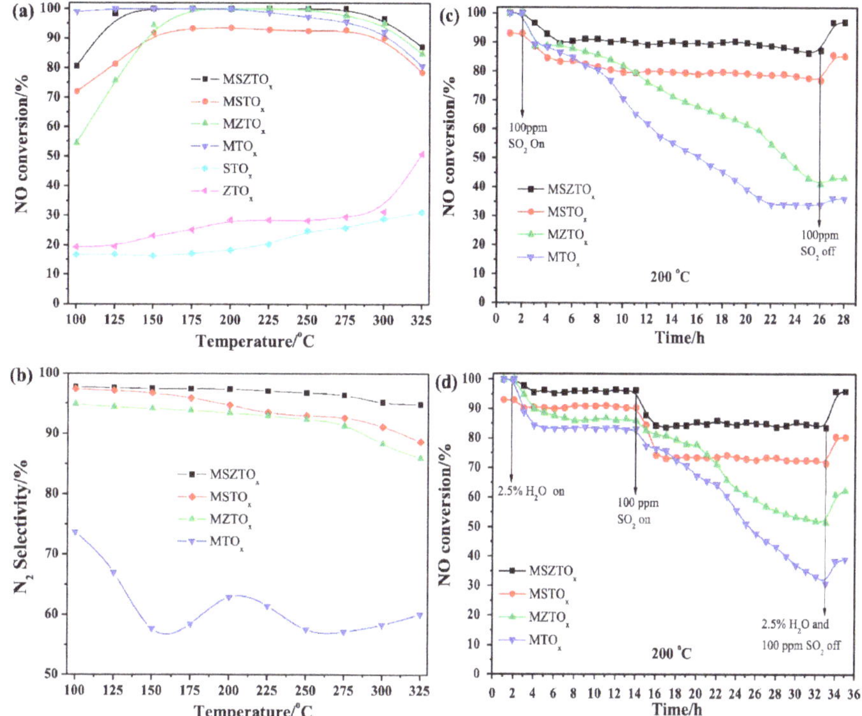

Figure 11. (**a**) NO conversions and (**b**) N_2 selectivities of the catalysts in the NH_3-SCR reaction as a function of temperature; (**c**) SO_2 (100 ppm) resistance tests, and (**d**) H_2O + SO_2 (2.5 vol%, 100 ppm) resistance tests at 200 °C over the catalysts. Reproduced from Reference [94]. Copyright 2018, with Permission from Elsevier.

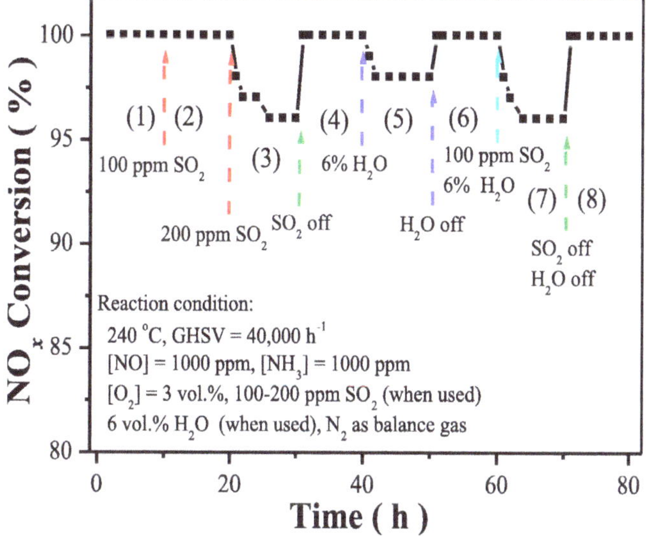

Figure 12. Effects of H_2O and/or SO_2 on NO_x conversion over the $Co_{0.2}Ce_{0.35}Mn_{0.45}Ti_{10}$ catalyst. Reproduced from Reference [96]. Copyright 2018, with Permission from Elsevier.

4. Supported Single Transition Metal-Based Catalysts

Support materials have proved to be highly beneficial for enhancing the activity and durability of catalysts as they possess high surface area and good thermal stability. In virtue of the fine dispersion of the active component on the surface of the support and the synergistic effect between active component and support, supported catalysts exhibits improved NH_3-SCR performance than the unsupported transition metal oxide. Therefore, a lot of attention has been focused to increase the de-NO_x efficiency by dispersing the transition metal oxides over different support materials such as TiO_2, Al_2O_3, SiO_2, carbon nanotubes (CNTs), etc. TiO_2 supported transition metal oxides, especially manganese oxides, have been widely reported as promising catalysts for NH_3-SCR reaction at low temperature [17,24,97–105]. Smirniotis and co-workers [17,106,107] first reported the transition metal oxides (V, Cr, Mn, Fe, Co, Ni, and Cu) supported on Hombikat TiO_2 for NH_3-SCR reaction at low-temperature. Among the investigated samples, the Mn/Hombikat TiO_2 catalyst found to exhibit the highest activity even in the presence of water. They also studied the effect of different supports on the NH_3-SCR performance and observed that the Mn/Hombikat TiO_2 (anatase, high surface area) had the best activity as compared to the Kemira TiO_2 (rutile), Degussa P25 TiO_2 (anatase, rutile), Aldrich TiO_2 (anatase, low surface area), Puralox γ-Al_2O_3, Aldrich SiO_2 supported Mn catalysts. It was concluded that the Lewis acidity, redox behavior, and a high surface concentration of MnO_2 could play a key role in improving the NH_3-SCR activity. Later, they investigated the effect of Mn loading on the NH_3-SCR performance of Mn/Hombikat TiO_2 and reported that the catalyst with 16.7 wt% Mn had optimal activity and excellent tolerance to H_2O during 10 days of the reaction [108]. In another work, they also proposed the NH_3-SCR reaction mechanism over the Mn/TiO_2 catalyst using transient isotopic labeled and in-situ FT-IR studies. As shown in Figure 13, the reaction proceeds via a Mars-van-Krevelen-like mechanism, in which NH_3 and NO species were first adsorbed onto the Mn^{4+} sites (Lewis acid sites), followed by the formation of nitrosamide and azoxy intermediate species. Finally, these intermediates converted into N_2 and H_2O products [24].

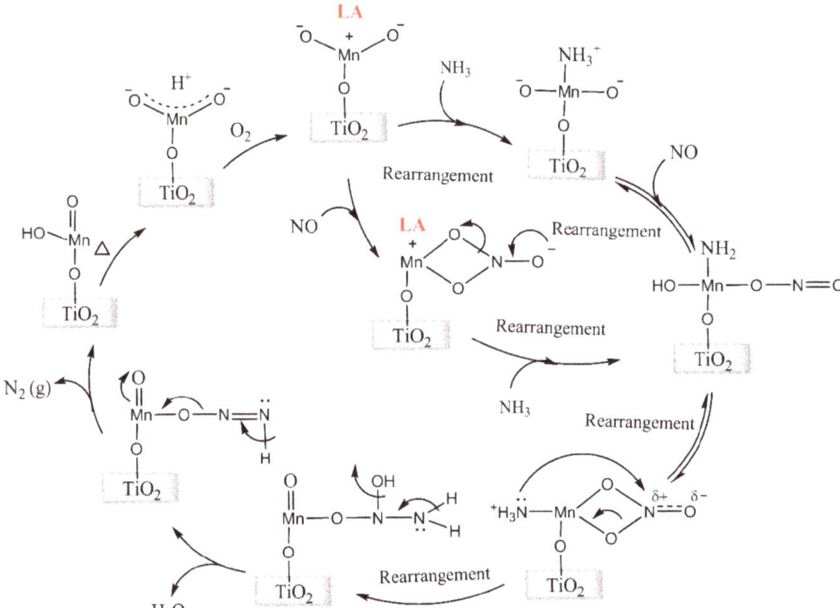

Figure 13. Plausible SCR mechanism over the surface of Mn/TiO_2 catalyst. Adapted from Reference [24]. Copyright 2012, with Permission from Elsevier.

In the aspect of catalyst structure design, Smirniotis and co-workers [26] developed a series of manganese confined titania nanotube (Mn/TNT-X) catalysts using different TiO_2 precursors (X = Ishihara (I), Kemira (K), Degussa P25 (P25), Sigma–Aldrich (SA), Hombikat (H), TiO_2 synthesized from titanium oxysulfate (TOS)) for low-temperature NH_3-SCR reaction. As can be observed from the Figure 14, all Mn/TNT-X catalysts exhibited an excellent NO_x conversion in broad temperature window, and especially, Mn(0.25)/TNT-H sample obtained the superior activity in the temperature range of 100–300 °C as compared to other catalysts. They believed that the better performance of Mn(0.25)/TNT-H catalyst was due to the high surface area (421 m^2/g) of the TNT-H support, and high dispersion of active components. The Mn(0.25)/TNT-H catalyst also showed greater catalytic performance than the conventional Mn-loaded titania nanoparticles (Mn/TiO_2), suggesting that the unique multiwall nanotube with open-ended structure could be advantageous to promote the reaction. Besides, the Mn (0.25)/TNT-H displayed outstanding tolerance to 10 vol% H_2O in the feed (Figure 15), which might be attributed to the preferential existence of highly active and redox potential pairs of Mn^{4+} and Mn^{3+} in the tubular framework.

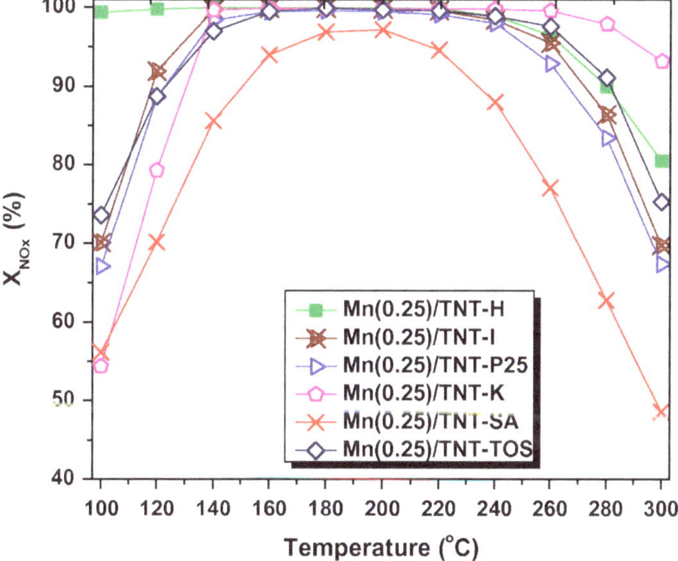

Figure 14. Catalytic evaluation of the Mn(0.25)/TNT-X (X = Hombikat, Ishihara, P25 Degussa, Kemira, Sigma–Aldrich, and Titania oxysulfate) family of catalyst for the SCR of NO_x by NH_3, in the presence of 900 ppm NO, 100 ppm NO_2, 1000 ppm NH_3, 10 vol% O_2 with He balance under a GHSV of 50,000 h^{-1} in the temperature range from 100–300 °C. Reproduced from Reference [26]. Copyright 2016, with Permission from Elsevier.

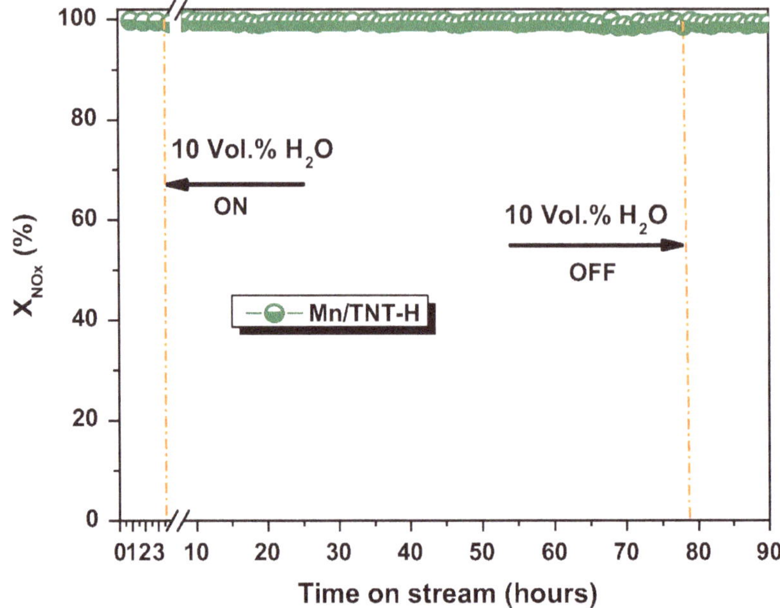

Figure 15. Influence of inlet water concentrations (10 vol%) on NO_x conversion in the SCR reaction over Mn(0.25)/TNT-H catalyst at 140 °C; feed: NO = 900 ppm, NO_2 = 100 ppm, NH_3/NO_x (ANR) = 1.0, O_2 = 10 vol%, He carrier gas, GHSV = 50,000 h^{-1}. Reproduced from Reference [26]. Copyright 2016, with Permission from Elsevier.

Recently, Boningari et al. [109] extended this work by comparing the NH_3-SCR activity of various metal oxide confined titania nanotubes M/TNT (M = Mn, Cu, Ce, Fe, V, Cr, and Co) based on the Hombikat TiO_2 support. As shown in Figure 16, the Mn-, V-, Cr-, and Cu-oxide confined titania nanotubes had excellent low-temperature activity and meanwhile, vanadium oxide confined titania nanotubes showed a broad operation temperature window for the NH_3-SCR reaction.

Sheng et al. [100] synthesized core-shell MnO_x/TiO_2 nanorod catalyst, showed high activity, stability, and N_2 selectivity in NH_3-SCR. They concluded that the abundant mesopores, Lewis-acid sites, and high redox capability could be beneficial to improve catalytic performance. Although the MnO_x/TiO_2 catalyst exhibited excellent resistance to H_2O, it was deactivated in the presence of SO_2 and SO_2/H_2O. Jia et al. [110] reported the low-temperature NH_3-SCR efficiency of MnO_x/TiO_2, MnO_x/ZrO_2, and MnO_x/ZrO_2-TiO_2 catalysts, and found that MnO_x/ZrO_2-TiO_2 obtained good activity at a temperature of 80–360 °C and excellent resistance to H_2O at 200 °C. However, all the catalysts showed poor tolerance to SO_2 and SO_2/H_2O that caused irreversible deactivation. Similar findings were also observed by Zhang et al. [111] over the Mn/Ti, Mn/Zr, and Mn/Ti-Zr catalysts, in which Mn/Ti-Zr sample exhibited an excellent NH_3-SCR performance in a wide temperature range due to its high surface area, Lewis acid sites, and surface Mn^{4+} ions.

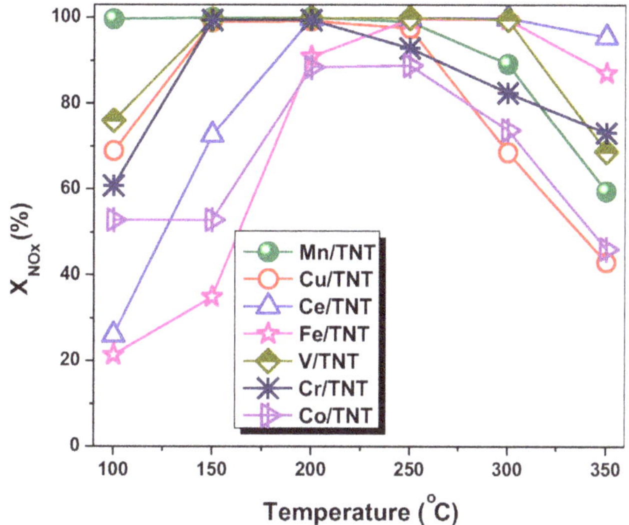

Figure 16. Catalytic activity evaluation of metal oxides confined titania (made of Hombikat titania) nanotube catalytic formulations M/TNT where M = Mn, Cu, Ce, Fe, V, Cr, and Co for the selective catalytic reduction of NO_x by NH_3 in the presence of 900 ppm NO, 100 ppm NO_2, 1000 ppm NH_3, 10 vol% O_2 in He balance, under a GHSV = 50,000 h^{-1}. Reprinted from Reference [109]. Copyright 2018, with Permission from Elsevier.

Carbon nanotubes (CNTs) have been reported as promising catalyst support for NH_3-SCR catalysis due to their excellent stability and unique electronic and structural properties [36,112–115]. Qu and co-workers [116] reported that the NH_3-SCR performance of Fe_2O_3 was dramatically enhanced when it supported on CNTs (Figure 17a). It was concluded that the large surface area, fine dispersion of Fe_2O_3, and interaction between Fe_2O_3 and CNTs were important factors to improve the NH_3-SCR activity. In addition, the Fe_2O_3/CNTs catalyst showed an excellent tolerance to H_2O/SO_2. Interestingly, SO_2 stream in the feed had promoting effect on the NO conversion (Figure 17b), which could be attributed to the increased acid sites for NH_3 adsorption and activation on the catalyst surface in presence of SO_2. Bai et al. [117] developed CNTs supported copper oxide catalysts, and found that the 10 wt% CuO/CNTs showed good NH_3-SCR activity and excellent stability at 200 °C. The 10 wt% CuO/CNTs also had greater performance in comparison to 10 wt% CuO/TiO_2. However, it exhibited poor resistance to SO_2 and moderate tolerance to H_2O.

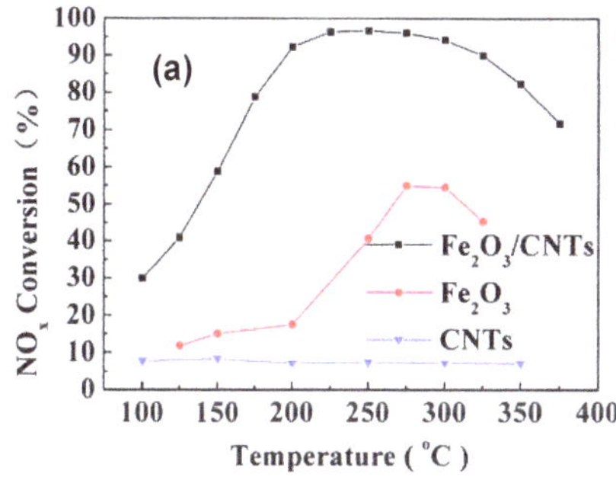

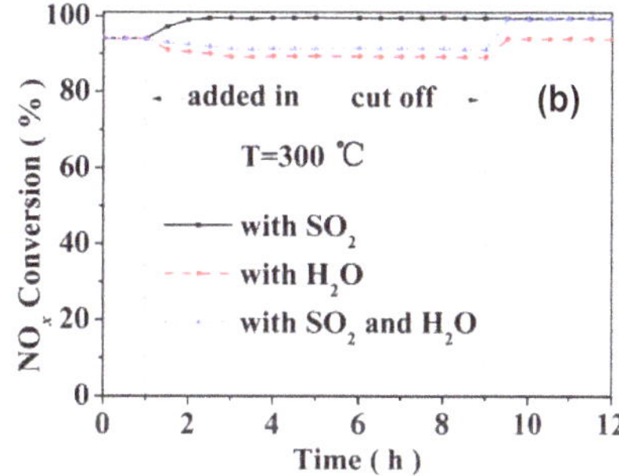

Figure 17. (**a**) NO_x conversion as a function of temperature over different catalysts and (**b**) SO_2/H_2O tolerance of the Fe_2O_3/CNTs catalyst. Reprinted from Reference [116]. Copyright 2015, with Permission from Royal Society of Chemistry.

5. Supported Binary and Multi Transition Metal-Based Catalysts

Given that the dispersion of two active components on support enhances the active sites further, researchers have been widely reported the supported binary transition metal-based oxides to improve the performance and SO_2/H_2O tolerance in NH_3-SCR reaction. With this perspective, several composites, such as MnCe/CNTs [118], Mn-Fe/TiO_2 [119], MnO_x-CeO_2/graphene [120], Mg-MnO_x/TiO_2 [121], CeO_x-MnO_x/TiO_2-graphene [122], Fe-Mn/Al_2O_3 [123], Mn-Fe/W-Ti [124], MnO_x-CeO_2/TiO_2-1%NG (NG = N-doped grapheme) [125], Mn-Ce/CeAPSO-34 [126], etc., were investigated for the NH_3-SCR reaction at low-temperature. Smirniotis et al. [22] studied the promotional effect of co-doped metals (Cr, Fe, Co, Ni, Cu, Zn, Ce, and Zr) on the NH_3-SCR performance of Mn/TiO_2. As shown in Figure 18, except Zn and Zr, all other co-doped metals had a positive impact on the activity of Mn/TiO_2, and particularly, the Mn-Ni/TiO_2 exhibited the highest NO conversion and N_2 selectivity among the other titania-supported bimetallic catalysts.

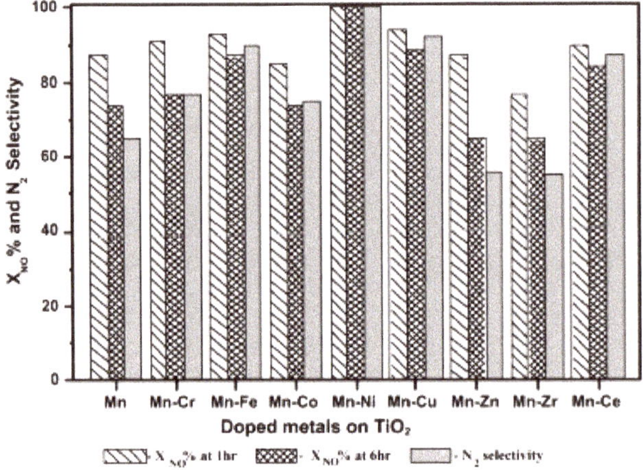

Figure 18. N_2 selectivity and catalytic performance of Mn-M′/TiO_2 anatage (M′ = Cr, Fe, Co, Ni, Cu, Zn, Zr, and Ce) catalysts: NH_3 = 400 ppm; NO = 400 ppm; O_2 = 2.0 vol%; GHSV = 50,000 h^{-1}; catalyst wt. = 0.1 g; reaction temperature = 200 °C. Reproduced from Reference [22]. Copyright 2011, with Permission from Elsevier.

They also investigated the influence of Ni loading on the activity of Mn/TiO_2 catalyst, and found that the 5wt%Mn-2wt%Ni/TiO_2 (Mn-Ni(0.4)/TiO_2, where Ni/Mn = 0.4) had the optimal activity with complete NO conversion at the temperature range of 200–250 °C (Figure 19a) and outstanding stability even in the presence of 10 vol% water (Figure 19b,c) [5,23]. The enhanced reducibility of manganese oxide and dominant phase of MnO_2 claimed to be responsible for the best activity and stability of Mn–Ni/TiO_2 catalyst [5,22,23]. In another study, they compared the de-NO_x performance of high surface texture hydrated titania and Hombikat TiO_2 supported Mn-Ce bimetallic catalysts, and observed that the Mn–Ce/TiO_2 (Hombikat) showed the better activity and excellent resistance to H_2O (Figure 20). The superior performance could be attributed to the enhancement in reduction potential of active components, broadening of acid sites distribution, and the promotion of Mn^{4+}/Mn^{3+}, Ce^{3+}/Ce^{4+} ratios including surface labile oxygen and small pore openings [25].

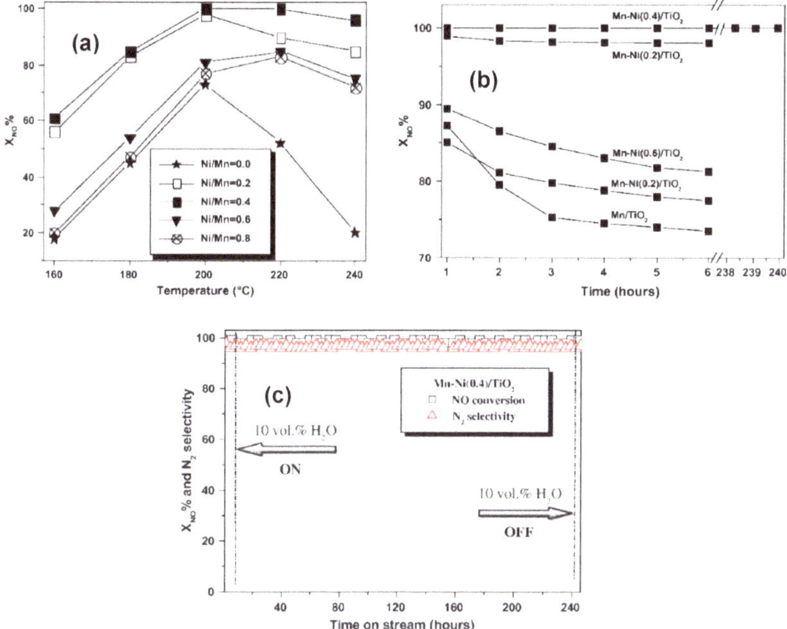

Figure 19. (**a**) Influence of Ni/Mn atomic ratio on NO conversion in the SCR reaction at a temperature range (160–240 °C) over Mn-Ni/TiO_2 catalysts (X_{NO}% = conversion of NO at 6 h on stream); (**b**) SCR of NO with NH_3 at 200 °C temperature over Mn/TiO_2 and Mn-Ni/TiO_2 catalysts; (**c**) Influence of inlet water concentrations (10 vol%) on NO conversion in the SCR reaction over Mn–Ni(0.4)/TiO_2 catalyst at 200 °C (GHSV = 50,000 h^{-1}; feed: NO = 400 ppm, NH_3 = 400 ppm, O_2 = 2 vol%, He carrier gas, catalyst = 0.1 g, total flow = 140 mL min^{-1}). Reprinted from Reference [5]. Copyright 2012, with Permission from Elsevier.

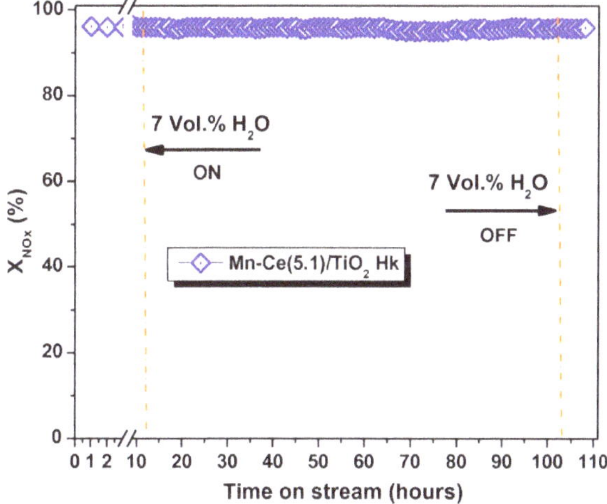

Figure 20. Influence of inlet water concentrations (7 vol%) on NO_x conversion in the SCR reaction over Mn–Ce(5.1)/TiO_2-Hk (Hombikat) catalyst at 175 °C; feed: NO = 900 ppm, NO_2 = 100 ppm NH_3/NO_x (ANR) = 1.0, O_2 = 10 vol%, He carrier gas, catalyst. 0.08 g, GHSV. 80,000 h^{-1}. Reprinted from Reference [25]. Copyright 2015, with Permission from Elsevier.

Xu and co-workers [127] reported Ce-Mn/TiO_2 catalysts with different Ce loadings, and the Ce(20)-Mn/TiO_2 found to show high activity with >90% NO conversion in the temperature range of 140–260 °C (Figure 21a). Their SO_2 tolerance results showed that the resistance ability was decreased in the order of Ce(20)-Mn/TiO_2 > Ce(30)-Mn/TiO_2 > Ce(10)-Mn/TiO_2 (Figure 21b). Although the Ce(20)-Mn/TiO_2 catalyst had reasonable resistance to 100 ppm SO_2 at different reaction temperatures (Figure 21c), it exhibited moderate tolerance to SO_2 poisoning when added higher than 100 ppm SO_2 to the reaction feed (Figure 21d). They ascribed the good SO_2 resistance of Ce(20)-Mn/TiO_2 to the widely distributed elements of Mn and Ce which in turn led to the inability of the sulfate material to remain on the surface.

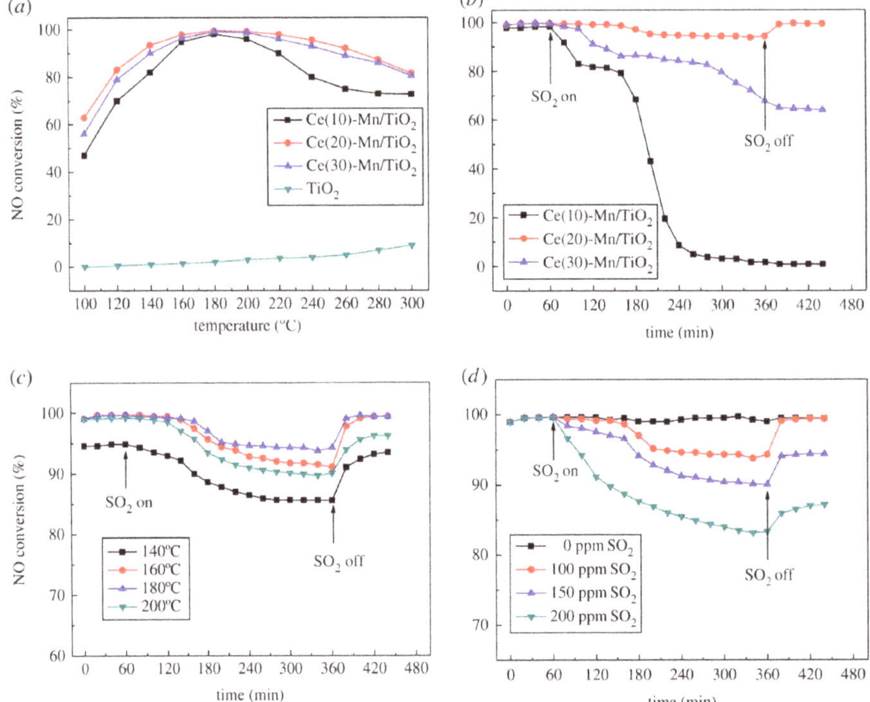

Figure 21. (**a**) Catalytic activity of Ce-Mn/TiO_2 catalyst for NH_3–SCR. Catalysts were loaded with 10%, 20% and 30% Ce and denoted as Ce(10), Ce(20) and Ce(30), respectively. Pure TiO_2 was also used for comparison; (**b**) The effect of various Ce concentrations using a Ce-Mn/TiO_2 catalyst on SO_2 resistance; (**c**) The effects of reaction temperature on NO conversion of the Ce(20)-Mn/TiO_2 catalyst in the presence of SO_2. The above three types of reactions were performed at: 500 ppm NO, 500 ppm NH_3, SO_2 100 ppm, 3% O_2, N_2 balance. gas, GHSV = 10 000 h^{-1}; and (**d**) the effects of SO_2 concentration on NO conversion of Ce(20)-Mn/TiO_2 catalysts (T = 180°C, 500 ppm NO, 500 ppm NH_3, 3%O_2, N_2 balance gas, GHSV = 10 000 h^{-1}). Adapted from Reference [127]. We thank the Royal Society Open Science for this Contribution.

Lin et al. [128] synthesized Me-Fe/TiO_2 (SD) catalyst via an aerosol-assisted deposition method, showing an excellent NH_3-SCR performance and good tolerance to SO_2/H_2O as compared to its counterparts prepared by co-precipitation and wet impregnation methods. The authors concluded that the enhanced surface reducibility and adsorption ability of NH_3/NO_x of Mn-Fe/TiO_2 (SD) catalyst could be responsible for its superior activity. Lee and co-workers [129] investigated the poisoning effect of SO_2 as metal sulfate and/or ammonium sulfate deposits on the low-temperature

NH_3-SCR activity of MnFe/TiO_2 catalysts. They found that the metal sulfates had a more serious deactivation effect than that of ammonium salts on the MnFe/TiO_2 catalysts. Their results showed that metal sulfates poisoning resulted in lower crystallinity, lower specific surface area, a lower ratio of Mn^{4+}/Mn^{3+}, higher surface acidity, and more chemisorbed oxygen, which in turn led to an adverse effect on the NH_3-SCR activity of the catalyst. Mu et al. [130] prepared Fe-Mn/Ti catalyst by ethylene glycol-assisted impregnation method, showing high NH_3-SCR efficiency over a broad temperature window (100−325 °C) and outstanding tolerance to sulfur poisoning. The formation of the Fe-O-Ti structure with strong interaction strengthened the electronic inductive effect and increased the ratio of surface chemisorption oxygen, thereby the enhancement of NO_x adsorption capacity and NO oxidation performance, which could be beneficial to improve the NH_3-SCR activity.

Liu and co-workers [131] reported that the addition of Eu had noticeably improved the NH_3-SCR performance of Mn/TiO_2 catalyst even after sulfation process under SCR conditions (Figure 22a). However, both the Mn/TiO_2 and MnEu/TiO_2 catalysts showed poor activity when they sulfated only with SO_2 + O_2 (Figure 22a). Further, the MnEu/TiO_2 catalyst found to show better SO_2 tolerance as compared to the Mn/TiO_2 (Figure 22b). Their results revealed that Eu modification could inhibit the formation of surface sulfate species on the Mn/TiO_2 catalyst during the NH_3-SCR in the presence of SO_2, which could be the reason for improved SO_2 resistance.

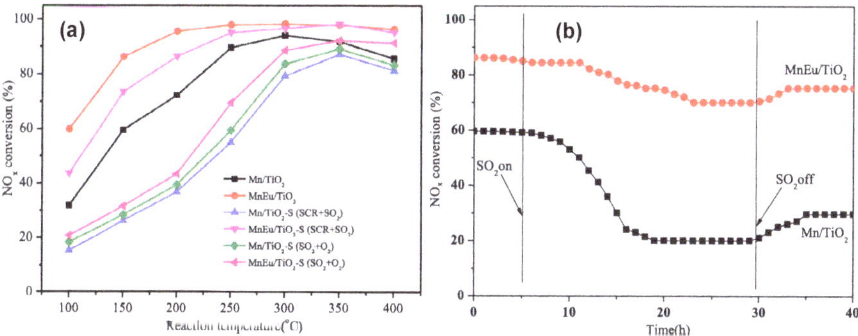

Figure 22. (**a**) NH_3-SCR activities of the fresh and sulfated catalysts; and (**b**) SO_2 tolerances of Mn/TiO_2 and MnEu/TiO_2 in NH_3-SCR reaction. Reprinted from Reference [131]. Copyright 2018, with Permission from Elsevier.

Sun et al. [132] investigated the NH_3-SCR activity over the Nb-doped Mn/TiO_2 catalysts with different Nb/Mn molar ratios, and found that the MnNb/TiO_2-0.12 (where Nb/Mn = 0.12) catalyst had optimal NO_x conversion and N_2 selectivity in the temperature range of 100–400 °C (Figure 23a,b). The optimal MnNb/TiO_2-0.12 catalyst also exhibited greater SO_2 resistance than Mn/TiO_2 catalyst (Figure 23c). The incorporation of Nb into Mn/TiO_2 catalyst led to increase surface acidity and reducibility as well as generate more surface Mn^{4+} and chemisorbed oxygen species along with more NO_2, which results in the better NH_3-SCR activity. In situ DRIFT studies over the Mn/TiO_2 and MnNb/TiO_2-0.12 catalysts disclosed that the NH_3-SCR took place through Eley–Rideal mechanism even in presence of SO_2, in which the reaction mainly occurred between adsorbed NO_2 and gaseous NH_3. Hence, it was concluded that the higher SO_2 tolerance of the MnNb/TiO_2-0.12 catalyst could be due to the existence of more adsorbed NO_2 on its surface.

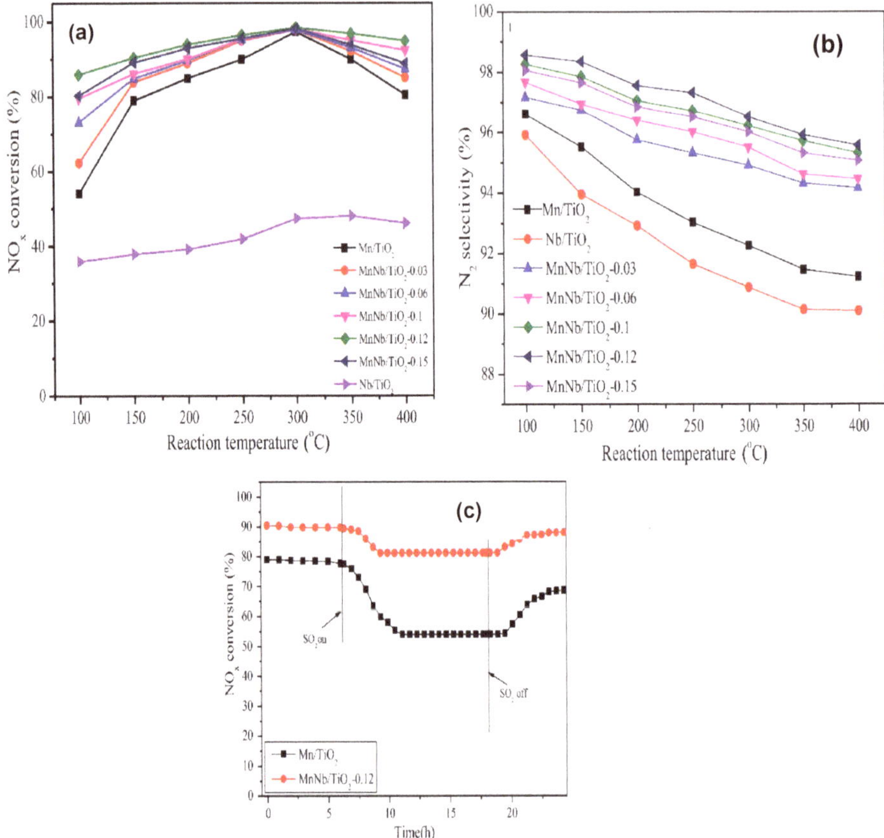

Figure 23. (**a**) SCR activities and (**b**) N_2 selectivities over different catalyst samples as a function of reaction temperature. Reaction conditions: [NO] = [NH_3] = 600 ppm, [O_2] = [H_2O] = 5%, balance Ar, GHSV = 108,000 h^{-1}; (c) Effect of SO_2 on the SCR activities over Mn/TiO_2 and MnNb/TiO_2-0.12 catalyst samples. Reaction conditions: [NO] = [NH_3] = 600 ppm, [O_2] = [H_2O] = 5%, [SO_2] = 100 ppm, balance Ar, GHSV = 108,000 h^{-1}, reaction temperature = 150 °C. Reproduced from Reference [132]. Copyright 2018, with Permission from Elsevier.

In another study, they reported the Mo-modified Mn/TiO_2 catalysts, exhibiting improved NH_3-SCR activity from 50 to 400 °C in comparison to Mn/TiO_2 catalyst. Particularly, the optimal MnMo/TiO_2-0.04 (where molar ratio of Mo/Mn = 0.04) catalyst better tolerance to SO_2 poisoning compared with Mn/TiO_2 (Figure 24) [133].

Fan and co-workers [134] fabricated ordered mesoporous titania supported CuO and MnO_2 composites (CuO/MnO_2-mTiO_2) through a facile acetic acid-assisted one-pot synthesis approach, showing high deNO$_x$ efficiency (>90% NO conversion) and N_2 selectivity (>95%) in a wide operating temperature range of 120–300 °C. They considered that the superior NH_3-SCR performance could be attributed to the unique structure and highly integrated mesoporous TiO_2 supported by the multicomponent system with high surface areas, accessible and homogenously dispersed CuO and MnO_2 with multivalent nature and good redox activity. Although the CuO/MnO_2-mTiO_2 catalyst had good tolerance to H_2O, the resistance to SO_2 and H_2O/SO_2 poisoning, as well as high space velocity (GHSV), still need to be enhanced for practical use. Li et al. [135] synthesized fly ash-derived SBA-15 mesoporous molecular sieves supported Fe and/or Mn catalysts, and reported that Fe-Mn/SBA-15 catalyst showed notably greater NH_3-SCR activity than Mn/SBA-15 or Fe/SBA-15 in the temperature range of 150–250 °C. Moreover, the Fe-Mn/SBA-15 catalyst exhibited good time-on-stream stability (200 h) and water tolerance at 200 °C. The high metal dispersion, Mn^{4+}/Mn^{3+} ratio, the concentration of adsorbed oxygen, and the redox activity are important features to enhance the NH_3-SCR performance of the Fe-Mn/SBA-15 catalyst. In their subsequent study, the authors investigated the mechanisms of NO reduction and N_2O formation using in-situ DRIFT and transient reaction studies and proposed a possible denitration mechanism over the Fe-Mn/SBA-15 catalyst which is shown in Figure 25. The NH_3-SCR reaction over the Fe-Mn/SBA-15 catalyst proceeded through Langmuir-Hinshelwood, Eley-Rideal, and Mars-van Krevelen mechanisms. Their results also revealed that a large amount of nitrate thereby N_2O being produced over the Fe-Mn/SBA-15 during the reaction due to its strong oxidation ability, low acidity, and high basicity, which resulted in the lower N_2 selectivity [136].

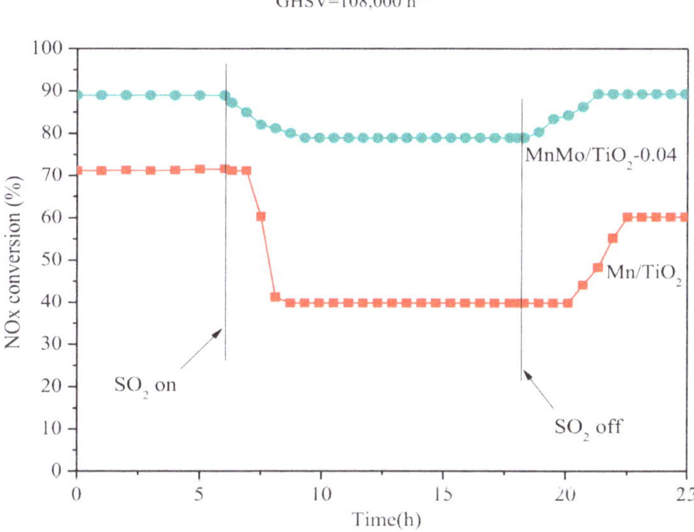

Figure 24. SO_2 tolerance of Mn/TiO_2 and $MnMo/TiO_2$-0.04 catalysts at 150 °C, Reaction conditions: 600 ppm NO, 600 ppm NH_3, 100 ppm SO_2, 5% O_2, balance Ar, GHSV = 108,000 h^{-1}. Reprinted from Reference [133]. Copyright 2018, with Permission from Elsevier.

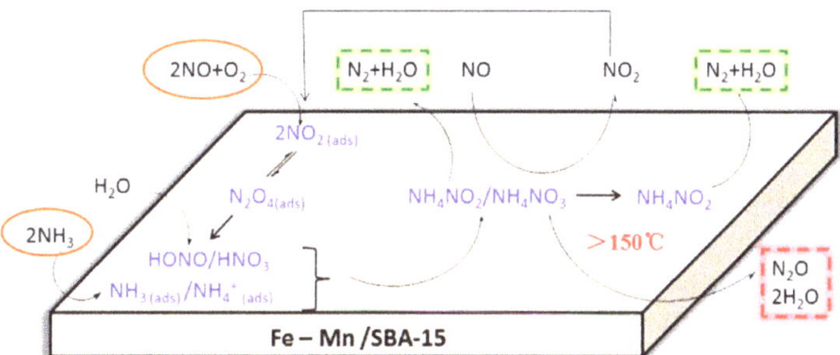

Figure 25. Low-temperature NH_3-SCR reaction mechanism on Fe-Mn/SBA-15 catalyst. Adapted from Reference [136]. Copyright 2018, with Permission from American Chemical Society.

Tang and co-workers [137] reported that Mn_2CoO_4/reduced graphene oxide (Mn_2CoO_4/rGO) catalyst with an optimal amount of $CoCl_2{\cdot}6H_2O$ of 0.3 (millimole) showed excellent NH_3-SCR activity and stability at low-temperature due to its large specific surface area, abundant Lewis acid sites, and special three-dimensional architecture. When 100 ppm SO_2 added to reaction feed, the NO_x conversion over the optimal catalyst decreased significantly (96% to 53%), and the NO_x conversion was recovered to original level by water-washing after stopping the supply of SO_2. However, the decreased activity (100% to 82% NO_x conversion) in the presence of H_2O was restored to the original level after removing H_2O from the feed gas. Wang et al. [138] investigated the honeycomb cordierite-based Mn-Ce/Al_2O_3 catalyst for NH_3-SCR reaction and found that it showed good activity and reasonable resistance to SO_2/H_2O. The catalyst deactivation in the presence of SO_2 was ascribed to the deposition of ammonium hydrogen sulfate and sulfated CeO_2 on the catalyst surface during the NH_3-SCR process.

Meng et al. [139] synthesized a novel $CuAlO_x$/CNTs (CNTs = carbon nanotubes) catalyst by facile one-step carbothermal reduction decomposition method for low-temperature NH_3-SCR. The $CuAlO_x$/CNTs catalyst was found to exhibit higher NO_x conversion (>80%) and N_2 selectivity (>90%) than the $CuAlO_x$ in the temperature range of 180–300 °C (Figure 26a). They concluded that more favorable formation of Cu^+ active sites, better dispersion of active CuO species and higher surface adsorbed oxygen were beneficial to enhance the NH_3-SCR activity of $CuAlO_x$/CNTs catalyst. As shown in Figure 26b, the $CuAlO_x$/CNTs catalyst displayed excellent resistance to SO_2/H_2O at 240 °C during the NH_3-SCR. The authors attributed this outstanding SO_2/H_2O tolerance to the presence of CNTs that could promote the reaction of NH_4HSO_4 and NO continuously to avoid the formation and accumulation of excess ammonium sulfate salts on the catalyst surface. Li group [140] reported a series of ultra-low content copper-modified TiO_2/CeO_2 catalysts and observed that the catalyst with a Cu/Ce molar ratio of 0.005 exhibited the high NH_3-SCR performance and good tolerance to SO_2. Their characterization results disclosed that the addition of Cu into TiO_2/CeO_2 lead to enhance the Brønsted acid sites, amount of surface adsorbed oxygen and Ce^{3+} species, redox, and surface acidic properties, which in turn improve the NH_3-SCR activity.

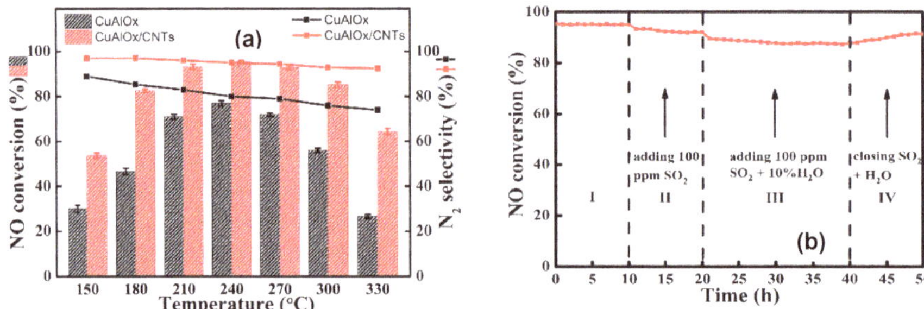

Figure 26. (**a**) NH_3-SCR activity and N_2 selectivity as a function of temperature from 150 °C to 330 °C; and (**b**) SO_2/H_2O resistance test of $CuAlO_x$/CNTs catalyst at 240 °C. Reaction conditions: 600 ppm NH_3, 600 ppm NO, 5.0 vol% O_2, 100 ppm SO_2 (when used), 10 vol% H_2O (when used) balanced by N_2 with a GHSV was 45,000 h^{-1}. Reprinted from Reference [139]. Copyright 2019, with Permission from Elsevier.

Recently, supported multi-metal oxide catalysts have been considered as the very promising candidates for low-temperature NH_3-SCR reaction because of the enlarged synergetic catalysis effects of different components as well as improved metal-support interactions [141–145]. Wang and co-workers [146] reported a series of Nb modified Cu-Ce-Ti mixed oxide (Nb_yCCT, where y represented the atomic ratio of Nb to Ti) catalysts for low-temperature NH_3-SCR reaction. It was found that Nb_yCCT catalysts demonstrated a better activity than the Cu-Ce-Ti (CCT) and Ce-Ti (CT) samples (Figure 27a). Among all the Nb_yCCT catalysts, $Nb_{0.05}CCT$ showed a higher NO conversion (>90%) in a broad temperature range of 180–360 °C under the GHSV of 40,000 h^{-1} (Figure 27a). Results indicated that the incorporation of Nb to Cu-Ce-Ti led to strong interactions among the active phases that increased the oxygen vacancies and inhibited the over-oxidation of NH_3, which in turn improved the NH_3-SCR activity and N_2 selectivity in a wide temperature window. DRIFTS studies revealed that the introduction of Nb promoted the generation of NO_2, which could improve the activity via "fast" SCR reaction process (Langmuir–Hinshelwood reaction pathway). As shown in Figure 27b, the optimal $Nb_{0.05}CCT$ catalyst exhibited higher resistance to SO_2/H_2O as compared to the Nb free catalyst. Li et al. [147] investigated the effect of Ho doping on the NH_3-SCR performance and the SO_2/H_2O resistance of Mn-Ce/TiO_2 catalyst. Among the catalysts tested, the catalyst with Ho/Ti of 0.1 ($Mn_{0.4}Ce_{0.07}Ho_{0.1}/TiO_2$) showed the best performance with >90% NO conversion in the temperature range of 150–220 °C, which was attributed to high concentration of chemisorbed oxygen, surface Mn^{4+}/Mn^{3+} ratio, and acidity, as well as large specific surface area. Although the $Mn_{0.4}Ce_{0.07}Ho_{0.1}/TiO_2$ showed higher resistance to SO_2 and H_2O than the $Mn_{0.4}Ce_{0.07}/TiO_2$ catalyst, it was deactivated some extent in presence of SO_2/H_2O which is irreversible.

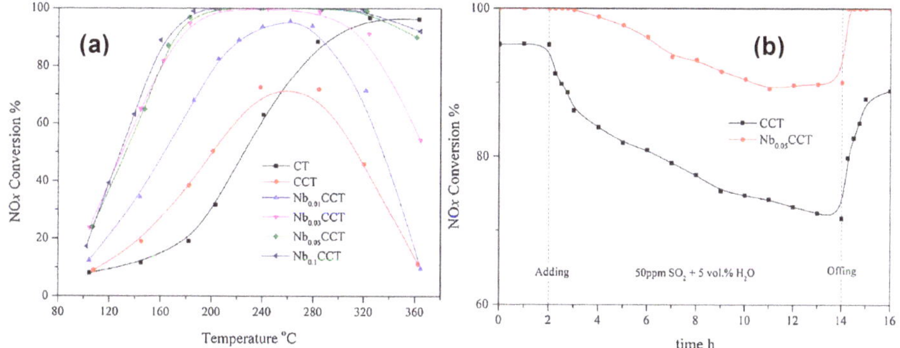

Figure 27. (**a**) NH_3-SCR activities of different catalysts {Reaction conditions: $[NH_3] = [NO] = 600$ ppm, $[O_2] = 3\%$, $[H_2O] = 5$ vol%, N_2 as balance. GHSV = 40 000 h^{-1}}; and (**b**) SO_2 and H_2O resistance of the catalysts at 250 °C in SCR reaction process {Reaction conditions: $[NH_3] = [NO] = 600$ ppm, $[O_2] = 3\%$, $[SO_2] = 50$ ppm, $[H_2O] = 5$ vol%, N_2 balance, GHSV = 40 000 h^{-1}}. Reproduced from Reference [146]. Copyright 2018, with Permission from Elsevier.

Lu group [148] synthesized a series of activated coke (AC) supported $Fe_xCo_yCe_zO_m$ catalysts for low-temperature NH_3-SCR, and found that the 3%$Fe_{0.6}Co_{0.2}Ce_{0.2}O_{1.57}$/AC catalyst had the best activity at 250–350 °C and good tolerance to H_2O/SO_2 at 250 °C. The superior performance of the catalyst was ascribed to the co-participation of Fe, Co, and Ce species with different valence states, high concentration of chemisorbed oxygen, well dispersed active components, increase of weak acid sites, good redox properties of metallic oxides, and abundant functional groups on the catalyst surface. Their mechanistic and kinetic studies also indicated that the enhanced active sites for the adsorption of NO and NH_3, and the redox cycle among Fe, Co and Ce were responsible for the improved activity. Zhao et al. [149] reported a series of Mn-Ce-V-WO_x/TiO_2 composite oxide catalysts, exhibiting greater NH_3-SCR activity than the TiO_2 supported single-component catalysts (Figure 28a,b). Particularly, the catalyst with a molar ratio of active components/TiO_2 = 0.2 showed the best performance (>90% NO conversion) from 150 to 400 °C (Figure 28a). As shown in Figure 28c, the optimal Mn-Ce-V-WO_x/TiO_2 (molar ratio of Mn-Ce-V-WO_x/TiO_2 = 0.2) showed excellent stability and outstanding tolerance to H_2O/SO_2 at 250 °C. The authors concluded that the better performance of Mn-Ce-V-WO_x/TiO_2 mainly attributed to the variety of valence states of the four active components and their high oxidation-reduction ability.

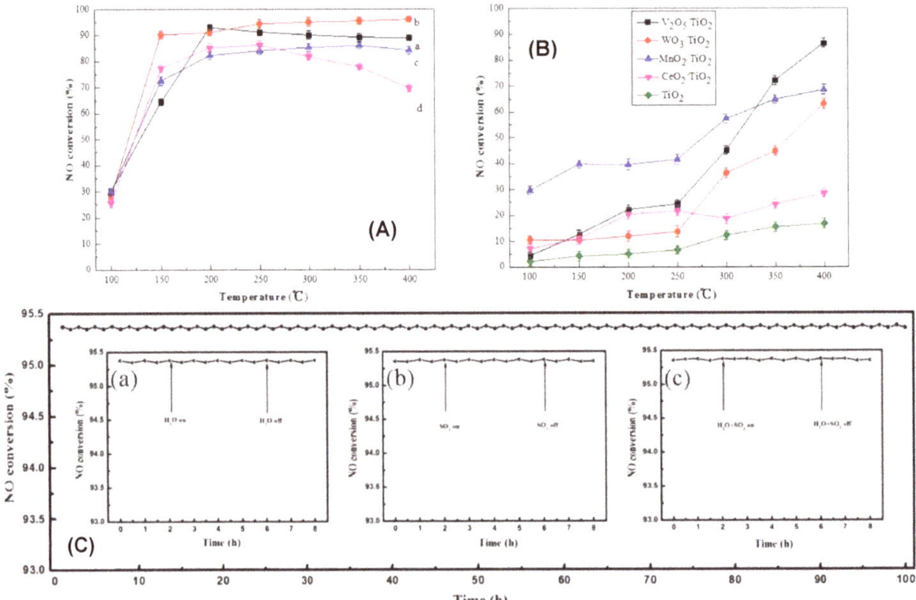

Figure 28. (**A**) Selective catalytic reduction (SCR) activity of Mn-Ce-V-WO_x/TiO_2 composite catalysts with molar ratio of active components/TiO_2 at different values; (a) 0.1; (b) 0.2; (c) 0.3; (d) 0.6; (**B**) SCR activity of V_2O_5/TiO_2, WO_3/TiO_2, MnO_2/TiO_2, CeO_2/TiO_2 and TiO_2. Reaction conditions: [NO] = [NH_3] = 1500 ppm, O_2 = 3%, gas hourly space velocity (GHSV) = 40,000 h^{-1}; and (**C**) the lifetime of Mn-Ce-V-WO_x/TiO_2 catalyst with molar ratio of 0.2 at 250 °C: inset (a-c) H_2O and SO_2 resistance at 250 °C. Reaction conditions: [NO] = [NH_3] = 1500 ppm, [O_2] = 3%, [H_2O] = 5%, [SO_2] = 100 ppm, GHSV = 40,000 h^{-1}. Adapted from Reference [149].

6. Conclusions

Since the emission standards for NO_x are becoming more stringent to keep our atmosphere clean, the widespread use of fossil fuel in automobiles and industries require advanced catalytic materials for NO_x emission control. The low-temperature SCR of NO_x with NH_3 would be a promising solution to mitigate the NO_x emissions from mobile and stationary sources. Hence, the development of efficient catalysts for the low-temperature NH_3-SCR with high deNO_x activity, N_2 selectivity, and high resistance toward SO_2/H_2O poisonings is the subject of increasing interest in the field of environmental catalysis. Transition metal-based oxide catalysts have drawn much attention for low-temperature NH_3-SCR due to their excellent redox properties, high activity, durability, and relatively low manufacturing costs. In this review, we have summarized the recent progress in the low-temperature NH_3-SCR technology over the various transition metal-based catalysts. Over the past decades, significant research efforts have been made to improve the de-NO_x efficiency and SO_2/H_2O tolerance of transition metal-based oxides in NH_3-SCR at low-temperatures. Various transition metal-based mixed oxides with and without support have been extensively studied for NH_3-SCR reaction and, particularly, MnO_x-based catalyst formulations have caught much attention because of their excellent de-NO_x efficiency at low-temperatures. The modification of transition metal oxides by doping with other metal oxides led to high redox ability and acidic sites, and consequently, better NH_3-SCR performance at low-temperature. The loading of single and multi-transition metal-based oxides on the surface of supports (TiO_2, TiO_2 nanotubes, carbon nanotubes, etc.) could also enhance the NO_x conversion and N_2 selectivity in NH_3-SCR reaction by the fine dispersion of active component/s and its/their strong interaction with the support. The choice of metal loading and the support could play a key role in the catalytic function of the supported transition metal-based catalysts. The synergistic redox interaction between the active components of mixed metal oxide/supported metal oxide catalysts was also found to be an important factor to design the efficient denitration catalysts. In spite of the significant progress on the SO_2/H_2O tolerance of the catalysts, the durability of catalysts in the presence of both SO_2 and H_2O still needs to be improved. Most transition metal-based catalysts suffered from low resistance when the reaction feed contains both SO_2 and H_2O streams simultaneously. Hence, researchers have continuously explored the different options of transition metal-based mixed oxides and active transition metal/s-support combinations in order to develop the better NH_3-SCR catalysts in terms of SO_2/H_2O tolerance at low-temperature. The understanding of the inhibition mechanism of SO_2 and H_2O could be a promising strategy to develop high SO_2/H_2O resistance catalysts for NH_3-SCR reaction. However, the SO_2/H_2O inhibition mechanism was not very clear that needs to be investigated deeply. Especially, the design of transition metal-based catalysts with a combination of high NO_x conversion and N_2 selectivity in a wide operation temperature window and good resistance to SO_2/H_2O have attracted paramount attention, but it is still challenging task. The scope of NH_3-SCR research is quite vast and a large number of improvements need to be achieved in the near future.

Author Contributions: The original draft was prepared by D.D., reviewed and edited by D.D., P.G.S., B.M.R, and P.R.E.

Funding: This research received no external funding.

Conflicts of Interest: The authors declare no conflict of interest.

References

1. Wan, Y.; Zhao, W.; Tang, Y.; Li, L.; Wang, H.; Cui, Y.; Gu, J.; Li, Y.; Shi, J. Ni-Mn bi-metal oxide catalysts for the low temperature SCR removal of NO with NH_3. *Appl. Catal. B* **2014**, *148–149*, 114–122. [CrossRef]
2. Boningari, T.; Smirniotis, P.G. Impact of nitrogen oxides on the environment and human health: Mn-based materials for the NO_x abatement. *Curr. Opin. Chem. Eng.* **2016**, *13*, 133–141. [CrossRef]
3. Xin, Y.; Li, Q.; Zhang, Z. Zeolitic materials for $DeNO_x$ selective catalytic reduction. *ChemCatChem* **2018**, *10*, 29–41. [CrossRef]
4. Damma, D.; Boningari, T.; Ettireddy, P.R.; Reddy, B.M.; Smirniotis, P.G. Direct decomposition of NO_x over TiO_2 supported transition metal oxides at low temperatures. *Ind. Eng. Chem. Res.* **2018**, *57*, 16615–16621. [CrossRef]
5. Thirupathi, B.; Smirniotis, P.G. Nickel-doped Mn/TiO_2 as an efficient catalyst for the low-temperature SCR of NO with NH_3: Catalytic evaluation and characterizations. *J. Catal.* **2012**, *288*, 74–83. [CrossRef]
6. Chen, L.; Si, Z.; Wu, X.; Weng, D. DRIFT study of $CuO-CeO_2-TiO_2$ mixed oxides for NO_x reduction with NH_3 at low temperatures. *ACS Appl. Mater. Interfaces* **2014**, *6*, 8134–8145. [CrossRef]
7. Chen, M.; Yang, J.; Liu, Y.; Li, W.; Fan, J.; Ran, X.; Teng, W.; Sun, Y.; Zhang, W.-X.; Li, G.; et al. TiO_2 interpenetrating networks decorated with SnO_2 nanocrystals: Enhanced activity of selective catalytic reduction of NO with NH_3. *J. Mater. Chem. A* **2015**, *3*, 1405–1409. [CrossRef]
8. Chen, Z.; Yang, Q.; Li, H.; Li, X.; Wang, L.; Tsang, S.C. Cr–MnO_x mixed-oxide catalysts for selective catalytic reduction of NO_x with NH_3 at low temperature. *J. Catal.* **2010**, *276*, 56–65. [CrossRef]
9. Boningari, T.; Koirala, R.; Smirniotis, P.G. Low-temperature selective catalytic reduction of NO with NH_3 over V/ZrO_2 prepared by flame-assisted spray pyrolysis: Structural and catalytic properties. *Appl. Catal. B* **2012**, *127*, 255–264. [CrossRef]
10. Chen, L.; Li, J.; Ge, M. DRIFT study on cerium-tungsten/titiania catalyst for selective catalytic reduction of NO_x with NH_3. *Environ. Sci. Technol.* **2010**, *44*, 9590–9596. [CrossRef]
11. Liu, F.; Asakura, K.; He, H.; Shan, W.; Shi, X.; Zhang, C. Influence of sulfation on iron titanate catalyst for the selective catalytic reduction of NO_x with NH_3. *Appl. Catal. B* **2011**, *103*, 369–377. [CrossRef]
12. Cai, S.; Zhang, D.; Zhang, L.; Huang, L.; Li, H.; Gao, R.; Shi, L.; Zhang, J. Comparative study of 3D ordered macroporous $Ce_{0.75}Zr_{0.2}M_{0.05}O_{2-\delta}$ (M = Fe, Cu, Mn, Co) for selective catalytic reduction of NO with NH_3. *Catal. Sci. Technol.* **2014**, *4*, 93–101. [CrossRef]

13. Wang, X.; Wen, W.; Su, Y.; Wang, R. Influence of transition metals (M = Co, Fe and Mn) on ordered mesoporous CuM/CeO_2 catalysts and applications in selective catalytic reduction of NO_x with H_2. *RSC Adv.* **2015**, *5*, 63135–63141. [CrossRef]
14. Liu, J.; Li, X.; Zhao, Q.; Ke, J.; Xiao, H.; Lv, X.; Liu, S.; Tadé, M.; Wang, S. Mechanistic investigation of the enhanced NH_3-SCR on cobalt-decorated Ce-Ti mixed oxide: In situ FTIR analysis for structure-activity correlation. *Appl. Catal. B* **2017**, *200*, 297–308. [CrossRef]
15. Schneider, H.; Maciejewski, M.; Köhler, K.; Wokaun, A.; Baiker, A. Chromia supported on titania, VI Properties of different chromium oxide phases in the catalytic reduction of NO by NH_3 studied by in situ diffuse reflectance FTIR spectroscopy. *J. Catal.* **1995**, *157*, 312–320. [CrossRef]
16. Chen, Z.; Wang, F.; Li, H.; Yang, Q.; Wang, L.; Li, X. Low-temperature selective catalytic reduction of NO_x with NH_3 over Fe-Mn mixed-oxide catalysts containing $Fe_3Mn_3O_8$ phase. *Ind. Eng. Chem. Res.* **2012**, *51*, 202–212. [CrossRef]
17. Smirniotis, P.G.; Peña, D.A.; Uphade, B.S. Low-temperature selective catalytic reduction (SCR) of NO with NH_3 by using Mn, Cr, and Cu oxides supported on hombikat TiO_2. *Angew. Chem. Int. Ed.* **2001**, *40*, 2479–2482. [CrossRef]
18. Wu, Z.; Jiang, B.; Liu, Y. Effect of transition metals addition on the catalyst of manganese/titania for low-temperature selective catalytic reduction of nitric oxide with ammonia. *Appl. Catal. B* **2008**, *79*, 347–355. [CrossRef]
19. Liu, F.; He, H.; Zhang, C.; Feng, Z.; Zheng, L.; Xie, Y.; Hu, T. Selective catalytic reduction of NO with NH_3 over iron titanate catalyst: Catalytic performance and characterization. *Appl. Catal. B* **2010**, *96*, 408–420. [CrossRef]
20. Yao, X.; Kong, T.; Chen, L.; Ding, S.; Yang, F.; Dong, L. Enhanced low-temperature NH_3-SCR performance of MnO_x/CeO_2 catalysts by optimal solvent effect. *Appl. Surf. Sci.* **2017**, *420*, 407–415. [CrossRef]
21. Sreekanth, P.M.; Peña, D.A.; Smirniotis, P.G. Titania supported bimetallic transition metal oxides for low-temperature SCR of NO with NH_3. *Ind. Eng. Chem. Res.* **2006**, *45*, 6444–6449. [CrossRef]
22. Thirupathi, B.; Smirniotis, P.G. Co-doping a metal (Cr, Fe, Co, Ni, Cu, Zn, Ce, and Zr) on Mn/TiO_2 catalyst and its effect on the selective reduction of NO with NH_3 at low-temperatures. *Appl. Catal. B* **2011**, *110*, 195–206. [CrossRef]
23. Thirupathi, B.; Smirniotis, P.G. Effect of nickel as dopant in Mn/TiO_2 catalysts for the low-temperature selective reduction of NO with NH_3. *Catal. Lett.* **2011**, *141*, 1399–1404. [CrossRef]
24. Ettireddy, P.R.; Ettireddy, N.; Boningari, T.; Pardemann, R.; Smirniotis, P.G. Investigation of the selective catalytic reduction of nitric oxide with ammonia over Mn/TiO_2 catalysts through transient isotopic labeling and in situ FT-IR studies. *J. Catal.* **2012**, *292*, 53–63. [CrossRef]
25. Boningari, T.; Ettireddy, P.R.; Somogyvari, A.; Liu, Y.; Vorontsov, A.; McDonald, C.A.; Smirniotis, P.G. Influence of elevated surface texture hydrated titania on Ce-doped Mn/TiO_2 catalysts for the low-temperature SCR of NO_x under oxygen-rich conditions. *J. Catal.* **2015**, *325*, 145–155. [CrossRef]
26. Pappas, D.K.; Boningari, T.; Boolchand, P.; Smirniotis, P.G. Novel manganese oxide confined interweaved titania nanotubes for the low-temperature selective catalytic reduction (SCR) of NO_x by NH_3. *J. Catal.* **2016**, *334*, 1–13. [CrossRef]
27. Yu, J.; Guo, F.; Wang, Y.; Zhu, J.; Liu, Y.; Su, F.; Gao, S.; Xu, G. Sulfur poisoning resistant mesoporous Mn-base catalyst for low-temperature SCR of NO with NH_3. *Appl. Catal. B* **2010**, *95*, 160–168. [CrossRef]
28. Cai, S.; Zhang, D.; Shi, L.; Xu, J.; Zhang, L.; Huang, L.; Li, H.; Zhang, J. Porous Ni–Mn oxide nanosheets in situ formed on nickel foam as 3D hierarchical monolith de-NO_x catalysts. *Nanoscale* **2014**, *6*, 7346–7353. [CrossRef] [PubMed]
29. Zhang, S.; Zhang, B.; Liu, B.; Sun, S. A review of Mn-containing oxide catalysts for low temperature selective catalytic reduction of NO_x with NH_3: Reaction mechanism and catalyst deactivation. *RSC Adv.* **2017**, *7*, 26226–26242. [CrossRef]
30. Zhang, L.; Shi, L.; Huang, L.; Zhang, J.; Gao, R.; Zhang, D. Rational design of high-performance $DeNO_x$ catalysts based on $Mn_xCo_{3-x}O_4$ nanocages derived from metal−organic frameworks. *ACS Catal.* **2014**, *4*, 1753–1763. [CrossRef]

31. Li, L.; Wu, Y.; Hou, X.; Chu, B.; Nan, B.; Qin, Q.; Fan, M.; Sun, C.; Li, B.; Dong, L.; et al. Investigation of two-phase intergrowth and coexistence in Mn−Ce−Ti−O catalysts for the selective catalytic reduction of NO with NH_3: Structure−activity relationship and reaction mechanism. *Ind. Eng. Chem. Res.* **2019**, *58*, 849–862. [CrossRef]
32. Gao, F.; Tang, X.; Yi, H.; Zhao, S.; Wang, J.; Gu, T. Improvement of activity, selectivity and H_2O&SO_2-tolerance of micromesoporous $CrMn_2O_4$ spinel catalyst for low-temperature NH_3-SCR of NO_x. *Appl. Surf. Sci.* **2019**, *466*, 411–424.
33. Li, X.J.; Du, Y.; Guo, X.M.; Wang, R.N.; Hou, B.H.; Wu, X. Synthesis of a novel NiMnTi mixed metal oxides from LDH precursor and its catalytic application for selective catalytic reduction of NO_x with NH_3. *Catal. Lett.* **2019**, *149*, 456–464. [CrossRef]
34. Yang, G.; Zhao, H.; Luo, X.; Shi, K.; Zhao, H.; Wang, W.; Chen, Q.; Fan, H.; Wu, T. Promotion effect and mechanism of the addition of Mo on the enhanced low temperature SCR of NO_x by NH_3 over MnO_x/γ-Al_2O_3 catalysts. *Appl. Catal. B* **2019**, *245*, 743–752. [CrossRef]
35. Wang, X.; Wu, W.; Chen, Z.; Wang, R. Bauxite-supported transition metal oxides: Promising low-temperature and SO_2-tolerant catalysts for selective catalytic reduction of NO_x. *Sci. Rep.* **2015**, *5*, 9766. [CrossRef]
36. Wang, Z.-Y.; Guo, R.-T.; Shi, X.; Pan, W.-G.; Liu, J.; Sun, X.; Liu, S.-W.; Liu, X.-Y.; Qin, H. The enhanced performance of Sb-modified Cu/TiO_2 catalyst for selective catalytic reduction of NO_x with NH_3. *Appl. Surf. Sci.* **2019**, *475*, 334–341. [CrossRef]
37. Hu, H.; Cai, S.; Li, H.; Huang, L.; Shi, L.; Zhang, D. Mechanistic aspects of deNO_x processing over TiO_2 supported Co−Mn oxide catalysts: Structure−activity relationships and in situ DRIFTs analysis. *ACS Catal.* **2015**, *5*, 6069–6077. [CrossRef]
38. Meng, B.; Zhao, Z.; Chen, Y.; Wang, X.; Li, Y.; Qiu, J. Low-temperature synthesis of Mn-based mixed metal oxides with novel fluffy structures as efficient catalysts for selective reduction of nitrogen oxides by ammonia. *Chem. Commun.* **2014**, *50*, 12396–12399. [CrossRef]
39. Qiu, M.; Zhan, S.; Zhu, D.; Yu, H.; Shi, Q. NH_3-SCR performance improvement of mesoporous Sn modified Cr-MnO_x catalysts at low temperatures. *Catal. Today* **2015**, *258*, 103–111. [CrossRef]
40. Shi, J.-W.; Gao, C.; Liu, C.; Fan, Z.; Gao, G.; Niu, C. Porous MnO_x for low-temperature NH_3-SCR of NO_x: The intrinsic relationship between surface physicochemical property and catalytic activity. *J. Nanopart. Res.* **2017**, *19*, 194–204. [CrossRef]
41. Husnain, N.; Wang, E.; Li, K.; Anwar, M.T.; Mehmood, A.; Gul, M.; Li, D.; Mao, J. Iron oxide-based catalysts for low-temperature selective catalytic reduction of NO_x with NH_3. *Rev. Chem. Eng.* **2019**, *35*, 239–264. [CrossRef]
42. Liu, F.; Shan, W.; Lian, Z.; Xie, L.; Yang, W.; He, H. Novel $MnWO_x$ catalyst with remarkable performance for low temperature NH_3-SCR of NO_x. *Catal. Sci. Technol.* **2013**, *3*, 2699–2707. [CrossRef]
43. Fan, G.; Shi, J.-W.; Gao, C.; Gao, G.; Wang, B.; Niu, C. Rationally designed porous MnO_x−FeO_x nanoneedles for low-temperature selective catalytic reduction of NO_x by NH_3. *ACS Appl. Mater. Interfaces* **2017**, *9*, 16117–16127. [CrossRef] [PubMed]
44. Gao, G.; Shi, J.-W.; Fan, G.; Gao, C.; Niu, C. MnM_2O_4 microspheres (M = Co, Cu, Ni) for selective catalytic reduction of NO with NH_3: Comparative study on catalytic activity and reaction mechanism via in-situ diffuse reflectance infrared Fourier transform spectroscopy. *Chem. Eng. J.* **2017**, *325*, 91–100. [CrossRef]
45. Gao, F.; Tang, X.; Yi, H.; Zhao, S.; Li, C.; Li, J.; Shi, Y.; Meng, X. A Review on selective catalytic reduction of NO_x by NH_3 over Mn–based catalysts at low temperatures: Catalysts, mechanisms, kinetics and DFT calculations. *Catalysts* **2017**, *7*, 199. [CrossRef]
46. Gao, C.; Shi, J.-W.; Fan, Z.; Gao, G.; Niu, C. Sulfur and water resistance of Mn-based catalysts for low-temperature selective catalytic reduction of NO_x: A review. *Catalysts* **2018**, *8*, 11. [CrossRef]
47. Hu, X.; Huang, L.; Zhang, J.; Li, H.; Zha, K.; Shi, L.; Zhang, D. Facile and template-free fabrication of mesoporous 3D nanosphere-like $Mn_xCo_{3-x}O_4$ as highly effective catalysts for low temperature SCR of NO_x with NH_3. *J. Mater. Chem. A* **2018**, *6*, 2952–2963. [CrossRef]
48. Meng, D.; Xu, Q.; Jiao, Y.; Guo, Y.; Guo, Y.; Wang, L.; Lu, G.; Zhan, W. Spinel structured $Co_aMn_bO_x$ mixed oxide catalyst for the selective catalytic reduction of NO_x with NH_3. *Appl. Catal. B* **2018**, *221*, 652–663. [CrossRef]

49. Yan, Q.; Nie, Y.; Yang, R.; Cui, Y.; O'Hare, D.; Wang, Q. Highly dispersed Cu_yAlO_x mixed oxides as superior low-temperature alkali metal and SO_2 resistant NH_3-SCR catalysts. *Appl. Catal. A* **2017**, *538*, 37–50. [CrossRef]
50. Meng, D.; Zhan, W.; Guo, Y.; Guo, Y.; Wang, L.; Lu, G. A Highly effective catalyst of Sm-MnO_x for the NH_3-SCR of NO_x at low temperature: Promotional role of Sm and its catalytic performance. *ACS Catal.* **2015**, *5*, 5973–5983. [CrossRef]
51. Sun, P.; Guo, R.-T.; Liu, S.-M.; Wang, S.-X.; Pan, W.-G.; Li, M.-Y. The enhanced performance of MnO_x catalyst for NH_3-SCR reaction by the modification with Eu. *Appl. Catal. A* **2017**, *531*, 129–138. [CrossRef]
52. Xin, Y.; Li, H.; Zhang, N.; Li, Q.; Zhang, Z.; Cao, X.; Hu, P.; Zheng, L.; Anderson, J.A. Molecular-level insight into selective catalytic reduction of NO_x with NH_3 to N_2 over a highly efficient bifunctional V_a-MnO_x catalyst at low temperature. *ACS Catal.* **2018**, *8*, 4937–4949. [CrossRef]
53. Han, Y.; Mu, J.; Li, X.; Gao, J.; Fan, S.; Tan, F.; Zhao, Q. Triple-shelled $NiMn_2O_4$ hollow spheres as an efficient catalyst for low-temperature selective catalytic reduction of NO_x with NH_3. *Chem. Commun.* **2018**, *54*, 9797–9800. [CrossRef] [PubMed]
54. Gao, F.; Tang, X.; Yi, H.; Zhao, S.; Wang, J.; Shi, Y.; Meng, X. Novel Co– or Ni–Mn binary oxide catalysts with hydroxyl groups for NH_3–SCR of NO_x at low temperature. *Appl. Surf. Sci.* **2018**, *443*, 103–113. [CrossRef]
55. Sun, W.; Li, X.; Zhao, Q.; Mu, J.; Chen, J. Fe–Mn mixed oxide catalysts synthesized by one-step urea-precipitation method for the selective catalytic reduction of NO_x with NH_3 at low temperatures. *Catal. Lett.* **2018**, *148*, 227–234. [CrossRef]
56. Kwon, D.W.; Nam, K.B.; Hong, S.C. The role of ceria on the activity and SO_2 resistance of catalysts for the selective catalytic reduction of NO_x by NH_3. *Appl. Catal. B* **2015**, *166–167*, 37–44. [CrossRef]
57. Fan, G.; Shi, J.-W.; Gao, C.; Gao, G.; Wang, B.; Wang, Y.; He, C.; Niu, C. Gd-modified MnO_x for the selective catalytic reduction of NO by NH_3: The promoting effect of Gd on the catalytic performance and sulfur resistance. *Chem. Eng. J.* **2018**, *348*, 820–830. [CrossRef]
58. Li, C.; Tang, X.; Yi, H.; Wang, L.; Cui, X.; Chu, C.; Li, J.; Zhang, R.; Yu, Q. Rational design of template-free MnO_x-CeO_2 hollow nanotube as de-NO_x catalyst at low temperature. *Appl. Surf. Sci.* **2018**, *428*, 924–932. [CrossRef]
59. Liu, F.; Yu, Y.; He, H. Environmentally-benign catalysts for the selective catalytic reduction of NO_x from diesel engines: Structure–activity relationship and reaction mechanism aspects. *Chem. Commun.* **2014**, *50*, 8445–8463. [CrossRef]
60. Zhan, S.; Qiu, M.; Yang, S.; Zhu, D.; Yu, H.; Li, Y. Facile preparation of MnO_2 doped Fe_2O_3 hollow nanofibers for low temperature SCR of NO with NH_3. *J. Mater. Chem. A* **2014**, *2*, 20486–20493. [CrossRef]
61. Li, Y.; Wan, Y.; Li, Y.; Zhan, S.; Guan, Q.; Tian, Y. Low-temperature selective catalytic reduction of NO with NH_3 over Mn_2O_3-doped Fe_2O_3 hexagonal microsheets. *ACS Appl. Mater. Interfaces* **2016**, *8*, 5224–5233. [CrossRef]
62. Yan, L.; Liu, Y.; Zha, K.; Li, H.; Shi, L.; Zhang, D. Scale-activity relationship of MnO_x-FeO_y nanocage catalysts derived from prussian blue analogues for low-temperature NO reduction: Experimental and DFT studies. *ACS Appl. Mater. Interfaces* **2017**, *9*, 2581–2593. [CrossRef] [PubMed]
63. Mu, J.; Li, X.; Sun, W.; Fan, S.; Wang, X.; Wang, L.; Qin, M.; Gan, G.; Yin, Z.; Zhang, D. Inductive effect boosting catalytic performance of advanced $Fe_{1-x}V_xO_\delta$ catalysts in low-temperature NH_3 selective catalytic reduction: Insight into the structure, interaction, and mechanisms. *ACS Catal.* **2018**, *8*, 6760–6774. [CrossRef]
64. Li, Y.; Han, X.; Hou, Y.; Guo, Y.; Liu, Y.; Cui, Y.; Huang, Z. Role of CTAB in the improved H_2O resistance for selective catalytic reduction of NO with NH_3 over iron titanium catalyst. *Chem. Eng. J.* **2018**, *347*, 313–321. [CrossRef]
65. Qiu, M.; Zhan, S.; Yu, H.; Zhu, D. Low-temperature selective catalytic reduction of NO with NH_3 over ordered mesoporous $Mn_xCo_{3-x}O_4$ catalyst. *Catal. Commun.* **2015**, *62*, 107–111. [CrossRef]
66. Qiu, M.; Zhan, S.; Yu, H.; Zhu, D.; Wang, S. Facile preparation of ordered mesoporous $MnCo_2O_4$ for low-temperature selective catalytic reduction of NO with NH_3. *Nanoscale* **2015**, *7*, 2568–2577. [CrossRef]
67. Hu, H.; Cai, S.; Li, H.; Huang, L.; Shi, L.; Zhang, D. In Situ DRIFTs investigation of the low-temperature reaction mechanism over Mn-doped Co_3O_4 for the selective catalytic reduction of NO_x with NH_3. *J. Phys. Chem. C* **2015**, *119*, 22924–22933. [CrossRef]

68. Qiao, J.; Wang, N.; Wang, Z.; Sun, W.; Sun, K. Porous bimetallic $Mn_2Co_1O_x$ catalysts prepared by a one-step combustion method for the low temperature selective catalytic reduction of NO_x with NH_3. *Catal. Commun.* **2015**, *72*, 111–115. [CrossRef]
69. Kang, M.; Park, E.D.; Kim, J.M.; Yie, J.E. Cu–Mn mixed oxides for low temperature NO reduction with NH_3. *Catal. Today* **2006**, *111*, 236–241. [CrossRef]
70. Guo, R.-T.; Zhen, W.-L.; Pan, W.-G.; Zhou, Y.; Hong, J.-N.; Xu, H.-J.; Jin, Q.; Ding, C.-G.; Guo, S.-Y. Effect of Cu doping on the SCR activity of CeO_2 catalyst prepared by citric acid method. *J. Ind. Eng. Chem.* **2014**, *20*, 1577–1580. [CrossRef]
71. Ali, S.; Chen, L.; Li, Z.; Zhang, T.; Li, R.; Bakhtiar, S.U.H.; Leng, X.; Yuan, F.; Niu, X.; Zhu, Y. Cu_x-$Nb_{1.1-x}$ (x = 0.45, 0.35, 0.25, 0.15) bimetal oxides catalysts for the low temperature selective catalytic reduction of NO with NH_3. *Appl. Catal. B* **2018**, *236*, 25–35. [CrossRef]
72. Yan, Q.; Chen, S.; Zhang, C.; Wang, Q.; Louis, B. Synthesis and catalytic performance of $Cu_1Mn_{0.5}Ti_{0.5}O_x$ mixed oxide as low-temperature NH_3-SCR catalyst with enhanced SO_2 resistance. *Appl. Catal. B* **2018**, *238*, 236–247. [CrossRef]
73. Liu, J.; Li, X.; Li, R.; Zhao, Q.; Ke, J.; Xiao, H.; Wang, L.; Liu, S.; Tadé, M.; Wang, S. Facile synthesis of tube-shaped Mn-Ni-Ti solid solution and preferable Langmuir-Hinshelwood mechanism for selective catalytic reduction of NO_x by NH_3. *Appl. Catal. A* **2018**, *549*, 289–301. [CrossRef]
74. Wang, C.; Yu, F.; Zhu, M.; Wang, X.; Dan, J.; Zhang, J.; Cao, P.; Dai, B. Microspherical MnO_2-CeO_2-Al_2O_3 mixed oxide for monolithic honeycomb catalyst and application in selective catalytic reduction of NO_x with NH_3 at 50–150 °C. *Chem. Eng. J.* **2018**, *346*, 182–192. [CrossRef]
75. Cheng, K.; Liu, B.; Song, W.; Liu, J.; Chen, Y.; Zhao, Z.; Wei, Y. Effect of Nb promoter on the structure and performance of iron titanate catalysts for the selective catalytic reduction of NO with NH_3. *Ind. Eng. Chem. Res.* **2018**, *57*, 7802–7810. [CrossRef]
76. Yan, Q.; Chen, S.; Qiu, L.; Gao, Y.; O'Hare, D.; Wang, Q. The synthesis of $Cu_yMn_zAl_{1-z}O_x$ mixed oxide as a low-temperature NH_3-SCR catalyst with enhanced catalytic performance. *Dalton Trans.* **2018**, *47*, 2992–3004. [CrossRef] [PubMed]
77. Geng, Y.; Shan, W.; Yang, S.; Liu, F. W-modified Mn−Ti mixed oxide catalyst for the selective catalytic reduction of NO with NH_3. *Ind. Eng. Chem. Res.* **2018**, *57*, 9112–9119. [CrossRef]
78. Gao, F.; Tang, X.; Yi, H.; Li, J.; Zhao, S.; Wang, J.; Chu, C.; Li, C. Promotional mechanisms of activity and SO_2 tolerance of Co- or Ni-doped MnO_x-CeO_2 catalysts for SCR of NO_x with NH_3 at low Temperature. *Chem. Eng. J.* **2017**, *317*, 20–31. [CrossRef]
79. Xiong, Z.-B.; Hu, Q.; Liu, D.-Y.; Wu, C.; Zhou, F.; Wang, Y.-Z.; Jin, J.; Lu, C.-M. Influence of partial substitution of iron oxide by titanium oxide on the structure and activity of iron–cerium mixed oxide catalyst for selective catalytic reduction of NO_x with NH_3. *Fuel* **2016**, *165*, 432–439. [CrossRef]
80. Wang, X.; Li, X.; Zhao, Q.; Sun, W.; Tade, M.; Liu, S. Improved activity of W-modified MnO_x–TiO_2 catalysts for the selective catalytic reduction of NO with NH_3. *Chem. Eng. J.* **2016**, *288*, 216–222. [CrossRef]
81. Zamudio, M.A.; Russo, N.; Fino, D. Low temperature NH_3 selective catalytic reduction of NO_x over substituted $MnCr_2O_4$ spinel-oxide catalysts. *Ind. Eng. Chem. Res.* **2011**, *50*, 6668–6672. [CrossRef]
82. Zhang, T.; Qiu, F.; Chang, H.; Peng, Y.; Li, J. Novel W-modified $SnMnCeO_x$ catalyst for the selective catalytic reduction of NO_x with NH_3. *Catal. Commun.* **2017**, *100*, 117–120. [CrossRef]
83. Ali, S.; Chen, L.; Yuan, F.; Li, R.; Zhang, T.; Bakhtiar, S.U.H.; Leng, X.; Niu, X.; Zhu, Y. Synergistic effect between copper and cerium on the performance of Cu_x-$Ce_{0.5-x}$-$Zr_{0.5}$ (x = 0.1–0.5) oxides catalysts for selective catalytic reduction of NO with ammonia. *Appl. Catal. B* **2017**, *210*, 223–234. [CrossRef]
84. Wei, Y.; Fan, H.; Wang, R. Transition metals (Co, Zr, Ti) modified iron-samarium oxide as efficient catalysts for selective catalytic reduction of NO_x at low-temperature. *Appl. Surf. Sci.* **2018**, *459*, 63–73. [CrossRef]
85. Gao, X.; Du, X.-S.; Cui, L.-W.; Fu, Y.-C.; Luo, Z.-Y.; Cen, K.-F. A Ce–Cu–Ti oxide catalyst for the selective catalytic reduction of NO with NH_3. *Catal. Commun.* **2010**, *12*, 255–258. [CrossRef]
86. Xiong, Z.B.; Wu, C.; Hu, Q.; Wang, Y.Z.; Jin, J.; Lu, C.M.; Guo, D.X. Promotional effect of microwave hydrothermal treatment on the low-temperature NH_3-SCR activity over iron-based catalyst. *Chem. Eng. J.* **2016**, *286*, 459–466. [CrossRef]
87. Cheng, K.; Song, W.; Cheng, Y.; Zheng, H.; Wang, L.; Liu, J.; Zhao, Z.; Wei, Y. Enhancing the low temperature NH_3-SCR activity of $FeTiO_x$ catalysts via Cu doping: A combination of experimental and theoretical study. *RSC Adv.* **2018**, *8*, 19301–19309. [CrossRef]

88. Fang, N.; Guo, J.; Shu, S.; Luo, H.; Chu, Y.; Li, J. Enhancement of low-temperature activity and sulfur resistance of $Fe_{0.3}Mn0_{.5}Zr_{0.2}$ catalyst for NO removal by NH_3-SCR. *Chem. Eng. J.* **2017**, *325*, 114–123. [CrossRef]
89. Guo, R.-T.; Sun, X.; Liu, J.; Pan, W.-G.; Li, M.-Y.; Liu, S.-M.; Sun, P.; Liu, S.-W. Enhancement of the NH_3-SCR catalytic activity of $MnTiO_x$ catalyst by the introduction of Sb. *Appl. Catal. A* **2018**, *558*, 1–8. [CrossRef]
90. Shi, J.-W.; Gao, G.; Fan, Z.; Gao, C.; Wang, B.; Wang, Y.; Li, Z.; He, C.; Niu, C. $Ni_yCo_{1-y}Mn_2O_x$ microspheres for the selective catalytic reduction of NO_x with NH_3: The synergetic effects between Ni and Co for improving low-temperature catalytic performance. *Appl. Catal. A* **2018**, *560*, 1–11. [CrossRef]
91. Wu, X.; Feng, Y.; Du, Y.; Liu, X.; Zou, C.; Li, Z. Enhancing $DeNO_x$ performance of CoMnAl mixed metal oxides in low-temperature NH_3-SCR by optimizing layered double hydroxides (LDHs) precursor template. *Appl. Surf. Sci.* **2019**, *467–468*, 802–810. [CrossRef]
92. Leng, X.; Zhang, Z.; Li, Y.; Zhang, T.; Ma, S.; Yuan, F.; Niu, X.; Zhu, Y. Excellent low temperature NH_3-SCR activity over $Mn_aCe_{0.3}TiO_x$ (a = 0.1–0.3) oxides: Influence of Mn addition. *Fuel Process. Technol.* **2018**, *181*, 33–43. [CrossRef]
93. Ali, S.; Li, Y.; Zhang, T.; Bakhtiar, S.U.H.; Leng, X.; Li, Z.; Yuan, F.; Niu, X.; Zhu, Y. Promotional effects of Nb on selective catalytic reduction of NO with NH_3 over Fe_x-$Nb_{0.5-x}$-$Ce_{0.5}$ (x = 0.45, 0.4, 0.35) oxides catalysts. *Mol. Catal.* **2018**, *461*, 97–107. [CrossRef]
94. Sun, C.; Liu, H.; Chen, W.; Chen, D.; Yu, S.; Liu, A.; Dong, L.; Feng, S. Insights into the Sm/Zr co-doping effects on N_2 selectivity and SO_2 resistance of a MnO_x-TiO_2 catalyst for the NH_3-SCR reaction. *Chem. Eng. J.* **2018**, *347*, 27–40. [CrossRef]
95. Yan, Q.; Chen, S.; Zhang, C.; O'Hare, D.; Wang, Q. Synthesis of $Cu_{0.5}Mg_{1.5}Mn_{0.5}Al_{0.5}O_x$ mixed oxide from layered double hydroxide precursor as highly efficient catalyst for low-temperature selective catalytic reduction of NO_x with NH_3. *J. Colloid Interface Sci.* **2018**, *526*, 63–74. [CrossRef] [PubMed]
96. Chen, L.; Yuan, F.; Li, Z.; Niu, X.; Zhu, Y. Synergistic effect between the redox property and acidity on enhancing the low temperature NH_3-SCR activity for NO_x removal over the $Co_{0.2}Ce_xMn_{0.8-x}Ti_{10}$ (x = 0–0.40) oxides catalysts. *Chem. Eng. J.* **2018**, *354*, 393–406. [CrossRef]
97. Peña, D.A.; Uphade, B.S.; Reddy, E.P.; Smirniotis, P.G. Identification of surface species on titania-supported manganese, chromium, and copper oxide low-temperature SCR catalysts. *J. Phys. Chem. B* **2004**, *108*, 9927–9936. [CrossRef]
98. Zhuang, K.; Qiu, J.; Tang, F.; Xu, B.; Fan, Y. The structure and catalytic activity of anatase and rutile titania supported manganese oxide catalysts for selective catalytic reduction of NO by NH_3. *Phys. Chem. Chem. Phys.* **2011**, *13*, 4463–4469. [CrossRef]
99. Boningari, T.; Pappas, D.K.; Ettireddy, P.R.; Kotrba, A.; Smirniotis, P.G. Influence of SiO_2 on M/TiO_2 (M = Cu, Mn, and Ce) formulations for low-temperature selective catalytic reduction of NO_x with NH_3: Surface properties and key components in relation to the activity of NO_x reduction. *Ind. Eng. Chem. Res.* **2015**, *54*, 2261–2273. [CrossRef]
100. Sheng, Z.; Ma, D.; Yu, D.; Xiao, X.; Huang, B.; Yang, L.; Wang, S. Synthesis of novel MnO_x@TiO_2 core-shell nanorod catalyst for low-temperature NH_3-selective catalytic reduction of NO_x with enhanced SO_2 tolerance. *Chin. J. Catal.* **2018**, *39*, 821–830. [CrossRef]
101. Cimino, S.; Totarella, G.; Tortorelli, M.; Lisi, L. Combined poisoning effect of K^+ and its counter-ion (Cl^- or NO^{3-}) on MnO_x/TiO_2 catalyst during the low temperature NH_3-SCR of NO. *Chem. Eng. J.* **2017**, *330*, 92–101. [CrossRef]
102. Fang, D.; Xie, J.; Hu, H.; Yang, H.; He, F.; Fu, Z. Identification of MnO_x species and Mn valence states in MnO_x/TiO_2 catalysts for low temperature SCR. *Chem. Eng. J.* **2015**, *271*, 23–30. [CrossRef]
103. Huang, J.; Huang, H.; Liu, L.; Jiang, H. Revisit the effect of manganese oxidation state on activity in low-temperature NO-SCR. *Mol. Catal.* **2018**, *446*, 49–57. [CrossRef]
104. Yang, S.; Qi, F.; Xiong, S.; Dang, H.; Liao, Y.; Wong, P.K.; Li, J. MnO_x supported on Fe–Ti spinel: A novel Mn based low temperature SCR catalyst with a high N_2 selectivity. *Appl. Catal. B* **2016**, *181*, 570–580. [CrossRef]
105. Liu, C.; Shi, J.-W.; Gao, C.; Niu, C. Manganese oxide-based catalysts for low-temperature selective catalytic reduction of NO_x with NH_3: A review. *Appl. Catal. A* **2016**, *522*, 54–69. [CrossRef]
106. Peña, D.A.; Uphade, B.S.; Smirniotis, P.G. TiO_2-supported metal oxide catalysts for low-temperature selective catalytic reduction of NO with NH_3, I. Evaluation and characterization of first row transition metals. *J. Catal.* **2004**, *221*, 421–431. [CrossRef]

107. Smirniotis, P.G.; Sreekanth, P.M.; Peña, D.A.; Jenkins, R.G. Manganese oxide catalysts supported on TiO_2, Al_2O_3, and SiO_2: A comparison for low-temperature SCR of NO with NH_3. *Ind. Eng. Chem. Res.* **2006**, *45*, 6436–6443. [CrossRef]
108. Ettireddy, P.R.; Ettireddy, N.; Mamedov, S.; Boolchand, P.; Smirniotis, P.G. Surface characterization studies of TiO_2 supported manganese oxide catalysts for low temperature SCR of NO with NH_3. *Appl. Catal. B* **2007**, *76*, 123–134. [CrossRef]
109. Boningari, T.; Pappas, D.K.; Smirniotis, P.G. Metal oxide-confined interweaved titania nanotubes M/TNT (M = Mn, Cu, Ce, Fe, V, Cr, and Co) for the selective catalytic reduction of NO_x in the presence of excess oxygen. *J. Catal.* **2018**, *365*, 320–333. [CrossRef]
110. Jia, B.; Guo, J.; Luo, H.; Shu, S.; Fang, N.; Li, J. Study of NO removal and resistance to SO_2 and H_2O of MnO_x/TiO_2, MnO_x/ZrO_2 and MnO_x/ZrO_2–TiO_2. *Appl. Catal. A* **2018**, *553*, 82–90. [CrossRef]
111. Zhang, B.; Zhang, S.; Liu, B.; Shen, H.; Li, L. High N_2 selectivity in selective catalytic reduction of NO with NH_3 over Mn/Ti–Zr catalysts. *RSC Adv.* **2018**, *8*, 12733–12741. [CrossRef]
112. Han, J.; Zhang, D.; Maitarad, P.; Shi, L.; Cai, S.; Li, H.; Huang, L.; Zhang, J. Fe_2O_3 nanoparticles anchored in situ on carbon nanotubes via an ethanol-thermal strategy for the selective catalytic reduction of NO with NH_3. *Catal. Sci. Technol.* **2015**, *5*, 438–446. [CrossRef]
113. Pourkhalil, M.; Moghaddam, A.Z.; Rashidi, A.; Towfighi, J.; Mortazavi, Y. Preparation of highly active manganese oxides supported on functionalized MWNTs for low temperature NO_x reduction with NH_3. *Appl. Surf. Sci.* **2013**, *279*, 250–259. [CrossRef]
114. Pan, S.; Luo, H.; Li, L.; Wei, Z.; Huang, B. H_2O and SO_2 deactivation mechanism of MnO_x/MWCNTs for low-temperature SCR of NO_x with NH_3. *J. Mol. Catal. A* **2013**, *377*, 154–161. [CrossRef]
115. Fang, C.; Zhang, D.; Shi, L.; Gao, R.; Li, H.; Ye, L.; Zhang, J. Highly dispersed CeO_2 on carbon nanotubes for selective catalytic reduction of NO with NH_3. *Catal. Sci. Technol.* **2013**, *3*, 803–811. [CrossRef]
116. Qu, Z.; Miao, L.; Wang, H.; Fu, Q. Highly dispersed Fe_2O_3 on carbon nanotubes for low-temperature selective catalytic reduction of NO with NH_3. *Chem. Commun.* **2015**, *51*, 956–958. [CrossRef] [PubMed]
117. Bai, S.; Li, H.; Wang, L.; Guan, Y.; Jiang, S. The properties and mechanism of CuO modified carbon nanotube for NO_x removal. *Catal. Lett.* **2014**, *144*, 216–221. [CrossRef]
118. Zhang, D.; Zhang, L.; Shi, L.; Fang, C.; Li, H.; Gao, R.; Huang, L.; Zhang, J. In situ supported MnO_x–CeO_x on carbon nanotubes for the low-temperature selective catalytic reduction of NO with NH_3. *Nanoscale* **2013**, *5*, 1127–1136. [CrossRef]
119. Putluru, S.S.R.; Schill, L.; Jensen, A.D.; Siret, B., Tabaries, F.; Fehrmann, R. Mn/TiO_2 and Mn–Fe/TiO_2 catalysts synthesized by deposition precipitation-promising for selective catalytic reduction of NO with NH_3 at low temperatures. *Appl. Catal. B* **2015**, *165*, 628–635. [CrossRef]
120. You, X.; Sheng, Z.; Yu, D.; Yang, L.; Xiao, X.; Wang, S. Influence of Mn/Ce ratio on the physicochemical properties and catalytic performance of graphene supported MnO_x-CeO_2 oxides for NH_3-SCR at low temperature. *Appl. Surf. Sci.* **2017**, *423*, 845–854. [CrossRef]
121. Fang, D.; He, F.; Liu, X.; Qi, K.; Xie, J.; Li, F.; Yu, C. Low temperature NH_3-SCR of NO over an unexpected Mn-based catalyst: Promotional effect of Mg doping. *Appl. Surf. Sci.* **2018**, *427*, 45–55. [CrossRef]
122. Lu, X.; Song, C.; Jia, S.; Tong, Z.; Tang, X.; Teng, Y. Low-temperature selective catalytic reduction of NO_X with NH_3 over cerium and manganese oxides supported on TiO_2–graphene. *Chem. Eng. J.* **2015**, *260*, 776–784. [CrossRef]
123. Wang, X.; Wu, S.; Zou, W.; Yu, S.; Gui, K.; Dong, L. Fe-Mn/Al_2O_3 catalysts for low temperature selective catalytic reduction of NO with NH_3. *Chin. J. Catal.* **2016**, *37*, 1314–1323. [CrossRef]
124. Shi, J.; Zhang, Z.; Chen, M.; Zhang, Z.; Shangguan, W. Promotion effect of tungsten and iron co-addition on the catalytic performance of MnO_x/TiO_2 for NH_3-SCR of NO_x. *Fuel* **2017**, *210*, 783–789. [CrossRef]
125. Zhao, C.; Wu, Y.; Liang, H.; Chen, X.; Tang, J.; Wang, X. N-doped graphene and TiO_2 supported manganese and cerium oxides on low-temperature selective catalytic reduction of NO_x with NH_3. *J. Adv. Ceram.* **2018**, *7*, 197–206. [CrossRef]
126. Fan, Y.; Ling, W.; Huang, B.; Dong, L.; Yu, C.; Xi, H. The synergistic effects of cerium presence in the framework and the surface resistance to SO_2 and H_2O in NH_3-SCR. *J. Ind. Eng. Chem.* **2017**, *56*, 108–119. [CrossRef]
127. Xu, Q.; Yang, W.; Cui, S.; Street, J.; Luo, Y. Sulfur resistance of Ce-Mn/TiO_2 catalysts for low-temperature NH_3–SCR. *R. Soc. Open Sci.* **2018**, *5*, 171846. [CrossRef] [PubMed]

128. Lin, L.-Y.; Lee, C.-Y.; Zhang, Y.-R.; Bai, H. Aerosol-assisted deposition of Mn-Fe oxide catalyst on TiO_2 for superior selective catalytic reduction of NO with NH_3 at low temperatures. *Catal. Commun.* **2018**, *111*, 36–41. [CrossRef]
129. Lee, T.; Bai, H. Metal Sulfate poisoning effects over MnFe/TiO_2 for selective catalytic reduction of NO by NH_3 at low temperature. *Ind. Eng. Chem. Res.* **2018**, *57*, 4848–4858. [CrossRef]
130. Mu, J.; Li, X.; Sun, W.; Fan, S.; Wang, X.; Wang, L.; Qin, M.; Gan, G.; Yin, Z.; Zhang, D. Enhancement of low-temperature catalytic activity over a highly dispersed Fe−Mn/Ti catalyst for selective catalytic reduction of NO_x with NH_3. *Ind. Eng. Chem. Res.* **2018**, *57*, 10159–10169. [CrossRef]
131. Liu, J.; Guo, R.-T.; Li, M.-Y.; Sun, P.; Liu, S.-M.; Pan, W.-G.; Liu, S.-W.; Sun, X. Enhancement of the SO_2 resistance of Mn/TiO_2 SCR catalyst by Eu modification: A mechanism study. *Fuel* **2018**, *223*, 385–393. [CrossRef]
132. Sun, P.; Huang, S.-X.; Guo, R.-T.; Li, M.-Y.; Liu, S.-M.; Pan, W.-G.; Fu, Z.-G.; Liu, S.-W.; Sun, X.; Liu, J. The enhanced SCR performance and SO_2 resistance of Mn/TiO_2 catalyst by the modification with Nb: A mechanistic study. *Appl. Surf. Sci.* **2018**, *447*, 479–488. [CrossRef]
133. Sun, X.; Guo, R.-T.; Liu, J.; Fu, Z.-G.; Liu, S.-W.; Pan, W.-G.; Shi, X.; Qin, H.; Wang, Z.-Y.; Liu, X.-Y. The enhanced SCR performance of Mn/TiO_2 catalyst by Mo modification: Identification of the promotion mechanism. *Int. J. Hydrogen Energy* **2018**, *43*, 16038–16048. [CrossRef]
134. Fan, J.; Lv, M.; Luo, W.; Ran, X.; Deng, Y.; Zhang, W.-X.; Yang, J. Exposed metal oxide active sites on mesoporous titania channels: A promising design for low-temperature selective catalytic reduction of NO with NH_3. *Chem. Commun.* **2018**, *54*, 3783–3786. [CrossRef] [PubMed]
135. Li, G.; Wang, B.; Wang, H.; Ma, J.; Xu, W.Q.; Li, Y.; Han, Y.; Sun, Q. Fe and/or Mn oxides supported on fly ash-derived SBA-15 for low-temperature NH_3-SCR. *Catal. Commun.* **2018**, *108*, 82–87. [CrossRef]
136. Li, G.; Wang, B.; Wang, Z.; Li, Z.; Sun, Q.; Xu, W.Q.; Li, Y. Reaction mechanism of low-temperature selective catalytic reduction of NO_x over Fe-Mn oxides supported on fly-ash-derived SBA-15 molecular sieves: Structure−activity relationships and in situ DRIFT analysis. *J. Phys. Chem. C* **2018**, *122*, 20210–20231. [CrossRef]
137. Tang, X.; Li, C.; Yi, H.; Wang, L.; Yu, Q.; Gao, F.; Cui, X.; Chu, C.; Li, J.; Zhang, R. Facile and fast synthesis of novel Mn_2CoO_4@rGO catalysts for the NH_3-SCR of NO_x at low temperature. *Chem. Eng. J.* **2018**, *333*, 467–476. [CrossRef]
138. Wang, C.; Zhang, C.; Zhao, Y.; Yan, X.; Cao, P. Poisoning effect of SO_2 on honeycomb cordierite-based Mn–Ce/Al_2O_3 catalysts for NO reduction with NH_3 at low temperature. *Appl. Sci.* **2018**, *8*, 95. [CrossRef]
139. Meng, H.; Liu, J.; Du, Y.; Hou, B.; Wu, X.; Xie, X. Novel Cu-based oxides catalyst from one-step carbothermal reduction decomposition method for selective catalytic reduction of NO with NH_3. *Catal. Commun.* **2019**, *119*, 101–105. [CrossRef]
140. Li, L.; Zhang, L.; Ma, K.; Zou, W.; Cao, Y.; Xiong, Y.; Tang, C.; Dong, L. Ultra-low loading of copper modified TiO_2/CeO_2 catalysts for low-temperature selective catalytic reduction of NO by NH_3. *Appl. Catal. B* **2017**, *207*, 366–375. [CrossRef]
141. Zhu, Y.; Zhang, Y.; Xiao, R.; Huang, T.; Shen, K. Novel holmium-modified Fe-Mn/TiO_2 catalysts with a broad temperature window and high sulfur dioxide tolerance for low-temperature SCR. *Catal. Commun.* **2017**, *88*, 64–67. [CrossRef]
142. Nam, K.B.; Kwon, D.W.; Hong, S.C. DRIFT study on promotion effects of tungsten-modified Mn/Ce/Ti catalysts for the SCR reaction at low-temperature. *Appl. Catal. A* **2017**, *542*, 55–62. [CrossRef]
143. Zhao, B.; Ran, R.; Guo, X.; Cao, L.; Xu, T.; Chen, Z.; Wu, X.; Si, Z.; Weng, D. Nb-modified Mn/Ce/Ti catalyst for the selective catalytic reduction of NO with NH_3 at low temperature. *Appl. Catal. A* **2017**, *545*, 64–71. [CrossRef]
144. Li, Y.; Li, G.; Lu, Y.; Hao, W.; Wei, Z.; Liu, J.; Zhang, Y. Denitrification performance of non-pitch coal-based activated coke by the introduction of MnO_x–CeO_x–M (FeO_x, CoO_x) at low temperature. *Mol. Catal.* **2018**, *445*, 21–28. [CrossRef]
145. Pan, Y.; Shen, Y.; Jin, Q.; Zhu, S. Promotional effect of Ba additives on $MnCeO_x$/TiO_2 catalysts for NH_3-SCR of NO at low temperature. *J. Mater. Res.* **2018**, *33*, 2414–2422. [CrossRef]
146. Wang, X.; Liu, Y.; Ying, Q.; Yao, W.; Wu, Z. The superior performance of Nb-modified Cu-Ce-Ti mixed oxides for the selective catalytic reduction of NO with NH_3 at low temperature. *Appl. Catal. A* **2018**, *562*, 19–27. [CrossRef]

147. Li, W.; Zhang, C.; Li, X.; Tan, P.; Zhou, A.; Fang, Q.; Chen, G. Ho-modified Mn–Ce/TiO_2 for low-temperature SCR of NO_x with NH_3: Evaluation and characterization. *Chin. J. Catal.* **2018**, *39*, 1653–1663. [CrossRef]
148. Lu, P.; Li, R.; Xing, Y.; Li, Y.; Zhu, T.; Yue, H.; Wu, W. Low temperature selective catalytic reduction of NO_X with NH_3 by activated coke loaded with $Fe_xCo_yCe_zO_m$: The enhanced activity, mechanism and kinetics. *Fuel* **2018**, *233*, 188–199. [CrossRef]
149. Zhao, X.; Mao, L.; Dong, G. Mn-Ce-V-WO_x/TiO_2 SCR catalysts: Catalytic activity, stability and interaction among catalytic oxides. *Catalysts* **2018**, *8*, 76. [CrossRef]

Review

Hydrogenation of Carbon Dioxide to Value-Added Chemicals by Heterogeneous Catalysis and Plasma Catalysis

Miao Liu [1], Yanhui Yi [1,2,*], Li Wang [3], Hongchen Guo [1] and Annemie Bogaerts [2]

1 State Key Laboratory of Fine Chemicals, School of Chemical Engineering, Dalian University of Technology, Dalian 116024, China; liumiao_dlut@163.com (M.L.); hongchenguo@163.com (H.G.)
2 Research Group PLASMANT, Department of Chemistry, University of Antwerp, Universiteitsplein 1, BE-2610 Wilrijk-Antwerp, Belgium; annemie.bogaerts@uantwerpen.be
3 College of Environmental Sciences and Engineering, Dalian Maritime University, Dalian 116026, China; liwang@dlmu.edu.cn
* Correspondence: yiyanhui@dlut.edu.cn; Tel.: +86-411-8498-6120

Received: 3 February 2019; Accepted: 8 March 2019; Published: 18 March 2019

Abstract: Due to the increasing emission of carbon dioxide (CO_2), greenhouse effects are becoming more and more severe, causing global climate change. The conversion and utilization of CO_2 is one of the possible solutions to reduce CO_2 concentrations. This can be accomplished, among other methods, by direct hydrogenation of CO_2, producing value-added products. In this review, the progress of mainly the last five years in direct hydrogenation of CO_2 to value-added chemicals (e.g., CO, CH_4, CH_3OH, DME, olefins, and higher hydrocarbons) by heterogeneous catalysis and plasma catalysis is summarized, and research priorities for CO_2 hydrogenation are proposed.

Keywords: carbon dioxide; hydrogenation; heterogeneous catalysis; plasma catalysis; value-added chemicals; methanol synthesis; methanation

1. Introduction

Climate changes are mostly induced by the greenhouse effect, and carbon dioxide (CO_2) accounts for a dominant proportion of this greenhouse effect. Indeed, one can barely ignore the connection between the emission of CO_2 and climate changes [1]. The CO_2 concentration in the atmosphere has actually climbed to 405 ppm in 2017, as shown in Figure 1a. As the global energy consumption is still mainly based on burning coal, oil, and natural gas, and this situation will last until the middle of the century (see Figure 1b), experts predict that the CO_2 concentration in the atmosphere will continue to rise to ~570 ppm by the end of the century if no measures are taken [2]. Hence, it is urgent to control the CO_2 emissions by taking effective measures to capture and utilize CO_2.

In principle, there are three strategies to reduce CO_2 emissions, i.e., reducing the amount of CO_2 produced, storage of CO_2, and utilization of CO_2 [3]. Recently, some comprehensive reviews have discussed the technological state-of-the-art of carbon capture and storage (CCS) [4–6]. Direct air capture technology (DAC) draws people's attention to mitigate climate change by taking advantage of chemical sorbents (e.g., basic solvents, supported amine and ammonium materials, etc.) [4]. In addition, Bui et al. also considered the economic and political obstacles in terms of the large-scale deployment of CCS [6]. Significantly reducing the amount of CO_2 produced is unrealistic in view of the current energy structure dominated by fossil energies, as shown in Figure 1b. CO_2 storage seems to be a potential approach, but there are some challenges, such as efficiency of capture and sequestration of CO_2, cutting down the operation costs for capture and separation of CO_2, and the long-term stability of underground storage [7,8]. Utilization of CO_2 is a promising approach, since CO_2 is a cheap and attractive carbon source, which can be used to yield a variety of industrial raw materials, which can be further converted into value-added chemicals and fuels. Interestingly, CO_2 can also be used as a desired trigger for stimuli-responsive materials. Darabi et al. summarized the synthesis, self-assembly, and applications of CO_2-responsive polymeric materials [9].

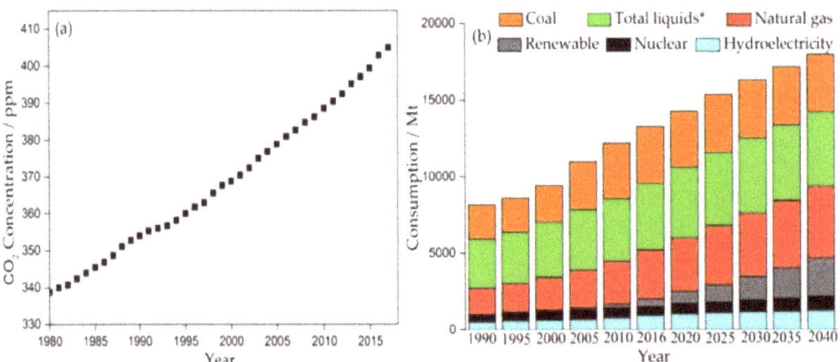

Figure 1. (**a**) Trends in atmospheric CO_2 concentrations (ppm). (**b**) Projected global energy consumption (Mt) from 1990 to 2040. Data from International Energy Agency.

One of the options to convert CO_2, is catalytic hydrogenation, as illustrated in Figure 2. The production of H_2, however, is also a vital problem to realize direct hydrogenation of CO_2 to oxygenates and hydrocarbons. Hydrogen production can be either based on renewable or non-renewable sources, such as electrical, thermal, photonic, and hybrid [10,11]. The main methods of hydrogen production include electrolysis, thermolysis, photo-electrolysis, and hybrid thermochemical cycles [10]. According to some evaluation criteria—such as global warming potential (GWP), social cost of carbon (SCC), acidification potential (AP), energy and exergy efficiencies, and production cost—the hybrid hydrogen production methods seems a promising route. On the other hand, the production cost evaluation shows that coal gasification ($0.92/kg H_2) and fossil fuel reforming ($0.75/kg H_2) are relatively low-cost methods compared to early R&D phase methods (e.g., photo-electrochemical: $10.36/kg H_2 from water dissociation). However, fossil fuel is non-renewable, being a major drawback. Therefore, reducing the cost of hydrogen production is an urgent and challenging issue, since we have to balance both the economy and sustainability.

Figure 2. Catalytic hydrogenation of carbon dioxide.

Various methods have been adopted—including photo-catalysis, electro-catalysis, heterogeneous catalysis, and plasma catalysis [12,13]—to realize hydrogenation of CO_2. Dalle et al. systematically presented the activity of the first-row transition metal complexes (i.e., Sc, Ti, V, Cr, Mn, Fe, Co, Ni, Cu, Zn) in CO_2 reduction by electro-catalysis, photo-catalysis and photoelectron-catalysis [14]. Li et al. discussed the important roles of co-catalysts (e.g., biomimetic, metal-based, metal-free, and multifunctional) in selective photo-catalytic CO_2 reduction [15]. Besides thermochemical approaches, Mota et al. made a detailed retrospective based on electrochemical and photo-chemical approaches for CO_2 hydrogenation to oxygenates and hydrocarbons [16]. In this review, we focus on the latter two methods. Heterogeneous catalysis has been widely studied, while plasma catalysis is still an emerging technology. Some excellent reviews were recently published for heterogeneous catalytic CO_2 hydrogenation [12,17,18]. Jadhav et al. focused on methanol production [17], Porosoff et al. on the synthesis of CO, CH_3OH, and hydrocarbons [18], while Alvarez et al. discussed the greener preparation of formates (formic acid), CH_3OH, and DME [12]. With regard to plasma catalysis for CO_2 hydrogenation, there are only a handful reports (as discussed below). Nevertheless, it exhibits great potential since plasma can operate at ambient temperature and atmospheric pressure.

In this review, we summarize the progress of mainly the last five years in CO_2 hydrogenation to value-added chemicals (e.g., CO, CH_4, CH_3OH, DME, olefins, and higher hydrocarbons) driven by both heterogeneous catalysis and plasma catalysis. The literature bibliography range is shown in Figure 3. Indeed, on the one hand, the insights obtained by heterogeneous catalysis can be useful for the further development of the emerging field of plasma catalysis. On the other hand, we also want to pinpoint the differences between heterogeneous and plasma catalysis, and thus the need for dedicated design of catalytic system tailored to the plasma environment.

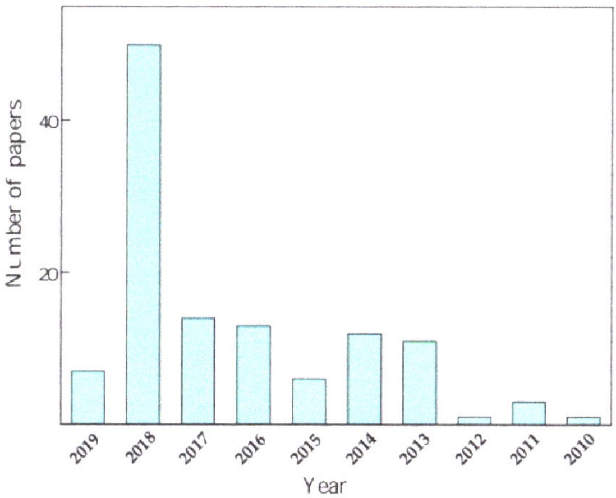

Figure 3. Literature bibliography included in this review.

2. Heterogeneous Catalysis

In heterogeneous catalysis, support materials, active metals, promoters, and the preparation methods of catalysts are major adjustable factors determining the catalytic activity. As far as the support materials are concerned, the catalytic performance is influenced by the metal-support interaction since it usually induces specific physicochemical properties for catalysis, e.g., active cluster, active oxygen vacancy, and acid-based property. Metal oxides and zeolites with special channel structures are usually selected. For the active metal components, research focuses on searching cheap and available metals to replace precious metals or to reduce the amount of precious metals in industrial heterogeneous catalysis. Promoters (structural-type and electron-type) can also significantly influence the catalytic activity by regulating the adsorption and desorption behavior of molecules (reactant, intermediate, and product) on the catalyst surface. Preparation methods usually determine the catalyst morphology including metal dispersion (particle size distribution), specific surface area, and channel structure, which also influences the catalytic performance [12]. In this section, we summarize recent progress of CO_2 hydrogenation to CO, CH_4, CH_3OH, and some other products in terms of rational design of heterogeneous catalysts.

2.1. CO_2 to CO

CO can be used as feedstock to produce liquid fuels and useful chemicals by Fischer–Tropsch (F–T) synthesis reaction. Therefore, the catalytic hydrogenation of CO_2 to CO via the reverse water–gas shift (RWGS) reaction, reaction (1), is promising to account for the shortage of energy. Accordingly, catalyst design for the RWGS reaction have attracted extensive attention.

$$CO_2 + H_2 \leftrightarrow CO + H_2O, \triangle H_{298K} = 41.2 \text{ kJ mol}^{-1} \quad (1)$$

Generally, noble metal catalysts (e.g., Pt, Ru, and Rh) exhibit effective ability towards H_2 dissociation, and thus precious metals have been investigated extensively for the RWGS reaction. However, these catalysts yield, methane as dominant product, mainly caused by the higher rate of C-H bond formation than CO desorption [19]. To change the product distribution, Bando et al. applied Li-promoted Rh ion-exchanged zeolites for CO_2 hydrogenation, achieving 87% CO selectivity for an atomic ratio of Li/Rh higher than 10/1 [19]. Although the catalytic performance of noble metals can be improved via promoters, the high price and instability of noble metals (aggregation of nano-particles) limit the industrial applicability. In order to properly decrease the operation cost and improve the life cycle of catalysts, non-noble metal carbides have been developed. Porosoff et al. synthesized a molybdenum carbide (Mo_2C) catalyst for the RWGS reaction [20], yielding 8.7% CO_2 conversion and 93.9% CO selectivity (at 573 K reaction temperature). The catalytic performance was much better than those of bimetallic catalysts (e.g., Pt-Co, Pt-Ni, Pd-Co, Pd-Ni) supported on CeO_2 (~5% CO_2 conversion and 83.3% CO selectivity at the same reaction temperature). The catalytic mechanism of Mo_2C in the RWGS reaction was further studied by ambient pressure X-ray photoelectron spectroscopy (AP-XPS) and in situ X-ray absorption near edge spectroscopy (XANES), which demonstrate that Mo_2C not only broke the C=O bond, but also dissociated hydrogen. Hence, Mo_2C has a dual functional and is an ideal catalytic material for the RWGS reaction. Interestingly, the authors also found that by modifying the catalyst with Co, the catalytic activity of the Co-Mo_2C catalyst was further improved (9.5% CO_2 conversion, 99% CO selectivity at 573 K), probably attributed to the existence of the $CoMoC_yO_z$ phase confirmed by X-ray Diffraction (XRD) (see Figure 4).

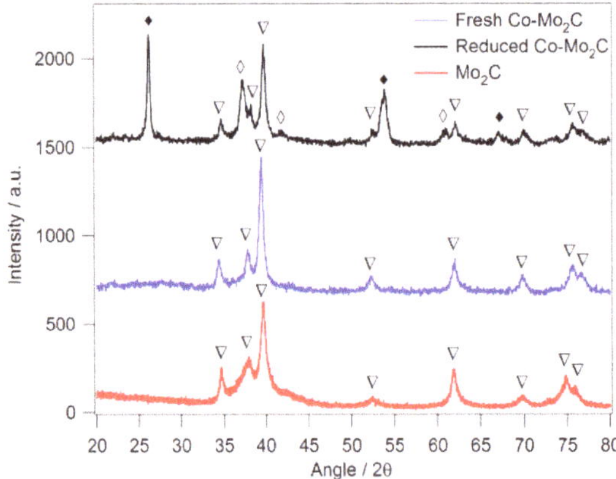

Figure 4. XRD pattern of fresh and reduced Co-Mo_2C with Mo_2C as a reference. The symbols correspond to the following: ▽—β-Mo_2C, ◊—$CoMoC_yO_z$, ♦—MoO_2, reproduced with permission from [20]. Copyright Wiley-VCH, 2014.

The support plays an important role in CO_2 hydrogenation to CO through the RWGS reaction. Kattel et al. prepared Pt, Pt/SiO_2 and Pt/TiO_2 materials as catalysts for the RWGS reaction [21], and showed that Pt nanoparticles alone cannot catalyze the RWGS reaction, while using SiO_2 and TiO_2 as support, the overall CO_2 conversion can be significantly improved, pointing towards a synergy effect between Pt and the oxide support. To reveal the synergy effect, they combined density functional theory (DFT), kinetic Monte Carlo (KMC) simulations and other experimental measurements. They found that the hydrogenation of CO_2 to CO is promoted once CO_2 is stabilized by the Pt-oxide interface. In the case of Pt nanoparticles (NP) alone, the conversion was close to 0, since the ability of Pt NP in binding CO_2 is weak. When SiO_2 and defected TiO_2 with oxygen vacancies served as supports, however, the CO_2 conversion was enhanced (to 3.35% and 4.51%, respectively) on the Pt-oxide interface. The synergy effect between Pt and oxide supports in activating and hydrogenating CO_2 is shown in Figure 5, and possible reaction pathway [21] for the RWGS reaction is illustrated in Scheme 1.

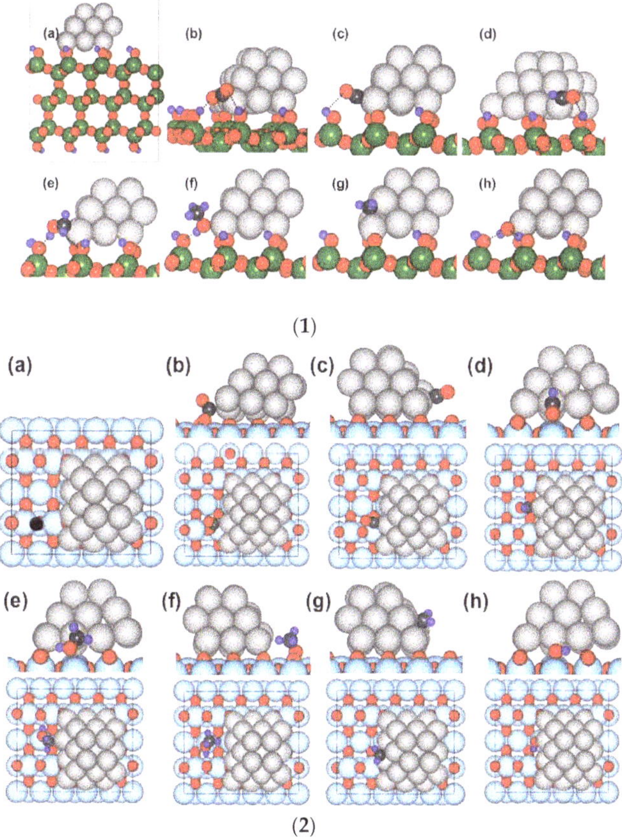

Figure 5. DFT optimized geometries. (**1**) (**a**) Pt_{25}/hydroxylated SiO_2 (111), and (**b**) *CO_2 species, (**c**) *CO species, (**d**) *HCO species, (**e**) *H_2COH species, (**f**) *CH_3OH species, (**g**) *CH_2 species, and (**h**) *OH species adsorbed on Pt/SiO_2 (111). The dashed lines show hydrogen bonds. (**2**) (**a**) Pt_{25}/TiO_2 (110) with oxygen vacancy, and side (**top**) and top (**bottom**) views of (**b**) *CO_2 species, (**c**) *CO species, (**d**) *HCO species, (**e**) *H_2COH species, (**f**) *CH_3OH species, (**g**) *CH_2 species, and (**h**) *OH species adsorbed on Pt/TiO_2 (110). The black circle in (a) depicts the position of oxygen vacancy on TiO_2 (110). Note: Si: green, Ti: light blue, Pt: light gray, C: dark gray, O: red, and H: blue, reprinted with permission from [21]. Copyright Elsevier, 2016.

$$\textbf{RWGS:}\quad CO_2(g) \longrightarrow {}^{*}CO_2 \xrightarrow{+H} {}^{*}HOCO \longrightarrow {}^{*}CO + {}^{*}OH \longrightarrow {}^{*}CO + {}^{*}H_2O$$

Scheme 1. Reaction pathways for the RWGS reaction, where '*X' represents species X adsorbed on a surface site, reproduced with permission from [21]. Copyright Elsevier, 2016.

Yan et al. investigated the effect of the Ru-Al_2O_3 interfaces on the catalytic activity of the RWGS reaction, and proposed Ru_{35}/Al_2O_3 and Ru_9/Al_2O_3 catalyst models to explain the experimental observations [22]. The product selectivity switched between CH_4 and CO over the Ru/Al_2O_3 catalyst, i.e., monolayer Ru sites favored the production of CO, while Ru nano-clusters preferred the production of CH_4, during CO_2 reduction reaction. Confirmed by kinetic analysis, characterization of the surface structures and real-time monitoring of the active intermediate species, the product selectivity of CH_4 and CO was regulated by the Ru sites and Ru-Al_2O_3 interfacial sites. Furthermore, based on the combination of theoretical calculations and isotope-exchange experimental results, the authors found that the O* species derived from the dissociative adsorption of CO_2 at interfacial Ru sites easily bridge with the Al sites from the γ-Al_2O_3 support, and new Ru-O-Al bonds are formed via the oxygen-exchange process. The interfacial O species existing in Ru-O-Al bonds was responsible for the CO_2 activation via oxygen-exchange with the O atoms of CO_2. Therefore, this experimental work is a good inspiration to further explore the influence of metal-support interfaces for the effective activation of CO_2.

Besides the widely used impregnation method, Yan et al. also reported a doping-segregation method for the preparation of Rh-doped $SrTiO_3$ [23]. Precursors with a molar ratio of Sr:Ti:Rh = 1.10:0.98:0.02. First, TiO_2 was suspended in deionized water, and then $Sr(OH)_2 \cdot 8H_2O$ and $Rh(NO_3)_3$ were introduced. Subsequently, the sample was poured in a stain steel acid digestion vessel, which was kept at 473 K for 24–48 h. Finally, the reaction product was dried at 343 K overnight after it was centrifuged, and washed via deionized water. As confirmed by in situ X-Ray-Diffraction (XRD) and X-ray Absorption Fine Structure (XAFS) measurements, the Rh-doped $SrTiO_3$ catalysts produce sub-nanometer Rh clusters, which are highly active for the conversion of CO_2 compared to the supported Rh/$SrTiO_3$ prepared by wetness impregnation. The better catalytic performance (7.9% CO_2 conversion and 95% CO selectivity at 573 K) could be ascribed to the cooperative effect between sub-nanometer Rh clusters and the reconstructed $SrTiO_3$ which is active for dissociation of H_2 and is favorable for adsorption/activation of CO_2. Therefore, the novel approach, doping-segregation method, maybe a novel strategy to tune the size of active metals and the physicochemical properties of supports for rational design of catalysts for the RWGS reaction.

Dai et al. studied CeO_2 catalysts which were prepared by the hard-template method (Ce-HT), the typical complex method (Ce-CA), and the typical precipitation method (Ce-PC) for the RWGS reaction [24]. The experimental results show that catalysts prepared by Ce-CA, Ce-HT, and Ce-PC methods exhibit a 100% CO selectivity and the CO_2 conversions were 9.3, 15.9, and 12.7% respectively at 853 K. Obviously, the hard-template (Ce-HT) method is beneficial for the RWGS reaction in view of the catalytic performance. The Ce-HT method comprises the following steps: (1) $Ce(NO_3)_3 \cdot 6H_2O$ was dissolved in ethanol, and KIT-6 mesoporous silica was added; (2) The mixture was stirred until a dry power was obtained; (3) The powder was calcined; (4) The obtained samples were treated with NaOH to remove the template. To reveal the relationship between the preparation method and the catalytic activity, the authors carried out XRD, Transmission Electron Microscopy (TEM) and Brunauer–Emmett–Teller (BET) characterization, and the characterization results show that CeO_2 catalysts which were prepared by the Ce-HT method have a porous structure and a high specific surface area, while CeO_2 catalysts prepared by the other methods (Ce-CA, Ce-PC) have an agglomerated structure (Ce-CA) and overlapped bulk structure (Ce-PC) with low porosity. Moreover, in the CeO_2 catalysts, oxygen vacancies as active sites were formed by H_2 reduction at 673 K, which were confirmed by in situ X-ray photoelectron spectroscopy (XPS) and H_2-temperature-programmed reduction (H_2-TPR). Therefore, the improvement of the catalytic activity for the RWGS reaction can be ascribed to the change of catalyst structure and oxygen vacancies.

Except for the above monometallic catalysts, some bimetallic catalysts—i.e., Pt-Co, Fe-Mo and Ni-Mo [25–27]—were also prepared and tested in the RWGS reaction. In general, the preparation method of bimetallic catalysts generally can be divided into two categories: (1) the bi-metal was first prepared and then supported on the carrier [25]; (2) the bi-metal as formed in the preparation progress [26,27]. Compared with pure Co catalyst, Pt-Co catalyst showed a better catalytic activity (mainly CO, close to 100%) for the RWGS reaction. Ambient pressure X-ray photoelectron spectroscopy (AP-XPS) and environmental transmission electron microscopy (ETEM) showed that Pt migrates on the catalyst surface, and Pt aids the reduction of Co to its metallic state under appropriate reaction conditions confirmed by near edge X-ray absorption fine structure (NEXAFS) spectroscopy [25]. Abolfazl et al. investigated Mo/Al_2O_3 and $Ni–Mo/Al_2O_3$ catalysts prepared by the impregnation method used for the RWGS reaction. The experimental results showed that $Ni–Mo/Al_2O_3$ catalyst is a promising catalyst (35% CO_2 conversion, i.e., close to the 38% equilibrium conversion) compared with Mo/Al_2O_3 catalyst (15% CO_2 conversion). As demonstrated by XPS, the electronic effect, which transfers electrons from Ni to Mo and leads to an electron-deficient state of the Ni species, is beneficial for CO_2 adsorption, and thus improves the CO_2 conversion. Furthermore, the XRD and H_2-TPR profiles indicated that the Ni–O–Mo structure crystallizes into the $NiMoO_4$ phase which can improves the adsorption and dissociation of H_2 on the $Ni-Mo/Al_2O_3$ catalyst [27]. Similarly, they also found that $Fe-Mo/Al_2O_3$ catalyst synthesized by the impregnation method is an efficient catalyst with high CO yield, almost no by-products and relatively stable (60 h) for the RWGS reaction. The enhancement of the catalytic activity may be attributed to better Fe dispersion with the addition of Mo and smaller particle size of active Fe species, which was confirmed by the BET method and scanning electron microscopy (SEM) [26].

Overall, to some extent, the reaction activity and stability of bimetallic catalysts for the direct hydrogenation of CO_2 to CO are excellent compared to mono-metallic catalysts. The reduction of the active metal, the formation of alloy metal, the dispersion and particle size of the catalyst are possible factors for the improvement of the catalytic performance in the RWGS reaction. Details of the conversion and product selectivity, along with the reaction conditions of several representative RWGS catalytic systems, are compared in Table 1. Obviously, precious metal catalysts (e.g., Pt, Rh, Ni) are still advantageous for the formation of CO compared to non-noble metal catalysts (e.g., Fe, Co, Mo), and upon increasing the gas hourly space velocity (GHSV), the selectivity of CO slightly decreases to some extent.

Table 1. Catalytic performance of several catalytic systems for CO_2 hydrogenation into CO, in terms of CO_2 conversion and CO selectivity, along with the reaction conditions

Catalyst	H_2:CO_2	GHSV	Temperature (°C)	Pressure (MPa)	CO_2 Conversion (%)	CO Selectivity (%)
Mo_2C [20]	3	36,000 [a]	300	0.1	8.7	93.9
Co-Mo_2C [20]	3	36,000 [a]	300	0.1	9.5	~99.0
Pt-TiO_2 [21]	1	119.7 [b]	300	0.1	4.5	99.1
Pt-SiO_2 [21]	1	24.7 [b]	300	0.1	3.3	100
Rh-$SrTiO_3$ [23]	1	12,000 [a]	300	N/A	7.9	95.4
Co-MCF-17 [25]	3	60,000 [b]	200-300	0.5	~5.0	~90.0
Pt-Co-MCF-17 [25]	3	60,000 [b]	200-300	0.5	~5.0	~99.0
Fe-Mo-Al_2O_3 [26]	1	30,000 [a]	600	1	~45.0	~100.0
Mo-Al_2O_3 [27]	1	30,000 [a]	600	0.1	16	N/A
Ni-Mo-Al_2O_3 [27]	1	30,000 [a]	600	0.1	34	N/A
La-Fe-Ni [28]	2	24,000 [a]	350	N/A	16.3	96.6

[a] $mL\ g_{cat}^{-1}\ h^{-1}$; [b] h^{-1}; N/A: not available.

2.2. CO_2 to CH_4

Methane, a high value carbon source, is used to produce syngas via steam reforming, and subsequently the syngas is usually converted into chemicals and/or fuels through F–T synthesis.

The direct hydrogenation of CO_2 to CH_4 (also called CO_2 methanation), reaction (2), is a feasible approach, if the production technology of H_2 becomes widespread and at low-cost.

$$CO_2 + 4H_2 \leftrightarrow CH_4 + 2H_2O, \triangle H_{298K} = -252.9 \text{ kJ mol}^{-1} \tag{2}$$

Ni-based [29], Co-based [30], and Ru-based [31] catalysts have been used in CO_2 methanation, and Ni-based catalysts are considered to be the most effective and the lowest cost alternative. However, coke formation is a serious problem for Ni-based catalytic systems [2]. Therefore, researchers are trying to seek appropriate promoters to improve the activity and stability of Ni-based catalytic systems for CO_2 methanation. Yuan et al. investigated the effect of Re on the catalytic activity of Ni-based catalysts in CO_2 methanation [32]. Based on DFT calculations, they found that, attributed to the strong affinity of Re to O (see Figure 6), the presence of Re markedly lowered the energy barrier of C-O bond cleavage, which benefits the activation of CO_2. Moreover, CH_4 selectivity can be enhanced owe to the presence of Re, which was confirmed by analysis of surface coverage of the adsorbed species on Ni (111) and Re@Ni (111). In addition, micro-kinetic analysis showed that, in addition to CO* and H*, a suitable amount of O_{ad} atoms were present on Re@Ni (111), and thus a possible reaction network of CO_2 methanation was proposed as shown in Scheme 2, in which a red line represents the preferable steps in each pathway.

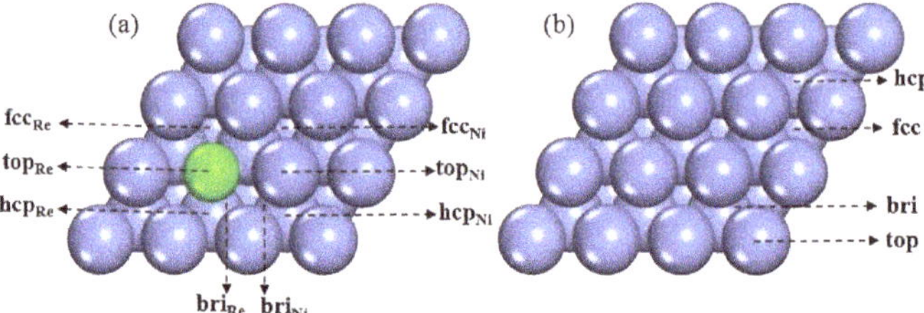

Figure 6. Adsorption sites existed in the (**a**) Re@Ni(111) and (**b**) Ni(111) surface. Ni: blue, Re: green, reprinted with permission from [32]. Copyright Elsevier, 2018.

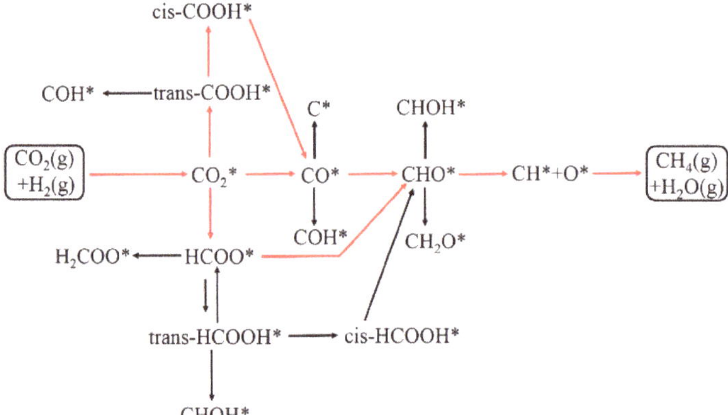

Scheme 2. Reaction network for CO_2 methanation. The preferable steps in each pathway are marked with a red line, reproduced with permission from [32]. Copyright Elsevier, 2018.

Besides Re, La is also an excellent promoter of Ni-based catalysts for CO_2 methanation. Quindimil et al. [29] applied Ni-La_2O_3/Na-BETA catalysts for CO_2 methanation. They found that the presence of La_2O_3 created more CO_2 adsorption sites and more hydrogenation sites, mainly attributed to the improved surface basicity and Ni dispersion by the co-catalyst effect of La. Under the optimized reaction conditions, 65% CO_2 conversion and nearly 100% CH_4 selectivity were achieved over a Ni-10%La_2O_3/Na-BETA catalyst with a good stability for more than 24 h at 593 K.

CeO_2, TiO_2, and SiO_2 have been used as the supports of methanation catalysts [33]. Reactions over Ni/CeO_2 catalyst performed full selectivity to CH_4 with higher TOF (up to forty-fold) compared to TiO_2 and SiO_2 supported Ni nanoparticles, and in CO_2 methanation, the catalytic stability of Ni/CeO_2 catalyst lasted for 50 h at 523 K. HRTEM analysis indicated that different supports induced distinctive crystal structure. Ni/CeO_2 catalyst presented hexagonal Ni nanocrystallites, while TiO_2 and SiO_2 favored the formation of pseudo-spherical Ni nanoparticles. Demonstrated by TPR, XPS, and UV Raman analysis, characterization results revealed partial reduction of the CeO_2 surface, and the partial reduction of the CeO_2 surface contributed to the generation of oxygen vacancies, which is beneficial for the formation of a strong metal-support interaction (SMSI) between Ni and CeO_2, while no SMSI was observed over Ni/SiO_2 and Ni-TiO_2 catalyst. Furthermore, pulse reaction by temporal analysis products (TAP) demonstrated the capacity of CO_2 adsorption following the order: Ni/SiO_2 < Ni/TiO_2 < Ni/CeO_2. Therefore, metal particle morphology and surface oxygen vacancies were used to anchor/stabilize Ni nanoparticles, and SMSI of Ni/CeO_2 catalyst contributed to the remarkable catalytic activity for CO_2 methanation.

Lin et al. investigated the influence of TiO_2 phase structure on the degree of dispersion of Ru nanoparticles [31]. Experiments showed that Ru/r-TiO_2 (rutile-type TiO_2) catalysts have a fast rate of CO_2 conversion, more than twice as fast as Ru/a-TiO_2 (anatase-type TiO_2) for CO_2 methanation. Meanwhile, compared to Ru/a-TiO_2 catalysts, Ru/r-TiO_2 catalysts exhibited a much higher thermal stability. High-angle annular dark-field scanning transmission electron microscopy (HAADF-STEM) images showed that r-TiO_2 supported Ru nanoparticles displayed a narrower particle size distribution (1.1 ± 0.2 nm) compared with a-TiO_2 supported Ru nanoparticles (4.0 ± 2.4 nm). As confirmed by XRD measurements and H_2-TPR experiments, a strong interaction existed in RuO_2 and r-TiO_2 contributed to the formation of the Ru–O–Ti bond. Experimental tests, revealed that the strong interaction between RuO_2 and r-TiO_2, not only promotes the highly dispersion of Ru nanoparticles, but also prevents nanoparticles' aggregation, which is responsible for the enhancement of the catalytic activity and thermal stability.

Furthermore, the influence of Al_2O_3, ZrO_2, SiO_2, KIT-6, and GO supports on Ni-based catalysts [30,34,35] has also been reported recently (Table 2). The specific area of support, the metal-support interaction and particle size of active sites are still mainly adjustable factors. Interestingly, Ni-SiO_2/GO-Ni-foam catalyst which was synthesized via intercalation of graphene oxide (GO) exhibited excellent activity compared with Ni-SiO_2/Ni-foam catalyst (see Figure 7) at 743 K for CO_2 methanation. Furthermore, the formation of nickel silicates on GO is responsible for the uniform dispersion of Ni active sites, which is favorable for inhibition of catalyst sintering.

Table 2. Catalytic performance of several catalysts for CO_2 methanation, in terms of CO_2 conversion and CH_4 selectivity, along with the reaction conditions

Catalyst	H_2:CO_2	GHSV	Temperature (°C)	CO_2 Conversion (%)	CH_4 Selectivity (%)
Ni-La/Na-BETA [29]	4	10,000 [b]	350	65	100
Co/meso-SiO_2 [30]	4.6	60,000 [a]	280	40	94.1
Co/KIT-6 [30]	4.6	22,000 [a]	280	48.9	100
Ru/BF_4/SiO_2 [31]	4	2400 [b]	250	70.5	N/A
Re-Ni(111) [32]	4	N/A	250	N/A	100
Ni/Al_2O_3-ZrO_2 [34]	4	40,000 [a]	400	~70.0	N/A
Ni-SiO_2/GO [35]	4	500 [b]	470	54.3	88
Ni-ZrO_2 [36]	4	75 [a]	350	~40.0	~95.0
Ni-Ce/USY [37]	4	N/A	350	65	95
Co/KIT-6 [38]	4	60,000 [a]	340	40	86.7
Co/SiO_2 [39]	4	60,000 [a]	360	44.3	86.5
Ni-Nb_2O_5 [40]	4	750 [b]	325	81	~99.0

[a] mL g_{cat}^{-1} h^{-1}; [b] h^{-1}; N/A: not available; pressure: 0.1 MPa.

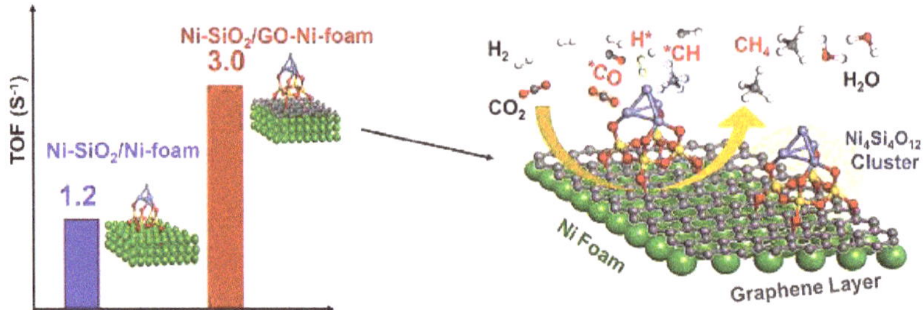

Figure 7. TOF of Ni-SiO_2/Ni-foam and Ni-SiO_2/GO-Ni-foam catalysts for CO_2 methanation, reprinted with permission from [35]. Copyright Elsevier, 2019.

Besides the traditional methods (co-impregnation method and deposition–precipitation method), some novel preparing methods, such as the sequential impregnation and a change of the preparation progress (e.g., the metal incorporation order, the reduction temperature, and the selection of different types of precursors), have been recently reported for preparing methanation catalysts.

Romero-Sáez et al. synthesized Ni-ZrO_2 catalysts supported on CNTs via the sequential and co-impregnation methods for CO_2 methanation [36]. The catalyst prepared by co-impregnation was apparently less active and selective to CH_4, compared with the catalyst synthesized by the sequential impregnation method. The preparation approach of the sequential impregnation method is as follows: (1) the appropriate amount of $ZrO(NO_3)_2$ xH_2O was dissolved in acetone; (2) the CNTs were added to the solution; (3) removal of the solvent, drying and heat treatment at 623 K; (4) in a rotatory evaporator, the same procedure for $Ni(NO_3)_2$ impregnation, followed drying and heat treatment was applied. As characterized by TEM analysis, NiO nanoparticles surrounded by ZrO_2 in core–shell structures were formed via co-impregnation method.The existence of core–shell structure reduced reactant access to Ni and Ni–ZrO_2 interface. However, when the catalyst was prepared via a sequential impregnation method, NiO nanoparticles were available and deposited either on the surface or next to the ZrO_2 nanoparticles, which improved the extent of the Ni–ZrO_2 interface. The ratio of Ni–O–Zr exposed species thus increases in the sequential impregnation method, and the interaction between H atoms (produced upon H_2 dissociation on the Ni surface) and the CO_2 molecule (activated by ZrO_2), can also be enhanced. The schematic representation is shown in Figure 8.

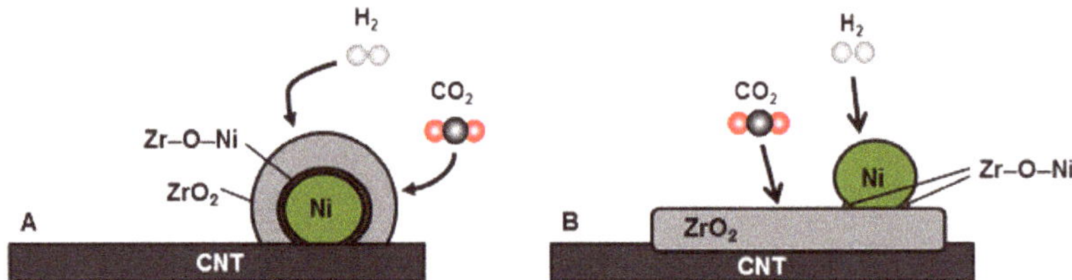

Figure 8. Schematic representation of the disposition of NiO, ZrO_2 and interface Ni-O-Zr at the surface of CNTs for (**A**) Ni-Zr-CNT-COI catalytic system, and (**B**) Ni-Zr-CNT-SEQ catalytic system, reproduced with permission from [36]. Copyright Elsevier, 2018.

Interestingly, Bacariza et al. investigated the influence of the metal incorporation order in the preparation of Ni-Ce/Y(USY) zeolite catalysts [37]. In their experiments, three ways were used for the preparation of catalysts: Ni before Ce (Ce/Ni), Ce before Ni (Ni/Ce), and co-impregnation (Ni-Ce). Experimental tests showed that the catalytic activity follows the order: Ce/Ni $\approx$ Ni/Ce < Ni-Ce. TEM and H-TPR characterizations demonstrated that the Ni^0 average size decreases to approximately 2.5 nm, and in terms of Ni/Ce or Ni-Ce catalysts, stronger interactions between Ni and Ce species are established. However, the CO_2 adsorption capacity is smaller for Ni/Ce catalyst. In contrast, even though larger Ni^0 particles (13.3 nm) are formed for Ce/Ni catalyst, CO_2 adsorption capacity can be enhanced. Finally, Ce-Ni was found as the best preparation method for CO_2 methanation using Ni-based catalysts supported on CeO_2.

In addition, it has been reported that the reduction temperature and a variety of different precursors (e.g., nitrate, chlorate, and oxalate) affect the number of active centers for CO_2 methanation [30,38]. As discussed above, the preparation methods showed beneficial influences on the catalytic performance, suggesting that reasonable adjustment of the preparation process is a strategy to optimize the experiments for the conversion of CO_2 to CH_4. Details of CO_2 conversion and product selectivity, along with the reaction conditions of several representative catalytic systems, are compared in Table 2. From Table 2, we can see that Ni-based catalysts are the main catalytic systems for the conversion of CO_2 to CH_4. Furthermore, the optimal temperature for CH_4 production is 300–400 °C.

2.3. CO_2 to CH_3OH

Methanol is not only an important industrial raw material, but also a stable hydrogen storage compound, which is convenient for transport and reserve. Recently, the concept of "methanol economy" advocated by the Nobel Laureate George Olah revealed the importance of methanol in energy structure [41]. Hence, a large amount of heterogeneous catalysts for the direct hydrogenation of CO_2 to CH_3OH, reaction (3), have emerged in recent years. Herein, we summarize some catalytic systems for the direct hydrogenation of CO_2 to CH_3OH (see Figure 9).

$$CO_2 + 3H_2 \leftrightarrow CH_3OH + H_2O,\ \triangle H_{298K} = -49.5\ \text{kJ mol}^{-1} \quad (3)$$

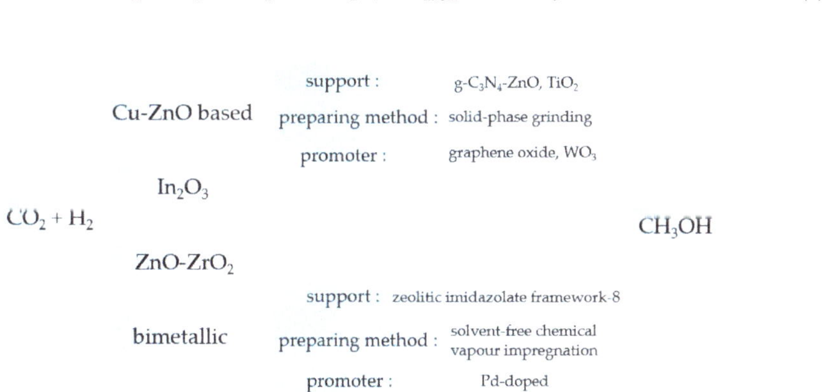

Figure 9. Catalytic system for the direct hydrogenation of CO_2 to CH_3OH in this review.

Cu-ZnO-based catalysts, developed by ICI (Imperial Chemical Industries) in the 1960s, are still widely used in CH_3OH synthesis from syngas (CO and H_2), and Cu^0 is considered to be the active site. Li et al. reported the synthesis of Cu/ZnO catalysts using a unique method, i.e., facile solid-phase grinding, using mixture of oxalic acid, copper nitrate, and zinc nitrate as raw materials [42]. Appropriate amount of these compounds were physically mixed, and then manually ground for 0.5 h. Subsequently, at 393 K, the obtained precursor was dried for 12 h, and at 623 K, calcined for 3 h in N_2 flow, followed by passivation in 1% O_2/N_2 flow for 5 h. In contrast, by the following steps: (1) at 623 K, calcining the dried precursor at for 3 h in air; (2) at 503 K, reducing the oxide for 10 h in 5% H_2/N_2 flow; and (3) at room temperature, passivating the catalyst in 1% O_2/N_2 flow for 5 h, the catalyst can be obtained via H_2 reduction. As demonstrated by XRD, H_2-TPR and thermal-gravity-differential thermal analysis (TG-DTA), in the facile solid-phase grinding process, the decomposition of oxalate complexes and the reduction of CuO took place simultaneously when the samples were calcined in N_2, which prevented the growth of active Cu^0 species and the aggregation of catalyst particles. In contrast, in the conventional H_2 reduction process, the growth of active species or the aggregation of catalyst particles were inevitable. Notably, Cu/ZnO catalysts prepared by a facile solid-phase grinding method achieved 29.2% CO_2 conversion (at 523 K and 3 MPa), which is even higher than the thermodynamic equilibrium conversion. Indeed, the equilibrium conversion of CO_2 at 473 K and 3 MPa is 25.78% according to the Benedict–Webb–Rubin equation [43], and it is even lower at 523 K, since CO_2 hydrogenation to CH_3OH is an exothermic reaction. Generally, tandem reaction can break the limit of thermodynamic on reaction result. Inspiring from which, the above unusual experimental results (higher than thermal equilibrium value) could be most likely achieved through tandem reactions, in which every stepwise reaction was accelerated by different metal sites. Therefore, the simple and solvent-free method based on solid-phase grinding, opened a new approach to synthesize bimetallic or multimetallic catalysts without further reduction.

To improve the catalytic activity of $Cu/ZnO/Al_2O_3$ catalysts, ZnO was replaced by g-C_3N_4-ZnO hybrid material, as reported by Deng et al. [44]. The experimental results showed that, at 12 bar and 523 K, the methanol space time yield (STY) reached 5.73 mmol h^{-1} $g_{Cu}{}^{-1}$ for Cu- g-C_3N_4-ZnO/Al_2O_3, which is superior to the methanol yield (5.45 mmol h^{-1} $g_{Cu}{}^{-1}$) of industrial catalyst (Cu-ZnO-Al_2O_3) under the same reaction pressure. As confirmed by time-resolved photoluminescence (TRPL) and electronic spin resonance (ESR), the electron-richness of ZnO was enhanced via the formation of type-II hetero-junction between g-C_3N_4 and ZnO. In addition, in the TPR curve, enhanced SMSI was observed, and the SMSI between electron-rich ZnO and Cu could boost the catalytic performance in CH_3OH production. The study provided a viable and economic method to modify traditional catalysts for improving CH_3OH production.

For CuO-ZnO-ZrO_2 catalysts, doping graphene oxide (GO) is a good strategy for improving the activity of CO_2 to CH_3OH [45]. The highest methanol selectivity reached up to 75.88% (473 K, 20 bar) at 1 wt % GO content, while under the same conditions, the methanol selectivity was found to be 68% over the GO-free catalyst. Furthermore, CuO-ZnO-ZrO_2-GO catalyst exhibited an almost constant space-time yield (STY) of methanol after 96 h time-on stream experiment. As demonstrated by H_2-TPD and CO_2-TPD, the CO_2 and H_2 adsorption capacity is enhanced over the CuO-ZnO-ZrO_2-GO catalysts. Additionally, to explain the improvement of catalytic activity, the authors proposed that GO nano-sheet can serve as a bridge between mixed metal oxides, which strengthens a hydrogen spillover (see Figure 10). That is, H species migrating from the copper surface to the carbon species are adsorbed on the isolated metal oxide particles.

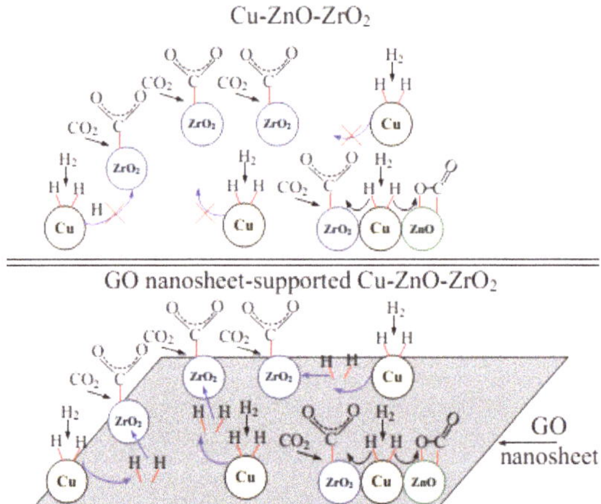

Figure 10. Graphene oxide (GO) nanosheet as a bridge promoting hydrogen spillover from the surface of copper to the surface of other metal oxides, reproduced with permission from [45]. Copyright Elsevier, 2018.

It is also reported that modification by small amounts (2 and 5 at %) of WO_3 can improve the CO_2 conversion and CH_3OH selectivity of the CuO-ZnO-ZrO_2 catalyst [46]. The optimal catalytic performance can be attributed to the specific surface area of metallic Cu, basic sites, and the reducibility of catalysts. However, if we want to effectively tune the catalytic reaction, a better understanding of the reaction mechanisms is essential. Up to now, the possible reaction pathways for the hydrogenation of CO_2 to methanol over Cu-ZnO-based catalysts are shown in Scheme 3.

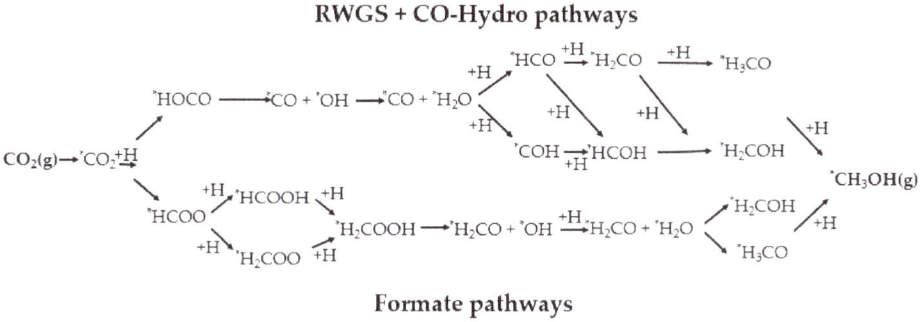

RWGS + CO-Hydro pathways

$CO_2(g) \rightarrow {}^*CO_2 \xrightarrow{+H}$ ${}^*HOCO \rightarrow {}^*CO + {}^*OH \rightarrow {}^*CO + {}^*H_2O$; ${}^*CO \xrightarrow{+H} {}^*HCO \xrightarrow{+H} {}^*H_2CO \xrightarrow{+H} {}^*H_3CO$; ${}^*CO \xrightarrow{+H} {}^*COH \xrightarrow{+H} {}^*HCOH \rightarrow {}^*H_2COH$; $\xrightarrow{+H} {}^*CH_3OH(g)$

$CO_2(g) \rightarrow {}^*CO_2 \xrightarrow{+H} {}^*HCOO \xrightarrow{+H} {}^*HCOOH \xrightarrow{+H} {}^*H_2COOH$; ${}^*HCOO \xrightarrow{+H} {}^*H_2COO \xrightarrow{+H} {}^*H_2COOH \rightarrow {}^*H_2CO + {}^*OH \xrightarrow{+H} {}^*H_2CO + {}^*H_2O$; ${}^*H_2COH$ / ${}^*H_3CO \xrightarrow{+H} {}^*CH_3OH(g)$

Formate pathways

Scheme 3. Possible reaction pathways for CO_2 hydrogenation to methanol, where '*X' represents species X adsorbed on a surface site, reprinted with permission from [21]. Copyright Elsevier, 2016.

Besides Cu-based catalysts, In_2O_3 catalyst with surface oxygen vacancies has attracted more and more attention from researchers. Using density functional theory (DFT) calculations, Ye et al. investigated a In_2O_3 catalyst for the hydrogenation of CO_2 to CH_3OH [47]. On a perfect In_2O_3 (110) surface, six possible surface oxygen vacancies (O_{v1} toO_{v6}) were investigated (see Figure 11a), and the D4 surface with the O_{v4} defective site was found to be most beneficial for CO_2 activation and further hydrogenation. Potential energy profiles of CO_2 hydrogenation and protonation on the D4 defective In_2O_3 (110) surfaces are shown in Figure 11b. In addition, the simulation results showed that the formation of CH_3OH replenishes the oxygen vacancy sites, while H_2 contributes to generate the vacancies, and this cycle between perfect and defective states of the surface is responsible for the formation of CH_3OH from CO_2 hydrogenation. To demonstrate this hypothesis, In_2O_3 catalyst was used for the hydrogenation of CO_2 to CH_3OH in practice, which showed the superior catalytic activity (7.1% CO_2 conversion, 39.7% CH_3OH selectivity at 603 K). This experimental result is better than for many other reported catalytic systems, which generally show low selectivity of CH_3OH at 603 K. Confirmed by thermo-gravimetric analysis (TGA), XRD, and HR-TEM, the authors found that In_2O_3 catalyst had satisfactory thermal and structural stability for CO_2 conversion to CH_3OH below 773 K [48]. Furthermore, to reveal the effect of oxygen vacancies, Martin et al. synthesized bulk In_2O_3 catalyst, and at a wide range of reaction conditions, the CH_3OH selectivity could be tuned up to 100%. XRD, H_2-TPR, CO_2-TPD, XPS, operando diffuse reflectance infrared Fourier transform spectroscopy (DRIFTS), and electron paramagnetic resonance (EPR) characterization confirmed that oxygen vacancies on the In_2O_3 catalyst are active sites for the reduction of CO_2. Additionally, to further improve the stability of In_2O_3 catalyst for the production of CH_3OH, the authors investigated a variety of supports (i.e., ZrO_2, TiO_2, ZnO_2, SiO_2, Al_2O_3, C, SnO_2, MgO). ZrO_2 showed the best catalytic performance since it prevented the sintering of the In_2O_3 phase, which was demonstrated by the enduring stability of In_2O_3/ZrO_2 catalyst over 1000 h [49]. Overall, In_2O_3 is a potential catalyst for CO_2 hydrogenation to CH_3OH with high selectivity.

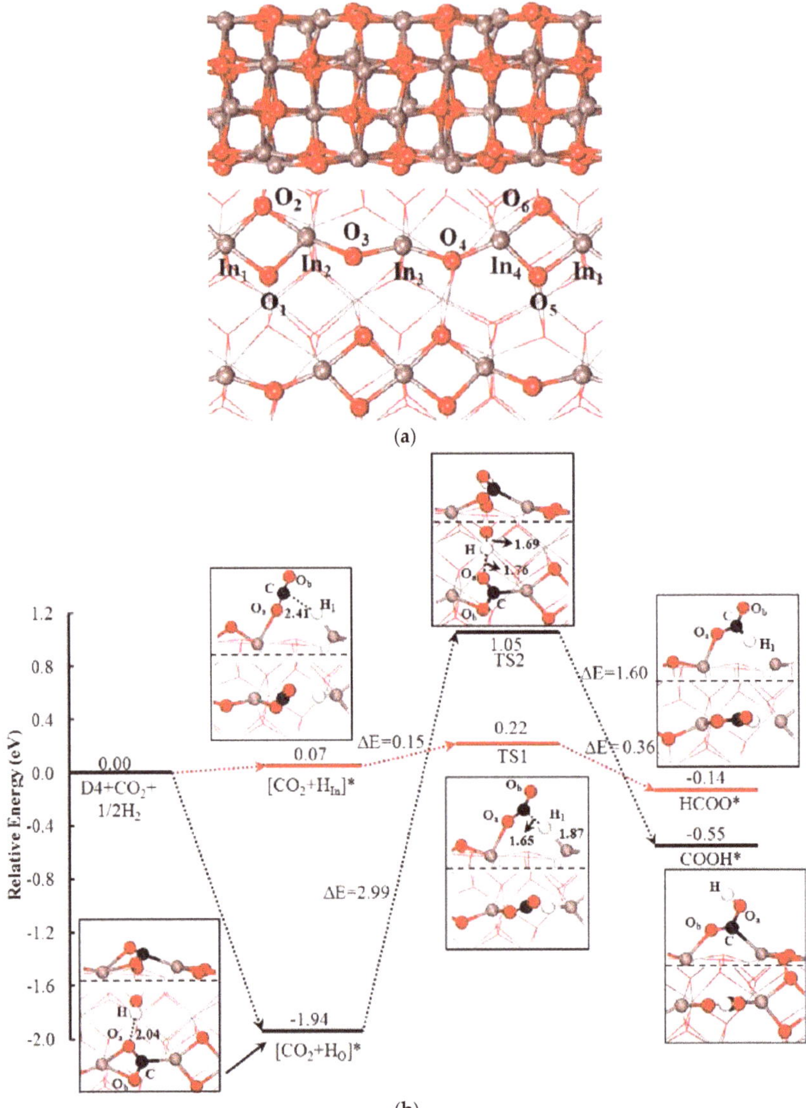

Figure 11. (**a**) Optimized structure of the In_2O_3 (110) surface. Red: O atoms; brown: In atoms. (**b**) Potential energy profiles of CO_2 hydrogenation and protonation on the D4 defective In_2O_3 (110) surfaces. Red line: hydrogenation; black line: protonation. A* represents the adsorption state of A on the surface, while [A+B]* represents the co-adsorption state of A and B on the surface, reproduced with permission from [47]. Copyright American Chemical Society, 2013.

Interestingly, Wang et al. synthesized a series of x% ZnO-ZrO_2 solid solution catalysts (x% represents the molar ratio of Zn) for CO_2 direct hydrogenation to CH_3OH [50]. Under the specified reaction conditions (5.0 MPa, H_2/CO_2 = 3:1 to 4:1, 593 K to 588 K), ZnO-ZrO_2 catalyst can achieve 86–91% methanol selectivity with single-pass CO_2 conversion more than 10%, better than the results reported by other researchers. Moreover, in the presence of 50 ppm SO_2 or H_2S in the reaction stream, no deactivation was observed. Experimental results confirmed that ZrO_2 and ZnO alone showed little activity in methanol synthesis, while the catalytic performance was significantly enhanced and CO_2 conversion reached the maximum value when the Zn/(Zn + Zr) molar percentage is close to 13%. Demonstrated by CO_2-TPD, most of the CO_2 adsorbed on the Zr sites of ZnO-ZrO_2 catalyst, and the H_2-D_2 exchange experiment indicated that ZnO had much higher activity than ZrO_2. Therefore, in ZnO-ZrO_2 solid solution catalyst, the synergetic effect between the Zn and Zr sites markedly promotes the activation of H_2 and CO_2. To explain the reaction mechanism on the solid solution catalyst (ZnO-ZrO_2), in situ diffuse reflectance infrared Fourier transform spectroscopy (DRIFTS) and density functional theory (DFT) calculations were conducted. It was concluded that CO_2 direct hydrogenation to CH_3OH is dominated via the formate pathway, and CH_3OH was formed by H_3CO^* protonation on the surface of ZnO-ZrO_2 catalyst.

Although Cu-ZnO-based catalysts are highly selective for CO_2 conversion to methanol, there are still some serious problems during industrial operation, such as deactivation of active sites at high temperature and agglomeration of catalytic particles under industrial reaction conditions. To overcome these issues, researchers designed a series of bimetallic catalysts for the direct hydrogenation of CO_2 to CH_3OH. Jiang et al. prepared a series of Pd-Cu bimetallic catalysts supported on SiO_2 with a wide range of total metal loading (i.e., 2.4–18.7 wt %) and evaluated the catalytic performance for the reduction of CO_2 to CH_3OH. With a decrease in metal loadings, the CO_2 conversion dropped stepwise, namely from 6.6 to 3.7%. However, the CH_3OH STY was slightly enhanced by 31% from 0.16 to 0.21 mmol mol^{-1} s^{-1} [51]. To unclose the reaction mechanisms, the authors carried out in situ diffuse reflectance FT-IR (DRIFTS) measurements on Pd(0.34)-Cu/SiO_2 catalyst [52]. The resulting spectra identified that the dominant species on a bimetallic surface were formate and carbonyl species on a bimetallic surface, which were dependent on the catalyst composition. Therefore, they proposed that on Pd-Cu catalysts, the surface coverage of formate species was correlated to the methanol promotion, implying its vital role in CH_3OH synthesis. Bahruji et al. prepared PdZn/TiO_2 bimetallic catalysts by a solvent-free chemical vapor impregnation method [53]. According to the order in which the metals were impregnated, the catalysts could be classified as ^{2}Pd-^{1}Zn-TiO_2, ^{2}Zn-^{1}Pd-TiO_2, and PdZn/TiO_2 (1 and 2 represent the order of sequential metal impregnations). The experimental results showed that PdZn/TiO_2 catalyst achieved optimal catalytic performance, 10.1% CO_2 conversion and 40% CH_3OH selectivity. The formation of $ZnTiO_3$ and PdZn nanoparticles was confirmed via XPS, XRD, and TEM. Combining the experimental results, the authors proposed that PdZn nanoparticles were beneficial for methanol formation and catalytic stability. Although it is hard to assign the formation of the PdZn alloy to the differences between the preparation methods, an interaction between $Pd(acac)_2$ and $Zn(acac)_2$ precursors might be responsible for the production of the PdZn active site in terms of PdZn/TiO_2 catalyst. Additionally, Yin et al. reported the preparation of Pd@zeolitic imidazolate framework-8 (ZIF-8) catalyst for the conversion of CO_2 to CH_3OH [54]. The optimal methanol yield can reach 0.65 g $g_{cat}{}^{-1}$ h^{-1} over a PdZn catalyst. As demonstrated by XPS, XRD, TEM and electron paramagnetic resonance (EPR), after H_2 reduction, PdZn alloy particles formed, and abundant oxygen defects existed on the ZnO surface. The experimental results revealed that the active site is a PdZn alloy rather than metallic Pd, in terms of CH_3OH formation.

Numerous studies suggested that promoters have an indispensable role in the catalytic performance of CuZn catalysts [55], although the mechanism of action still remains unclear. Pd-doped CuZn catalysts were prepared to evaluate the surface modification effect. An interesting observation was made from the volcano-shaped relationship between CH_3OH STY and Pd loading, which implied that an appropriate amount of Pd loading is beneficial for methanol synthesis. The chemisorption analyses further revealed a strong interaction between Pd and Cu surface, where a hydrogen spillover has taken place, generating more activated Cu sites. Hence, the CH_3OH STY and the methanol turnover frequency (TOF) improved a lot with Pd loading of 1 wt % [55]. In addition, a variety of bimetallic catalysts (e.g., Cu-Fe, Cu-Co, Cu-Ni, Co-Ga, Ni-Ga) have been evaluated by many researchers [56–59]. A possible catalytic reaction mechanism of CO_2 hydrogenation over Cu−Ni/CeO_2−NT is shown in

Figure 12. Details of the CO_2 conversion and product selectivity, along with reaction conditions of several representative catalytic systems are compared in Table 3. In terms of hydrogenation of CO_2 to CH_3OH, In-based and Pd-based catalysts have gradually drawn researchers' attention, besides traditional Cu-based catalysts. Additionally, many researchers demonstrated that 250 °C is the optimal temperature to produce CH_3OH with the high selectivity. According to the current experimental results, increasing the reaction pressure is also a good strategy to improve CO_2 conversion, but a high pressure means a high cost of operation and equipment.

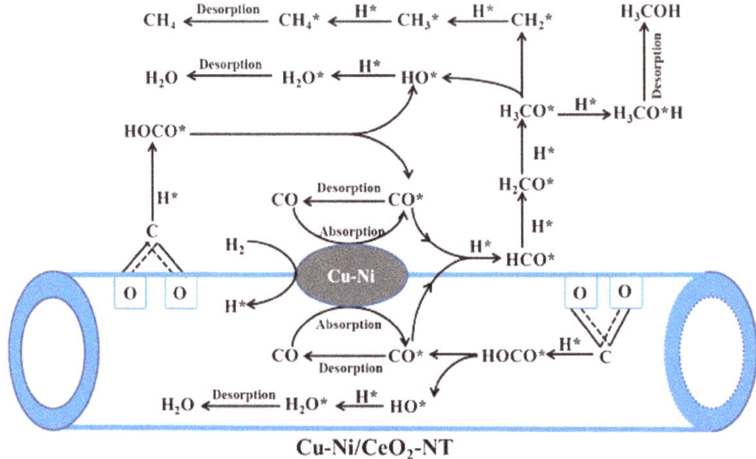

Figure 12. Possible catalytic mechanism of CO_2 hydrogenation over Cu−Ni/CeO_2−NT, reproduced with permission from [59]. Copyright American Chemical Society, 2018.

Table 3. Catalytic performance of several catalytic systems for CO_2 hydrogenation into CH_3OH, in terms of CO_2 conversion and CH_3OH selectivity, along with the reaction conditions.

Catalyst	H_2:CO_2	GHSV	Temperature (°C)	Pressure (MPa)	CO_2 Conversion (%)	CH_3OH Selectivity (%)
Cu-ZnO [42]	2.9	2160 ^{a}	250	3	29.2	83.6
Cu/g-C_3N_4-Zn/Al_2O_3 [44]	3	6800 ^{a}	250	1.2	~7.0	~55.0
CuO-ZnO-ZrO_2-GO [45]	3	15,600 ^{a}	240	2	N/A	75.8
W-Cu-Zn-Zr [46]	2.7	2400 ^{a}	240	3	19.7	49.3
In_2O_3 [48]	3	15,000 ^{a}	270	4	1.1	54.9
In_2O_3 [48]	3	15,000 ^{a}	330	4	7.1	3
In_2O_3 [49]	4	16,000–48,000 ^{b}	300	5	N/A	100
ZnO-ZrO_2 [50]	3-4	24,000 ^{a}	315-320	5	>10	86–91
Pd-Zn-TiO_2 [53]	3	916 ^{b}	250	2	10.3	61
Pd-Zn-ZIF-8 [54]	3	21,600 ^{a}	270	4.5	~22.0	~50.0
Pd-Cu-Zn [55]	3	10,800 ^{a}	270	4.5	~8.0	~65.0
Co_5Ga_3 [58]	3	N/A	250	3	1	63
$CuNi_2$/CeO_2-NT [59]	3	6000 ^{b}	260	3	17.8	78.8
Cu-Zn-SiO_2 [60]	3	2000 ^{a}	220	3	14.1	57.3
Ni_5Ga_3/SiO_2/Al_2O_3/Al [61]	3	3000 ^{a}	210	0.1	~1.0	86.7
Cu-Zr-SiO_2 [62]	3	N/A	230	5	N/A	77
Cu/Mg/Al [63]	2.8	2000 ^{b}	200	2	3.6	31
Cu-Ce-Zr [64]	3	7500 ^{a}	250	3	14.3	53.8
Cu-TiO_2 [65]	3	3600 ^{a}	260	3	N/A	64.7
Cu-Zn-Mn-KIT-6 [66]	3	120,000 ^{a}	180	4	8.2	>99.0
Cu-SBA-15 [67]	3	N/A	210	2.2	13.9	91.3
Au-CuO/SBA-15 [68]	3	3600 ^{b}	250	3	24.2	13.5
Cu-Zr-SBA-15 [69]	3	N/A	250	3.3	15	N/A
Pd/In_2O_3 [70]	4	>21,000 ^{a}	300	5	>20.0	>70.0

^{a} mL g_{cat}^{-1} h^{-1}; ^{b} h^{-1}; N/A: not available.

2.4. CO_2 to Other Products

Besides CO, CH_4, and CH_3OH, dimethyl ether (DME), light olefins [71–74], alcohol [75], isoparaffins [76], and aromatics [77,78] have also been produced by CO_2 hydrogenation. Clearly, in terms of CO_2 conversion, the coupling of C-C bond to produce higher hydrocarbons and oxygenates is a technological barrier. However, taking advantage of tandem catalysis to realize one-step synthesis of hydrocarbons via hydrogenation of CO_2 is feasible. The extensive studies can be mainly categorized into two categories: methanol-mediated and non-methanol-mediated reactions [79,80]. In the methanol-mediated approach, DME is usually produced as main product, while for the non-methanol-mediated approach, alkenes and alkanes are generally produced as main products. The possible reaction mechanism for CO_2 hydrogenation to DME and light olefins is shown in Scheme 4.

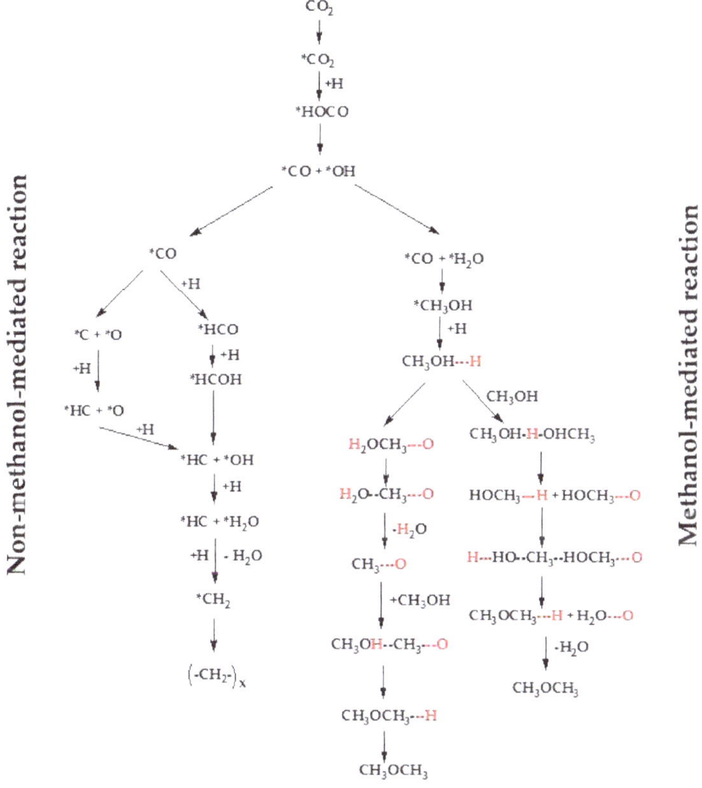

Scheme 4. Possible reaction pathways for CO_2 hydrogenation to DME and light olefins. Note: ---H, ---O derived from H-ZSM5 zeolite, reprinted with permission from [21]. Copyright Elsevier, 2016.

Cu-based catalysts are in practice used for methanol synthesis, and HZSM-5 zeolites are widely employed for methanol dehydration due to its solid acid catalysis. Therefore, a variety of studies have been reported with regard to the bi-functional catalysts consisting of Cu and HZSM-5 used for the direct hydrogenation of CO_2.

Zhang et al. reported a Cu-ZrO_2/HZSM-5 catalyst promoted by Pd/CNT for direct synthesis of DME from CO_2/H_2 [79]. Although a minor amount of the Pd-decorated CNTs into the CuZr/HZSM-5 catalytic system caused little change in the activation energy for CO_2 conversion compared with the CuZr/HZSM-5 catalyst, the former created a micro-environment including higher concentration of active H species and adsorbed CO_2 species on the surface of the catalyst system. Owing to the increase of hydrogenation reactions, under reaction conditions of 5 MPa and 523 K, the specific rate of CO_2 hydrogenation-conversion was 1.22 times higher than the pure CuZr/HZSM-5 catalyst. Furthermore, Zhang et al. reported a series of V-modified CuO-ZnO-ZrO_2/HZSM-5 catalysts which were prepared via an oxalate co-precipitation method for the synthesis of DME from CO_2/H_2 [81]. The catalytic performances of the catalysts were strongly dependent on the content of V, which was confirmed by XRD, N_2O chemisorption, and X-ray photoelectron spectroscopy (XPS). Similarly, the influence of the promoter—i.e., La and W—has also been examined. The optimal catalytic activity was obtained (43.8% CO_2 conversion and 71.2% DME selectivity) when the amount of La was 2 wt %. In contrast, the Cu-ZnO-ZrO_2 catalyst admixed with WO_x obtained 18.9% CO_2 conversion and 15.3% DME selectivity. Obviously, La is a better promoter compared with W, and this can be explained by the fact that the reducibility and dispersion of bi-functional catalysts (CuO-ZnO-Al_2O_3-La_2O_3/HZSM-5) were largely dependent on the modification of La, while the hybrid catalyst (Cu-ZnO-ZrO_2-WO_x/Al_2O_3) modified by W strongly adsorbed water molecules, resulting in a lower catalytic performance. Moreover, as shown in Figure 13, the STY of DME of different WO_x/Al_2O_3 catalysts (i.e., on supports with different pore sizes) as a function of W surface density shows a volcanic curve relation. Details of CO_2 conversion and DME selectivity reported recently are shown in Table 4 [82,83]. The above studies indicate that the current catalysts used for DME preparation from CO_2/H_2 are still dominated by Cu-based-HZSM-5 catalysts. Some other zeolite catalysts (SBA-15, SAPO-5, SUZ-4) with similar acid properties (acid content and acid strength) as HZSM-5 zeolite may be a direction of future research.

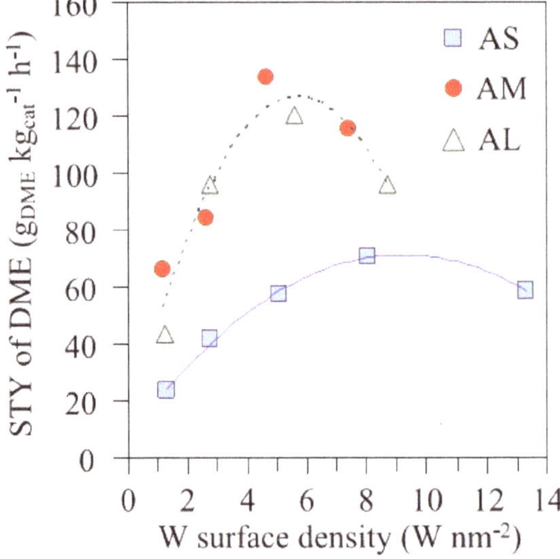

Figure 13. STY of DME of different WO_x/Al_2O_3 catalysts, with different support pore size as a function of W surface density. AS, AM, and AL refer to the mean pore size of alumina supports with small, medium and large pores, respectively, reproduced with permission from [83]. Copyright Elsevier, 2018.

Among the large-scale technologies with industrial potential, the conversion of CO_2 to DME is promising and relatively mature [84]. Korea Gas Corporation (KOGAS) has developed a DME plant, with CO_2 as a raw material. The schematic diagram of the KOGAS tri-reforming process is shown in Figure 14 [85,86]. Kansai Electric Power Co. and Mitsubishi Heavy Industries have realized a bench-scale (100 cm^3 catalyst loading) experiment for DME synthesis [87,88]. However, during the reaction process, water formation decreased the yield of DME [89–91]. Catizzone et al. and Bonura et al. evidenced that zeolites or ferrierite could effectively mitigate the influence of water and avoid catalyst sintering [89,90]. In a fixed bed reactor, kinetic modeling confirmed the negative effect of water (formed during the reaction process) on DME production, which is consistent with the experimental results. Therefore, Falco et al. suggested that hydrophilic membranes could be promising for industrial production of DME [92]. Recently, Fang et al. advocated CO_2 capture and conversion by using a membrane reactor system, where a high-temperature mixed electronic and carbonate-ion conductor (MECC) membrane was used for CO_2 capture and a solid oxide electrolysis cell (SOEC) was used for CO_2 reduction [93]. Furthermore, based on membrane technology, Sofia et al. carried out a techno-economic analysis project of power and hydrogen co-production via an integrated gasification combined cycle (IGCC) plant with CO_2 capture [94]. The development of membrane technology would be an advantageous factor for the capture and utilization of CO_2.

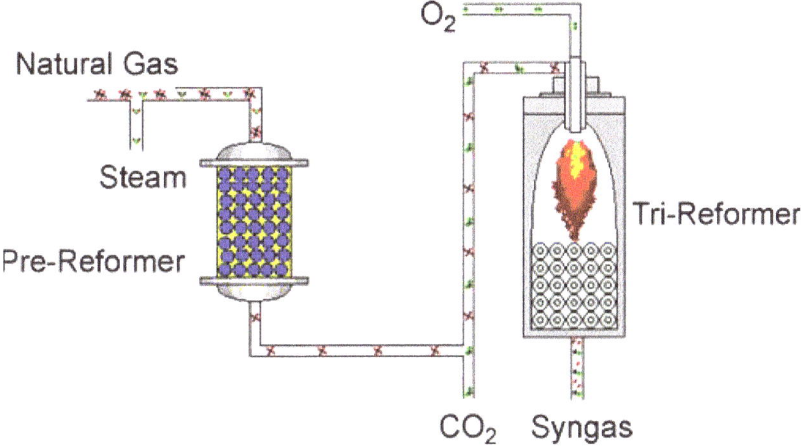

Figure 14. Schematic diagram of KOGAS tri-reforming process, reproduced with permission from [86]. Copyright Elsevier, 2009.

Li et al. synthesized a novel ZnZrO/SAPO tandem catalyst, combined by a ZnO ZrO_2 solid solution and a Zn-modified SAPO-34 zeolite, which achieved 80–90% olefin selectivity among the hydrocarbon products [95]. Based on the surface reaction kinetics, they proposed that the tandem reaction process proceeded as follows: (1) generation of CH_xO species on ZnZrO via CO_2 reduction; (2) olefins production from the derived CH_xO species which migrate/transfer onto SAPO zeolite pore structure. Moreover, experimental results confirmed that the excellent selectivity can be ascribed to the effective synergy between ZnZrO and SAPO for the tandem catalyst. In addition, this catalyst (ZnZrO and SAPO-34) shows an excellent stability toward thermal and sulfur treatments, indicating the potential value for industrial application.

Recently, CO_2 was converted to aromatics with a selectivity up to 73%, at 14% CO_2 conversion over ZnZrO/HZSM-5 catalyst [77]. Demonstrated by operando infrared (IR) characterization, Li et al. proposed that CH_xO, as an intermediate species, transformed from the ZnZrO surface into the pore structure of HZSM-5, which is responsible for C-C bond formation for aromatics production. The reaction scheme is shown in Figure 15. Interestingly, the presence of H_2O and CO_2 markedly suppressed the generation of polycyclic aromatics, consequently, enhanced the stability (100 h in the reaction stream) of the tandem catalyst, which showed potential industrial application prospects. Similarly, Ni et al. synthesized a tandem catalyst of $ZnAlO_x$ and HZSM-5, which yields 73.9% aromatics selectivity and 0.4% CH_4 selectivity among the carbon products without CO [78]. Confirmed by XRD, SEM, TEM, and element distribution analysis, ZnAlOx was formed and uniformly dispersed in the tandem catalyst. Furthermore, demonstrated by 2,6-di-tert-butyl-pyridine absorption (DTBPy-FTIR), the external Brønsted acid of HZSM-5 can be shielded by $ZnAlO_x$, which is beneficial to aromatization. According to the operando diffuse reflectance infrared Fourier transform spectroscopy (DRIFTS) study, they proposed the following possible reaction mechanism: MeOH and DME, produced by hydrogenation of formate species, are transmitted to HZSM-5 and then converted into olefins and finally aromatics.

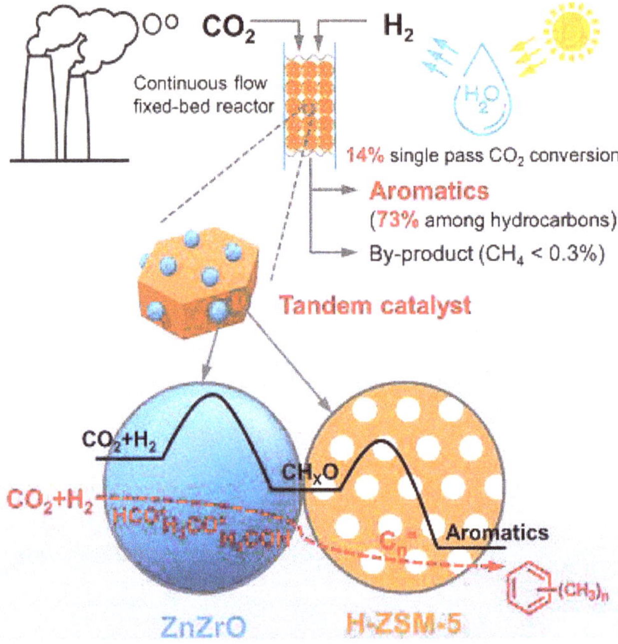

Figure 15. Reaction scheme of CO_2 hydrogenation to aromatics, reproduced with permission from [77]. Copyright Elsevier, 2019.

Non-methanol-mediated reactions, i.e., the direct hydrogenation of CO_2 to light olefins is even more significant than CH_3OH synthesis from CO_2 hydrogenation, since a large proportion of methanol is used for the synthesis of olefins in industry, through the methanol-to-olefins (MTO) reaction by using SAPO-34 zeolite catalysts. CO_2 to lower olefins can be realized by coupling of the RWGS reaction and F–T synthesis, as shown in Schemes 1 and 3.

Iron-based catalysts has been considered as an excellent option for the synthesis of light olefins from CO_2/H_2, mainly because of their excellent activity, high selectivity, and low price. Using honeycomb-structure graphene (HSG) as the support and K as a promoter, Wu et al. prepared an iron-based catalyst [80]. They found out that the confinement effect of the porous HSG was beneficial for the sintering of the Fe active sites, and within 120 h stability test, no significant deactivation occurred. In addition, as confirmed by CO_2 temperature programmed desorption (CO_2-TPD) and H_2-TPD, they found out that potassium as a promoter effectively enhanced the chemisorption and activation of the reactants CO_2 and H_2. Moreover, as revealed by ^{57}Fe Mossbauer absorption spectroscopy, for the Fe/HSG catalyst, there were two doublets with isomer shift (IS) of 0.96 mm s^{-1} and quadrupole splitting (QS) of 0.28 mm s^{-1} and IS of 1.21 mm s^{-1} and QS of 1.82 mm s^{-1}, which corresponded to the Fe (II) species in low- and high-coordination environments. Instead, for the FeK1.5/HSG catalyst, only one doublet with IS of 0.31 mm s^{-1} and QS of 1.05 mm s^{-1} ascribed to the Fe (III) species, which implies that K is capable of stabilizing high valence-state iron (Fe (III)) during CO_2 hydrogenation to light olefins (CO_2–FTO). The above mentioned three factors contributed to the excellent catalytic performance, i.e., 56% olefins selectivity and a 120-h stability testing experiment (as shown in Figure 16).

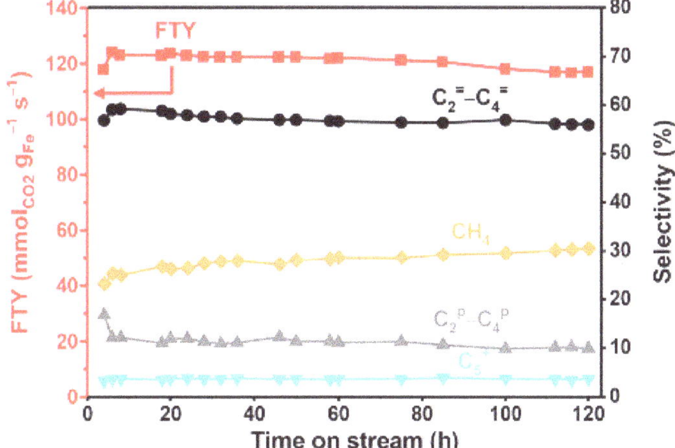

Figure 16. CO_2−FTO results over the FeK1.5/HSG catalyst during 120 h on stream (Fe time yield to hydrocarbons, termed as FTY). Reaction conditions: 0.15 g catalyst, T = 613 K, P = 20 bar, H_2/CO_2 = 3 by volume, and the space velocity of 26 L h^{-1} g^{-1}. The CO selectivity is in the range of 39−43%, reproduced with permission from [80]. Copyright American Chemical Society, 2018.

Zhang et al. used an impregnation method to prepare Fe-Zn-K catalysts, in which K was used as a promoter [71]. The experimental results showed that the Fe-Zn-K catalyst with H_2/CO reduction showed the best catalytic activity, with 51.03% CO_2 conversion and 53.58% C_2–C_4 olefins selectivity, at the reaction conditions of 593 K and 0.5 MPa. XRD, H_2-TPR, and XPS characterization results revealed that, in the Fe-Zn-K catalyst, $ZnFe_2O_4$ spinel phase and ZnO phase were formed. Among them, $ZnFe_2O_4$ spinel phase strengthens the interaction between iron and zinc, and changed the reduction and CO_2 adsorption behaviors. In addition, the H_2-TPR profiles show that the catalyst modified by K contributed to a slight shift of the initial reduction peak to higher temperature, indicating the reduction of Fe_2O_3 and formation of Fe phase was inhibited by K modification.

You et al. investigated the catalytic activity of non-supported Fe catalysts (bulk Fe catalysts) modified by alkali metal ions (i.e., Li, Na, K, Rb, Cs) for the conversion of CO_2 to light olefins [72]. By calcining ammonium ferric citrate, non-supported Fe catalyst was prepared. Compared with Fe catalyst without modification (5.6% CO_2 conversion, 0% olefins), the modification of the Fe catalyst with an alkali metal ion markedly enhanced the catalytic activity for CO_2 conversion to light olefins. For Fe catalyst modified by K and Rb, the conversion of CO_2 and the olefin selectivity (based on only the hydrocarbon compounds, without CO) increased to about 40% and 50%, respectively, and the yield of light (C_2–C_4) olefins can reach 10–12%. Further investigation via XRD showed that over the alkali-metal-ion-modified Fe catalysts, Fe_5C_2 was formed, while in the unmodified catalyst, iron carbide species were not observed after the reaction. Accordingly, they proposed that in the presence of an alkali metal ion, the generation of iron carbide species was one possible reason for the enhanced catalytic activity. Therefore, modification by alkali metal ions remains a good strategy to tune the product distribution.

Additionally, Wei et al. reported a highly efficient, stable and multifunctional Na–Fe_3O_4/HZSM-5 catalytic system [76], for direct hydrogenation of CO_2 to gasoline-range (C_5–C_{11}), in which the selectivity of C_5–C_{11} hydrocarbons reached 78% (based on total hydrocarbons), while under industrial conditions only reached 4% methane selectivity at a 22% CO_2 conversion. Characterization by high resolution transmission electron microscopy (HRTEM), X-ray diffraction (XRD), and Mossbauer spectroscopy showed that two different types of iron phase—i.e., Fe_3O_4 and x-Fe_5C_2—were discerned in the spent Na–Fe_3O_4 catalyst, which cooperatively catalyzed a tandem reaction (RWGS and FT).

The possible reaction scheme for CO_2 hydrogenation to gasoline-range hydrocarbons is shown in Figure 17. Furthermore, acid sites existing in the HZSM-5 zeolite were favorable for acid-catalyzed reactions (oligomerization, isomerization, and aromatization). Notably, the multifunctional catalyst exhibited a significant stability for 1000 h on stream, showing the potential as promising industrial catalyst material for CO_2 conversion to liquid fuels. Similarly, Gao et al. investigated a bi-functional catalytic system composed of In_2O_3 and HZSM-5, which can achieve 78.6% C_{5+} selectivity with only 1% CH_4 at a 13.1% CO_2 conversion [73]. As demonstrated by Ye et al. [47], CO_2 and H_2 can be activated in the oxygen vacancies on the In_2O_3 surfaces, and catalyzed CH_3OH formation. Subsequently, C−C coupling occurred inside the zeolite pores structure (HZSM-5) to synthesize gasoline-range hydrocarbons with a relatively high octane number (C_{5+}).

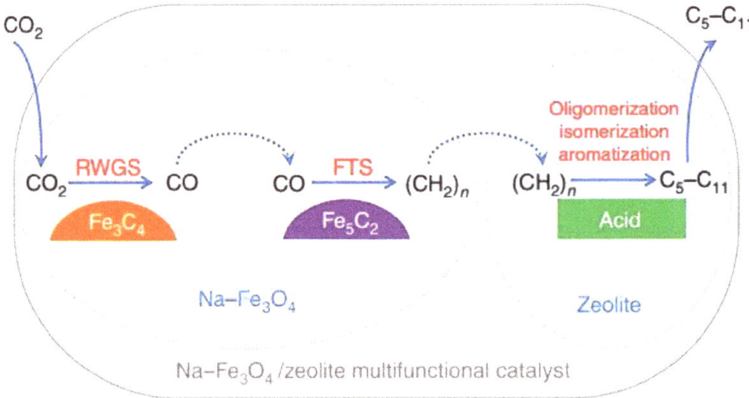

Figure 17. Reaction scheme for CO_2 hydrogenation to gasoline-range hydrocarbons, reproduced with permission from [76]. Copyright Nature Publishing Group, 2017.

Details of the conversion and selectivity for the hydrogenation of CO_2 into other products, along with the reaction conditions of several representative catalytic systems, are compared in Table 4. Fe-based catalysts are beneficial for the production of olefins, while Cu-based + HZSM-5 is still a dominant catalytic system for CO_2 conversion to DME. Similarly, 250 °C is the optimal temperature for the production of DME based on the current experimental results, and 300–400 °C, which is close to industrial production conditions, seems to be better for the direct conversion of CO_2 to olefins. On the other hand, the selectivity of DME can be maintained upon increasing GHSV (>10,000 mL g_{cat}^{-1} h^{-1}), but the high selectivity of olefins generally relies on a relatively low GHSV (<5000 mL g_{cat}^{-1} h^{-1}).

Table 4. Catalytic activity of several catalysts for CO_2 hydrogenation into other products (e.g., DME, olefins, alcohol, isoparaffins, gasoline, aromatics), in terms of CO_2 conversion and product selectivity, along with the reaction conditions.

Catalyst	H_2:CO_2	GHSV	Temperature (°C)	Pressure (MPa)	CO_2 Conversion (%)	Selectivity (%)
Fe-Zn-K [71]	3	1000 [b]	320	0.5	51.03	olefins (53.58)
K-Fe [72]	3	1200 [a]	340	2	38	light olefins (78)
In/HZSM-5 [73]	3	9000 [a]	340	3	13.1	liquid fuels (78.6)
Ce-Pt@mSi-Co [74]	3	N/A	250	N/A	~3.0	C_2–C_4 (60)
K/Cu-Zn-Fe [75]	3	5000 [b]	300	6	42.3	alcohol (56.43)
Na-Fe/HMCM-22 [76]	2	4000 [a]	320	3	26	isoparaffins (74)
ZnZrO-HZSM-5 [77]	N/A	1200 [a]	320	4	14	aromatic (73)
ZnAlOx-HZSM-5 [78]	3	2000 [a]	320	3	9.1	aromatics (74)
Cu-Zr-Pd/HZSM-5 [79]	3	25,000 [a]	250	5	18.9	DME (51.8)
Fe-K/HSG [80]	3	26,000 [a]	340	20	N/A	olefins (59)

Table 4. *Cont.*

Catalyst	H_2:CO_2	GHSV	Temperature (°C)	Pressure (MPa)	CO_2 Conversion (%)	Selectivity (%)
V-Cu-Zn-Zr/HZSM-5 [81]	3	4200 [b]	270	3	32.5	DME (58.8)
Cu-Zn-Al-La/HZSM-5 [82]	3	3000 [b]	250	3	43.8	DME (71.2)
ZnO-ZrO_2-SAPO-34 [95]	N/A	3600 [a]	380	2	12.6	olefins (80–90)
Cu-ZnO-Al_2O_3 [96]	3	10 [b]	270	5	9	DME (31)
In-Zr/SAPO-34 [97]	3	9000 [a]	380	3	26.2	C_{2+} (36.1)
Cu-Zn-kaolin-SAPO-34 [98]	3	1800 [a]	400	3	50.4	C_2–C_4 (65.3)
Fe/C-Bio [99]	3	N/A	320	1	31	C_{4-18} alkenes (50.3)
Cu-Zn-Zr/HZSM-5 [100]	3	10,000 [a]	220	3	9.6	DME (46.6)
Cu-Zn-Zr/HZSM-5 [101]	3	9000 [a]	240	3	~30	DME (~35)
Cu-Zn-Al/HZSM-5 [102]	3	4200 [b]	270	3	30.6	DME (49.02)
Cu-Zn-Al/HZSM-5 [103]	3	1800 [a]	262	3	46.2	DME (45.2)
Cu-Zn-Zr-zeolite [104]	3	10,000 [b]	240	3	24	DME (38.5)
Cu-Zn-Al/HZSM-5 [105]	3	1800 [a]	270	3	48.3	DME (48.5)
Fe-K/HPCMs-1 [106]	3	3600 [a]	400	3	33.4	olefins (47.6)
Na-Fe/HZSM-5 [107]	3	4000 [a]	320	3	22	C_5–C_{11} (78)
Fe_5C_2-/a-Al_2O_3 [108]	3	3600 [a]	400	3	31.5	C_{2+} (69.2)
Fe-Zr-Ce-K [109]	3	1000 [b]	320	2	57.34	C_2–C_4 (55.67)
Fe/C+K [110]	3	24,000 [a]	320	30	24	C_2–C_6 (36)
Co-Cu/TiO_2 [111]	3	3000 [a]	250	3	18.4	C_{5+} (42.1)

[a] mL g_{cat}^{-1} h^{-1}; [b] h^{-1}; N/A: not available.

2.5. Opportunities of Heterogeneous Catalysis for CO_2 Conversion

The relationship between products selectivity and CO_2 conversion by heterogeneous catalytic hydrogenation is shown in Figure 18. It is obvious that, in the RWGS reaction, although the CO selectivity reaches up to nearly 100%, the CO_2 conversion is low. In the methanation reaction of CO_2, the CH_4 selectivity is high enough, and the CO_2 conversion also exceeds 50%. With regard to the synthesis of CH_3OH, DME, and light olefins, the relationship between conversion and selectivity is less clear.

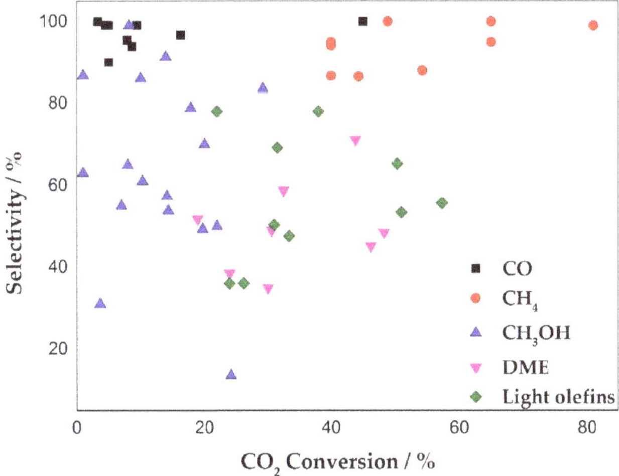

Figure 18. Selectivity and conversion distribution of different products according to recent reports in heterogeneous catalysis.

In the RWGS reaction, CO is mainly produced over Fe-, Co-, Mo-, and Pt-based catalysts (Table 1), while CO_2 methanation is typically carried out on Co- and Ni-based catalysts (Table 2). Although CO is a main component of syngas and CH_4 is an important energy resource, neither is the best product for CO_2 conversion from a relatively economic point of view. Furthermore, the conversion and utilization of CO and CH_4 is also an important research field, which also implies that CO and CH_4 are not the optimal choice as end-products of CO_2 conversion. Therefore, the coupling reaction is a potential direction for CO_2 conversion to some higher value products, i.e., CH_3OH and light olefins.

Cu-ZnO-based catalyst is still the main catalytic material for direct conversion of CO_2 to CH_3OH under industrial reaction condition, while highly innovative methodologies are awaited to achieve low-pressure and low-temperature CH_3OH synthesis processes. Considering the industrial application value of methanol, exploring novel catalytic systems and designing rational reactors to improve the catalytic activity should be targeted in the future. DME, up to now, is synthesized via Cu-based and H-ZSM5 catalyst (Table 4), and this process essentially remains a two-step tandem reaction. Hence, the bottleneck of DME production could be the deactivation of CH_3OH dehydration catalyst, due to water poisoning. The formation of coke is also the major reason of catalyst deactivation, especially in a relatively long-term operation process. To maintain high activity of catalyst for the production of DME, water- and coke-resistant catalysts need to be developed for CH_3OH dehydration.

As shown in Table 4, the direct hydrogenation of CO_2 to light olefins is mainly catalyzed by Fe-based catalysts. However, the starting point of current research work is mainly the coupling of RWGS, F–T and/or methanol-to-olefins (MTO) reactions. Therefore, suitable catalytic materials are sought for the coupling of C-C bonds, which is an effective strategy for improving the catalytic activity.

3. Plasma Catalysis

Plasma, the 'fourth state of matter', consists of electrons, neutral species (i.e., molecules, radicals, and excited species) and ions. Plasma can be in so-called thermal equilibrium or not, based on which it is subdivided into 'thermal plasma' and 'non-thermal plasma' (NTP) [13,112,113]. In non-thermal plasma, the gas temperature remains near room temperature, while electrons temperature is extremely high, usually in the range of 1–10 eV (~10,000–100,000 K). The latter is enough to activate stable gas molecules into reactive species (e.g., radicals, excited atoms, molecules, and ions). These reactive species, especially the radicals, can trigger reactions at low temperature. That is, NTP offers a unique approach to enable thermodynamically unfavorable chemical reactions to proceed at low temperature by breaking thermodynamic limits. Nevertheless, the control of selectivity of desired products in plasma is extremely difficult, since the reactions in plasma are mainly triggered through nonselective collision between active species (radicals, molecules, atoms, and ions).

To improve the desired product selectivity of the reactions in plasma, the combination of catalysts with plasma technology (so-called plasma catalysis) is a promising strategy, since catalysts usually have a special feature of regulating product distribution.

NTP can be generated through various types of discharges—i.e., microwave discharges, glow discharges, gliding arc discharges, dielectric barrier discharges (DBD), etc.—but DBD are the best option to be used in plasma catalysis. Indeed, DBDs are usually operated at atmospheric pressure and ambient temperature, and the integration of DBD plasma with catalysts has the advantages of simple operation and low cost. Thus plasma-driven direct hydrogenation of CO_2 is mostly based on DBD plasmas. The possible plasma/catalyst synergism is illustrated in Figure 19 [13].

A lot of research has been performed for pure CO_2 splitting, in various types of plasma reactors, including DBD, microwave discharge and gliding arc discharge, without catalysts [114–117]. A typical experimental set-up of a DBD plasma for CO_2 decomposition is shown in Figure 20. Paulussen et al. studied the conversion of CO_2 to CO and oxygen in DBD [114], and they found that the gas flow rate is the most crucial parameter affecting the CO_2 conversion. At 0.05 L min^{-1} flow rate, 14.75 W cm^{-3} power density and 60 kHz discharge frequency, 30% CO_2 conversion was achieved. The performance might be further enhanced by optimizing the discharge parameters (i.e., power, frequency, dielectric material) or by implementing parallel reactors.

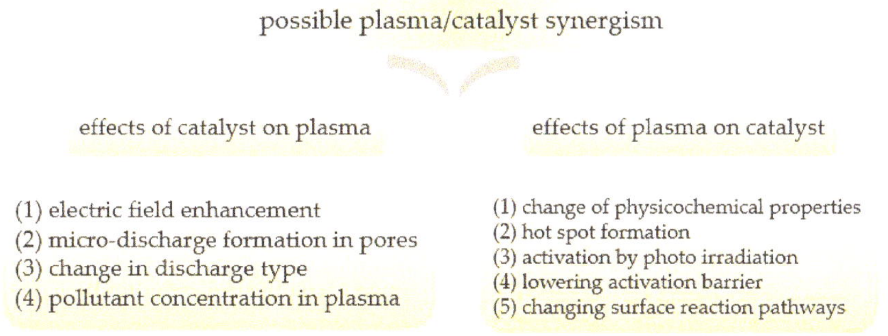

possible plasma/catalyst synergism
effects of catalyst on plasma
effects of plasma on catalyst
(1) electric field enhancement
(2) micro-discharge formation in pores
(3) change in discharge type
(4) pollutant concentration in plasma
(1) change of physicochemical properties
(2) hot spot formation
(3) activation by photo irradiation
(4) lowering activation barrier
(5) changing surface reaction pathways

Figure 19. Overview of the possible effects of the catalyst on the plasma and vice versa, possibly leading to synergism in plasma-catalysis, reproduced with permission from [13]. Copyright Royal Society of Chemistry, 2017.

AC electrode (rod)
CO2
CO
O, O2
Dielectric tube
(alumina)
Grounded
electrode (mesh)
OES
MS
CO2
200 mL.min-1
HV electrode
Grounded
electrode
V1
V2
Power
supply
10 nF
Rogowski
coil
CH1 CH2
Oscilloscope

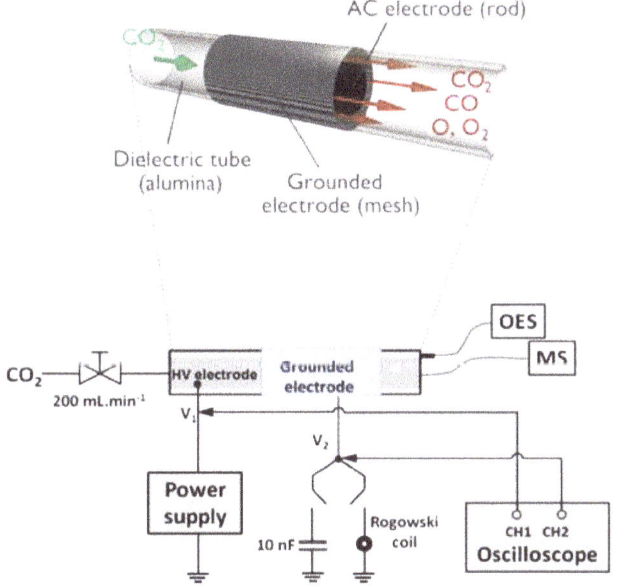

Figure 20. Schematic diagram of the experimental set-up used in the decomposition of CO_2, reproduced with permission from [116]. Copyright IOP Publishing, 2016.

Dry reforming of methane (DRM), i.e., the combined conversion of CH_4 and CO_2 by plasma and/or plasma catalysis, has attracted extensive attention in recent years. For instance, Li et al. studied CO_2 reforming of CH_4 by taking advantage of atmospheric pressure glow discharge plasma [118]. Liu et al. reported high-efficient conversion of CO_2 and CH_4 in AC-pulsed tornado gliding arc plasma [119]. Kolb et al. investigated DRM in a DBD reactor [120]. In the above-mentioned studies [118–120], syngas (CO and H_2) was produced as the main product. Recently, Wang et al. reported a novel one-step reforming of CO_2 and CH_4 into liquid oxygenate products, dominated by acetic acid, at room temperature by the coupling of DBD plasma and catalysts [121]. They examined the effect of CH_4/CO_2 molar ratio and of various catalysts (γ-Al_2O_3, Cu-γ-Al_2O_3, Au-γ-Al_2O_3, Pt-γ-Al_2O_3). Interestingly, compared with plasma-only mode, the coupling of plasma and catalysts tuned the selectivity of liquid chemicals, and oxygenates selectivity was achieved up to approximately 60% when Cu catalyst was used. Details about the effects of operating mode and catalysts on the CO_2 conversion reaction results are shown in Figure 21. Although much efforts need to be made to reveal the unknown mechanisms, the results are attractive and show an excellent application prospect.

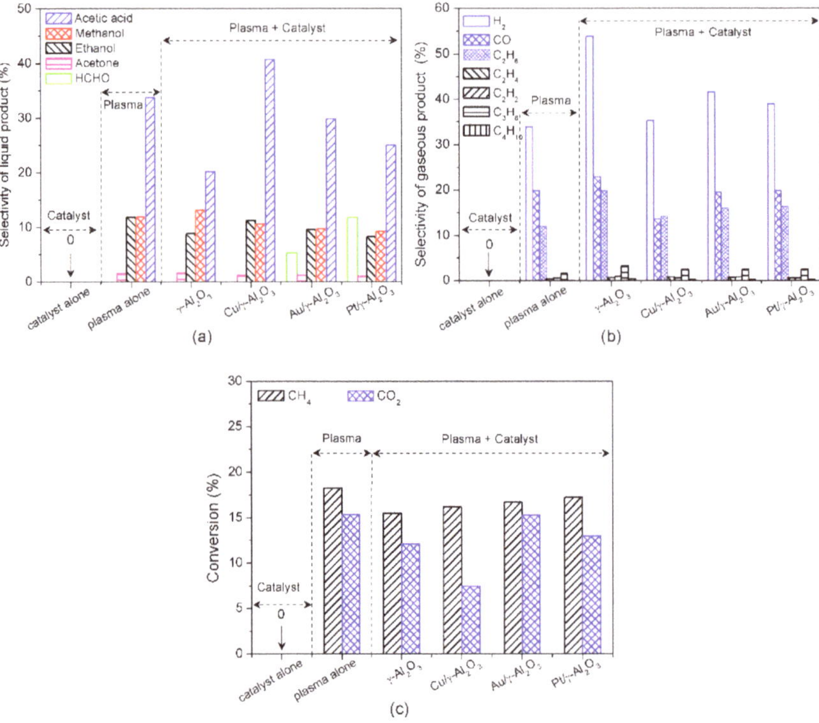

Figure 21. Effect of operating modes and catalysts on the reaction: (**a**) selectivity of liquid oxygenates, (**b**) selectivity of gaseous products, (**c**) conversion of CH_4 and CO_2 (total flow rate 40 mL min^{-1}, discharge power 10 W, catalyst ca. 2 g), reproduced with permission from [121]. Copyright Wiley online library, 2017.

Besides the above-mentioned metal catalysts, zeolites have also been used in plasma catalytic DRM. Zhang et al. studied the catalytic performance of zeolite catalysts (i.e., NaA, NaY, and HY) for the direct conversion of CH_4 and CO_2 at relatively low temperature range and ambient pressure via DBD plasma [122], and the products were dominated by syngas and C_4 hydrocarbons. Form the investigated catalysts (NaA, NaY, and HY), HY zeolite catalyst exhibited the best performance (26.7% CO_2 conversion and 52.1% C_4 hydrocarbon selectivity), which was mainly attributed to the appropriate pore size and electrostatic properties of HY zeolite.

Although direct hydrogenation of CO_2 to CH_3OH is an exothermic reaction, studies of heterogeneous catalysis for CO_2 hydrogenation to CH_3OH were usually operated at high temperature and high pressure, mainly caused by the high stability of CO_2 and the low equilibrium constant at atmospheric pressure. Plasma catalysis, however, is a promising approach to enable CO_2 conversion to CH_3OH at ambient conditions, and has gradually attracted more and more interest. For instance, Eliasson et al. reported the direct hydrogenation of CO_2 to CH_3OH by coupling of DBD plasma and a discharge-activated catalyst (CuO-ZnO-Al_2O_3) [123]. By comparison of the experiments, with catalyst only, discharge only, and discharge + catalyst, they found that DBD plasma effectively lowered the optimal reaction temperature corresponding to the best catalytic performance. Indeed, the optimal reaction temperature was 493 K for catalyst only, while in terms of discharge only and discharge + catalyst, the optimal reaction temperature for CO_2 conversion was 373 K. The maximum selectivity for the methanol formation (10%) was achieved at a temperature of 373 K, which implies that the plasma improved the catalytic activity of CuO-ZnO-Al_2O_3 catalyst at low temperature. Additionally, the authors found that low input power and high pressure are beneficial for the improvement of the methanol selectivity.

Zeng et al. studied CO_2 hydrogenation by combining various catalysts (i.e., Cu/γ-Al_2O_3, Mn/γ-Al_2O_3, and Cu–Mn/γ-Al_2O_3) with DBD plasma in a coaxial packed-bed [124]. The experimental results showed that the addition of catalysts in the reactor improved the conversion of CO_2. At the same time, with the increase of the H_2/CO_2 molar ratio, the CO_2 conversion was improved, and the CO yield was also enhanced. It is also worth mentioning that, compared with plasma only experiments, the energy efficiency was enhanced by adding catalysts, although the synergetic mechanism between catalysts and plasma is still unknown.

Recently, Wang et al. examined the influence of plasma reactor structure and catalysts on CO_2 conversion and CH_3OH selectivity for plasma catalytic CO_2 hydrogenation [125], and a schematic diagram of the experimental setup and images of the H_2/CO_2 discharge are shown in Figure 22. They investigated three kinds of reactors, i.e., a cylindrical reactor (aluminum foil sheet as ground electrode), a double dielectric barrier discharge reactor (water as ground electrode), and a single dielectric barrier discharge reactor (water as a ground electrode). The single DBD reactor equipped with a special water-electrode showed the optimal reaction performance (21.2% CO_2 conversion and 53.7% CH_3OH selectivity). In addition, they tested the catalytic performance of Pt/γ-Al_2O_3 catalysts and Cu/γ-Al_2O_3 catalysts in the optimized reactor for direct conversion of CO_2 to CH_3OH. As shown in Figure 23, both Pt/γ-Al_2O_3 catalysts and Cu/γ-Al_2O_3 catalysts improved not only the CO_2 conversion, but also the CH_3OH selectivity, and Cu/γ-Al_2O_3 catalysts showed a better catalytic performance (21.2% CO_2 conversion and 53.7% methanol selectivity). That is, a strong synergistic effect between the plasma and Cu/γ-Al_2O_3 catalysts promoted the hydrogenation of CO_2 to methanol, although the reaction temperature in the optimized reactor remained near room temperature.

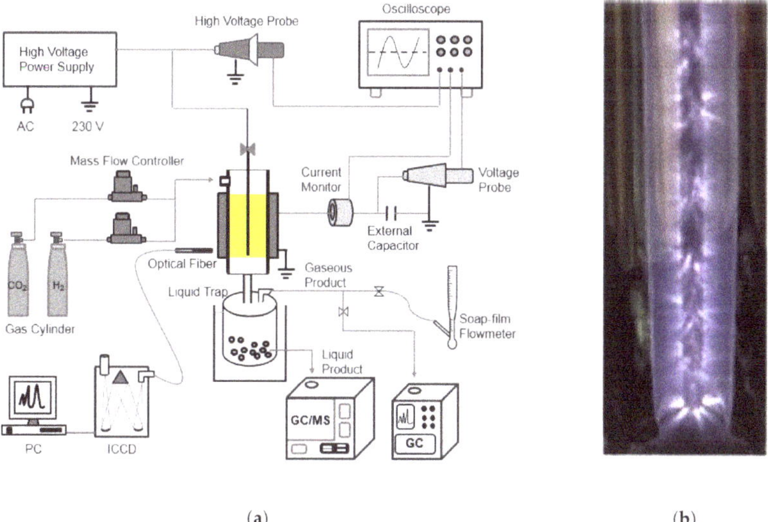

(a) (b)

Figure 22. (**a**) Schematic diagram of the experimental setup of a DBD plasma catalytic reactor. (**b**) Images of H_2/CO_2 discharge generated in DBD reactor without catalyst, reprinted with permission from [125]. Copyright American Chemical Society, 2017.

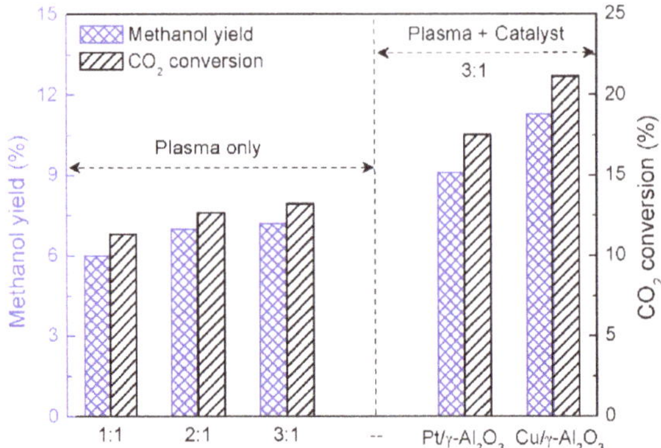

Figure 23. Effect of H_2/CO_2 molar ratio and catalysts on the reaction performance of the plasma hydrogenation process, reprinted with permission from [125]. Copyright American Chemical Society, 2017.

It is obvious that the combination of the plasma and catalysts can enhance the catalytic reaction at room temperature and atmospheric pressure. The maximum methanol selectivity of 53.7% was achieved with a CO_2 conversion of 21.2% via the plasma-catalysis process (see Figure 23). All the above studies show a strong synergistic effect between plasma and catalysts.

Using one-dimensional fluid modeling [126], De Bie et al. proposed a chemical reaction network in the plasma (i.e., without catalysts) for conversion of CO_2 to value-added chemicals, i.e., CO, CH_4, CH_2O, CH_3OH, and hydrocarbons. The simulation results indicated the dominant reaction pathways for the conversion of CO_2 and H_2, as illustrated in Scheme 5. According to the model, the combination between H atoms and CHO radicals is the most important reaction to form CO, while this reaction is counterbalanced by the reorganization of H with CO into CHO radicals. Therefore, the most effective net CO formation reaction is dissociation of CO_2 influenced by electrons. The production of CH_4 was generally driven by two reactions, i.e., three-body recombination reaction between CH_3 and H radicals, and charge transfer reaction between CH_5^+ and H_2O. However, the latter reaction is partly balanced by the loss of CH_4, resulting from a charge transfer reaction with H_3^+. The production of CH_2O is closely related to the initial CO_2 fraction in the gas mixture. At a low initial fraction of CO_2, the reaction between CO_2 and CH_2 radicals seem to be the most important channel for the formation of CH_2O, while at higher initial CO_2 fractions, CH_2O is also produced out of two CHO radicals to some extent. In addition, as predicted by the model, the most important channel for the formation of CH_3OH is the three-body reaction between CH_3 and OH radicals, while the three-body reaction between CH_2OH and H radicals is also an effective production channel for CH_3OH. Furthermore, they also found that a higher density of CH_3 and CH_2 radicals would be essential to tune the distribution of end products. Therefore, it can be predicted that the degree of hydrogenation in the reaction has a significant influence for the targeted products.

Figure 24 summarizes the selectivity of CO, CH_4, or CH_3OH, as a function of the CO_2 conversion, obtained from the recent reports discussed above. Several main trends are clear. Firstly, the main products are CO and CH_4 for direct hydrogenation of CO_2 in plasma catalysis, while the selectivity of CH_3OH is relatively low. Moreover, it is obvious that researchers have paid more attention to the reduction of CO_2 to CO and CH_4 up to now. However, the production of other hydrocarbons, such as olefins and gasoline hydrocarbons, would also be a promising direction, to achieve the maximum utilization of CO_2 by plasma catalysis. Therefore, some insights from heterogeneous catalysis, especially the combination of metal catalysts, metal oxide catalysts, and zeolite catalysts could be helpful to develop a methodology for plasma catalytic hydrogenation of CO_2. On the other hand, there is no guarantee that good catalysts in thermal processes would also perform well in plasma catalysis, because of the clearly different operating conditions (e.g., lower temperature, abundance of reactive species, excited species, charges, and electric field present in plasma).

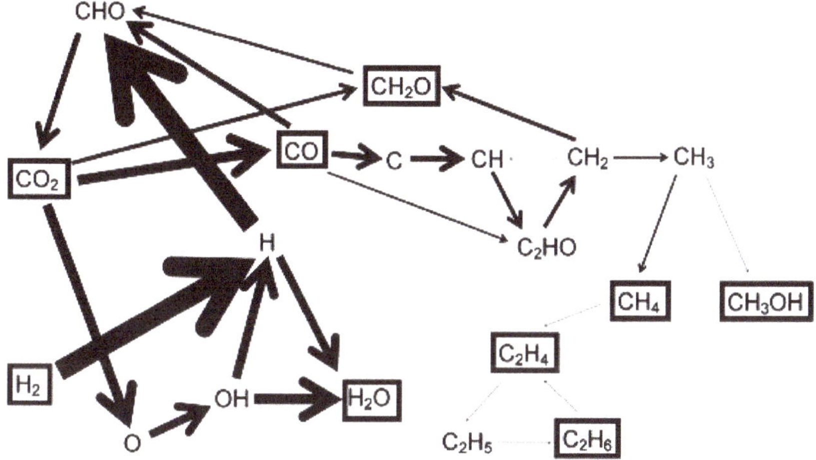

Scheme 5. Dominant reaction pathways for the plasma-based conversion (without catalysts) of CO_2 and H_2 into various products, in a 50/50 CO_2/H_2 gas mixture. The thickness of the arrow lines is proportional to the rates of the net reactions. The stable molecules are indicated with black rectangles, reprinted with permission from [126]. Copyright American Chemical Society, 2016.

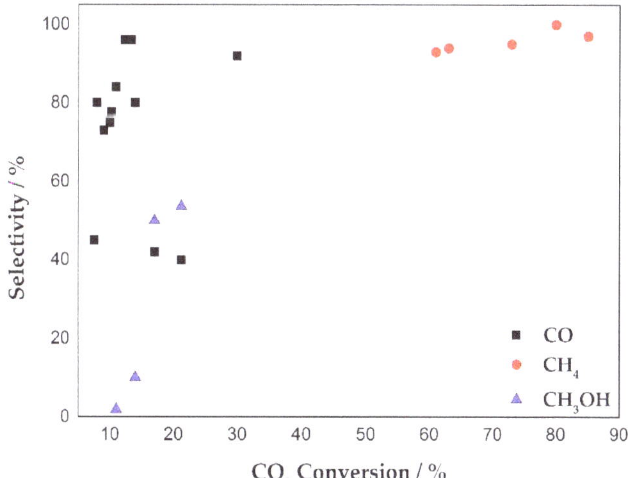

Figure 24. Overview of selectivity into CO, CH_4, and CH_3OH, as a function of CO_2 conversion, based on all reports available in literature about plasma catalysis (all in DBD reactors). The references where these data are adopted from are discussed in the text.

Up to now, in view of the complexity of this interdisciplinary field, the development of plasma catalysis still needs major research efforts. On one hand, plasma catalysis has potential for industrial application, because it drives the CO_2 conversion reaction at ambient temperature and atmospheric pressure, breaking thermodynamic equilibrium to make full use of feedstock. On the other hand, the excessive energy consumption—caused by high input power and heat loss—is an important disadvantage, but it can be mitigated with the further development of renewable energy (i.e., wind, solar, and tidal energy). Additionally, to improve the reaction activity for CO_2 conversion, it is crucial to explore the reaction mechanisms by in situ characterization and computer modeling, to improve the synergistic effect between plasma and catalysts. Finally, we need to search for suitable catalytic materials to strength the reaction performance and decrease the production costs.

4. Outlook and Conclusions

Currently, fuels and base chemicals are nearly all produced from non-renewable fossil energy (oil, natural gas, and coal), and CO_2 is generally the end product (e.g., upon burning fossil fuels) or a waste product in chemical industry. This indicates that we should use CO_2 as the main carbon source when fossil energy would get depleted in the future. Thus, in the long term, hydrogenation of CO_2 (as well as CO_2 conversion with other H-sources) into value-added chemicals and fuels is very significant, since it can close the carbon cycle, as shown in Figure 25. However, some crucial issues should be addressed in advance for application of CO_2 hydrogenation in industry.

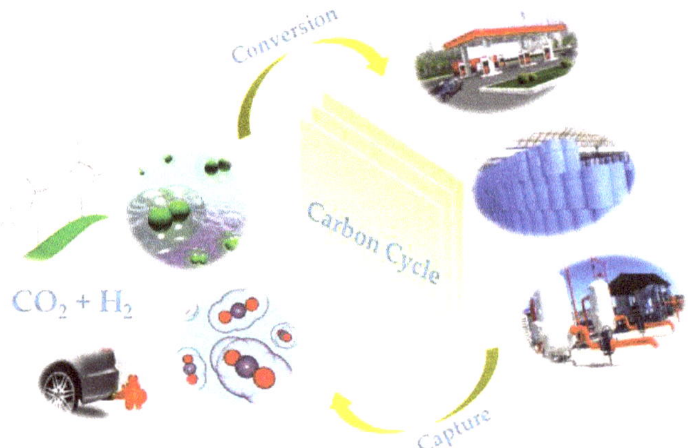

Figure 25. CO_2 conversion into fuels, which release CO_2 again upon burning, aiming a closed carbon cycle.

From a relatively economical point of view, the direct hydrogenation of CO_2 is not yet a viable approach. On one hand, CO_2 of a certain purity generally depends on the development of capture and separation techniques. On the other hand, compared with the price of the feedstock (H_2, 10,000 \$/ton), the price of the main products (i.e., liquefied natural gas, 770 \$/ton, and CH_3OH, 340 \$/ton) obtained by CO_2 hydrogenation is too low to make this an economically viable process. However, once the production of H_2 can be realized with fully-fledged solar power technology, the production cost for CO_2 hydrogenation will be greatly reduced. Meanwhile, the CO_2 conversion driven by solar energy (artificial photosynthesis) is also a promising routing.

Technologically, the direct hydrogenation of CO_2 to CO is an endothermic reaction ($\Delta H = 41.2$ kJ mol^{-1}), and a higher reaction temperature is beneficial for the production of CO according to Le Chatelier's principle. Based on the available experimental results (Table 1), the CO selectivity can reach nearly 100% under conditions of 873 K and 1 MPa. On the other hand, the production of other products (i.e., CH_4, CH_3OH, DME, and light olefins) from CO_2 is exothermic, and in theory a low temperature favors the equilibrium conversion. However, the inert CO_2 molecule generally needs relatively high reaction temperature to be activated. The competition between these two factors makes the reaction performance not very satisfactory. Up to now, from the perspective of technology readiness level, methanol synthesis is a far more advanced production process, with a high selectivity up to 80–90% with Cu-based catalyst (Table 3). However, there are some limitations in heterogeneous catalysis for direct hydrogenation of CO_2, such as catalyst sintering at high temperature, the influence of water produced in the reaction process and low CO_2 conversion caused by thermo-dynamical restriction in terms of the formation of CH_3OH and hydrocarbons. Hence, we should explore novel catalytic materials and reactors to break the thermodynamic equilibrium, as well as design bi-functional and/or multi-functional catalysts (metal, metal oxides, and zeolites) to couple the chemical reactions (RWGS, methanation, methanol synthesis, and/or MTO).

Plasma catalysis, a new field of catalysis, has attracted sufficient attention in recent years, due to its simple operating conditions (ambient temperature and atmospheric pressure) and unique advantages in activating inert molecules. However, the energy efficiency is still too high for commercial exploitation. On the other hand, renewable energy (e.g., wind, solar, and tidal energy) will further develop in the near future, and it will be a perfect match with plasma catalysis (because plasma is generated by electricity and can simply be switched on/off—allowing storage of fluctuating energy), hence making the problem of limited energy efficiency less dramatic. Nevertheless, significant efforts must be devoted to elucidate the reaction mechanisms in plasma catalysis, to improve not only the energy efficiency, but certainly also the selectivity towards value-added products. Indeed, plasma catalysis is complicated from chemistry and physics point of view, but the potential of plasma technology for industrial applications deserves major research efforts. A combination of computer simulations with experiments will be needed for an in-depth understanding of the reaction mechanisms, responsible for the synergy between plasma and catalysts. Although the development process in plasma catalysis may be slow due to its complex character, it would bring great benefits to human society, once it is developed mature enough, and once appropriate catalysts for plasma catalysis can be designed. Therefore, future research should definitely emphasize on a better understanding and rational screening of highly active catalysts.

Compared with the numerous studies on CO_2 hydrogenation by heterogeneous catalysis, much more research should be carried out in the field of plasma catalysis, to improve the CO_2 conversion and target products selectivity (and even to exploit new products, such as olefins, gasoline hydrocarbons, and aromatics), since the results achieved by plasma catalysis (Figure 24) are far from those of heterogeneous catalysis (Figure 18). To achieve this goal, the advantage of plasma should be exploited in full, and at the same time, insights from heterogeneous catalysis (e.g., catalyst combination, reaction combination, active sites design, etc.) can help to further improve the potential of this promising field of catalysis.

Author Contributions: Conceptualization: M.L., Y.Y., L.W. and A.B.; Validation: M.L., Y.Y., L.W., H.G. and A.B.; Formal analysis: M.L., Y.Y., L.W., H.G. and A.B.; Resources: Y.Y. and A.B.; Data curation: Y.Y., L.W. and A.B.; Writing—original draft preparation: M.L. and Y.Y.; Writing—review and editing: Y.Y. and A.B.; Supervision: Y.Y., L.W. and A.B.; Funding acquisition: Y.Y. and A.B.

Funding: This research was funded by the Fundamental Research Funds for the Central Universities of China (DUT18JC42), the National Natural Science Foundation of China (21503032), PetroChina Innovation Foundation (2018D-5007-0501), and the TOP research project of the Research Fund of the University of Antwerp (32249).

Conflicts of Interest: The authors declare no conflict of interest.

References

1. Ingwersen, W.W.; Garmestani, A.S.; Gonzalez, M.A.; Templeton, J.J. A systems perspective on responses to climate change. *Clean Technol. Environ. Policy* **2014**, *16*, 719–730. [CrossRef]
2. Wang, W.; Wang, S.; Ma, X.; Cong, J. Recent advances in catalytic hydrogenation of carbon dioxide. *Chem. Soc. Rev.* **2011**, *40*, 3703–3727. [CrossRef] [PubMed]
3. Irish, J.L.; Sleath, A.; Cialone, M.A.; Knutson, T.R.; Jensen, R.E. Simulations of Hurricane Katrina (2005) under sea level and climate conditions for 1900. *Clim. Chang.* **2014**, *122*, 635–649. [CrossRef]

4. Sanz-perez, E.S.; Murdock, C.R.; Didas, S.A.; Jones, C.W. Direct Capture of CO_2 from Ambient Air. *Chem. Rev.* **2016**, *116*, 11840–11876. [CrossRef] [PubMed]
5. Boot-Handford, M.E.; Abanades, J.C.; Anthony, E.J.; Blunt, M.J.; Brandani, S.; Dowell, N.M. Carbon capture and storage update. *Energy Environ. Sci.* **2014**, *7*, 130–189. [CrossRef]
6. Bui, M.; Adjiman, C.S.; Bardow, A.; Anthony, E.J.; Boston, A.; Brown, S.; Fennell, P.S.; Fuss, S. Carbon capture and storage (CCS): The way forward. *Energy Environ. Sci.* **2018**, *11*, 1062–1176. [CrossRef]
7. Bobicki, E.R.; Liu, Q.; Xu, Z.; Zeng, H. Carbon capture and storage using alkaline industrial wastes. *Prog. Energy Combust. Sci.* **2012**, *38*, 302–320. [CrossRef]
8. Rahman, F.A.; Aziz, M.M.A.; Saidur, R.; Wan, W.A.; Hainin, M.R.; Putrajava, R.; Hassan, N.A. Pollution to solution: Capture and sequestration of carbon dioxide (CO_2) and its utilization as a renewable energy source for a sustainable future. *Renew. Sustain. Energy Rev.* **2017**, *71*, 112–126. [CrossRef]
9. Darabi, A.; Jessop, P.G.; Cunningham, M.F. CO_2-responsive polymeric materials: Synthesis, self-assembly, and functional applications. *Chem. Soc. Rev.* **2016**, *45*, 4391–4436. [CrossRef]
10. Dincer, I.; Acar, C. Review and evaluation of hydrogen production methods for better sustainability. *Int. J. Hydrogen Energy* **2015**, *40*, 11094–11111. [CrossRef]
11. Centi, G.; Quadrelli, E.A.; Perathoner, S. Catalysis for CO_2 conversion: A key technology for rapid introduction of renewable energy in the value chain of chemical industries. *Energy Environ. Sci.* **2013**, *6*, 1711–1731. [CrossRef]
12. Álvarez, A.; Bansode, A.; Urakawa, A.; Bavykina, A.V.; Wezendonk, T.A.; Makkee, M.; Gascon, J.; Kapteijn, F. Challenges in the Greener Production of Formates/Formic Acid, Methanol, and DME by Heterogeneously Catalyzed CO_2 Hydrogenation Processes. *Chem. Rev.* **2017**, *117*, 9804–9838. [CrossRef]
13. Snoeckx, R.; Bogaerts, A. Plasma technology—A novel solution for CO_2 conversion? *Chem. Soc. Rev.* **2017**, *46*, 5805–5863. [CrossRef]
14. Dalle, K.E.; Warnan, J.; Leung, J.J.; Reuillard, B.; Karmel, I.S.; Reisner, E. Electro- and Solar-Driven Fuel Synthesis with First Row Transition Metal Complexes. *Chem. Rev.* **2018**. [CrossRef]
15. Li, X.; Yu, J.G.; Jaroniec, M.; Chen, X.B. Cocatalysts for Selective Photoreduction of CO_2 into Solar Fuels. *Chem. Rev.* **2018**. [CrossRef]
16. Mota, F.M.; Kim, D.H. From CO_2 methanation to ambitious long-chain hydrocarbons: Alternative fuels paving the path to sustainability. *Chem. Soc. Rev.* **2019**, *48*, 205–259. [CrossRef]
17. Jadhav, S.G.; Vaidya, P.D.; Bhanage, B.M.; Joshi, J.B. Catalytic carbon dioxide hydrogenation to methanol: A review of recent studies. *Chem. Eng. Res. Des.* **2014**, *92*, 2557–2567. [CrossRef]
18. Porosoff, M.D.; Yan, B.; Chen, J.G. Catalytic reduction of CO_2 by H_2 for synthesis of CO, methanol and hydrocarbons: Challenges and opportunities. *Energy Environ. Sci.* **2016**, *9*, 62–73. [CrossRef]
19. Bando, K.K.; Soga, K.; Kunimori, K.; Ichikuni, N.; Okabe, K.; Kusama, H.; Sayama, K.; Arakawea, H. CO_2 hydrogenation activity and surface structure of zeolite-supported Rh catalysts. *Appl. Catal. A* **1998**, *173*, 47–60. [CrossRef]
20. Porosoff, M.D.; Yang, X.; Boscoboinik, J.A.; Chen, J.G. Molybdenum Carbide as Alternative Catalysts to Precious Metals for Highly Selective Reduction of CO_2 to CO. *Angew. Chem. Int. Ed.* **2014**, *53*, 6705–6709. [CrossRef]
21. Kattel, S.; Yan, B.; Chen, J.G.; Liu, P. CO_2 hydrogenation on Pt, Pt/SiO_2 and Pt/TiO_2: Importance of synergy between Pt and oxide support. *J. Catal.* **2016**, *343*, 115–126. [CrossRef]
22. Yan, Y.; Wang, Q.; Jiang, C.; Yao, Y.; Lu, D.; Zheng, J.; Dai, Y.; Wang, H.; Yang, Y. Ru/Al_2O_3 catalyzed CO_2 hydrogenation: Oxygen-exchange on metal-support interfaces. *J. Catal.* **2018**, *367*, 194–205. [CrossRef]
23. Yan, B.; Wu, Q.; Cen, J.; Timoshenko, J.; Frenkel, A.I.; Su, D.; Chen, X.; Parise, J.B.; Stach, E.; Orlov, A.; et al. Highly active subnanometer Rh clusters derived from Rh-doped $SrTiO_3$ for CO_2 reduction. *Appl. Catal. B* **2018**, *237*, 1003–1011. [CrossRef]
24. Dai, B.; Cao, S.; Xie, H.; Zhou, G.; Chen, S. Reduction of CO_2 to CO via reverse water-gas shift reaction over CeO_2 catalyst. *Korean J. Chem. Eng.* **2018**, *35*, 421–427. [CrossRef]
25. Alayoglu, S.; Beaumont, S.K.; Zheng, F.; Pushkarev, V.V.; Zheng, H.; Iablokov, V.; Liu, Z.; Guo, J.; Kruse, N.; Somorjai, G.A. CO_2 Hydrogenation Studies on Co and CoPt Bimetallic Nanoparticles Under Reaction Conditions Using TEM, XPS and NEXAFS. *Top. Catal.* **2011**, *54*, 778–785. [CrossRef]
26. Kharaji, A.G.; Shariati, A.; Takassi, M.A. A Novel γ-Alumina Supported Fe-Mo Bimetallic Catalyst for Reverse Water Gas Shift Reaction. *Chin. J. Chem. Eng.* **2013**, *21*, 1007–1014. [CrossRef]

27. Kharaji, A.G.; Shariati, A.; Ostadi, M. Development of Ni–Mo/Al_2O_3 Catalyst for Reverse Water Gas Shift (RWGS) Reaction. *J. Nanosci. Nanotechnol.* **2014**, *14*, 6841–6847. [CrossRef]
28. Zhao, B.; Yan, B.; Jiang, Z.; Yao, S.; Liu, Z.; Wu, Q.; Ran, R.; Senanayake, S.D.; Weng, D.; Chen, J.G. High selectivity of CO_2 hydrogenation to CO by controlling the valence state of nickel using perovskite. *Chem. Commun.* **2018**, *54*, 7354–7357. [CrossRef]
29. Quindimil, A.; De-La-Torre, U.; Pereda-Ayo, B.; González-Marcos, J.A.; González-Velasco, J.R. Ni catalysts with La as promoter supported over Y- and BETA-zeolites for CO_2 methanation. *Appl. Catal. B* **2018**, *238*, 393–403. [CrossRef]
30. Zhou, G.; Wu, T.; Xie, H.; Zheng, X. Effects of structure on the carbon dioxide methanation performance of Co-based catalysts. *Int. J. Hydrogen Energy* **2013**, *38*, 10012–10018. [CrossRef]
31. Lin, Q.; Liu, X.Y.; Jiang, Y.; Wang, Y.; Huang, Y.; Zhang, T. Crystal phase effects on the structure and performance of ruthenium nanoparticles for CO_2 hydrogenation. *Catal. Sci. Technol.* **2014**, *4*, 2058–2063. [CrossRef]
32. Yuan, H.; Zhu, X.; Han, J.; Wang, H.; Ge, Q. Rhenium-promoted selective CO_2 methanation on Ni-based catalyst. *J. CO_2 Util.* **2018**, *26*, 8–18. [CrossRef]
33. Li, M.; Amari, H.; van Veen, A.C. Metal-oxide interaction enhanced CO_2 activation in methanation over ceria supported nickel nanocrystallites. *Appl. Catal. B* **2018**, *239*, 27–35. [CrossRef]
34. Zhan, Y.; Wang, Y.; Gu, D.; Chen, C.; Jiang, L.; Takehira, K. Ni/Al_2O_3-ZrO_2 catalyst for CO_2 methanation: The role of γ-(Al, Zr)$_2O_3$ formation. *Appl. Surf. Sci.* **2018**, *459*, 74–79. [CrossRef]
35. Ma, H.; Ma, K.; Ji, J.; Tang, S.; Liu, C.; Jiang, W.; Yue, H.; Liang, B. Graphene intercalated Ni-SiO_2/GO-Ni-foam catalyst with enhanced reactivity and heat-transfer for CO_2 methanation. *Chem. Eng. Sci.* **2019**, *194*, 10–21. [CrossRef]
36. Romero-Sáez, M.; Dongil, A.B.; Benito, N.; Espinoza-González, R.; Escalona, N.; Gracia, F. CO_2 methanation over nickel-ZrO_2 catalyst supported on carbon nanotubes: A comparison between two impregnation strategies. *Appl. Catal. B* **2018**, *237*, 817–825. [CrossRef]
37. Bacariza, M.C.; Graça, I.; Lopes, J.M.; Henriques, C. Ni-Ce/Zeolites for CO_2 Hydrogenation to CH_4: Effect of the Metal Incorporation Order. *ChemCatChem* **2018**, *10*, 2773–2781. [CrossRef]
38. Liu, H.; Xu, S.; Zhou, G.; Huang, G.; Huang, S.; Xiong, K. CO_2 hydrogenation to methane over Co/KIT-6 catalyst: Effect of reduction temperature. *Chem. Eng. J.* **2018**, *351*, 65–73. [CrossRef]
39. Zhou, G.; Liu, H.; Xing, Y.; Xu, S.; Xie, H.; Xiong, K. CO_2 hydrogenation to methane over mesoporous Co/SiO_2 catalysts: Effect of structure. *J. CO_2 Util.* **2018**, *26*, 221–229. [CrossRef]
40. Gnanakumar, E.S.; Chandran, N.; Kozhevnikov, I.V.; Grau-Atienza, A.; Ramos Fernández, E.V.; Sepulveda-Escribano, A.; Shiju, N.R. Highly efficient nickel-niobia composite catalysts for hydrogenation of CO_2 to methane. *Chem. Eng. Sci.* **2019**, *194*, 2–9. [CrossRef]
41. Olah, G.A.; Goeppert, A.; Surya Prakash, G.K. Beyond oil and gas: The methanol Economy. *Angew. Chem. Int. Ed.* **2005**, *44*, 2636–2639. [CrossRef]
42. Li, W.; Lu, P.; Xu, D.; Tao, K. CO_2 hydrogenation to methanol over Cu/ZnO catalysts synthesized via a facile solid-phase grinding process using oxalic acid. *Korean J. Chem. Eng.* **2018**, *35*, 110–117. [CrossRef]
43. Cao, F.H.; Liu, D.H.; Hou, Q.S.; Fang, D.Y. Thermodynamic Analysis of CO_2 Direct Hydrogenation Reactions. *J. Nat. Gas Chem.* **2001**, *10*, 24–33.
44. Deng, K.; Hu, B.; Lu, Q.; Hong, X. Cu/g-C_3N_4 modified ZnO/Al_2O_3 catalyst: Methanol yield improvement of CO_2 hydrogenation. *Catal. Commun.* **2017**, *100*, 81–84. [CrossRef]
45. Witoon, T.; Numpilai, T.; Phongamwong, T.; Donphai, W.; Boonyuen, C.; Warakulwit, C.; Chareonpanich, M.; Limtrakul, J. Enhanced activity, selectivity and stability of a CuO-ZnO-ZrO_2 catalyst by adding graphene oxide for CO_2 hydrogenation to methanol. *Chem. Eng. J.* **2018**, *334*, 1781–1791. [CrossRef]
46. Wang, G.; Mao, D.; Guo, X.; Yu, J. Enhanced performance of the CuO-ZnO-ZrO_2 catalyst for CO_2 hydrogenation to methanol by WO_3 modification. *Appl. Surf. Sci.* **2018**, *456*, 403–409. [CrossRef]
47. Ye, J.; Liu, C.; Mei, D.; Ge, Q. Active Oxygen Vacancy Site for Methanol Synthesis from CO_2 Hydrogenation on In_2O_3(110): A DFT Study. *ACS Catal.* **2013**, *3*, 1296–1306. [CrossRef]
48. Sun, K.; Fan, Z.; Ye, J.; Yan, J.; Ge, Q.; Li, Y.; He, W.; Yang, W.; Liu, C. Hydrogenation of CO_2 to methanol over In_2O_3 catalyst. *J. CO_2 Util.* **2015**, *12*, 1–6. [CrossRef]

49. Martin, O.; Martin, A.J.; Mondelli, C.; Mitchell, S.; Segawa, T.F.; Hauert, R.; Drouilly, C.; Curulla-Ferre, D.; Perez-Ramirez, J. Indium Oxide as a Superior Catalyst for Methanol Synthesis by CO_2 Hydrogenation. *Angew. Chem. Int. Ed.* **2016**, *55*, 6261–6265. [CrossRef]
50. Wang, J.; Li, G.; Li, Z.; Tang, C.; Feng, Z.; An, H.; Liu, H.; Liu, T.; Li, C. A highly selective and stable ZnO-ZrO_2 solid solution catalyst for CO_2 hydrogenation to methanol. *Sci. Adv.* **2017**, *3*, e1701290. [CrossRef]
51. Jiang, X.; Jiao, Y.; Moran, C.; Nie, X.; Gong, Y.; Guo, X.; Walton, K.S.; Song, C. CO_2 hydrogenation to methanol on Pd-Cu bimetallic catalysts with lower metal loadings. *Catal. Commun.* **2019**, *118*, 10–14. [CrossRef]
52. Jiang, X.; Wang, X.; Nie, X.; Koizumi, N.; Guo, X.; Song, C. CO_2 hydrogenation to methanol on Pd-Cu bimetallic catalysts: H_2/CO_2 ratio dependence and surface species. *Catal. Today* **2018**, *316*, 62–70. [CrossRef]
53. Bahruji, H.; Esquius, J.R.; Bowker, M.; Hutchings, G.; Armstrong, R.D.; Jones, W. Solvent Free Synthesis of PdZn/TiO_2 Catalysts for the Hydrogenation of CO_2 to Methanol. *Top. Catal.* **2018**, *61*, 144–153. [CrossRef]
54. Yin, Y.; Hu, B.; Li, X.; Zhou, X.; Hong, X.; Liu, G. Pd@zeolitic imidazolate framework-8 derived PdZn alloy catalysts for efficient hydrogenation of CO_2 to methanol. *Appl. Catal. B* **2018**, *234*, 143–152. [CrossRef]
55. Hu, B.; Yin, Y.; Liu, G.; Chen, S.; Hong, X.; Tsang, S.C.E. Hydrogen spillover enabled active Cu sites for methanol synthesis from CO_2 hydrogenation over Pd doped CuZn catalysts. *J. Catal.* **2018**, *359*, 17–26. [CrossRef]
56. Studt, F.; Sharafutdinov, I.; Abild-Pedersen, F.; Elkjær, C.F.; Hummelshøj, J.S.; Dahl, S.; Chorkendorff, I.; Nørskov, J.K. Discovery of a Ni-Ga catalyst for carbon dioxide reduction to methanol. *Nat. Chem.* **2014**, *6*, 320–324. [CrossRef]
57. Marcos, F.C.F.; Assaf, J.M.; Assaf, E.M. CuFe and CuCo supported on pillared clay as catalysts for CO_2 hydrogenation into value-added products in one-step. *Mol. Catal.* **2018**, *458*, 297–306. [CrossRef]
58. Joseph, A.S.; Can, A.; Julia, S.; Wang, T.; Jens, K.N.; Frank, A.P.; Stacey, F.B. Theoretical and Experimental Studies of CoGa Catalysts for the Hydrogenation of CO_2 to Methanol. *Catal. Lett.* **2018**, *148*, 3583–3591.
59. Tan, Q.; Shi, Z.; Wu, D. CO_2 Hydrogenation to Methanol over a Highly Active Cu–Ni/CeO_2–Nanotube Catalyst. *Ind. Eng. Chem. Res.* **2018**, *57*, 10148–10158. [CrossRef]
60. Jiang, Y.; Yang, H.; Gao, P.; Li, X.; Zhang, J.; Liu, H.; Wang, H.; Wei, W.; Sun, Y. Slurry methanol synthesis from CO_2 hydrogenation over micro-spherical SiO_2 support Cu/ZnO catalysts. *J. CO_2 Util.* **2018**, *26*, 642–651. [CrossRef]
61. Chen, P.; Zhao, G.; Liu, Y.; Lu, Y. Monolithic $Ni_5Ga_3/SiO_2/Al_2O_3$/Al-fiber catalyst for CO_2 hydrogenation to methanol at ambient pressure. *Appl. Catal. A* **2018**, *562*, 234–240. [CrossRef]
62. Lam, E.; Larmier, K.; Wolf, P.; Tada, S.; Safonova, O.V.; Copéret, C. Isolated Zr Surface Sites on Silica Promote Hydrogenation of CO_2 to CH_3OH in Supported Cu Catalysts. *J. Am. Chem. Soc.* **2018**, *140*, 10530–10535. [CrossRef]
63. Dasireddy, V.D.B.C.; Stefancic, N.S.; Likozar, B. Correlation between synthesis pH, structure and Cu/MgO/Al_2O_3 heterogeneous catalyst activity and selectivity in CO_2 hydrogenation to methanol. *J. CO_2 Util.* **2018**, *28*, 189–199. [CrossRef]
64. Shi, Z.; Tan, Q.; Wu, D. Ternary copper-cerium-zirconium mixed metal oxide catalyst for direct CO_2 hydrogenation to methanol. *Mater. Chem. Phys.* **2018**, *219*, 263–272. [CrossRef]
65. Bao, Y.; Huang, C.; Chen, L.; Zhang, Y.D.; Liang, L.; Wen, J.; Fu, M.; Wu, J.; Ye, D. Highly efficient Cu/anatase TiO_2 {001}-nanosheets catalysts for methanol synthesis from CO_2. *J. Energy Chem.* **2018**, *27*, 381–388. [CrossRef]
66. Koh, M.K.; Khavarian, M.; Chai, S.P.; Mohamed, A.R. The morphological impact of siliceous porous carriers on copper-catalysts for selective direct CO_2 hydrogenation to methanol. *Int. J. Hydrogen Energy* **2018**, *43*, 9334–9342. [CrossRef]
67. Tasfy, S.; Zabidi, N.A.M.; Shaharum, M.S.; Subbarao, D. Methanol production via CO_2 hydrogenation reaction: Effect of catalyst support. *Int. J. Nanotechnol.* **2017**, *14*, 410–421. [CrossRef]
68. Li, Y.; Na, W.; Wang, H.; Gao, W. Hydrogenation of CO_2 to methanol over Au–CuO/SBA-15 catalysts. *J. Porous Mater.* **2017**, *24*, 591–599. [CrossRef]
69. Atakan, A.; Mäkie, P.; Söderlind, F.; Keraudy, J.; Björk, E.M.; Odén, M. Synthesis of a Cu-infiltrated Zr-doped SBA-15 catalyst for CO_2 hydrogenation into methanol and dimethyl ether. *Phys. Chem. Chem. Phys.* **2017**, *19*, 19139–19149. [CrossRef]
70. Rui, N.; Wang, Z.; Sun, K.; Ye, J.; Ge, Q.; Liu, C. CO_2 hydrogenation to methanol over Pd/In_2O_3: Effects of Pd and oxygen vacancy. *Appl. Catal. B* **2017**, *218*, 488–497. [CrossRef]

71. Zhang, J.; Lu, S.; Su, X.; Fan, S.; Ma, Q.; Zhao, T. Selective formation of light olefins from CO_2 hydrogenation over Fe–Zn–K catalysts. *J. CO_2 Util.* **2015**, *12*, 95–100. [CrossRef]
72. You, Z.; Deng, W.; Zhang, Q.; Wang, Y. Hydrogenation of carbon dioxide to light olefins over non-supported iron catalyst. *Chin. J. Catal.* **2013**, *34*, 956–963. [CrossRef]
73. Gao, P.; Li, S.; Bu, X.; Dang, S.; Liu, Z.; Wang, H.; Zhong, L.; Qiu, M.; Yang, C.; Cai, J.; et al. Direct conversion of CO_2 into liquid fuels with high selectivity over a bifunctional catalyst. *Nat. Chem.* **2017**, *9*, 1019–1024. [CrossRef] [PubMed]
74. Xie, C.; Chen, C.; Yu, Y.; Su, J.; Li, Y.; Somorjai, G.A.; Yang, P. Tandem Catalysis for CO_2 Hydrogenation to C_2–C_4 Hydrocarbons. *Nano Lett.* **2017**, *17*, 3798–3802. [CrossRef]
75. Li, S.; Guo, H.; Luo, C.; Zhang, H.; Xiong, L.; Chen, X.; Ma, L. Effect of Iron Promoter on Structure and Performance of K/Cu–Zn Catalyst for Higher Alcohols Synthesis from CO_2 Hydrogenation. *Catal. Lett.* **2013**, *143*, 345–355. [CrossRef]
76. Wei, J.; Yao, R.; Ge, Q.; Wen, Z.; Ji, X.; Fang, C.; Zhang, J.; Xu, H.; Sun, J. Catalytic Hydrogenation of CO_2 to Isoparaffins over Fe-Based Multifunctional Catalysts. *ACS Catal.* **2018**, *8*, 9958–9967. [CrossRef]
77. Li, Z.; Qu, Y.; Wang, J.; Liu, H.; Li, M.; Miao, S.; Li, C. Highly Selective Conversion of Carbon Dioxide to Aromatics over Tandem Catalysts. *Joule* **2019**, *3*, 1–14. [CrossRef]
78. Ni, Y.; Chen, Z.; Fu, Y.; Liu, Y.; Zhu, W.; Liu, Z. Selective conversion of CO_2 and H_2 into aromatics. *Nat. Commun.* **2018**, *9*, 3457. [CrossRef] [PubMed]
79. Zhang, M.H.; Liu, Z.M.; Lin, G.D.; Zhang, H.B. Pd/CNT-promoted $CuZrO_2$/HZSM-5 hybrid catalysts for direct synthesis of DME from CO_2/H_2. *Appl. Catal. A* **2013**, *451*, 28–35. [CrossRef]
80. Wu, T.; Lin, J.; Cheng, Y.; Tian, J.; Wang, S.; Xie, S.; Pei, Y.; Yan, S.; Qiao, M.; Xu, H.; et al. Porous Graphene-Confined Fe–K as Highly Efficient Catalyst for CO_2 Direct Hydrogenation to Light Olefins. *ACS Appl. Mater. Interfaces* **2018**, *10*, 23439–23443. [CrossRef]
81. Zhang, Y.; Li, D.; Zhang, Y.; Cao, Y.; Zhang, S.; Wang, K.; Ding, F.; Wu, J. V-modified CuO–ZnO–ZrO_2/HZSM-5 catalyst for efficient direct synthesis of DME from CO_2 hydrogenation. *Catal. Commun.* **2014**, *55*, 49–52. [CrossRef]
82. Gao, W.; Wang, H.; Wang, Y.; Guo, W.; Jia, M. Dimethyl ether synthesis from CO_2 hydrogenation on La-modified CuO-ZnO-Al_2O_3/HZSM-5 bifunctional catalysts. *J. Rare Earths* **2013**, *31*, 470–476. [CrossRef]
83. Suwannapichat, Y.; Numpilai, T.; Chanlek, N.; Faungnawakij, K.; Chareonpanich, M.; Limtrakul, J.; Witoon, T. Direct synthesis of dimethyl ether from CO_2 hydrogenation over novel hybrid catalysts containing a Cu ZnO ZrO_2 catalyst admixed with WO_x/Al_2O_3 catalysts: Effects of pore size of Al_2O_3 support and W loading content. *Energy Convers. Manag.* **2018**, *159*, 20–29. [CrossRef]
84. Michailos, S.; McCord, S.; Sick, V.; Stokes, G.; Styring, P. Dimethyl ether synthesis via captured CO_2 hydrogenation within the power to liquids concept: A techno-economic assessment. *Energy Convers. Manag.* **2019**, *184*, 262–276. [CrossRef]
85. Quadrelli, E.A.; Centi, G.; Duplan, J.L.; Perathoner, S. Carbon Dioxide Recycling: Emerging Large-Scale Technologies with Industrial Potential. *ChemSusChem* **2011**, *4*, 1194–1215. [CrossRef] [PubMed]
86. Cho, W.; Song, T.; Mitsos, A.; McKinnon, J.T.; Ko, G.H.; Tolsma, J.E.; Denholm, D.; Park, T. Optimal design and operation of a natural gas tri-reforming reactor for DME synthesis. *Catal. Today* **2009**, *139*, 261–267. [CrossRef]
87. Hirano, M.; Tatsumi, M.; Yasutake, T.; Kuroda, K. Dimethyl ether synthesis via reforming of steam/carbon dioxide and methane. *J. Jpn. Pet. Inst.* **2005**, *48*, 197–203. [CrossRef]
88. Hirano, M.; Yasutake, T.; Kuroda, K. Dimethyl ether synthesis from carbon dioxide by catalytic hydrogenation (Part 3) direct synthesis using hybrid by recycling process. *J. Jpn. Pet. Inst.* **2007**, *50*, 34–43. [CrossRef]
89. Bonura, G.; Migliori, M.; Frusteri, L.; Cannilla, C.; Catizzone, E.; Giordano, G.; Frusteri, F. Acidity control of zeolite functionality on activity and stability of hybrid catalysts during DME production via CO_2 hydrogenation. *J. CO_2 Util.* **2018**, *24*, 398–406. [CrossRef]
90. Catizzone, E.; Migliori, M.; Purita, A.; Giordano, G. Ferrierite vs. γ-Al_2O_3: The superiority of zeolites in terms of water-resistance in vapour-phase dehydration of methanol to dimethyl ether. *J. Energy Chem.* **2019**, *30*, 162–169. [CrossRef]
91. Catizzone, E.; Bonura, G.; Migliori, M.; Frusteri, F.; Giordano, G. CO_2 Recycling to Dimethyl Ether: State-of-the-Art and Perspectives. *Molecules* **2018**, *23*, 31. [CrossRef] [PubMed]

92. De Falco, M.; Capocelli, M.; Centi, G. Dimethyl ether production from CO_2 rich feedstocks in a one-step process: Thermodynamic evaluation and reactor simulation. *Chem. Eng. J.* **2016**, *294*, 400–409. [CrossRef]
93. Fang, J.; Jin, X.F.; Huang, K. Life cycle analysis of a combined CO_2 capture and conversion membrane reactor. *J. Membr. Sci.* **2018**, *549*, 142–150. [CrossRef]
94. Sofia, D.; Giuliano, A.; Poletto, M.; Barletta, D. Techno-economic analysis of power and hydrogen co-production by an IGCC plant with CO_2 capture based on membrane technology. *Comput. Aided Process Eng.* **2015**, *37*, 1373–1378.
95. Li, Z.; Wang, J.; Qu, Y.; Liu, H.; Tang, C.; Miao, S.; Feng, Z.; An, H.; Li, C. Highly Selective Conversion of Carbon Dioxide to Lower Olefins. *ACS Catal.* **2017**, *7*, 8544–8548. [CrossRef]
96. Da Silva, R.J.; Pimentel, A.F.; Monteiro, R.S.; Mota, C.J.A. Synthesis of methanol and dimethyl ether from the CO_2 hydrogenation over Cu·ZnO supported on Al_2O_3 and Nb_2O_5. *J. CO_2 Util.* **2016**, *15*, 83–88. [CrossRef]
97. Dang, S.; Gao, P.; Liu, Z.; Chen, X.; Yang, C.; Wang, H.; Zhong, L.; Li, S.; Sun, Y. Role of zirconium in direct CO_2 hydrogenation to lower olefins on oxide/zeolite bifunctional catalysts. *J. Catal.* **2018**, *364*, 382–393. [CrossRef]
98. Wang, P.; Zha, F.; Yao, L.; Chang, Y. Synthesis of light olefins from CO_2 hydrogenation over (CuO-ZnO)-kaolin/SAPO-34 molecular sieves. *Appl. Clay Sci.* **2018**, *163*, 249–256. [CrossRef]
99. Guo, L.; Sun, J.; Ji, X.; Wei, J.; Wen, Z.; Yao, R.; Xu, H.; Ge, Q. Directly converting carbon dioxide to linear α-olefins on bio-promoted catalysts. *Commun. Chem.* **2018**, *1*, 11. [CrossRef]
100. Bonura, G.; Cordaro, M.; Spadaro, L.; Cannilla, C.; Arena, F.; Frusteri, F. Hybrid Cu–ZnO–ZrO_2/H-ZSM5 system for the direct synthesis of DME by CO_2 hydrogenation. *Appl. Catal. B* **2013**, *140–141*, 16–24. [CrossRef]
101. Bonura, G.; Cordaro, M.; Cannilla, C.; Mezzapica, A.; Spadaro, L.; Arena, F.; Frusteri, F. Catalytic behaviour of a bifunctional system for the one step synthesis of DME by CO_2 hydrogenation. *Catal. Today* **2014**, *228*, 51–57. [CrossRef]
102. Zhang, Y.; Li, D.; Zhang, S.; Wang, K.; Wu, J. CO_2 hydrogenation to dimethyl ether over CuO–ZnO–Al_2O_3/HZSM-5 prepared by combustion route. *RSC Adv.* **2014**, *4*, 16391–16396. [CrossRef]
103. Zha, F.; Tian, H.; Yan, J.; Chang, Y. Multi-walled carbon nanotubes as catalyst promoter for dimethyl ether synthesis from CO_2 hydrogenation. *Appl. Surf. Sci.* **2013**, *285*, 945–951. [CrossRef]
104. Frusteri, F.; Bonura, G.; Cannilla, C.; Drago Ferrante, G.; Aloise, A.; Catizzone, E.; Migliori, M.; Giordano, G. Stepwise tuning of metal-oxide and acid sites of CuZnZr-MFI hybrid catalysts for the direct DME synthesis by CO_2 hydrogenation. *Appl. Catal. B* **2015**, *176–177*, 522–531. [CrossRef]
105. Liu, R.; Tian, H.; Yang, A.; Zha, F.; Ding, J.; Chang, Y. Preparation of HZSM-5 membrane packed CuO–ZnO–Al_2O_3 nanoparticles for catalysing carbon dioxide hydrogenation to dimethyl ether. *Appl. Surf. Sci.* **2015**, *345*, 1–9. [CrossRef]
106. Dai, C.; Zhang, A.; Liu, M.; Li, J.; Song, F.; Song, C.; Guo, X. Facile one-step synthesis of hierarchical porous carbon monoliths as superior supports of Fe-based catalysts for CO_2 hydrogenation. *RSC Adv.* **2016**, *6*, 10831–10836. [CrossRef]
107. Wei, J.; Ge, Q.; Yao, R.; Wen, Z.; Fang, C.; Guo, L.; Xu, H.; Sun, J. Directly converting CO_2 into a gasoline fuel. *Nat. Commun.* **2017**, *8*, 16170. [CrossRef]
108. Liu, J.; Zhang, A.; Jiang, X.; Liu, M.; Zhu, J.; Song, C.; Guo, X. Direct Transformation of Carbon Dioxide to Value-Added Hydrocarbons by Physical Mixtures of Fe_5C_2 and K-Modified Al_2O_3. *Ind. Eng. Chem. Res.* **2018**, *57*, 9120–9126. [CrossRef]
109. Zhang, J.; Su, X.; Wang, X.; Ma, Q.; Fan, S.; Zhao, T.S. Promotion effects of Ce added Fe–Zr–K on CO_2 hydrogenation to light olefins. *React. Kinet. Mech. Catal.* **2018**, *124*, 575–585. [CrossRef]
110. Ramirez, A.; Gevers, L.; Bavykina, A.; Ould-Chikh, S.; Gascon, J. Metal Organic Framework-Derived Iron Catalysts for the Direct Hydrogenation of CO_2 to Short Chain Olefins. *ACS Catal.* **2018**, *8*, 9174–9182. [CrossRef]
111. Shi, Z.; Yang, H.; Gao, P.; Chen, X.; Liu, H.; Zhong, L.; Wang, H.; Wei, W.; Sun, Y. Effect of alkali metals on the performance of CoCu/TiO_2 catalysts for CO_2 hydrogenation to long-chain hydrocarbons. *Chin. J. Catal.* **2018**, *39*, 1294–1302. [CrossRef]
112. Bogaerts, A.; Neyts, E.; Gijbels, R.; van der Mullen, J. Gas discharge plasmas and their applications. *Spectrochim. Acta Part B* **2002**, *57*, 609–658. [CrossRef]
113. Bogaerts, A.; Neyts, E.C. Plasma Technology: An Emerging Technology for Energy Storage. *ACS Energy Lett.* **2018**, *3*, 1013–1027. [CrossRef]

114. Paulussen, S.; Verheyde, B.; Tu, X.; De Bie, C.; Martens, T.; Petrovic, D.; Bogaerts, A.; Sels, B. Conversion of carbon dioxide to value-added chemicals in atmospheric pressure dielectric barrier discharges. *Plasma Sources Sci. Technol.* **2010**, *19*, 034015. [CrossRef]
115. Silva, T.; Britun, N.; Godfroid, T.; Snyders, R. Optical characterization of a microwave pulsed discharge used for dissociation of CO_2. *Plasma Sources Sci. Technol.* **2014**, *23*, 25009. [CrossRef]
116. Ozkan, A.; Dufour, T.; Silva, T.; Britun, N.; Snyders, R.; Reniers, F.; Bogaerts, A. DBD in burst mode: Solution for more efficient CO_2 conversion? *Plasma Sources Sci. Technol.* **2016**, *25*, 055005. [CrossRef]
117. Wang, W.; Berthelot, A.; Kolev, S.; Tu, X.; Bogaerts, A. CO_2 conversion in a gliding arc plasma: 1D cylindrical discharge model. *Plasma Sources Sci. Technol.* **2016**, *25*, 065012. [CrossRef]
118. Li, D.; Li, X.; Bai, M.; Tao, X.; Shang, S.; Dai, X.; Yin, Y. CO_2 reforming of CH_4 by atmospheric pressure glow discharge plasma: A high conversion ability. *Int. J. Hydrogen Energy* **2009**, *34*, 308–313. [CrossRef]
119. Liu, J.L.; Park, H.W.; Chung, W.J.; Park, D.W. High-Efficient Conversion of CO_2 in AC-Pulsed Tornado Gliding Arc Plasma. *Plasma Chem. Plasma Process.* **2016**, *36*, 437–449. [CrossRef]
120. Kolb, T.; Voigt, J.H.; Gericke, K.H. Conversion of Methane and Carbon Dioxide in a DBD Reactor: Influence of Oxygen. *Plasma Chem. Plasma Process.* **2013**, *33*, 631–646. [CrossRef]
121. Wang, L.; Yi, Y.; Wu, C.; Guo, H.; Tu, X. One-Step Reforming of CO_2 and CH_4 into High-Value Liquid Chemicals and Fuels at Room Temperature by Plasma-Driven Catalysis. *Angew. Chem. Int. Ed.* **2017**, *56*, 13679–13683. [CrossRef]
122. Zhang, K.; Eliasson, B.; Kogelschatz, U. Direct Conversion of Greenhouse Gases to Synthesis Gas and C_4 Hydrocarbons over Zeolite HY Promoted by a Dielectric-Barrier Discharge. *Ind. Eng. Chem. Res.* **2002**, *41*, 1462–1468. [CrossRef]
123. Eliasson, B.; Kogelschatz, U.; Xue, B.; Zhou, L.M. Hydrogenation of Carbon Dioxide to Methanol with a Discharge-Activated Catalyst. *Ind. Eng. Chem. Res.* **1998**, *37*, 3350–3357. [CrossRef]
124. Zeng, Y.; Tu, X. Plasma-Catalytic CO_2 Hydrogenation at Low Temperatures. *IEEE Trans. Plasma Sci.* **2016**, *44*, 405–411. [CrossRef]
125. Wang, L.; Yi, Y.; Guo, H.; Tu, X. Atmospheric Pressure and Room Temperature Synthesis of Methanol through Plasma-Catalytic Hydrogenation of CO_2. *ACS Catal.* **2018**, *8*, 90–100. [CrossRef]
126. De Bie, C.; van Dijk, J.; Bogaerts, A. CO_2 Hydrogenation in a Dielectric Barrier Discharge Plasma Revealed. *J. Phys. Chem. C* **2016**, *120*, 25210–25224. [CrossRef]

 catalysts

Article

H_2O and/or SO_2 Tolerance of Cu-Mn/SAPO-34 Catalyst for NO Reduction with NH_3 at Low Temperature

Guofu Liu [1,†], Wenjie Zhang [1,†], Pengfei He [2], Shipian Guan [2], Bing Yuan [3], Rui Li [3], Yu Sun [3] and Dekui Shen [1,*]

1 Key Laboratory of Energy Thermal Conversion and Control of Ministry of Education, School of Energy and Environment, Southeast University, Nanjing 210000, China; qzliuguofu@163.com (G.L.); z282780680@163.com (W.Z.)

2 Jiangsu Frontier Electric Power Technology Co., Ltd., Nanjing 211102, China; hpf66805@163.com (P.H.); parallel79@163.com (S.G.)

3 Jiangsu Power Design Institute Co., Ltd., Nanjing 210008, China; letfighting@163.com (B.Y.); chongba_js@163.com (R.L.); sunyunjpdi@163.com (Y.S.)

* Correspondence: 101011398@seu.edu.cn; Tel.: +86-025-83794744

† These two authors contribute equally to the manuscript.

Received: 2 February 2019; Accepted: 19 March 2019; Published: 21 March 2019

Abstract: A series of molecular sieve catalysts (Cu–Mn/SAPO-34) with different loadings of Cu and Mn components were prepared by the impregnation method. The deNO$_x$ activity of the catalyst was investigated during the selective catalytic reduction (SCR) of NO with NH_3 in the temperature range of 120 °C to 330 °C, including the effects of H_2O vapors and SO_2. In order to understand the poisoning mechanism by the injection of H_2O and/or SO_2 into the feeding gas, the characteristics of the fresh and spent catalyst were identified by means of Brunner−Emmet−Teller (BET), X-ray Diffraction (XRD), Scanning Electronic Microscopy (SEM) and Thermal Gravity- Differential Thermal Gravity (TG-DTG). The conversion of NO by the catalyst can achieve at 72% under the reaction temperature of 120 °C, while the value reached more than 90% under the temperature between 180 °C and 330 °C. The deNO$_x$ activity test shows that the H_2O has a reversible negative effect on NO conversion, which is mainly due to the competitive adsorption of H_2O and NH_3 on Lewis acid sites. When the reaction temperature increases to 300 °C, the poisoning effect of H_2O can be negligible. The poisoning effect of SO_2 on deNO$_x$ activity is dependent on the reaction temperature. At low temperature, the poisoning effect of SO_2 is permanent with no recovery of deNO$_x$ activity after the elimination of SO_2. The formation of (NH4)$_2$SO$_4$, which results in the plug of active sites and a decrease of surface area, and the competitive adsorption of SO_2 and NO should be responsible for the loss of deNO$_x$ activity over Cu/SAPO-34.

Keywords: SCR; Catalyst; $(NH_4)_2SO_4$; deNO$_x$; H_2O and SO_2 poisoning

1. Introduction

Nitrogen oxides were estimated as one of the major air pollutants released from the combustion of fossil fuels (especially coal), being hazardous for the ecological and environmental system [1–4]. Currently, selective catalytic reduction (SCR) was regarded as a widely-used deNO$_x$ technology for the purification of flue gas, where the performance of the catalyst plays a significant role in the process [5]. Most of the commercial catalysts (e.g., V-W-Ti system) exhibited the effective activity with temperature located in a narrow and relatively high window as 300–400 °C, accelerating the deactivation of catalyst through sintering and occlusion of salts produced from H_2O or SO_2. The high deNO$_x$ activity of SCR catalyst at relatively low temperature is highly required without the formation of salts from H_2O and

SO_2 in the flue gas. The stability of the air-preheating system can be improved along with the secure low-temperature SCR system, leading to a full-time deNO$_x$ for the power plant under different power loadings. Therefore, it is of great significance to develop efficient and stable low-temperature SCR catalyst [6].

Among all these catalysts, transition metal loaded on zeolite materials with CHA structure have been widely focused due to the broad operating temperature range and the high deNO$_x$ activity [7–9]. A number of research works have been conducted on the deNO$_x$ performance of molecular sieve catalysts, such as the ZSM-5, BEA, USY, SSZ-13 and so on [10–12]. Kim [13] prepared Mn–Fe/ZSM-5 by the impregnation method, exhibiting the NO_x conversion ratio as high as 95% at 175 °C, and the conversion of NO_x nearly to 100% during the temperature between 200 and 350 °C. SAPO-34 possesses pear-shaped cages with 8-membered ring (8MR) openings, and double 6-membered ring (D6MR) units linked by 4-membered ring (4MR) units, while most of the P is replaced by Si to generate a Si–O–Al linkage, resulting in remarkable SCR performance. Compared with Fe-zeolites and vanadia-zeolites, Cu-zeolites exhibited a superior deNO$_x$ activity and high N_2 selectivity [14,15]. Wang [16,17], Ye [18] and Deka [19] prepared Cu/SAPO-34 by different methods, exhibiting an excellent low-temperature SCR activity in the temperature range from 150 °C to 500 °C. Among all these researches, Cu-SAPO-34 based catalysts showed a remarkable deNO$_x$ activity compared to other zeolites [8,20].

However, Cu-SAPO-34 catalyst is proved to be sensitive to SO_2 poisoning due to the strong chemical binding strength and the oxidative conditions in the gas and the deactivation effect is more pronounced at low temperatures [21]. Zhang et al. [7] have used DRIFT and Temperature Programmed Desorption (TPD) method to study the poisoning effect of SO_2 over Cu/SAPO-34 catalyst, they reported that low-temperature deactivation is caused by the formation of ammonium sulfates. Shen et al. [22] observed no obvious sulfur species on the Cu/SAPO-34 catalyst and they concluded that the reduction of the number of isolated Cu^{2+} caused by SO_2 might induce the loss of SCR activity. Wijayanti et al. [23] studied the SO_2 poisoning effects and found that the main reason for the deactivation is the formation of copper sulfates, resulting in the loss of redox properties. Jangjou et al. [24] reported the S species formed on Cu^{2+} at 6MR by DRIFT study. On the basis of such observations, the formation of ammonium sulfates in a complex with Cu is claimed as the main mechanism for the loss of low-temperature deNO$_x$ activity. In general, different SO_2 poisoning mechanisms have been proposed by different researchers.

In addition, H_2O is also one of the main components in the flue gas and often causes the loss of low-temperature deNO$_x$ activity [3]. In this work, for a better and specific understanding of the SO_2 poisoning mechanism and the synergistic effect of H_2O and SO_2 over Cu/SAPO-34 catalyst, the influence of SO_2 or/and H_2O with different concentrations at different reaction temperatures on deNO$_x$ activity and physicochemical properties over Cu/SAPO-34 catalyst was studied. A series of Cu-Mn/SAPO-34 catalysts were prepared through the impregnation method. The low-temperature deNO$_x$ activity of the catalysts was estimated by means of a self-designed apparatus, where the effects of H_2O and SO_2 on the reaction activity were also investigated. BET, XRD, SEM and TG-DTG were employed to determine the characteristics of the fresh and spent catalyst, in order to understand the poisoning mechanism of the catalyst by H_2O and SO_2 during the low-temperature SCR process.

2. Results and Discussion

2.1. DeNOx Activity of the Catalysts Without H_2O and SO_2

Figure 1a illustrates the deNO$_x$ activity of Cu/SAPO-34, Mn/SAPO-34, Cu–Mn/SAPO-34 catalysts under different temperatures without H_2O and SO_2. The deNO$_x$ activity of Cu–Mn/SAPO-34 catalyst (bimetallic composite molecular sieve) was much higher than that of Cu/SAPO-34 and Mn/SAPO-34 catalysts (monomeric molecular sieve). Even when the reaction temperature is lower than 100 °C the NO conversion by Cu–Mn/SAPO-34 catalyst can be achieved at about 60%, while for the other two catalysts was around 20%.

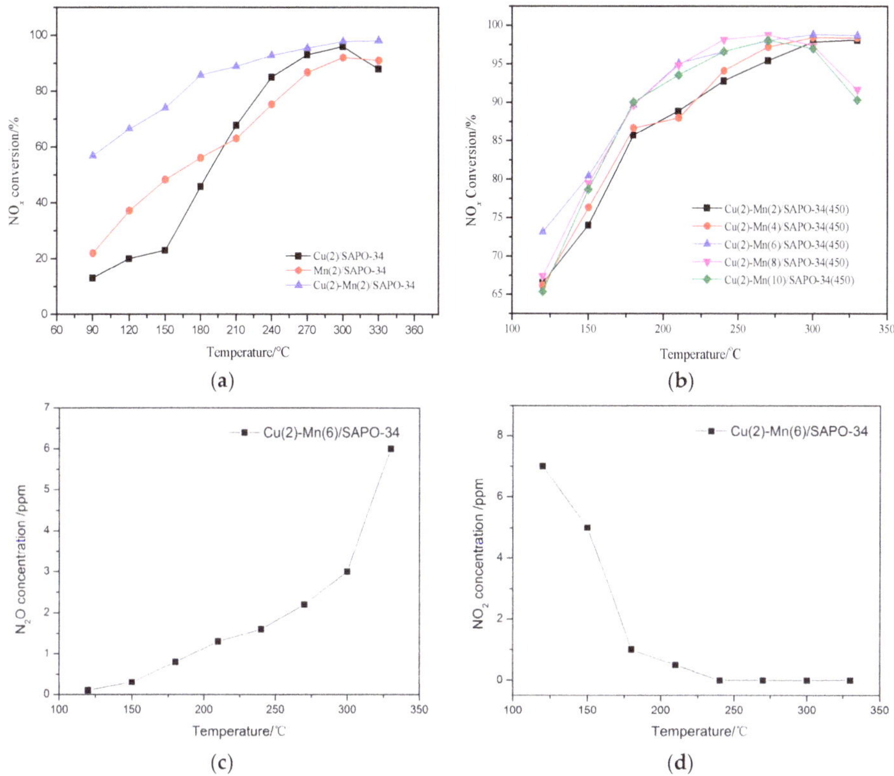

Figure 1. DeNO$_x$ performance of mono-component and multi-component catalysts (**a**); deNO$_x$ performance of molecular catalysts with different *Mn* loadings (**b**); outlet N_2O concentration over Cu(2)-Mn(6)/SAPO-34 catalyst (**c**); outlet NO_2 concentration over Cu(2)-Mn(6)/SAPO-34 catalyst (**d**).

The NO conversion over Cu–Mn/SAPO-34 catalyst could reach 90% at around 180 °C, compared to that of 220 °C for Cu/SAPO-34 and 270 °C for Mn/SAPO-34. It can be concluded that the Cu–Mn/SAPO-34 catalyst gives a better deNO$_x$ performance under the relatively low-temperature range from 100 to 200 °C. This can be attributed to the promotion of NH_3 adsorption on the surface of the catalyst by the bimetallic system on SAPO-34 [25–27]. In addition, the deNO$_x$ reaction energy might be declined by the bimetallic interaction on the catalyst, improving the low-temperature activity of the catalyst and broadening its SCR temperature range.

The effect of metal loading on the activity of Cu-Mn/SAPO-34 catalyst is shown in Figure 1b. The deNO$_x$ activity of Cu(2)–Mn(6)/SAPO-34 performed as the best one among the catalysts at the temperature from 120 to 210 °C and from 270 to 330 °C, while Cu(2)–Mn(8)/SAPO-34 performs the best one at the temperature from 210 to 270 °C. It needs to be noted that the NO conversion over Cu–Mn/SAPO-34 with the Mn content more than 8% was notably declined when the reaction temperature is higher than 270 °C. Under high temperatures, the oxidation of the catalyst can be enhanced with the increased loading of Mn [28]. This leads to the oxidation of NH_3 to NO and then the decline of the NO conversion (Figure 1b). In addition, the nonselective catalytic reduction (NSCR) reaction and catalytic oxidation reaction (i.e., the C–O reaction) happens simultaneously during the NH_3-SCR reaction. The outlet N_2O and NO_2 concentration over Cu(2)–Mn(6)/SAPO-34 catalyst are shown in Figure 1c,d. It can be seen that the N_2O concentration increases with the increase of reaction temperature, which may contribute to the decrease of deNO$_x$ activity over Cu(2)–Mn(6)/SAPO-34 catalyst at 330 °C. With the increase of reaction temperature, the NO_2 concentration decreases. These results demonstrate that N_2 is the main product of NH_3-SCR reaction over Cu(2)–Mn(6)/SAPO-34 catalyst according to the "standard SCR" reaction [29].

2.2. Effect of H_2O on the deNO$_x$ Activity of the Catalyst

The effect of the injected H_2O concentration into feeding gas on the deNO$_x$ activity of Cu(2)–Mn(6)/SAPO-34 catalyst at a reaction temperature of 240 °C was shown in Figure 2a. The NO conversion decreased with an increased H_2O concentration. When the volume concentration of 2% water vapor was injected into the original feeding gas after 8 h, the deNO$_x$ activity of the catalyst was declined from 93% to 91%. While the volume concentration of the vapor injected into the feeding gas was increased to 10%, after 8 h, the activity of the catalyst decreased to 86%. It needs to be noted that the activity of catalyst after the 8-hour injection of vapor was recovered to its original level after a while of the vapor cut-off in spite of the concentration of vapor. This indicates that the poisoning of the catalyst by the water is due to the competitive adsorption between H_2O vapor and NH_3 or NO, which is reversible. When the water vapor concentration was increased, the activity of the spent catalyst (after the cut-off of vapor) was a little bit lower than that of the original catalyst. It is demonstrated that the hydroxyl may be created due to the adsorption and decomposition of H_2O on the surface of Cu/SAPO-34 with the increase of H_2O concentration, resulting in the irreversible loss of deNO$_x$ activity [3].

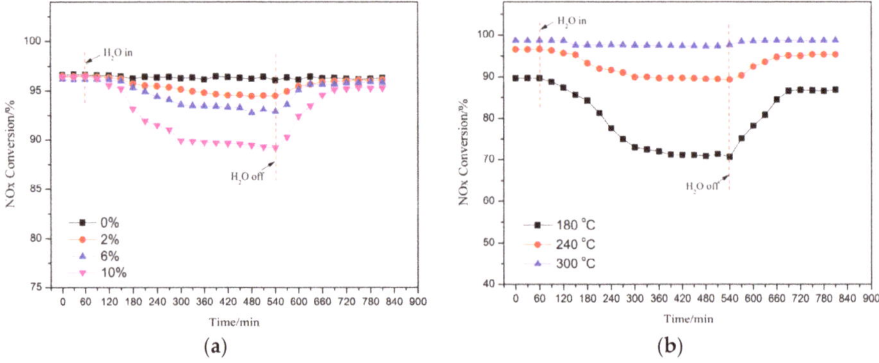

Figure 2. Effect of the injected H_2O concentration on the deNO$_x$ activity of the catalyst (240 °C) (**a**); effect of H_2O on the deNO$_x$ activity of catalyst under different temperatures (10% H_2O) (**b**).

Figure 2b shows the effect of H_2O on the deNO$_x$ activity of the catalyst Cu(2)–Mn(6)/SAPO-34 under different temperatures with the concentration of water vapor as 10%. The deNO$_x$ activity of the catalyst was significantly influenced at low reaction temperature with the presence of H_2O. A sharp decline of the deNO$_x$ activity of catalyst (from 86% to 68%) can be observed under the reaction temperature of 180 °C after the 4-hour injection of vapor into the feeding gas. However, no significant change in the activity of the catalyst can be found for the reaction under 300 °C in the presence of vapor. After 8 h injection of the vapor under 180 °C, the activity of the catalyst was all recovered to its original level in 2 h, while the activity recovery duration for the catalyst was decreased with the increased reaction temperature after the cut-off of vapor. The adsorption capacity of H_2O on the surface of the catalyst can be enhanced under the lower temperatures, occupying more active sites than that of NH_3, NO and other reaction gases [30]. It can be also concluded that poisoning performance of water on the activity of the catalyst can be ignored while the reaction temperature is increased over 300 °C.

2.3. Effect of SO_2 on the deNO$_x$ Activity of the Catalyst

The effect of the injected SO_2 concentration (from 500 ppm to 1500 ppm) on the deNO$_x$ activity of the catalyst Cu(2)–Mn(6)/SAPO-34 under the reaction temperature of 240 °C was shown in Figure 3a.

It is obvious that the deNO$_x$ activity of the catalyst was remarkably and abruptly decreased with the injection of SO_2 for the three different concentrations. After 2 h injection of SO_2, the deNO$_x$ activity of catalyst was reduced from 97% to 72%, 68%, 63% for the SO_2 concentration of 500, 1000 and 1500 ppm. Compared to that of the injected H_2O, the high concentration of the injected SO_2 could accelerate the decline of the activity of the catalyst. The formation of SO_3 can be promoted by the high concentration of the injected SO_2, consequently enhancing the formation of ammonium sulfate with NH_3 covering the catalyst active sites on the surface. After the cut-off (8 h) of the injected SO_2, the activity of the catalyst was increased, but much lower than its initial activity. It is indicated that the competitive adsorption between SO_2 and NH_3 or NO is not the main reason for the loss of deNO$_x$ activity in the presence of SO_2. Part of active sites on the catalyst was occupied by SO_2 over NH_3 and NO leading to the temporary poisoning, while a great number of active sites was covered by the formed sulfate salts (such as ammonium sulfate) for the permanent deactivation of the catalyst [31,32].

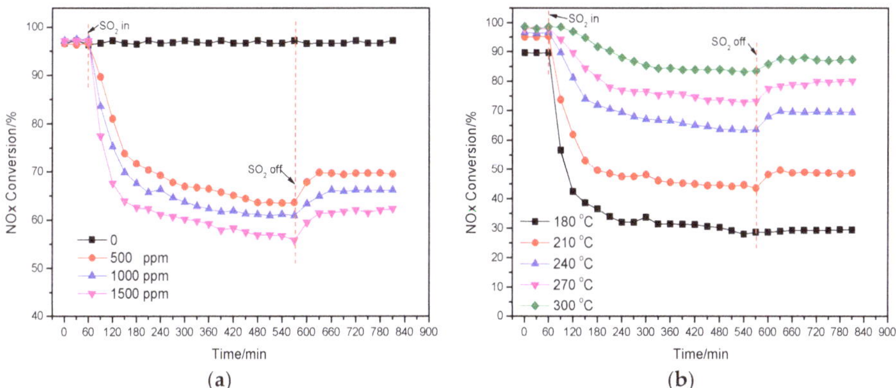

Figure 3. Effect of the injected SO_2 concentration of on deNO$_x$ activity of catalyst (240 °C) (**a**); Effect of the injected SO_2 (500 ppm) on deNO$_x$ activity of catalyst under different temperatures (**b**).

The effect of the injected SO_2 on the deNO$_x$ activity of the catalyst Cu(2)–Mn(6)/SAPO-34 under different reaction temperatures was shown in Figure 3b. The effect of the injected SO_2 on the activity of the catalyst was greatly influenced by the reaction temperature. It needs to be noted that the activity of the catalyst was decreased from 90% to 29% after 8-hour injection of 500 ppm SO_2 under 180 °C. Comparatively, under the reaction temperature of 300 °C, the activity of the catalyst was reduced from 99% to 88% after 8-hour injection of 500 ppm SO_2. This result demonstrates that the effect of SO_2 on deNO$_x$ activity is greatly related to the reaction temperature and a much more rapid decrease of deNO$_x$ activity happened with the addition of SO_2 at a lower temperature. It should be noted that no obvious recovery of deNO$_x$ activity at the reaction temperature of 180 °C can be observed. The more detailed mechanism of SO_2 poisoning over Cu/SAPO-34 at different reaction temperatures will be further discussed.

2.4. *Effect of Both H_2O and SO_2 Injection on the deNO$_x$ Activity of the Catalyst*

Figure 4 shows the effect of injection of H_2O and SO_2 on the catalytic activity of Cu(2)–Mn(6)/SAPO-34. NO conversion was decreased at the presence of H_2O and SO_2, compared to the performance the individual injection of H_2O or SO_2. Moreover, the co-existence of H_2O and SO_2 in the feeding gas gave a much more serious decline on the deNO$_x$ activity of the catalyst compared to the sum of the effect of H_2O and SO_2 respectively, which is labeled as the estimated value as shown in Figure 4. The existence of H_2O could enhance the deactivation of the catalyst by SO_2 through two ways: (1) SO_2 in gas phase reacted with the H_2O adsorbed on active site of catalyst to generate sulfuric acid or sulfurous acid which is easily reacted with NH_3; (2) the thermal decomposition of the formed ammonium sulfate on the active sites was confined at the presence of H_2O.

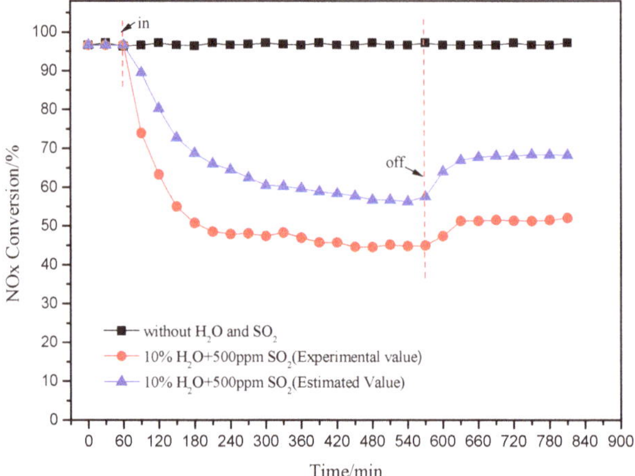

Figure 4. Effect of the injected H_2O and SO_2 on the deNO$_x$ activity of catalyst (240 °C).

2.5. Mechanism of the Catalyst Poisoning by H_2O and/or SO_2

In order to identify the change of crystalline phases for the fresh and spent Cu(2)–Mn(6)/SAPO-34 catalyst after SO_2 and/or H_2O poisoning. Powder X-ray diffraction (XRD) measurements were carried out and the patterns are shown in Figure 5. For the fresh Cu/SAPO-34 catalyst, sharp diffraction peaks corresponding to CHA phases are obtained. After poisoned with SO_2 or H_2O, the intensity of each peak decreases, especially for the catalyst after SO_2 + H_2O treatment, indicating the obvious skeleton damage of SAPO-34 after adding SO_2 + H_2O for 8 h [33]. No obvious crystalline change can be observed in the presence of 10% H_2O, this indicates that the phase poisoned by H_2O existed either in amorphous form or in the particle beyond the limited size of XRD detection [34]. For the catalyst after SO_2 poisoning for 8 h, A number of new diffraction peaks at 2θ = 11.9°, 29.8°and 40.7° can be observed. The peaks at 11.9° and 29.8° can be attributed to NH_4HSO_4 and $(NH_4)_2SO_4$, while the latter peak at 40.7° is assigned to $MnSO_4$. For the catalyst after SO_2 + H_2O poisoning for 8h, new diffraction peaks at 2θ = 29.8° and 34.1° can be observed, which can be assigned to $(NH_4)_2SO_4$. In addition, the new peak at 2θ = 21.1° can be assigned to the formation of $CuSO_4$. It can be deduced that the newly formed NH_4HSO_4, $(NH_4)_2SO_4$ and $MnSO_4$ result in the loss of deNO$_x$ activity in the presence of SO_2. When H_2O and SO_2 were added simultaneously, SO_2 in gas phase may react with adsorbed H_2O to generate sulfuric acid or sulfurous acid, which is easily reacted with NH_3 to form $(NH_4)_2SO_4$. Thus, the formation of $(NH_4)_2SO_4$ and $CuSO_4$ may cause the loss of deNO$_x$ activity in the presence of SO_2 + H_2O.

In order to determine the morphology before and after SO_2 or/and H_2O poisoning, the fresh Cu/SAPO-34 and poisoned catalyst by SO_2 or/and H_2O were characterized by SEM method shown in Figure 6. No significant change of the surface morphology can be found between the catalysts before and after the H_2O poisoning. The active particles on the surface of fresh catalyst were replaced by the bulks of ammonium sulfate after the poisoning by individual SO_2, inhibiting the deNO$_x$ activity of the catalyst. Similar phenomena can be observed for the spent catalyst in presence of H_2O and SO_2, where the agglomeration of ammonium sulfate lead to the decline in the surface active sites of the catalyst and thus the activity of the catalyst.

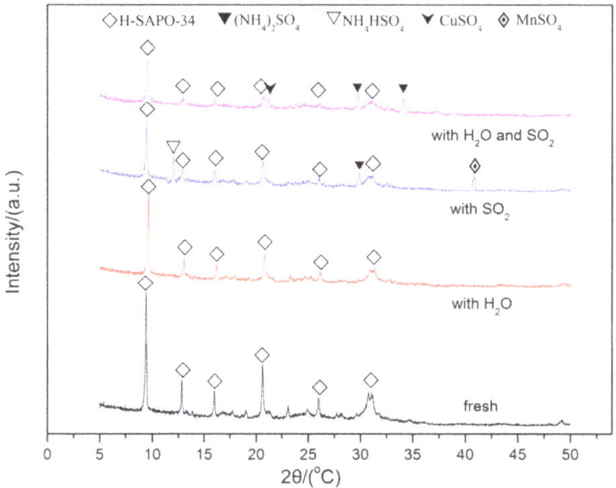

Figure 5. X-ray diffraction spectra of the fresh and spent Cu(2)–Mn(6)/SAPO-34.

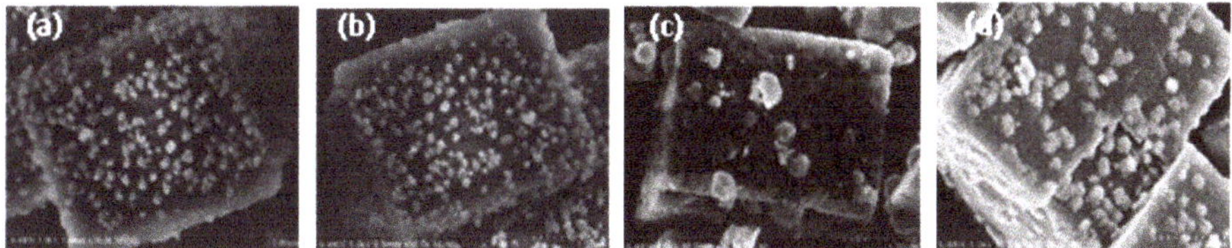

Figure 6. SEM micrographs of fresh and spent Cu(2)-Mn(6)/SAPO-34: (**a**) fresh; (**b**) 10% H_2O 240 °C 8 h; (**c**) 500 ppm SO_2 240 °C 8 h; (**d**) 10% H_2O and 500 ppm SO_2 240 °C 8 h.

The TG curves of the poisoned catalyst in the presence of individual SO_2 and both H_2O and SO_2 were shown in Figure 7. Three mass loss stages can be observed for the two kinds of the poisoned catalyst: the first mass loss in the temperature between 50 °C and 100 °C was attributed to the dehydration of the catalyst; the second stage between 200 °C and 400 °C can be designated to the thermal decomposition of NH_4HSO_4 and $(NH_4)_2SO_4$ is 200 °C [35]; the third stage between 600 °C and 800 °C can be attributed to the decomposition of the sulfate metal salts [36]. The peak value of the second mass loss stage of the poisoned catalyst in the presence of H_2O and SO_2 is notably higher than that of the poisoned catalyst in presence of individual SO_2, indicating that more content of $(NH_4)SO_4$ deposited on the surface of the catalyst. This confirms that the formation of ammonium sulfate can be facilitated and promoted with the injection of H_2O in the feeding gas, which is consistent with the XRD and TG-DTG analysis. While the peak value of the third mass loss stage of the poisoned catalyst in the presence of H_2O and SO_2 is lower than that in the presence of SO_2. It might be because the formation of $(NH_4)SO_4$ can inhibit the formation of sulfate metal salts due to the consumption of SO_2.

The specific surface area, pore volume and pore size of fresh and spent Cu(2)–Mn(6)/SAPO-34 were determined by N_2 adsorption and summarized in Table 1. The specific surface area and pore volume of Cu(2)–Mn(6)/SAPO-34 catalyst was decreased after the all different NO reduction experiments with or without the injection of H_2O and/or SO_2. It can be seen that the damage of the surface area is related to the concentration of SO_2 or H_2O and the reaction temperature. With the increase of reaction temperature, the damage of the surface area caused by SO_2 is lightening. The low reaction temperature could confine the thermal decomposition of the formed NH_4HSO_4 and $(NH_4)_2SO_4$ on the surface, resulting in a decline in surface area and the blockage of pore channel [37]. With the increase of the concentration of SO_2, a change of which (from 331 m^2/g to 320 m^2/g) can be observed for different SO_2 concentrations (from 500 ppm to 1500 ppm) under the temperature of

240 °C. For the poisoning effect of H_2O, with the increase of reaction temperature, the existence of H_2O has a more severe effect on the surface area of Cu/SAPO-34, which may result from the inhibited adsorption capacity of H_2O. With the increase of the concentration of H_2O, the surface area decreases from 425 m^2/g to 408 m^2/g as the concentration of H_2O increased from 2% to 10% as well as the pore volume of catalyst. When H_2O and SO_2 added into the gas stream simultaneously for 8 h, the surface area decreases from 457 m^2/g to 276 m^2/g, which is a dramatically decrease compared to the injection of SO_2 for 8 h at 240 °C (from 457 m^2/g to 331 m^2/g) and the injection of H_2O for 8 h at 240 °C (from 457 m^2/g to 408 m^2/g). It is demonstrated that the synergistic poisoning effect of SO_2 and H_2O is enhanced, ascribed to the large amount of deposited $(NH_4)_2SO_4$ on the pore channel of Cu/SAPO-34, which is consistent with the TG-DTG results.

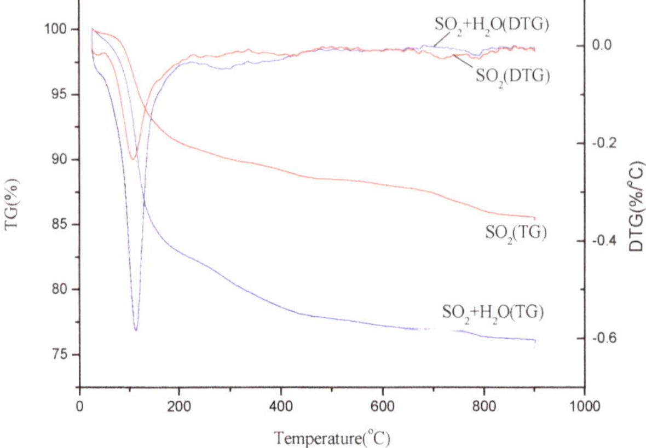

Figure 7. TG-DTG curves of the thermal decomposition for the Cu(2)–Mn(6)/SAPO-34 catalyst after 8 h poisoning in presence of 500 ppm SO_2 and/or 10% H_2O.

Table 1. Brunner−Emmet−Teller (BET) analysis of the fresh and spent Cu(2)–Mn(6)/SAPO-34 catalyst.

Catalyst	BET Surface Area (m^2g^{-1})	Pore Volume (cm^3g^{-1})	Pore Diameter (nm)
Fresh catalyst	457	0.23	2.08
Catalyst exposed to 500 ppm SO_2 at 180 °C for 8 h	307	0.14	2.42
Catalyst exposed to 500 ppm SO_2, at 210 °C for 8 h	314	0.16	2.29
Catalyst exposed to 500 ppm SO_2, at 240 °C for 8 h	331	0.16	2.15
Catalyst exposed to 500 ppm SO_2, at 270 °C for 8 h	367	0.19	2.09
Catalyst exposed to 500 ppm SO_2, at 300 °C for 8 h	372	0.18	2.09
Catalyst exposed to 1000 ppm SO_2, at 240 °C for 8 h	321	0.16	2.14
Catalyst exposed to 1500 ppm SO_2, at 240 °C for 8 h	320	0.16	2.28
Catalyst exposed to 10% H_2O, at 180 °C for 8 h	383	0.18	2.03
Catalyst exposed to 10% H_2O, at 240 °C for 8 h	408	0.19	2.02
Catalyst exposed to 10% H_2O, at 300 °C for 8 h	441	0.21	2.04
Catalyst exposed to 2% H_2O, at 240 °C for 8 h	425	0.20	2.03
Catalyst exposed to 6% H_2O, at 240 °C for 8 h	413	0.19	1.95
Catalyst exposed to 10% H_2O+500 ppmSO_2, at 240 °C for 8 h	276	0.17	2.62

Based on the activity tests and characterizations, the poisoning effect of SO_2 or/and H_2O can be described as follows. The poisoning effect of H_2O, especially at low temperatures, can be ascribed to the competitive adsorption between H_2O and NH_3 on Lewis acid sites by occupying the metal sits [38], which can be recovered to nearly the original activity after H_2O was removed. The XRD patterns of the catalyst after H_2O poisoning did not show ant obvious changes, indicating no change with the active metal sites.

While for the poisoning effect of SO_2 over Cu/SAPO-34 catalyst, the deNO$_x$ activity decreases from 90% to 29% at 180 °C in a short time of SO_2 injection. In addition, the deNO$_x$ activity could not be recovered after the elimination of SO_2, indicating the permanent deactivation of SO_2 on Cu/SAPO-34 at 180 °C. The poisoning mechanism could be summarized as three aspects. Firstly, SO_2 in the gas may be oxidized to SO_3 and further react with NH_3 to form NH_4HSO_4, which is a drying powdery decomposed at 280 °C [39]. The formation of NH_4HSO_4 causes the plug of active sites of the catalyst and the decline in surface area. Ammonium sulfate crystallite is observed on the XRD spectra and the weight loss peak ascribed to the decomposition of ammonium sulfates is also observed from the TG curves. Secondly, the active sites (i.e., MnO_2 or CuO species) may react with SO_2 or SO_3 to form $MnSO_4$ or $CuSO_4$, which inhibited the redox properties [23]. The diffraction peaks assigned to $MnSO_4$ and $CuSO_4$ are observed in XRD spectra. Thirdly, the competitive adsorption of SO_2 and NO on metal sites may be a part of the reason for the loss of $deNO_x$ activity at the reaction temperature of above 180 °C [40]. With regards to the synergistic effect of SO_2 + H_2O, the $deNO_x$ activity tests and characterizations show that the existence of H_2O could enhance the deactivation of the catalyst by SO_2 through two ways: (1) SO_2 in gas phase reacted with the H_2O adsorbed on active site of catalyst to generate sulfuric acid or sulfurous acid which is easily reacted with NH_3 to form large amount of $(NH_4)_2SO_4$; (2) the thermal decomposition of the formed ammonium sulfate on the active sites was confined at the presence of H_2O. Both of the two explanations can be assigned to the deposition of $(NH_4)_2SO_4$, which further plug the pore channel of catalyst and cause the rapid decrease of surface area as shown in Table 1.

3. Materials and Methods

3.1. Catalyst Preparation

The molecular sieve was modified by impregnation method. A certain amount of zeolite molecular sieve H-SAPO-34 ($n(P_2O_5)$:$n(SiO_2)$:$n(Al_2O_3)$ = 1:1:1), provided by the catalyst factory of Nankai University, was weighed and dried in a drying oven at 105 °C for 30 min. A certain amount of $Cu(NO_3)_2 \bullet 3H_2O$ powder was mixed with manganese nitrate solution with the 50 wt.% in the beaker of 200 mL, and then added 50 mL of deionized water into the immersion liquid. The beaker with a magnetic stirrer inside was immersed in a water bath at a constant temperature of 40 °C. After 12 h of immersion, the solution was thoroughly mixed and heated until the moisture was completely evaporated. The powder was then put into a dry oven at about 100 °C for 12 h. The dried powder was ground and sieved by the 40 to 60 mesh. The obtained powders were placed in a tube furnace and calcined at 450 °C for 6 h to obtain the catalyst sample for the experiment. The catalyst sample Cu(2)-Mn(6)/SAPO-34(450) indicated that the 2 wt.% Cu and 6 wt.% Mn were loaded on the molecular sieve SAPO-34 with the calcination temperature of 450 °C.

3.2. DeNOx Activity Measurements

$DeNO_x$ activity measurements using NH_3 were carried out in a fixed-bed stainless steel tubular flow reactor (Inner diameter: 16 mm) (Figure 8). The tube furnace of the reactor can be heated from the room temperature to 800 °C The gas feeding system was composed of five gas feeding pipes controlled by the mass flow meter (0–1.5 L) and one liquid feeder a microinjection pump (0.001 μL/min–127 mL/min), equipped with the reactor for adjusting the composition of the feeding gas. The reaction temperature in the experiments was set to be changed from 90 to 330 °C.

Two milliliter catalyst was placed on the holder of the reactor, and the feeding gas consisted of 350 ppm NO, 350 ppm NH_3, 3 vol.% O_2 and N_2 as balanced. H_2O (0–10 vol.%) and/or 500–1500 ppm SO_2 would be injected into the feeding gas for investigating the effect of H_2O and/or SO_2 on the catalyst poisoning. The GHSV was set to be 15,000h^{-1}. The NO and NO_2 concentration of the reactor inlet and outlet was collected by an airbag and analyzed by the flue gas analyzer (Testo 350, Testo, Inc., Lenzkirch, Germany). The outlet N_2O concentration is collected by the N_2O analyzer (Medi-Gas G200,

Bedfont Scientific Ltd., Bedfont, United Kingdom). The NO conversion was obtained from the equation as follows:

$$\eta = \frac{C_{NO}^{in} - C_{NO}^{out}}{C_{NO}^{in}} \times 100\%$$

where η, C_{NO}^{in}, C_{NO}^{out} represented the NO conversion, inlet and outlet NO concentration.

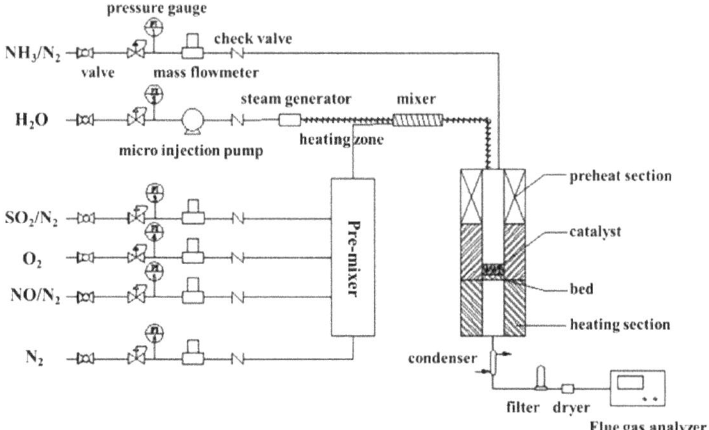

Figure 8. Schematics for the fixed-bed tubular flow reactor.

3.3. Catalyst Characterization

A micromeritics ASAP 2010M micropore size analyzer was used to measure the N_2 adsorption isotherms of the catalyst sample at liquid N_2 temperature (−196 °C. Specific surface area, pore volume and pore diameter can be determined by N_2 adsorption using the BET and BJH methods.

The XRD measurement for the catalyst was carried out on a Rigaku D/Smartlab(III) system (Rigaku, Neu-Isenburg, Germany) with Cu Ka radiation. The X-ray source was operated at 40kV and 40mA. The diffraction patterns were taken in the 2θ range of 5–50° at a scan speed of 10° min^{-1} and a resolution of 0.02°.

SEM was performed using a SIRION-50 scanning electron microscope from Field Electron and Ion Company, the Netherlands, with a resolution of 150 eV.

Thermal gravimetric analysis of the catalyst samples was performed using TGA-101 type produced by the Nanjing Exhibition Electrical and Mechanical Technology Company (Nanjing, China). The accuracy of the instrument is 0.2 μg. For the TG experiments, the spent catalyst samples were measured under the temperature from room temperature to 900 °C at the heating rate of 15 °C min^{-1}.

4. Conclusions

Cu-Mn/SAPO-34 with the loading of 2 wt.% Cu and 6 wt.% Mn exhibited remarkable low-temperature NO reduction activity among the prepared Cu-Mn/SAPO-34 catalysts. The NO conversion could achieve as high as 72% under the reaction temperature of 120 °C, while the value reached more than 90% under the temperature between 180 °C and 330 °C. The reversibly poisoning effect of H_2O is mainly due to the competitive adsorption between H_2O and NH_3 on Lewis acid sites by occupying the metal sits. With the increase of reaction temperature, the poisoning effect is less important. The poisoning effect of SO_2 on deNO$_x$ activity is dependent on the reaction temperature. At low temperature, the poisoning effect of SO_2 is permanent with no recovery of deNO$_x$ activity after the elimination of SO_2. XRD, SEM and BET analysis suggested that the deposition of $(NH_4)_2SO_4$ on active sites may be the main reason for the loss of deNO$_x$ activity. TG-DTG analysis shows that some metal sulfates are formed on the surface of Cu/SAPO-34 catalyst, which may inhibit the redox properties and cut off the redox cycle during the low-temperature SCR reaction. The addition of H_2O into the SO_2-containing atmosphere promotes the formation of $(NH_4)_2SO_4$ on the surface of the catalyst, causing the damage of surface area and the rapid decrease of deNO_x activity. At a higher reaction temperature, the formation of ammonium sulfate might be inhibited by the high reaction temperature, or the formed ammonium sulfate from SO_2 and NH_3 can be easily decomposed under the high temperature.

Author Contributions: Funding acquisition, D.S.; Investigation, S.G., B.Y., R.L. and Y.S.; Methodology, G.L., W.Z., P.H., S.G., B.Y., R.L., Y.S. and D.S.; Supervision, D.S.; Writing–original draft, G.L., W.Z. and P.H.; Writing–review & editing, G.L., W.Z. and P.H.

Funding: This work was supported by National Natural Science Foundation of China [grant number 51676047], the international collaboration project from Department of Science and Technology of Jiangsu Province [grant number BZ2017014], Key Technology R and D Program of Jiangsu Province [grant number BE2015677], and the Science and Technology Project of Jiangsu Power Design Institute Co., Ltd. [grant number 32-JK-2019-017].

Conflicts of Interest: The authors declare no conflict of interest.

References

1. Tie, X.; Huang, R.-J.; Dai, W.; Cao, J.; Long, X.; Su, X.; Zhao, S.; Wang, Q.; Li, G. Effect of heavy haze and aerosol pollution on rice and wheat productions in China. *Sci. Rep.* **2016**, *6*, 29612. [CrossRef] [PubMed]
2. Maher, B.A.; Ahmed, I.A.M.; Karloukovski, V.; MacLaren, D.A.; Foulds, P.G.; Allsop, D.; Mann, D.M.A.; Torres-Jardón, R.; Calderon-Garciduenas, L. Magnetite pollution nanoparticles in the human brain. *Natl. Acad. Sci.* **2016**, *113*, 10797–10801. [CrossRef] [PubMed]
3. Li, J.; Chang, H.; Ma, L.; Hao, J.; Yang, R.T. Low-temperature selective catalytic reduction of NO_x with NH_3 over metal oxide and zeolite catalysts—A review. *Catal. Today.* **2011**, *175*, 147–156. [CrossRef]
4. Boningari, T.; Smirniotis, P.G. Impact of nitrogen oxides on the environment and human health: Mn-based materials for the NO_x abatement. *Curr. Opin. Chem. Eng.* **2016**, *13*, 133–141. [CrossRef]
5. Forzatti, P.; Nova, I.; Tronconi, E.; Kustov, A.; Thøgersen, J.R. Effect of operating variables on the enhanced SCR reaction over a commercial V_2O_5–WO_3/TiO_2 catalyst for stationary applications. *Catal. Today* **2012**, *184*, 153–159. [CrossRef]
6. Zhang, X.P.; Shen, B.X.; Chen, J.H.; Cai, J.; He, C.; Wang, K. $Mn_{0.4}/Co_{0.1}Ce_{0.45}Zr_{0.45}O_X$, high performance catalyst for selective catalytic reduction of NO by ammonia. *J. Energy Inst.* **2013**, *86*, 119–124. [CrossRef]
7. Zhang, L.; Wang, D.; Liu, Y.; Kamasamudram, K.; Li, L.; Epling, W. SO_2 poisoning impact on the NH_3-SCR reaction over a commercial Cu-SAPO-34 SCR catalyst. *Appl. Catal. B Environ.* **2014**, *156–157*, 371–377. [CrossRef]
8. Ma, L.; Cheng, Y.; Cavataio, G.; McCabe, R.W.; Fu, L.; Li, L. Characterization of commercial Cu-SSZ-13 and Cu-SAPO-34 catalysts with hydrothermal treatment for NH_3-SCR of NO_x in diesel exhaust. *Chem. Eng. J.* **2013**, *225*, 323–330. [CrossRef]
9. Blakeman, P.G.; Burkholder, E.M.; Chen, H.-Y.; Collier, J.E.; Fedeyko, J.M.; Jobson, H.; Rajaram, R.R. The role of pore size on the thermal stability of zeolite supported Cu SCR catalysts. *Catal. Today* **2014**, *231*, 56–63. [CrossRef]
10. Bates, S.A.; Delgass, W.N.; Ribeiro, F.H.; Miller, J.T.; Gounder, R. Methods for NH_3 titration of Brønsted acid sites in Cu-zeolites that catalyze the selective catalytic reduction of NO_x with NH_3. *J. Catal.* **2014**, *312*, 26–36. [CrossRef]
11. Bates, S.A.; Verma, A.A.; Paolucci, C.; Parekh, A.A.; Anggara, T.; Yezerets, A.; Schneider, W.F.; Miller, J.T.; Delgass, W.N.; Ribeiro, F.H. Identification of the active Cu site in standard selective catalytic reduction with ammonia on Cu-SSZ-13. *J. Catal.* **2014**, *312*, 87–97. [CrossRef]
12. Ettireddy, P.R.; Kotrba, A.; Boningari, T.; Smirniotis, P. Low temperature SCR catalysts optimized for cold-start and low-load engine exhaust conditions. *SAE Technical Paper* **2015**, *01*, 1026.
13. Kim, Y.J.; Kwon, H.J.; Heo, I.; Nam, I.S.; Cho, B.K.; Jin, W.C.; Cha, M.S.; Yeo, G.K. Mn–Fe/ZSM5 as a low-temperature SCR catalyst to remove NOx from diesel engine exhaust. *Appl. Catal. B Environ.* **2012**, *126*, 9–21. [CrossRef]
14. Nova, I.; Tronconi, E. *Urea-SCR Technology for deNOx after Treatment of Diesel Exhausts*; Springer: New York, NY, USA, 2014.

15. Chen, H.-Y.; Wei, Z.; Kollar, M.; Gao, F.; Wang, Y.; Szanyi, J.; Peden, C.H. A comparative study of N_2O formation during the selective catalytic reduction of NOx with NH_3 on zeolite supported Cu catalysts. *J. Catal.* **2015**, *329*, 490–498. [CrossRef]
16. Wang, J.; Yu, T.; Wang, X.; Qi, G.; Xue, J.; Shen, M.; Li, W. The influence of silicon on the catalytic properties of Cu/SAPO-34 for NO_x reduction by ammonia-SCR. *Appl. Catal. B Environ.* **2012**, *127*, 137–147. [CrossRef]
17. Xue, J.; Wang, X.; Qi, G.; Wang, J.; Shen, M.; Li, W. Characterization of copper species over Cu/SAPO-34 in selective catalytic reduction of NO_x with ammonia: Relationships between active Cu sites and de-NO_x performance at low temperature. *J. Catal.* **2013**, *297*, 56–64. [CrossRef]
18. Ye, Q.; Wang, L.; Yang, R.T. Activity, propene poisoning resistance and hydrothermal stability of copper exchanged chabazite-like zeolite catalysts for SCR of NO with ammonia in comparison to Cu/ZSM-5. *Appl. Catal. A Gen.* **2012**, *427–428*, 24–34. [CrossRef]
19. Deka, U.; Lezcano-Gonzalez, I.; Warrender, S.J.; Picone, A.L.; Wright, P.A.; Weckhuysen, B.M.; Beale, A.M. Changing active sites in Cu–CHA catalysts: deNO_x selectivity as a function of the preparation method. *Microporous Mesoporous Mater.* **2013**, *166*, 144–152. [CrossRef]
20. Gao, F.; Kwak, J.H.; Szanyi, J.; Peden, C.H. Current Understanding of Cu-Exchanged Chabazite Molecular Sieves for Use as Commercial Diesel Engine DeNO_x Catalysts. *Top. Catal.* **2013**, *56*, 1441–1459. [CrossRef]
21. Liu, X.; Wu, X.; Weng, D.; Si, Z.; Ran, R. Migration, reactivity, and sulfur tolerance of copper species in SAPO-34 zeolite toward NO_x reduction with ammonia. *RSC Adv.* **2017**, *7*, 37787–37796. [CrossRef]
22. Shen, M.; Wen, H.; Hao, T.; Yu, T.; Fan, D.; Wang, J.; Li, W.; Wang, J. Deactivation mechanism of SO_2 on Cu/SAPO-34 NH_3-SCR catalysts: structure and active Cu^{2+}. *Catal. Sci. Technol.* **2015**, *5*, 1741–1749. [CrossRef]
23. Wijayanti, K.; Andonova, S.; Kumar, A.; Li, J.; Kamasamudram, K.; Currier, N.M.; Yezerets, A.; Olsson, L. Impact of sulfur oxide on NH_3-SCR over Cu-SAPO-34. *Appl. Catal. B: Environ.* **2015**, *166–167*, 568–579. [CrossRef]
24. Jangjou, Y.; Wang, D.; Kumar, A.; Li, J.; Epling, W.S. SO_2 Poisoning of the NH_3-SCR Reaction over Cu-SAPO-34: Effect of Ammonium Sulfate versus Other S-Containing Species. *ACS Catal.* **2016**, *6*, 6612–6622. [CrossRef]
25. Liang, J.; Li, H.; Zhao, S.; Guo, W.; Wang, R.; Ying, M. Characteristics and performance of SAPO-34 catalyst for methanol-to-olefin conversion. *Appl. Catal.* **1990**, *64*, 31–40. [CrossRef]
26. Xu, L.; Du, A.; Wei, Y.; Wang, Y.; Yu, Z.; He, Y.; Zhang, X.; Liu, Z. Synthesis of SAPO-34 with only Si(4Al) species: Effect of Si contents on Si incorporation mechanism and Si coordination environment of SAPO-34. *Microporous Mesoporous Mater.* **2008**, *115*, 332–337. [CrossRef]
27. Buchholz, A.; Wang, W.; Xu, M.; Arnold, A.; Hunger, M. Thermal stability and dehydroxylation of Brønsted acid sites in silicoaluminophosphates H-SAPO-11, H-SAPO-18, H-SAPO-31, and H-SAPO-34 investigated by multi-nuclear solid-state NMR spectroscopy. *Microporous Mesoporous Mater.* **2002**, *56*, 267–278. [CrossRef]
28. Wang, J.; Huang, Y.; Yu, T.; Zhu, S.; Shen, M.; Li, W.; Wang, J. The migration of Cu species over Cu–SAPO-34 and its effect on NH_3 oxidation at high temperature. *Catal. Sci. Technol.* **2014**, *4*, 3004–3012. [CrossRef]
29. Busca, G.; Lietti, L.; Ramis, G.; Berti, F. Chemical and mechanistic aspects of the selective catalytic reduction of NO_x by ammonia over oxide catalysts: A review. *Appl. Catal. B Environ.* **1998**, *18*, 1–36. [CrossRef]
30. Huang, Z.; Liu, Z.; Zhang, X.; Liu, Q. Inhibition effect of H_2O on V_2O_5/AC catalyst for catalytic reduction of NO with NH_3 at low temperature. *Appl. Catal. B Environ.* **2006**, *63*, 260–265. [CrossRef]
31. Shen, B.; Zhang, X.; Ma, H.; Yao, Y.; Liu, T. A comparative study of Mn/CeO_2, Mn/ZrO_2 and Mn/Ce-ZrO_2 for low temperature selective catalytic reduction of NO with NH_3 in the presence of SO_2 and H_2O. *J. Environ. Sci.* **2013**, *25*, 791–800. [CrossRef]
32. Notoya, F.; Su, C.; Sasaoka, E.; Nojima, S. Effect of SO_2 on the Low-Temperature Selective Catalytic Reduction of Nitric Oxide with Ammonia over TiO_2, ZrO_2, and Al_2O_3. *Ind. Eng. Chem. Res.* **2001**, *40*, 3732–3739. [CrossRef]
33. Yan, C.; Cheng, H.; Yuan, Z.; Wang, S. The role of isolated Cu^{2+} location in structural stability of Cu-modified SAPO-34 in NH_3-SCR of NO. *Environ. Technol.* **2015**, *36*, 169–177. [CrossRef]
34. Zhang, R.; Alamdari, H.; Kaliaguine, S. SO_2 poisoning of $LaFe_{0.8}Cu_{0.2}O_3$ perovskite prepared by reactive grinding during NO reduction by C_3H_6. *Appl. Catal. A Gen.* **2008**, *340*, 140–151. [CrossRef]
35. Fan, Y.Z.; Cao, F.H. Thermal Decomposition Kinetics of Ammonium Sulfate. *J. Chem. Eng. Chin. Univ.* **2011**, *25*, 341–346.

36. Shen, B.X.; Liu, T. Deactivation of MnO_x-CeO_x/ACF Catalysts for Low-Temperature NH_3-SCR in the Presence of SO_2. *Acta Physico-Chimica Sinica* **2010**, *26*, 3009–3016.
37. Xu, W.; He, H.; Yu, Y. Deactivation of a Ce/TiO_2 catalyst by SO_2 in the selective catalytic reduction of NO by NH_3. *J. Phys. Chem. C* **2009**, *113*, 4426–4432. [CrossRef]
38. Pan, S.; Luo, H.; Li, L.; Wei, Z.; Huang, B. H_2O and SO_2 deactivation mechanism of MnO_x/MWCNTs for low-temperature SCR of NO_x with NH_3. *J. Mol. Catal. A Chem.* **2013**, *377*, 154–161. [CrossRef]
39. Kiyoura, R.; Urano, K. Mechanism, Kinetics, and Equilibrium of Thermal Decomposition of Ammonium Sulfate. *Ind. Eng. Chem. Proc. Dev.* **1970**, *9*, 489–494. [CrossRef]
40. Jiang, B.Q.; Wu, Z.B.; Liu, Y.; Lee, S.C.; Ho, W.K. DRIFT Study of the SO_2 Effect on Low-Temperature SCR Reaction over Fe−Mn/TiO_2. *J. Phys. Chem. C* **2010**, *114*, 4961–4965. [CrossRef]

Article

Comprehensive Comparison between Nanocatalysts of Mn−Co/TiO_2 and Mn−Fe/TiO_2 for NO Catalytic Conversion: An Insight from Nanostructure, Performance, Kinetics, and Thermodynamics

Yan Gao [1,2,*], Tao Luan [3], Shitao Zhang [2], Wenchao Jiang [2], Wenchen Feng [3] and Haolin Jiang [3]

1. Key Laboratory of Renewable Energy Building Application Technology of Shandong Province, Shandong Jianzhu University, Jinan 250101, China
2. Department of Thermal Engineering, Shandong Jianzhu University, Jinan 250101, China; shitao_zhang@hotmail.com (S.Z.); jiang_wc@126.com (W.J.)
3. Engineering Laboratory of Power Plant Thermal System Energy Saving of Shandong Province, Shandong University, Jinan 250061, China; prof.luantao@gmail.com (T.L.); wcfeng18@126.com (W.F.); haolin_jiang@hotmail.com (H.J.)

* Correspondence: gaoyan.sdu@hotmail.com; Tel.: +86-138-6415-4887

Received: 15 January 2019; Accepted: 9 February 2019; Published: 13 February 2019

Abstract: The nanocatalysts of Mn−Co/TiO_2 and Mn−Fe/TiO_2 were synthesized by hydrothermal method and comprehensively compared from nanostructures, catalytic performance, kinetics, and thermodynamics. The physicochemical properties of the nanocatalysts were analyzed by N_2 adsorption, transmission electron microscope (TEM), X-ray diffraction (XRD), H_2-temperature-programmed reduction (TPR), NH_3-temperature-programmed desorption (TPD), and X-ray photoelectron spectroscopy (XPS). Based on the multiple characterizations performed on Mn−Co/TiO_2 and Mn−Fe/TiO_2 nanocatalysts, it can be confirmed that the catalytic properties were decidedly dependent on the phase compositions of the nanocatalysts. The Mn−Co/TiO_2 sample presented superior structure characteristics than Mn−Fe/TiO_2, with the increased surface area, the promoted active components distribution, the diminished crystallinity, and the reduced nanoparticle size. Meanwhile, the Mn^{4+}/Mn^{n+} ratios in the Mn−Co/TiO_2 nanocatalyst were higher than Mn−Fe/TiO_2, which further confirmed the better oxidation ability and the larger amount of Lewis acid sites and Bronsted acid sites on the sample surface. Compared to Mn−Fe/TiO_2 nanocatalyst, Mn−Co/TiO_2 nanocatalyst displayed the preferable catalytic property with higher catalytic activity and stronger selectivity in the temperature range of 75–250 °C. The results of mechanism and kinetic study showed that both Eley-Rideal mechanism and Langmuir-Hinshelwood mechanism reactions contributed to selective catalytic reduction of NO with NH_3 (NH_3-SCR) over Mn−Fe/TiO_2 and Mn−Co/TiO_2 nanocatalysts. In this test condition, the NO conversion rate of Mn−Co/TiO_2 nanocatalyst was always higher than that of Mn−Fe/TiO_2. Furthermore, comparing the reaction between doping transition metal oxides and NH_3, the order of temperature−Gibbs free energy under the same reaction temperature is as follows: Co_3O_4 < CoO < Fe_2O_3 < Fe_3O_4, which was exactly consistent with nanostructure characterization and NH_3-SCR performance. Meanwhile, the activity difference of MnO_x exhibited in reducibility properties and Ellingham Diagrams manifested the promotion effects of cobalt and iron dopings. Generally, it might offer a theoretical method to select superior doping metal oxides for NO conversion by comprehensive comparing the catalytic performance with the insight from nanostructure, catalytic performance, reaction kinetics, and thermodynamics.

Keywords: NH_3-SCR; nanostructure; kinetics; thermodynamics; manganese oxides

---

1. Introduction

Nitrogen oxides (NO_x) generated from fossil fuels are regarded as the primary reason for acid rain, ozone depletion, photochemical smog, and greenhouse effects [1]. Selective catalytic reduction of NO_x with NH_3 or urea as a reductant (NH_3-SCR) is proposed to be the most efficient method for eliminating NO_x from stationary source and mobile source [2]. In recent decades, the commercial catalyst of V_2O_5-WO_3(MoO_3)/TiO_2 used for NH_3-SCR process exhibited an excellent catalytic property in the typical standard SCR reaction within the temperature range of 300–400 °C [3,4].

$$4NO + 4NH_3 + O_2 \rightarrow 4N_2 + 6H_2O \quad (1)$$

Nowadays, for the purpose of reducing the inhibiting effects of ash and SO_2 over the SCR catalysts, there is an intense interest in developing catalysts for NO reduction at lower temperature [5]. However, the $V_2O_5-WO_3(MoO_3)/TiO_2$ catalysts demand a strict temperature window, which limits their arrangement flexibility. The vanadium-based catalysts cannot reach satisfactory efficiency of eliminating NO_x as the reaction temperature lower than 250 °C. Hence, the catalysts appropriate to low temperature SCR are strongly desired, which could be placed at the downstream of electrostatic precipitator and desulfurizer [6]. A great deal of catalysts comprised of different transition metal oxides on various supports were analyzed to improve the catalytic ability for low-temperature deNO_x. The typical transition metals, such as Mn [7], Co [8], Fe [9], Ni [10], Zn [11], and Cr [9], were well-reported to display satisfactory properties at the low temperature. Among the various transition metal elements applied in the catalysts for NO_x reduction, manganese oxides display superior activity especially at the low temperature, which can be attributed to the various types of labile oxygen and high mobility of valence states [12]. Meanwhile, it was found that cobalt and iron species can combine with manganese to produce mixed nanoparticle-oxides and exhibit high SCR activity and excellent N_2 selectivity with a wide temperature window from 100—300 °C [13,14]. These mixed nanoparticle-oxides contain abundant oxygen vacancies on the catalyst surface, forming strong interaction bands at atomic scale, such as Mn-O-Fe [15,16] and Mn-O-Co [17]. Meanwhile, the active metal species of CoO_x and FeO_x are also regarded as the typical promoters for NO_x conversion, which serve as core catalyst components of active metal oxides, supplying surface oxygen to accelerate NO_x elimination [17,18].

Another crucial field of the investigation on NH_3-SCR process is the reaction mechanism on the active sites with a better understanding of the surface chemistry and developed kinetic models to evaluate the reactions. It was widely accepted that the main reaction of low temperature SCR complies with both the Eley-Rideal mechanism and Langmuir-Hinshelwood mechanism, which is further depended on the reaction temperature and the catalyst components [19,20]. According to the Eley-Rideal mechanism, the reaction occurred between NO and NH_3 (adsorbed) to generate an activated transition state and further decomposed into H_2O and N_2 [20]. Based on the Langmuir-Hinshelwood mechanism, the reaction happened between adsorbed NO and adsorbed NH_3 on the adjacent sites to form H_2O and N_2 [21]. Meanwhile, it was widely accepted that the first step in SCR was an oxidative abstraction of the hydrogen from adsorbed ammonia [22]. Therefore, oxidative NH_3 played an important part in the mechanism of NH_3-SCR, which could be evaluated by thermodynamics in details [7]. However, the comprehensive comparison of NO elimination over the nanocatalysts of $MnCoO_x$ and $MnFeO_x$ on TiO_2 support, especially with an insight from structure, performance, kinetics, and thermodynamics, has not been explored clearly.

In this study, the nanocatalysts of $Mn-Co/TiO_2$ and $Mn-Fe/TiO_2$ were synthesized by hydrothermal method. The physicochemical properties of the obtained samples were researched by SEM, BET, XRD, H2-TPR, NH3-TPD, and XPS. In the meantime, the catalytic performance comparation of $Mn-Co/TiO_2$ and $Mn-Fe/TiO_2$ nanocatalysts was investigated using kinetic and thermodynamic analysis. The purpose of this work was mean to explore a comprehensive perspective for optimizing multimetals SCR catalysts, especially for the Mn-based bimetals nanocatalysts.

2. Results and Discussions

2.1. Physicochemical Properties of the Nanocatalysts

2.1.1. TEM Analysis

The topographic characteristics and the porous structures of the Mn−Co/TiO_2 and Mn−Fe/TiO_2 nanocatalysts were collected by TEM test. As shown in Figure 1a, the most part of Mn−Co/TiO_2 nanocatalyst was formed with uniform elliptic nanoparticles with glabrous surfaces and well-proportioned size distribution and without apparent agglomeration, although there was a spot of tightly aggregated TiO_2 nanoparticles interfused into the smaller regular particles. Therefore, a distinct and unbroken mesh structure of micropore was form in Mn−Co/TiO_2 sample. However, according to the TEM images of Mn−Fe/TiO_2 nanocatalyst from Figure 1b, a noticeable augment in the particle size was observed. The nanoparticles were irregular, lots of which stacked on the nanocatalyst surface. According to Figure 1(c)(l), captured at a larger scope, it could be found that all active elements contained in Mn−Co/TiO_2 nanocatalyst were well dispersed without obvious irregular stacked particles. On the surface of Mn−Fe/TiO_2 nanocatalyst, the element of manganese appeared slight regional accumulation, which might be caused by the abundant micropore structure collapsing [23].

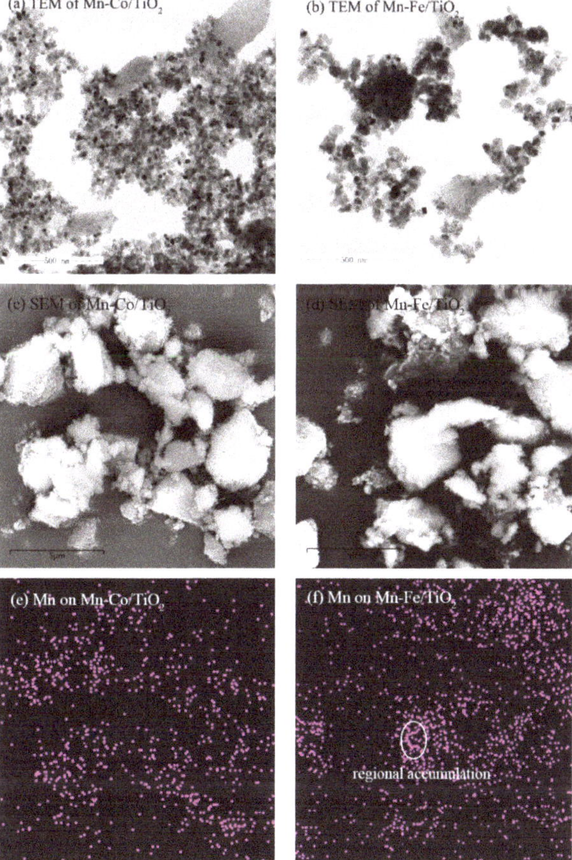

Figure 1. *Cont.*

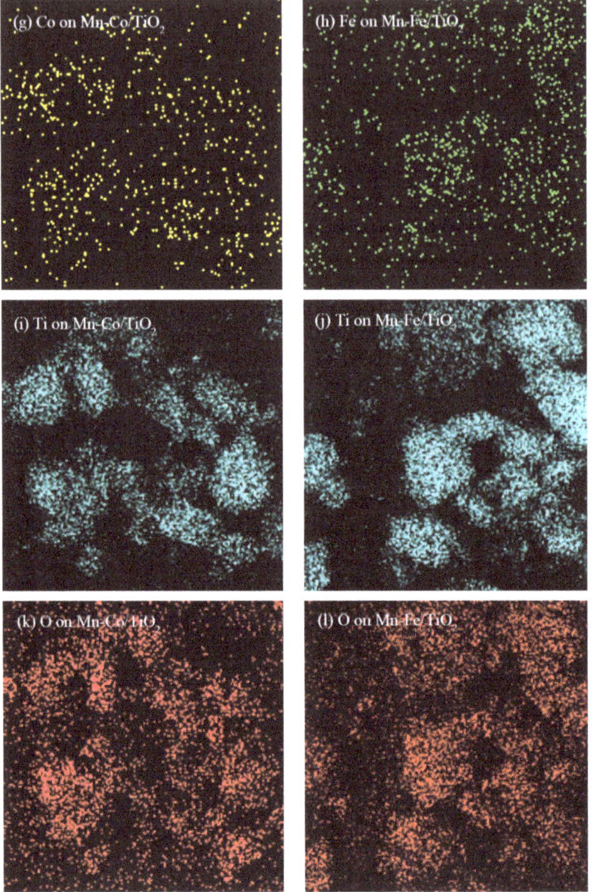

Figure 1. TEM, SEM, and Mapping of Mn−Co/TiO_2 and Mn−Co/TiO_2 nanocatalysts. (**a**) TEM of Mn−Co/TiO_2; (**b**) TEM of Mn−Fe/TiO_2; (**c**) SEM of Mn−Co/TiO_2; (**d**) SEM of Mn−Fe/TiO_2; (**e**) Mapping of Mn on Mn−Co/TiO_2; (**f**) Mapping of Mn on Mn−Fe/TiO_2; (**g**) Mapping of Co on Mn−Co/TiO_2; (**h**) Mapping of Fe on Mn−Fe/TiO_2; (**i**) Mapping of Ti on Mn−Co/TiO_2; (**j**) Mapping of Ti on Mn−Fe/TiO_2; (**k**) Mapping of O on Mn−Co/TiO_2; (**l**) Mapping of O on Mn−Fe/TiO_2.

2.1.2. BET Analysis

The structure parameters of Mn−Co/TiO_2 and Mn−Co/TiO_2 nanocatalysts, such as the specific surface areas, total pore volumes, and average pore diameters, were investigated by nitrogen-adsorption-desorption, and the test data was summarized in Table 1. Mn−Co/TiO_2 nanocatalyst attained the larger specific surface areas than Mn−Fe/TiO_2, which was possible due to the improved dispersion of $MnCoO_x$ species on the nanocatalyst surface. In the meantime, the average pore diameters increased obviously, from 33.06 nm in the Mn−Co/TiO_2 sample to 54.85 nm in the Mn−Fe/TiO_2 sample. Therefore, the $MnCoO_x$ species were more likely to promote nanocatalyst to form micropores compared with $MnFeO_x$ species [24,25]. Nevertheless, the difference of total pore volumes between Mn−Co/TiO_2 and Mn−Fe/TiO_2 nanocatalysts was not conspicuous. The total pore volumes reduced slightly from 0.531 $cm^3 \cdot g^{-1}$ in Mn−Co/TiO_2 sample to 0.424 $cm^3 \cdot g^{-1}$ in Mn−Fe/TiO_2 sample, which was probable due to the mesoporosity generation, which blocked the micropore formation, resulting in a small decrease of total pore volume. Considering the comprehensive test results, the Mn−Co/TiO_2 nanocatalyst exhibited superior physical properties than the Mn−Fe/TiO_2 sample with higher specific surface area, larger total pore volume, and smaller average pore diameter, which were contributed to abundant micropores structures.

Table 1. Physical properties of Mn−Co/TiO_2 and Mn−Fe/TiO_2 nanocatalysts.

Samples	Specific Surface Aarea ($m^2 \cdot g^{-1}$)	Total Pore Volume ($cm^3 \cdot g^{-1}$)	Average Pore Diameter (nm)
Mn−Co/TiO_2	189.9	0.531	33.06
Mn−Fe/TiO_2	104.6	0.424	54.85

2.1.3. Components Analysis

The XRD patterns of Mn−Co/TiO_2 and Mn−Fe/TiO_2 nanocatalysts were displayed in Figure 2. The diffraction peaks for TiO_2 support contained in the nanocatalysts were preserved entirely, with the strong and distinguished reflections appearing at 2θ = 25.3°, 37.8°, 48.0°, 53.9°, 62.7°, 68.8°, 70.3°, 75.1°, and 82.7°. These accorded with the XRD pattern of anatase TiO_2 (ICDD PDF card # 71-1166) [26], with the diffraction angles of the matching peaks shifted at tiny degrees. For Mn−Co/TiO_2 nanocatalyst, each corresponding peak of anatase TiO_2 appeared at the lower diffraction angle, which indicated the interaction between $MnCoO_x$ and anatase TiO_2 was stronger than that of $MnFeO_x$ and anatase TiO_2. Meanwhile, the diffraction peaks of anatase TiO_2 in the Mn−Co/TiO_2 sample were broader and weaker than that in Mn−Fe/TiO_2 sample, which demonstrated the crystalline of anatase TiO_2 reduced with $MnCoO_x$ loading.

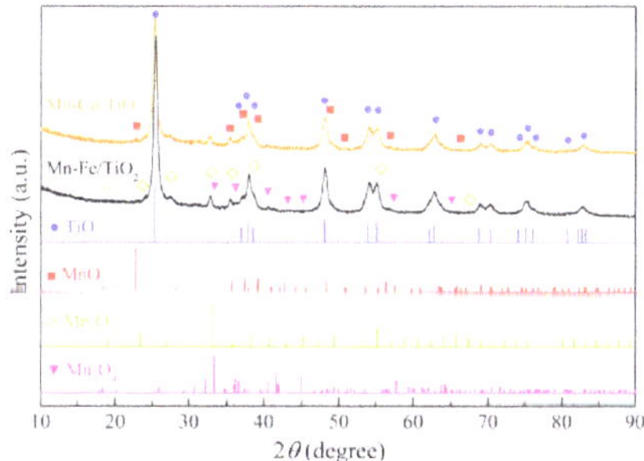

Figure 2. XRD patterns of Mn−Co/TiO_2 and Mn−Fe/TiO_2 nanocatalysts.

The diffraction peaks corresponding to MnO_x contained in Mn−Co/TiO_2 and Mn−Fe/TiO_2 nanocatalysts were very complex due to the transformation among the incomplete crystallization of manganese oxides, such as MnO_2, Mn_2O_3, Mn_3O_4, and MnO. The characterization reflections for MnO_x in the Mn−Co/TiO_2 sample were obviously weaker than that in Mn−Fe/TiO_2 sample, which manifested the active species in the Mn−Co/TiO_2 nanocatalyst, which were superior dispersed on the nanocatalyst surface or the active species incorporated into TiO_2 lattice better [27]. The diffraction peaks fitted MnO_2 (ICDD PDF card # 82-2169) precisely at about 2θ values of 22.10°, 35.19°, 36.96°, 38.72°, 47.86°, and 57.166°, which were coinciding with the crystallographic plane reflections of (110), (310), (201), (111), (311), and (420), respectively [12]. Meanwhile, both Mn−Co/TiO_2 and Mn−Fe/TiO_2 nanocatalysts exhibited the diffraction peaks corresponding to Mn_2O_3 and Mn_3O_4, apparently. The Mn_2O_3 (ICDD PDF card # 78-0390) was well-matched with intensive distinct signals at 2θ = 23.08°, 26.72°, 32.87°, and 56.89°, attributed to the crystallographic plane reflections of (211), (220), (222), and (433), respectively. The Mn_3O_4 (ICDD PDF card # 75-0765) was fitted to characteristic peaks at 36.28°, 40.67°, 41.80°, 57.73° and 64.17°, ascribed to the crystallographic plane reflections of (112), (130), (131), (115), and (063), respectively [23].

Comparing the XRD patterns of the Mn−Co/TiO_2 and Mn−Fe/TiO_2 nanocatalysts, it could be observed that the diffraction peaks of both Mn_2O_3 and Mn_3O_4 were visibly weakened in the Mn−Co/TiO_2 sample. In the meantime, the diffraction peaks of anatase TiO_2 in the Mn−Co/TiO_2 sample were also notably lower than that in Mn−Fe/TiO_2 sample. It was believed that the doping of cobalt into MnO_x obtained superior enhancement than iron on reducing the crystallization of MnO_x and TiO_2 simultaneously. Moreover, in the XRD patterns of the Mn−Co/TiO_2 nanocatalyst, there were no distinguished diffraction peaks for CoO_x, which manifested the doping of cobalt not only improved MnO_x dispersion, but also enhanced CoO_x dispersing on the nanocatalyst surface completely. The similar results were received over Mn−Fe/TiO_2 sample. Based on the phase characteristics comparation between Mn−Co/TiO_2 and Mn−Fe/TiO_2 nanocatalysts, it was believed that the Mn−Co/TiO_2 sample exhibited more excellent properties with smaller crystallinities of chemical compounds and better distribution of the active species, which were conducive to catalytic reactions.

2.1.4. Reducibility Properties

The oxidation states of the active compounds contained in nanocatalysts of Mn−Co/TiO_2 and Mn−Fe/TiO_2 were tested by H_2-TPR. The H_2 consumption curve was fitted by reduction peak separation with Gaussian function, as shown in Figure 3. The details of H_2 consumptions and reduction temperatures for each peak were listed in Table 2. In Mn−Co/TiO_2 and Mn−Fe/TiO_2 nanocatalysts, the anatase TiO_2 support did not bring remarkable reduction peaks within the test temperature range. Hence, all H_2 consumption could be attributed to the reduction process of MnO_x, CoO_x, and FeO_x. It was proposed that the typical reduction process of MnO_x contained in the nanocatalysts of Mn−Co/TiO_2 and Mn−Fe/TiO_2 complied with the following order: $MnO_2 \rightarrow Mn_2O_3$ (Mn_3O_4) $\rightarrow$MnO [28]. Mn−Co/TiO_2 nanocatalyst exhibited four reduction peaks with temperature rising from 50—850 °C, as shown in Figure 3b. The first reduction peak centered, at about 218 °C, was attributed to the high oxidation state of manganese ion converting from MnO_2 to Mn_2O_3 [29]. The dominant reduction peak (Peak 3) was caused by two sequential processes of Mn_2O_3 reducing to Mn_3O_4 and Mn_2O_3 reducing to MnO, as reported in previous literatures [12,28]. The reduction process of Mn_2O_3 to Mn_3O_4 was more liable to happen over the original amorphous Mn_2O_3 [30], while reduction process of Mn_2O_3 to MnO preferred to occur at relatively higher temperature [31]. For the Mn−Co/TiO_2 sample, the typical reduction process of cobalt oxides commonly displayed two separated peaks, the one of Co_2O_3 reducing to CoO appeared at around 327 °C, presenting as an asymmetrical reduction peak (Peak 2), the other one of CoO reducing to Co^0 occurred at about 418 °C and overlapped with MnO_x reduction peaks in whole or partly (Peak 3) [8,32]. Therefore, the fourth medium reduction peak in the Mn−Co/TiO_2 nanocatalyst was related to the reduction processes of Mn_3O_4 to MnO.

Table 2. H_2-TPR quantitative analysis of Mn−Co/TiO_2 and Mn−Fe/TiO_2 nanocatalysts.

Samples	Temperature (°C) / H_2 Consumption ($mmol{\cdot}g^{-1}$)				
	Peak 1	Peak 2	Peak 3	Peak 4	Total
Mn−Co/TiO_2	218/1.14	327/0.21	418/2.12	517/0.96	- /4.43
Mn−Fe/TiO_2	276/0.83	387/0.41	436/0.54	501/1.59	- /3.37

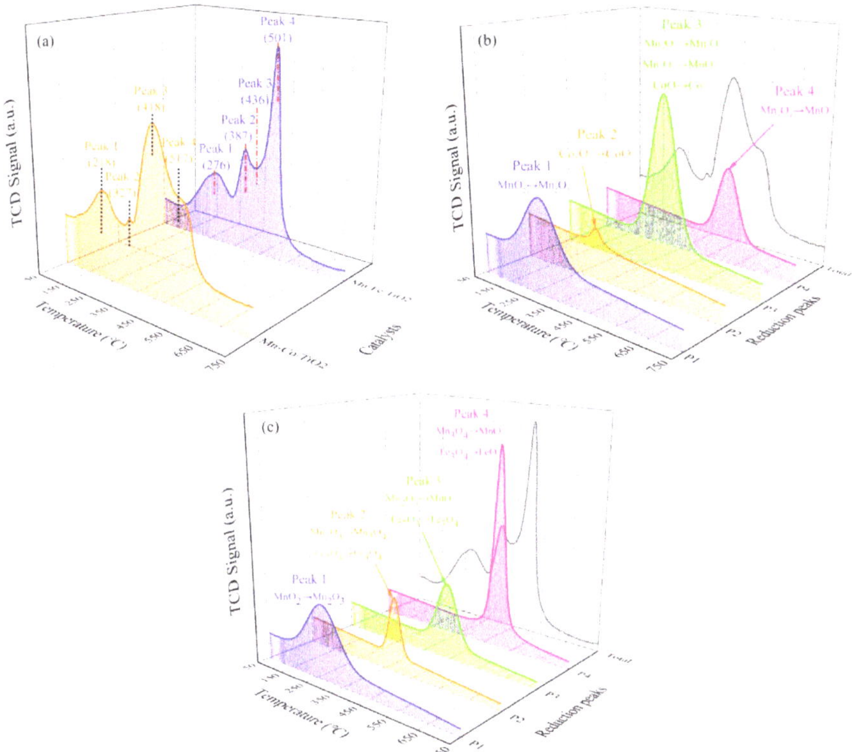

Figure 3. H_2-TPR profiles of Mn−Co/TiO_2 and Mn−Fe/TiO_2 nanocatalysts: (**a**) Total H_2-TPR curves; (**b**) Multi-peaks Gaussian fitting for Mn−Co/TiO_2 nanocatalyst; (**c**) Multi-peaks Gaussian fitting for Mn−Fe/TiO_2 nanocatalyst.

In comparison, the peak intensions and the reduction temperatures of the Mn−Fe/TiO_2 nanocatalyst were evidently different from that of the Mn−Co/TiO_2 sample. The peak of MnO_2 reducing to Mn_2O_3 weakened remarkably and shifted to higher temperature (276 °C). There were wide joint peaks (Peak 2 and Peak 3) at temperature of 330–530 °C, which were ascribed to the coinstantaneous reduction processes of Mn_2O_3 to Mn_3O_4 and Fe_2O_3 to Fe_3O_4. It was proposed that the majority of Fe_2O_3 was reduced at around 380 °C, which located at effortlessly reducible sites in the form of oligomeric clusters, nanoparticles or isolated ions [33]. After the majority of Fe_2O_3 reduction, there was a small quantity of residual Fe_2O_3 reducing to Fe_3O_4 at the higher temperature [34]. The dominating peak (Peak 4), at around 501 °C, was assigned to the overlapped peaks of Mn_3O_4 to MnO and Fe_3O_4 to FeO. Figure 4 provided a graphical representation of the reduction process, in which each active component embodied qualitatively [1,35]. In these two kinds of nanocatalysts, Mn−Co/TiO_2 sample displayed the higher low-temperature reducibility and exhibited a noticeable medium temperature reduction peak at the same time, which manifested the higher oxidation states of manganese ion (Mn^{4+} and Mn^{3+}) constituted the dominating phase [31].

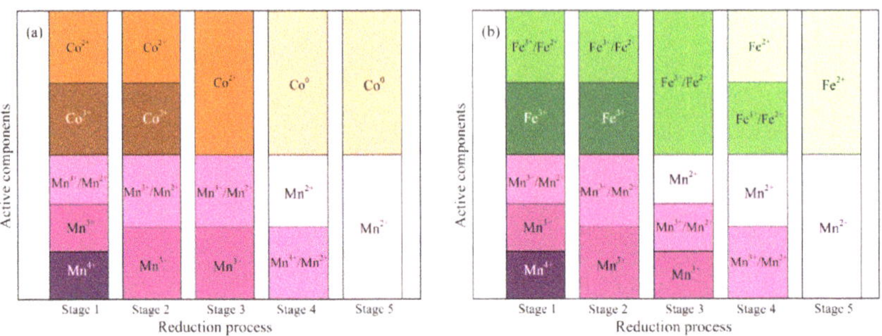

Figure 4. The qualitative component states of active elements during reduction process.

According to Table 2, the total H_2 consumption of Mn−Co/TiO_2 nanocatalyst was 4.43 mmol·g^{-1} much larger than that of Mn−Fe/TiO_2 sample. Meanwhile, the starting reduction peak temperature of Mn−Co/TiO_2 nanocatalyst was at 218 °C lower than that of Mn−Fe/TiO_2, which was regarded as an important factor influencing the reducibility. It was believed that the catalyst performed superior catalytic ability at low temperatures as the reduction peaks arising in lower temperature regions [1].

2.1.5. Ammonia Adsorption Properties

The acid capacity of the Mn−Co/TiO_2 and Mn−Fe/TiO_2 nanocatalysts was measured by NH_3 TPD test, which was another crucial factor effecting the catalyst performance in SCR process [36]. The results were exhibited in Figure 5 and Table 3, respectively. The NH_3-TPD curves for the Mn−Co/TiO_2 and Mn−Fe/TiO_2 nanocatalysts were ascribed to four desorption peaks of chemisorbed NH_3 with the temperature rising from 150 °C to 750 °C. For Mn−Fe/TiO_2 nanocatalyst, the first NH_3 desorption peak occurred at around 209 °C, attributed to NH_3 desorption from weak acid sites, which was too feeble to bound NH_3 steadily in the gas mixture during SCR process [36]. Peak 2 and peak 3 were combined from 398 °C to 585 °C, which were attributed to the medium strong acid sites on the nanocatalyst surface. Peak 4, at around 649 °C, was assigned to the strong acid sites, which were regarded as abundant Lewis acid sites adsorbing a great deal of strongly bound NH_3 [37]. In comparation, the NH_3 desorption results of the Mn−Co/TiO_2 nanocatalyst exhibited better acidity capacity at medium and high temperatures. However, there was an undesired temperature shift to higher temperature occurred in the meantime. For Mn−Co/TiO_2 sample, the desorption temperature of weak acid sites, medium strong acid sites and strong acid sites were 276 °C (Peak 1), 402 °C (Peak 2), 587 °C (Peak 3), and 653 °C (Peak 4), respectively. It was obvious that the medium strong acid sites and the strong acid sites were enriched in the Mn−Co/TiO_2 sample, which were positive to form more abundant Brønsted acid sites and Lewis acid sites promoting NH_3 adsorption on the nanocatalyst surface [38,39].

Table 3. Quantitative analysis of NH_3-TPD profiles.

Samples	Temperature (°C) / NH_3 composition (mmol·g^{-1})				
	Peak 1	Peak 2	Peak 3	Peak 4	Total
Mn−Co/TiO_2	276/0.11	402/0.49	587/0.52	653/0.32	−/1.44
Mn−Fe/TiO_2	209/0.14	398/0.24	585/0.39	649/0.20	−/0.97

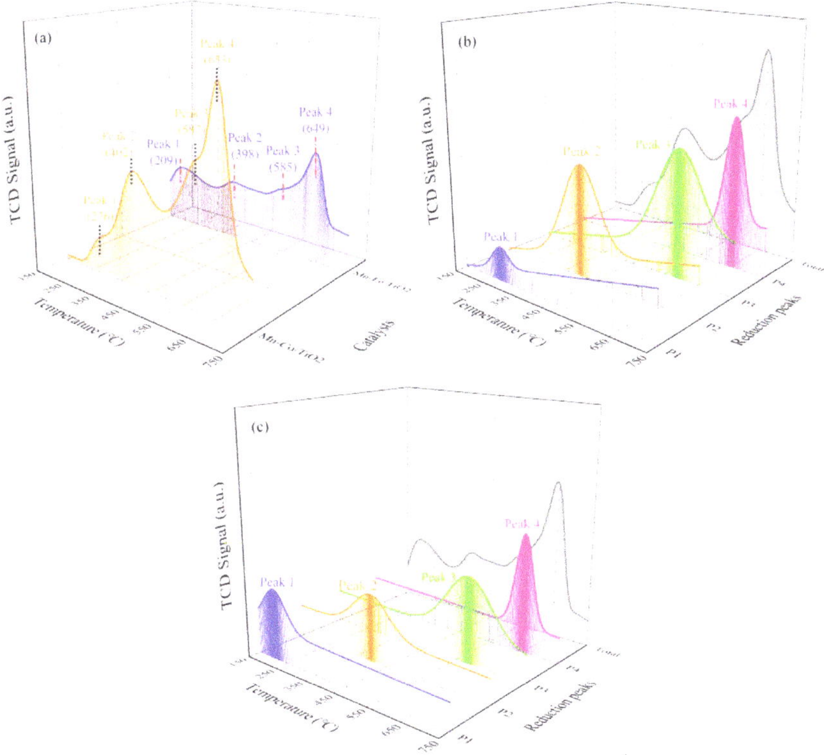

Figure 5. NH_3-TPD profiles of Mn−Co/TiO_2 and Mn−Fe/TiO_2 nanocatalysts: (**a**) Total NH_3-TPD curves; (**b**) Multi-peaks Gaussian fitting for Mn−Co/TiO_2 nanocatalyst; (**c**) Multi-peaks Gaussian fitting for Mn−Fe/TiO_2 nanocatalyst.

The quantitative comparation of the total acid capacity between the Mn−Co/TiO_2 and Mn−Fe/TiO_2 nanocatalysts was performed and summarized in Table 3. The Mn−Co/TiO_2 sample achieved larger total NH_3 desorption of 1.44 mmol·g^{-1} than Mn−Fe/TiO_2, which further confirmed the superior enhancement of $MnCoO_x$ on the surface acidity than $MnFeO_x$. Although the Mn−Fe/TiO_2 sample generated weak acid sites at the lower temperature, the NH_3 desorption on the weak acid sites was only 0.14 mmol·g^{-1}, which was too little to offer enough NH_3 for SCR process. However, it was reported that NH_3 could restrain the adsorption and activation of NO onto the active sites on catalyst surface via the undesirable electron transfer between the adsorbed NH_3 and the metal sites [40]. Consequently, the excessive adsorbed NH_3 could inhibit SCR reactions. Therefore, the surface acidity properties of the Mn−Co/TiO_2 and Mn−Fe/TiO_2 nanocatalysts were bound up with their reducibility properties.

2.1.6. Oxidation States of Active Species

In order to better understand the oxidation states and the atomic concentrations of active species on the surface of Mn−Co/TiO_2 and Mn−Fe/TiO_2 nanocatalysts, the XPS spectra of Ti 2*p*, Mn 2*p*, Co 2*p*, Fe 2*p*, and O 1 *s* were tested and numerically analyzed by Gaussian fitting, respectively, as shown in Figure 6. The binding energy and atomic concentration of each element was summarized in Table 4.

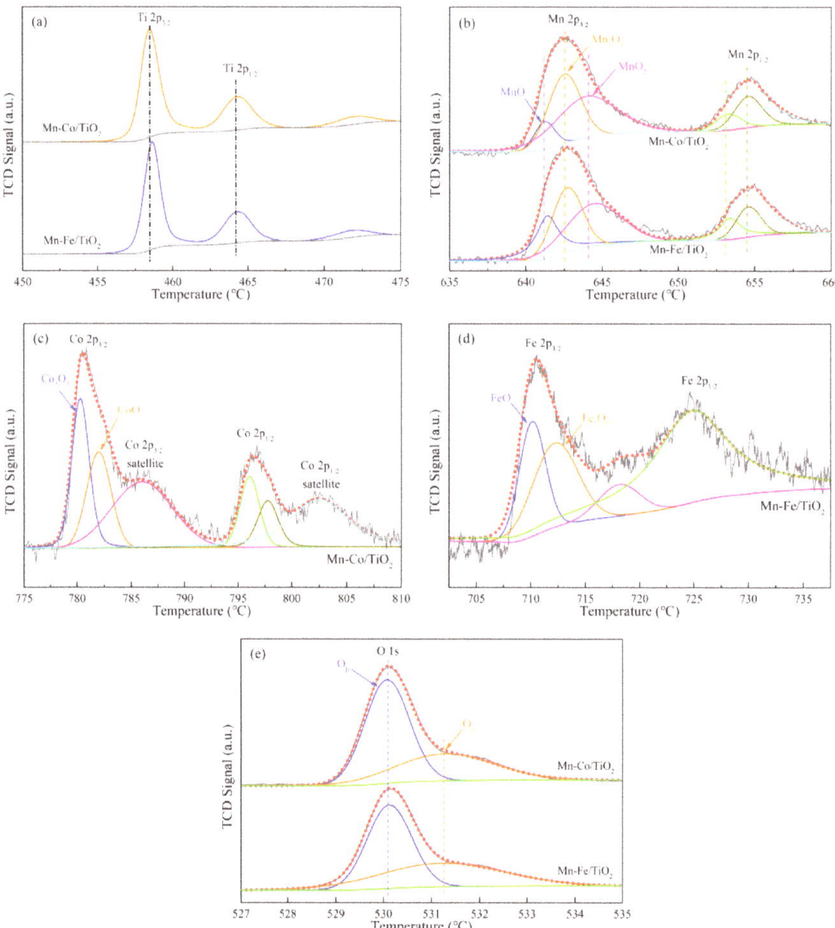

Figure 6. XPS analysis of Mn−Co/TiO_2 and Mn−Fe/TiO_2 nanocatalysts: (**a**) XPS spectra for Ti 2*p*; (**b**) XPS spectra for Mn 2*p*; (**c**) XPS spectra for Co 2*p*; (**d**) XPS spectra for Fe 2*p*; (**e**) XPS spectra for O 1*s*.

Table 4. Surface atomic compositions of the catalysts determined by XPS.

Samples	Binding Energy (eV) / Atomic Composition (%)								
		Mn		Fe		Co		O	
	Mn^{2+}	Mn^{3+}	Mn^{4+}	Fe^{2+}	Fe^{3+}	Co^{2+}	Co^{3+}	O_α	O_β
Mn−Co/TiO_2	641.2/ 13.8	642.6/ 39.4	644.1/ 46.8	-/-	-/-	779.6 39.3	782.1/ 60.7	531.4/ 33.7	530.3/ 66.3
Mn−Fe/TiO_2	641.4/ 21.3	642.8/ 38.5	644.6/ 40.2	709.6/ 43.4	711.7/ 56.6	-/-	-/-	531.6/ 28.2	530.3/ 71.8

As shown in Figure 8a, the XPS spectrum of Ti 2*p* was consisted of two characteristic peaks, attributed to Ti $2p_{1/2}$ at around 464.3 eV and Ti $2p_{3/2}$ at about 458.7 eV, respectively [41]. It was obvious that Ti^{4+} ion was stabilized on the catalyst surface and appeared as the dominating state in the nanocatalysts of Mn−Co/TiO_2 and Mn−Fe/TiO_2. The XPS spectra of Mn 2*p* was composed of Mn $2p_{1/2}$ peak at about 653 eV and Mn $2p_{3/2}$ peak at around 642 eV [42], as exhibited in Figure 6b. The Mn $2p_{3/2}$ peak could be further split into three peaks for Mn^{2+} at 641.2 ± 0.3 eV, Mn^{3+} at 642.6 ± 0.2 eV, and Mn^{4+} at 644.1 ± 0.5 eV, respectively, according to the multi-peaks Gaussian fitting results [43]. It is evidently difficult to distinguish the three valence states of manganese in MnO_x with the binding energy difference value less than 3.7 eV. For accurate comparison of the atomic concentration of Mn^{n+} on the surface of $Mn-Co/TiO_2$ and $Mn-Fe/TiO_2$ nanocatalysts, a quantitative analysis was performed according to the integral area under every divided peak, as exhibited in Table 4. It was widely accepted that the catalytic ability of MnO_x ranked as: $MnO_2 > Mn_2O_3 > Mn_3O_4$ [12,22,30]. The abundant MnO_2 generated on the nanocatalyst surface was beneficial to SCR reactions [5]. As listed in Table 4, the major amount of manganese in both $Mn-Co/TiO_2$ and $Mn-Fe/TiO_2$ nanocatalysts was Mn^{4+}. Meanwhile, the Mn^{4+}/Mn^{n+} atomic composition was 46.8 % in $Mn-Co/TiO_2$ sample higher than that of 40.2 % in $Mn-Fe/TiO_2$ sample, generating more lattice oxygen and plenty of oxygen vacancy on the catalyst surface [1], which was regarded as the main reason for higher reducibility properties of $Mn-Co/TiO_2$ sample as discussed above.

In Co 2*p* spectrum of $Mn-Co/TiO_2$ nanocatalyst, the two individual peaks were attributed to Co $2p_{1/2}$ at about 796.5 eV and Co $2p_{3/2}$ at around 780.6 eV, respectively. Both of these two main peaks had satellite peaks at 803.1 eV and 786.8 eV, correspondingly, as shown in Figure 6c. The gentler and broader satellite peaks appeared at relatively higher binding energy, which were introduced by the metal-to-ligand electron transfer or the shakeup process of cobalt in its high spin state. However, this shakeup process was only observed with the high spin state of Co^{2+} ion, but did not appear with the diamagnetic low-spin Co^{3+} ion [13]. The intense peak of Co $2p_{3/2}$ was composed of two overlapped peaks, one attributed to Co^{3+} seated at about 780.0 eV and the other ascribed Co^{2+} and located at around 781.6 eV. These two distinguishing peaks indicated the co-occurrence of cobalt in +2 and +3 valence states on the $Mn-Co/TiO_2$ nanocatalyst surface. There was a dynamic equilibrium sustained on the nanocatalyst surface with the electron transfer between Mn and Co ions during the catalytic oxidation process, expressed as $Co^{3+} + Mn^{3+} \leftrightarrow Co^{2+} + Mn^{4+}$ [44]. Moreover, Co^{3+} species were exhibited as the major atomic concentration of 60.7%, which were at the comparatively higher valence state of Co^{n+} generating vast anionic defects, forming abundant surface oxygen and facilitating adsorption and oxidation during SCR process [45].

In Fe 2*p* spectra of the $Mn-Fe/TiO_2$ nanocatalyst, as shown in Figure 6d, the two main peaks were ascribed to Fe $2p_{1/2}$ at about 724 eV and Fe $2p_{3/2}$ at around 710 eV. There was a satellite peak at 718.3 eV assigned to Fe^{3+} in Fe_2O_3 [46]. The XPS spectra of Fe $2p_{3/2}$ was composed of two characteristic peaks, ascribed to Fe^{3+} at around 711.6 eV and Fe^{2+} at about 709.6 eV, which meant these two kinds of Fe^{n+} coexisted on the surface of the $Mn-Fe/TiO_2$ nanocatalyst. It was proposed that the promotion effect between manganese and iron was attributed to the electron transfer in the redox reaction: $Mn^{3+} + Fe^{3+} \leftrightarrow Mn^{4+} + Fe^{2+}$ [47].

The O 1*s* spectrum of $Mn-Co/TiO_2$ and $Mn-Fe/TiO_2$ nanocatalysts were compared in Figure 6e, which was composed of the chemisorbed oxygen peak at 531.2–531.6 eV (O_α) and the lattice oxygen peak at 530.2–530.3 eV (O_β). According to the curve-fitting results, the binding energy of O_α shifted to higher values from 531.4 eV in $Mn-Co/TiO_2$ sample to 531.6 eV in $Mn-Fe/TiO_2$ sample. Meanwhile, the similar changes happened to the binding energy of O_β. It was believed that the chemisorbed oxygen (O_α) was the most active oxygen species owing to its high mobility [48]. On the surface of $Mn-Co/TiO_2$ nanocatalyst, the atomic concentration of O_α reached 33.7% much higher than that of $Mn-Fe/TiO_2$ sample, which was regarded as another reason for its superior redox ability.

2.2. Catalytic Performance

Figure 7 exhibited the NO conversion and catalytic selectivity over $Mn-Co/TiO_2$ and $Mn-Fe/TiO_2$ nanocatalysts. It can be seen that the NO conversion improved conspicuously with the temperature increasing from 25 °C to 250 °C. For both $Mn-Co/TiO_2$ and $Mn-Fe/TiO_2$ nanocatalysts, the satisfactory conversion (>90%) was obtained above 150 °C. Meanwhile, the $Mn-Ce/TiO_2$ nanocatalyst exhibited higher catalytic activity than the $Mn-Fe/TiO_2$ sample within the temperature range of 50–175 °C. Furthermore, the catalytic selectivity of $Mn-Ce/TiO_2$ nanocatalyst was higher than that of Mn−Fe/TiO_2 within 75–250 °C. Therefore, both the catalytic activity and the catalytic selectivity over Mn−Co/TiO_2 were remarkably improved compared to Mn−Fe/TiO_2, potentially due to the interaction between manganese and cobalt stronger than that between manganese and iron, which enhanced the generation of Brønsted acid sites and Lewis acid sites on the nanocatalyst surface, resulting in promoting Langmuir–Hinshelwood mechanism reactions and Eley-Rideal mechanism reactions, simultaneously [49].

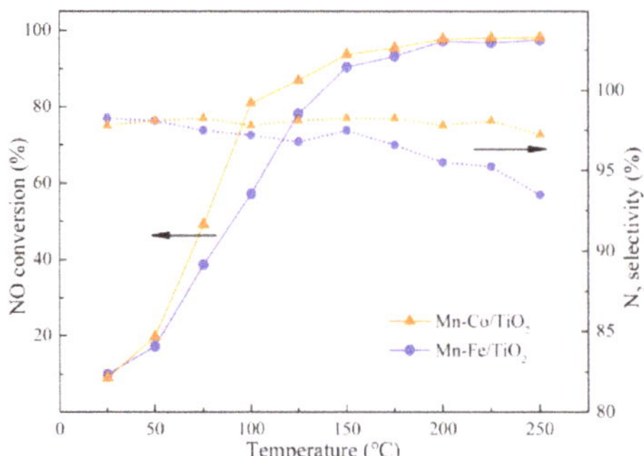

Figure 7. Catalytic performance of Mn−Co/TiO_2 and Mn−Fe/TiO_2 nanocatsalysts.

2.3. Reaction Kinetic Study

It was proposed that both the Langmuir–Hinshelwood mechanism and the Eley–Rideal mechanism contributed to the SCR reactions over nanostructured Fe-Mn oxides [36]. According to Langmuir-Hinshelwood mechanism, the SCR reactions over Mn−Co/TiO_2 and Mn−Fe/TiO_2 nanocatsalysts could be approximately described as:

$$M^{n+} = O + NO\,(ad) \rightarrow M^{(n-1)+}{-}O - NO \quad (2)$$

$$M^{(n-1)+}{-}O - NO + NH_3(ad) \rightarrow M^{(n-1)+}{-}O - NO - NH_3 \rightarrow M^{(n-1)+}{-}OH + N_2 + H_2O \quad (3)$$

$$M^{(n-1)+}{-}O - NO + O_2 \rightarrow M^{(n-1)+}{-}O - NO_2 \quad (4)$$

$$M^{(n-1)+}{-}O - NO + NH_3(ad) \rightarrow M^{(n-1)+}{-}O - NO_2 - NH_3 \rightarrow M^{(n-1)+}{-}OH + N_2O + H_2O \quad (5)$$

$$M^{(n-1)+}{-}OH + O_2 \rightarrow M^{n+}{=}\ O + H_2O \quad (6)$$

The adsorbed NO(ad) was oxidized by M^{n+} (Mn^{4+}, Mn^{3+}, Co^{3+}, Fe^{3+}) on the nanocatalyst surface to form NO_2^- (Reaction 2). The NO_2^- reacted with adsorbed NH_3(ad) to generate NH_4NO_2 and decomposed to N_2 and H_2O, subsequently (Reaction 2). In the meantime, an undesired oxidization reaction occurred on NO_2^- which was oxidized to NO_3^- under O_2 (Reaction 4). The formed NO_3^- reacted with adsorbed NH_3(ad) to generate NH_4NO_3 and further decomposed to N_2O and H_2O (Reaction 5). Finally, the active species $M^{(n-1)+} - OH$ was oxidized by O_2 to regenerate $M^{n+}= O$ (Reaction 6).

According to the Eley-Rideal mechanism, the SCR reactions over Mn−Co/TiO_2 and Mn−Fe/TiO_2 nanocatsalysts could be described as:

$$M^{n+}{=}\ O + NH_3(ad) \rightarrow M^{(n-1)+}{-}OH + NH_2(ad) \quad (7)$$

$$NH_2(ad) + NO \rightarrow N_2 + H_2O \quad (8)$$

The adsorbed NH_3(ad) on the nanocatalyst surface was activated by M^{n+} to generate NH_2 (Reaction 7). Then, the NH_2 reacted with NO to form N_2 and H_2O (Reaction 8).

However, in addition to the main Eley-Rideal mechanism reactions, undesirable side reactions occurred on the active species of $M^{n+}= O$.

$$M^{n+}{=}O + NH_2(ad) \rightarrow M^{(n-1)+}{-}OH + NH(ad) \tag{9}$$

$$M^{n+}{=}O + NH(ad) + NO \rightarrow M^{(n-1)+}{-}OH + N_2O \tag{10}$$

$$M^{n+}{=}O + NH(ad) + O_2 \rightarrow M^{(n-1)+}{-}OH + NO \tag{11}$$

The activated NH_2(ad) formed in Reaction 7 was further oxidized to NH(ad) over $M^{n+}{=}O$. Then NH(ad) reacted with NO to form N_2O or react with O_2 to generate NO.

According to Reactions 3 and 8, the kinetic equations of N_2 formation rates via the Langmuir-Hinshelwood mechanism and the Eley-Rideal mechanism could be described as:

$$\begin{aligned} k_{scr} &= \frac{d[N_2]}{dt} = \frac{d[N_2]}{dt}\Big|_{LH} + \frac{d[N_2]}{dt}\Big|_{ER} = -\frac{d\left[M^{(n-1)+}{-}O{-}NO{-}NH_3\right]}{dt}\Big|_{LH} - \frac{d[NH_2(ad)]}{dt}\Big|_{ER} \\ &= k_3\left[M^{(n-1)+}{-}O-NO-NH_3\right] + k_8[NH_2(ad)][NO] \end{aligned} \tag{12}$$

where k_3 and k_8 were the kinetic constants of Reactions 3 and 8, $\left[M^{(n-1)+}{-}O-NO-NH_3\right]$, $[NH_2(ad)]$, and [NO] were the concentrations of NH_4NO_2, NH_2(ad), and NO, respectively. When the reactions above reached stable conditions, the concentrations of NH_4NO_2 and NH_2(ad) on the surface of Mn−Co/TiO_2 and Mn−Fe/TiO_2 nanocatalysts could be regarded as constants without connecting with the gaseous concentrations of NH_3 and NO [50]. Therefore, the SCR reaction rate was in a linear relationship with NO concentration as exhibited in Equation (13), the slope and the intercept could be obtained from the linear regression in Figure 8, as listed in Table 5.

$$k_{scr} = k_{SCR-LH} + k_{SCR-ER}[NO] \tag{13}$$

$$k_{SCR-LH} = k_3\left[M^{(n-1)+}{-}O-NO-NH_3\right] \tag{14}$$

$$k_{SCR-ER} = k_8[NH_2(ad)] \tag{15}$$

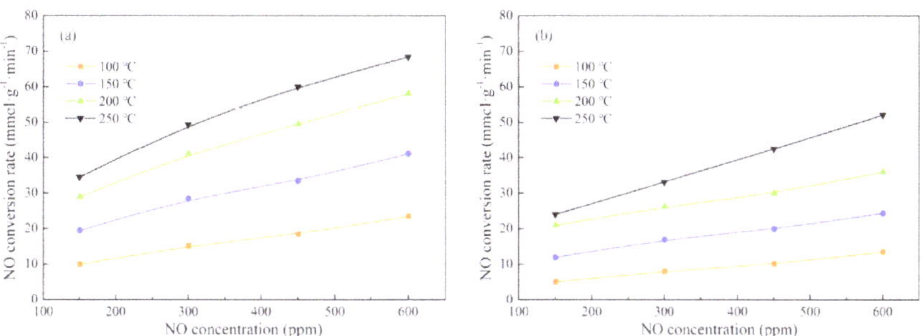

Figure 8. Effect of NO concentration on selective catalytic reduction (SCR) reaction rate.

Table 5. Reaction kinetic constants of NO reduction.

Sample	Temperature (°C)	$k_{SCR-ER}/10^6$	k_{SCR-LH}	R^2	k_{SCR} [a]
$Mn-Co/TiO_2$	100	0.029	5.75	0.993	23.5
	150	0.046	13.25	0.994	41.1
	200	0.064	20.50	0.988	58.2
	250	0.075	25.01	0.984	68.4
$Mn-Fe/TiO_2$	100	0.018	2.35	0.989	13.5
	150	0.027	8.25	0.990	24.4
	200	0.033	16.01	0.991	36.0
	250	0.062	14.50	0.998	52.1

[a] $[NO] = [NH_3] = 600$ ppm.

2.4. Thermodynamic Calculation

The standard thermodynamic characteristics of the chemical individual species of M_xO_y and M in crystalline states, as well as NH_3, N_2 and H_2O in gaseous states, were listed in Table 6. The oxidative capacities of M_xO_y were evident according to the reliable standard thermodynamic parameters, which was a convincing support for investigating the roles of M_xO_y in NH_3-SCR process. The Gibbs free energies for the reactions between M_xO_y species and ammonia were calculated for estimating their oxidative capacities according to Equation (16), displayed in Table 7. In order to discuss the reaction mechanism conveniently, simplifying assumptions were introduced as follows. Only the kind of main oxidation state was considered in the thermodynamic calculation, although there were several kinds of oxidation states for each M_xO_y [51]. Meanwhile, it was assumed that every M_xO_y was reduced into metal, although the metallic oxide in the high-oxidation state was reduced step-by-step in the actual process. It had been proved that the entropy of metals would vary drastically as the metal physical states transforming [52]. Furthermore, the transformation could bring about the changes of the straight slope of the temperature–Gibbs free energy curve which was named Ellingham diagrams. The Ellingham diagrams were a particular graphical form of the principle that the thermodynamic feasibility of a reaction depends on the Gibbs free energy change ($\Delta_f G^0$), which could be calculated using Equation (16), where $\Delta_f H^0$ was the enthalpy change and ΔS^0 was the entropy change under standard state.

$$\Delta_f G^0 = \Delta_f H^0 - T\Delta S^0 \quad (16)$$

Table 6. Standard thermodynamic properties of M_xO_y [7,53–55].

	C_p J mol^{-1} K	$\Delta_f H^0$ kJ mol^{-1}	S^0 J mol^{-1} K	$\Delta_f G^0$ kJ mol^{-1}
Mn	26.3	0.0	32.0	-
MnO	45.4	−385.2	59.7	−362.9
Mn_3O_4	139.7	−1387.8	155.6	-1283.2
Mn_2O_3	107.7	−959.0	110.5	−881.1
MnO_2	54.1	−520.0	53.1	−465.1
Co	24.8	0.0	30.0	-
CoO	55.2	−237.9	53.0	−214.2
Co_3O_4	123.4	−891.0	102.5	−774.0
Fe	25.1	0.0	27.3	-
FeO	-	−272.0	-	-
Fe_3O_4	143.4	−1118.4	146.4	−1015.4
Fe_2O_3	103.9	−824.2	87.4	−742.2
NH_3	-	45.9	192.8	-
N_2	29.1	0.0	191.6	-
H_2O (g)	33.6	−241.8	188.8	−228.6

Table 7. Gibbs free energy for the reaction between M_xO_y and NH_3 under different temperatures.

Substance	$\triangle_fG$ /kJ					
	298 K	**323 K**	**373 K**	**423 K**	**473 K**	**523 K**
MnO_2	−46.25	−49.09	−54.77	−60.45	−66.12	−71.80
Mn_3O_4	42.00	39.27	33.81	28.34	22.88	17.42
MnO	84.06	81.65	76.83	72.01	67.19	62.37
Co_3O_4 [a]	−85.71	−88.74	−94.79	−100.84	−106.89	−112.94
CoO	−64.64	−67.17	−72.22	−77.28	−82.34	−87.39
Fe_2O_3	−31.40	−34.23	−39.89	−45.55	−51.21	−56.87
Fe_3O_4	−24.99	−27.69	−33.09	−38.49	−43.89	−49.29

[a] Co_3O_4 was regard as the compound of Co_2O_3 and CoO.

The Ellingham diagram plotted Δ_fG for each oxidation reaction as a function of temperature. In order to compare different reactions, all Δ_fG values referred to the same quantity of oxygen, chosen as one mole O (1/2 mol O_2) [53], as shown in Equation (17)

$$\frac{1}{y}M_xO_y + \frac{2}{3}NH_3 \rightarrow \frac{x}{y}M + \frac{1}{3}N_2 + H_2O \tag{17}$$

According to Equations (16) and (17), the redox reactions that happened between M_xO_y and NH_3 were discussed under thermodynamic calculation, assuming that the enthalpy, entropy, and Gibbs free energy remained unchanged with temperature. The Δ_fG in Table 7 for each kind of M_xO_y was calculated in order to estimate its oxidative capacities and gain semiquantitative conclusions. Meanwhile, this research just focused on the catalytic performance and reaction mechanism of cobalt and iron doped in the Mn/TiO_2 nanocatalysts, so that only one M_xO_y was different at a time. The negative Δ_fG values for Co_3O_4 (the compound of Co_2O_3 and CoO) and Fe_2O_3 indicated M_xO_y was effective for the oxidative abstraction of the hydrogen from adsorbed ammonia [7].

The Ellingham diagrams for the Gibbs free energy of reactions between M_xO_y and NH_3 with the reaction temperature rising was displayed in Figure 9. As the $\triangle_fG$ was negative ($\triangle_fG < 0$), the reaction took place more easily. On the contrary, the reaction would happen at higher temperature or would not happen if the $\triangle_fG$ was positive ($\triangle_fG > 0$). Under the same reaction temperature, the order of temperature−Gibbs free energy for the reaction between M_xO_y and NH_3 is as follows: Co_3O_4 (compound of Co_2O_3 and CoO) < CoO < Fe_2O_3 < Fe_3O_4 (only considering the doping transition metals). These results were exactly consistent with the NH_3-SCR performance and almost coincided with the data of the nanostructure characterization, as discussed above, which demonstrated the simplifying assumptions used in the thermodynamic calculations was reasonable. However, the activity of MnO_x exhibited in Ellingham Diagrams was not in well accord with the reducibility properties, which might be on account of positive kinetics or simplifying assumptions during the thermodynamic calculations [51,53]. It manifested that MnO_x, as the dominate phase in nonacatalysts of Mn−Co/TiO_2 and Mn−Fe/TiO_2, was activated by the doping of cobalt or iron. A large number of researches had been carried out focusing on optimizing catalysts via doping transition metals in order to enhance NO catalytic conversion remarkably. Based on the thermodynamic study, Ellingham diagrams would provide a novel insight for developing catalysts doped proper metal oxides with higher NH_3-SCR activity.

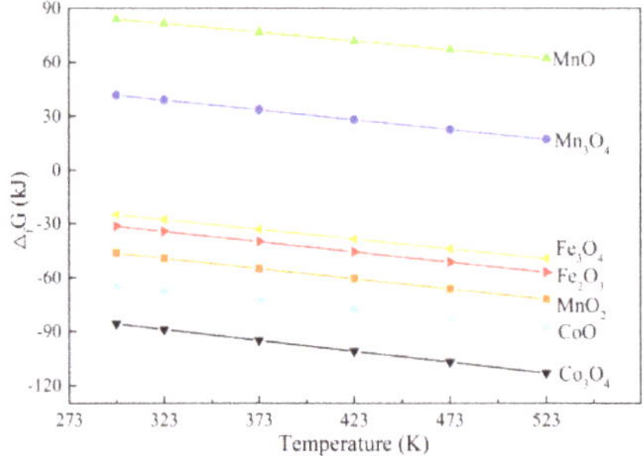

Figure 9. Temperature–Gibbs free energy curve for the reaction between M_xO_y and NH_3.

3. Materials and Methods

3.1. Catalysts Preparation

The noncatalysts of Mn−Co/TiO_2 and Mn−Fe/TiO_2 were prepared by hydrothermal method. $Co(CH_3COO)_2{\cdot}4H_2O$ (analytical pure 99.9%, Kermel, Tianjin, China) and $Fe(NO_3)_3{\cdot}9H_2O$ (analytical pure 99.9%, Sinopharm, Shanghai, China) were used as the precursors of CoO_x and FeO_x, respectively. The aqueous solution of $Mn(NO_3)_2$ (analytical pure 50.0%, Sinopharm, Shanghai, China) was used as the precursors of MnO_x. The tetrabutyl titanate was introduced as the precursors of anatase TiO_2 to support the bimetals oxides. The precursors were added into deionized water at room temperature. One-hundred fifty mL glycol was added into the above mixture with magnetic stirring continuously to regulate the reaction rate of hydrothermal reaction. Then the homogeneous solution was heat at 180 °C for 8 h in a Teflon-lined stainless steel autoclave. After that, tetrabutyl titanate was added into the solution and reheated in the autoclave at 180 °C for 3 h. The precipitate was obtained by reduplicative centrifugation and wash. In the end, the nanocatalysts were dried at 150 °C for 12 h and calcined in air at 500 °C for 4 h. The nanocatalyst was denoted as Mn−Co/TiO_2 and Mn−Fe/TiO_2 with the molar ratios of Mn:M:Ti = 2:1:7.

3.2. Catalysts Characterization

The advanced microstructural image data and the surface element contents of the nanocatalysts was achieved by a high resolution transmission electron microscope JEOL JEM-2010 (Japan electronics corporation, Tokyo, Japan) combined with EDS. The Maxon Tristar II 3020 micropore-size analyzer (Maxon, Chicago, IL, USA) was used for testing N_2 adsorption isotherms of the prepared nanocatalysts at −196 °C. The surface areas and the pore-size distributions of the nanocatalysts were measured after the nanocatalysts degassing in vacuum at 350 °C for 10 h. BET plot linear portion was used to determine the nanocatalysts specific surface areas, and the desorption branch with Barrett–Joyner–Halenda (BJH) formula was introduced to calculate the pore-size distributions. The XRD data was captured by a Bruker D8 advance analyzer (Bruker, Frankfurt, Germany) with Mo K_α radiation, diffraction intensity from 10° to 90°, point counting time of 1 s and point counting step of 0.02°. The element phases contained in the nanocatalysts were distinguished by comparing characteristic peaks presented in the XRD patterns with the International Center for Diffraction Data (ICDD). H_2-TPR and NH_3-TPD tests were performed with a Micromeritics Autochem II 2920 chemical adsorption instrument (Micromeritics, Houston, TX, USA). During H_2-TPR experiment, nanocatalysts were pretreatment in He at 400 °C for 1 h, and then cooled to environment temperature in H_2 and He gas mixture at 30 mL/min. The test temperature range of H_2 consumptions was from 50 °C to 850 °C with the heating rate of 10 °C/min. The operating process of NH_3-TPD test was similar to that of H_2-TPR test with NH_3 replacing H_2. XPS analysis was performed on a Thermo ESCALAB 250XI (Thermo Fisher, Boston, MA, USA) with pass energy 46.95 eV, Al K_{α} radiation 1486.6 eV, X-ray source 150 W, and binding energy precision ±0.3 eV. The C 1 *s* line at 284.6 eV was measured as a reference.

3.3. Catalytic Performance Tests

The catalytic properties of the Mn−Co/TiO_2 and Mn−Fe/TiO_2 nanocatalysts were tested on a fixed-bed quartz tube reactor including a tube furnace and a temperature control unit, as shown in Figure 10. The mass of the catalyst was 250 mg with the particle diameter of 40–60 mesh. During NO conversion test, 5 mL nanocatalyst with the particle diameter of 40–60 mesh was used under the total flow rate of 2000 mL/min, corresponding to the gas hourly space velocity (GHSV) of 24,000 h^{-1}. The feed gas contained 300 ppm NO, 300 ppm NH_3, 5% O_2, and N_2 as balance gas. During reaction kinetic study, the reaction conditions were NO 150–600 ppm, NH_3 600 ppm, O_2 5%, catalyst mass of 200 mg, total flow rate at 4000 mL/min, and GHSV at 1 200 000 $cm^3{\cdot}g^{-1}{\cdot}h^{-1}$. The concentrations of NO, N_2O, and NH_3 in the outlet were continually monitored using German MRU MGA-5 analyzer (MRU, Berlin, Germany) joint with an external special detector for N_2O and NH_3. The NO conversion was calculated according to Equation (18). The N_2 selectivity was calculated by the concentrations of N_2O and NO, as show in Equation (19). Each experiment was repeated three times to confirm the results accuracy.

$$\text{NO conversion} = \left(\frac{[\text{NO}]_{\text{in}} - [\text{NO}]_{\text{out}}}{[\text{NO}]_{\text{in}}}\right) \times 100\% \tag{18}$$

$$\text{N}_2\text{ selectivity} = 1 - \frac{2[\text{N}_2\text{O}]_{\text{out}}}{[\text{NO}]_{\text{in}} - [\text{NO}]_{\text{out}}} \times 100\% \tag{19}$$

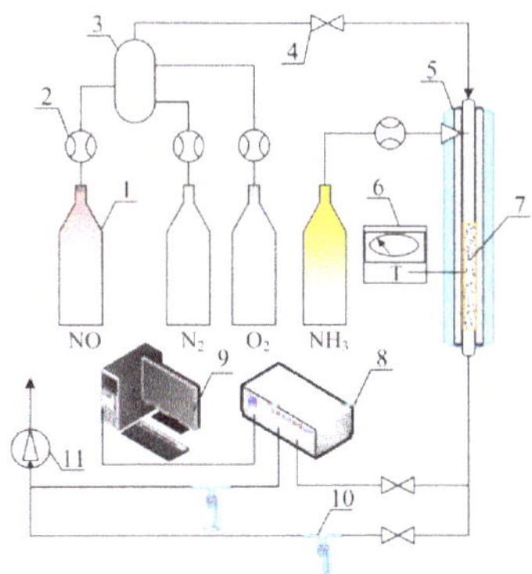

Figure 10. The schematic diagram of SCR experiments. 1, standard gas; 2, mass flowmeter; 3, gas mixer; 4, shutdown valve; 5, resistance furnace; 6, temperature controller; 7, nanocatalysts; 8, flue gas analyzer; 9, record system; 10, gas washing bottle; 11, induced draft fan.

4. Conclusions

The multiple characterizations performed on the Mn−Co/TiO_2 and Mn−Fe/TiO_2 nanocatalysts indicated that the Mn−Co/TiO_2 sample obtained the superior structure characteristics than Mn−Fe/TiO_2 with the surface area increased by 85.3 $m^2{\cdot}g^{-1}$, the nanoparticle size reduced by 21.79 nm, the active components distribution improved, and the crystallinity decreased. Meanwhile, these results further confirmed the catalytic property was highly dependent on the phase compositions of the catalysts. The ratios of Mn^{4+}/Mn^{n+} in the Mn−Co/TiO_2 sample was 46.8% higher than that of 40.2% in the Mn−Fe/TiO_2 sample, which demonstrated the better oxidation ability and the larger amount of Bronsted acid sites and Lewis acid sites on the sample surface. Both Mn−Co/TiO_2 and Mn−Fe/TiO_2 nanocatalysts exhibited the satisfactory conversion (>90%) above 150 °C. When compared to the

Mn−Fe/TiO_2 nanocatalyst, the Mn−Ce/TiO_2 sample displayed the preferable catalytic property with higher catalytic activity and stronger selectivity in the temperature range of 75—250 °C. In the meantime, the mechanism and kinetic study showed that Langmuir-Hinshelwood mechanism and Eley-Rideal mechanism reactions contributed to NH_3-SCR catalysis, simultaneously. Under the same test conditions, the NO conversion rate of Mn−Co/TiO_2 nanocatalyst was always higher than that of Mn−Fe/TiO_2 sample. Furthermore, Ellingham Diagrams of Mn−Co/TiO_2 and Mn−Fe/TiO_2 nanocatalysts were in accordance with their catalytic performances, based on the Gibbs free energy calculation of the reactions between ammonia and each kind of active metal oxide. Generally, Mn−Co/TiO_2 nanocatalyst exhibited the better catalytic properties than Mn−Fe/TiO_2 sample according to the comprehensive comparison in this research. The comparison with the insight from structure, performance, kinetics, and thermodynamics might supply a theoretical method to select superior metal oxides for NH_3-SCR.

Author Contributions: Conceptualization, Y.G.; Funding acquisition, Y.G., T.L.; Methodology, Y.G., T.L. Project administration, Y.G.; Writing—original draft, Y.G.; Writing—review & editing, T.L.; Data curation, S.Z., W.J., W.F., H.J.

Funding: This work was supported by Shandong Provincial Science and Technology Development Plan (2011GSF11716), Doctoral research fund project of Shandong Jianzhu university (XNBS1835), Shandong Jianzhu university open experimental project (2018yzkf023, 2018wzkf013), and the Shandong electric power engineering consulting institute science and technology project (37-K2014-33).

Conflicts of Interest: The authors declare no conflict of interest.

References

1. France, L.J.; Yang, Q.; Li, W.; Chen, Z.; Guang, J.; Guo, D.; Wang, L.; Li, X. Ceria modified $FeMnO_x$ —Enhanced performance and sulphur resistance for low-temperature SCR of NO_x. *Appl. Catal. B Environ.* **2017**, *206*, 203–215. [CrossRef]
2. Gillot, S.; Tricot, G.; Vezin, H.; Dacquin, J.-P.; Dujardin, C.; Granger, P. Induced effect of tungsten incorporation on the catalytic properties of $CeVO_4$ systems for the selective reduction of NO_x by ammonia. *Appl. Catal. B Environ.* **2018**, *234*, 318–328. [CrossRef]
3. Liu, J.; Li, X.; Zhao, Q.; Ke, J.; Xiao, H.; Lv, X.; Liu, S.; Tadé, M.; Wang, S. Mechanistic investigation of the enhanced NH_3-SCR on cobalt-decorated Ce-Ti mixed oxide: In situ FTIR analysis for structure-activity correlation. *Appl. Catal. B Environ.* **2017**, *200*, 297–308. [CrossRef]
4. Gao, Y.; Luan, T.; Lu, T.; Cheng, K.; Xu, H. Performance of V_2O_5-WO_3-MoO_3/TiO_2 Catalyst for Selective Catalytic Reduction of NO_x by NH_3. *Chin. J. Chem. Eng.* **2013**, *21*, 1–7. [CrossRef]
5. Huang, J.; Huang, H.; Jiang, H.; Liu, L. The promotional role of Nd on Mn/TiO_2 catalyst for the low-temperature NH_3 -SCR of NO_x. *Catal. Today* **2018**, *411*, 338–346. [CrossRef]
6. Zhang, R.; Yang, W.; Luo, N.; Li, P.; Lei, Z.; Chen, B. Low-temperature NH_3-SCR of NO by lanthanum manganite perovskites: Effect of A-/B-site substitution and TiO_2/CeO_2 support. *Appl. Catal. B Environ.* **2014**, *146*, 94–104. [CrossRef]
7. Fang, D.; He, F.; Mei, D.; Zhang, Z.; Xie, J.; Hu, H. Thermodynamic calculation for the activity and mechanism of Mn/TiO_2 catalyst doped transition metals for SCR at low temperature. *Catal. Commun.* **2014**, *52*, 45–48. [CrossRef]
8. Qiu, M.; Zhan, S.; Yu, H.; Zhu, D. Low-temperature selective catalytic reduction of NO with NH_3 over ordered mesoporous $Mn_xCo_{3-x}O_4$ catalyst. *Catal. Commun.* **2015**, *62*, 107–111. [CrossRef]
9. Zhou, C.; Zhang, Y.; Wang, X.; Xu, H.; Sun, K.; Shen, K. Influence of the addition of transition metals (Cr, Zr, Mo) on the properties of MnO_x –FeO_x catalysts for low-temperature selective catalytic reduction of NO_x by Ammonia. *J. Colloid Interface Sci.* **2013**, *392*, 319–324. [CrossRef]
10. Wang, R.; Wu, X.; Zou, C.; Li, X.; Du, Y. NOx Removal by Selective Catalytic Reduction with Ammonia over a Hydrotalcite-Derived NiFe Mixed Oxide. *Catalysts* **2018**, *8*, 384. [CrossRef]
11. Li, W.; Guo, R.-T.; Wang, S.-X.; Pan, W.-G.; Chen, Q.-L.; Li, M.-Y.; Sun, P.; Liu, S.-M. The enhanced Zn resistance of Mn/TiO_2 catalyst for NH_3 -SCR reaction by the modification with Nb. *Fuel Process. Technol.* **2016**, *154*, 235–242. [CrossRef]

12. Gao, F.; Tang, X.; Yi, H.; Chu, C.; Li, N.; Li, J.; Zhao, S. In-situ DRIFTS for the mechanistic studies of NO oxidation over α-MnO_2, β-MnO_2 and γ-MnO_2 catalysts. *Chem. Eng. J.* **2017**, *322*, 525–537. [CrossRef]
13. Hu, H.; Cai, S.; Li, H.; Huang, L.; Shi, L.; Zhang, D. Mechanistic Aspects of $deNO_x$ Processing over TiO_2 Supported Co–Mn Oxide Catalysts: Structure–Activity Relationships and In Situ DRIFTs Analysis. *ACS Catal.* **2015**, *5*, 6069–6077. [CrossRef]
14. Yang, S.; Wang, C.; Li, J.; Yan, N.; Ma, L.; Chang, H. Low temperature selective catalytic reduction of NO with NH_3 over Mn–Fe spinel: Performance, mechanism and kinetic study. *Appl. Catal. B Environ.* **2011**, *110*, 71–80. [CrossRef]
15. Zhang, Y.; Zheng, Y.; Wang, X.; Lu, X. Preparation of Mn–FeO_x/CNTs catalysts by redox co-precipitation and application in low-temperature NO reduction with NH_3. *Catal. Commun.* **2015**, *62*, 57–61. [CrossRef]
16. Gao, Y.; Jiang, W.; Luan, T.; Li, H.; Zhang, W.; Feng, W.; Jiang, H. High-Efficiency Catalytic Conversion of NO_x by the Synergy of Nanocatalyst and Plasma: Effect of Mn-Based Bimetallic Active Species. *Catalysts* **2019**, *9*, 103. [CrossRef]
17. Qiu, L.; Pang, D.; Zhang, C.; Meng, J.; Zhu, R.; Ouyang, F. In situ IR studies of Co and Ce doped Mn/TiO_2 catalyst for low-temperature selective catalytic reduction of NO with NH_3. *Appl. Surf. Sci.* **2015**, *357*, 189–196. [CrossRef]
18. Chen, Y.; Wang, J.; Yan, Z.; Liu, L.; Zhang, Z.; Wang, X. Promoting effect of Nd on the reduction of NO with NH_3 over CeO_2 supported by activated semi-coke: An in situ DRIFTS study. *Catal. Sci. Technol.* **2015**, *5*, 2251–2259. [CrossRef]
19. Wang, X.; Shi, Y.; Li, S.; Li, W. Promotional synergistic effect of Cu and Nb doping on a novel Cu/Ti-Nb ternary oxide catalyst for the selective catalytic reduction of NO_x with NH_3. *Appl. Catal. B Environ.* **2018**, *220*, 234–250. [CrossRef]
20. Yang, S.; Fu, Y.; Liao, Y.; Xiong, S.; Qu, Z.; Yan, N.; Li, J. Competition of selective catalytic reduction and non selective catalytic reduction over MnO_x/TiO_2 for NO removal: the relationship between gaseous NO concentration and N_2O selectivity. *Catal. Sci. Technol.* **2014**, *4*, 224–232. [CrossRef]
21. Wei, L.; Cui, S.; Guo, H.; Ma, X. Study on the role of Mn species in low temperature SCR on MnO_x/TiO_2 through experiment and DFT calculation. *Mol. Catal.* **2018**, *445*, 102–110. [CrossRef]
22. Kapteijn, F.; Singoredjo, L.; Andreini, A.; Moulijn, J.A. Activity and selectivity of pure manganese oxides in the selective catalytic reduction of nitric oxide with ammonia. *Appl. Catal. B Environ.* **1994**, *3*, 173–189. [CrossRef]
23. Cao, Y.; Luan, T.; Zhang, W.; Li, H. The promotional effects of cerium on the catalytic properties of Al_2O_3-supported $MnFeO_x$ for NO oxidation and fast SCR reaction. *Res. Chem. Intermed.* **2019**, *45*, 663–686. [CrossRef]
24. Liu, Z.; Yi, Y.; Zhang, S.; Zhu, T.; Zhu, J.; Wang, J. Selective catalytic reduction of NO_x with NH_3 over Mn-Ce mixed oxide catalyst at low temperatures. *Catal. Today* **2013**, *216*, 76–81. [CrossRef]
25. Li, X.; Zhang, S.; Jia, Y.; Liu, X.; Zhong, Q. Selective catalytic oxidation of NO with O_2 over Ce-doped MnO_x/TiO_2 catalysts. *J. Nat. Gas Chem.* **2012**, *21*, 17–24. [CrossRef]
26. Ma, Z.; Yang, H.; Li, Q.; Zheng, J.; Zhang, X. Catalytic reduction of NO by NH_3 over Fe–Cu–O_x/CNTs-TiO_2 composites at low temperature. *Appl. Catal. A Gen.* **2012**, *427–428*, 43–48. [CrossRef]
27. Shi, Y.; Chen, S.; Sun, H.; Shu, Y; Quan, X. Low-temperature selective catalytic reduction of NO_x with NH_3 over hierarchically macro-mesoporous Mn/TiO_2. *Catal. Commun.* **2013**, *42*, 10–13. [CrossRef]
28. Gong, P.; Xie, J.; Fang, D.; Han, D.; He, F.; Li, F.; Qi, K. Effects of surface physicochemical properties on NH_3-SCR activity of MnO_2 catalysts with different crystal structures. *Chin. J. Catal.* **2017**, *38*, 1925–1934. [CrossRef]
29. Boningari, T.; Pappas, D.K.; Ettireddy, P.R.; Kotrba, A.; Smirniotis, P.G. Influence of SiO_2 on M/TiO_2 (M = Cu, Mn, and Ce) Formulations for Low-Temperature Selective Catalytic Reduction of NO_x with NH_3: Surface Properties and Key Components in Relation to the Activity of NO_x Reduction. *Ind. Eng. Chem. Res.* **2015**, *54*, 2261–2273. [CrossRef]
30. Luo, S.; Zhou, W.; Xie, A.; Wu, F.; Yao, C.; Li, X.; Zuo, S.; Liu, T. Effect of MnO_2 polymorphs structure on the selective catalytic reduction of NO_x with NH_3 over TiO_2–Palygorskite. *Chem. Eng. J.* **2016**, *286*, 291–299. [CrossRef]
31. Jiang, H.; Wang, C.; Wang, H.; Zhang, M. Synthesis of highly efficient MnO_x catalyst for low-temperature NH_3-SCR prepared from Mn-MOF-74 template. *Mater. Lett.* **2016**, *168*, 17–19. [CrossRef]

32. Meng, D.; Xu, Q.; Jiao, Y.; Guo, Y.; Guo, Y.; Wang, L.; Lu, G.; Zhan, W. Spinel structured CoaMnbOx mixed oxide catalyst for the selective catalytic reduction of NO_x with NH_3. *Appl. Catal. B Environ.* **2018**, *221*, 652–663. [CrossRef]
33. Stanciulescu, M.; Caravaggio, G.; Dobri, A.; Moir, J.; Burich, R.; Charland, J.P.; Bulsink, P. Low-temperature selective catalytic reduction of NO_x with NH_3 over Mn-containing catalysts. *Appl. Catal. B Environ.* **2012**, *123–124*, 229–240. [CrossRef]
34. Kong, F.; Qiu, j.; Liu, H.; Zhao, R.; Zeng, H. Effect of NO/SO_2 on Elemental Mercury Adsorption by Nano-Fe_2O_3. *Korean J. Chem. Eng.* **2010**, *30*, 43–48. [CrossRef]
35. Gao, Y.; Luan, T.; Zhang, M.; Zhang, W.; Feng, W. Structure–Activity Relationship Study of Mn/Fe Ratio Effects on Mn−Fe−Ce−O_x/γ-Al_2O_3 Nanocatalyst for NO Oxidation and Fast SCR Reaction. *Catalysts* **2018**, *8*, 642. [CrossRef]
36. Zhang, C.; Chen, T.; Liu, H.; Chen, D.; Xu, B.; Qing, C. Low temperature SCR reaction over Nano-Structured Fe-Mn Oxides: Characterization, performance, and kinetic study. *Appl. Surf. Sci.* **2018**, *457*, 1116–1125. [CrossRef]
37. Jin, R.; Liu, Y.; Wu, Z.; Wang, H.; Gu, T. Low-temperature selective catalytic reduction of NO with NH_3 over MnCe oxides supported on TiO_2 and Al_2O_3: A comparative study. *Chemosphere* **2010**, *78*, 1160–1166. [CrossRef]
38. Xiong, Z.-b.; Liu, J.; Zhou, F.; Liu, D.-y.; Lu, W.; Jin, J.; Ding, S.-f. Selective catalytic reduction of NO_x with NH_3 over iron-cerium-tungsten mixed oxide catalyst prepared by different methods. *Appl. Surf. Sci.* **2017**, *406*, 218–225. [CrossRef]
39. Liu, F.; He, H.; Zhang, C.; Shan, W.; Shi, X. Mechanism of the selective catalytic reduction of NO_x with NH_3 over environmental-friendly iron titanate catalyst. *Catal. Today* **2011**, *175*, 18–25. [CrossRef]
40. Fan, X.; Qiu, F.; Yang, H.; Tian, W.; Hou, T.; Zhang, X. Selective catalytic reduction of NO_X with ammonia over Mn–Ce–O_X/TiO_2-carbon nanotube composites. *Catal. Commun.* **2011**, *12*, 1298–1301. [CrossRef]
41. Guan, D.S.; Wang, Y. Synthesis and growth mechanism of multilayer TiO_2 nanotube arrays. *Nanoscale* **2012**, *4*, 2968–2977. [CrossRef] [PubMed]
42. Lu, X.; Shen, C.; Zhang, Z.; Barrios, E.; Zhai, L. Core–Shell Composite Fibers for High-Performance Flexible Supercapacitor Electrodes. *ACS Appl. Mater. Interfaces* **2018**, *10*, 4041–4049. [CrossRef]
43. Wang, C.; Yu, F.; Zhu, M.; Wang, X.; Dan, J.; Zhang, J.; Cao, P.; Dai, B. Microspherical MnO_2-CeO_2-Al_2O_3 mixed oxide for monolithic honeycomb catalyst and application in selective catalytic reduction of NO_x with NH_3 at 50–150 °C. *Chem. Eng. J.* **2018**, *346*, 182–192. [CrossRef]
44. Deng, L.; Ding, Y.; Duan, B.; Chen, Y.; Li, P.; Zhu, S.; Shen, S. Catalytic deep combustion characteristics of benzene over cobalt doped Mn-Ce solid solution catalysts at lower temperatures. *Mol. Catal.* **2018**, *446*, 72–80. [CrossRef]
45. Li, K.; Tang, X.; Yi, H.; Ning, P.; Kang, D.; Wang, C. Low-temperature catalytic oxidation of NO over Mn–Co–Ce–O_x catalyst. *Chem. Eng. J.* **2012**, *192*, 99–104. [CrossRef]
46. Zhang, W.; Wang, F.; Li, X.; Liu, Y.; Liu, Y.; Ma, J. Fabrication of hollow carbon nanospheres introduced with Fe and N species immobilized palladium nanoparticles as catalysts for the semihydrogenation of phenylacetylene under mild reaction conditions. *Appl. Surf. Sci.* **2017**, *404*, 398–408. [CrossRef]
47. Wang, T.; Wan, Z.; Yang, X.; Zhang, X.; Niu, X.; Sun, B. Promotional effect of iron modification on the catalytic properties of Mn-Fe/ZSM-5 catalysts in the Fast SCR reaction. *Fuel Process. Technol.* **2018**, *169*, 112–121. [CrossRef]
48. Marbán, G.; Valdés-Solis, T.; Fuertes, A.B. Mechanism of low-temperature selective catalytic reduction of NO with NH_3 over carbon-supported Mn_3O_4: Role of surface NH_3 species: SCR mechanism. *J. Catal.* **2004**, *226*, 138–155. [CrossRef]
49. Liu, Z.; Zhu, J.; Li, J.; Ma, L.; Woo, S.I. Novel Mn–Ce–Ti Mixed-Oxide Catalyst for the Selective Catalytic Reduction of NO_x with NH_3. *ACS Appl. Mater. Interfaces* **2014**, *6*, 14500–14508. [CrossRef] [PubMed]
50. Xiong, S.; Liao, Y.; Xiao, X.; Dang, H.; Yang, S. Novel Effect of H_2O on the Low Temperature Selective Catalytic Reduction of NO with NH_3 over MnO_x-CeO_2: Mechanism and Kinetic Study. *J. Phys. Chem. C* **2015**, *119*, 4180–4187. [CrossRef]
51. Lu, J.J.; Murray, N.V.; Rosenthal, E. Signed formulas and annotated logics. In *Proceedings of the [1993] Proceedings of the Twenty-Third International Symposium on Multiple-Valued Logic*; Institute of Electrical and Electronics Engineers (IEEE): Piscatville, NJ, USA, 2002; pp. 48–53.

52. Lee, J.D. *A New Concise Inorganic Chemistry*; Van Nostrand Reinhold: Berkshire, UK, 1991; ISBN 0-412-40290-4.
53. Cooksy, A. *Physical Chemistry: Thermodynamics, Statistical Mechanics, and Kinetics*; PEARSON EDUCATION: Boston, MA, USA, 2012; ISBN 0-321-81415-0.
54. Lide, D.R. *CRC Handbook of Chemistry and Physics*, 90th ed.; CRC Press: Boca Raton, FL, USA, 2010; ISBN 978-1439820773.
55. Koebel, M.; Elsener, M. Selective catalytic reduction of NO over commercial $DeNO_x$-catalysts: Experimental determination of kinetic and thermodynamic parameters. *Chem. Eng. Sci.* **1998**, *53*, 657–669. [CrossRef]

Article

Mechanism and Performance of the SCR of NO with NH_3 over Sulfated Sintered Ore Catalyst

Wangsheng Chen [1,2], Fali Hu [1,2], Linbo Qin [1,2], Jun Han [1,2,*], Bo Zhao [1,2], Yangzhe Tu [1,2] and Fei Yu [3]

1 School of Resource and Environmental Engineering, Wuhan University of Science and Technology, Wuhan 430081, China; chenwangsheng@wust.edu.cn (W.C.); hufali@hotmail.com (F.H.); qinlinbo@wust.edu.cn (L.Q.); zhaobo87@wust.edu.cn (B.Z.); tuyangzhe123@hotmail.com (Y.T.)

2 Industrial Safety Engineering Technology Research Center of Hubei Province, Wuhan University of Science and Technology, Wuhan 430081, China

3 Department of Agricultural and Biological Engineering, Mississippi State University, Mississippi State, MS 39762, USA; fyu@abe.msstate.edu

* Correspondence: hanjun@wust.edu.cn; Tel.: +86-27-6886-2880

Received: 25 December 2018; Accepted: 14 January 2019; Published: 16 January 2019

Abstract: A sulfated sintered ore catalyst (SSOC) was prepared to improve the denitration performance of the sintered ore catalyst (SOC). The catalysts were characterized by X-ray Fluorescence Spectrometry (XRF), Brunauer–Emmett–Teller (BET) analyzer, X-ray diffraction (XRD), X-ray photoelectron spectroscopy (XPS) and diffuse reflectance infrared spectroscopy (DRIFTS) to understand the NH_3-selective catalytic reduction (SCR) reaction mechanism. Moreover, the denitration performance and stability of SSOC were also investigated. The experimental results indicated that there were more Brønsted acid sites at the surface of SSOC after the treatment by sulfuric acid, which lead to the enhancement of the adsorption capacity of NH_3 and NO. Meanwhile, Lewis acid sites were also observed at the SSOC surface. The reaction between $-NH_2$, NH_4^+ and NO (E-R mechanism) and the reaction of the coordinated ammonia with the adsorbed NO_2 (L-H mechanism) were attributed to NO_x reduction. The maximum denitration efficiency over the SSOC, which was about 92%, occurred at 300 °C, with a 1.0 NH_3/NO ratio, and 5000 h^{-1} gas hourly space velocity (GHSV).

Keywords: sintered ore catalyst; sulfate; In-situ DRIFTS; SCR

1. Introduction

Nitrogen oxide (NO_x) is one of the major atmospheric pollutants and is mainly generated from the combustion of fossil fuel, which has serious harmful effects on human health and the ecological environment. In 2016, the total emission of NOx from the iron and steel industry was about 1.04 million tonnes [1]. The sintering process in iron-making plants uses coal or coke as fuel, which is a major emission source of NOx. About 35–50% of the total NOx emission from the iron and steel industry is attributed to the sintering process [2–4]. The new ultra-low emission standard of air pollutants for the iron and steel industry will be issued by Chinese government and require that NOx concentration in the sintering flue gas should be below 50 mg/m^3. Hence, it is urgent to address the treatment of sintering flue gas.

Selective catalytic reduction (SCR) over V_2O_5-WO_3 (MoO_3)/TiO_2 catalyst has been widely applied in power stations because of its high denitration efficiency [5–7]. However, there are still shortcomings such as the toxicity of the catalyst, and the high operation costs [8]. In particular, the reaction temperature of the current commercial catalysts is higher than the temperature of the sintering flue gas [9]. Therefore, the sintering flue gas must be heated to the reaction temperature of the catalyst by using additional fuels.

Many researchers paid more attention to developing the low temperature catalysts or the catalysts prepared by the non-noble metals for sintering flue gas. Zhang et al. reported that the Fe–based catalysts exhibited a high catalytic activity for NO reduction [10]. Wang [11] also stated that the Fe_2O_3 particles had a good performance during NO_x elimination, where the highest NO_x conversion reached 95%. Yang et al. [12] claimed that α–Fe_2O_3 had a poor SCR activity, while γ–Fe_2O_3 had an excellent SCR activity at 200–350 °C. Meanwhile, it was also found that the further increase reaction temperature from 350 to 500 °C would suppress NOx conversion over γ–Fe_2O_3 [13].

It was well known that the sintered ore was one of the raw materials in the sintering plant, its main component was Fe_2O_3. In our previous study, Han et al. and Chen et al. [14,15] proposed that the hot sintered ore was used as catalysts for NO_x removal from the sintering flue gas. The experimental results illustrated that the hot sintered ore had a good denitration performance, the denitration efficiency was about 60% at 300 °C, with a 1.0 NH_3/NO ratio, and 1000 h^{-1} GHSV. However, the denitration efficiency was too low and the NOx concentration from the sintering flue gas could reach the limit after SCR over the sintered ore.

Ciambelli [16] found that introduction of SO_4^{2-} into SCR catalysts can promote the surface acidity of the catalysts. Thus, the catalytic activity was promoted. Khodayari [17] and Xu [18] thought that the sulfation of the catalysts would decrease the oxidization ability of Fe^{3+}, and separated the active sites of adsorbing –NH_2 and the active sites of oxidizing –NH_2. Therefore, the catalytic oxidization of NH_3 over γ–Fe_2O_3 was suppressed. This process resulted in an obvious promotion of NOx conversion. Zhang [19] also investigated the sulfation of CeO_2-ZrO_2, and their experiment results demonstrated that the sulfated CeO_2-ZrO_2 provided more surface acidities and acidic sites, and Brønsted acid sites were also increased.

In order to improve the denitration efficiency, SSOC was prepared and the influence of the acidification on SCR performance was investigated. At the same time, the reaction mechanisms of SCR over SSOC were also discussed.

2. Results and Discussion

2.1. Characterizations

The main components of the two catalysts were shown in Table 1. It presented that the main components of both two catalysts were Fe_2O_3 and CaO. After the introduction of sulfuric acid, the proportion of iron oxide in the catalyst was decreased and sulphate content was significantly increased.

Table 1. Chemical compositions of catalysts (%).

	Fe_2O_3	CaO	SiO_2	Al_2O_3	MgO	MnO_2	TiO_2	P_2O_5	SO_x
SOC	75.42	12.54	4.22	1.19	1.02	0.19	0.20	0.09	0.14
SSOC	48.92	11.52	2.82	0.79	0.85	0.13	0.09	0.05	29.92

The BET surface area, pore-size and pore volume of the catalysts were presented in Table 2. The specific surface area of the SOC was 3.684 m^2/g, the total pore volume was 0.00714 cm^3/g, and the average pore diameter was 6.3 nm. The BET surface area, pore size and pore volume of the SSOC were 4.734 m^2/g, 5.9 nm and 0.00964 cm^3/g, respectively. It was indicated that the introduction of sulfuric acid would open some micro-pores, which lead to the increase of the total pore volume. At the same time, the sulphate would block some pores. The BET of SSOC was higher than that of SCO, which meant that the sulfation had a positive effect on the pore structure of SOC.

Table 2. Textural properties of the catalysts.

Catalysts	BET Surface Area ($m^2 \cdot g^{-1}$)	Total Pore Volume ($cm^3 \cdot g^{-1}$)	Average Pore Diameter (nm)
SOC	3.684	0.00714	6.3
SSOC	4.734	0.00964	5.9

The XRD patterns of the catalysts were demonstrated in Figure 1. Several sharp diffraction peaks at 24.1°, 33.2°, 35.6°, 49.6°, 54.2°, 57.1° and 62.6° were observed, which were assigned to α-Fe_2O_3 (JCPDS NO. 33-0664), and the characteristic peaks of α-Fe_3O_4 at 30.1°, 33.2°, 49.6°, 54.2° and 57.1° were also detected in Figure 1. The results indicated that the main components of SOC were α-Fe_2O_3 and a small amount of α-Fe_3O_4. After the acidification with sulfuric acid, it could be seen that the peak of α-Fe_2O_3 was decreased. Meanwhile, the characteristic peaks (25.4°, 31.4°, 38.7°, 52.2°) of $Fe_2(SO_4)_3$ were detected, and a small amount of $FeSO_4$ and $CaSO_4$ also appeared.

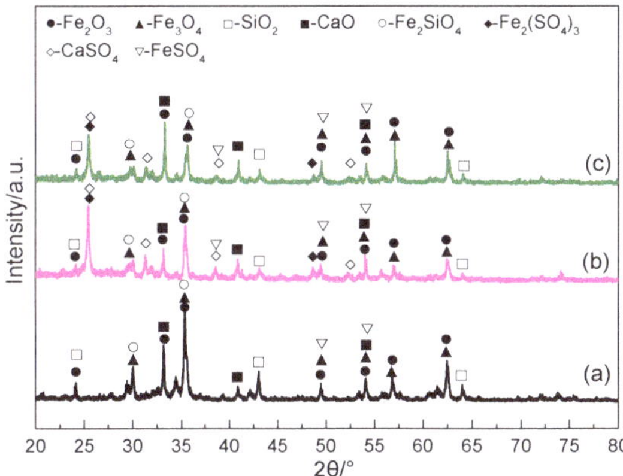

Figure 1. XRD patterns of the catalysts: SOC (**a**), SSOC (**b**) and SSOC after tested (**c**).

X-ray photoelectron spectroscopy (XPS) was used to study the valence states of the catalysts, as shown in Figure 2. It could be seen from Figure 2 (I) that there were obvious peaks of C (1s), O (1s), and Fe (2p) in both of the two catalysts. Moreover, there were obvious S (2p) peaks after the introduction of sulfuric acid. As shown in Figure 2 (II), the binding energies of Fe 2p 2/3 and Fe 2p 1/2 on SOC were mainly centered at about 710.8 and 724.3 eV, which were indications that the iron species were mainly in the form of Fe^{3+} in the SOC [20]. At the same time, a small number of Fe^{2+} existed, which was also consistent with the XRD results. The binding energies of Fe 2p 2/3 and Fe 2p 1/2 of SSOC were higher than those of SOC. The peak of Fe 2p 2/3 appeared at 711.8 eV and the peak of Fe 2p 1/2 appeared at 724.9 eV. After the acidification with sulfuric acid, both $Fe_2(SO_4)_3$ and $FeSO_4$ appeared in the SSOC, which contributed to Fe 2p peaks shifting to the higher position. In addition, in Figure 2 (III), the peak of S 2p appeared at 169.3 eV, indicating that sulfur compounds existed in the form of SO_4^{2-} [21].

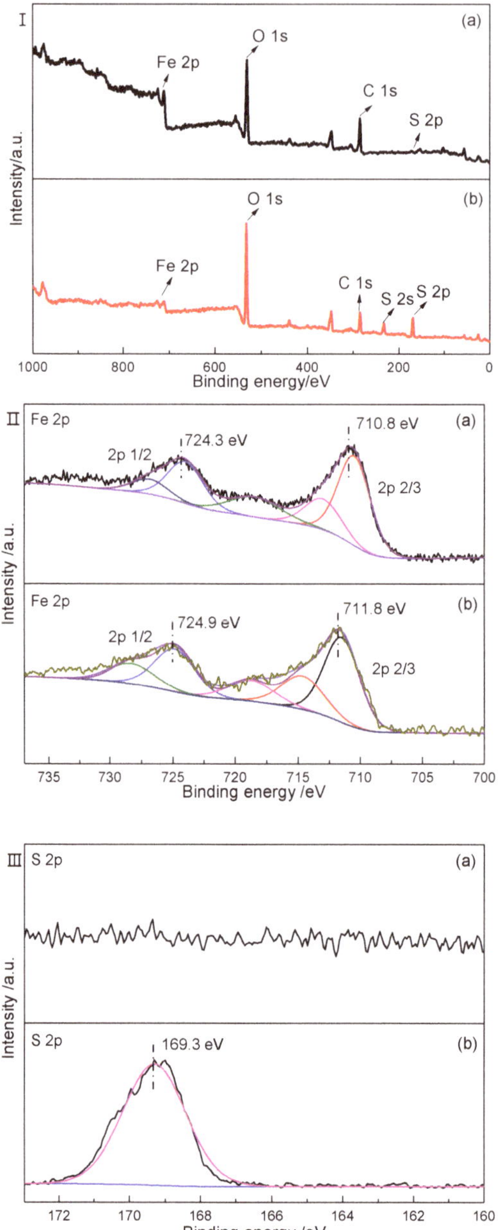

Figure 2. XPS spectra of the catalysts: SOC (**a**) and SSOC (**b**).

2.2. *In-Situ DRIFTS Studies*

2.2.1. Adsorption of NH_3

Figure 3 showed the FTIR spectra of NH_3 adsorption over the SSOC with 1000 ppm NH_3/Ar under different temperatures (250–350 °C). Thirty minutes after NH_3/Ar introduction, the bands at

1690, 1525, 1460 and 1408 cm^{-1} were observed. The bands at 1690, 1460 and 1408 cm^{-1} were symmetric bending vibration of NH_4^+ species at Brønsted acid sites [20,22]. The bands at 1525 and 3434 cm^{-1} were assigned to the intermediate species, such as ammonium or amide species (–NH_2) [20]. Meanwhile, the bands at, 3242 and 3120 cm^{-1} could be ascribed to the N–H stretching vibration of coordination NH_3, and the bands at 1259 cm^{-1} (symmetric bending vibration of NH_3 on Lewis acid sites) and 1102 cm^{-1} (NH_3 species adsorbed on Lewis acid sites) also appeared [20,23,24]. The intensities of the peaks of NH_4^+ species (1690, 1460 and 1408 cm^{-1}) and –NH_2 species (1525 cm^{-1}) first increased and then decreased in the temperature range of 250–350 °C. There were peaks at 960 and 930 cm^{-1} (NH_3 of gaseous state or weak adsorption state) under 300 °C. Yu et al. [25] stated that the following reactions probably took place in the reaction:

$$NH_3(g) \rightarrow NH_3(a) \tag{1}$$

$$NH_3(a) + O(a)/O^{2-} \rightarrow NH_2(a) + OH(a) \tag{2}$$

$$NH_3(g) + H^+ \rightarrow NH_4^+ \tag{3}$$

According to Equations (1)–(3), the more functional groups formed at the surface of the catalyst, the higher the reaction activity. Thus, the optimum adsorption activity of NH_3 over the SSOC was 300 °C, as shown in Figure 3.

The dependence of FTIR spectra over the SOC and SSOC on the reaction time at 1000 ppm NH_3/Ar and 300 °C was presented in Figure 4. Figure 4a demonstrated that there were 6 bands in the range of 1690–1100 cm^{-1}. The bands at 1690, 1405 and 1454 cm^{-1} were related to the symmetric bending vibration of NH_4^+ species, and the band at 1525 cm^{-1} was ascribed to the intermediate products of ammonium or –NH_2 species. Moreover, the band at 3242 cm^{-1} (N–H stretching vibration of coordinated NH_3) appeared at 10 min. With the feed of NH_3, the adsorption band at 1259 cm^{-1} (symmetric bending vibration of NH_3 on Lewis acid sites) appeared at 30 min, and the band at 1107 cm^{-1} (NH_3 of gaseous state or weak adsorption state) appeared at 60 min. Hence, there were both Lewis acid sites and Brønsted acid sites on the surface of SOC. The FTIR spectra over SSOC dependence of the reaction time was presented in Figure 4b, and the bands of NH_3 adsorption were basically the same as those of SOC. Moreover, the bands at 3434 cm^{-1} (–NH_2 groups) and 3128 cm^{-1} (coordination NH_3 on Lewis acid sites) were observed because the surface acidity of SSOC was strengthened. There were two new weak bands that appeared at 965 and 927 cm^{-1}, where NH_3 was in a gaseous state or weak adsorption state. These weakly adsorbed or gaseous NH_3 could rapidly adsorbed on the acid sites once the adsorbed NH_3 species was consumed. It could also be seen that the strength of NH_3 adsorption peak was enhanced after SOC acidification with sulfuric acid, which was the reason that the surface acidity of SSOC was enhanced after the acidification. The influence of the acidity of SSCO on the dentiration efficiency is presented in Figure 5. The different acidity of SSOC was achieved by sulfating with 1, 3 and 5 mol/L sulfuric acid. In this experimental run, the mass of catalyst and the volume of sulfuric acid were the same. Only the concentration of sulfuric acid was varied. It was demonstrated the maximum denitration efficiency occurred at the catalyst treated by 5 mol/L sulfuric acid, and its denitration efficiency was 92.3% at 300 °C. At the same time, the denitration efficiency of the catalysts sulfated by 1 and 3 mol/L was 56.6% and 68.5% at the same reaction temperature. Moreover, the experimental results proved that the optimum reaction temperature for all catalysts was 300 °C. These sulfates on the surface of SSOC provided more Brønsted acid sites (Peaks at 1690, 1525 cm^{-1} were increased), which promoted the adsorption capacity for NH_3. Hence, the catalyst denitration performance was also improved [26,27].

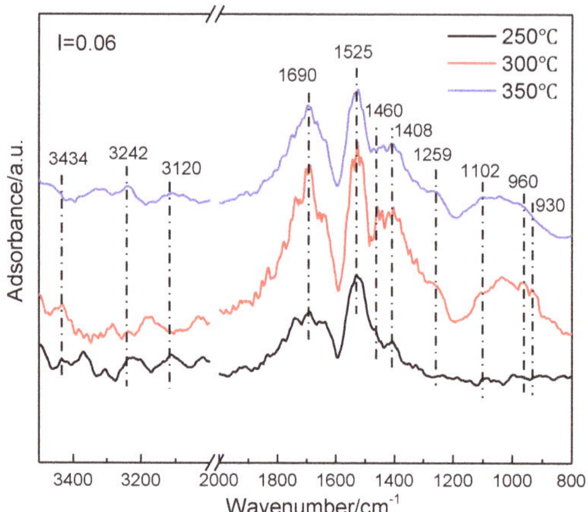

Figure 3. DRIFT spectra of SSOC in the condition of 1000 ppm NH_3 at 250–350 °C for 30 min.

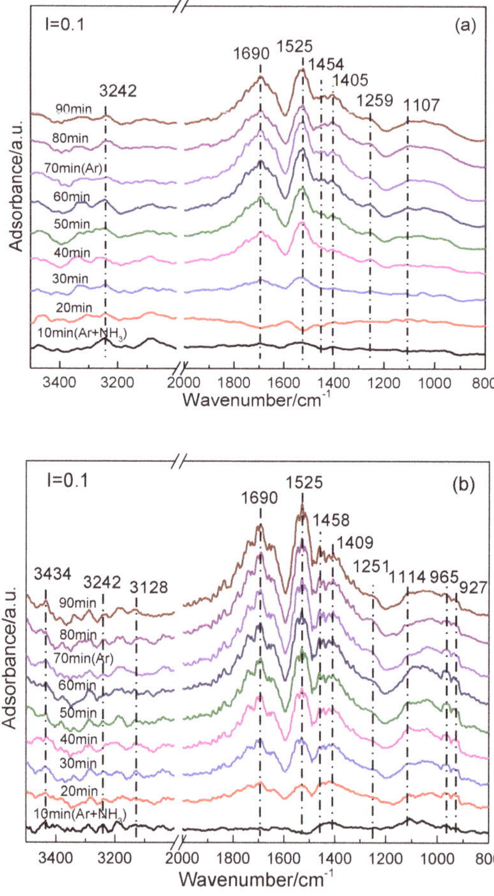

Figure 4. DRIFT spectra of SOC (**a**) and SSOC (**b**) exposed to 1000 ppm NH_3 at 300 °C at different time.

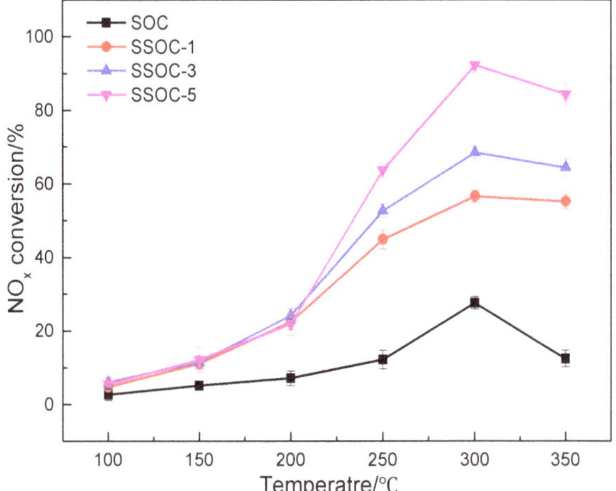

Figure 5. The influence of acidity of SSOC on SCR performance.

2.2.2. Co-Adsorption of NO and O_2

Figure 6 indicated the DRIFTS spectra of co–adsorption of NO and O_2 under 250–350 °C, and 1000 ppm NO + 15% O_2/Ar. After 30 min, there were 4 bands appeared in the range of 1620–1290 cm^{-1} and a weak band appeared at 1014 cm^{-1}. The band at 1607 cm^{-1} was ascribed to bridged nitrate species and adsorbed NO_2 molecules [28,29]. The bands at 1485 and 1417 cm^{-1} were assigned to bidentate nitrates and monodentate nitrates respectively [30–32]. In addition, the bands at 1293 and 1014 cm^{-1} were related to nitro compounds [31]. The variation trend of the intensities of these bands were same as NH_3 adsorption in the temperature range of 250–350 °C. The adsorption process can be explained by the following formula:

$$NO(g)NO + O(a) \leftrightarrow NO_2(a) - \text{Bridgednitrites} \tag{4}$$

$$NO(g) + O(a) \leftrightarrow NO_2(a) - \text{Monodentatednnitrites} \tag{5}$$

$$NO_2(g) + O(a) \leftrightarrow NO_3(a) - \text{Bidentatenitrites} \tag{6}$$

$$NH_2(a) + NO(g) \rightarrow NH_2NO \rightarrow N_2 + H_2O \tag{7}$$

$$NH_4^+(a) + NO(g) \rightarrow \{NH_3NO \rightarrow NH_2NO + H_2O\} \rightarrow N_2 + H_2O \tag{8}$$

$$NO_2(a) + NH_3(a) \rightarrow NO_2[NH_3]_2(a) + NO \rightarrow 2N_2 + 3H_2O \tag{9}$$

$$NO_2(a) + 2NH_4^+(a) \rightarrow NO_2NH_4^+(a) + NO \rightarrow N_2 + 2H_2 \tag{10}$$

According to Equations (4)–(10), all the peaks in Figure 6 were important to the SCR reaction. The intensity of peaks at 300 °C was the highest, which mean that the optimum adsorption temperature for NO was also 300 °C.

Figure 7 showed the DRIFTS spectra over SOC and SSOC at different reaction times under 300 °C, 1000 ppm NO + 15% O_2/Ar. Similar with the spectra in Figure 6, after 10 min, there were 4 bands that appeared in the range of 1620–1300 cm^{-1} and a weak band appeared at 1022 cm^{-1}. The bands intensities were gradually increased with the adsorption time. It could be seen that in Figure 7b, the adsorption intensity of SSOC was obviously higher than that of SOC, especially for the nitro compounds (1290 cm^{-1}) and the nitrate species (1490 cm^{-1}). The results indicated that the adsorption capacity of NO was improved after the acidification with sulfuric acid. It had been shown that the introduction of SO_4^{2-} enhanced the adsorption of NO [26].

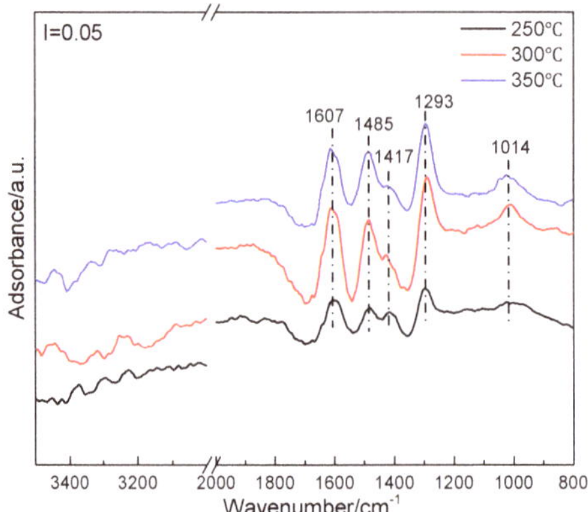

Figure 6. DRIFT spectra of SSOC in the condition of 1000 ppm NO and 15% O_2 at 250–350 °C for 30 min.

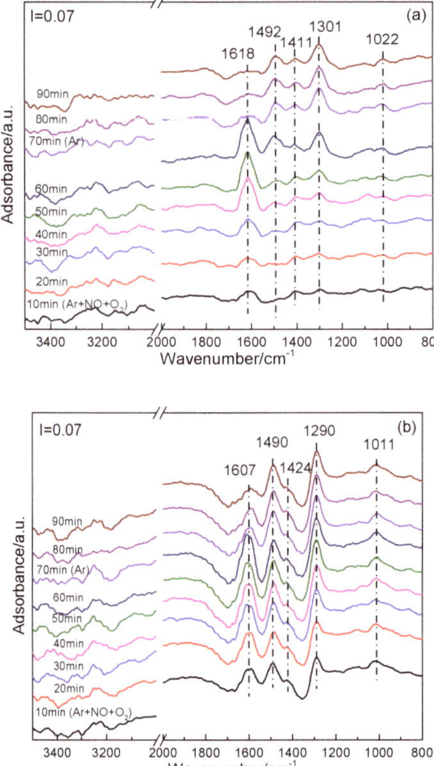

Figure 7. DRIFT spectra of SOC (**a**) and SSOC (**b**) exposed to 1000 ppm NO and 15% O_2 at 300 °C for a different time.

The results showed that the peak intensities of the adsorbed species related to NO_2 molecules and bridged nitrate species (1607 and 1618 cm^{-1}) were decreased obviously with purging by Ar, which indicated NO_2 molecules and bridged nitrate species absorbed at the surface of catalysts were unstable.

At the same time, it was found that the intensities of bidentate nitrates, monodentate nitrates and nitro compounds were stable even after an Ar purge.

2.2.3. Reaction between NH_3 and NO

The reaction at the surface of catalyst by introducing NH_3 and NO + O_2 into an in-situ reactor was also studied. Figure 8 shows that NH_4^+ species (1695 cm^{-1}), amide species(–NH_2) or ammonium (1533 cm^{-1}), and several weak adsorption bands at 1252, and 3242 cm^{-1} (NH_3 adsorbed on Lewis acid site) were detected on the surface of SOC after NH_3 introduction at 30 min. On the surface of SSOC catalyst, NH_4^+ species at 1695, 1427 and 1454 cm^{-1} and weakly adsorbed NH_3 or gaseous NH_3 (966, 925 cm^{-1}) were also detected. There were two bands at 3434 cm^{-1} (–NH_2 groups) and 3127 cm^{-1} (coordination NH_3 on Lewis acid sites) that appeared at the same time. After switching to NO + O_2, the adsorbed species of NH_3 over SOC and SSOC gradually disappeared and the adsorbed species of NO_x appeared. These bands were ascribed to NO_2 molecules and bridged nitrate species (1602 and 1618 cm^{-1}), bidentate nitrates (1488 and 1498), monodentate nitrates (1413 and 1419 cm^{-1}) and nitro compounds (1290, 1295, 1020 and 1011 cm^{-1}). Comparing Figure 8a with Figure 8b, it could be seen that there were more nitrite species and NH_3 species adsorbed at SSOC than those at SOC. This could be explained that the SSOC contained more active sites, which resulted from the sulfates on SSOC, and Equations (7)–(10) occurred [33]. The reaction of amide species (–NH_2) and NH_4^+ species on the surface of catalysts with the gaseous NO was E-R mechanism, and the reaction between adsorbed state NO_2 and adsorbed NH_3 followed L-H mechanism. Therefore, the reaction between NO and NH_3 over SOC and SSOC had two mechanisms.

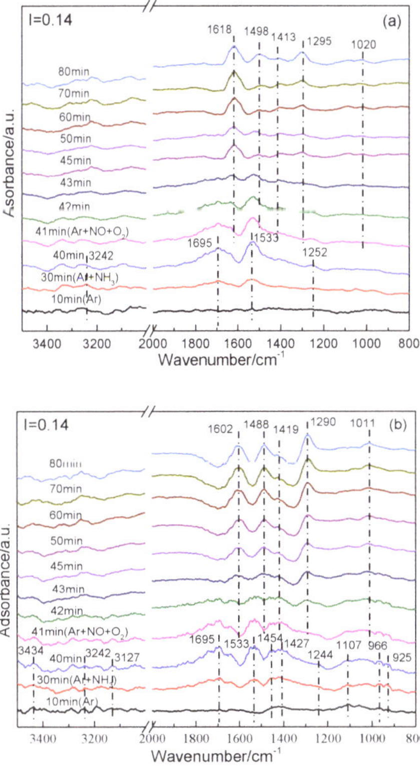

Figure 8. DRIFT spectra of SOC (a) and SSOC (b) successively exposed to 1000 ppm NH_3, 1000 ppm NO + 15% O_2 for different time under 300 °C.

2.2.4. SCR Performance

SCR performance were carried out in a fixed-bed reactor, which was made of quartz with an inner diameter of 20 mm and a length of 1000 mm. In this experiment, the mass of the catalysts samples was 13.49 g, the flow rate of the flue gas which simulated sintering flue gas was about 600 mL/min, the simulated sintering flue gas contained 300 ppm NO, 15% O_2 and balance N_2. The temperature in the reactor was kept at 100 °C–350 °C, with the condition of a 0.5–1.0 NH_3/NO ratio and 5000 h^{-1} GHSV. The NO_x concentrations in simulated flue gas at the inlet and outlet of the reactor were continuously recorded by a gas analyzer (PG-350, Horiba, Kyoto, Japan) with an accuracy of ±1.0%. The NO_x conversion was calculated according to the following equations:

$$NO_x \text{ conversion} = \frac{[NO_x]_{in} - [NO_x]_{out}}{[NO_x]_{in}} \times 100\% \quad (11)$$

Figure 9 presented the denitration performance of SOC and SSOC in the temperature range of 100 °C–350 °C. It was found that the reaction temperature had a great effect on the NH_3-SCR denitration performance of SOC and SSOC. The optimum reaction temperature was 300 °C, which conformed well with the in-situ DRIFTS results. The NO_x conversion of SOC was only 27% at 300 °C, 1.0 NH_3/NO ratio and 5000 h^{-1} GHSV, and the NO_x conversion of SSOC was 92% at the same condition. It was found that the denitration performance of SSOC was greatly improved and the optimum reaction temperature was 300 °C.

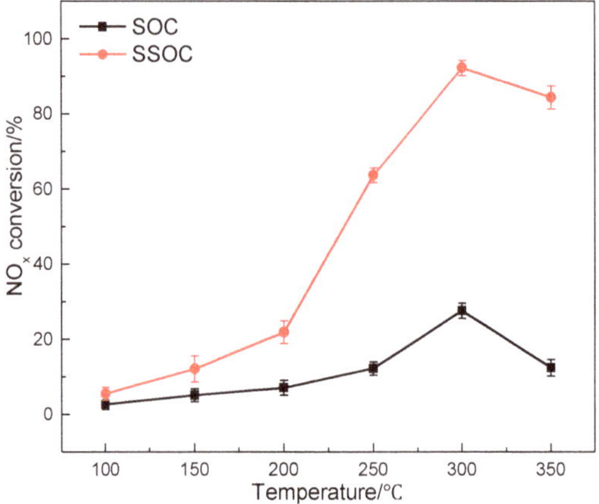

Figure 9. Effect of temperature on SCR activity on SOC and SSOC. 300 ppm NO, 15% O_2, Ar as balance gas, NH_3/NO ratio: 1.0, GHSV: 5000 h^{-1}.

The stability test of SSOC was shown at Figure 10. The reaction continued for 24 h at 300 °C, 1.0 NH_3/NO ratio, GHSV = 5000 h^{-1}. It could be seen that the NO_x conversion was stable at about 92%, which indicated that SSOC has a good denitration stability. Compared with the SOC, the denitration performance had been greatly improved after the acidification with sulfuric acid. The adsorption of NH_3 and NO on SSOC was obviously improved, which subsequently promoted the NH_3-SCR reaction.

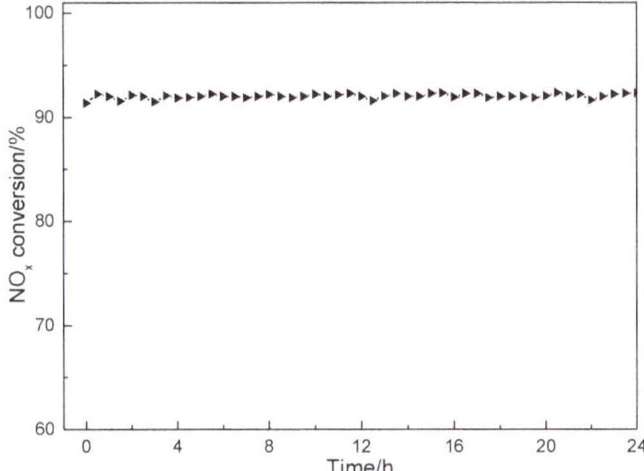

Figure 10. The stability test of SSOC. 300 ppm NO, 15% O_2, Ar as balance gas, NH_3/NO ratio: 1.0, GHSV: 5000 h^{-1}.

3. Experimental Process

3.1. Catalyst Preparation

In this experiment, the sintered ore was sampled from a sintering workshop in Wuhan. The sintered ore was dried, milled and sieved to 0.15–0.25 mm, which was denoted as SOC. The SSOC was prepared by using an impregnation method. Firstly, 100 g sintered ore (0.15–0.25 mm) was weighed and put into a beaker, then 50 mL sulfuric acid solution with a concentration of 5 mol/L was added and stirred simultaneously for 30 min. After filtration and washing with the deionized water, the mixture was dried at 105 °C and then calcined at 500 °C for 3 h in the air atmosphere. Finally, the prepared SSOC were naturally cooled to the room temperature, then crushed and sieved to 0.15–0.25 mm.

3.2. Catalyst Characterization

The main chemical composition of the SOC and SSOC were analyzed by X-ray fluorescence spectroscopy (XRF) (ARL SMS-XY, Thermo Fisher Scientific Corp., Waltham, MA, USA). The specific surface area, pore volume and pore size distribution of the catalysts were measured by an automated adsorption analyzer (Micromeritics ASAP 2020, Micromeritics Corp., Norcross, GA, USA). The catalysts samples were firstly degassed at 240 °C for 4 h before the test, and the adsorption medium was liquid nitrogen. The nitrogen adsorption and desorption were analyzed at −196 °C using BET analyzer (Micromeritics ASAP 2020, Micromeritics Corp., Norcross, GA, USA). The surface area was calculated by using the BET method according to nitrogen adsorption data in the relative pressure (P/P0) range of 0.01–1. X-ray diffraction (XRD; Rigaku RINT2000, Tokyo, Japan) was performed using CuKα radiation (λ = 1.54056 Å) to detect the crystalline phases of the samples. The analysis of XRD was referred to International Centre for Diffraction Data (ICDD). X-ray photoelectron spectroscopy (XPS, AXIS Ultra DLD, Shimazu Corp., Kyoto, Japan) was used to determine the valence states of the surface atoms of the catalysts with Al Kα radiation.

In-situ DRIFTS experiments were carried out in a FTIR spectrometer (FT-IR, Bruker Tensor II, Bruker optics Corp., Karlsruhe, Germany) equipped with an in-situ cell and a mercury-cadmium-telluride detector [34–37]. For the adsorption of NH_3 (or NO + O_2), the catalyst was exposed to a 20 mL/min NH_3 (or NO + O_2), which resulted in the variation with adsorption time of the DRIFT spectra, and argon purging was subsequently performed. In the reaction mechanism studies, the catalyst was pretreated in a flow of 20 mL/min NH_3 for 40 min, then was shifted NO + O_2 at 300 °C to get the DRIFT spectra. All spectra were recorded by accumulating 100 scans at a spectra resolution of 4 cm^{-1}.

4. Conclusions

In this paper, SSOC displayed an excellent catalytic denitration activity. There were some sulfates that appeared in SSOC after acidification with sulfuric acid solution, which provided more Brønsted acid sites. In-situ DRIFTS results demonstrated that there were Lewis and Brønsted acid sites simultaneously on the surfaces of SOC and SSOC. The acidification contributed to the increase of the Brønsted acid sites at SSOC, which improved the adsorption capacity of NH_3 and NO. NH_3 and NO were adsorbed on the surface of catalysts to form amide species ($-NH_2$), NH_4^+ species, NO_2 molecules in gaseous or weakly adsorbed state and nitrate species. Meanwhile, it could be seen that the NH_3-SCR process of SOC and SSOC followed E-R and L-H mechanisms. Moreover, the optimum reaction temperature of catalysts was 300 °C and maximum NO_x conversion fraction over SSOC was 92% at 1.0 NH_3/NO ratio and 5000 h^{-1} GHSV. It was observed that the NO_x conversion could be steadily maintained.

Author Contributions: Data curation, F.H.; Investigation, Y.T.; Methodology, L.Q.; Project administration, J.H.; Software, B.Z.; Writing—original draft, W.C.; Writing—review & editing, F.Y.

Funding: The present work was partly supported by the National Natural Science Foundation of China (Grant No. 51476118 and 51576146).

Conflicts of Interest: The authors declare no conflict of interest. The funders had no role in the design of the study; in the collection, analyses, or interpretation of data; in the writing of the manuscript, or in the decision to publish the results.

References

1. Zhou, H.; Chen, J.; Zhou, M.; Cen, K. Experimental investigation on the mixing performance of heating gas into the low temperature sintering flue gas selective catalyst reaction facilities. *Appl. Therm. Eng.* **2017**, *115*, 378–392. [CrossRef]
2. Ren, S.; Guo, F.; Yang, J.; Yao, L.; Zhao, Q.; Kong, M. Selection of carbon materials and modification methods in low-temperature sintering flue gas denitrification. *Chem. Eng. Res. Des.* **2017**, *126*, 278–285. [CrossRef]
3. Mousavi, S.; Panahi, P. Modeling and optimization of NH_3-SCR performance of MnO_x/γ-alumina nanocatalysts by response surface methodology. *J. Taiwan Inst. Chem. Eng.* **2016**, *69*, 68–77. [CrossRef]
4. Han, J.; Liang, Y.; Hu, J.; Qin, L.; Street, J.; Lu, Y.; Yu, F. Modeling downdraft biomass gasification process by restricting chemical reaction equilibrium with Aspen Plus. *Energy Convers. Manag.* **2017**, *153*, 641–648. [CrossRef]
5. Ogidiama, O.; Shamim, T. Performance Analysis of Industrial Selective Catalytic Reduction (SCR) Systems. *Energy Procedia* **2014**, *61*, 2154–2157. [CrossRef]
6. Chen, C.; Cao, Y.; Liu, S.; Chen, J.; Jia, W. Review on the latest developments in modified vanadium-titanium-based SCR catalysts. *Chin. J. Catal.* **2018**, *39*, 1347–1365. [CrossRef]
7. Liu, F.; Asakura, K.; He, H.; Shan, W.; Shi, X.; Zhang, C. Influence of sulfation on iron titanate catalyst for the selective catalytic reduction of NO_x, with NH_3. *Appl. Catal. B Environ.* **2011**, *103*, 369–377. [CrossRef]
8. Zhang, Q.; Fan, J.; Ning, P.; Song, Z.; Liu, X.; Wang, L.; Wang, J.; Wang, H.; Long, K. In situ DRIFTS investigation of NH_3-SCR reaction over CeO_2/zirconium phosphate catalyst. *Appl. Surf. Sci.* **2018**, *435*, 1037–1045. [CrossRef]
9. Chen, W.; Luo, J.; Qin, L.; Han, J. Selective autocatalytic reduction of NO from sintering flue gas by the hot sintered ore in the presence of NH_3. *J. Environ. Manag.* **2015**, *164*, 146–150. [CrossRef]
10. Zhang, C.; Chen, T.; Liu, H.; Chen, D.; Xu, B.; Qing, C. Low temperature SCR reaction over Nano-Structured Fe-Mn Oxides: Characterization, performance, and kinetic study. *Appl. Surf. Sci.* **2018**, *457*, 1116–1125. [CrossRef]
11. Wang, X.; Gui, K. Fe_2O_3 particles as superior catalysts for low temperature selective catalytic reduction of NO with NH_3. *J. Environ. Sci.* **2013**, *25*, 2469–2475. [CrossRef]
12. Yang, S.; Liu, C.; Chang, H.; Ma, L.; Qu, Z.; Yan, N.; Wang, C.; Li, J. Improvement of the activity of γ-Fe_2O_3 for the selective catalytic reduction of NO with NH_3 at high temperatures: No reduction versus NH_3 oxidization. *Ind. Eng. Chem. Res.* **2013**, *52*, 5601–5610. [CrossRef]
13. Shi, X.; He, H.; Xie, L. The effect of Fe species distribution and acidity of Fe-ZSM-5 on the hydrothermal stability and SO_2 and hydrocarbons durability in NH_3-SCR reaction. *Chin. J. Catal.* **2015**, *36*, 649–656. [CrossRef]
14. Han, J.; He, X.; Qin, L.; Chen, W.; Yu, F. NO_x removal coupled with energy recovery in sintering plant. *Ironmak. Steelmak.* **2013**, *41*, 350–354. [CrossRef]

15. Chen, W.; Li, Z.; Hu, F.; Qin, L.; Han, J.; Wu, G. In-situ DRIFTS investigation on the selective catalytic reduction of NO with NH_3 over the sintered ore catalyst. *Appl. Surf. Sci.* **2018**, *439*, 75–81. [CrossRef]
16. Ciambelli, P.; Fortuna, M.; Sannino, D.; Baldacci, A. The influence of sulphate on the catalytic properties of V_2O_5-TiO_2 and WO_3-TiO_2 in the reduction of nitric oxide with ammonia. *Catal. Today* **1996**, *29*, 161–164. [CrossRef]
17. Khodayari, R.; Odenbrand, C. Regeneration of commercial SCR catalysts by washing and sulphation: Effect of sulphate groups on the activity. *Appl. Catal. B Environ.* **2001**, *33*, 277–291. [CrossRef]
18. Xu, T.; Wu, X.; Liu, X.; Cao, L.; Lin, Q.; Weng, D. Effect of barium sulfate modification on the SO_2 tolerance of V_2O_5/TiO_2 catalyst for NH_3-SCR reaction. *J. Environ. Sci.* **2017**, *57*, 110–117. [CrossRef] [PubMed]
19. Zhang, H.; Zou, Y.; Peng, Y. Influence of sulfation on CeO_2-ZrO_2 catalysts for NO reduction with NH_3. *Chin. J. Catal.* **2017**, *38*, 160–167. [CrossRef]
20. Liu, F.; He, H.; Ding, Y.; Zhang, C. Effect of manganese substitution on the structure and activity of iron titanate catalyst for the selective catalytic reduction of NO with NH_3. *Appl. Catal. B Environ.* **2009**, *93*, 194–204. [CrossRef]
21. Ma, L.; Seo, C.; Nahata, M.; Chen, X.; Li, J.; Schwank, J. Shape dependence and sulfate promotion of CeO_2 for selective catalytic reduction of NO_x with NH_3. *Appl. Catal. B Environ.* **2018**, *232*, 246–259. [CrossRef]
22. Liu, J.; Guo, R.; Li, M.; Sun, P.; Liu, S.; Pan, W. Enhancement of the SO_2 resistance of Mn/TiO_2 SCR catalyst by Eu modification: A. mechanism study. *Fuel* **2018**, *223*, 385–393. [CrossRef]
23. Ramis, G.; Larrubia, M. An FT-IR study of the adsorption and oxidation of N-containing compounds over Fe_2O_3/Al_2O_3 SCR catalysts. *J. Mol. Catal. A Chem.* **2004**, *215*, 161–167. [CrossRef]
24. Kumar, P.; Jeong, Y.; Gautam, S.; Ha, H.; Lee, K.; Chae, K. XANES and DRIFTS study of sulfated Sb/V/Ce/TiO_2 catalysts for NH_3-SCR. *Chem. Eng. J.* **2015**, *275*, 142–151. [CrossRef]
25. Yu, C.; Huang, B.; Dong, L.; Chen, F.; Liu, X. In situ FT-IR study of highly dispersed MnO_x/SAPO-34 catalyst for low-temperature selective catalytic reduction of NOx by NH_3. *Catal. Today* **2017**, *281*, 610–620. [CrossRef]
26. Zhang, L.; Qu, H.; Du, T.; Ma, W.; Zhong, Q. H_2O and SO_2 tolerance, activity and reaction mechanism of sulfated Ni-Ce-La composite oxide nanocrystals in NH_3-SCR. *Chem. Eng. J.* **2016**, *296*, 122–131. [CrossRef]
27. Chen, L.; Niu, X.; Li, Z.; Dong, Y.; Zhang, Z.; Yuan, F.; Zhu, Y. Promoting catalytic performances of Ni-Mn spinel for NH_3-SCR by treatment with SO_2 and H_2O. *Catal. Commun.* **2016**, *85*, 48–51. [CrossRef]
28. Li, X.; Li, K.; Peng, Y.; Li, X.; Zhang, Y.; Wang, D.; Chen, J.; Li, J. Interaction of phosphorus with a $FeTiO_x$ catalyst for selective catalytic reduction of NO_x with NH_3: Influence on surface acidity and SCR mechanism. *Chem. Eng. J.* **2018**, *347*, 173–183. [CrossRef]
29. Liu, N.; Wang, J.; Wang, F.; Liu, J. Promoting effect of tantalum and antimony additives on $deNO_x$ performance of $Ce_3Ta_3SbO_x$ for NH_3-SCR reaction and DRIFT studies. *J. Rare Earths* **2018**, *36*, 594–602. [CrossRef]
30. Cao, L.; Chen, L.; Wu, X.; Ran, R.; Xu, T.; Chen, Z.; Weng, D. TRA and DRIFTS studies of the fast SCR reaction over CeO_2/TiO_2 catalyst at low temperatures. *Appl. Catal. A Gen.* **2018**, *557*, 46–54. [CrossRef]
31. Pen, D.; Uphade, B.; Reddy, E.; Smirniotis, P. Identification of Surface Species on Titania-Supported Manganese, Chromium, and Copper Oxide Low-Temperature SCR Catalysts. *J. Phys. Chem. B* **2004**, *108*, 9927–9936. [CrossRef]
32. Han, J.; Zhang, L.; Zhao, B.; Qin, L.; Wang, Y.; Xing, F. The N-doped activated carbon derived from sugarcane bagasse for CO_2 adsorption. *Ind. Crop. Prod.* **2019**, *128*, 290–297. [CrossRef]
33. Wang, H.; Qu, Z.; Dong, S.; Tang, C. Mechanism study of FeW mixed oxides to the selective catalytic reduction of NO_x with NH_3: In situ DRIFTS and MS. *Catal. Today* **2018**, *307*, 35–40. [CrossRef]
34. Han, J.; Li, W.; Liu, D.; Qin, L.; Chen, W.; Xing, F. Pyrolysis characteristic and mechanism of waste tyre: A thermogravimetry-mass spectrometry analysis. *J. Anal. Appl. Pyrolysis* **2018**, *129*, 1–5. [CrossRef]
35. Han, J.; Zhang, L.; Lu, Y.; Hu, J.; Cao, B.; Yu, F. The effect of syngas composition on the Fischer Tropsch synthesis over three-dimensionally ordered macro-porous iron based catalyst. *Mol. Catal.* **2017**, *440*, 175–183. [CrossRef]
36. Lu, Y.; Yan, Q.; Han, J.; Cao, B.; Street, J.; Yu, F. Fischer-Tropsch synthesis of olefin-rich liquid hydrocarbons from biomass-derived syngas over carbon-encapsulated iron carbide/iron nanoparticles catalyst. *Fuel* **2017**, *193*, 369–384. [CrossRef]
37. Huang, Z.; Qin, L.; Xu, Z.; Chen, W.; Xing, F.; Han, J. The effects of Fe_2O_3 catalyst on the conversion of organic matter and bio-fuel production during pyrolysis of sewage sludge. *J. Energy Inst.* **2018**. [CrossRef]

Article

Promotional Effect of Cerium and/or Zirconium Doping on Cu/ZSM-5 Catalysts for Selective Catalytic Reduction of NO by NH_3

Ye Liu, Chonglin Song *, Gang Lv, Chenyang Fan and Xiaodong Li

State Key Laboratory of Engines, Tianjin University, Tianjin 300072, China; liuyeskle@tju.edu.cn (Y.L.); lvg@tju.edu.cn (G.L.); fanchenyang@tju.edu.cn (C.F.); linahaifeng@tju.edu.cn (X.L.)

* Correspondence: songchonglin@tju.edu.cn; Tel.: +86-22-27406840-8020

Received: 7 July 2018; Accepted: 24 July 2018; Published: 28 July 2018

Abstract: The cerium and/or zirconium-doped Cu/ZSM-5 catalysts ($CuCe_xZr_{1-x}O_y$/ZSM-5) were prepared by ion exchange and characterized by X-ray diffraction (XRD), transmission electron microscopy (TEM), X-ray photoelectron spectroscopy (XPS), temperature-programmed reduction by hydrogen (H_2-TPR). Activities of the catalysts obtained on the selective catalytic reduction (SCR) of nitric oxide (NO) by ammonia were measured using temperature programmed reactions. Among all the catalysts tested, the $CuCe_{0.75}Zr_{0.25}O_y$/ZSM-5 catalyst presented the highest catalytic activity for the removal of NO, corresponding to the broadest active window of 175–468 °C. The cerium and zirconium addition enhanced the activity of catalysts, and the cerium-rich catalysts exhibited more excellent SCR activities as compared to the zirconium-rich catalysts. XRD and TEM results indicated that zirconium additions improved the copper dispersion and prevented copper crystallization. According to XPS and H_2-TPR analysis, copper species were enriched on the ZSM-5 grain surfaces, and part of the copper ions were incorporated into the zirconium and/or cerium lattice. The strong interaction between copper species and cerium/zirconium improved the redox abilities of catalysts. Furthermore, the introduction of zirconium abates N_2O formation in the tested temperature range.

Keywords: selective catalytic reduction; nitric oxide; ammonia; Cu/ZSM-5; cerium; zirconium

1. Introduction

Metal-based ZSM-5 catalyst has received considerable attention due to its excellent performance in the low-temperature selective catalytic reduction (SCR) reaction [? ? ? ?]. As a crystalline inorganic polymer, ZSM-5 consists of a three-dimensional network of SiO_4 and AlO_4 tetrahedra linked by interconnecting oxygen ions. When transition metal is introduced in ZSM-5, the catalytic performance can be enhanced through different types of active species, including isolated ions on the exchange sites within the ZSM-5 structure, metal oxide clusters, and metal oxide particles in the appearance of ZSM-5 [? ?]. Among all the relevant catalysts studied, Cu/ZSM-5 is of particular interest because it exhibits excellent SCR activity in NO abatement [? ?]. However, it has been reported that the catalytic activity of Cu/ZSM-5 catalysts is decreased, after an initial increase, with increasing copper content [?], which suggests that it is difficult to improve catalytic performance by merely increasing the copper content.

Fabrication of copper catalysts can be implemented through incorporating more reducible promoters, such as ceria, lanthana, and zirconia [? ? ?]. Cerium-containing materials are promising candidates for the application of NO abatement due to their high oxygen storage capacity and unique redox properties [? ?]. Nevertheless, a major drawback of cerium lies in its poor thermal and hydrothermal stability, especially under a practical diesel exhaust atmosphere [? ?]. It is well known that the incorporation of zirconia in cerium-based materials via high-temperature calcination can enhance the thermal stability and dispersion of the active component on the support surface [? ?]. Furthermore, the addition of zirconia to ceria introduces structural defects through substitution of Ce^{4+} by Zr^{4+}, which further enhances the oxygen storage capacity of ceria, the oxygen mobility in the lattice, the redox property, and thermal resistance [? ? ?]. Consequently, cerium and zirconium materials have the potential to improve the catalytic activity of Cu/ZSM-5 catalyst. Based on this background, the present work attempts to address the effects of the cerium and/or zirconium addition on Cu/ZSM-5 catalysts for SCR of NO by NH_3. A series of $CuCe_xZr_{1-x}O_y$/ZSM-5 catalysts (x = 0, 0.25, 0.50, 0.75 and 1) were synthesized using a conventional ion-exchange method, and the effects of adding cerium/zirconium metal ions into Cu/ZSM-5 catalysts on the SCR reaction were investigated using X-ray diffraction (XRD), transmission electron microscopy (TEM), X-ray photoelectron spectroscopy (XPS), and temperature-programmed reduction by hydrogen (H_2-TPR). The correlation between the structural characteristics, dispersion, reduction, and activity for the SCR process are discussed. The purpose of this work is to establish the relationships between structure and catalytic performance, which will be beneficial for the design and rationalization of the practical diesel catalysts.

2. Results and Discussion

2.1. Structure and Morphology

Figure ?? shows the XRD patterns of ZSM-5, Cu/ZSM-5, and $CuCe_xZr_{1-x}$/ ZSM-5 (x = 0, 0.25, 0.5, 0.75 and 1). As expected, the catalysts still retain the ordered microstructure after the addition of copper, and cerium and/or zirconium because the inherent MFI structure of ZSM-5 (2θ = 7.8°, 8.7°, 24.5°, 24.9°, PDF 44-003) can be observed for all the catalysts. However, the incorporation of copper, and cerium and/or zirconium leads to a decrease of the intensity of the ZSM-5 principal diffraction peaks, which can be attributed to the higher absorption coefficient of metal compounds for X-ray radiation [?]. Crystallized CuO (2θ = 35.5°, 38.6°, PDF = 02-1041) nanoparticles are detected for the Cu/ZSM-5 catalysts. For the $CuZr_1$/ZSM-5, $CuCe_{0.25}Zr_{0.75}$/ZSM-5, $CuCe_{0.5}Zr_{0.5}$/ZSM-5, and $CuCe_{0.75}Zr_{0.25}$/ZSM-5 catalysts, no diffraction peaks for metal or metal oxide clusters were observed, indicating that the copper, zirconium, and cerium oxides are well dispersed as amorphous metal species, or aggregated into mini-crystals that are too small (<4 nm) to be detected by XRD [?]. For $CuCe_1$/ZSM-5, the observation of a small peak of CeO_2 (2θ = 28.2°, PDF = 34-0394) suggests that the extra-framework cerium is prone to agglomerating into cerium oxide clusters.

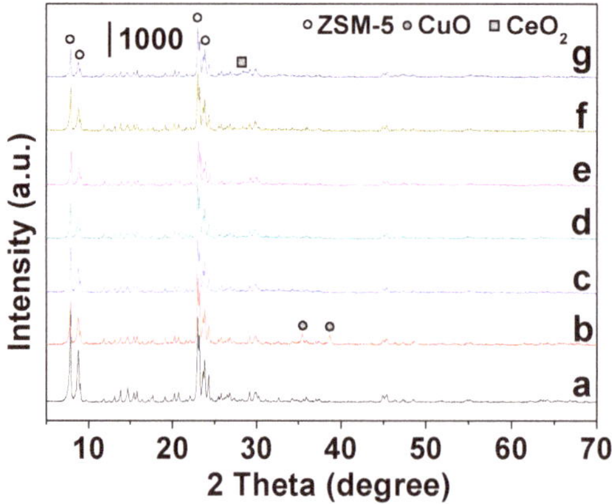

Figure 1. XRD patterns of $CuCe_xZr_{1-x}$/ZSM-5 catalysts with different loading ratios: (a) ZSM-5, (b) Cu/ZSM-5, (c) $CuZr_1$/ZSM-5, (d) $CuCe_{0.25}Zr_{0.75}$/ZSM-5, (e) $CuCe_{0.5}Zr_{0.5}$/ZSM-5, (f) $CuCe_{0.75}Zr_{0.25}$/ ZSM-5, and (g) $CuCe_1$/ZSM-5.

To validate the XRD results, for each catalyst sample, more than 200 metal oxide particles from different TEM images were randomly chosen to determine the size distribution of the metal oxide particles. The size distributions of the metal oxide particles for $CuCe_{0.25}Zr_{0.75}$/ZSM-5, $CuCe_{0.5}Zr_{0.5}$/ZSM-5, $CuCe_{0.75}Zr_{0.25}$/ZSM-5, and $CuCe_1$/ZSM-5 are shown in Figure ??. The metal oxide particles in the d, e and f samples are less than 4 nm in size, while most metal oxide particles in the g sample are larger than 4 nm in size. These results are consistent with the XRD data.

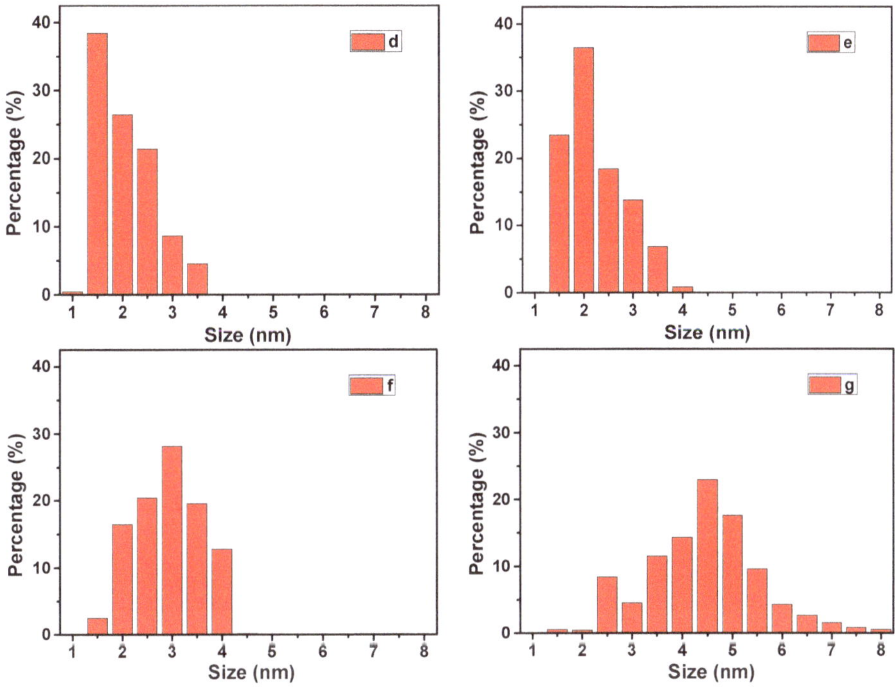

Figure 2. Size distributions of metal oxide particles for (**d**) $CuCe_{0.25}Zr_{0.75}$/ZSM-5, (**e**) $CuCe_{0.5}Zr_{0.5}$/ZSM5, (**f**) $CuCe_{0.75}Zr_{0.25}$/ZSM-5, and (**g**) $CuCe_1$/ZSM-5.

The typical TEM images of pure ZSM-5 and $CuCe_xZr_{1-x}$/ZSM-5 catalysts (x = 0, 0.5 and 1) are shown in Figure ??. The typical interference fringes of the ZSM-5 crystal structure are found in Figure ??a. Upon zirconium addition, small aggregates are observed for the $CuZr_1$/ZSM-5 catalyst (see Figure ??b), and most of them are less than 2 nm in size. The subsequent energy-dispersive X-ray (EDX) analysis for Point 1 in Figure ??b indicates a copper, zirconium, and oxygen-rich phase for these small aggregates. With increasing cerium content, the sizes of aggregates become larger, as shown in Figure ??c for $CuCe_{0.5}Zr_{0.5}$/ZSM-5 and Figure ??d for $CuCe_1$/ZSM-5. These results reveal that the addition of cerium decreases the dispersion of metal ions. The EDX analysis for Point 2 in Figure ??d indicates copper and cerium enrichment for the large black aggregates.

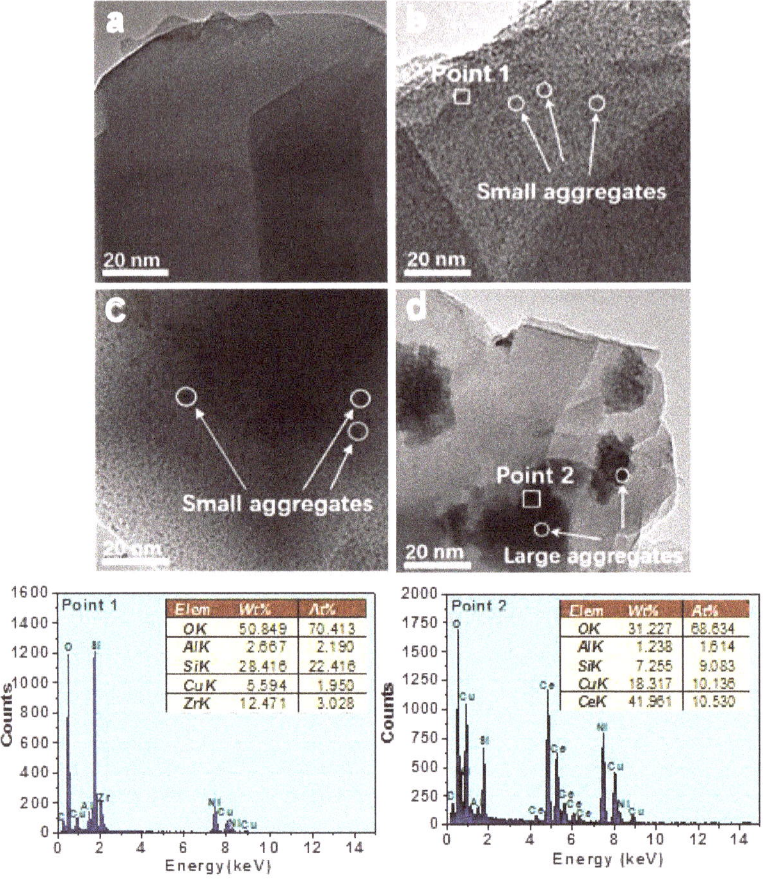

Figure 3. TEM images and energy-dispersive X-ray (EDX) quantitative analyses for (**a**) ZSM-5, (**b**) $CuZr_1$/ZSM-5, (**c**) $CuCe_{0.5}Zr_{0.5}$/ZSM-5, and (**d**) $CuCe_1$/ZSM-5.

2.2. XPS Analysis

The chemical state and surface composition of the elements in the catalysts were characterized by XPS. The Cu 2p spectra of the $CuCe_xZr_{1-x}$/ZSM-5 catalysts (x = 0, 0.5 and 1) in Figure **??**a show two main peaks: a peak at 933.3 eV that can be attributed to Cu $2p_{3/2}$, and a peak at about 953 eV that can be attributed to Cu $2p_{1/2}$. The intense satellite peak located at approximately 943 eV confirms the existence of divalent copper. Peak deconvolution and fitting to experimental data indicate that the Cu $2p_{3/2}$ peak can be well fitted by two peaks at 932.5 and 933.6 eV, corresponding to the Cu^+ and Cu^{2+} ions, respectively [**? ? ?**].

The XPS spectra of O 1s for the $CuCe_xZr_{1-x}$/ZSM-5 catalysts (x = 0, 0.5, and 1) in Figure **??**b show two primary peaks. The peak at a higher binding energy (BE) of 531.9 eV can be assigned to regular lattice oxygen from the ZSM-5 zeolite structure [**? ? ?**], while the shoulder peak at about 529.6 eV corresponds to characteristic lattice oxygen bound to metal (copper, cerium and zirconium) cations. As the cerium content increases by 6.5% in $CuCe_1$/ZSM-5, the shoulder peak corresponding to the lattice oxygen slightly shifts to a lower BE value (529.3 eV). Considering the findings from XRD that no metal oxide crystals are observed except for the $CuCe_1$/ZSM-5 catalyst, the existence of lattice oxygen implies that the oxides are well dispersed on the ZSM-5 support as the mini-crystal form.

The XPS doublet spectra of Zr 3d obtained from the $CuCe_xZr_{1-x}$/ZSM-5 catalysts (x = 0 and 0.5) are shown in Figure ??c. The peaks centered at 181.9 and 184.2 eV correspond to Zr $3d_{5/2}$ and Zr $3d_{3/2}$, respectively. The BE of Zr $3d_{5/2}$ is about 0.9 eV, higher than that reported for zirconium metal and 0.5 eV lower than that reported for zirconia [?]. This phenomenon probably originates from the contribution of electrons from the surrounding species. Wang et al. [?] also found that the zirconium cations in Cu/ZrO_2 catalysts possess a lower BE of Zr $3d_{5/2}$ than zirconia, and they ascribed it to copper oxides located near oxygen vacancies on the exterior of the zirconia.

Figure ??d shows the XPS spectra of Ce 3d for the $CuCe_xZr_{1-x}$/ZSM-5 catalysts (x = 0.5 and 1). The complex spectrum of Ce 3d with respect to Ce $3d_{3/2}$ and Ce $3d_{5/2}$ ionization features are deconvoluted into eight components. The peaks at about 882.5, 889.3, and 898.5 eV are ascribed to the peak of Ce $3d_{5/2}$ and two "shake-up" satellite peaks, respectively. The peaks at approximately 901.1, 908.1, and 916.6 eV can be assigned to Ce $3d_{3/2}$ and two shake-up satellite peaks, respectively [? ?]. These features are viewed as the fingerprints for the existence of Ce^{4+}. The inconspicuous peaks at about 885.5 and 904.3 eV represent the initial state of Ce^{3+} [? ?]. It is evident from Figure ??d that cerium is mostly in a quadrivalent oxidation state, and a small quantity of Ce^{3+} co-exists. The presence of Ce^{3+} can be attributed to the relative homogeneous $Ce_xZr_{1-x}O_2$ or the substitution of Ce^{4+} by Zr^{4+} ions [?].

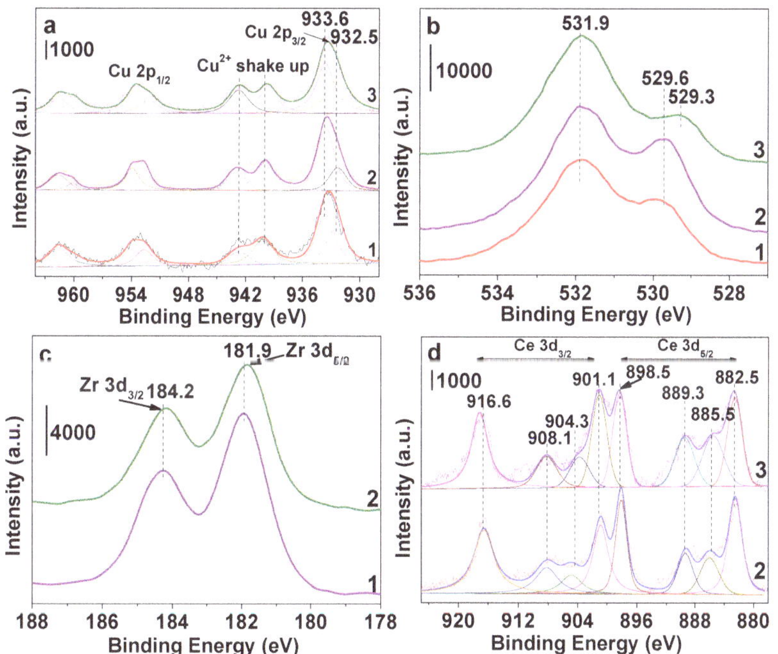

Figure 4. X-ray photoelectron spectra for (**a**) Cu 2p, (**b**) O 1s, (**c**) Zr 3d, and (**d**) Ce 3d over (1) $CuZr_1$/ZSM-5, (2) $CuCe_{0.5}Zr_{0.5}$/ZSM-5 and (3) $CuCe_1$/ZSM-5.

For the catalysts, the atomic ratios of Cu/Si, Ce/Si, and Zr/Si obtained from AAS and XPS are listed in Table ??. The atomic ratios determined from XPS are larger than those obtained from AAS, suggesting that the metal oxides become enriched on the surface of ZSM-5 grains. The atomic ratios of Cu^+/Cu^{2+} and Ce^{3+}/Ce^{4+} calculated from XPS spectra are also listed in Table ??. As the cerium content increases from 0% for $CuZr_1$/ZSM-5 to 3.4% for $CuCe_{0.5}Zr_{0.5}$/ZSM-5, the Cu^+/Cu^{2+} ratio decreases from 0.64 to 0.35, while the Ce^{3+}/Ce^{4+} ratio increases from 0 to 0.198. Upon further increasing the cerium content, the Cu^+/Cu^{2+} ratio increases to 0.71 for the $CuCe_1$/ZSM-5 catalyst, while the Ce^{3+}/Ce^{4+} ratio decreases after an initial increase.

Table 1. Surface compositions of $CuCe_xZr_{1-x}$/ZSM-5 catalysts (x = 0, 0.25, 0.5, 0.75, 1) obtained from X-ray photoelectron spectroscopy (XPS) analysis.

Sample	Cu^+/Cu^{2+}	Ce^{3+}/Ce^{4+}	Cu/Si (Atomic)		Ce/Si (Atomic)		Zr/Si (Atomic)	
			AAS	XPS	AAS	XPS	AAS	XPS
$CuZr_1$/ZSM-5	0.64	/	0.025	0.118	0	0	0.026	0.804
$CuCe_{0.25}Zr_{0.75}$/ZSM-5	0.498	0.187	0.026	0.116	0.005	0.015	0.018	0.723
$CuCe_{0.5}Zr_{0.5}$/ZSM-5	0.35	0.198	0.025	0.119	0.013	0.029	0.014	0.643
$CuCe_{0.75}Zr_{0.25}$/ZSM-5	0.672	0.298	0.024	0.117	0.020	0.034	0.007	0.312
$CuCe_1$/ZSM-5	0.71	0.247	0.026	0.116	0.025	0.066	0	0

2.3. H_2-TPR

Figure ?? shows the H_2-TPR profiles of ZrO_2, CeO_2, and the Cu/ZSM-5 and $CuCe_{0.75}Zr_{0.25}$/ZSM-5 catalysts, and the relevant data are listed in Table ??. Three main reduction peaks are evident for the catalysts: the α-peak (182–213 °C) is ascribed to the reduction of the copper species dispersed on the ZSM-5 support, the β-peak (260–264 °C) is associated with the reduction of the copper oxide adhering to the external surface of zeolite crystallites, and the γ-peak (406 °C) is generally considered to be due to the reduction of bulk and unsupported copper oxide. Some of the copper species have incorporated into the vacant sites of cerium and/or zirconium oxides to form a coordinated surface structure with capping oxygen [?], corresponding to the δ peak (164–178 °C) and the ε peak (315–342 °C). As the cerium content increases, the δ peak shifts from 171 °C for $CuZr_1$/ZSM-5 to 164 °C for $CuCe_1$/ZSM-5, indicating that the composite oxides of copper and cerium are more readily reduced than those of copper and zirconium.

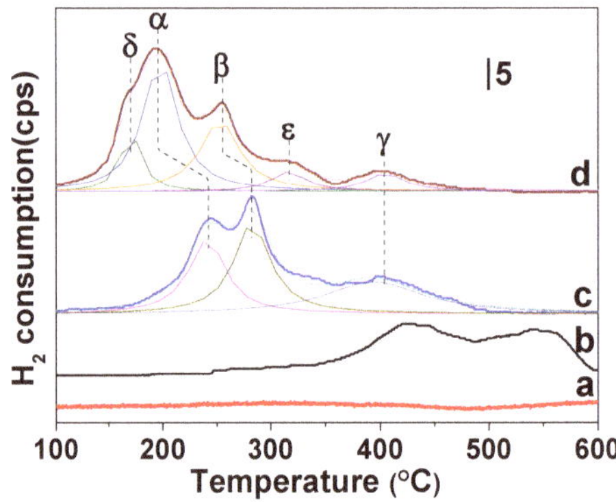

Figure 5. Temperature-programmed reduction by hydrogen (H_2-TPR) profiles of (a) ZrO_2, (b) CeO_2, (c) Cu/ZSM-5 and (d) $CuCe_{0.75}Zr_{0.25}$/ZSM-5.

Table 2. H_2-TPR results from Cu/ZSM-5 and $CuCe_xZr_{1-x}$/ZSM-5 catalysts (x = 0, 0.25, 0.5, 0.75, 1).

Sample	δ Peak		α Peak		β Peak		ε Peak		γ Peak		Total H_2 Consumption (μmol/g_{cat})
	T (°C)	H_2 Consumption (μmol/g_{cat})	T (°C)	H_2 Consumption (μmol/g_{cat})	T (°C)	H_2 Consumption (μmol/g_{cat})	T (°C)	H_2 Consumption (μmol/g_{cat})	T (°C)	H_2 Consumption (μmol/g_{cat})	
Cu/ZSM-5	/	/	236	56.4	278	68.23	/	/	393	63.96	188.59
$CuZr_1$/ZSM-5	171	15.49	198	45.06	261	76.22	342	25.12	398	28.53	190.42
$CuCe_{0.25}Zr_{0.75}$/ZSM-5	175	16.92	209	66.36	259	68.49	332	17.51	407	23.54	192.82
$CuCe_{0.5}Zr_{0.5}$/ZSM-5	178	22.49	213	91.39	263	61.40	338	3.24	399	18.66	197.18
$CuCe_{0.75}Z_{0.25}$/ZSM-5	170	24.97	192	91.07	255	58.10	315	14.16	402	14.09	202.39
$CuCe_1$/ZSM-5	164	18.55	182	67.15	260	82.83	322	10.37	405	23.33	202.23

From the H_2-TPR quantitative data in Table **??**, it can be seen that as the cerium content increases from 0% for $CuZr_1$/ZSM-5 to 5.2% for $CuCe_{0.75}Zr_{0.25}$/ZSM-5, the hydrogen consumption for the δ and α peak monotonously increases, while the hydrogen consumption for the β and γ peak gradually decreases. As mentioned above, the δ and α peak are related to highly dispersed copper species. Therefore, this increase in hydrogen consumption for the δ and α peak indicates that the addition of cerium improves the copper dispersion. For the $CuCe_1$/ZSM-5 catalyst, although the reduction temperatures of the δ and α peaks are lower than those for the other catalysts, the H_2 consumption of the δ and α peaks is not the maximum. This can be attributed to the formation of ceria, which is evidenced by the above XRD and TEM results. The ceria formed over the catalyst leads to the reduction not only extending deeply into the bulk of crystalline ceria but also being confined to its surface, so that the $CuCe_1$/ZSM-5 catalyst may consume less hydrogen than the $CuCe_{0.75}Zr_{0.25}$/ZSM-5 catalyst. Moreover, because the total H_2 consumption shows an increase with the cerium content increasing from 0% ($CuZr_1$/ZSM-5) to 5.2% ($CuCe_{0.75}Zr_{0.25}$/ZSM-5), the hydrogen uptake can be attributed not only to the copper reduction but also to a partial reduction of the cerium surface for the cerium-containing catalysts.

2.4. Catalytic Activity Test

Figure **??** shows the NO conversion in the SCR reaction on all the catalysts from 50 to 600 °C. All of the catalysts exhibit excellent SCR activity. For the Cu/ZSM-5 catalyst, the temperature range for 95% NO conversion is 209–406 °C. After the addition of cerium and/or zirconium to Cu/ZSM-5, the temperature range for 95% NO conversion extends toward both lower and higher temperatures, corresponding to active window broadening. The temperature ranges for 95% NO conversion are 190–436 °C for the $CuZr_1$/ZSM-5, 202–456 °C for $CuCe_{0.25}Zr_{0.75}$/ZSM-5, 203–460 °C for $CuCe_{0.5}Zr_{0.5}$/ZSM-5, 175–468 °C for $CuCe_{0.75}Zr_{0.25}$/ZSM-5, and 179–435 °C for $CuCe_1$/ZSM-5. Among the catalysts tested, the $CuCe_{0.75}Zr_{0.25}$/ZSM-5 catalyst has the highest catalytic activity for NO conversion. The improvements in the SCR activity after the addition of cerium and/or zirconium can be accounted for by three factors:

(1) The active copper species are well dispersed on the surface of the catalysts. The XRD and TEM results show that the introduction of zirconium promotes surface copper enrichment and prevents copper crystallization.
(2) The addition of cerium results in high Cu^+/Cu^{2+} ratios, as evidenced from the XPS results. In the SCR reaction, NO preferentially chemisorbs on Cu^+, and the formation of intermediate Cu^+–NO adsorbed sites is crucial for the activation of the NO molecule [? ?]. Thus, a high percentage of Cu^+ species favors catalytic NO reduction.
(3) Partial substitution of cerium and zirconium ions by copper ions contributes to the formation of copper/cerium or copper/zirconium composite oxides. As shown in the H_2-TPR experiment, these composite oxides greatly increase the reactive lattice oxygen content and reducibility, and thus improve the low-temperature activity [?].

Furthermore, as the temperature increases from 406 to 600 °C, a clear decrease of NO conversion for all the tested catalysts is observed in Figure **??**. This decrease is mainly due to the occurrence of non-selective NH_3 oxidation by the reaction $4NH_3 + 5O_2 \rightarrow 4NO + 6H_2O$ at high temperatures [? ? ?]. The consumption of NH_3, together with the production of new NO, limits NO conversion.

The SCR process induced by all the catalysts inevitably produces N_2O and NO_2 as byproducts. In the whole temperature range considered in this study, the NO_2 concentration is below 3 ppm for all the catalysts (not shown). Figure **??** shows the yield of N_2O plotted against reaction temperature. For all the catalysts, the yield of N_2O increases with increasing temperature up until a certain temperature and then starts to decrease. The N_2O produced in the low temperature range mainly arises from side reactions between NO and NH_3: $4NO + 4NH_3 + 3O_2 \rightarrow 4N_2O + 6H_2O$ [? ?]. Upon further increase of the reaction temperature, the formation of N_2O is mainly from NH_3 oxidation via the reaction $2NH_3 + 2O_2 \rightarrow N_2O + 3H_2O$ [? ?], and the amount of N_2O rapidly increases with the maximum yield at reaction temperatures <450 °C. In the high temperature range, non-selective NH_3 oxidation takes place by the reaction $4NH_3 + 5O_2 \rightarrow 4NO + 6H_2O$ [? ? ?], and thus the amount of N_2O rapidly decreases as the temperature increases. Moreover, close inspection of Figure **??** shows that the zirconium-containing catalysts yield less N_2O than the Cu/ZSM-5 and $CuCe_1$/ZSM-5 catalysts, which means that the introduction of zirconium abates N_2O formation in the tested temperature range.

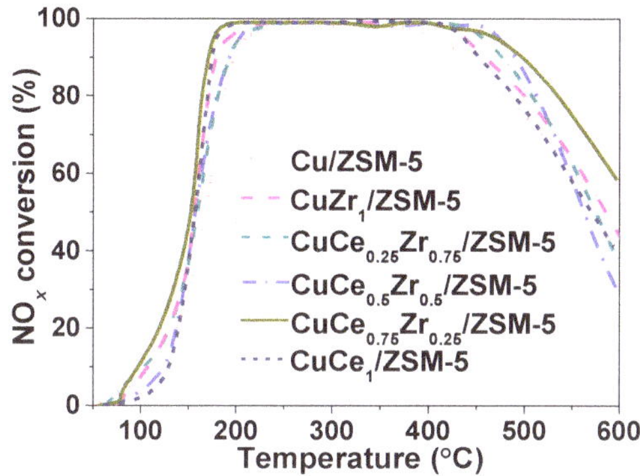

Figure 6. Catalytic activities for NO_x reduction by NH_3 for Cu/ZSM-5 and $CuCe_xZr_{1-x}$/ZSM-5 catalysts (x = 0, 0.25, 0.5, 0.75 and 1).

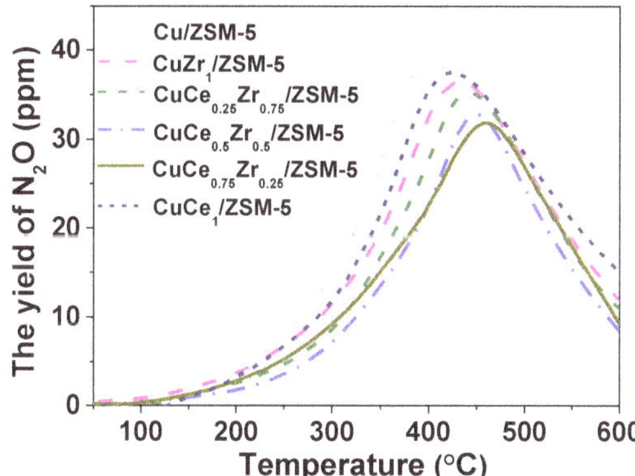

Figure 7. Yield of N_2O during the selective catalytic reduction (SCR) process as a function of reaction temperature for Cu/ZSM-5 and$CuCe_xZr_{1-x}$/ZSM-5 catalysts (x = 0, 0.25, 0.5, 0.75 and 1).

3. Experimental

3.1. Synthesis of Catalysts

H/ZSM-5 with an atomic Si/Al ratio of 25 was provided by Nankai University, Tianjin, China. A series of $CuCe_xZr_{1-x}O_y$/ZSM-5 catalysts (x = 0, 0.25, 0.5, 0.75, and 1) were synthesized by an aqueous ion-exchange technique. A desired amount of copper acetate, and zirconium and cerium nitrate was added to deionized water and mixed with 20 g of H/ZSM-5 powder at room temperature. The resulting solution was stirred at 80 °C for 24 h. The samples were filtered and dried by evaporation in air and then calcined at 550 °C for 4 h. The copper content of the $CuCe_xZr_{1-x}O_y$/ZSM-5 catalysts was fixed at 3 wt. %, and the molar ratio of Cu/(Ce + Zr) was 1:1. These catalysts are denoted as $CuZr_1$/ZSM-5, $CuCe_{0.25}Zr_{0.75}$/ZSM-5, $CuCe_{0.5}Zr_{0.5}$/ZSM-5, $CuCe_{0.75}Zr_{0.25}$/ZSM-5, and $CuCe_1$/ZSM-5. The copper, cerium, and zirconium contents of each calcined catalyst were determined by atomic absorption spectroscopy (AAS) using a PerkinElmer AAnalyst 300 spectrometer, and the results are shown in Table ??. To evaluate the catalytic activity, the catalysts were compressed by a pressure of 20 MPa, then granulated and screened to the size of a 20–40 mesh.

Table 3. Element compositions of all catalysts.

Sample	Element Composition Measured by AAS (wt. %)			
	Cu	Ce	Zr	Si
$CuZr_1/ZSM-5$	3.0	0	4.4	52.4
$CuCe_{0.25}Zr_{0.75}/ZSM-5$	3.1	1.3	3.1	52.2
$CuCe_{0.5}Zr_{0.5}/ZSM-5$	2.9	3.4	2.3	52.3
$CuCe_{0.75}Zr_{0.25}/ZSM-5$	2.8	5.2	1.2	52.4
$CuCe_1/ZSM-5$	3.1	6.5	0	52.2

3.2. Characterization

Powder X-ray diffraction (XRD) was performed using a Rigaku D/MAC/max 2500 v/pc instrument with Cu Kα radiation (40 kV, 200 mA, λ = 1.5418 Å) (Jananese Science, Tokyo, Japan). The scan was performed with a 2θ rate of 0.02°/min from 5–80°. Transmission electron microscopy (TEM) images of the catalysts were obtained using a Philips Tecnai G2 F20 microscope operating at 200 kV equipped with an Oxford-1NCA energy-dispersive X-ray detector (EDX) (FEI, Hillsboro, OR, USA). Prior to measurement, the catalysts were dispersed in pure ethanol through sonication, and then mounted on nickel grids with a carbon film. X-ray photoelectron spectroscopy (XPS) was recorded using a Perkin-Elmer PHI-1600 ESCA spectrometer with a Mg Kα X-ray source (Perkin Elmer, Wellesley, MA, USA). The binding energies (BEs) were calibrated by referencing the C 1s at 284.8 eV. A ChemBet Pulsar system was used to perform temperature-programmed reduction by hydrogen (H_2-TPR) (Quantachrome Instruments, Boynton Beach, FL, USA). For the H_2-TPR experiments, 100 mg of the catalyst was exposed to pure Ar at a flow rate of 30 mL/min for 1 h at 300 °C. Then, the same gas was used to cool the catalyst down to 50 °C. The H_2-TPR measurements were carried out by increasing the temperature to 700 °C at a heating rate of 10 °C/min in 5% hydrogen at a constant flow rate of 30 mL/min. The hydrogen consumption was monitored and quantified using a thermal conductivity detector.

3.3. Catalytic Activity

The activities of the catalysts were measured in a continuous flow apparatus at atmosphere pressure. Before each test run, the catalyst powder was first pressed into a wafer and sieved into 20–40 meshes, and then 0.5 g of catalyst was set into a fix-bed reactor made of a quartz tube with an internal diameter of 10 mm. The reaction was carried out in the temperature range 50–600 °C, and a K-type thermocouple was located inside the catalyst bed to monitor the reaction temperature. The feed gas was controlled by calibrated electronic mass flow controllers, and contained 1000 ppm NO, 1000 ppm NH_3, and 10% O_2 and N_2 as the balance gas. The space velocity was set at 15,000 h^{-1}. An online mass spectrometry (Dycor LC-D200) was used to monitor the effluent NO, NO_2, N_2O, and NH_3. From the concentration of the gases at steady state, the NO_x conversion was defined as:

$$NO_x \text{ conversion } (\%) = \frac{[NO_x]_{in} - [NO_x]_{out}}{[NO_x]_{in}} \times 100, \ [NOx] = [NO] + [NO_2]$$

4. Conclusions

The present study has highlighted the effects of cerium and/or zirconium incorporation into Cu/ZSM-5 on the SCR activity in NO abatement. The $CuCe_xZr_{1-x}O_y/ZSM-5$ catalysts used in this study present a more than 95% NO conversion rate in a wide temperature range (175–468 °C), which is an improvement relative to the Cu/ZSM-5 catalyst (209–406 °C) in terms of SCR activity. The zirconium doping improves copper dispersion, and the introduction of zirconium promotes surface copper enrichment and prevents copper crystallization, which favors catalytic NO reduction. Partial substitution of cerium and zirconium ions by copper ions increases the reactive lattice oxygen content and reducibility, and thus improves the low-temperature activity. Because of the existence of the Ce^{3+}/Ce^{4+} redox couple in the cerium-containing catalysts, the SCR activities of cerium-rich catalysts are higher than those of zirconium-rich catalysts. Moreover, formation of N_2O is suppressed in the range of test temperatures due to the presence of zirconium in the catalysts.

Author Contributions: This study was conducted through contributions of all authors. C.S. and G.L. designed the study and wrote the manuscript. Y.L., C.F. and X.L. designed and performed the experiments and analyzed the data. Y.L. checked and corrected the manuscript.

Funding: This research received no external funding.

Acknowledgments: This study was supported by the National Natural Science Foundation of China (No. 51476116).

Conflicts of Interest: The authors declare no conflict of interest.

References

1. Landi, G.; Lisi, L.; Pirone, R.; Russo, G.; Tortorelli, M. Effect of water on NO adsorption over Cu-ZSM-5 based catalysts. *Catal. Today* **2012**, *191*, 138–141. [CrossRef]
2. Ji, Y.; Yang, H.; Yan, W. Strategies to enhance the catalytic performance of ZSM-5 zeolite in hydrocarbon cracking: A review. *Catalysts* **2017**, *7*, 367. [CrossRef]
3. Wang, T.; Wan, Z.; Yang, X.; Zhang, X.; Niu, X.; Sun, B. Promotional effect of iron modification on the catalytic properties of Mn-Fe/ZSM-5 catalysts in the Fast SCR reaction. *Fuel Process. Technol.* **2018**, *169*, 112–121. [CrossRef]
4. Xu, W.J.; Zhang, G.X.; Chen, H.W.; Zhang, G.M.; Han, Y.; Chang, Y.C.; Gong, P. Mn/beta and Mn/ZSM-5 for the low-temperature selective catalytic reduction of NO with ammonia: Effect of manganese precursors. *Chin. J. Catal.* **2018**, *39*, 118–127. [CrossRef]
5. De Lucas, A.; Valverde, J.L.; Dorado, F.; Romero, A.; Asencio, I. Influence of the ion exchanged metal (Cu, Co, Ni and Mn) on the selective catalytic reduction of NO_x over mordenite and ZSM-5. *J. Mol. Catal.* **2005**, *225*, 47–58. [CrossRef]
6. Ma, T.; Imai, H.; Yamawaki, M.; Terasaka, K.; Li, X. Selective synthesis of gasoline-ranged hydrocarbons from syngas over hybrid catalyst consisting of metal-loaded ZSM-5 coupled with copper-zinc oxide. *Catalysts* **2014**, *4*, 116–128. [CrossRef]
7. Abu-Zied, B.M.; Schwieger, W.; Unger, A. Nitrous oxide decomposition over transition metal exchanged ZSM-5 zeolites prepared by the solid-state ion-exchange method. *Appl. Catal. B* **2008**, *84*, 277–288. [CrossRef]
8. Chen, P.; Simböck, J.; Schönebaum, S.; Rauch, D.; Simons, T.; Palkovits, R.; Simon, U. Monitoring NH_3 storage and conversion in Cu-ZSM-5 and Cu-SAPO-34 catalysts for NH_3-SCR by simultaneous impedance and DRIFT spectroscopy. *Sens. Actuators B* **2016**, *236*, 1075–1082. [CrossRef]
9. Jiang, L.; Zhu, H.; Razzaq, R.; Zhu, M.; Li, C.; Li, Z. Effect of zirconium addition on the structure and properties of CuO/CeO_2 catalysts for high-temperature water-gas shift in an IGCC system. *Int. J. Hydrogen Energy* **2012**, *37*, 15914–15924. [CrossRef]
10. Seo, C.K.; Choi, B.; Kim, H.; Lee, C.H.; Lee, C.B. Effect of ZrO_2 addition on de-NO_x performance of Cu-ZSM-5 for SCR catalyst. *Chem. Eng. J.* **2012**, *191*, 331–340. [CrossRef]
11. Zhang, R.; Teoh, W.Y.; Amal, R.; Chen, B.; Kaliaguine, S. Catalytic reduction of NO by CO over $Cu/Ce_xZr_{1-x}O_2$ prepared by flame synthesis. *J. Catal.* **2010**, *272*, 210–219. [CrossRef]
12. Fu, M.; Yue, X.; Ye, D.; Ouyang, J.; Huang, B.; Wu, J.; Liang, H. Soot oxidation via CuO doped CeO_2 catalysts prepared using coprecipitation and citrate acid complex-combustion synthesis. *Catal. Today* **2010**, *153*, 125–132. [CrossRef]
13. Łamacz, A.; Krztoń, A.; Djéga-Mariadassou, G. Catalytic decomposition of nitrogen oxides from coal combustion flue gases on $CeZrO_2$ supported Cu catalysts. *Catal. Today* **2011**, *176*, 126–130. [CrossRef]

14. Si, R.; Zhang, Y.W.; Li, S.J.; Lin, B.X.; Yan, C.H. Urea-Based Hydrothermally derived homogeneous nanostructured $Ce_{1-x}Zr_xO_2$ ($x = 0-0.8$) solid solutions: A strong correlation between oxygen storage capacity and lattice strain. *J. Phys. Chem. B* **2004**, *108*, 12481–12488. [CrossRef]
15. Qi, G.; Yang, R.T. Selective catalytic oxidation (SCO) of ammonia to nitrogen over Fe/ZSM-5 catalysts. *Appl. Catal. A* **2005**, *287*, 25–33. [CrossRef]
16. Ohtsuka, H.; Tabata, T.; Okada, O.; Sabatino, L.M.; Bellussi, G. A study on selective reduction of NO_x by propane on Co-Beta. *Catal. Lett.* **1997**, *44*, 265–270. [CrossRef]
17. Oliveira, M.L.M.; Silva, C.M.; Moreno-Tost, R.; Farias, T.L.; Jiménez-López, A.; Rodríguez-Castellón, E. A study of copper-exchanged mordenite natural and ZSM-5 zeolites as SCR-NO_x catalysts for diesel road vehicles: Simulation by neural networks approach. *Appl. Catal. B* **2009**, *88*, 420–429. [CrossRef]
18. Jones, S.D.; Neal, L.M.; Everett, M.L.; Hoflund, G.B.; Hagelin-Weaver, H.E. Characterization of ZrO_2-promoted Cu/ZnO/nano-Al_2O_3 methanol steam reforming catalysts. *Appl. Surf. Sci.* **2010**, *256*, 7345–7353. [CrossRef]
19. Correa, C.M.; Castrillón, F.C. Supported bimetallic Pd-Co catalysts: characterization and catalytic activity. *J. Mol. Catal. A Chem.* **2005**, *228*, 267–273. [CrossRef]
20. Wu, Z.; Jin, R.; Liu, Y.; Wang, H. Ceria modified MnO_x/TiO_2 as a superior catalyst for NO reduction with NH_3 at low-temperature. *Catal. Commun.* **2008**, *9*, 2217–2220. [CrossRef]
21. Hwang, I.C.; Kim, D.H.; Woo, S.I. Role of oxygen on NO_x SCR catalyzed over Cu/ZSM-5 studied by FTIR, TPD, XPS and micropulse reaction. *Catal. Today* **1998**, *44*, 47–55. [CrossRef]
22. Liu, L.; Yao, Z.; Liu, B.; Dong, L. Correlation of structural characteristics with catalytic performance of $CuO/Ce_xZr_{1-x}O_2$ catalysts for NO reduction by CO. *J. Catal.* **2010**, *275*, 45–60. [CrossRef]
23. Wang, L.C.; Liu, Q.; Chen, M.; Liu, Y.M.; Cao, Y.; He, H.Y.; Fan, K.N. Structural evolution and catalytic properties of nanostructured Cu/ZrO_2 catalysts prepared by oxalate gel-coprecipitation technique. *J. Phys. Chem. C* **2007**, *111*, 16549–16557. [CrossRef]
24. Nelson, A.E.; Schulz, K.H. Surface chemistry and microstructural analysis of $Ce_xZr_{1-x}O_{2-y}$ model catalyst surfaces. *Appl. Surf. Sci.* **2003**, *210*, 206–221. [CrossRef]
25. Deo, G.; Wachs, I.E. Reactivity of supported vanadium oxide catalysts: The partial oxidation of methanol. *J. Catal.* **1994**, *146*, 323–334. [CrossRef]
26. Ansell, G.; Diwell, A.; Golunski, S.; Hayes, J.; Rajaram, R.; Truex, T.; Walker, A. Mechanism of the lean NO_x reaction over Cu/ZSM-5. *Appl. Catal. B.* **1993**, *2*, 81–100. [CrossRef]
27. Curtin, T.; Grange, P.; Delmon, B. The effect of pretreatments on different copper exchanged ZSM-5 for the decomposition of NO. *Catal. Today* **1997**, *36*, 57–64. [CrossRef]
28. Liu, Q.; Liu, Z.; Li, C. Adsorption and activation of NH_3 during selective catalytic reduction of NO by NH_3. *Chin. J. Catal.* **2006**, *27*, 636–646. [CrossRef]
29. Koebel, M.; Elsener, M.; Madia, G. Reaction pathways in the selective catalytic reduction process with NO and NO_2 at low temperatures. *Ind. Eng. Chem. Res.* **2001**, *40*, 52–59. [CrossRef]
30. Sirdeshpande, A.R.; Lighty, J.S. Kinetics of the selective catalytic reduction of NO with NH_3 over CuO/ γ-Al_2O_3. *Ind. Eng. Chem. Res.* **2000**, *39*, 1781–1787. [CrossRef]

Article

A CeO_2/ZrO_2-TiO_2 Catalyst for the Selective Catalytic Reduction of NO_x with NH_3

Wenpo Shan [1,2], Yang Geng [3], Yan Zhang [1,2], Zhihua Lian [1] and Hong He [1,2,4,*]

1 Center for Excellence in Regional Atmospheric Environment, Institute of Urban Environment, Chinese Academy of Sciences, Xiamen 361021, China; wpshan@iue.ac.cn (W.S.); yzhang3@iue.ac.cn (Y.Z.); zhlian@iue.ac.cn (Z.L.)

2 Ningbo Urban Environment Observation and Research Station-NUEORS, Institute of Urban Environment, Chinese Academy of Sciences, Ningbo 315800, China

3 School of Environmental and Biological Engineering, Nanjing University of Science and Technology, Nanjing 210094, China; yanggeng_njust@163.com

4 State Key Joint Laboratory of Environment Simulation and Pollution Control, Research Center for Eco-Environmental Sciences, Chinese Academy of Sciences, Beijing 100085, China

* Correspondence: honghe@rcees.ac.cn; Tel./Fax: +86-10-62849123

Received: 30 September 2018; Accepted: 27 November 2018; Published: 30 November 2018

Abstract: In this study, $CeZr_{0.5}Ti_aO_x$ (with a = 0, 1, 2, 5, 10) catalysts were prepared by a stepwise precipitation approach for the selective catalytic reduction of NO_x with NH_3. When Ti was added, all of the Ce-Zr-Ti oxide catalysts showed much better catalytic performances than the $CeZr_{0.5}O_x$. Particularly, the $CeZr_{0.5}Ti_2O_x$ catalyst showed excellent activity for broad temperature range under high space velocity condition. Through the control of pH value and precipitation time during preparation, the function of the $CeZr_{0.5}Ti_2O_x$ catalyst could be controlled and the structure with highly dispersed CeO_2 (with redox functions) on the surface of ZrO_2-TiO_2 (with acidic functions) could be obtained. Characterizations revealed that the superior catalytic performance of the catalyst is associated with its outstanding redox properties and adsorption/activation functions for the reactants.

Keywords: Ce-based catalyst; stepwise precipitation; selective catalytic reduction; diesel exhaust; nitrogen oxides abatement

---

1. Introduction

NO_x (mainly NO and NO_2) in the atmosphere plays critical roles in the formation of severe air pollution problems, such as haze, acid rain, and photochemical smog. In the last few decades, great efforts have been devoted to the development of NO_x emission control technologies [1–3]. Selective catalytic reduction of NO_x with NH_3 (NH_3 SCR) has been widely applied for the removal of NO_x generated from stationary sources for many years, and it has also been used for the control of NO_x emission from diesel vehicles [2,4].

Catalysts play an important role in the development of NH_3-SCR technology [5,6]. Vanadium-based catalyst (especially V_2O_5-WO_3/TiO_2), with excellent SO_2 resistance, is the most widely used NH_3-SCR catalyst for NO_x emission control from power plants, and it was also applied on diesel vehicles as the first generation of SCR catalyst [4]. However, this catalyst system still has some problems, including the toxicity of active V_2O_5, narrow temperature window, and low thermal stability [2].

There has been strong interest in developing a vanadium-free catalyst that can be used on diesel vehicles [5–11]. Ce is a key component in three-way catalysts for emission control in automobiles for gasoline. CeO_2 provides an oxygen storage function through redox cycling between Ce^{3+} and

Ce^{4+}. In recent years, Ce has also attracted great attention for applications as a support [12,13], promoter [14–18], or main active component [19–26] for NH_3-SCR catalysts.

Pure Ce oxide is not suitable for use as an NH_3-SCR catalyst [27,28]. When Zr oxide was introduced into Ce oxide, the thermal stability and the oxygen storage capacity of the oxide could be significantly improved. Therefore, Ce-Zr oxide was investigated for NH_3-SCR [12,13,29–34]. In the NH_3-SCR reaction, both redox functions and acidic functions of the catalyst are needed [4,35]. Therefore, a high dispersion of active sites and close coupling of redox with acid sites is the way to design a highly efficient NH_3-SCR catalyst.

In this study, starting from a preparation of Ce-Zr oxide by the co-preparation method, we developed a Ce-Zr-Ti oxide catalyst using a stepwise precipitation approach, under the theoretical guidance of the close combination of the Ce-Zr oxide with strong redox functions and Ti oxide with excellent acid properties [4,5]. This obtained catalyst showed superior catalytic performance for NH_3-SCR.

2. Results and Discussion

2.1. NH_3-SCR Activity

Figure 1A presents the NO_x conversion over the catalysts with different Ti contents under a relatively high gas hourly space velocity (GHSV) of 200,000 h^{-1}. The $CeZr_{0.5}O_x$ just exhibited over 50% NO_x conversion in a narrow temperature range of 350–425 °C. When Ti was introduced, all of the Ce-Zr-Ti oxide catalysts exhibited much better activities. With the increase in Ti content, the low temperature firstly increased and then decreased. As a result, the $CeZr_{0.5}Ti_2O_x$ catalyst presented the best activity in a low temperature range, together with a high NO_x conversion in a wide temperature range. On the other hand, the variation in high temperature activity with Ti content was contrary to that of low temperature activity, with the activity of $CeZr_{0.5}Ti_2O_x$ slightly lower than those of the other Ce-Zr-Ti oxide catalysts in a high temperature range. In addition, adding Ti to the catalyst also enhanced the N_2 selectivity, and the Ce-Zr-Ti oxide catalysts all presented higher N_2 selectivity than $CeZr_{0.5}O_x$ (Figure 1B).

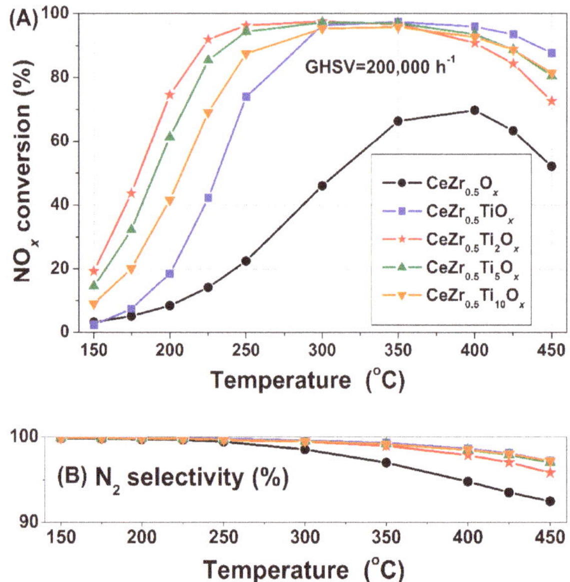

Figure 1. (**A**) NO_x conversions and (**B**) N_2 selectivity over the $CeZr_{0.5}O_x$ and Ce-Zr-Ti oxide catalysts. Reaction conditions: [NO] = [NH_3] = 500 ppm, [O_2] = 5 vol.%, N_2 balance, and GHSV = 200,000 h^{-1}.

The influences of H_2O and space velocity on the NO_x conversion over $CeZr_{0.5}Ti_2O_x$ were tested and the results are shown in Figure 2. The existence of 5% H_2O in the flow gas decreased the low temperature activity, but enhanced the high temperature activity. As a result, over 80% NO_x conversion could still be achieved from 250 to 450 °C. When the GHSV was decreased from 200,000 h^{-1} to 100,000 h^{-1}, the activity of the catalyst at low temperatures was obviously improved.

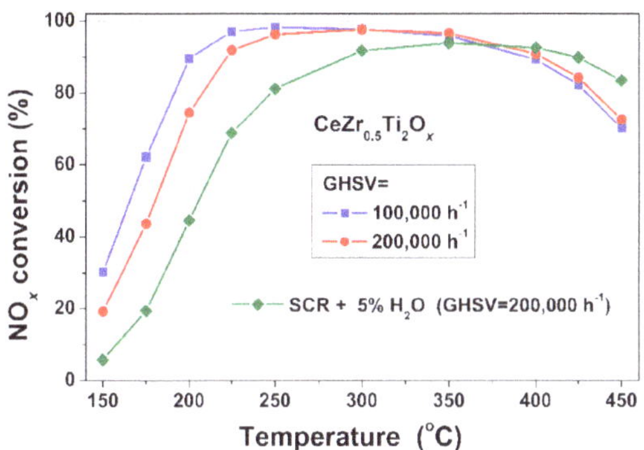

Figure 2. NO_x conversion over $CeZr_{0.5}Ti_2O_x$ catalyst under different reaction conditions. Reaction conditions: [NO] = [NH_3] = 500 ppm, [O_2] = 5 vol.%, [H_2O] = 5 vol.% (when used), N_2 balance, and GHSV = 100,000 or 200,000 h^{-1}.

2.2. Separated NO/NH$_3$ Oxidation

To analyze the effects of Ti on the catalyst, separated NO oxidation and NH_3 oxidation tests were carried out for the $CeZr_{0.5}O_x$ and $CeZr_{0.5}Ti_2O_x$ (Figure 3). The NO_2 production during NO oxidation over the $CeZr_{0.5}Ti_2O_x$ was clearly higher than that over $CeZr_{0.5}O_x$ at a low temperature. Since the presentation of NO_2 in the reaction gas could promote the SCR reaction at a low temperature by accelerating the fast SCR process ($2\,NH_3 + NO + NO_2 \rightarrow 2N_2 + 3H_2O$), the enhanced low-temperature activity by the introduction of Ti should be associated with the promoted oxidation of NO to NO_2 over $CeZr_{0.5}Ti_2O_x$ [10,35]. In addition, the introduction of Ti also promoted NH_3 oxidation over the catalyst at a high temperature. The NH_3-SCR reaction route at a high temperature mainly follows the Eley-Rideal mechanism, and the activation of NH_3 to form NH_2 species by oxidation plays the key role for the reaction with NO to form N_2 and H_2O, owing to $NH_2 + NO(g) \rightarrow N_2 + H_2O$. Therefore, promoted NH_3 oxidation would be beneficial for the improvement of high temperature activity.

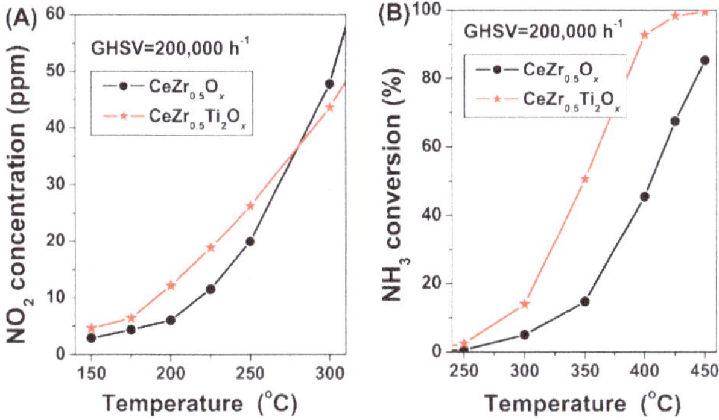

Figure 3. (**A**) NO_2 productions during separate NO oxidation reaction and (**B**) NH_3 conversions during separate NH_3 oxidation reaction over the $CeZr_{0.5}O_x$ and $CeZr_{0.5}Ti_2O_x$ catalysts. Reaction conditions: (A) [NO] _ENREF_30= 500 ppm, (B) [NH_3] = 500 ppm, [O_2] = 5 vol.%, N_2 balance and GHSV = 200,000 h^{-1}.

2.3. XRD

The X-ray diffraction (XRD) results of the $CeZr_{0.5}O_x$ and Ce-Z-Ti oxide catalysts are presented in Figure 4. Both CeO_2 and ZrO_2 were detected in $CeZr_{0.5}O_x$. With the increase of Ti, the peaks for CeO_2 and ZrO_2 became more and more weak, and only anatase TiO_2 was observed for $CeZr_{0.5}Ti_{10}O_x$. Only weak peaks for CeO_2 with cubic fluorite structures (PDF# 43-1002) were observed in the $CeZr_{0.5}Ti_2O_x$, indicating that the introduction of Ti had induced the structural change of the $CeZr_{0.5}O_x$, and the crystallizations of Ce, Zr and Ti oxides in $CeZr_{0.5}Ti_2O_x$ were significantly inhibited. As a result, the $CeZr_{0.5}Ti_2O_x$ (165.1 m^2/g) showed a higher Brunauer–Emmett–Teller (BET) surface area than $CeZr_{0.5}O_x$ (113.5 m^2/g).

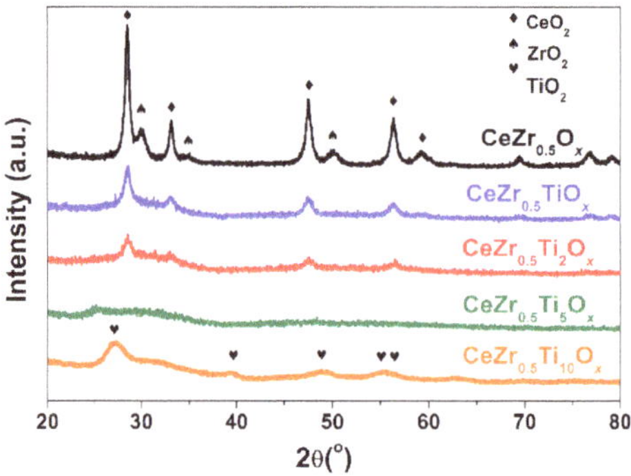

Figure 4. XRD patterns of the $CeZr_{0.5}O_x$ and Ce-Z-Ti oxide catalysts.

2.4. H_2-TPR

The H_2 temperature-programmed reduction (H_2-TPR) profiles of $CeZr_{0.5}O_x$ and $CeZr_{0.5}Ti_2O_x$ are presented in Figure 5. The $CeZr_{0.5}O_x$ exhibited two peaks at 496 and 755 °C due to the surface and bulk reductions of CeO_2 (as detected by XRD), respectively [31,36–38]. During the test, coordinatively unsaturated surface oxygen anions are easily reduced by H_2 in the low temperature region, while the bulk oxygen species are reduced only after the transportation to the surface [39]. With the addition of Ti, a sharp H_2 consumption peak appeared at 567 °C, which indicates that another type of Ce species might be formed. Considering the XRD results, this sharp peak might be associated with the reduction of the highly dispersed Ce species from Ce^{4+} to Ce^{3+} [22,34]. In addition, the H_2 consumption of $CeZr_{0.5}Ti_2O_x$ was much higher than that of $CeZr_{0.5}O_x$ at a low temperature. The H_2-TPR results clearly indicated the enhancement of redox functions for $CeZr_{0.5}Ti_2O_x$.

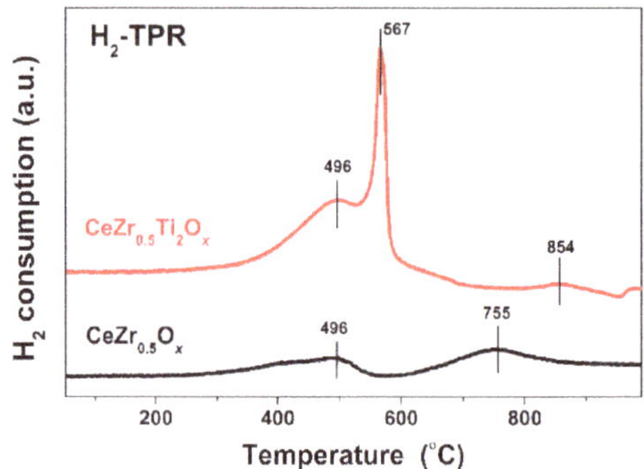

Figure 5. H_2-TPR profiles of the $CeZr_{0.5}O_x$ and $CeZr_{0.5}Ti_2O_x$ catalysts.

Previous studies have indicated that the redox properties of NH_3-SCR catalyst play a dominant role in the low temperature activity [35,40,41]. Therefore, the enhanced redox function of $CeZr_{0.5}Ti_2O_x$ would beneficial for low temperature activity.

2.5. NO_x/NH_3-TPD

To investigate the NO_x and NH_3 adsorption/desorption properties of $CeZr_{0.5}O_x$ and $CeZr_{0.5}Ti_2O_x$, NO_x temperature-programmed desorption (NO_x-TPD) and NH_3 temperature-programmed desorption (NH_3-TPD) were performed for the catalysts (Figure 6).

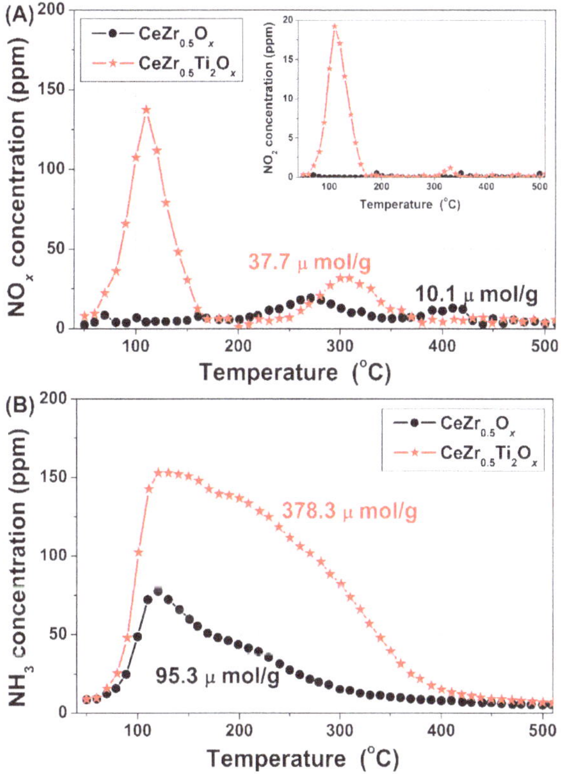

Figure 6. (**A**) NO_x-TPD and (**B**) NH_3-TPD profiles of the $CeZr_{0.5}O_x$ and $CeZr_{0.5}Ti_2O_x$ catalysts.

The NO_x-TPD profiles are presented in Figure 6A. The first NO_x peak of $CeZr_{0.5}Ti_2O_x$ was at ca. 110 °C, mainly due to the desorption of physisorbed NO_x, while the other NO_x peak was at ca. 300 °C and was associated with the decomposition of chemsorbed NO_x species [42,43]. On the other hand, two weak peaks were observed for $CeZr_{0.5}O_x$ at ca. 270 °C and ca. 410 °C, respectively, which were due to the decomposition of different types of chemsorbed NO_x species. With the addition of Ti, the adsorbed NO_x on $CeZr_{0.5}Ti_2O_x$ was obviously more than that of $CeZr_{0.5}O_x$. Particularly, the desorbed NO_2 of $CeZr_{0.5}Ti_2O_x$ was much higher, owing to the enhanced low-temperature activity for NO oxidation (as shown by the separated NO oxidation results), which could facilitate the conversion of NO_x in NH_3-SCR.

Surface acidity plays a dominate role in the high-temperature SCR activity due to its effects on the adsorption and activation of NH_3 [35,41]. Previous studies have revealed that Ti species of NH_3-SCR catalysts mainly act as acid sites in the reaction for NH_3 adsorption [4]. Therefore, the adsorbed NH_3

of $CeZr_{0.5}Ti_2O_x$ was much more than that of $CeZr_{0.5}O_x$, which might be an important reason for the better NH_3-SCR activity of $CeZr_{0.5}Ti_2O_x$ at high temperatures.

2.6. XPS

The X-ray photoelectron spectroscopy (XPS) results for Ce 3d of the $CeZr_{0.5}O_x$ and $CeZr_{0.5}Ti_2O_x$ are shown in Figure 7. The sub-bands labeled with u′/v′ and u^0/v^0 represent the $3d^{10}4f^1$ initial electronic state corresponding to Ce^{3+} and the $3d^94f^2$ state of Ce^{3+}, respectively [44]. The sub-bands labeled with u‴ and v‴ represent the $3d^{10}4f^0$ state of Ce^{4+}, and the sub-bands labeled with u, u″, v and v″ represent the $3d^94f^1$ state corresponding to Ce^{4+} [44]. The presence of Ce^{3+} would induce a charge imbalance, which could lead to unsaturated chemical bonds and oxygen vacancies. The calculated Ce^{3+} ratio of $CeZr_{0.5}Ti_2O_x$ (36.0%) was higher than that of $CeZr_{0.5}O_x$ (33.8%), indicating that more surface oxygen vacancies presented in $CeZr_{0.5}Ti_2O_x$. In addition, the Ce^{3+} ratio of the catalyst could influence the redox ability and reactant adsorption and activation functions, and thereby contribute to NH_3-SCR performance.

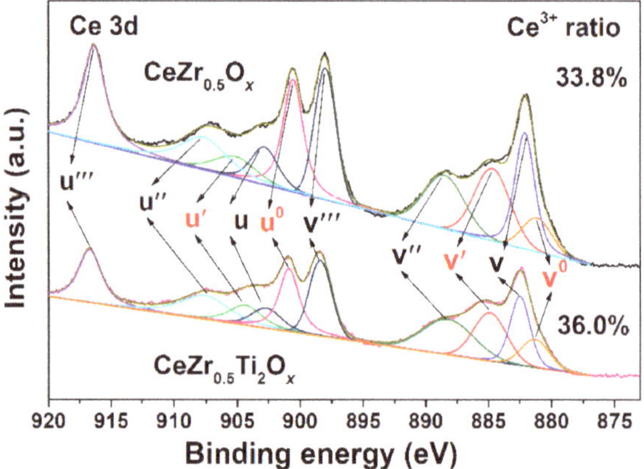

Figure 7. XPS results of Ce 3d of the $CeZr_{0.5}O_x$ and $CeZr_{0.5}Ti_2O_x$ catalysts.

The surface oxygen vacancies of the catalysts might generate weakly-adsorbed oxygen species or additional chemisorbed oxygen on the surface of the catalyst [27,45]. The XPS results of O 1s of the $CeZr_{0.5}O_x$ and $CeZr_{0.5}Ti_2O_x$ are shown in Figure 8. The O 1s peak was fit into two sub-bands. The sub-bands at 531.2–531.5 eV and 529.1–529.6 eV were assigned to the surface adsorbed oxygen (O_α), such as the O_2^{2-} and O^- belonging to defect-oxide or a hydroxyl-like group, and the lattice oxygen O^{2-} (O_β), respectively [46]. The O_α ratios of the catalysts were calculated by $O_\alpha/(O_\alpha + O_\beta)$, and the $CeZr_{0.5}Ti_2O_x$ showed higher O_α ratio than $CeZr_{0.5}O_x$. The results confirmed that the addition of Ti indeed induced more surface-adsorbed oxygen, which would facilitate NO oxidation to NO_2 (as shown by the separated NO oxidation and NO_x-TPD results), and thus facilitates the conversion of NO by fast SCR effects.

2.7. Formation Process Analysis of the $CeZr_{0.5}Ti_2O_x$ Catalyst

Figure 9 shows the pH variations of the mixed solutions for the preparation of the $CeZr_{0.5}O_x$ and $CeZr_{0.5}Ti_2O_x$ catalysts. During the preparation of $CeZr_{0.5}O_x$, the initial pH value of the solution was 1.6. With the hydrolysis of urea, the pH increased gradually to be 7.6 after heating for 12 h. Due to the increase in pH, suspended particles began to appear in the solution in the second hour. The particles with the precipitation time of 2 h, 4 h, 6 h, and 12 h were collected and then calcined to be catalyst samples. The activity tests of these samples showed similar NO_x conversions with each other.

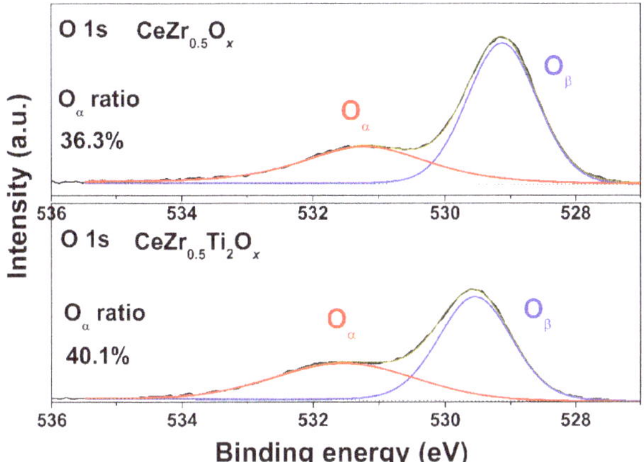

Figure 8. XPS results of O 1s of the $CeZr_{0.5}O_x$ and $CeZr_{0.5}Ti_2O_x$ catalysts.

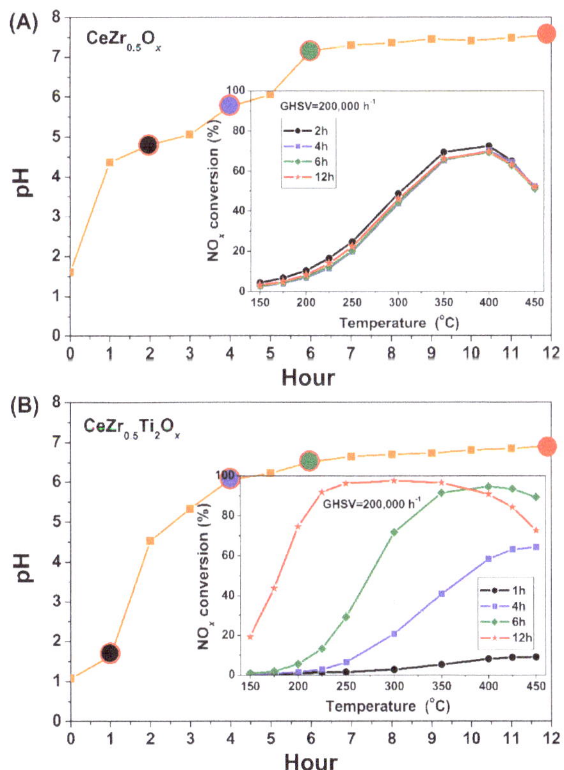

Figure 9. The pH variation of the mixed solution during the preparation of the (**A**) $CeZr_{0.5}O_x$ and (**B**) $CeZr_{0.5}Ti_2O_x$ catalysts, and the NO_x conversions of the obtained samples at different precipitation time.

Due to the acidity induced by the added $Ti(SO_4)_2$, the initial pH value of the mixed solution during the preparation of $CeZr_{0.5}Ti_2O_x$ dropped to be 1.1. With the hydrolysis of urea, the pH increased gradually after heating, and some white particles generated in the first hour and suspended in the solution. With the increase of time, the particles gradually turned yellow. The pH reached ca. 7.0 after 12 h of reaction. The particles with the precipitation times of 1 h, 4 h, 6 h, and 12 h were collected and then calcined to be catalyst samples. Interestingly, the activity test showed a remarkable enhancement of NO_x conversions for the four samples with the increase in precipitation time.

The surface metal atomic concentrations of the $CeZr_{0.5}Ti_2O_x$ samples with different precipitation times were analyzed using XPS, and the variations in Ce, Zr, and Ti concentrations with precipitation time are shown in Figure 10. For the 1-h precipitation sample, only Ti and Zr, without Ce, were detected. With the increase in precipitation time, surface Ce concentration increased gradually in the samples. At the same time, Ti and Zr concentrations gradually decreased with the increase in precipitation time. A TEM-EDS mapping image showed that Ce was highly dispersed in the $CeZr_{0.5}Ti_2O_x$ catalyst (Figure 11).

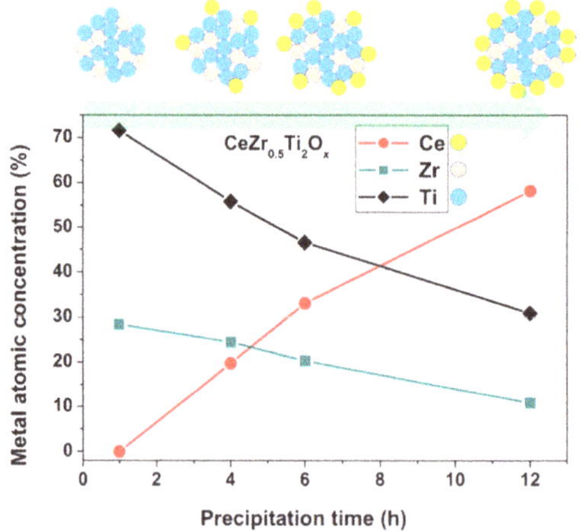

Figure 10. Surface metal atomic concentrations of the $CeZr_{0.5}Ti_2O_x$ samples with different precipitation times.

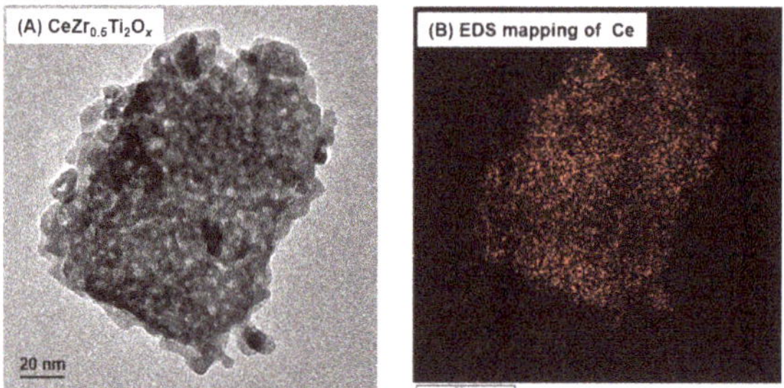

Figure 11. TEM image (A) and the corresponding EDS mapping (B) for the Ce of the $CeZr_{0.5}Ti_2O_x$ catalyst.

Considering the variations in the solution pH value when preparing the $CeZr_{0.5}Ti_2O_x$, the formation process of the catalyst can be proposed as follows: The Ti and Zr species were first co-precipitated with the increase in solution pH. Then, the Ce species uniformly precipitated onto the precipitated Zr-Ti species with the further increase in pH. Finally, a $CeZr_{0.5}Ti_2O_x$ catalyst with a higher surface Ce concentration than Ti and Zr was obtained. Through control of the hydrolysis of urea, the variations in the solution pH can be controlled, and then we can control the precipitation process of the catalyst, which is very important for the formation of highly-dispersed CeO_2 on ZrO_2-TiO_2. Thus, the obtained catalyst can present excellent NH_3-SCR performance.

3. Experimental Section

3.1. Catalyst Preparation and Activity Test

The $CeZr_{0.5}Ti_aO_x$ (a = Ti/Ce molar ratio = 0, 1, 2, 5, 10), with a Zr/Ce molar ratio fixed to be 0.5, was prepared using a precipitation method. Desired precursors of $Ce(NO_3)_3{\cdot}6H_2O$ (>99%, Sinopharm Chemical Reagent Co., Ltd., Shanghai, China), $Zr(NO_3)_4{\cdot}5H_2O$ (>99%, Sinopharm Chemical Reagent Co., Ltd., Shanghai, China) and $Ti(SO_4)_2$ (>98%, Sinopharm Chemical Reagent Co., Ltd., Shanghai, China) were dissolved together in distilled water, and urea (>99%, Sinopharm Chemical Reagent Co., Ltd., Shanghai, China) was added to the mixed solution as a slowly-releasing precipitator. Then, the solution was heated to 90 °C to facilitate the release of NH_3 and thereby raise the pH value gradually. The temperature of the mixed solution was held at 90 °C for 12 h under vigorous stirring (some samples with shorter precipitation times were also prepared). After that, the precipitated powders were collected via filtration, washed using distilled water, and dried for 12 h at 100 °C. Finally, the catalyst was obtained after calcination at 500 °C for 5 h.

The SCR activity of the catalysts (40–60 mesh) were tested in a fixed-bed quartz flow reactor. The reaction conditions were controlled as follows: 500 ppm NO, 500 ppm NH_3, 5 vol.% O_2, N_2 balance, and 400 mL/min total flow rate. Different gas hourly space velocities (GHSVs) were obtained by changing the volume of catalysts, i.e., 0.24 mL catalyst for a GHSV = 100,000 h^{-1} and 0.12 mL catalyst for a GHSV = 200,000 h^{-1}. The concentrations of effluent N-containing gases (NO, NH_3, NO_2 and N_2O) were continuously measured by an online FTIR gas analyzer (Nicolet Antaris IGS analyzer, Thermo-Fisher Scientific, Waltham, MA, USA). NO_x conversion and N_2 selectivity were calculated using the following equations, respectively:

$$NO_x \text{ conversion} = (1 - \frac{[NO]_{out} + [NO_2]_{out}}{[NO]_{in} + [NO_2]_{in}}) \times 100\%$$

$$N_2 \text{ selectivity} = (1 - \frac{2[N_2O]_{out}}{[NO_x]_{in} + [NH_3]_{in} - [NO_x]_{out} - [NH_3]_{out}}) \times 100\%$$

3.2. Characterizations

X-ray diffraction (XRD) measurements were carried out on a computerized AXS D8 diffractometer (Bruker, GER), with Cu Kα (λ = 0.15406 nm) radiation, from 20 to 80° at 8°/min.

Surface areas were tested using an ASAP 2020 (Micromeritics, Norcross, GA, USA) at −196 °C by N_2 adsorption/desorption and calculated using a BET equation in the 0.05–0.35 partial pressure range.

The X-ray photoelectron spectroscopy (XPS) results of Ce 3d and O 1s were measured on an ESCALAB 250Xi Scanning X-ray Microprobe (Thermo-Fisher Scientific, Waltham, MA, USA) using Al Ka radiation (1486.7 eV) and a C 1 s peak, with BE = 284.8 eV as the calibration standard.

The transmission electron microscopy (TEM) image and energy-dispersive X-ray spectroscopy (EDS) mapping of Ce were obtained using a JEM-2100F equipment (JEOL, Tokyo, Japan), combined with a specimen tilting beryllium holder for energy dispersive spectroscopy. The accelerating voltage was 200 kV.

The H_2 temperature-programmed reduction (H_2-TPR) was tested using an AutoChem_II_2920 chemisorption analyzer (Micromeritics, Norcross, GA, USA), and the temperature-programmed desorption of NH_3 and NO_x (NO_x-TPD and NH_3-TPD) were tested using the same reaction system as the activity tests. Experiment details can be found in Reference [42].

4. Conclusions

A series of Ce-Zr-Ti oxide catalysts were prepared using a stepwise precipitation approach for NH_3-SCR. $CeZr_{0.5}O_x$ without Ti just showed a relatively low NO_x conversion. When Ti was introduced, Ce-Zr-Ti catalysts showed much better activities and N_2 selectivity. A $CeZr_{0.5}Ti_2O_x$ catalyst, which contains moderate Ti amounts, showed the best performance, which is associated with its optimal ratios for the redox (CeO_x) and acidic (TiO_2) components.

$CeZr_{0.5}O_x$ and $CeZr_{0.5}Ti_2O_x$ catalysts were characterized using various methods and the formation process during preparation was investigated. $CeZr_{0.5}Ti_2O_x$ catalyst showed superior redox properties (by H_2-TPR), good adsorption and NO_x/NH_3 activation functions (by NO_x-TPD and NH_3-TPD, respectively), and enhanced charge imbalance (by XPS).

During preparation, the Ti and Zr species were first co-precipitated with an increase in solution pH. Then, the Ce species uniformly precipitated onto the precipitated Zr-Ti species with the further increase in pH. As a result, $CeZr_{0.5}Ti_2O_x$ catalyst with a surface Ce concentration higher than those of Ti and Zr was obtained. This preparation process resulted in the formation of highly-dispersed CeO_2 on ZrO_2-TiO_2, and thus the catalyst can present excellent NH_3-SCR performance.

Author Contributions: W.S. and H.H. conceived the project; Y.G. and Y.Z. performed the experiments; W.S. and Z.L. carried out the data analysis; W.S. and Y.G. wrote the paper; H.H. supervised the study.

Funding: This work was supported by the National Key R&D Program of China (2017YFC0212502, 2017YFC0211101), the National Natural Science Foundation of China (201637005), and the Key Research Program of the Chinese Academy of Sciences (ZDRW-ZS-2017-6-2-3).

Conflicts of Interest: The authors declare no conflicts of interest.

References

1. Parvulescu, V.I.; Granger, P.; Delmon, B. Catalytic removal of NO. *Catal. Today* **1998**, *46*, 233–316. [CrossRef]
2. Granger, P.; Parvulescu, V.I. Catalytic NO_x abatement systems for mobile sources: From three-way to lean burn after-treatment technologies. *Chem. Rev.* **2011**, *111*, 3155–3207. [CrossRef] [PubMed]
3. Liu, Z.; Woo, S.I. Recent advances in catalytic $DeNO_x$ science and technology. *Catal. Rev.* **2006**, *48*, 43–89. [CrossRef]
4. Busca, G.; Lietti, L.; Ramis, G.; Berti, F. Chemical and mechanistic aspects of the selective catalytic reduction of NO_x by ammonia over oxide catalysts: A review. *Appl. Catal. B* **1998**, *18*, 1–36. [CrossRef]
5. Liu, F.; Yu, Y.; He, H. Environmentally-benign catalysts for the selective catalytic reduction of NO_x from diesel engines: Structure-activity relationship and reaction mechanism aspects. *Chem. Commun.* **2014**, *50*, 8445–8463. [CrossRef] [PubMed]
6. Shan, W.; Song, H. Catalysts for the selective catalytic reduction of NO_x with NH_3 at low temperature. *Catal. Sci. Technol.* **2015**, *5*, 4280–4288. [CrossRef]
7. Guan, B.; Zhan, R.; Lin, H.; Huang, Z. Review of state of the art technologies of selective catalytic reduction of NO_x from diesel engine exhaust. *Appl. Therm. Eng.* **2014**, *66*, 395–414. [CrossRef]
8. Gao, F.; Kwak, J.; Szanyi, J.; Peden, C.F. Current understanding of Cu-exchanged chabazite molecular sieves for use as commercial diesel engine $DeNO_x$ catalysts. *Top. Catal.* **2013**, *56*, 1441–1459. [CrossRef]
9. Brandenberger, S.; Kröcher, O.; Tissler, A.; Althoff, R. The state of the art in selective catalytic reduction of NO_x by ammonia using metal-exchanged zeolite catalysts. *Catal. Rev.* **2008**, *50*, 492–531. [CrossRef]
10. Li, J.; Chang, H.; Ma, L.; Hao, J.; Yang, R.T. Low-temperature selective catalytic reduction of NO_x with NH_3 over metal oxide and zeolite catalysts-a review. *Catal. Today* **2011**, *175*, 147–156. [CrossRef]
11. Deka, U.; Lezcano-Gonzalez, I.; Weckhuysen, B.M.; Beale, A.M. Local environment and nature of Cu active sites in zeolite-based catalysts for the selective catalytic reduction of NO_x. *ACS Catal.* **2013**, 413–427. [CrossRef]
12. Li, Y.; Cheng, H.; Li, D.; Qin, Y.; Xie, Y.; Wang, S. WO_3/CeO_2-ZrO_2, a promising catalyst for selective catalytic reduction (SCR) of NO_x with NH_3 in diesel exhaust. *Chem. Commun.* **2008**, 1470–1472. [CrossRef] [PubMed]

13. Can, F.; Berland, S.; Royer, S.; Courtois, X.; Duprez, D. Composition-dependent performance of $Ce_xZr_{1-x}O_2$ mixed-oxide-supported WO_3 catalysts for the NO_x storage reduction–selective catalytic reduction coupled process. *ACS Catal.* **2013**, *3*, 1120–1132. [CrossRef]
14. Long, R.Q.; Yang, R.T. Superior Fe-ZSM-5 catalyst for selective catalytic reduction of nitric oxide by ammonia. *J. Am. Chem. Soc.* **1999**, *121*, 5595–5596. [CrossRef]
15. Chen, L.; Li, J.; Ge, M. Promotional effect of Ce-doped V_2O_5-WO_3/TiO_2 with low vanadium loadings for selective catalytic reduction of NO_x by NH_3. *J. Phys. Chem. C* **2009**, *113*, 21177–21184. [CrossRef]
16. Wu, Z.; Jin, R.; Liu, Y.; Wang, H. Ceria modified MnO_x/TiO_2 as a superior catalyst for NO reduction with NH_3 at low-temperature. *Catal. Commun.* **2008**, *9*, 2217–2220. [CrossRef]
17. Niu, C.; Shi, X.; Liu, K.; You, Y.; Wang, S.; He, H. A novel one-pot synthesized CuCe-SAPO-34 catalyst with high NH_3-SCR activity and H_2O resistance. *Catal. Commun.* **2016**, *81*, 20–23. [CrossRef]
18. Carja, G.; Delahay, G.; Signorile, C.; Coq, B. Fe-Ce-ZSM-5 a new catalyst of outstanding properties in the selective catalytic reduction of NO with NH_3. *Chem. Commun.* **2004**, 1404–1405. [CrossRef] [PubMed]
19. Gao, X.; Jiang, Y.; Fu, Y.; Zhong, Y.; Luo, Z.; Cen, K. Preparation and characterization of CeO_2/TiO_2 catalysts for selective catalytic reduction of NO with NH_3. *Catal. Commun.* **2010**, *11*, 465–469. [CrossRef]
20. Liu, Z.; Yi, Y.; Li, J.; Woo, S.I.; Wang, B.; Cao, X.; Li, Z. A superior catalyst with dual redox cycles for the selective reduction of NO_x by ammonia. *Chem. Commun.* **2013**, *49*, 7726–7728. [CrossRef] [PubMed]
21. Liu, Z.; Zhang, S.; Li, J.; Ma, L. Promoting effect of MoO_3 on the NO_x reduction by NH_3 over CeO_2/TiO_2 catalyst studied with in situ DRIFTS. *Appl. Catal. B* **2014**, *144*, 90–95. [CrossRef]
22. Peng, Y.; Qu, R.; Zhang, X.; Li, J. The relationship between structure and activity of MoO_3-CeO_2 catalysts for NO removal: Influences of acidity and reducibility. *Chem. Commun.* **2013**, *49*, 6215–6217. [CrossRef] [PubMed]
23. Qu, R.; Gao, X.; Cen, K.; Li, J. Relationship between structure and performance of a novel cerium-niobium binary oxide catalyst for selective catalytic reduction of NO with NH_3. *Appl. Catal. B* **2013**, *142*, 290–297. [CrossRef]
24. Shan, W.; Liu, F.; He, H.; Shi, X.; Zhang, C. Novel cerium-tungsten mixed oxide catalyst for the selective catalytic reduction of NO_x with NH_3. *Chem. Commun.* **2011**, *47*, 8046–8048. [CrossRef] [PubMed]
25. Krishna, K.; Seijger, G.B.F.; Bleek, C.M.; Calis, H.P.A. Very active CeO_2-zeolite catalysts for NO_x reduction with NH_3. *Chem. Commun.* **2002**, *18*, 2030–2031. [CrossRef]
26. Shan, W.; Liu, F.; Yu, Y.; He, H. The use of ceria for the selective catalytic reduction of NO_x with NH_3. *Chin. J. Catal.* **2014**, *35*, 1251–1259. [CrossRef]
27. Gu, T.; Liu, Y.; Weng, X.; Wang, H.; Wu, Z. The enhanced performance of ceria with surface sulfation for selective catalytic reduction of NO by NH_3. *Catal. Commun.* **2010**, *12*, 310–313. [CrossRef]
28. Zhang, L.; Pierce, J.; Leung, V.L.; Wang, D.; Epling, W.S. Characterization of ceria's interaction with NO_x and NH_3. *J. Phys. Chem. C* **2013**, *117*, 8282–8289. [CrossRef]
29. Si, Z.; Weng, D.; Wu, X.; Yang, J.; Wang, B. Modifications of CeO_2–ZrO_2 solid solutions by nickel and sulfate as catalysts for NO reduction with ammonia in excess O_2. *Catal. Commun.* **2010**, *11*, 1045–1048. [CrossRef]
30. Si, Z.; Weng, D.; Wu, X.; Ran, R.; Ma, Z. NH_3-SCR activity, hydrothermal stability, sulfur resistance and regeneration of $Ce_{0.75}Zr_{0.25}O_2$–PO_4^{3-} catalyst. *Catal. Commun.* **2012**, *17*, 146–149. [CrossRef]
31. Shen, B.; Wang, Y.; Wang, F.; Liu, T. The effect of Ce-Zr on NH_3-SCR activity over MnOx(0.6)/$Ce_{0.5}Zr_{0.5}O_2$ at low temperature. *Chem. Eng. J.* **2014**, *236*, 171–180. [CrossRef]
32. Ko, J.H.; Park, S.H.; Jeon, J.-K.; Kim, S.-S.; Kim, S.C.; Kim, J.M.; Chang, D.; Park, Y.-K. Low temperature selective catalytic reduction of NO with NH_3 over Mn supported on $Ce_{0.65}Zr_{0.35}O_2$ prepared by supercritical method: Effect of Mn precursors on NO reduction. *Catal. Today* **2012**, *185*, 290–295. [CrossRef]
33. Liu, Z.; Su, H.; Li, J.; Li, Y. Novel MoO_3/CeO_2–ZrO_2 catalyst for the selective catalytic reduction of NO_x by NH_3. *Catal. Commun.* **2015**, *65*, 51–54. [CrossRef]
34. Ding, S.P.; Liu, F.D.; Shi, X.Y.; Liu, K.; Lian, Z.H.; Xie, L.J.; He, H. Significant promotion effect of Mo additive on a novel Ce-Zr mixed oxide catalyst for the selective catalytic reduction of NO_x with NH_3. *ACS Appl. Mater. Interfaces* **2015**, *7*, 9497–9506. [CrossRef] [PubMed]
35. Shan, W.; Liu, F.; He, H.; Shi, X.; Zhang, C. A superior Ce-W-Ti mixed oxide catalyst for the selective catalytic reduction of NO_x with NH_3. *Appl. Catal. B* **2012**, *115–116*, 100–106. [CrossRef]
36. Chen, A.; Zhou, Y.; Ta, N.; Li, Y.; Shen, W. Redox properties and catalytic performance of ceria-zirconia nanorods. *Catal. Sci. Technol.* **2015**, *5*, 4184–4192. [CrossRef]

37. Liu, Z.; Zhang, S.; Li, J.; Zhu, J.; Ma, L. Novel V_2O_5-CeO_2/TiO_2 catalyst with low vanadium loading for the selective catalytic reduction of NO_x by NH_3. *Appl. Catal. B* **2014**, *158–159*, 11–19. [CrossRef]
38. Vuong, T.H.; Radnik, J.; Rabeah, J.; Bentrup, U.; Schneider, M.; Atia, H.; Armbruster, U.; Grünert, W.; Brückner, A. Efficient $VO_x/Ce_{1-x}Ti_xO_2$ catalysts for low-temperature NH_3-SCR: Reaction mechanism and active sites assessed by in situ/operando spectroscopy. *ACS Catal.* **2017**, *7*, 1693–1705. [CrossRef]
39. Wang, Z.; Qu, Z.; Quan, X.; Wang, H. Selective catalytic oxidation of ammonia to nitrogen over ceria–zirconia mixed oxides. *Appl. Catal. A* **2012**, *411–412*, 131–138. [CrossRef]
40. Topsøe, N.-Y. Mechanism of the selective catalytic reduction of nitric oxide by ammonia elucidated by in situ on-line Fourier Transform Infrared Spectroscopy. *Science* **1994**, *265*, 1217–1219. [CrossRef] [PubMed]
41. Lietti, L.; Forzatti, P.; Berti, F. Role of the redox properties in the SCR of NO by NH_3 over V_2O_5-WO_3/TiO_2 catalyst. *Catal. Lett.* **1996**, *41*, 35–39. [CrossRef]
42. Geng, Y.; Huang, H.; Chen, X.; Ding, H.; Yang, S.; Liu, F.; Shan, W. The effect of Ce on a high-efficiency CeO_2/WO_3-TiO_2 catalyst for the selective catalytic reduction of NO_x with NH_3. *RSC Adv.* **2016**, *6*, 64803–64810. [CrossRef]
43. Liu, F.; He, H. Structure-activity relationship of iron titanate catalysts in the selective catalytic reduction of NO_x with NH_3. *J. Phys. Chem. C* **2010**, *114*, 16929–16936. [CrossRef]
44. Bêche, E.; Charvin, P.; Perarnau, D.; Abanades, S.; Flamant, G. Ce 3d XPS investigation of cerium oxides and mixed cerium oxide ($Ce_xTi_yO_z$). *Surf. Interface Anal.* **2008**, *40*, 264–267. [CrossRef]
45. Liu, C.; Chen, L.; Chang, H.; Ma, L.; Peng, Y.; Arandiyan, H.; Li, J. Characterization of CeO_2-WO_3 catalysts prepared by different methods for selective catalytic reduction of NO_x with NH_3. *Catal. Commun.* **2013**, *40*, 145–148. [CrossRef]
46. Shan, W.; Liu, F.; He, H.; Shi, X.; Zhang, C. An environmentally-benign CeO_2-TiO_2 catalyst for the selective catalytic reduction of NO_x with NH_3 in simulated diesel exhaust. *Catal. Today* **2012**, *184*, 160–165. [CrossRef]

Article

Gas-Phase Phosphorous Poisoning of a Pt/Ba/Al_2O_3 NO_x Storage Catalyst

Rasmus Jonsson [1], Oana Mihai [1], Jungwon Woo [1], Magnus Skoglundh [1], Eva Olsson [1], Malin Berggrund [2] and Louise Olsson [1,*]

1 Competence Centre for Catalysis, Chalmers University of Technology, SE-412 96 Gothenburg, Sweden; rasjon@chalmers.se (R.J.); oana.mihai@chalmers.se (O.M.); jungwon@chalmers.se (J.W.); skoglund@chalmers.se (M.S.); eva.olsson@chalmers.se (E.O.)

2 Volvo Car Corporation, SE-405 31 Gothenburg, Sweden; malin.berggrund@volvocars.com

* Correspondence: louise.olsson@chalmers.se; Tel.: +46-31-772-4390

Received: 9 March 2018; Accepted: 7 April 2018; Published: 11 April 2018

Abstract: The effect of phosphorous exposure on the NO_x storage capacity of a Pt/Ba/Al_2O_3 catalyst coated on a ceramic monolith substrate has been studied. The catalyst was exposed to phosphorous by evaporating phosphoric acid in presence of H_2O and O_2. The NO_x storage capacity was measured before and after the phosphorus exposure and a significant loss of the NO_x storage capacity was detected after phosphorous exposure. The phosphorous poisoned samples were characterized by X-ray photoelectron spectroscopy (XPS), environmental scanning electron microscopy (ESEM), N_2-physisorption and inductive coupled plasma atomic emission spectroscopy (ICP-AES). All characterization methods showed an axial distribution of phosphorous ranging from the inlet to the outlet of the coated monolith samples with a higher concentration at the inlet of the samples. Elemental analysis, using ICP-AES, confirmed this distribution of phosphorous on the catalyst surface. The specific surface area and pore volume were significantly lower at the inlet section of the monolith where the phosphorous concentration was higher, and higher at the outlet where the phosphorous concentration was lower. The results from the XPS and scanning electron microscopy (SEM)-energy dispersive X-ray (EDX) analyses showed higher accumulation of phosphorus towards the surface of the catalyst at the inlet of the monolith and the phosphorus was to a large extent present in the form of P_4O_{10}. However, in the middle section of the monolith, the XPS analysis revealed the presence of more metaphosphate (PO_3^-). Moreover, the SEM-EDX analysis showed that the phosphorous to higher extent had diffused into the washcoat and was less accumulated at the surface close to the outlet of the sample.

Keywords: LNT; NSR; NO_x storage; phosphorous; deactivation; poisoning

1. Introduction

In diesel and lean burn gasoline engines, the engine operates with a large excess of air, and this results in an increased fuel economy, and thereby reduced emissions of CO_2, which is a greenhouse gas. However, it is critical to remove NO_x, which can be done by different catalytic aftertreatment techniques. Emission standards have therefore been implemented in large parts of the world and the emission levels have been significantly reduced over the years [1].

Since the traditional Three-way catalyst (TWC), requires stoichiometric air:fuel ratio, the Lean NO_x Trap (LNT) was introduced as a catalytic concept for exhaust gas treatment in diesel vehicles [2–5]. In the LNT, usually barium is used as a storage component, and the nitrogen oxides during lean conditions are stored on barium as barium nitrates. Moreover, noble metals, such as platinum, play a key role to oxidize the NO to NO_2, which facilities the NO_x storage [2,6–10]. Under short periods of time the engine is switched from operating under lean to rich conditions. Under rich conditions, barium nitrates decompose and are reduced over the noble metal sites to mainly N_2 and H_2O. The reducing agents in the exhaust gas are hydrocarbons (HC), carbon monoxide (CO) and hydrogen (H_2), which react with the NO_x to produce CO_2, H_2O and N_2. Nitrogen is the desired product from the reduction of stored nitrates. However, there are other possible bi-products from the reduction process such as ammonia (NH_3) and nitrous oxide (N_2O) [3].

A vehicle is expected to be in service for a long time, which requires the catalyst to be durable. Over time, the catalytic properties of the catalyst deteriorate. There are a few mechanisms that contribute to degradation of catalysts, such as sintering of catalytic particles as a result of high-temperature exposure [11]. The catalytic washcoat can also undergo mechanical tear and poisoning of active sites on the catalyst caused by chemisorption or reactions of catalytically active sites with poisons in the exhaust gas [12]. The poisons are often introduced to the system through the gasoline or diesel fuel, which is the case for SO_2, or through the lubricant of the engine. Lubricants, such as zinc dialkyldithiophosphate (ZDDP), are commonly used in vehicles and are a source for accumulation of both sulfates, zinc and phosphorus in the catalyst [13]. The deactivation effect of sulfur has extensively been studied over LNTs [12,14–16], and also how to regenerate the catalyst from sulfur [17].

There are only a few studies available in the open literature where phosphorus poisoning has been studied over noble metal catalysts used for emission control. Bunting et al. [13] studied a diesel oxidation catalyst (DOC) connected to an engine rig where the fuel was doped with ZDDP, which showed accumulation of foreign compounds such as phosphates, zinc and sulfates on the catalyst. Also, results from characterization of field-aged TWCs showed accumulation of elements found from ZDDP poisoning of catalysts [18–20]. Moreover, it has been shown that the distribution of phosphorus as it accumulates in the catalytic washcoat exhibits a gradient, with higher concentration at the inlet and lower along the washcoat length [19–23]. Compared to sulfur, which has a more even distribution over the washcoat and penetrates deep into the washcoat, phosphorus tends to be more concentrated close to the washcoat surface [13]. A study performed by Galisteo et al. [24], where a $Pt/Ba/Al_2O_3$ catalyst was impregnated with $(NH_4)_2HPO_4$, showed that the amount of phosphorus in the catalyst correlated with the loss of NO_x storage capacity.

However, there are, to our knowledge, no studies in the open literature that have examined phosphorus poisoning of LNT catalysts using vapour phase exposure that is closer to realistic poisoning conditions, which is the objective of the present work. In this study, we have examined a $Pt/Ba/Al_2O_3$ catalyst exposed to phosphorous in the gas phase using different phosphorous concentrations. The catalyst samples were studied in a flow reactor to determine changes in NO_x storage capacity, NO_x reduction activity and bi-product formation. The phosphorous-exposed samples were also characterized by X-ray photoelectron spectroscopy (XPS), nitrogen-physisorption, Inductive coupled plasma atomic emission spectroscopy (ICP-AES) and Environmental Scanning Electron Microscopy (ESEM) with energy dispersive X-ray (EDX) analysis.

2. Results and Discussion

2.1. Activity Measurements

Lean/rich cycling experiments were conducted in which the system was exposed to 400 vol.-ppm NO, 5 vol.-% CO_2, 5 vol.-% H_2O and 8 vol.-% O_2 during the lean phase. In this phase, NO is first oxidized over the platinum sites, and thereafter storage of NO_x species occurs as nitrates through a disproportionation mechanism, but also via nitrite formation at low temperature [5]. The NO oxidation capacity of the catalyst is clearly visible by the NO_2 formation seen in Figure 1. In the initial part of the lean phase, all NO_x is stored and after a certain amount of time, breakthrough of NO_x, due to saturation of the storage sites, is seen. Most NO_x is stored onto the barium sites in form of $Ba(NO_3)_2$, but some NO_x is also stored onto alumina sites [3]. At lower temperature, storage of NO_x occurs over both the barium and alumina sites, but at higher temperature the barium sites dominate the NO_x storage [25], due to the lower stability of alumina nitrates.

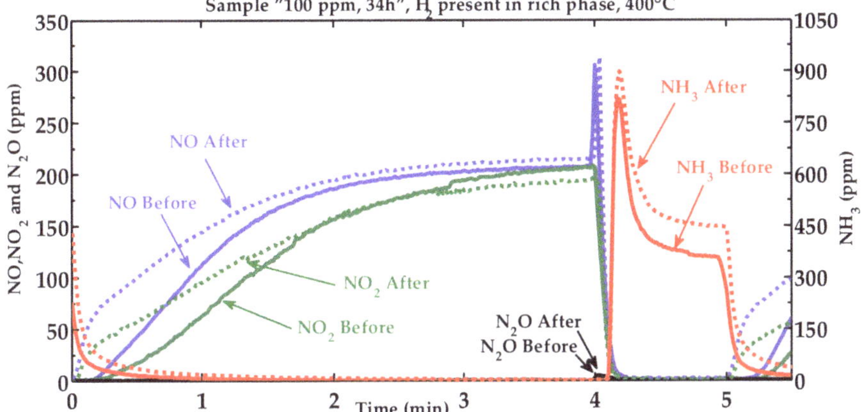

Figure 1. Measured outlet NO, NO_2, N_2O and NH_3 concentrations for the Pt/Ba/Al_2O_3 sample "100 ppm, 34 h" before and after exposure to phosphorous (100 vol.-ppm H_3PO_4, 8 vol.-% O_2, 5 vol.-% H_2O at 200 °C) in a flow-reactor experiment at 400 °C. Lean phase; 400 vol.-ppm NO, 5 vol.-% H_2O, 5 vol.-% CO_2 and 8 vol.-% O_2. Rich phase; 400 vol.-ppm NO, 5 vol.-%H_2O, 5 vol.-% CO_2 and 1 vol.-% H_2.

As the system is switched from lean to rich phase, a NO_x peak is visible, which is related to release of stored NO_x in the catalytic material [26]. Directly after the switch from the lean to the rich (1 vol.-% H_2, 5 vol.-% CO_2 and 5 vol.-% H_2O) phase, minor formation of N_2O is seen and it is during this short period of time, that most of the N_2 formation is suggested to occur [27–29]. After the N_2O peak a large ammonia peak is observed, which is consistent with the work by Lindholm et al. [25]. Most of the formed NH_3, early in the rich phase, is consumed by a Selective Catalytic Reduction (SCR) reaction with the stored NO_x, according to Lietti et al. [27] and Lindholm et al. [6]. Hence, the NH_3 peak is delayed and will dominate more as the rich phase proceeds and this is also seen in Figure 1. Indeed, spatially resolved MS measurements performed by Partridge et al. [10] showed a clear axial consumption of ammonia.

The effect of phosphorus exposure was studied by exposing the Pt/Ba/Al_2O_3 catalyst to 100 vol.-ppm H_3PO_4, 8 vol.-% O_2 and 5 vol.-% H_2O at 200 °C for 34 h and thereafter repeating the same experiment, see Figure 1. In addition, a second catalyst was used, which was exposed to 50 vol.-ppm H_3PO_4 instead. These catalysts were studied under lean/rich cycling experiments, which were conducted both at 300 and 400 °C, using 400 vol.-ppm NO, 5 vol.-% CO_2, 8 vol.-% O_2, and 5 vol.-% H_2O for the lean conditions and 1 vol.-% H_2 or 1000 vol.-ppm C_3H_6, 400 vol.-ppm NO, 5 vol.-% CO_2 and 5 vol.-% H_2O during the rich conditions, and the resulting outlet NO_x concentrations are shown in Figure 2. The results show that exposure to phosphorus causes a significant loss of NO_x storage capacity for the Pt/Ba/Al_2O_3 catalyst. The loss of NO_x storage capacity is summarised in Table 1. It is clear that the loss in storage capacity is similar for both poisoning levels and, moreover, that the effect is more pronounced after exposure at 300 compared to 400 °C.

Our results are in line with those in the study by Galisteo et al. [24], where the effect of phosphorous on the NO_x storage capacity of Pt/Ba/Al_2O_3 was studied by impregnation of the catalyst with $(NH_3)_3PO_4$ dissolved in water. The authors also observed a decreased NO_x storage capacity after phosphorous exposure. At 300 °C, the NO_x reduction properties of C_3H_6 are poor compared to 400 °C (compare graph B and D in Figure 2), which previously has been observed by

Olsson et al. [30]. However, loss in NO_x storage capacity is still observed for this case after phosphorus exposure. For exposure at 400 °C, the phosphorus poisoning is similar for both H_2 and C_3H_6, which indicates that as long as the reductant can remove the NO_x, the phosphorus poisoning is related to poisoning of the reactions under lean conditions. For phosphorous exposure at 300 °C on the other hand, the reduction with propene is poor, which results in low NO_x storage and therefore the effect of phosphorus poisoning is not seen equally clear.

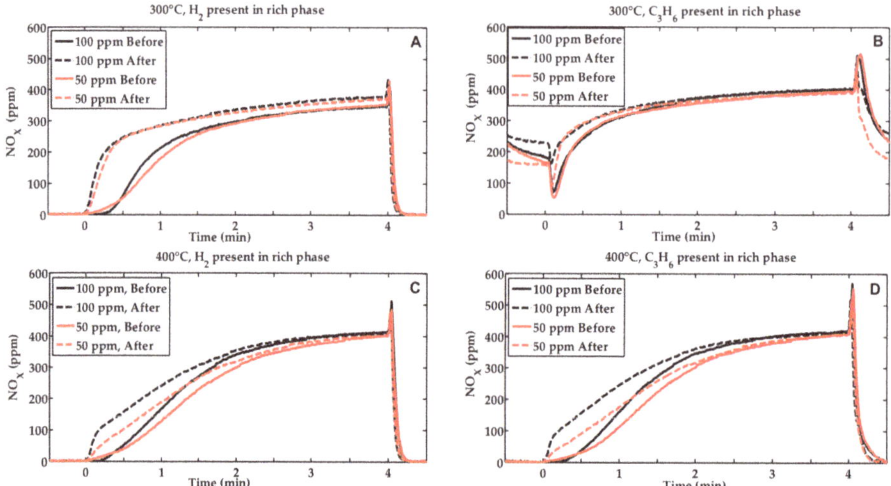

Figure 2. Measured outlet NO_x before and after exposing Pt/Ba/Al_2O_3 to 50 or 100 vol.-ppm H_3PO_4, 8 vol.-% O_2 and 5 vol.-% H_2O for 34 h. The lean phase (400 vol.-ppm NO, 5 vol.-% H_2O, 5 vol.-% CO_2 and 8 vol.-% O_2) starts at time = 0. The rich phase consist of 400 vol.-ppm NO, 5 vol.-% H_2O, 5 vol.-% CO_2 and H_2 or C_3H_6, where (**A**) 300 °C with 1 vol.-% H_2 in the rich phase, (**B**) 300 °C with 0.1 vol.-% C_3H_6 in the rich phase, (**C**) 400 °C with 1 vol.-% H_2 in the rich phase, and (**D**) 400 °C with 0.1 vol.-% C_3H_6 in the rich phase.

Table 1. NOx storage capacity for fresh and phosphorous exposed samples.

Condition	NO_x Storage Capacity (μmol)	Loss in NO_x Storage Capacity (μmol)
NO+H_2, 300 °C, fresh	92.4	-
NO+H_2, 300 °C, 50 ppm P	56.2	36.2
NO+H_2, 300 °C, fresh	88.8	-
NO+H_2, 300 °C, 100 ppm P	51.4	37.4
NO+H_2, 400 °C, fresh	85.9	-
NO+H_2, 400 °C, 50 ppm P	70.6	15.3
NO+H_2, 400 °C, fresh	70.3	-
NO+H_2, 400 °C, 100 ppm P	51.5	18.8

The resulting N_2O formation during these experiments is shown in Figure 3, and it is observed that the selectivity towards N_2O is higher at low temperatures, which is in accordance with the literature [31]. Most of the N_2O is formed in the first part of the rich period, but also a peak in the beginning of the lean phase is observed. Choi et al. [32] found that the second N_2O peak relates to reactions with NH_3. When shifting from rich to lean phase, the formation of NH_3 ceases due to the replacement of H_2 with O_2 in the feed. Therefore, it is plausible that the second N_2O peak relates to consumption of stored NH_3 to form N_2O.

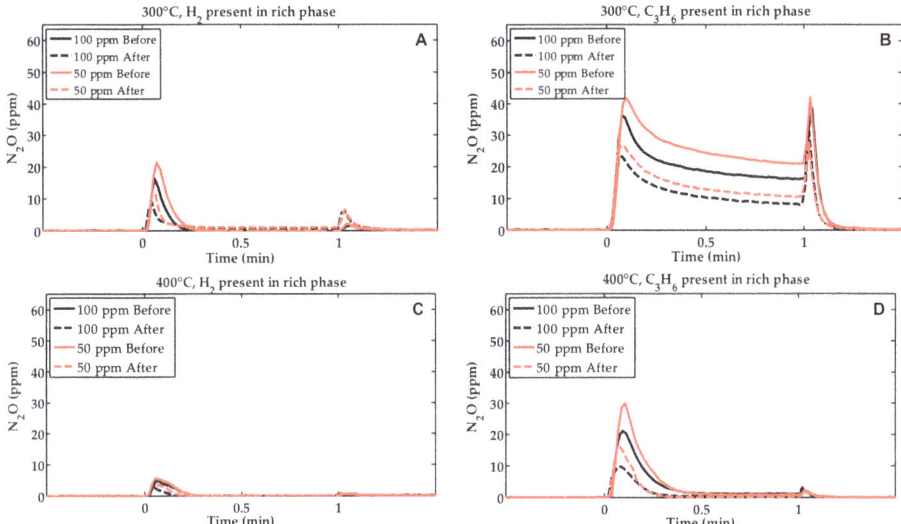

Figure 3. Measured outlet N_2O before and after exposing Pt/Ba/Al_2O_3 to 50 or 100 vol.-ppm H_3PO_4, 8 vol.-% O_2 and 5 vol.-% H_2O for 34 h. The rich phase (400 vol.-ppm NO, 5 vol.-% H_2O, 5 vol.-% CO_2 and H_2 or C_3H_6) starts at time = 0. The lean phase consist of 400 vol.-ppm NO, 5 vol.-% H_2O, 5 vol.-% CO_2 and 8 vol.-% O_2, where (**A**) 300 °C with 1 vol.-% H_2 in the rich phase, (**B**) 300 °C with 0.1 vol.-% C_3H_6 in the rich phase, (**C**) 400 °C with 1 vol.-% H_2 in the rich phase, and (**D**) 400 °C with 0.1 vol.-% C_3H_6 in the rich phase.

In addition, more N_2O is formed when using propene as the reducing agent compared to hydrogen. The N_2O formation was previously studied by Olsson et al. [30] where a kinetic model for the reduction of NO_x by C_3H_6 was developed. The reason for the N_2O formation is the reaction between the hydrocarbon and NO_x to form N_2O [33]. Furthermore, in the beginning of the lean phase after propene reduction, there is an increase in the N_2O formation. This is due to reaction between NO_x and HC species, stored during the rich phase. In addition, the results clearly show that the formation of N_2O in the rich phase decreases after exposure to phosphoric acid. The decrease in N_2O formation occurs at both 300 and 400 °C for both H_2 and C_3H_6 as the reducing agent. When comparing the ratio of (N_2O formed)/(NO_x stored) at 300 °C with propene as reductant, the ratio clearly decreases after phosphorous exposure (0.015, 0.0084, 0.020 and 0.011 for first sample before and after exposure for 100 vol.-ppm P, and for second sample before and after exposure for 50 vol.-ppm P, respectively). Thus, it is clear that the N_2O formation not decreases only due that less NO_x is stored, but also due to that the selectivity changes because of the poisoning. De Abreu Goes et al. [11] studied thermal exposure of LNT catalysts, which resulted in sintering of the catalytic particles that in turn resulted in decreased formation of N_2O. Since the catalyst in the present study was not exposed to high temperature again after the degreening step, sintering as a cause for the loss of N_2O formation is not likely. Therefore, the reduced N_2O formation after phosphorous exposure is more likely to relate to poisoning of the platinum sites.

With H_2 as the reducing agent, the formation of NH_3 is higher compared to C_3H_6 as reductant, which can be seen in Figure 4. At 300 °C, the formation of NH_3 is even negligible for the C_3H_6 case. For both samples, the NH_3 formation is slightly higher after the exposure to phosphorus when H_2 is used as the reducing agent. The ammonia formed over the LNT depends on several reactions occurring simultaneously: (i) ammonia production from inlet NO_x; (ii) ammonia production from stored NO_x; and (iii) reaction of the formed ammonia with stored NO_x in an SCR process [6]. An increase in the ammonia production has also been observed for field-aged, oven-aged and sulphur-poisoned LNT catalysts [11,34]. The reason for the increase in ammonia formation after phosphorus exposure could be related to the loss of NO_x storage capacity depicted in Figure 2. Since less NO_x is available in the form of nitrates during the lean phase, less ammonia can in turn react in the SCR reaction with the stored nitrates [6], and thereby the ammonia release is higher. However, less ammonia will also be produced from the stored NO_x, thus this means that the SCR reaction is more influenced by the phosphorous poisoning compared to the ammonia production. The increased ammonia production during rich conditions can be beneficial for the cases where an SCR catalyst is placed downstream of the LNT [8], in the so-called passive SCR technique.

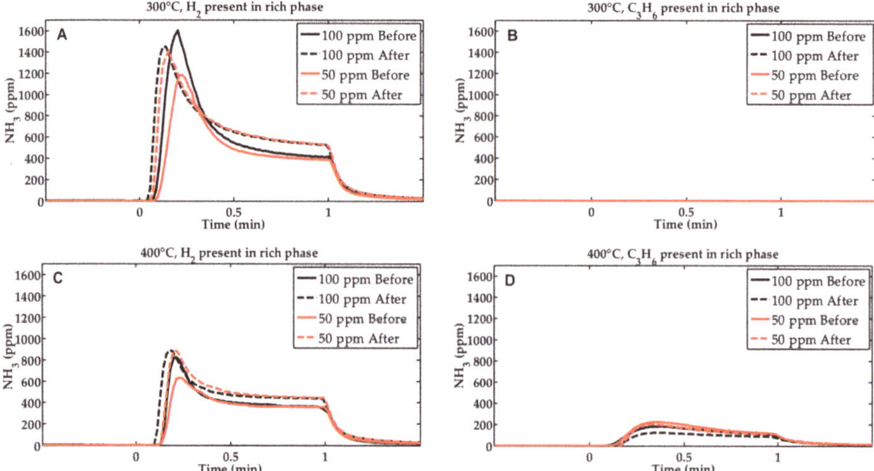

Figure 4. Measured outlet NH_3 concentration before and after exposing $Pt/Ba/Al_2O_3$ to 50 or 100 vol.-ppm H_3PO_4, 8 vol.-% O_2 and 5 vol.-% H_2O for 34 h. The rich phase (400 vol.-ppm NO, 5 vol.-% H_2O, 5 vol.-% CO_2 and H_2 or C_3H_6) starts at time = 0. The lean phase consist of 400 vol.-ppm NO, 5 vol.-% H_2O, 5 vol.-% CO_2 and 8 vol.-% O_2, where (**A**) 300 °C with 1 vol.-% H_2 in the rich phase, (**B**) 300 °C with 0.1 vol.-% C_3H_6 in the rich phase, (**C**) 400 °C with 1 vol.-% H_2 in the rich phase, and (**D**) 400 °C with 0.1 vol.-% C_3H_6 in the rich phase.

The outlet N_2 concentration was not measured due to IR spectroscopy was used for the gas phase analysis. By assuming that all nitrogen leaving the catalyst are either in form of NO, NO_2, NH_3, N_2O or N_2, the remaining N_2 can be determined. The quantities of NO_x, NO, NO_2 NH_3 and N_2O were estimated by integration of the data from the FTIR analysis. The results from these calculations are shown in Figure 5 and it is clear that the formed amount of N_2 is quite low. The reason for this is that we intentionally used long lean and rich cycles to come closer to saturation and also more complete regeneration. In a real application, the cycle times would be different, giving a high amount of N_2. Moreover, the results clearly show a minor N_2O formation in all conditions except with C_3H_6 as reducing agent at 300 °C. Hence, the main difference between the results from the different experimental conditions is the formation of NH_3, N_2 and release of NO_x. The phosphorous-exposed samples tend to release higher amounts of NO_x and NH_3 at the cost of formation of the desired product; N_2. According to the ICP-AES results shown in Table 2, the amount of accumulated phosphorous is similar for both H_3PO_4 concentrations. Thus the H_3PO_4 concentration does not seem to be such a critical factor for the poisoning at these levels. Therefore, the difference before and after phosphorous exposure between the two cases appears similar. For H_2 as the reducing agent, exposure to phosphorus has a more pronounced effect at 300 than 400 °C. The reason for this could be that since the ammonia production is higher at lower temperature, and is to a large extent influenced by the phosphorous poisoning, it results in a more pronounced effect on the N_2 selectivity. However, for C_3H_6 as the reducing agent, there is no clear trend regarding the selectivities between the fresh and phosphorous exposed samples at 300 and 400 °C. The reason for this could be that the amounts of NH_3 and N_2O formed in these experiments are low, so most of the species are NO_x and N_2.

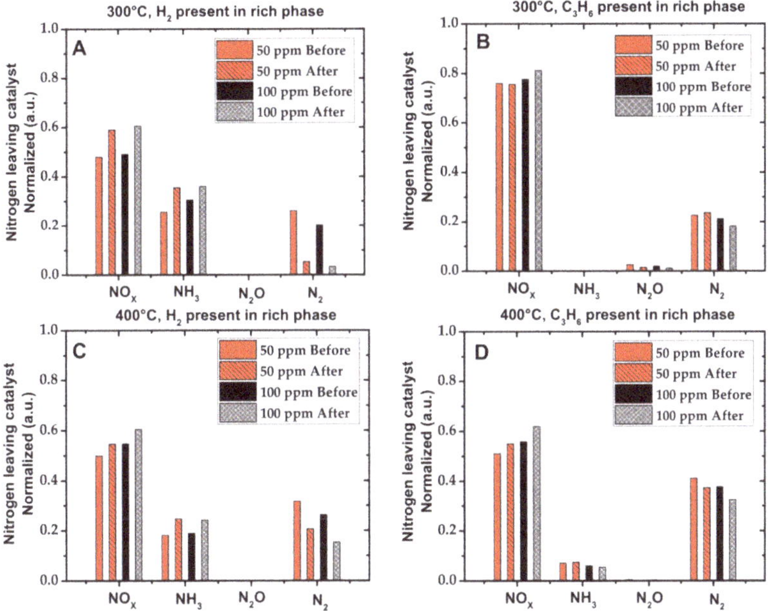

Figure 5. Balance over outlet nitrogen atoms over a lean/rich cycle before and after exposing $Pt/Ba/Al_2O_3$ to 50 and 100 vol.-ppm H_3PO_4 for 34 h, assuming that all nitrogen leaving the catalyst is either in form of NO_x, NH_3, N_2O or N_2. (**A**) 300 °C with 1 vol.-% H_2 in the rich phase, (**B**) 300 °C with 0.1 vol.-% C_3H_6 in the rich phase, (**C**) 400 °C with 1 vol.-% H_2 in the rich phase, (**D**) 400 °C with 0.1 vol.-% C_3H_6 in the rich phase.

2.2. *Characterization of the Phosphorous Exposed Samples*

One channel of the phosphorous-exposed monolith sample "50 ppm, 34 h" was cut out and the surface composition in the inlet, middle, and in the outlet section of the channel was analyzed by XPS. The overall spectra (see Figure 6A), show that the magnitude of the phosphorus peaks is highest from the surface of the inlet section and lowest from the outlet section of the phosphorous exposed sample. The P 2p binding energy region measured with higher resolution for the inlet and outlet sections for the phosphorous exposed sample are shown in Figure 6B. Due to different signal intensities from the two samples, the spectra are normalized to facilitate the interpretation of the peak positions when comparing different spectra with each other. The intensity of the P 2p peak from the middle section of the sample is significantly lower compared to the peak from the inlet section. In addition, the results show that the position of the P 2p peak shifts towards lower binding energy for the middle section compared to the inlet section of the sample. Deconvolution of the P 2p peak for the inlet and middle section of the sample can be seen in Figure 7. For the inlet section of the sample the deconvolution shows a higher presence of P_4O_{10} with a binding energy ranging between 135.0 and 135.5 eV [35]. However, for the middle region of the sample, the deconvolution shows that metaphosphates, PO_3^-, dominate, which have a binding energy ranging between 134.0 and 134.5 eV [35]. Formation of metaphosphates has previously been observed after exposing Cu/BEA [36] as well as Fe/BEA [37] for phosphoric acid in gas phase. In addition, formation of P_4O_{10} was also found on Fe/BEA [37], which is in line with our observations. Our results show that the phosphorus-containing surface species mainly are in the form of P_4O_{10} in the inlet section of the sample, where the phosphorus concentration and deposition is higher, while in the middle section more metaphosphates are seen.

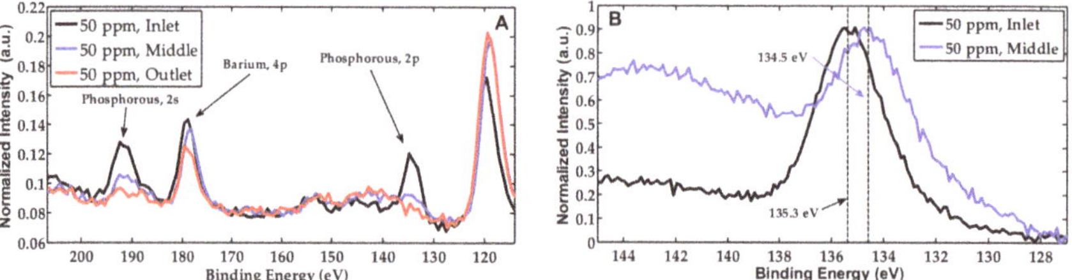

Figure 6. (**A**) Normalized X-ray photoelectron spectroscopy (XPS) spectra from the inlet, middle and outlet section of the $Pt/Ba/Al_2O_3$ sample exposed to 50 vol.-ppm H_3PO_4 for 34 h. (**B**) Corresponding XPS spectra for the P 2p region from the inlet and middle section of the sample.

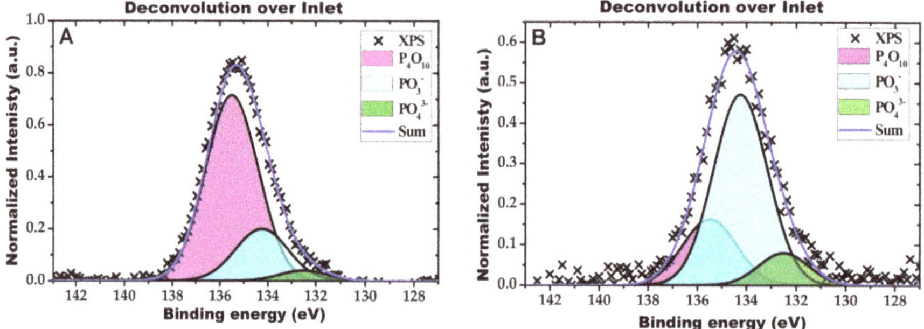

Figure 7. Deconvolution of the P 2p XPS spectra from the inlet (**A**) and middle (**B**) section of the $Pt/Ba/Al_2O_3$ sample exposed to 50 vol.-ppm H_3PO_4 for 34 h. The considered phosphorous species are P_4O_{10}, PO_3^- and PO_4^{3-}.

In order to achieve further insight into the effect of phosphorous exposure in the axial direction of the $Pt/Ba/Al_2O_3$ catalyst, the specific surface area and pore volume, and elemental composition of the inlet, middle and outlet sections of the samples exposed for 100 and 50 vol.-ppm phosphorous for 34 h were analysed together with a degreened sample. In Table 2, the phosphorus content from ICP-AES analysis is presented along with the specific surface area and pore volume. The results show that phosphorus has a gradient distribution, from the inlet to the outlet section of the two phosphorous-exposed samples, which is in line with other studies of both field-aged catalysts and experiments in engine rigs [13,19–23,38,39]. Moreover, it is found that the effect of the phosphorus concentration is minor, which is in line with the flow reactor experiments. Glisteo et al. [24] impregnated a $Pt/Ba/Al_2O_3$ catalyst with aqueous solutions of phosphorous of different concentrations. Their results showed that the specific surface area decreased more by higher phosphorous concentration accumulated on the catalyst, which is in line with our results. We find that the phosphorus content is the highest, and the surface area and pore volume have lowest values, for the inlet section of both catalysts exposed to phosphorous. Close to the outlet section, the phosphorus content is much lower for both samples, whereas the specific surface area and pore volume are higher and more similar to the degreened sample.

Table 2. Phosphorous content, specific surface area and pore volume of the fresh Pt/Ba/Al_2O_3 sample and of the Pt/Ba/Al_2O_3 samples exposed to 50 and 100 vol.-ppm phosphorous for 34 h. The phosphorus contents are measured by inductive coupled plasma atomic emission spectroscopy (ICP-AES).

	P-content, (wt.-%)	S_{BET} (m^2/(g washcoat)) [1]	V_P (cm^3/(g washcoat)) [1]
Fresh	-	134	0.46
50 ppm, 34 h, inlet	2.2	98	0.36
50 ppm, 34 h, middle	0.40	130	0.45
50 ppm, 34 h, outlet	0.11	134	0.45
100 ppm, 34 h, inlet	2.0	113	0.35
100 ppm, 34 h, middle	0.38	123	0.39
100 ppm, 34 h, outlet	0.07	132	0.39

[1] Estimated based on washcoat amount.

The results from ESEM and EDX analyses acquired from two different positions; 2 mm from the front and 2 mm from the back of the monolith sample exposed for 50 vol.-ppm phosphorous for 34 h are presented in Figures 8 and 9, respectively. Besides phosphorus, other elements (Al and Ba) that are present in the washcoat are depicted in the images. Moreover, silicon, which is one of the major compounds of the cordierite substrate, is also shown. The analysis 2 mm from the front of the sample (Figure 8) indicates that phosphorus is located towards the surface of the washcoat, which previously has been shown for catalysts exposed to ZDDP doped fuel in engine rigs [13]. Closer to the outlet of the sample, less phosphorus is present and the distribution of the phosphorus is more uniform in the washcoat. Accumulation of phosphorus towards the front of the catalyst has been shown for field-aged catalysts [21]. The ESEM-EDX results in the present study can be related to the XPS results (Figure 7), which showed more P_4O_{10} in the inlet section of the sample, while more metaphosphates in the middle section of the phosphorous exposed sample. Finally, in the outlet section of the monolith sample, only small amounts of phosphorus are observed with EDX, but it should be noted that phosphorus is still visible, which means that in a catalytic aftertreatment system it is possible that a catalyst placed downstream the NO_x storage catalyst also can be exposed to phosphorus.

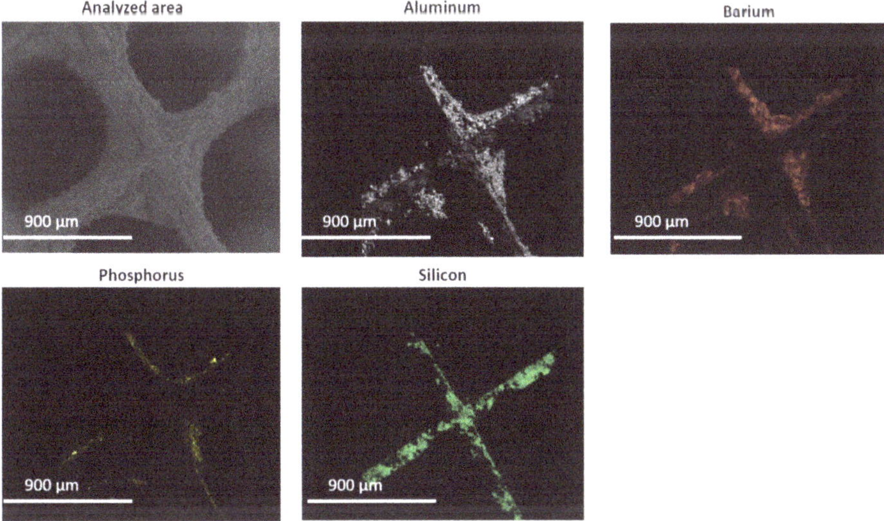

Figure 8. Environmental scanning electron microscopy (ESEM) and energy dispersive X-ray (EDX) images from the cross-section 2 mm from the front of the Pt/Ba/Al_2O_3 sample exposed to 50 vol.-ppm H_3PO_4 for 34 h.

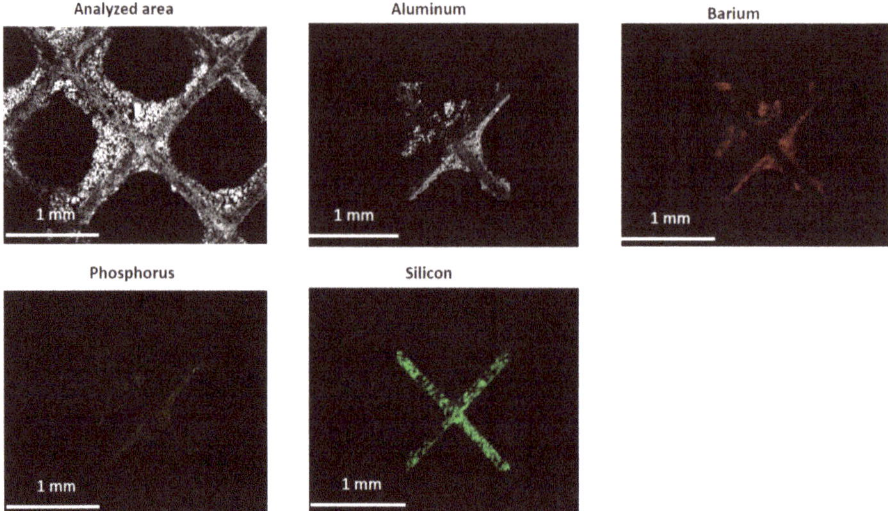

Figure 9. ESEM and EDX images from the cross-section 2 mm from the back of the Pt/Ba/Al_2O_3 sample exposed to 50 vol.-ppm H_3PO_4 for 34 h.

To summarize, the combined XPS, EDX, BET and ESEM-EDX data suggest that during phosphorus exposure in the vapour phase more phosphorus is deposited in the inlet of the monolith sample and this phosphorous is mainly located at the surface of the washcoat and contains a large fraction of P_4O_{10}. While the middle section of the monolith contains less phosphorus and has a higher fraction of metaphosphates. Moreover, closer to the outlet the phosphorus is more evenly distributed.

3. Materials and Methods

3.1. Catalyst Synthesis

The Pt/BaO/Al_2O_3 catalyst was synthesized using incipient wetness impregnation. To enhance the NO_x storage capacity of the chosen model catalyst, the impregnation of platinum on the support material was performed before incorporation of barium, which previously has shown beneficial to achieve high NO_x storage capacity of the catalyst [25]. Prior to the impregnation, the γ-alumina (Sigma-Aldrich) support was calcined for 2 h at 600 °C. Thereafter the γ-alumina support was impregnated with a platinum nitrate precursor (Heraeus platinum (II) nitrate; solution type K, batch: 12207), diluted with Milli-Q water (18 MΩ·cm) of the estimated volume targeting 2 wt.-% Pt. The impregnated support material was dried in air at 100 °C for 2 h and subsequently calcined in air at 550 °C for 2 h, starting from room temperature with a temperature increase of 5 °C/min. The calcined Pt/Al_2O_3 sample was then impregnated with a barium precursor (Sigma-Aldrich barium acetate, 99% A.C.S reagent, batch: 11415KA), dissolved in Milli-Q water in two steps also by incipient wetness impregnation, targeting a total of 16 wt.-% of BaO. The catalytic material was dried and calcined after each impregnation step under same conditions as for the impregnation of platinum.

3.2. Monolith Preparation

The catalytic material was coated on honeycomb-shaped ceramic monolith substrates (20 mm in length, 21 mm in diameter, a wall thickness of 0.106 mm and with 400 cpsi). To improve the adhesiveness, 10 wt.-% boehmite (Sasol disperal P2, product code: 538116) was mixed with the catalyst in a solution of 50 vol.-% Milli-Q water and 50 vol.-% ethanol. The monolith substrates were stepwise dipped in the solution and dried at 90 °C until the target washcoat weight of 700 mg per monolith sample was reached. Thereafter, the catalysts were calcined at 550 °C for 2 h, starting from room temperature with a temperature increase of 5 °C/min.

3.3. *Flow-Reactor Experiments*

The flow-reactor experiments were carried out in a synthetic gas bench (SGB) reactor. The coated monolith samples were placed in a quartz tube, which was 750 mm in length and with an inner diameter of 22 mm. To prevent by-pass, quartz wool was wrapped around the monoliths. The gas flow and gas concentrations were regulated using mass flow controllers from Bronkhorst and water vapor was dosed using a Bronkhorst Controlled evaporation system (CEM) system. The quartz tube was placed in an electric heating coil and covered with insulation material, and the temperatures were measured by two thermocouples, one measuring and regulating the gas temperature before the catalyst and one measuring the catalyst temperature in a center channel of the monolith. An MKS Multigas 2030 FTIR spectrometer provided by MKS instruments, Andover, MA, USA was used to monitor the gas phase concentrations of NO, NO_2, N_2O, NH_3 and H_2O, and the total flow was 3500 mL/min, corresponding to 30,300 h^{-1} space velocity, with argon as carrier gas for all measurements. All lines and connections prior and after the reactor were heated to 200 °C, to prevent condensation.

To avoid sintering of platinum particles under the experiments, the coated monoliths were first H_2-treated in 1 vol.-% H_2, 5 vol.-% CO_2 and 5 vol.-% H_2O for 20 min at 500 °C and thereafter degreened at 600 °C, according to the following procedure. First the catalyst was exposed to the rich phase (400 vol.-ppm NO, 5 vol.-% CO_2, 1 vol.-% H_2 and 5 vol.-% H_2O with argon as carrier gas) for 60 min, followed by exposure to the lean phase (400 vol.-ppm NO, 5 vol.-% CO_2, 8 vol.-% O_2, and 5 vol.-% H_2O in argon) for 15 min. Subsequently, the catalyst was reduced in 1 vol.-% H_2, 5 vol.-% CO_2 and 5 vol.-% H_2O in Ar for 20 min at 600 °C.

The NO_x storage and reduction performance of the catalysts was studied at 300 and 400 °C using lean/rich cycling. At each temperature, two cycle segments consisting of five cycles in each segment were conducted. Cycle number 4 for all sequences is shown in the Figures. The gas composition of the lean phase was 400 vol.-ppm NO, 5 vol.-% CO_2, 8 vol.-% O_2, and 5 vol.-% H_2O. In the first lean/rich phase cycle segment, hydrogen was used as the reductant (1 vol.-% H_2, 400 vol.-ppm NO, 5 vol.-% CO_2 and 5 vol.-% H_2O) and in the second cycle segments propene was used as the reducing agent (0.1 vol.-% C_3H_6, 400 vol.-ppm NO, 5 vol.-% CO_2 and 5 vol.-% H_2O). For all cycles, the duration of the lean and rich phase were 4 and 1 min, respectively. Prior to the experiment at each temperature, the catalyst was pre-treated with 1 vol.-% H_2, 5 vol.-% CO_2 and 5 vol.-% H_2O for 15 min at 400 °C. This pre-treatment was also conducted for the phosphorus-exposed catalysts.

3.4. *Phosphorous Exposure*

The NO_x storage capacity of the coated monoliths was measured in the flow reactor before and after the exposure to phosphorus, using the procedure described in the previous section. The same reactor as described above was used for the phosphorous exposure, however using a separate reactor tube. In order to reduce surfaces for the evaporated phosphoric acid to stick to, the monolith was not wrapped in quartz wool during the phosphorous exposure. Phosphoric acid was introduced to the gaseous flow by a syringe pump connected to a Teflon pipe disposing the acid inside the quarts tube. The composition of the gas phase during the phosphorous exposure was 50 or 100 vol.-ppm evaporated phosphoric acid, 5 vol.-% H_2O, 8 vol.-% O_2 in argon as carrier gas, and the temperature was 200 °C. Two different phosphorous exposure conditions were studied in this work; in the first case the catalyst was exposed to 50 vol.-ppm phosphoric acid for 34 h, and in the second case the sample was exposed to 100 vol.-ppm for phosphoric acid for 34 h. This corresponds to an exposure of 1.7 and 3.4 g P/l catalyst, respectively. Note that high amounts of phosphorus passed through the catalyst without adsorption. These samples are denoted "50 ppm, 34 h" and "100 ppm, 34 h", respectively. Note that new monolith samples were used for each poisoning condition.

3.5. XPS

To evaluate the oxidation state of phosphorus accumulated on the catalytic washcoat, an entire channel from the coated monolith was placed on a holder to measure the surface species of the washcoat. The use of a whole channel also enabled the identification of the axial position where measurements took place. The instrument used for the XPS analysis was a Perkin Elmer PHI 5000 ESCA system provided by PerkinElmer, Waltham, MA, USA equipped with an EDS elemental mapping system. The X-ray source for the XPS measurements was monochromatic Al Kα radiation at 1486.6 eV. Correction for charging was performed by normalizing the spectra using the C 1s peak at 284.6 eV [40] as reference. Furthermore, since normalizing will cause the largest peak being equal to 1, all spectra were divided by 1.1 for better visualization. Deconvolution of the P 2p peak from each sample was performed by fitting a Gaussian function to the experimental data, with subtraction of a linear baseline under both peaks. The peak positions were optimized according to the same procedure for both samples to achieve the lowest standard deviation and found to be 134.25, 135.5 and 132.5 eV for PO_3^-, P_4O_{10} and PO_4^{3-}, respectively.

3.6. N_2-Physisorption and ICP-AES

The specific surface area and pore volume of the samples were measured by N_2-physisorption isotherms at −195 °C and were collected using a TriStar 3000 gas adsorption analyzer. The inlet-, middle- and outlet section of the coated monoliths were crushed and grinded to powder form. Approximately 300 mg sample from each section was thermally dried at 90 °C in N_2 gas flow for 4 h and used for specific surface area and pore-volume measurements.

Powder samples from each section were used for elemental analysis by inductive coupled plasma atomic emission spectroscopy (ICP-AES), and the measurements were performed by ALS Scandinavia AB, Luleå, Sweden. The concentration of the measured phosphorus is reported as weight percentages.

3.7. ESEM

Environmental scanning electron microscopy (ESEM) was used to acquire an overview of the location of different elements over the catalytic washcoat. The instrument used to gain these images was a Quanta200 ESEM FEG from FEI, Hillsboro, OR, USA. An energy dispersive X-ray (EDX) system from Oxford Inca was equipped to the instrument to map the elements over scanned areas.

4. Conclusions

In the present study, a $Pt/Ba/Al_2O_3$ NO_x storage catalyst coated on a ceramic monolith substrate was exposed to two different phosphorus containing atmospheres for 34 h at 200 °C; 50 vol.-ppm H_3PO_4 + 8 vol.-% O_2 + 5 vol.-% H_2O and 100 vol.-ppm H_3PO_4 + 8 vol.-% O_2 + 5 vol.-% H_2O. The catalysts were characterized by N_2-physisorption, XPS, ESEM and ICP-AES, and the NO_x storage capacity was measured in flow reactor experiments.

The specific surface area and pore volume measurements, together with the elemental analysis by ICP-AES, showed a clear gradient distribution of phosphorous along the axial direction of the monolith. The phosphorus concentrations close to the inlet section of the sample were significantly higher compared to the middle and outlet section. The specific surface area and pore volume were found to be lower in the sections where the concentration of phosphorus was higher. At the outlet section, both the specific surface area and the pore volume were found to be comparable with those for the fresh sample, and the accumulated amount of phosphorus was only about 0.1 wt.-%, which can be compared with 2 wt.-% at the inlet section of the phosphorous exposed catalyst. Moreover, there was no significant difference between the different phosphorous concentrations, which indicates that the phosphorus concentration is not limiting the deposition of phosphorous on the catalyst, due to long exposure time.

Flow-reactor experiments showed that the NO_x storage capacity of $Pt/BaO/Al_2O_3$ decreased with presence of phosphorous species on the catalyst. The results also showed that the formation of N_2 decreased after the exposure to phosphorous, in favour of ammonia production. Furthermore, the formation of N_2O decreased after the phosphorus exposure. The increased ammonia production can be explained by the fact that the outlet ammonia concentration is dependent on three reactions over the catalyst: (i) the formation of ammonia from NO in the inlet gas; (ii) formation of ammonia from stored NO_x; and (iii) the consumption of ammonia through the selective catalytic reduction by the reaction with the stored nitrates. Since the amount of stored nitrates is lower for the phosphorus exposed sample, it means that less ammonia is produced in reaction (ii), but simultaneously less ammonia is consumed in reaction (iii). Since there is more ammonia in total, it means that the SCR reaction is more influenced by the phosphorus poisoning than the ammonia formation from the stored nitrates.

The XPS analysis showed that phosphorus exposure resulted in the formation of different species depending on the axial location of the washcoat. At the inlet section of the catalyst, where most of the phosphorus was accumulated, the dominating phosphorous species was found to be P_4O_{10} while further along the axial direction metaphosphates (PO_3^-) was the dominating species. From the ESEM and EDX analysis it could be concluded that most phosphorus, close to the inlet of the catalyst, was located at the washcoat surface. At the outlet, only a small amount of phosphorus could be detected by SEM-EDX, however, it was more diffused into the washcoat. To summarize, the results from XPS and SEM indicate that the accumulation of phosphorus differs between the front and the back of the monolith catalyst.

Acknowledgments: This work was conducted at the Competence Centre for Catalysis (KCK) hosted by Chalmers University of Technology and financially supported by the Swedish Energy Agency, Chalmers and the member companies AB Volvo, ECAPS AB, Haldor Topsøe A/S, Scania CV AB, Volvo Car Corporation AB, and Wärtsilä Finland Oy. KCK is greatly acknowledged for their financial support. Lars Ilver is gratefully acknowledged for the assistance with XPS measurements.

Author Contributions: R.J. and L.O. conceived and designed the experiments; R.J. performed the experiments; R.J., O.M., M.S., E.O., M.B. and L.O. analyzed the data; L.I. and J.W. contributed analysis tools; R.J. wrote the paper.

Conflicts of Interest: The authors declare no conflict of interest.

References

1. Guerreiro, C.; Gonzalez Ortiz, A.; de Leeuw, F.; Viana, M.; Horalek, J. *Air Quality in Europe—2016 Report*; European Environment Agency (EEA): Copenhagen, Denmark, 2016; ISBN 9789292138240.
2. Takahashi, N.; Shinjoh, H.; Iijima, T.; Suzuki, T.; Yamazaki, K.; Yokota, K.; Suzuki, H.; Miyoshi, N.; Matsumoto, S.; Tanizawa, T.; et al. The new concept 3-way catalyst for automotive lean-burn engine: NO_x storage and reduction catalyst. *Catal. Today* **1996**, *27*, 63–69. [CrossRef]
3. Epling, W.S.; Campbell, L.E.; Yezerets, A.; Currier, N.W.; Parks, J.E. Overview of the fundamental reactions and degradation mechanisms of NO_x storage/reduction catalysts. *Catal. Rev. Sci. Eng.* **2004**, *46*, 163–245. [CrossRef]
4. Matsumoto, S. Recent advances in automobile exhaust catalysts. *Catal. Today* **2004**, *90*, 183–190. [CrossRef]
5. Fridell, E.; Skoglundh, M.; Westerberg, B.; Johansson, S.; Smedler, G. NO_x Storage in Barium-Containing Catalysts. *J. Catal.* **1999**, *183*, 196–209. [CrossRef]
6. Lindholm, A.; Currier, N.W.; Li, J.; Yezerets, A.; Olsson, L. Detailed kinetic modeling of NO_x storage and reduction with hydrogen as the reducing agent and in the presence of CO_2 and H_2O over a Pt/Ba/Al catalyst. *J. Catal.* **2008**, *258*, 273–288. [CrossRef]
7. Lindholm, A.; Currier, N.W.; Fridell, E.; Yezerets, A.; Olsson, L. NO_x storage and reduction over Pt based catalysts with hydrogen as the reducing agent. Influence of H_2O and CO_2. *Appl. Catal. B Environ.* **2007**, *75*, 78–87. [CrossRef]
8. Lindholm, A.; Sjövall, H.; Olsson, L. Reduction of NO_x over a combined NSR and SCR system. *Appl. Catal. B Environ.* **2010**, *98*, 112–121. [CrossRef]

9. Nova, I.; Castoldi, L.; Lietti, L.; Tronconi, E.; Forzatti, P.; Prinetto, F.; Ghiotti, G. NO_x adsorption study over Pt-Ba/alumina catalysts: FT-IR and pulse experiments. *J. Catal.* **2004**, *222*, 377–388. [CrossRef]
10. Partridge, W.P.; Choi, J.S. NH_3 formation and utilization in regeneration of Pt/Ba/Al_2O_3 NO_x storage-reduction catalyst with H_2. *Appl. Catal. B Environ.* **2009**, *91*, 144–151. [CrossRef]
11. De Abreu Goes, J.E.; Olsson, L.; Berggrund, M.; Kristoffersson, A.; Gustafson, L.; Hicks, M. Performance Studies and Correlation between Vehicle- and Rapid-Aged Commercial Lean NO_x Trap Catalysts. *SAE Int. J. Engines* **2017**, *10*. [CrossRef]
12. Bartholomew, C.H. Mechanism of catalyst deactivation. *Appl. Catal. A Gen.* **2001**, *212*, 17–60. [CrossRef]
13. Bunting, B.G.; More, K.; Lewis, S.; Toops, T. Phosphorous Poisoning and Phosphorous Exhaust Chemistry with Diesel Oxidation Catalysts. *SAE Tech. Pap.* **2005**. [CrossRef]
14. Sedlmair, C.; Seshan, K.; Jentys, A.; Lercher, J.A. Studies on the deactivation of NO_x storage-reduction catalysts by sulfur dioxide. *Catal. Today* **2002**, *75*, 413–419. [CrossRef]
15. Huang, H.Y.; Long, R.Q.; Yang, R.T. A highly sulfur resistant Pt-Rh/TiO_2/Al_2O_3 storage catalyst for NO_x reduction under lean-rich cycles. *Appl. Catal. B Environ.* **2001**, *33*, 127–136. [CrossRef]
16. Umeno, T.; Hanzawa, M.; Hayashi, Y.; Hori, M. Development of New Lean NO_x Trap Technology with High Sulfur Resistance. *SAE Tech. Pap.* **2014**. [CrossRef]
17. Le Phuc, N.; Corbos, E.C.; Courtois, X.; Can, F.; Marecot, P.; Duprez, D. NO_x storage and reduction properties of Pt/$CexZr_{1-x}O_2$ mixed oxides: Sulfur resistance and regeneration, and ammonia formation. *Appl. Catal. B Environ.* **2009**, *93*, 12–21. [CrossRef]
18. Darr, S.T.; Choksi, R.A.; Hubbard, C.P.; Johnson, M.D.; Mccabe, R.W.; Co, F.M. Effects of Oil-Derived Contaminants on Emissions from TWC-Equipped Vehicles. *SAE Tech. Pap.* **2000**. [CrossRef]
19. Sumida, H.; Koda, Y.; Sadai, A.; Ichikawa, S.; Kyogoku, M.; Takato, M.; Miwa, Y. Analysis of Phosphorus Poisoning on Exhaust Catalysts from Compact-Class Vehicle. *SAE Tech. Pap.* **2004**. [CrossRef]
20. Rokosz, M.J.; Chen, A.E.; Lowe-Ma, C.K.; Kucherov, A.V.; Benson, D.; Paputa Peck, M.C.; McCabe, R.W. Characterization of phosphorus-poisoned automotive exhaust catalysts. *Appl. Catal. B Environ.* **2001**, *33*, 205–215. [CrossRef]
21. Angove, D.E.; Cant, N.W. Position dependent phenomena during deactivation of three-way catalytic converters on vehicles. *Catal. Today* **2000**, *63*, 371–378. [CrossRef]
22. Ball, D.J.; Mohammed, A.G.; Schmidt, W.A. Application of Accelerated Rapid Aging Test (RAT) Schedules with Poisons: The Effects of Oil Derived Poisons, Thermal Degradation and Catalyst Volume on FTP Emissions. *SAE Tech. Pap.* **1997**. [CrossRef]
23. Guevremont, J.M.; Guinther, G.; Jao, T.; Herlihy, T.; White, R.; Howe, J. Total Phosphorus Detection and Mapping in Catalytic Converters. *Powertrain Fluid Syst. Conf. Exhib.* **2007**. [CrossRef]
24. Cabello Galisteo, F.; López Granados, M.; Martín Alonso, D.; Mariscal, R.; Fierro, J.L.G. Loss of NO_x storage capacity of Pt-Ba/Al_2O_3 catalysts due to incorporation of phosphorous. *Catal. Commun.* **2008**, *9*, 327–332. [CrossRef]
25. Lindholm, A.; Currier, N.W.; Dawody, J.; Hidayat, A.; Li, J.; Yezerets, A.; Olsson, L. The influence of the preparation procedure on the storage and regeneration behavior of Pt and Ba based NO_x storage and reduction catalysts. *Appl. Catal. B Environ.* **2009**, *88*, 240–248. [CrossRef]
26. Sharma, M.; Harold, M.P.; Balakotaiah, V. Analysis of periodic storage and reduction of NO_x in catalytic monoliths. *Ind. Eng. Chem. Res.* **2005**, *44*, 6264–6277. [CrossRef]
27. Lietti, L.; Nova, I.; Forzatti, P. Role of ammonia in the reduction by hydrogen of NO_x stored over Pt-Ba/Al_2O_3 lean NO_x trap catalysts. *J. Catal.* **2008**, *257*, 270–282. [CrossRef]
28. Nova, I.; Castoldi, L.; Lietti, L.; Tronconi, E.; Forzatti, P. How to control the selectivity in the reduction of NO_x with H_2 over Pt-Ba/Al_2O_3 Lean NO_x Trap catalysts. *Top. Catal.* **2007**, *42–43*, 21–25. [CrossRef]
29. Tonkyn, R.G.; Disselkamp, R.S.; Peden, C.H.F. Nitrogen release from a NO_x storage and reduction catalyst. *Catal. Today* **2006**, *114*, 94–101. [CrossRef]
30. Olsson, L.; Fridell, E.; Skoglundh, M.; Andersson, B. Mean field modelling of NO_x storage on Pt/BaO/Al_2O_3. *Catal. Today* **2002**, *73*, 263–270. [CrossRef]
31. Clayton, R.D.; Harold, M.P.; Balakotaiah, V. Selective catalytic reduction of NO by H_2 in O_2 on Pt/BaO/Al_2O_3 monolith NO_x storage catalysts. *Appl. Catal. B Environ.* **2008**, *81*, 161–181. [CrossRef]

32. Choi, J.-S.; Partridge, W.P.; Pihl, J.A.; Kim, M.-Y.; Kočí, P.; Daw, C.S. Spatiotemporal distribution of NO_x storage and impact on NH_3 and N_2O selectivities during lean/rich cycling of a Ba-based lean NO_x trap catalyst. *Catal. Today* **2012**, *184*, 20–26. [CrossRef]
33. Bamwenda, G. The role of the metal during NO_2 reduction by C_3H_6 over alumina and silica-supported catalysts. *J. Mol. Catal. A Chem.* **1997**, *126*, 151–159. [CrossRef]
34. Ji, Y.; Easterling, V.; Graham, U.; Fisk, C.; Crocker, M.; Choi, J.S. Effect of aging on the NO_x storage and regeneration characteristics of fully formulated lean NO_x trap catalysts. *Appl. Catal. B Environ.* **2011**, *103*, 413–427. [CrossRef]
35. Moulder, J.F.; Stickle, W.; Sobol, P.E.; Bomben, K.D. *Handbook of X-ray Photoelectron Spectroscopy*; Perkin Elmer Corporation: Eden Prairie, MN, USA, 1992.
36. Andonova, S.; Vovk, E.; Sjöblom, J.; Ozensoy, E.; Olsson, L. Chemical deactivation by phosphorous under lean hydrothermal conditions over Cu/BEA NH_3-SCR catalysts. *Appl. Catal. B Environ.* **2014**, *147*, 251–263. [CrossRef]
37. Shwan, S.; Jansson, J.; Olsson, L.; Skoglundh, M. Chemical deactivation of Fe-BEA as NH_3-SCR catalyst-Effect of phosphorous. *Appl. Catal. B Environ.* **2014**, *147*, 111–123. [CrossRef]
38. Andersson, J.; Antonsson, M.; Eurenius, L.; Olsson, E.; Skoglundh, M. Deactivation of diesel oxidation catalysts: Vehicle- and synthetic aging correlations. *Appl. Catal. B Environ.* **2007**, *72*, 71–81. [CrossRef]
39. Beck, D.; Monroe, D.; Di Maggio, C.; Sommers, J. Impact of Oil-Derived Catalyst Poisons on FTP Performance of LEV Catalyst Systems. *SAE Trans.* **1997**. [CrossRef]
40. Yamashita, T.; Hayes, P. Analysis of XPS spectra of Fe^{2+} and Fe^{3+} ions in oxide materials. *Appl. Surf. Sci.* **2008**, *254*, 2441–2449. [CrossRef]

Article

"PdO vs. PtO"—The Influence of PGM Oxide Promotion of Co_3O_4 Spinel on Direct NO Decomposition Activity

Gunugunuri K. Reddy *, Torin C. Peck and Charles A. Roberts *

Toyota Research Institute of North America, Ann Arbor, MI 48105, USA; torin.peck@toyota.com
* Correspondence: krishna.gunugunuri@toyota.com (G.K.R.); charles.roberts@toyota.com (C.A.R.)

Received: 17 November 2018; Accepted: 5 January 2019; Published: 9 January 2019

Abstract: Direct decomposition of NO into N_2 and O_2 ($2NO \rightarrow N_2 + O_2$) is recognized as the "ideal" reaction for NOx removal because it needs no reductant. It was reported that the spinel Co_3O_4 is one of the most active single-element oxide catalysts for NO decomposition at higher reaction temperatures, however, activity remains low below 650 °C. The present study aims to investigate new promoters for Co_3O_4, specifically PdO vs. PtO. Interestingly, the PdO promoter effect on Co_3O_4 was much greater than the PtO effect, yielding a 4 times higher activity for direct NO decomposition at 650 °C. Also, Co_3O_4 catalysts with the PdO promoter exhibit higher selectivity to N_2 compared to PtO/Co_3O_4 catalysts. Several characterization measurements, including X-ray diffraction (XRD), X-ray photoelectron spectroscopy (XPS), H_2-temperature programmed reduction (H_2-TPR), and *in situ* FT-IR, were performed to understand the effect of PdO vs. PtO on the properties of Co_3O_4. Structural and surface analysis measurements show that impregnation of PdO on Co_3O_4 leads to a greater ease of reduction of the catalysts and an increased thermal stability of surface adsorbed NOx species, which contribute to promotion of direct NO decomposition activity. In contrast, rather than remaining solely as a surface species, PtO enters the Co_3O_4 structure, and it promotes neither redox properties nor NO adsorption properties of Co_3O_4, resulting in a diminished promotional effect compared to PdO.

Keywords: direct NO decomposition; PGM oxide promotion; PdO vs. PtO; in-situ FT-IR; NO adsorption properties; redox properties

---

1. Introduction

Nitrogen oxides (NOx) formed by combustion from fixed and mobile sources cause severe detrimental environmental problems, such as acid rain and photochemical smog [1–3]. Effectively controlling the emission of NOx is the topic of much research and has led to the introduction of many new catalyst technologies, such as three-way catalysts (TWC), NOx storage-reduction (NSR), and selective catalytic reduction (SCR) for NOx gas removal from mobile sources, and SCR and selective non-catalytic reduction (SNCR) for NOx gas removal from fixed sources [4–6]. Among various deNOx strategies, direct decomposition of NO ($NO \rightarrow 1/2O_2 + 1/2N_2$) has been considered to be the most desirable method because this reaction is thermodynamically favorable at low temperatures and does not need any reductants, such as NH_3, H_2, CO, or hydrocarbons. However, kinetic studies have indicated that the reaction needs to overcome a large activation energy (~335 kJ mol^{-1}) barrier [4–15]. Accordingly, there is an apparent need for a suitable catalyst to decompose NOx at a given temperature, and therefore, significant research has been undertaken towards development of active and stable catalysts.

Since the pioneering work of Jellinek on the catalytic decomposition of NO in 1906, much research has been reported on NO direct decomposition over several materials, including perovskites, rare earth oxides, and Cu-zeolites [2–7]. Numerous metal oxides have also been examined as candidates for NO decomposition catalysts [16] and Co_3O_4 is often recognized as a significant component in many active catalysts at higher reaction temperatures [17–21]. However, Haneda et al. recently reported that NO decomposition takes place slowly, if at all, over pure Co_3O_4 at temperatures below 650 °C [18]. They reported that the presence of small amounts of alkali metals were essential to activate NO decomposition over Co_3O_4 oxide by enhancing NO adsorption [18–20]. This interesting effect of alkali metals, particularly Na, was also reported by Kung et al. [21], but dependence on alkali metals is not feasible for practical applications due to their volatile nature at temperatures above 600 °C.

Metal oxide supported platinum group metals (PGM metals) were also one of the earliest types of NO decomposition catalysts studied, and the results have been widely reported; mainly Au, Pt, Pd, and Ir at temperatures higher than 700 °C [22,23]. Suzuki et al. [24] synthesized a porous $CaZrO_3$/MgO/Pt composite and found that this catalyst could obtain a NO conversion rate of about 52% at 900 °C in the absence of O_2. Haneda et al. [25] found that the addition of Pt improved the direct NO decomposition performance of rare earth oxides. They [26,27] also compared the activity of $[Pd(NH_3)_4]$ $(NO_3)_2$, $Pd(NO_3)_2$, $Pd(CH_3COO)_2$, and $(NH_4)_2$-$[PdCl_4]$ as palladium precursors for NO decomposition in a Pd/Al_2O_3 catalyst at 700 °C, and the activity was found to decrease in the order of $Pd(NO_3)_2 > [Pd(NH_3)_4]$ $(NO_3)_2 > Pd(CH_3COO)_2 >> (NH_4)_2[PdCl_4]$. Almusaiteer et al. [28] reported that compared to Pd/Al_2O_3, the Pd/C (activated carbon) catalyst was found to be more beneficial for O_2 desorption, but both have similar activity. Oliveira et al. [29] investigated the catalytic performance of palladium and copper catalysts loaded in mordenite (MOR) and found that these catalysts were more active for NO decomposition than alumina supported catalysts. However, the reports on supported PtO catalysts for direct NO decomposition at temperatures below 700 °C are very limited in the literature. Similarly, metallic Pd catalysts always deactivate over time due to oxidation of Pd metal to PdO at temperatures below 650 °C [30].

To the best of our knowledge, Co_3O_4 supported PGM catalysts have never been explored for direct NO decomposition, likely due to the inactivity of the individual components at lower temperatures. However, the need for enhanced NO adsorption on Co_3O_4 suggests that the addition of PGM promotion can lead to increased low temperature activity. Hence, the present study aims to investigate the promotional effect of PdO vs. PtO on the Co_3O_4 for direct NO decomposition. The activity measurements show that the optimum PdO/Co_3O_4 catalyst exhibits 4 times higher activity than PtO/Co_3O_4 catalysts. Several characterization techniques such as X-ray diffraction (XRD), X-ray photoelectron spectroscopy (XPS), H_2-Temperature programmed reduction (H_2-TPR), and *in situ* FT-IR are employed to understand the influence of PdO and PtO on the structural and surface properties of Co_3O_4.

2. Results and Discussion

2.1. Direct NO Decomposition Activity Measurements

To qualify as direct NO decomposition, the catalyst must decompose NO into just two products: N_2 and O_2. The possibility for unwanted N_2O and NO_2 formation as side products cannot be neglected during the analysis of this reaction. Therefore, high NO conversion is desired, but not sufficient; it is also very important to maximize selectivity towards N_2 rather than N_2O or NO_2. Considering all possible products, the reaction can be written as:

$$2\,NO \rightarrow N_2 + O_2$$
$$4\,NO \rightarrow 2N_2O + O_2$$
$$NO + [O] \rightarrow NO_2 \text{ ([O]: catalyst lattice oxygen)}$$

The selectivity to N_2 can be defined as:

$$N_2 \text{ selectivity } (\%) = 2 \times [N_2]/(2 \times [N_2] + 2[N_2O] + [NO_2])$$

Also, when NO dissociates on the catalyst surface, it is possible for N_2 production to occur, but without simultaneous release of the stoichiometric amount of the O_2 product. This situation can occur quite frequently and is the topic of a previous study from some of us [30]. Rather than desorb as the O_2 product, oxygen atoms either remain strongly adsorbed on the surface or chemically react with catalyst material, oxidizing both surface and bulk, and changing the catalyst composition. In either case, the catalyst deactivates over time. Hence, it is very important to confirm that the catalyst releases both N_2 and O_2 as products.

Direct NO decomposition measurements were performed over the pure spinel oxide Co_3O_4 and over the PdO- and PtO-promoted Co_3O_4 catalysts, denoted PdO/Co_3O_4 and PtO/Co_3O_4, respectively. Activity was measured for 2 h at 400, 450, 550, and 650 °C. The direct NO decomposition activity to N_2 of PdO/Co_3O_4 and PtO/Co_3O_4 with varying PGM loading is presented in Figure 1 as a function of temperature and compared to the pure Co_3O_4 spinel. All numeric values of NO conversion and N_2, N_2O, NO_2 ppm product concentrations of the Co_3O_4, PdO/Co_3O_4, and PtO/Co_3O_4 catalysts at various reaction temperatures are presented in Tables S1 and S2. The raw NO conversion profiles (NO converted to all the products) of Co_3O_4, $3PdO/Co_3O_4$, and 4 PtO/Co_3O_4 during the steady state direct NO decomposition obtained from FT-IR detector in the temperature region 400 to 650 °C are presented in Figure S1. High values for NO conversion at lower temperatures may seem counterintuitive, however, most of the NO conversion at low temperature is simply due to thermodynamically favorable NO oxidation to NO_2. For example, as shown in Table S1, the NO conversion values of Co_3O_4, 3 PdO/Co_3O_4, are 3 PtO/Co_3O_4 are 3.15, 2.92, and 7.66, respectively, giving the impression that 3 PtO/Co_3O_4 is the most active catalyst at 400 °C. However, for 3 PtO/Co_3O_4, the NO conversion specifically to N_2 is lower than that of 3 PdO/Co_3O_4 because most of the NO conversion is due to NO oxidation. Hence, for NO decomposition to N_2, it is more instructive to consider NO converted into N_2 rather than total NO conversion. The activity values were calculated in this manner and are presented in Figure 1 in units of micromoles of NO converted to N_2 per gram per second.

NO and Ar partial pressure values obtained by mass spectrometry (MS) during the steady state measurements are presented in Figure S2 and compared with data obtained in the absence of the catalyst, which serves as a baseline. Inert Ar gas was introduced as a tracer to monitor for potential systematic variation in signal intensity during the experiment. As shown in Figure S2, no change in Ar signal intensity was observed during the experiment, however, the intensities of NO signal changed based on catalyst identity and reaction temperature. These measurements confirm the change in the NO signal is due to catalytic conversion of NO and not due to artifacts. Also, the MS signal for NO (m/z = 30) tracks with the conversion reported by the FT-IR measurements (Figure S1). Figures S3–S5 present the MS partial pressure values of the N_2, N_2O, and NO_2 products, respectively, during direct NO decomposition and are also compared to the MS partial pressures obtained in the absence of the catalyst. As shown in Figure S3, the N_2 and O_2 signal intensities are higher compared to the background signals, which confirms the simultaneous release of N_2 and O_2 as expected for NO decomposition. Furthermore, the MS intensity of the N_2 signal qualitatively tracks with the N_2 concentrations calculated by nitrogen mass balance from the FTIR measurements (Tables S1 and S2), lending additional confidence in the activity results. Similarly, good correlation between the FT-IR detection of N_2O and NO_2 and the MS signals was observed (compare Tables S1 and S2 to Figures S3 and S4). Finally, the release of oxygen as a product (Figure S3) and stable NO conversion (Figures S1 and S2) during the steady state measurements suggest that the catalysts were not poisoned by the irreversible chemical adsorption of oxygen. Thus, the good correlation between the FT-IR data and MS signals suggests that the calculation of N_2 production from the FT-IR is reliable and can be used for the calculation of activity from NO conversion to N_2. As shown in Figure 1b, NO decomposition activity to N_2 increases slightly with temperature up to 550 °C for the pure Co_3O_4 catalyst. Further increase in the temperature to 650 °C results in decreased activity. This result suggests that the Co_3O_4 spinel is not a good catalyst for NO decomposition at temperatures below 650 °C, as indicated by Hamada et al. [21]. The addition of PdO and PtO to the Co_3O_4 spinel improves the direct NOx decomposition activity of

Co_3O_4. Direct NOx activity increases with temperature for all palladium and platinum loadings, but unlike the pure Co_3O_4, deactivation was not observed above 550 °C for the any of PdO or PtO catalysts.

The direct NO decomposition activity of PdO/Co_3O_4 and PtO/Co_3O_4 to N_2 at various temperatures is presented as a function of the weight percent loading of the respective PGMs in Figure 2. For PdO/Co_3O_4, activity increases with palladium loading up to 3 wt%, but further increase in the palladium loading leads to decreased activity (Figure 2a). For PtO/Co_3O_4, activity increases with PtO loading up to 4 wt%, however, the overall effect on activity is significantly diminished compared to PdO/Co_3O_4 (Figure 2b). The optimum loading of each PGM was found to be 3 wt% PdO/Co_3O_4 and 4wt% PtO/Co_3O_4. Interestingly, PdO-promoted catalysts exhibit higher activity than the PtO-promoted catalysts. At 650 °C, the optimum PdO/Co_3O_4 catalyst exhibits 4 times higher activity compared to the optimum PtO/Co_3O_4 catalyst.

To confirm the reaction is indeed direct NO decomposition to N_2 rather than the unwanted production of N_2O or NO_2, the selectivity to N_2 was calculated. The N_2 selectivity is presented as a function of PGM loading from 400–650 °C for PdO/Co_3O_4 and PtO/Co_3O_4 in Figure 3a,b, respectively. As expected, pure Co_3O_4 (0 wt% PGM loading) exhibited relatively low selectivity to N_2 ($\leq$20%) at 400 and 450 °C. The N_2 selectivity increased to 80% at 550 °C and to 100% at 650 °C. Formation of N_2O was not observed and only N_2 and NO_2 products are detected during direct NO decomposition over the PtO- and PdO-promoted Co_3O_4 catalysts (Tables S1 and S2). Thus, the product distribution measurements suggest NO oxidation (NO_2 formation) is more favorable at lower reaction temperatures ($\leq$450 °C), and at higher reaction temperatures, NO decomposition (N_2 formation) is predominant. Remarkably, the addition of 1 wt% PdO to the Co_3O_4 improves the N_2 selectivity from 1 to 40% at 400 °C and from 21 to 70% at 450 °C. The N_2 selectivity further increased to 50% at 400 °C and 75% at 450 °C with increasing PdO loading to 3 wt%. Increasing the PdO loading from 3 wt% to 4 wt% lead to only a slight decrease in the N_2 selectivity. When also considering the activity measurement in Figure 2, the N_2 selectivity measurements confirm 3 wt% PdO as the optimum loading on Co_3O_4 for direct NO decomposition.

Regarding N_2 selectivity as a function of PtO loading (Figure 3b), the addition of 1 wt% PtO yielded less improvement at 400 °C (1 to 16%) compared to the addition of 1 wt% PdO (1 to 40%). Furthermore, the N_2 selectivity decreases with increasing PtO loading, dropping to 9% for the 4wt% PtO/Co_3O_4 catalyst. Moreover, there is no improvement in the selectivity observed at reaction temperatures at and above 450 °C. Therefore, the direct NO decomposition measurements show that the addition of PdO to Co_3O_4 improves the decomposition activity and N_2 selectivity and 3wt% is the optimum Pd loading over Co_3O_4, whereas PtO loaded on Co_3O_4 leads to only slight improvement in the activity and almost no influence on the N_2 selectivity.

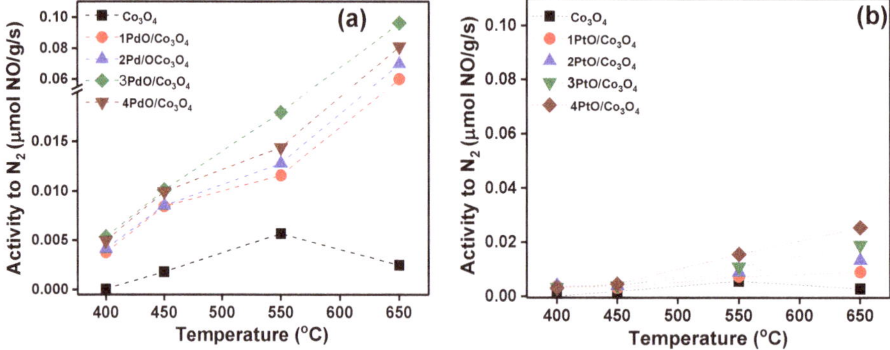

Figure 1. Direct NO decomposition activity to N_2 as a function of reaction temperature for (**a**) PdO- and (**b**) PtO-promoted Co_3O_4 catalysts with varying PGM loading. The pure Co_3O_4 support is included for comparison. (Gas hourly space velocity (GHSV) = 2100 h^{-1}, 1% NO/He).

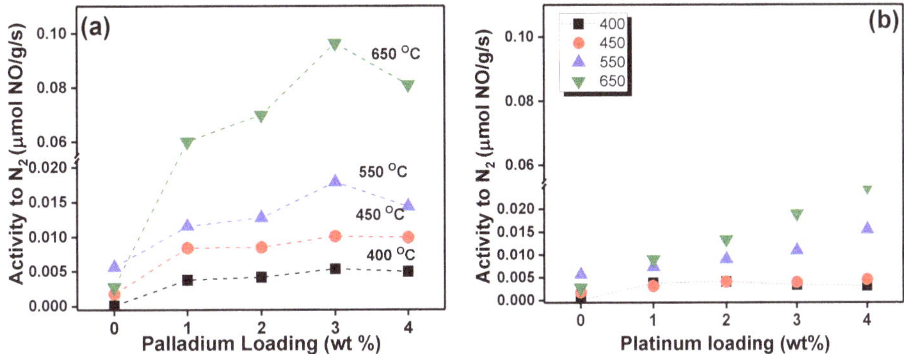

Figure 2. Direct NO decomposition activity to N_2 as a function of Pd and Pt loading for (**a**) PdO- and (**b**) PtO-promoted Co_3O_4 catalysts at varying temperature. The pure Co_3O_4 support is included for comparison. (GHSV = 2100 h^{-1}, 1% NO/He).

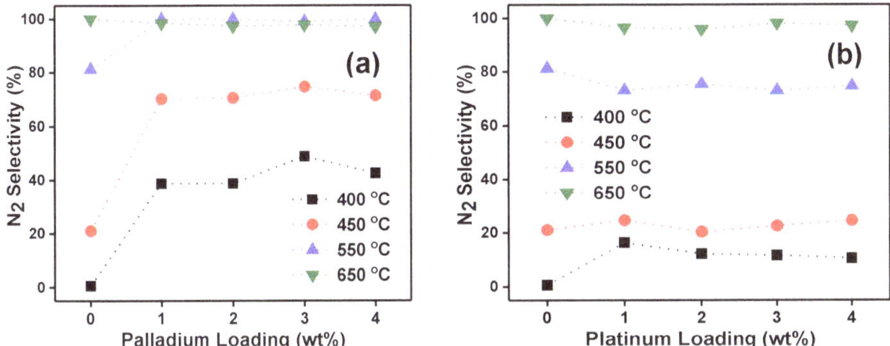

Figure 3. N_2 selectivity as a function of PdO and PtO loading for (**a**) PdO- and (**b**) PtO-promoted Co_3O_4 catalysts at varying temperature. The pure Co_3O_4 support is included for comparison. (GHSV = 2100 h^{-1}, 1% NO/He).

2.2. Catalyst Characterization

2.2.1. Structural and Textural Properties

These catalysts have been evaluated using several characterization techniques, like XRD, XPS, H_2-TPR, BET surface area, and *in-situ* FT-IR during NO adsorption, to understand the influence of PdO and PtO on the structural and surface properties of Co_3O_4 and to explain the greater promoter effect of PdO on Co_3O_4 compared to PtO during NO decomposition. The palladium and platinum loadings of the studied catalysts prepared by impregnation were verified with XRF spectrometry. For the nominal 1.0, 2.0, 3.0, and 4.0 wt% of PdO, the experimental values were 0.83, 1.94, 2.80, and 4.15, respectively (Table 1). The experimental values for PtO doped Co_3O_4 catalysts were 0.93, 2.12, 3.23, and 4.02, respectively. Differences may be due to surface heterogeneity or incomplete precursor dispersion during the impregnation procedure or the inherent uncertainty related to the employed XRF method, which did not utilize a standard material to aid the data analysis.

The BET surface area values of PdO/Co_3O_4 and PtO/Co_3O_4 catalysts and the pure Co_3O_4 are presented in Table 1. The pure Co_3O_4 catalyst exhibits a BET surface area of 36 m^2/g. Little change in the surface area is observed after impregnating Co_3O_4 with 1, 2, and 3 wt% Pd. However, increasing Pd loading from 3 to 4 wt% on to Co_3O_4 lead to a decrease in the surface area from 33 to 26 m^2/g. These values suggest that PdO dispersed very well on the surface of Co_3O_4 until 3wt% and further increase in the loading to 4wt% likely leads to surface agglomeration, which can block access to the active surface. The BET surface area measurements are corroborated by the activity measurements, which showed that the activity of Co_3O_4 increases with increasing palladium doping only until 3 wt%. Further increase in the Pd loading to 4 wt% lead to a decrease in the activity. Little change in surface area was observed for the Co_3O_4 with platinum impregnation (Table 1), as only a slight decrease in the surface area was observed at the highest loading. As suggested above, the blockage of the active surface may be the cause of the decrease in the activity for the 4 wt% PdO/Co_3O_4 catalyst. It is hypothesized that the formation of PdO crystallites is responsible for this behavior, and this hypothesis will be investigated below.

Table 1. Co/M (M = Pd, Pt) and BET surface area values of PdO/Co_3O_4 and PtO/Co_3O_4 catalysts.

Catalyst Loading	PGM Loading (wt%) *		BET Surface Area (m^2/g)	
	Pd wt%	Pt wt%	PdO/Co_3O_4	PtO/Co_3O_4
0	-	-	36	36
1	0.83	0.93	36	35
2	1.94	2.12	35	39
3	2.8	3.23	33	34
4	4.15	4.02	26	33

* As measured by XRF.

The X-ray diffraction (XRD) patterns of the fresh PdO/Co_3O_4 and PtO/Co_3O_4 catalysts are shown in Figure 4a,b, with the pattern of the pure Co_3O_4 for reference. The X-ray diffraction lines characteristic of the cubic cobalt spinel structure were indexed within the Fd3m space group (JCPDS card no. 01-080-1533) in the case of the pure Co_3O_4 catalyst [31]. As shown in Figure 4a, all the PdO/Co_3O_4 catalysts exhibit peaks due to Co_3O_4. The diffractograms provide evidence that the spinel structure was preserved after Pd impregnation, revealing no observable structural changes compared to the pure Co_3O_4 carrier. Diffraction peaks related to Pd or PdO were not detected on samples up to a nominal Pd loading of 3 wt%. For the 4 wt% Pd sample, a low-intensity diffraction peak indicative of tetragonal PdO (0 0 2) (JCPDS card no. 75-584) was visible at 2θ of 33.6° [32]. These measurements suggest that PdO is well-dispersed on the Co_3O_4 support up to 3 wt%, and above this loading, crystalline PdO forms on Co_3O_4 surface. As stated above, the BET surface decreases from 33 to 26 m^2/g with increasing Pd loading from 3 wt% to 4 wt%. XRD measurements confirm that the decrease in the surface area is due to the formation of crystalline PdO on the surface of Co_3O_4 and blocking of the active surface. The X-ray diffraction patterns of PtO/Co_3O_4 catalysts are shown in Figure 4b and only peaks due to Co_3O_4 are present at all Pt loadings, suggesting the absence of bulk metallic Pt or PtO with long-range order. However, unlike PdO/Co_3O_4, the XRD peaks of the PtO/Co_3O_4 samples were shifted to higher values relative to the pure Co_3O_4 spinel for all Pt loadings. The shift in the peak position to higher values indicates that Pt is likely incorporating into the Co_3O_4 spinel structure in contrast to PdO/Co_3O_4, where the absence of the peak shift indicates that Pd remained dispersed on the Co_3O_4 surface.

Figure 5a,b display X-ray diffraction patterns of the spent PdO/Co_3O_4 and PtO/Co_3O_4 catalysts after direct NO decomposition. As shown in Figure 5a, the spent Co_3O_4 exhibits only peaks due to spinel structure. There are no peaks due to either CoO or metallic Co, suggesting the Co_3O_4 spinel is structurally stable during direct NO decomposition. In addition to the Co_3O_4 spinel peaks, the spent PdO/Co_3O_4 catalysts with 1, 2, and 3 wt% Pd loading also exhibit peaks at 2θ values of 40.3, 46.79, and 68.4°. These peaks are due to the (1 1 1), (2 0 0), and (2 2 0) facets of Pd metal (JCPDS no: 46-1043). The XRD measurements of the spent catalysts show that the dispersed PdO reduced to Pd metal during direct NO decomposition. Similar to the XRD pattern for the fresh 4 wt% PdO/Co_3O_4, the spent XRD pattern also exhibits peaks due to PdO along with the Pd metal and Co_3O_4 peak, which suggests that the crystalline PdO remains even after direct NO decomposition. The X-ray diffraction patterns of the spent PtO/Co_3O_4 catalysts after direct NO decomposition are displayed in Figure 5b. The spent

PtO/Co_3O_4 catalysts exhibit peaks due only to Co_3O_4 after direct NO decomposition. In contrast to the metallic phase observed in the spent PdO/Co_3O_4 catalysts, there are no peaks due to metallic Pt observed in the spent PtO/Co_3O_4 catalysts.

The XRD measurements show that in the case of PdO/Co_3O_4 catalysts, the PdO reduced to metallic Pd during direct NO decomposition and promotes the activity of Co_3O_4 catalysts. On the other hand, no metallic Pt formation occurred in the PtO/Co_3O_4 catalysts, leading to a greatly diminished promoter effect compared to the PdO/Co_3O_4 catalysts. Also, the NO decomposition measurements show that the catalytic activity decreases with increasing Pd loading from 3 wt% to 4 wt%. The formation of crystalline PdO and decrease in the surface area (blocking of the active surface) explains the lower activity of 4 wt% sample compared to 3 wt% sample. Hence, XRD and BET surface area measurements corroborate with the activity measurements. The spent PtO/Co_3O_4 catalysts also exhibit a shift in the peak positions to higher 2θ values compared to the pure Co_3O_4 catalyst, which suggests the incorporation of Pt into the Co_3O_4 spinel structure even after direct NO decomposition.

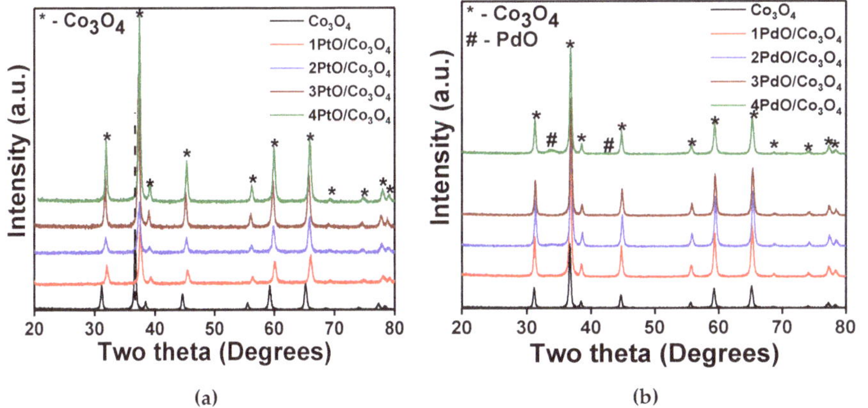

Figure 4. X-ray diffraction patterns of fresh (**a**) PdO- and (**b**) PtO-promoted Co_3O_4 catalysts. The pattern for the fresh pure Co_3O_4 support is included for reference.

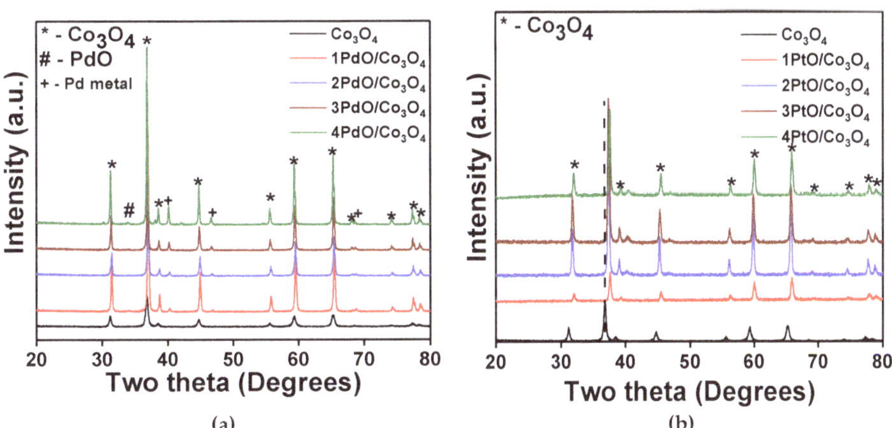

Figure 5. X-ray diffraction patterns of spent (**a**) PdO- and (**b**) PtO-promoted Co_3O_4 catalysts. The pattern for the spent pure Co_3O_4 support is included for reference.

2.2.2. Redox Properties

The influence of PdO and PtO on the redox properties of Co_3O_4 are investigated using H_2-temperature programmed reduction (H_2-TPR) measurements. The H_2-TPR profiles of PdO/Co_3O_4 and PtO/Co_3O_4 catalysts are presented in Figure 6a,b, along with that of the pure Co_3O_4 for comparison. Several authors reported that the reduction behavior of Co_3O_4 is strongly dependent on the preparation method, catalyst composition, and dispersion on a support [33,34]. The reduction behavior of Co_3O_4 was widely accepted as a stepwise process, including the reduction of Co^{3+} to Co^{2+} and Co^{2+} to metallic Co. There are three well-defined reduction peaks in the TPR profile of Co_3O_4 (Figure 6). The peak at 235 °C is attributed to the reduction of surface oxygen species. The other two peaks are for the stepwise reduction of Co_3O_4 to metallic cobalt. According to the literature, the second reduction peak centered at 275 °C is due to the reduction of Co_3O_4 to CoO, and the third peak at the region of 305 °C is due to the reduction of CoO to metallic cobalt [33,34].

$$Co_3O_4 + H_2 \rightarrow 3CoO + H_2O$$

$$CoO + H_2 \rightarrow Co + H_2O$$

The addition of 1 wt% PdO to the Co_3O_4 leads to a drastic change in the redox profile of Co_3O_4 (Figure 6). No peaks were observed in the 250–310 °C temperature region. Both PdO and Co_3O_4 were reduced at much lower temperature and all reduction events completed below 150 °C. These measurements show that PdO promotes the reduction of Co_3O_4. Two reduction peaks were observed in the 1 wt% PdO/Co_3O_4 H_2-TPR profile at 79 and 104 °C. The first reduction peak at 79 °C is due to the reduction of PdO to metallic Pd and the second reduction peak is due to the reduction of Co_3O_4. The H_2-TPR profiles for 1, 2, and 3 wt% PdO/Co_3O_4 were all very similar (Figure 6). The promotion of Co_3O_4 reduction by Pd observed in H_2-TPR is possibly ascribed to hydrogen spillover and the synergistic effect between Pd species and Co_3O_4. The synergistic effect can weaken the Co-O bond. Chen et al. [35] also reported a similar promotional effect for PdO impregnated on Co_3O_4 catalysts with different morphologies, and the synergistic effect between Pd and Co existed, regardless of Co_3O_4 morphology. In the present study, the intensity of the first reduction peak increases with increasing PdO loading from 1 to 3 wt%, and the increase is accompanied by a slight shift in the peak temperature from 79 to 85 °C. This may be due to the increase in the PdO loading on the Co_3O_4 surface. The reduction profile of 4 wt% PdO/Co_3O_4 is slightly different from the PdO promoted catalysts of lower loading. Along with the peaks due to Co_3O_4 and PdO, a small additional peak is observed at 220 °C. Given the identification of crystalline PdO in the XRD pattern of the 4 wt% PdO/Co_3O_4, it is reasonable to assign this peak to reduction of crystalline PdO.

The H_2-TPR profiles of PtO/Co_3O_4 catalysts are presented in Figure 6b. Two types of reduction features were observed in the case of PtO promoted Co_3O_4 catalysts, one from 130 to 190 °C and another from 200 to 325 °C. The first feature corresponds to reduction of PtO to metallic Pt and the second is reduction of Co_3O_4. Unlike PdO/Co_3O_4, little to no shift in the Co_3O_4 reduction temperature of PtO/Co_3O_4 catalysts was observed relative to the pure Co_3O_4. The reduction of Co^{3+} to Co^{2+} occurred in the 260–275 °C temperature region for Co_3O_4 and PtO/Co_3O_4 catalysts, irrespective of PtO loading, and the reduction of PtO occurred separately at a distinctly lower temperature. Yang et al. [36] observed similar behavior for Pt promoted Co_3O_4/Al_2O_3 catalysts, wherein both PtO and Co_3O_4 reduced separately in distinct temperature regions. Even though the Co_3O_4 reduction shifted to slightly lower temperatures at higher Pt loadings in their study, a synergistic effect, similar to Pd and Co in PdO/Co_3O_4, was not observed between Pt and Co. This lack of synergistic effect by Pt on the reduction of Co_3O_4 is consistent with the smaller promotional effect of Pt on direct NO decomposition activity compared to Pd promotion. Conversely, the decreased reduction temperature of Co_3O_4 observed in H_2-TPR measurements of PdO/Co_3O_4 illustrates how Pd can promote direct NO decomposition by enhancing the reducibility of the catalyst.

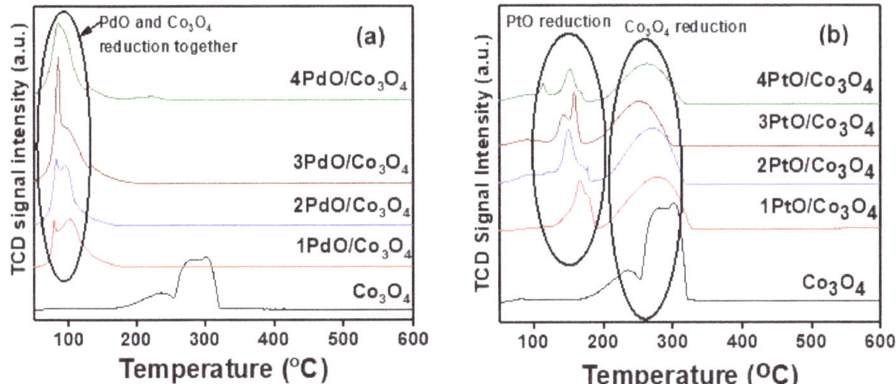

Figure 6. H_2- Temperature programmed reduction profiles of fresh (**a**) PdO- and (**b**) PtO-promoted Co_3O_4 catalysts. The pure Co_3O_4 profile is shown for comparison.

2.2.3. Surface Properties

The X-ray photoelectron spectroscopy (XPS) was used to investigate the surface elemental compositions, metal oxidation states, and adsorbed oxygen species of the as-prepared and spent samples. The *O1s* XPS spectra of fresh PdO- and PtO-promoted Co_3O_4 catalysts are presented in Figure 7, with that of the pure Co_3O_4 for comparison. The pure Co_3O_4 exhibits two peaks in the *O1s* spectra. The large peak at lower binding energy (BE = 530.2–530.7 eV) is attributed to the surface lattice oxygen in Co_3O_4 (denoted as O_{lat}) [37]. The shoulder at higher BE (532.0–532.7 eV) is associated with oxygen atoms present as surface adsorbed oxygen or surface hydroxyl groups or defect oxide (denoted as O_{ad}). The PdO- and PtO-promoted Co_3O_4 samples also exhibit two peaks in their *O1s* spectra due to the O_{lat} and O_{ad} species., however, little difference in the peak energies is observed. This is may be due to lower loadings of promoters.

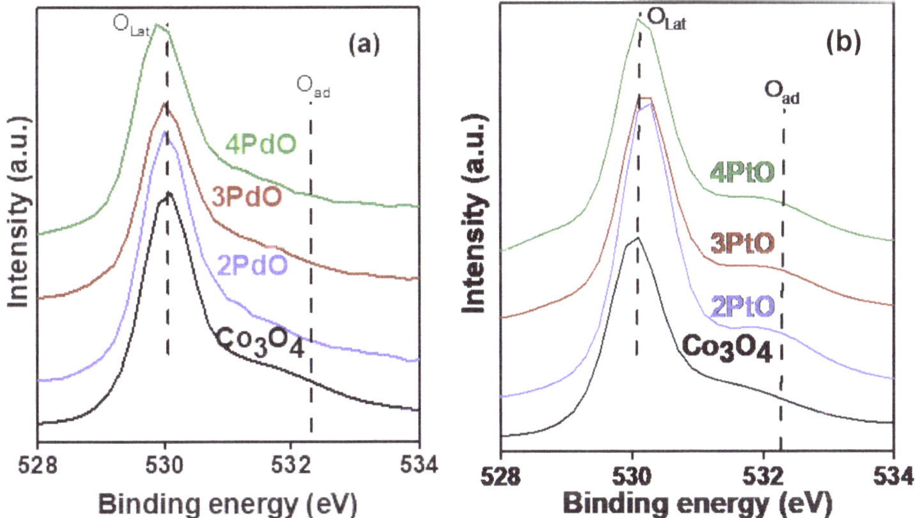

Figure 7. *O1s* XPS spectra of fresh (**a**) PdO- and (**b**) PtO-promoted Co_3O_4 catalysts. Co_3O_4 spectrum is shown for reference.

The fitted Co2*p* XPS spectra of the fresh PdO- and PtO-promoted Co_3O_4 catalysts are presented in Figure 8, with that of the pure Co_3O_4 for comparison. In the pure Co_3O_4 XPS spectrum, the main peak in the BE range of 780.7–782.2 eV is assigned to Co2$p_{3/2}$, and the shoulder at 795.9–797.9 eV is attributed to Co2$p_{1/2}$. Pure Co_3O_4 exhibits peaks due to both Co^{3+} and Co^{2+} and their satellites. The main Co2$p_{3/2}$ feature can be further resolved into two components, with BE values centered at 778.7–780.4 eV and 779.8–781.6 eV, and corresponding to Co^{3+} and Co^{2+}, respectively [38]. Furthermore, the presence of the satellite peaks also confirms the presence of Co^{2+} in the catalysts. As expected for samples containing Co_3O_4 spinel, all catalysts exhibited peaks and satellites due to both Co^{3+} and Co^{2+}, irrespective of Pd or Pt promoter loading. Also, no significant change in the position of the peaks was observed upon impregnation of Co_3O_4 with PdO or PtO. The spent catalysts also exhibit peaks due to the Co^{3+} and Co^{2+} ions, irrespective of promoter identity or loading. The Co2*p* XPS measurements show that the Co_3O_4 spinel is very stable during direct NO decomposition, which agrees with XRD measurements.

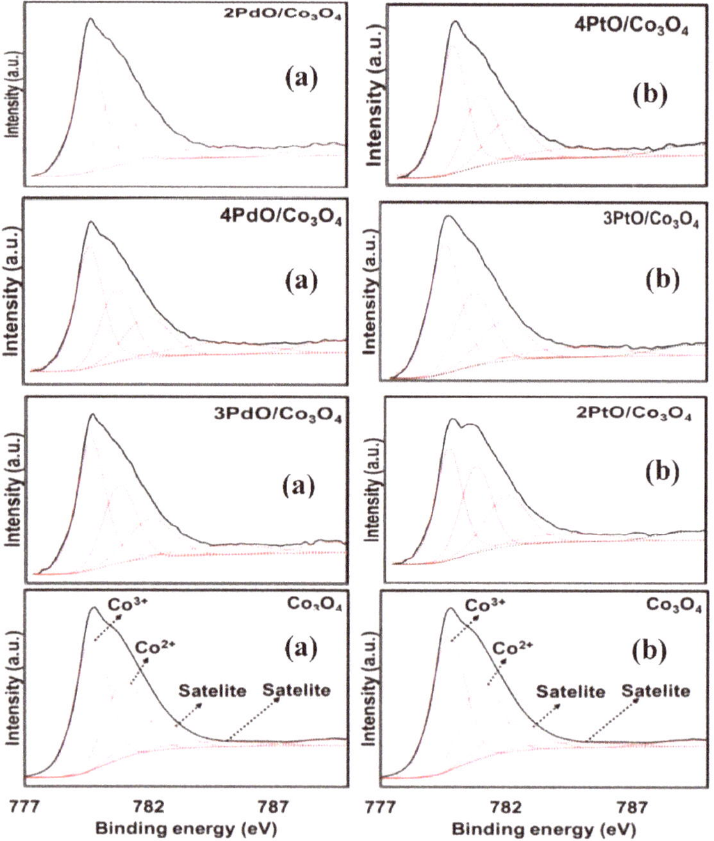

Figure 8. Co2*p* XPS profiles of fresh (**a**) PdO- and (**b**) PtO-promoted Co_3O_4. The pure Co_3O_4 spectra is presented for reference.

The Pd3*d* XPS spectra of the fresh and spent 2, 3, and 4 wt% PdO/Co_3O_4 are presented in Figure 9a,b. In general, Pd may exist as Pd^0 (335.1–335.4 eV [39]), Pd^{2+} (336.8.1–337.2 eV or 336.3–336.8 eV [40–44]), Pd^{4+} (337.8–339.3 eV), or a combination thereof. All the PdO/Co_3O_4 catalysts exhibit peaks due to the Pd^{2+} and Pd^{4+} after calcination at all PdO loadings. However,

XRD measurements show no peaks corresponding to PdO or PdO_2 up to 3 wt% loading, which indicates that the PdO present on Co_3O_4 is in amorphous form and dispersed very well on the surface. In agreement with the above XRD analysis of the spent samples (see Figure 6a), XPS indicates PdO/Co_3O_4 catalysts exhibit peaks due to PdO and metallic Pd after direct NOx decomposition. These results suggest that some of the PdO reduced to metallic Pd during direct NOx decomposition, which corroborates the evidence from H_2-TPR and XRD of the promotional effect of Pd on the activity of Co_3O_4 spinel catalysts. The intensity of the metallic Pd increases with increasing PdO loading from 2 to 3 wt%, however, the intensity of the metallic Pd peak decreases drastically with further PdO loading from 3 to 4 wt%. This is due to the formation of the separate bulk PdO phase in the spent 4 wt% PdO/Co_3O_4 sample, which is clearly observed in the spent 4 wt% PdO/Co_3O_4 Pd3*d* spectrum. The formation of a separate PdO phase leads to less reduction of PdO to metallic Pd during direct NOx decomposition and is the likely cause of the lower activity compared to the 3 wt% PdO/Co_3O_4 catalyst. These results agree with the conclusions made based on the spent XRD patterns (see Figure 5a), further strengthening the evidence that 3 wt% Pd is the optimum loading for promoting activity.

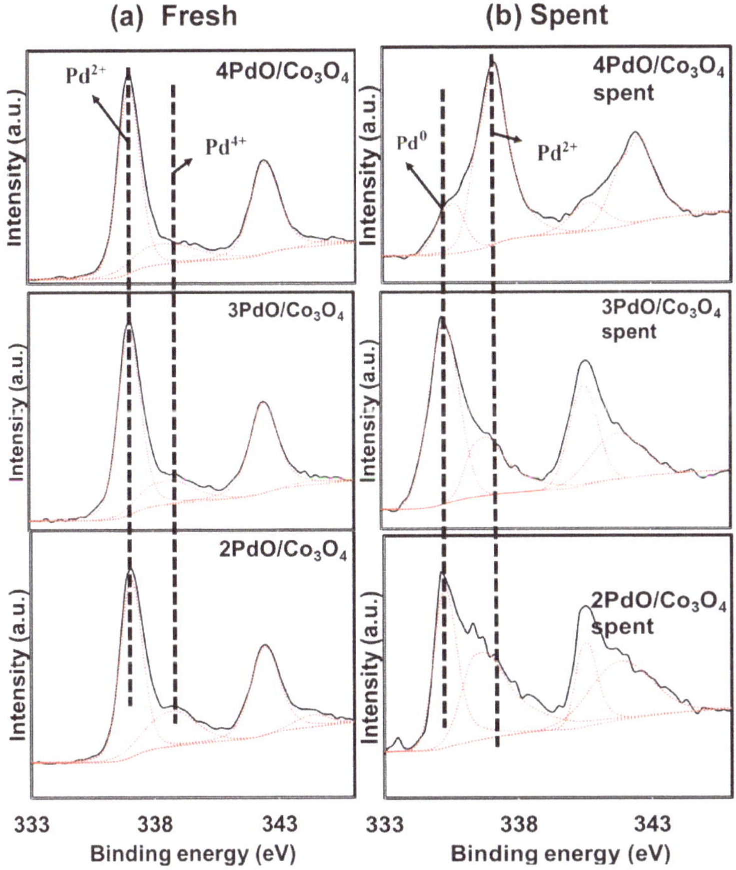

Figure 9. Pd3d XPS spectra of (**a**) fresh and (**b**) spent PdO-promoted Co_3O_4 catalysts.

Figure 10a,b present the Pt4*f* XPS spectra of the fresh and spent 2, 3, and 4 wt% PdO/Co_3O_4. All fresh and spent PtO/Co_3O_4 catalysts only exhibit peaks due to Pt^{2+} (72.3, 74.1 eV) at all Pt loadings [45]. There are no observed peaks due to either Pt^{4+} or metallic Pt^0 in contrast to the

PdO/Co_3O_4 catalysts, wherein the PGM underwent significant changes in oxidation state with exposure to reaction conditions. Overall, XPS measurements show that the support Co_3O_4 is very stable during the promoter impregnation, as well as during the direct NO decomposition. PdO reduced to metallic Pd during the direct NO decomposition and improves the direct NO decomposition activity of Co_3O_4. On the other hand, PtO stays in an oxidized state (no metallic Pt formation) during the direct NO decomposition and exhibits less promotional effects compared to PdO.

2.2.4. NO Adsorption Properties

The adsorption of NO and formation of surface intermediates is essential to establishing activity for direct NO decomposition. In situ FT-IR spectroscopy was performed on pure Co_3O_4 and the PdO- and PtO-promoted Co_3O_4 catalysts to understand the interaction of NO with the catalyst during adsorption. In situ FT-IR measurements were collected during NO adsorption over the catalysts at 300 °C. Before NO adsorption, all the catalysts were pretreated at 350 °C in the presence of 10% O_2 in a helium balance and cooled to 300 °C in the presence of helium. All the spectra collected were normalized with respect to the gas phase NO peak at 1874 cm^{-1}. The in situ FTIR spectra of Co_3O_4, PdO- and PtO-promoted Co_3O_4 catalysts during NO adsorption are presented in Figure 11a,b. As shown in Figure 11a, little to no NO adsorption occurs over the pure Co_3O_4 spinel oxide at 300 °C, as no clear peaks were present relative to the noise level. Interestingly, impregnating PdO over Co_3O_4 leads to a formation of chelating surface nitrate intermediates (1577 and 1254 cm^{-1}) during the NO adsorption [46]. The intermediate formation was observed for all the catalysts irrespective of loading and the intensity of the peak increases with PdO loading. On the other hand, PtO-promoted Co_3O_4 catalysts do not produce spectroscopically relevant amounts of intermediates during the NO adsorption and exhibit spectra similar to the pristine Co_3O_4 catalyst. The catalysts in the current study exhibit activity at temperatures ≥400 °C, however, at these temperatures, no spectroscopically relevant surface NOx species were observed by in situ FTIR (not shown). This observation indicates that neither the surface chelating nitrate nor any other surface NOx species is the most abundant reactive intermediate in the direct NO decomposition mechanism. The in situ FTIR results at 300 °C are, therefore, interpreted as a probe of the affinity of the NO reactant molecule to interact with the catalyst surface. In this interpretation, it is concluded that the presence of PdO improves the affinity of the catalyst to interact with NO compared to PtO or the pure Co_3O_4 support.

The direct NO decomposition measurements show that PdO promotes direct NO decomposition activity of Co_3O_4 much better compared to the PtO. The Co_3O_4 is a normal spinel with AB_2O_4 formula, where A (T_d) sites are occupied by Co^{2+} ions and B (Oh) sites are occupied by Co^{3+} ions. According to the general mechanism proposed by Haneda et al. [18], initially NO adsorbs on the surface and decomposes into N and O. The oxygen atoms adsorb on Co^{2+} ions and are oxidized to Co^{3+}. Then, Co^{3+} ions reduce back to Co^{2+} upon release of the oxygen as a product. Hence, NO adsorption and oxygen release (redox) properties are very important for direct NO decomposition. The *in-situ* FT-IR results reveal that PdO increases the affinity of the catalyst to form surface NOx species compared to PtO or a pure Co_3O_4 support. The H_2-TPR studies in our work show that the Co_3O_4 reduction temperature is significantly decreased by the presence of dispersed PdO, thus suggesting a more facile reduction of Co^{3+} to Co^{2+} to release O_2 during direct NOx decomposition is possible. The improvement in the NOx adsorption properties and ease of cobalt reduction explains the better direct NO decomposition activity of PdO catalysts compared to PtO catalysts.

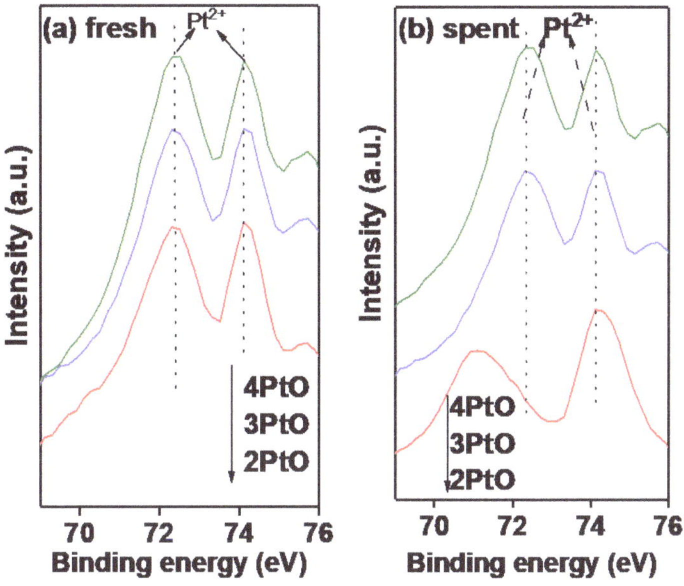

Figure 10. Pt4*f* XPS profiles of (**a**) fresh and (**b**) spent PtO-promoted Co_3O_4 catalysts.

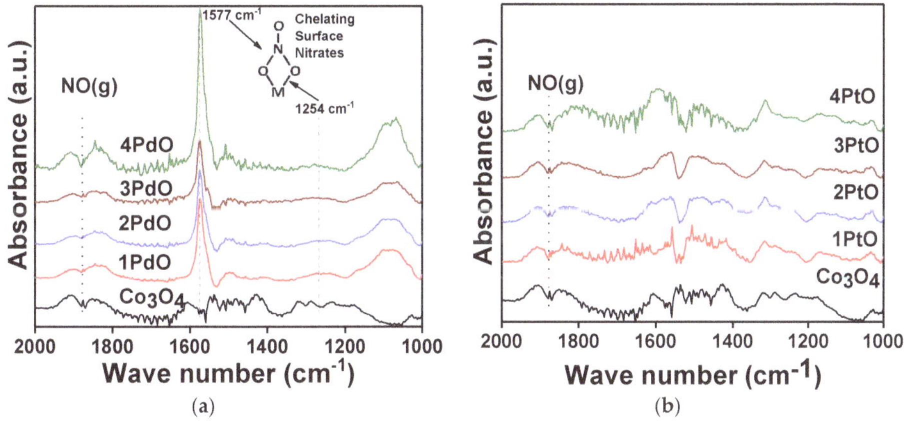

Figure 11. *In-situ* FT-IR spectra of Co_3O_4, PdO (**a**), and PtO (**b**) promoted Co_3O_4 catalysts during NO adsorption at 300 °C.

3. Materials and Methods

3.1. Catalyst Synthesis

Palladium and platinum promoted Co_3O_4 catalysts were synthesized using the wet impregnation method. Commercial Co_3O_4 was purchased from Sigma-Aldrich (St. Louis, MO, USA) (99.5% trace metal basis) and used as received without any further modification for the synthesis. In a typical synthesis procedure, 5 g of commercial Co_3O_4 were mixed with 50 mL of water. Then, the required quantity of palladium nitrate hydrate (Sigma-Aldrich), or tetraamine platinum (II) nitrate (Sigma-Aldric, 99.995% trace metal basis) was dissolved separately in deionized water and combined with the Co_3O_4 suspension. The mixture was heated to 80 °C with continuous stirring. The obtained powder was then dried in an oven at 120 °C for 12 h under air. Finally, the catalyst was calcined at 400 °C for 4 h with a 1 °C/min ramp. Different loadings of palladium and platinum on Co_3O_4 (nominally 1 to 4 wt%) catalysts were prepared by varying the amount of palladium nitrate or platinum nitrate. For reference, the commercial Co_3O_4 support was also calcined at 400 °C for 4 h.

3.2. Catalyst Characterization

X-ray diffraction: X-ray powder diffraction (XRD) patterns were obtained using a Rigaku SmartLab X-ray diffractometer (Rigaku, The Woodlands, TX, USA) using Cu Kα radiation (1.5405 A). A glass holder was used to support the sample. The scanning range was from 10 to 80° (2θ) with a step size of 0.02° and a step time of 1 s. The XRD phases present in the samples were identified with the help of ICDD-JCPDS [31] data files.

BET Surface Area Measurements: The surface area of the PdO and PtO promoted Co_3O_4 materials were measured using a Micromeritics 3Flex surface characterization instrument (Micromeritics, Atlanta, GA, USA). N_2 physisorption isotherms was conducted at −196 °C, and the surface area was measured by the BET method. Prior to the analyses, the samples were outgassed at 300 °C under vacuum (5×10^{-3} Torr) for 3 h.

X-ray Fluorescence Measurements: XRF was collected using a Rigaku ZSX, primus II X-ray spectrometer (Rigaku, The Woodlands, TX, USA). Impurities in the crystals were gained by X-ray fluorescence in operation of spectrometer in standard fewer modes with coverage of a full element. The amount of any elements and oxides particles was detected by the XRF experiment.

H_2-Temperature Programmed Reduction (H_2-TPR) Measurements: The redox properties of the PtO/Co_3O_4 and PdO/Co_3O_4 catalysts were studied using H_2 temperature programmed reduction (H_2-TPR) experiments. H_2-TPR experiments were performed using a Micromeritics 3Flex surface characterization instrument (Micromeritics, Atlanta, GA, USA) equipped with a thermal conductivity detector. Before the experiment, the catalysts were preheated to 300 °C in the presence of 20% O_2/He (30 mL/min). After the pretreatment, the temperature was decreased to 50 °C. The H_2-TPR measurements were performed by heating the catalyst from 50 to 600 °C in the presence of 10% H_2/Ar (30 mL/min).

X-ray Photo Electron Spectroscopy: The XPS measurements were performed using a PHI 5000 Versa Probe II X-ray photoelectron spectrometer (Physical Electronics, East Chanhassen, MN, USA) using an Al Kα source. Charging of the catalyst samples was corrected by setting the binding energy of the adventitious carbon (C*1s*) to 284.6 eV [47]. The XPS analysis was performed at ambient temperature and at pressures typically on the order of 10^7 Torr. Prior to the analysis, the samples were outgassed under vacuum for 30 min.

In Situ FTIR Spectroscopy Measurements during NO Adsorption: The NO adsorption properties were measured using in situ Fourier transform infrared (FTIR) spectroscopy. The Harrick High Temperature Cell with environmental (gas flow) and temperature control was used for in situ diffuse-reflectance FTIR spectroscopy. Spectra were recorded using a Thermo Scientific Nicolet 8700 Research FT-IR Spectrometer (Thermo Scientific Fidher, Waltham, MA USA) equipped with a liquid N_2 cooled MCT detector. Spectra were obtained with a resolution of 2 cm^{-1} and by averaging 64 scans. In situ diffuse-reflectance FTIR spectra were collected during NO adsorption at 300 °C. Prior to NO adsorption, the sample was first pretreated at 350 °C in 30 mL/min of 10% O_2/He. The background spectrum (64 scans) was of the catalyst after cooling to 300 °C in 30 mL/min UHP He. Adsorption of NO was achieved by flowing 30 mL/min of 10,000 ppm NO over the catalyst for 25 min. Adsorption of NO proceeded for 25 min, while spectra were obtained every minute using a series collection. To compare peak intensities among different catalyst samples, the adsorption spectra were normalized to the NO gas phase peak at ~1876 cm^{-1}.

3.3. Direct NO Decomposition Measurements

Direct NO decomposition measurements were performed in a fixed bed flow reactor. A gas mixture of ~1% NO in helium balance was used with a gas hourly space velocity of ~2100 h^{-1} and in the temperature region of 450–650 °C. Before the reaction, catalysts were pretreated at 500 °C in the presence of 20% O_2/He. After pretreatment, the bed temperature was decreased to 400 °C and direct NO decomposition measurements were collected. The measurements were performed at 400, 450, 550, and 650 °C, with 2 h of steady state at each temperature. The NO, N_2O, and NO_2 concentrations were analyzed with a FTIR detector (CAI 600 SC FTIR California Analytical Instruments, Inc., Orange County, CA, USA). The N_2 concentration was calculated by mass balance of the total nitrogen species. The raw NO conversion (NO converted to all the products) during the steady state measurements are presented in Figure S1 of the supporting information and activity to N_2 was reported in Figure 1 of the manuscript. The steady state direct NO decomposition measurements were also performed in a reactor, which was equipped with the mass spectroscopy (MKs, Cirrus 2). The changes in the NO, N_2, O_2, N_2O, and NO_2 signal intensities were monitored during the reaction (Figures S2–S5). The inert Ar gas was introduced as a tracer to monitor for potential systematic variation in signal intensity during the experiment (Figure S2).

4. Conclusions

The direct NO decomposition measurements show that PdO promotes the activity of Co_3O_4 and is 4 times more active compared to PtO at 650 °C. Also, the activity increases with increasing PdO loading until 3 wt% and further increase in the loading leads to a decrease in the activity. On the other hand, only a slight increase in the activity was observed with increasing PtO loading up to 4 wt%. Surface area measurements indicated that both PdO and PtO have little to no influence on the surface area of Co_3O_4, except for a decrease in surface area for 4 wt% PdO/Co_3O_4. The X-ray diffraction measurements show that Pt incorporated into the Co_3O_4 structure during the synthesis and PdO stays mostly on the surface. The diffraction measurements also suggested that PdO is in an amorphous form up to 3 wt% over Co_3O_4 surface and crystalline PdO forms at 4 wt% loading, whereas PtO mostly stays as amorphous from or incorporated into Co_3O_4 structure until 4 wt%. Due to the synergistic effect between Pd species and Co_3O_4, an improvement in the redox properties of Co_3O_4 was observed in the case of PdO/Co_3O_4 catalysts. Conversely, PtO do not have any influence on the redox properties of Co_3O_4. The X-ray photo electron spectroscopic measurements reveal that PdO reduced to Pd metal during the direct NO decomposition reaction and Pt was in 2+ oxidation state before and after the direct NO decomposition reaction. In situ NO adsorption measurements show that PdO improve the NO adsorption properties of Co_3O_4 by forming the nitrate ion intermediates, whereas PtO/Co_3O_4 do not form any intermediates during the NO adsorption at 300 °C. Overall, PdO ease the redox properties of Co_3O_4 and forms surface adsorbed species during NO adsorption and improves the direct NO decomposition activity of Co_3O_4. On the other hand, PtO do not have any influence on the redox or NO adsorption properties of Co_3O_4 and exhibits lesser promotional effects compared to PdO. For PdO/Co_3O_4 catalysts, the PdO remains in amorphous form until 3PdO/Co_3O_4 and improves the activity of Co_3O_4 with loading. However, further increase in the loading to 4 wt% leads to formation of crystalline PdO, which reduces separately during H_2-TPR and inhibits the PdO reduction to metallic Pd during direct NO decomposition and exhibits lesser activity compared to 3 wt% PdO/Co_3O_4.

Supporting information: The NO conversion, and N_2, N_2O, and NO_2 ppm values of various PdO/Co_3O_4 and PtO/Co_3O_4 catalysts are presented in Tables S1 and S2. The total NO conversion profiles of the Co_3O_4, 3PdO/Co_3O_4, and 4PtO/CoO calculated from the FT-IR detector during the steady state direct NO decomposition measurements are presented in Figure S2. NO, Ar, N_2, O_2, N_2O, and NO_2 partial pressure values (obtained from mass spectroscopy) of Co_3O_4, 3 wt% PdO/Co_3O_4, and 3 wt% PtO/Co_3O_4 during steady state direct NO decomposition in the 400 to 650 °C temperature region are presented in Figures S3 and S4.

Supplementary Materials: The following are available online at http://www.mdpi.com/2073-4344/9/1/62/s1. Table S1: NO conversion, N_2, N_2O, NO_2 ppm values of the Co_3O_4 and various PdO/Co_3O_4 catalysts in the temperature region 400–650 °C. Table S2: NO conversion, N_2, N_2O, NO_2 ppm values of the Co_3O_4 and various PtO/Co_3O_4 catalysts in the temperature region 400–650 °C. Figure S1: Steady state NO conversion values of the Co_3O_4, $3PdO/Co_3O_4$, and $4PtO/Co_3O_4$ catalysts during the direct NO decomposition in the temperature region 400 to 650 °C. Figure S2: NO and Ar M.S. partial pressures of the (**a**) Co_3O_4, (**b**) $3PdO/Co_3O_4$, and (**c**) $3PtO/Co_3O_4$ catalysts during the steady state direct NO decomposition in the temperature region 400 to 650 °C. Figure S3: N_2 and O_2 M.S. partial pressures of the (**a**) Co_3O_4, (**b**) $3PdO/Co_3O_4$, and (**c**) $3PtO/Co_3O_4$ catalysts during the steady state direct NO decomposition in the temperature region 400 to 650 °C. Figure S4: N_2O M.S. partial pressures of the (**a**) Co_3O_4, (**b**) $3PdO/Co_3O_4$, and (**c**) $3PtO/Co_3O_4$ catalysts during the steady state direct NO decomposition in the temperature region 400 to 650 °C. Figure S5: NO_2 M.S. partial pressures of the (**a**) Co_3O_4, (**b**) $3PdO/Co_3O_4$, and (**c**) $3PtO/Co_3O_4$ catalysts during the steady state direct NO decomposition in the temperature region 400 to 650 °C.

Author Contributions: This study was conducted through contributions of all authors. G.K.R. Reddy designed the study, performed the experiments, and wrote the manuscript. T.C.P. was involved in performing the experiments. C.A.R. checked and corrected the manuscript.

Funding: This research received no external funding.

Acknowledgments: The authors thank Hongfei Jia from Toyota Research Institute—North America and Naoto Nagata from Toyota Motor Corp. for their support.

Conflicts of Interest: The authors declare no conflict of interest.

References

1. Irfan, M.F.; Goo, J.H.; Kim, S.D. Effects of NO_2, CO, O_2, and SO_2 on oxidation kinetics of NO over Pt-WO_3/TiO_2 catalyst for fast selective catalytic reduction process. *Int. J. Chem. Kinet.* **2006**, *38*, 613–620. [CrossRef]
2. Masui, T.; Uejima, S.; Tsujimoto, S.; Nagai, R.; Imanaka, N. Direct NO decomposition over C-type cubic Y_2O_3–Pr_6O_{11}–Eu_2O_3 solid solutions. *Catal. Today* **2015**, *242*, 338–342. [CrossRef]
3. Hong, Z.; Wang, Z.; Li, X.B. Catalytic oxidation of nitric oxide (NO) over different catalysts: An overview. *Catal. Sci. Technol.* **2017**, *7*, 3440–3452. [CrossRef]
4. Imanaka, N.; Masui, T. Advances in direct NO decomposition catalysts. *Appl. Catal. A* **2012**, *431–432*, 1–8. [CrossRef]
5. Haneda, M.; Hamada, H. Recent progress in catalytic NO decomposition. *C. R. Chim.* **2016**, *19*, 1254–1265. [CrossRef]
6. Locci, C.; Vervisch, L.; Farcy, B.; Domingo, P.; Perret, N. Selective Non-Catalytic Reduction (SNCR) of nitrogen oxide emissions: A perspective from numerical modeling. *Flow Turbul. Combust.* **2018**, *100*, 301–340. [CrossRef]
7. Jellinek, K. About decomposition rate of nitric oxide and dependence of the same on the temperature. *J. Inorg. Chem.* **1906**, *49*, 229–276.
8. Glick, H.S.; Klein, J.J.; Squire, W. Single-Pulse Shock Tube Studies of the Kinetics of the Reaction $N_2 + O_2 \rightleftarrows$ 2NO between 2000–3000 °K. *J. Chem. Phys.* **1957**, *27*, 850–857. [CrossRef]
9. Zhu, J.; Thomas, A. Perovskitetype mixed oxides as catalytic material for NO removal. *Appl. Catal. B* **2009**, *92*, 225–233. [CrossRef]
10. Royer, S.; Duprez, D.; Can, F.; Courtois, X.; Batiot-Dupeyrat, C.; Laassiri, S.; Alamdari, H. Perovskites as Substitutes of Noble Metals for Heterogeneous Catalysis: Dream or Reality. *Chem. Rev.* **2014**, *114*, 10292–10368. [CrossRef]
11. Tofan, C.; Klvana, D.; Kirchnerova, J. Direct decomposition of nitric oxide over perovskite-type catalysts: Part I. Activity when no oxygen is added to the feed. *Appl. Catal. A* **2002**, *223*, 275–286. [CrossRef]
12. Yokoi, Y.; Uchida, H. The influence of palladium on the structure and catalytic activity of lanthanum based mixed oxide. *Catal. Today* **1998**, *42*, 167–174. [CrossRef]
13. Zhao, Z.; Yang, X.G.; Wu, Y. Comparative study of Nickel-based perovskite-like mixed oxide catalysts for direct decomposition of NO. *Appl. Catal. B* **1996**, *8*, 281–297. [CrossRef]
14. Zhu, J.J.; Xiao, D.H.; Li, J.; Yang, X.G.; Wu, Y. Effect of Ce on NO direct decomposition in the absence/presence of O_2 over $La_{1-x}Ce_xSrNiO_4$ ($0 \leq x \leq 0.3$). *J. Mol. Catal. A Chem.* **2005**, *234*, 99–105. [CrossRef]
15. Zhu, J.J.; Xiao, D.H.; Li, J.; Yang, X.G.; Wu, Y.; Wei, K. Effect of Ce and MgO on NO decomposition over La_{1-x}–Cex–Sr–Ni–O/MgO. *Catal. Commun.* **2006**, *7*, 432–435. [CrossRef]
16. Winter, E.R.S. The catalytic decomposition of NO by metallic oxides. *J. Catal.* **1971**, *22*, 158. [CrossRef]

17. Boreskov, G.K. Forms of oxygen bonds on the surface of oxidation catalysts Discuss. *Faraday Soc.* **1966**, *41*, 263. [CrossRef]
18. Haneda, M.; Kintaichi, Y.; Bion, N.; Hamada, H. Alkali metal-doped cobalt oxide catalysts for NO decomposition. *Appl. Catal. B* **2003**, *46*, 473–482. [CrossRef]
19. Haneda, M.; Kintaichi, Y.; Hamada, H. Reaction mechanism of NO decomposition over alkali metal-doped cobalt oxide catalysts. *Appl. Catal. B* **2005**, *55*, 169–175. [CrossRef]
20. Haneda, M.; Nakamura, I.; Fujitani, T.; Hamada, H. Catalytic Active Site for NO Decomposition Elucidated by Surface Science and Real Catalyst. *Catal. Surv. Asia* **2005**, *9*, 207–215. [CrossRef]
21. Park, P.W.; Kil, J.K.; Kung, H.H.; Kung, M.C. NO decomposition over Na promoted cobalt oxide. *Catal. Today* **1998**, *42*, 51–60. [CrossRef]
22. Amirnazmi, A.; Benson, J.E.; Boudart, M. Oxygen inhibition in the decomposition of NO on metal oxides and platinum. *J. Catal.* **1973**, *30*, 55–65. [CrossRef]
23. Behm, R.J.; Brundle, C.R. Decomposition of NO on Ag(111) at low temperatures. *J. Vac. Sci. Technol. A* **1984**, *2*, 1040–1041. [CrossRef]
24. Suzuki, Y.; Hwang, H.J.; Kondo, N.; Ohji, T. In Situ Processing of a Porous Calcium Zirconate/Magnesia Composite with Platinum Nanodispersion and Its Influence on Nitric Oxide Decomposition. *J. Am. Ceram. Soc.* **2001**, *84*, 2713–2715. [CrossRef]
25. Haneda, M.; Kintaichi, Y.; Hamada, H. Surface reactivity of prereduced rare earth oxides with nitric oxide: new approach for NO decomposition. *Phys. Chem. Chem. Phys.* **2002**, *4*, 3146–3151. [CrossRef]
26. Haneda, M.; Kintaichi, Y.; Nakamura, I.; Fujitani, T.; Hamada, H. Comprehensive study combining surface science and real catalyst for NO direct decomposition. *Chem. Commun.* **2002**, *21*, 2816–2817. [CrossRef]
27. Haneda, M.; Kintaichi, Y.; Nakamura, I.; Fujitani, T.; Hamada, H. Effect of surface structure of supported palladium catalysts on the activity for direct decomposition of nitrogen monoxide. *J. Catal.* **2003**, *218*, 405–410. [CrossRef]
28. Almusaiteer, K.; Krishnamurthy, R.; Chuang, S.S.C. In situ infrared study of catalytic decomposition of NO on carbon-supported Rh and Pd catalysts. *Catal. Today* **2000**, *55*, 291–299. [CrossRef]
29. De Oliveira, A.M.; Crizel, L.E.; da Silveira, R.S.; Pergher, S.B.C.; Baibich, I.M. NO decomposition on mordenite-supported Pd and Cu catalysts. *Catal. Commun.* **2007**, *8*, 1293–1297. [CrossRef]
30. Reddy, G.K.; Ling, C.; Peck, T.; Jia, H. Understanding the chemical state of palladium during the direct NO decomposition—Influence of pretreatment environment and reaction temperature. *RSC Adv.* **2017**, *7*, 19645–19655. [CrossRef]
31. Klug, H.P.; Alexander, L.E. *X-ray Diffraction Procedures for Polycrystalline and Amorphous Materials*, 2nd ed.; Wiley: New York, NY, USA, 1974.
32. Ercolino, G.; Grzybek, G.; Stelmachowski, P.; Specchia, S.; Kotarba, A.; Specchia, V. Pd/Co_3O_4-based catalysts prepared by solution combustion synthesis for residual methane oxidation in lean conditions. *Catal. Today* **2015**, *257*, 66–71. [CrossRef]
33. Bahlawane, N.; Rivera, E.F.; Ho¨inghaus, K.K.; Brechling, A.; Kleineberg, U. Characterization and tests of planar Co_3O_4 model catalysts prepared by chemical vapor deposition. *Appl. Catal. B* **2004**, *53*, 245. [CrossRef]
34. Lin, H.Y.; Chen, Y.W. The mechanism of reduction of cobalt by hydrogen. *Mater. Chem. Phys.* **2004**, *85*, 171. [CrossRef]
35. Chen, Z.; Wanga, S.; Dinga, Y.; Zhanga, L.; Lva, L.; Wanga, M.; Wanga, S. Pd catalysts supported on Co_3O_4 with the specified morphologies in CO and CH_4 oxidation. *Appl. Catal. A Gen.* **2017**, *532*, 95–104. [CrossRef]
36. Yang, H.; Deng, J.; Liu, Y.; Xie, S.; Xu, P.; Dai., H. Pt/Co_3O_4/3DOM Al_2O_3: Highly effective catalysts for toluene combustion. *Chin. J. Catal.* **2016**, *37*, 934–946. [CrossRef]
37. Voogt, E.H.; Mens, A.J.M.; Gijzeman, O.L.J.; Geus, J.W. XPS analysis of palladium oxide layers and particles. *Surf. Sci.* **1996**, *350*, 21–31. [CrossRef]
38. Zhang, C.; Zhang, L.; Xu, G.C.; Ma, X.; Li, Y.H.; Zhang, C.; Jia, D. Metal organic framework-derived Co_3O_4 microcubes and their catalytic applications in CO oxidation. *New J. Chem.* **2017**, *41*, 1631–1636. [CrossRef]
39. Gnanamani, M.K.; Jacobs, G.; Hamdeh, H.H.; Shafer, W.D.; Liu, F.; Hopps, S.D.; Thomas, G.A.; Davis, B.H. Hydrogenation of Carbon Dioxide over Co–Fe Bimetallic Catalysts. *ACS Catal.* **2016**, *6*, 913–927. [CrossRef]
40. Brun, M.; Berthet, A.; Bertolini, J.C. XPS, AES and Auger parameter of Pd and PdO. *J. Electron Spectrosc. Relat. Phenom.* **1999**, *104*, 55–60. [CrossRef]

41. Suhonen, S.; Valden, M.; Pessa, M.; Savimaki, A.; Harkonen, M.; Hietikko, M.; Pursiainen, J.; Laitinen, R. Characterization of alumina supported Pd catalysts modified by rare earth oxides using X-ray photoelectron spectroscopy and X-ray diffraction: Enhanced thermal stability of PdO in Nd/Pd catalysts. *Appl. Catal. A Gen.* **2001**, *207*, 113–120. [CrossRef]
42. Mirkelamoglu, B.; Karakas, G. The role of alkali-metal promotion on CO oxidation over PdO/SnO_2 catalysts. *Appl. Catal. A Gen.* **2006**, *299*, 84–94. [CrossRef]
43. Mucalo, M.R.; Cooney, R.P.; Metson, J.B. Platinum and palladium hydrosols: Characterisation by X-ray photoelectron spectroscopy and transmission electron microscopy. *Colloids Surf.* **1991**, *60*, 175–197. [CrossRef]
44. Kibis, L.S.; Titkov, A.I.; Stadnichenko, A.I.; Koscheev, S.V.; Boronin, A.I. X-ray photoelectron spectroscopy study of Pd oxidation by RF discharge in oxygen. *Appl. Surf. Sci.* **2009**, *255*, 9248–9254. [CrossRef]
45. Hegdea, M.S.; Bera., P. Noble metal ion substituted CeO_2 catalysts: Electronic interaction between noble metal ions and CeO_2 lattice. *Catal. Today* **2015**, *253*, 40–50. [CrossRef]
46. Hadjiivanov, K.I. Identification of Neutral and Charged N_xO_y Surface Species by IR Spectroscopy. *Catal. Rev. Sci. Eng.* **2000**, *42*, 71–144. [CrossRef]
47. Wagner, C.D.; Riggs, W.M.; Davis, L.E.; Moulder, J.F. *Handbook of X-ray Photoelectron Spectroscopy*; Muilenberg, G.E., Ed.; Perkin-Elmer Corp.: Waltham, MA, USA, 1978.

Article

Investigation of Various Pd Species in Pd/BEA for Cold Start Application

Beibei Zhang [1], Meiqing Shen [1,2,3], Jianqiang Wang [1], Jiaming Wang [4] and Jun Wang [1,*]

1 Key Laboratory for Green Chemical Technology of State Education Ministry, School of Chemical Engineering & Technology, Tianjin University, Tianjin 300072, China; zhang_525@tju.edu.cn (B.Z.); mqshen@tju.edu.cn (M.S.); jianqiangwang@tju.edu.cn (J.W.)

2 State Key Laboratory of Engines, Tianjin University, Tianjin 300072, China

3 Collaborative Innovation Center of Chemical Science and Engineering, Tianjin 300072, China

4 Wuxi Weifu Lida Catalytic Converter Co., Ltd., Wuxi 214028, China; jiaming.wang@weifu.com.cn

* Correspondence: wangjun@tju.edu.cn; Tel.: +86-22-2740-7002

Received: 16 February 2019; Accepted: 4 March 2019; Published: 7 March 2019

Abstract: A series of Pd/BEA catalysts with various Pd loadings were synthesized. Two active Pd^{2+} species, Z^{-}-Pd^{2+}-Z^{-} and Z^{-}-$Pd(OH)^{+}$, on exchanged sites of zeolites, were identified by in situ FTIR using CO and NH_3 respectively. Higher NO_x storage capacity of Z^{-}-Pd^{2+}-Z^{-} was demonstrated compared with that of Z^{-}-$Pd(OH)^{+}$, which was caused by the different resistance to H_2O. Besides, lower Pd loading led to a sharp decline of Z^{-}-$Pd(OH)^{+}$, which was attributed to the 'exchange preference' for Z^{-}-Pd^{2+}-Z^{-} in BEA. Based on this research, the atom utilization of Pd can be improved by decreasing Pd loading.

Keywords: Pd/BEA; Cold start; Pd species; NOx abatement

1. Introduction

The exhaust regulation on NO_x emissions is getting more stringent [1]. Presently, NH_3 selective catalytic reduction (SCR) [2] and NO_x storage reduction (NSR) [3] are widely used for NO_x removal. However, standard NO_x aftertreatment technologies fail to function efficiently at low temperatures, which results in a large proportion of the tailpipe NO_x emission [4]. Meanwhile, high efficiency internal combustion engines require new and/or improved technologies which specifically address their low exhaust temperatures. In response to difficulties of low temperature emissions control, numerous efforts are underway to develop catalysts that light-off at temperatures below 150 °C. Passive NO_x adsorbers (PNAs) could play a critical role in enabling high efficiency advanced combustion systems.

Recently, Pd/zeolite serving as passive NO_x adsorbers (PNAs) was first proposed by Chen et al. [4]. This catalyst is emerging as effective passive NO_x adsorbent technology because of its NO_x storage/release capabilities, resistance to sulfur poisoning and hydrothermal deactivation [4–8]. Due to these excellent characteristics, Pd/zeolite has attracted great interest recently and has been further optimized [6–18].

Several recent studies show that isolated Pd ions in Pd/zeolite are the main active sites for NO_x trapping [6,7,11,13,16]. It is reported that there are nine skeletal T sites with various chemical environments in BEA [19]. So, isolated Pd ions located in various positions of zeolite framework are likely to be formed. As reported by Gao et al. and Giordanino et al. [20–22], two kinds of isolated Cu species (Cu^{2+} and $[Cu(OH)]^{+}$) are identified on various framework positions of Cu/SSZ-13, which indicates that species of isolated cations may be influenced by their locations in zeolites. They also mentioned that Cu species were significantly affected by Cu loading. So, various isolated Pd species may co-exist in Pd/BEA, and Pd loading may be capable of influencing the content of them. Actually, two kinds of active isolated Pd ions (bare Pd^{2+} and $Pd(OH)^{+}$) have been observed by Zheng et al. [10]

and Khivantsev et al. [12]. However, they did not point out the difference in adsorption between these two species.

Therefore, these two kinds of isolated Pd ions were further studied in this research. A series of in situ FTIR experiments in CO and NH_3 were carried out, and a semiquantitative method was adopted to distinguish these two species. Based on this method, the NO_x storage capacity of each isolated Pd species was compared. Further, by increasing the proportion of isolated Pd species with higher NO_x storage capacities, the atom utilization of Pd can be improved.

2. Results

2.1. Ex-Situ FTIR

Ex-situ FTIR spectra of each sample are exhibited in Figure 1. Compared with 0-Pd (Figure 1a), an extra vibration at 926 cm^{-1} is observed on the FTIR spectrum of 1-Pd. This peak is attributed to the vibration distortion of the skeletal T–O–T bond due to the strong interaction of Pd ions [4]. This is an obvious piece of evidence for the existence of Pd ions on the exchange sites of zeolites. Peaks at 926 cm^{-1} appear on spectra of 0.2-Pd, 0.5-Pd and 2-Pd, too (Figure 1b). So, the existence of Pd ions on exchange sites in these samples is confirmed.

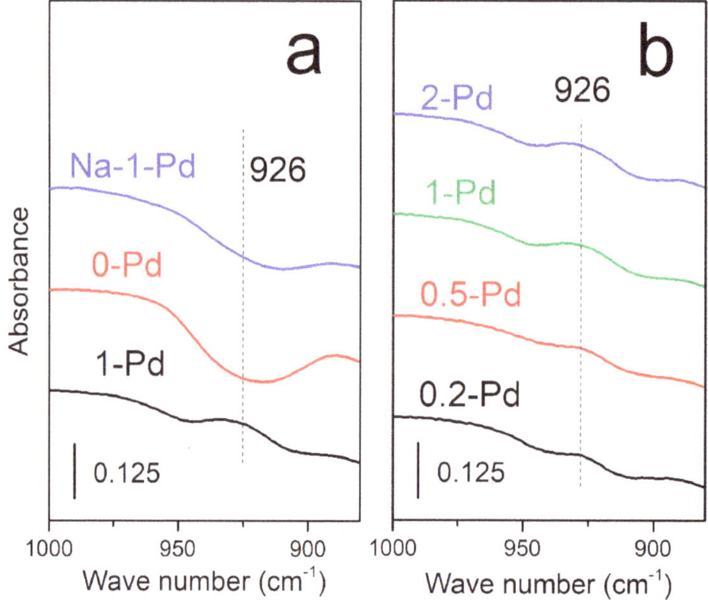

Figure 1. (**a**) Ex-situ FTIR spectra of 0-Pd, 1-Pd, Na-1-Pd; (**b**) Ex-situ FTIR spectra of 2-Pd, 1-Pd, 0.5-Pd, 0.2-Pd (Temperature: 200 °C; Flow: N_2, 1 L/min).

Compared with 1-Pd (Figure 1a), the peak at 926 cm^{-1} disappears in the spectrum of Na-1-Pd, which indicates the elimination of isolated Pd ions on exchange sites. This is firm evidence of which isolated Pd ions in Na-1-Pd have been completely exchanged by Na^+ ions through the titration process. So, Pd loaded on Na-1-Pd mainly exists in form of Pd oxidations.

2.2. Na^+ Titration

Na^+ titration was adopted to measure the content of isolated Pd ions in each sample [23], and the proportion of isolated Pd ions in total Pd loading (marked as Isolated Pd/Pd loading in Table 1) was further calculated. Considering that Pd is sensitive to Cl [24,25], the titration process was carried out in a $NaNO_3$ solution. Besides, this measurement is believed to have no negative effect on zeolite structures since the titration process was carried out in mild conditions and no calcination was done. As shown in Table 1, lower Pd loading leads to less isolated Pd^{2+} content, whereas the proportion of isolated Pd ions in total Pd loading becomes larger. This phenomenon indicates that the increase of Pd loading leads to the formation of more Pd oxidations (PdOx), which is consistent with Jaeha Lee et al.'s study [11].

Table 1. The content of Pd ions on exchange sites of each sample.

Catalyst	0-Pd	0.2-Pd	0.5-Pd	1-Pd	2-Pd
Isolated Pd (wt %)	0.00	0.22	0.53	0.87	0.91
Isolated Pd/Pd loading (%)	0.00	95.7	96.4	77.7	42.7

2.3. CO In Situ FTIR

Pd species were further probed by CO using in situ FTIR, and the result is displayed in Figure 2. Peaks below 2100 cm^{-1} are attributed to CO signals on metallic Pd (Pd^0) formed via CO reduction [10,26], among which Pd^0–CO (atop) are found at 2098 cm^{-1}, 2076 cm^{-1} and $Pd_2{}^0$–CO (bridging) are found at 1951 cm^{-1}. Besides, the peak at 2117 cm^{-1} is attributed to the C-O vibration on Pd^+ [10,27]. As reported by Vu et al. [9], Pd^+ is formed by the CO reduction of ion-exchanged Pd species [9]. CO signals on isolated Pd^{2+} species are observed above 2100 cm^{-1}. The peak at 2152 cm^{-1} with a shoulder peak at 2137 cm^{-1} is attributed to the C-O vibration on isolated Pd^{2+} bonded with the hydroxy of zeolites (marked as Z^--Pd^{2+}-Z^-) [10]. The peak at 2179 cm^{-1} is attributed to the vibration of C-O adsorbed by another kind of isolated Pd^{2+} [10], which was first determined as Z^--$Pd(OH)^+$ by Okumura et al. [28].

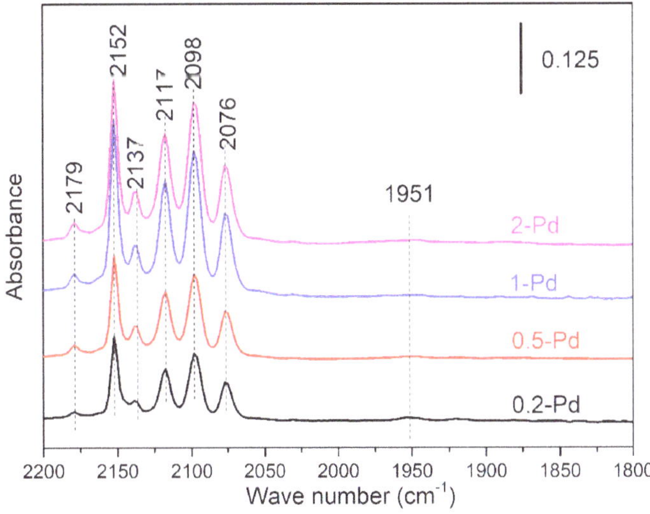

Figure 2. CO in situ FTIR spectra of 2-Pd, 1-Pd, 0.5-Pd, 0.2-Pd (Temperature: 80 °C; Flow: 1000 ppm CO, balanced with N_2, 500 mL/min).

In short, two kinds of isolated Pd^{2+} (Z^--Pd^{2+}-Z^- and Z^--$Pd(OH)^+$) were identified in 0.2-Pd, 0.5-Pd, 1-Pd and 2-Pd. Note that all spectra in Figure 2 were obtained when the steady state had been achieved, and corresponding peaks of Z^--Pd^{2+}-Z^- and Z^--$Pd(OH)^+$ were still observed. So, these two isolated Pd^{2+} species cannot be completely reduced by CO at this temperature, which indicated that the reduction reaction is a reversible one. As shown in Figure S1, the existence of Z^{-}-Pd^{2+}-Z^{-} in 1-Pd-80 is also demonstrated by CO whereas no Z^{-}-$Pd(OH)^{+}$ is observed. Since the complete reduction of Z^{-}-$Pd(OH)^{+}$ cannot be achieved by CO, there is only one isolated Pd^{2+} species, Z^{-}-Pd^{2+}-Z^{-}, in 1-Pd-80 (see Figure S1).

2.4. Catalyst Evaluation

Profiles of the NO adsorption stage are shown in Figure 3. NO_x storage capacities are calculated by the integration of negative peaks on these profiles, and the dead volume has been subtracted. The result is displayed in Table 2. Note that samples with the same Pd loading as much as 1 wt % (see Figure S2) exhibit entirely different NO_x storage capacities (53.3 μmol/gcat and 9.9 μmol/gcat respectively), which indicates that only part of Pd loaded on samples is efficient. As shown in Figure 3, NO_x storage capacities of 0-Pd and Na-1-Pd are very low. So, Brønsted hydroxyl group and PdOx are inefficient active centers for NO_x storage, whereas they do trap NO_x in this condition as reported [10,27]. Since the NO_x storage capacity of 1-Pd is much higher, isolated Pd^{2+} ions are likely to be the main active sites for NO trapping.

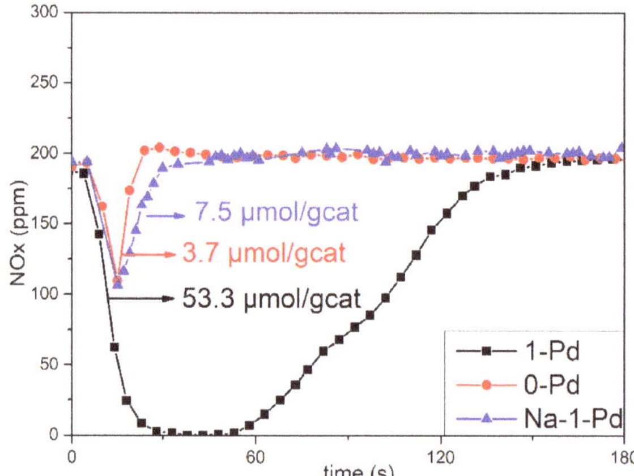

Figure 3. NO_x adsorption profiles (Temperature: 80 °C; Flow: NO_x, CO, H_2O, O_2, CO_2, balanced with N_2, 1 L/min).

Table 2. NO_x storage capacities in NO_x, CO, H_2O, O_2 and CO_2.

Catalyst	0-Pd	0.2-Pd	0.5-Pd	1-Pd	2-Pd
NO_x storage capacity (μmol/gcat)	3.7	21.6	47.5	53.3	50.1

2.5. NH_3 In Situ FTIR

Acidity over samples is probed by NH_3 with in situ FTIR, and the result is exhibited in Figure 4. The peak at 1463cm^{-1} observed in both spectra of 1-Pd and 0-Pd (Figure 4a) is attributed to the vibration of NH^{4+} in Brønsted hydroxyl groups (NH^{4+}-B) [29]. In the spectrum of 1-Pd, two additional peaks at 1625 cm^{-1} and 1313cm^{-1} corresponding to the vibration of NH_3 on Lewis acid (NH_3-L) [30,31] is observed. Nevertheless, these two peaks do not exist in Na-1-Pd in which isolated Pd^{2+} ions (Z^{-}-Pd^{2+}-Z^{-} and Z^{-}-$Pd(OH)^{+}$) are replaced by Na^{+} completely. So, it is reasonable to believe that Lewis acid is generated from Z^{-}-Pd^{2+}-Z^{-} and Z^{-}-$Pd(OH)^{+}$ species.

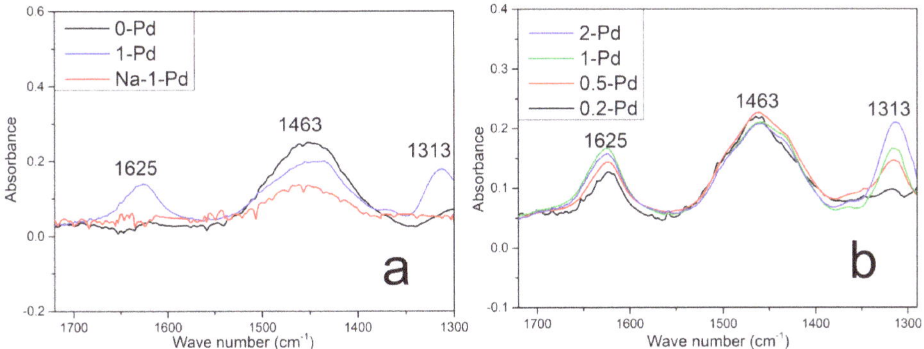

Figure 4. (**a**) NH_3 in situ FTIR spectra of 0-Pd, 1-Pd, Na-1-Pd; (**b**) NH_3 in situ FTIR spectra of 2-Pd, 1-Pd, 0.5-Pd, 0.2-Pd (Temperature: 80 °C; Flow: 500 ppm NH_3, balanced with N_2, 500 mL/min).

As shown in Figure S3, there is only one peak at 1625cm^{-1} corresponding to the vibration of NH_3-L observed in the spectrum of 1-Pd-80. Since there is only one kind of isolated Pd^{2+}, Z^--Pd^{2+}-Z^-, in this sample as discussed above, the peak at 1625cm^{-1} should be assigned to the vibration of NH_3-L originated from Z^--Pd^{2+}-Z^-. In this case, the peak at 1313cm^{-1} should be attributed to the vibration of NH_3-L originating from Z^--$Pd(OH)^+$.

Since Z^--Pd^{2+}-Z^- and Z^--$Pd(OH)^+$ are capable of being probed by NH_3, the ratio of the height of the peaks at 1625 cm^{-1} ($P_{Z^-\text{-}Pd^{2+}\text{-}Z^-}$) and 1313 cm^{-1} ($P_{Z^-\text{-}Pd(OH)^+}$) in Figure 4b can be defined as the relative content between these two isolated Pd^{2+} species. Considering that extinction coefficients for NH_3 molecules adsorbed on Z^--Pd^{2+}-Z^- and Z^--$Pd(OH)^+$ are both constants, they have no effect on the tendency of peak height ratios. So, the extinction coefficient is not considered in this part [32–34]. The result is shown in Figure 5a. It is worth nothing that $P_{Z^-\text{-}Pd(OH)^+}/P_{Z^-\text{-}Pd^{2+}\text{-}Z^-}$ rises in parallel with the increase of isolated Pd^{2+}, which means that the higher content of isolated Pd^{2+} leads to a much more obvious increase of Z^--$Pd(OH)^+$ than that of Z^--Pd^{2+}-Z^-. So, isolated Pd^{2+} in 0.2-Pd mainly exists in the form of Z^--Pd^{2+}-Z^-, whereas a large amount of Z^--$Pd(OH)^+$ ions are formed in 2-Pd.

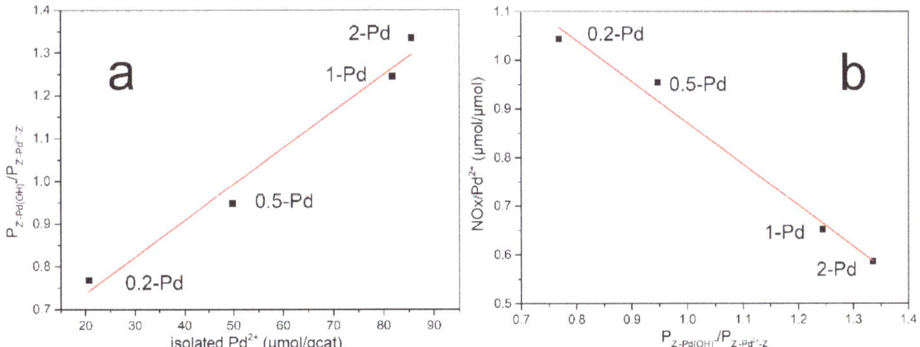

Figure 5. (**a**) tendency between isolated Pd^{2+} and $P_{Z^-\text{-}Pd(OH)^+}/P_{Z^-\text{-}Pd^{2+}\text{-}Z^-}$; (**b**) tendency between $P_{Z^-\text{-}Pd(OH)^+}/P_{Z^-\text{-}Pd^{2+}\text{-}Z^-}$ and NO_x/Pd^{2+}.

Mole ratios of NO_x adsorbed to isolated Pd^{2+} for each sample (marked as NO/Pd^{2+}) are calculated via data in Tables 1 and 2. Figure 5b is plotted by $P_{Z^-\text{-}Pd(OH)^+}/P_{Z^-\text{-}Pd^{2+}\text{-}Z^-}$ on the horizontal axis and NO_x/Pd^{2+} on the vertical. As reported, NO_x is believed to be trapped in the form of $Pd^{\delta+}$-NO [12,15], so the maximum NO_x/Pd^{2+} ratio is 1 in theory. However, the downward trend between $P_{Z^--Pd(OH)^+}/P_{Z^--Pd^{2+}-Z^-}$ and NO_x/Pd^{2+} means that higher $P_{Z^--Pd(OH)^+}/P_{Z^--Pd^{2+}-Z^-}$ leads to lower NO_x storage capacity of isolated Pd^{2+}, which indicates that the NO_x storage capacity of Z^--Pd(OH)$^+$ is obviously lower than that of Z^--Pd^{2+}-Z^- in this condition.

A further discussion is given below so as to give a reasonable explanation from the perspective of the structure-function relationship.

In BEA zeolite, nine skeletal T sites with various chemical environments are determined [19]. Among these T sites, T5 and T6 are demonstrated to have the lowest Al substitution energy [35]. Thus, Al substitution on these two sites (marked as Al_{T5} and Al_{T6}) is preferential, and the amount of Al_{T5} and Al_{T6} is larger than that of Al atoms on the other T sites in BEA. Besides, exchange sites formed on these two Al atoms are more active due to the lowest deprotonation energy [36] of them. Meanwhile, Al_{T5} and Al_{T6} are located in the meta-position of the same five-membered ring [37], and the co-ion-exchange of protons on these two exchange sites formed on Al_{T5} and Al_{T6} is feasible. These characteristics of BEA can explain the 'exchange preference' for Z^--Pd^{2+}-Z^- formed in Pd/BEA with lower content of isolated Pd ions to some extent. Besides, with the increase of Pd loading, more exchange sites are taken up. Limited by the amount of protons which are capable of being exchanged, Z^--Pd(OH)$^+$ ions, which take up less exchange sites than Z^--Pd^{2+}-Z^-, are preferred to be formed. So, higher content of isolated Pd^{2+} leads to much more obvious increase of Z^--Pd(OH)$^+$ than that of Z^--Pd^{2+}-Z^-.

As discussed above, the maximum NO_x/Pd^{2+} ratio is 1 in theory. As reported by Khivantsev et al. [12], H_2O molecules can occupy NO_x storage sites due to strong hydration of isolated Pd^{2+} species. Since isolated Pd^{2+} is the main active site for NO_x storage, the NO_x/Pd^{2+} ratio of Pd/zeolite-based PNA materials will significantly smaller than 1 in the presence of H_2O. However, as shown in Figure 5b, NO_x/Pd^{2+} of 0.2-Pd is as much as 1, which indicates that isolated Pd^{2+} in this sample is not occupied by H_2O. As discussed above, isolated Pd^{2+} in 0.2-Pd mainly exists in the form of Z^--Pd^{2+}-Z^-. Thus Z^--Pd^{2+}-Z^- may be insensitive to H_2O in this condition. Furthermore, Z^--Pd(OH)$^+$ probably tends to hydrate due to the hydrogen bond interaction between H_2O and hydroxy. Accordingly, the difference in NO_x storage capacities of Z^--Pd(OH)$^+$ and Z^--Pd^{2+}-Z^- is probably caused by the different resistance to H_2O in this condition.

The atom utilization of Pd can be represented by mole ratios of NO_x adsorbed to total Pd loading. According to Figure 6, the Pd utilization of 0.2-Pd is as much as 100% whereas that of 2-Pd is only 25%. So, it is obvious that lower Pd loading benefits to the increase of Pd utilization. Besides, Khivantsev et al. [13] have reported that the Pd utilization of the 1 wt% Pd/SSZ-13 (Si/Al ratio = 6) sample is 100%, which is much higher than that of 1-Pd (Si/Al ratio = 16) using BEA as a support. Accordingly, zeolite structure and Si/Al ratio can also influence the Pd utilization. In this part, only the influence of Pd loading will be discussed.

As discussed above, lower Pd loading leads to lower content of isolated Pd^{2+}, whereas the proportion of isolated Pd ions in total Pd loading becomes larger. On the on hand, a larger proportion of isolated Pd ions in total Pd loading is beneficial for the improvement of Pd utilization, since isolated Pd ions are identified as the main active center for NO_x storage. On the other hand, less isolated Pd^{2+} results in a preference for the formation of Z^--Pd^{2+}-Z^- which exhibits higher NO_x storage capacity. Thus, the atom utilization of Pd can be improved by decreasing Pd loading.

Nevertheless, lower Pd loading also leads to less NO_x storage capacities of unit mass of Pd/BEA. Note that the coating amount of catalysts is limited in order to obtain acceptable back pressure. Accordingly, to determine optimum Pd loading for low temperature NO_x adsorption, the NO_x storage capacity and the atom utilization of Pd should be both considered. As shown in Figure 6, 0.2-Pd and 0.5-Pd exhibits much higher Pd atom utilization than that of the other two samples. Meanwhile, Pd atom utilization of 0.5-Pd is 92%, which is only slightly lower than that of 0.2-Pd. However, the NO_x storage capacity of 0.5-Pd is larger than twice of that of 0.2-Pd. So, 0.5wt% should be determined as the optimum Pd loading for Pd/BEA (Si/Al ratio = 16) served as PNA material.

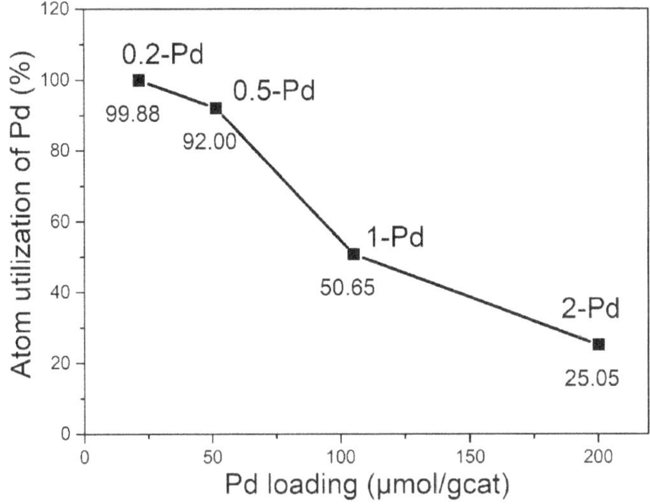

Figure 6. Tendency between Pd loading and the atom utilization of Pd.

3. Materials and Methods

3.1. Catalyst Preparation

H-BEA zeolite (Si/Al = 16.2) was supplied by Novel Chemistry. Pd/BEA samples with various Pd loading were prepared by incipient wetness impregnation, and $Pd(NO_3)_2$ solution (15.47wt% Pd, Heraeus Materials Technology Shanghai Ltd., Shanghai, China) was used. Then, the powders obtained were dried under ambient temperature followed by a 4-h calcination at 550 °C in air. Finally, samples were stabilized at 750 °C for 12 h in air with 10% H_2O. Pd loading of each sample was detected by inductively coupled plasma (ICP) analysis (USA Agilent 5100 ICP-OES, Santa Clara, CA, USA). Si/Al ratios were measured by X-ray fluorescence (XRF, ThermoFisher PERFORM'X, Waltham, MA, USA) analysis. Detailed information is exhibited in Table 3. In the following text, samples are abbreviated as *x*-Pd, where "*x*" represents the Pd loading of corresponding samples. Besides, the sample treated by Na^+ titration was named as Na-1-Pd.

Table 3. Pd loading of samples.

Catalyst	0-Pd	0.2-Pd	0.5-Pd	1-Pd	2-Pd	Na-1-Pd
ICP Pd (wt %)	0	0.23	0.55	1.12	2.13	0.25

3.2. Catalyst Characterization

Na^+ titration was used to quantify isolated Pd ions as reported by Ogura et al. [23]. Each sample was mixed with $NaNO_3$ (Purity above 99.0%, Tianjin Jiangtian Chemical Technology Co., Ltd., Tianjin, China) solution (0.1M) at 80 °C. The solution was stirred for 4 h followed by suction filtration. At last, the powders obtained were washed by deionized water and dried at 100 °C for 6 h. The whole process above was repeated three times. By analyzing the change of Pd loading in each sample after the titration process, the content of isolated Pd can be measured.

NO_x storage capacities were measured by standard catalyst evaluation tests carried out in a plug flow reactor system. 0.25 g sample (60–80 mesh) mixed with 0.75 g quartz (60–80 mesh) was loaded in a quartz reactor with a thermocouple. The sample was oxidized in the flow of 10% O_2/N_2 for 30 min at 500 °C and was cooled to 80 °C in N_2. Then the flow (200 ppm NO_x, 200 ppm CO, 5% CO_2, 5%H_2O, 10% O_2, balanced with N_2) mixed in the bypass in advance was introduced into the reactor. The NO_x

adsorption stage continued for 3 min. After that, the sample was heated to 500 °C with a ramping rate of 10 °C/min in the flow of N_2. Gas concentrations were measured by an online MKS MultiGas 2030 FTIR gas analyzer in the whole process. Besides, a space velocity of 28,800 h^{-1} was adopted.

Ex-situ FTIR was carried out on a Nicolet iS10 FTIR spectrometer equipped with a liquid N_2 cooled mercury cadmium telluride (MCT) detector to identify the existence of Pd ions. The sample (20 mg) was pressed into a self-supporting wafer with a diameter of 13 mm and was inserted into a cell sealed with ZnSe windows connected with a gas manifold. KBr was used to obtain background spectra. All samples, as well as KBr, were oxidized in the flow of 10% O_2/N_2 for 30 min at 500 °C in advance. Spectra were obtained at 200 °C in N_2.

In situ FTIR was carried out on a Nicolet iS10 FTIR spectrometer, too. Test temperatures and feed compositions varied according to the needs of different experiments, and detailed information will be reported below.

4. Conclusions

In this work, two isolated Pd^{2+} ions, Z^--Pd^{2+}-Z^- and Z^--$Pd(OH)^+$, on exchange sites of zeolites are confirmed as the main active sites for NO trapping in cold-start applications. Lower Pd loading leads to a lower content of isolated Pd^{2+} whereas the proportion of isolated Pd ions in total Pd loading becomes larger. Lower content of isolated Pd^{2+} further leads to a sharp decline of Z^--$Pd(OH)^+$, which is attributed to the 'exchange preference' for Z^--Pd^{2+}-Z^- in BEA. Besides, the higher NO_x storage capacity of Z^--Pd^{2+}-Z^- is demonstrated compared with that of Z^--$Pd(OH)^+$, which is caused by the different resistance to H_2O. In conclusion, the atom utilization of Pd can be improved by using lower Pd loading. 0.5wt% should be determined as the optimum Pd loading for Pd/BEA (Si/Al ratio = 16) serving as PNA material among all samples.

Supplementary Materials: The following are available online at http://www.mdpi.com/2073-4344/9/3/247/s1, Table S1: Detailed information of 1-Pd-80, Figure S1: CO in situ FTIR spectra of 1-Pd-80, Figure S2: NO_x adsorption profiles of 1-Pd and 1-Pd-80, Figure S3: NH_3 in situ FTIR spectra of 1-Pd-80.

Author Contributions: Conceptualization, B.Z.; methodology, J.W. (Jun Wang), M.S., and J.W. (Jianqiang Wang); software, B.Z.; validation, J.W. (Jianqiang Wang); formal analysis, J.W. (Jianqiang Wang); investigation, B.Z and J.W. (Jiaming Wang).; writing—original draft preparation, B.Z.; writing—review and editing, B.Z.; visualization, B.Z.; supervision, J.W. (Jun Wang); resources, M.S.; project administration, M.S.; funding acquisition, M.S.

Funding: This research was funded by National Key Research and Development Program, grant number 2017YFC0211002; State Key Laboratory of Advanced Technologies for Comprehensive Utilization of Platinum Metals, grant number SKL-SPM-2018017.

Conflicts of Interest: The authors declare no conflict of interest.

References

1. Johnson, T.; Joshi, A. Review of vehicle engine efficiency and emissions. *SAE Int. J. Engines* **2018**, *11*, 1307–1330.
2. Gao, F.; Tang, X.; Yi, H.; Zhao, S.; Li, C.; Li, J.; Shi, Y.; Meng, X. A Review on Selective Catalytic Reduction of NOx by NH3 over Mn–Based Catalysts at Low Temperatures: Catalysts, Mechanisms, Kinetics and DFT Calculations. *Catalysts* **2017**, *7*, 199. [CrossRef]
3. Kubiak, L.; Matarrese, R.; Castoldi, L.; Lietti, L.; Daturi, M.; Forzatti, P. Study of N_2O Formation over Rh- and Pt-Based LNT Catalysts. *Catalysts* **2016**, *6*, 36. [CrossRef]
4. Chen, H.-Y.; Collier, J.E.; Liu, D.; Mantarosie, L.; Durán-Martín, D.; Novák, V.; Rajaram, R.R.; Thompsett, D. Low temperature NO storage of zeolite supported Pd for low temperature diesel engine emission control. *Catal. Lett.* **2016**, *146*, 1706–1711. [CrossRef]
5. Chen, H.-Y.; Mulla, S.; Weigert, E.; Camm, K.; Ballinger, T.; Cox, J.; Blakeman, P. Cold Start Concept (CSC™) A Novel Catalyst for Cold Start Emission Control. *SAE Int. J. Fuels Lubr.* **2013**, *6*, 372–381. [CrossRef]
6. Ryou, Y.; Lee, J.; Cho, S.J.; Lee, H.; Kim, C.H.; Kim, D.H. Activation of Pd/SSZ-13 catalyst by hydrothermal aging treatment in passive NO adsorption performance at low temperature for cold start application. *Appl. Catal. B* **2017**, *212*, 140–149. [CrossRef]

7. Lee, J.; Ryou, Y.; Hwang, S.; Kim, Y.; Cho, S.J.; Lee, H.; Kim, C.H.; Kim, D.H. Comparative study of the mobility of Pd species in SSZ-13 and ZSM-5, and its implication for their activity as passive NOx adsorbers (PNAs) after hydro-thermal aging. *Catal. Sci. Technol.* **2019**, *9*, 163–173. [CrossRef]
8. Ryou, Y.; Lee, J.; Lee, H.; Kim, C.H.; Kim, D.H. Effect of various activation conditions on the low temperature NO adsorption performance of Pd/SSZ-13 passive NOx adsorber. *Catal. Today* **2019**, *320*, 175–180. [CrossRef]
9. Vu, A.; Luo, J.; Li, J.; Epling, W.S. Effects of CO on Pd/BEA Passive NO_x Adsorbers. *Catal. Lett.* **2017**, *147*, 745–750. [CrossRef]
10. Zheng, Y.; Kovarik, L.; Engelhard, M.H.; Wang, Y.; Wang, Y.; Gao, F.; Szanyi, J. Low-Temperature Pd/Zeolite Passive NO_x Adsorbers: Structure, Performance, and Adsorption Chemistry. *J. Phys. Chem. C* **2017**, *121*, 15793–15803. [CrossRef]
11. Lee, J.; Ryou, Y.; Cho, S.J.; Lee, H.; Kim, C.H.; Kim, D.H. Investigation of the active sites and optimum Pd/Al of Pd/ZSM–5 passive NO adsorbers for the cold-start application: Evidence of isolated-Pd species obtained after a high-temperature thermal treatment. *Appl. Catal. B* **2018**, *226*, 71–82. [CrossRef]
12. Khivantsev, K.; Gao, F.; Kovarik, L.; Wang, Y.; Szanyi, J. Molecular Level Understanding of How Oxygen and Carbon Monoxide Improve NO_x Storage in Palladium/SSZ-13 Passive NO_x Adsorbers: The Role of NO+ and Pd (II)(CO)(NO) Species. *J. Phys. Chem. C* **2018**, *122*, 10820–10827. [CrossRef]
13. Khivantsev, K.; Jaegers, N.R.; Kovarik, L.; Hanson, J.C.; Tao, F.; Tang, Y.; Zhang, X.; Koleva, I.Z.; Aleksandrov, H.A.; Vayssilov, G.N.; et al. Achieving Atomic Dispersion of Highly Loaded Transition Metals in Small-Pore Zeolite SSZ-13: High-Capacity and High-Efficiency Low-Temperature CO and Passive NOx Adsorbers. *Angew. Chem.* **2018**, *130*, 16914–16919. [CrossRef]
14. Mei, D.; Gao, F.; Szanyi, J.; Wang, Y. Mechanistic insight into the passive NOx adsorption in the highly dispersed Pd/HBEA zeolite. *Appl. Catal. A* **2019**, *569*, 181–189. [CrossRef]
15. Khivantsev, K.; Jaegers, N.R.; Kovarik, L.; Prodinger, S.; Derewinski, M.A.; Wang, Y.; Gao, F.; Szanyi, J. Palladium/Beta zeolite passive NOx adsorbers (PNA): Clarification of PNA chemistry and the effects of CO and zeolite crystallite size on PNA performance. *Appl. Catal. A Gen.* **2019**, *569*, 141–148. [CrossRef]
16. Ryou, Y.; Lee, J.; Kim, Y.; Hwang, S.; Lee, H.; Kim, C.H.; Kim, D.H. Effect of reduction treatments (H2 vs. CO) on the NO adsorption ability and the physicochemical properties of Pd/SSZ-13 passive NOx adsorber for cold start application. *Appl. Catal. A* **2019**, *569*, 28–34. [CrossRef]
17. Porta, A.; Pellegrinelli, T.; Castoldi, L.; Matarrese, R.; Morandi, S.; Dzwigaj, S.; Lietti, L. Low Temperature NOx Adsorption Study on Pd-Promoted Zeolites. *Top. Catal.* **2018**, *61*, 2021–2034. [CrossRef]
18. Theis, J.R.; Lambert, C.K. Mechanistic assessment of low temperature NOx adsorbers for cold start NOx control on diesel engines. *Catal. Today* **2019**, *320*, 181–195. [CrossRef]
19. Newsam, J.; Treacy, M.M.; Koetsier, W.; De Gruyter, C. Structural characterization of zeolite beta. *Proc. R. Soc. Lond. A* **1988**, *420*, 375–405. [CrossRef]
20. Gao, F.; Washton, N.M.; Wang, Y.; Kollár, M.; Szanyi, J.; Peden, C.H. Effects of Si/Al ratio on Cu/SSZ-13 NH 3-SCR catalysts: Implications for the active Cu species and the roles of Brønsted acidity. *J. Catal.* **2015**, *331*, 25–38. [CrossRef]
21. Giordanino, F.; Vennestrøm, P.N.; Lundegaard, L.F.; Stappen, F.N.; Mossin, S.; Beato, P.; Bordiga, S.; Lamberti, C. Characterization of Cu-exchanged SSZ-13: A comparative FTIR, UV-Vis, and EPR study with Cu-ZSM-5 and Cu-β with similar Si/Al and Cu/Al ratios. *Dalton Trans.* **2013**, *42*, 12741–12761. [CrossRef] [PubMed]
22. Gao, F.; Peden, C.H.F. Recent Progress in Atomic-Level Understanding of Cu/SSZ-13 Selective Catalytic Reduction Catalysts. *Catalysts* **2018**, *8*, 140.
23. Ogura, M.; Hayashi, M.; Kage, S.; Matsukata, M.; Kikuchi, E. Determination of active palladium species in ZSM-5 zeolite for selective reduction of nitric oxide with methane. *Appl. Catal. B* **1999**, *23*, 247–257. [CrossRef]
24. Kumar, A.; Buttry, D.A. Influence of Halide Ions on Anodic Oxidation of Ethanol on Palladium. *Electrocatalysis* **2016**, *7*, 201–206. [CrossRef]
25. Yuan, N.; Pascanu, V.; Huang, Z.; Valiente, A.; Heidenreich, N.; Leubner, S.; Inge, A.K.; Gaar, J.; Stock, N.; Persson, I. Probing the evolution of palladium species in Pd@ MOF catalysts during the Heck coupling reaction: An operando X-ray absorption spectroscopy study. *J. Am. Chem. Soc.* **2018**. [CrossRef] [PubMed]
26. Palazov, A.; Chang, C.; Kokes, R. The infrared spectrum of carbon monoxide on reduced and oxidized palladium. *J. Catal.* **1975**, *36*, 338–350. [CrossRef]

27. Aylor, A.W.; Lobree, L.J.; Reimer, J.A.; Bell, A.T. Investigations of the Dispersion of Pd in H-ZSM-5. *J. Catal.* **1997**, *172*, 453–462. [CrossRef]
28. Okumura, K.; Amano, J.; Yasunobu, N.; Niwa, M. X-ray absorption fine structure study of the formation of the highly dispersed PdO over ZSM-5 and the structural change of Pd induced by adsorption of NO. *J. Phys. Chem. B* **2000**, *104*, 1050–1057. [CrossRef]
29. Barzetti, T.; Selli, E.; Moscotti, D.; Forni, L. Pyridine and ammonia as probes for FTIR analysis of solid acid catalysts. *J. Chem. Soc. Faraday Trans.* **1996**, *92*, 1401–1407. [CrossRef]
30. Pawelec, B.; Campos-Martin, J.; Cano-Serrano, E.; Navarro, R.; Thomas, S.; Fierro, J. Removal of PAH compounds from liquid fuels by Pd catalysts. *Environ. Sci. Technol.* **2005**, *39*, 3374–3381. [CrossRef] [PubMed]
31. Centeno, M.; Carrizosa, I.; Odriozola, J. NO–NH3 coadsorption on vanadia/titania catalysts: Determination of the reduction degree of vanadium. *Appl. Catal. B* **2001**, *29*, 307–314. [CrossRef]
32. Niwa, M.; Nishikawa, S.; Katada, N. IRMS–TPD of ammonia for characterization of acid site in β-zeolite. *Microporous Mesoporous Mater.* **2005**, *82*, 105–112. [CrossRef]
33. Datka, J.; Gil, B.; Kubacka, A. Acid properties of NaH-mordenites: Infrared spectroscopic studies of ammonia sorption. *Zeolites* **1995**, *15*, 501–506. [CrossRef]
34. Ding, K.; Gulec, A.; Johnson, A.M.; Schweitzer, N.M.; Stucky, G.D.; Marks, L.D.; Stair, P.C. Identification of active sites in CO oxidation and water-gas shift over supported Pt catalysts. *Science* **2015**, *350*, 189. [CrossRef] [PubMed]
35. Bare, S.R.; Kelly, S.D.; Sinkler, W.; Low, J.J.; Modica, F.S.; Valencia, S.; Corma, A.; Nemeth, L.T. Uniform catalytic site in Sn-β-Zeolite determined using x-ray absorption fine structure. *J. Am. Chem. Soc.* **2005**, *127*, 12924–12932. [CrossRef] [PubMed]
36. Jones, A.J.; Iglesia, E. The strength of Brønsted acid sites in microporous aluminosilicates. *ACS Catal.* **2015**, *5*, 5741–5755. [CrossRef]
37. Higgins, J.; LaPierre, R.B.; Schlenker, J.; Rohrman, A.; Wood, J.; Kerr, G.; Rohrbaugh, W. The framework topology of zeolite beta. *Zeolites* **1988**, *8*, 446–452. [CrossRef]

Article

Optimization of Ammonia Oxidation Using Response Surface Methodology

Marek Inger *, Agnieszka Dobrzyńska-Inger, Jakub Rajewski and Marcin Wilk

New Chemical Syntheses Institute, Al. Tysiąclecia Państwa Polskiego 13a, 24-100 Puławy, Poland; agnieszka.dobrzynska-inger@ins.pulawy.pl (A.D.-I.); jakub.rajewski@ins.pulawy.pl (J.R.); marcin.wilk@ins.pulawy.pl (M.W.)

* Correspondence: marek.inger@ins.pulawy.pl; Tel.: +48-(81)-473-1415

Received: 22 December 2018; Accepted: 4 March 2019; Published: 9 March 2019

Abstract: In this paper, the design of experiments and response surface methodology were proposed to study ammonia oxidation process. The following independent variables were selected: the reactor's load, the temperature of reaction and the number of catalytic gauzes, whereas ammonia oxidation efficiency and N_2O concentration in nitrous gases were assumed as dependent variables (response). Based on the achieved results, statistically significant mathematical models were developed which describe the effect of independent variables on the analysed responses. In case of ammonia oxidation efficiency, its achieved value depends on the reactor's load and the number of catalytic gauzes, whereas the temperature in the studied range (870–910 °C) has no effect on this dependent variable. The concentration of nitrous oxide in nitrous gases depends on all three parameters. The developed models were used for the multi-criteria optimization with the application of desirability function. Sets of parameters were achieved for which optimization assumptions were met: maximization of ammonia oxidation efficiency and minimization of the N_2O amount being formed in the reaction.

Keywords: ammonia oxidation; response surface methodology; desirability function; Box-Behnken design

---

1. Introduction

Nitric acid is mainly used for producing nitrogen fertilizers: ammonium nitrate (AN) and calcium ammonium nitrate (CAN) which constitute 75–80% of its entire production. The remaining amount of nitric acid is used in other industrial applications for example as a nitration agent for the production of explosives and other semi-organic products (aliphatic nitro compounds and aromatic nitro compounds) for the production of adipic acid, for metallurgy (etching steel) [1].

The industrial production of nitric acid is based on Ostwald process [1] which involves three basic stages: the catalytic oxidation of ammonia to nitrogen oxide (NO) with the use of oxygen from air, oxidation of nitrogen oxide (NO) to nitrogen dioxide (NO_2) and absorption of nitrogen oxides in water with the formation of HNO_3.

Ammonia consumption depends on the selectivity of the applied ammonia oxidation catalyst and on the process conditions. Among numerous catalysts [2–8], packages of gauzes made of noble metal alloys such as platinum and rhodium are most commonly applied in industrial practice [7–10]. Properly selected catalyst package allows to obtain ammonia conversion to main product (NO) in the range of 90–98% depending mainly on oxidation pressure [1,7,8]. Oxidation pressure has an inversely proportional effect on ammonia oxidation efficiency. In order to alleviate this effect, the temperature of reaction should be higher. However, this leads to the increased platinum losses and as a consequence, shortens the lifetime of the catalytic gauzes. For example, platinum losses are six times higher after increasing the temperature of reaction from 820 to 920 °C [1,2]. Therefore, both these aspects should be taken into account to determine the temperature of reaction.

The application of medium pressure in the oxidation unit (0.35–0.55 MPa) and high pressure in the absorption unit (0.8–1.5 MPa) is optimal for specific ammonia consumption and efficient energy use. Therefore, modern nitric acid plants are dual-pressure ones. The average pressure in the oxidation unit is a kind of trade-off between the capacity that is possible to achieve per 1 m^2 of catalytic gauzes, oxidation efficiency, number of gauzes in package, lifetime of gauzes and noble metals losses during exploitation [1,2].

In the context of global warming and climate changes, a very important issue related to ammonia oxidation process is the amount of the by-product formed that is nitrous oxide (N_2O). In Kyoto Protocol, N_2O was qualified as a greenhouse gas with a very high global warming potential, about 300 times higher than CO_2 [11]. At room temperature, N_2O is a colourless, non-flammable gas with a delicate pleasant smell and sweet taste [12]. Since it was isolated at the end of 17th century and because of its pain-relieving and anaesthetic properties, it has been widely applied in dentistry and surgery. Currently, due to some concerns, there is an ongoing discussion on its safe use which has the effect of decreasing the N_2O application in medicine [12,13]. At the same time, the increasing trend of its use for recreational purposes is observed. Inhaling the 'laughing gas' causes euphoria and hallucinations [13].

Microbial nitrification and de-nitrification in land and aqueous eco-systems are the natural sources of N_2O in environment. The anthropogenic sources are cultivated soils fertilized intensely with nitrogen fertilizers and industrial processes such as burning fossil fuels and biomass as well as the production of adipic acid and nitric acid with the last one being regarded as the biggest source of N_2O in the chemical industry [14,15]. Nitrous oxide formed in nitric acid plant does not undergo any conversions and it is released to atmosphere. Currently, the emission of this gas is monitored and industry is obliged to reduce it. Pursuant to BAT requirements, concentration of this gas in outlet gases cannot exceed 20–300 ppm depending on the type of nitric acid plants [16,17]. However, due to the battle against climate change and global warming, further restrictions in emission limits can be expected.

There are several methods of limiting N_2O emissions from nitric acid plants [14]. Generally, they can be classified as primary and secondary methods. Primary methods involve preventing the formation of N_2O during ammonia oxidation. They include modification of catalytic gauzes (so-called low-emissions systems) and parameters optimization of ammonia oxidation process. Secondary methods involve the removal of N_2O. At the temperature over 800 °C, thermal decomposition of N_2O occurs but the efficient decomposition requires ensuring adequately long residence time at high temperatures [14,15].

The achievement of low level of N_2O emissions requires the application of the catalytic methods such as high temperature N_2O decomposition from nitrous gases, low- or middle temperature N_2O decomposition or reduction from tail gases. High temperature method is more common. In some cases, the combination of primary method (application of modified catalytic gauzes packages and/or optimization of ammonia oxidation parameters) and high temperature N_2O decomposition ensures meeting the emission standards.

Optimization of production process requires extensive knowledge and understanding the effect of particular parameters on the process. Until recently, the most commonly applied approach of researchers to study simple and complex processes was 'one-factor-at-a-time' (*OFAT*), which is time consuming and ineffective method for processes with multiple complicated dependencies between parameters. Over the last years, mathematical and statistical methods for design of experiments and parameters optimization have been applied more frequently [18]. Because of its usability, this method is applied for the design, improvement and optimization of production processes and products [19–21]. It is a widely applied method in research of various processes [22–27] and approx. 50% of all applications is in medicine, engineering, biochemistry, physics and computer science [28]. In this method, reaction kinetics equations and process mechanism are not taken into account and they are regarded as a 'black-box' [29] (Figure 1).

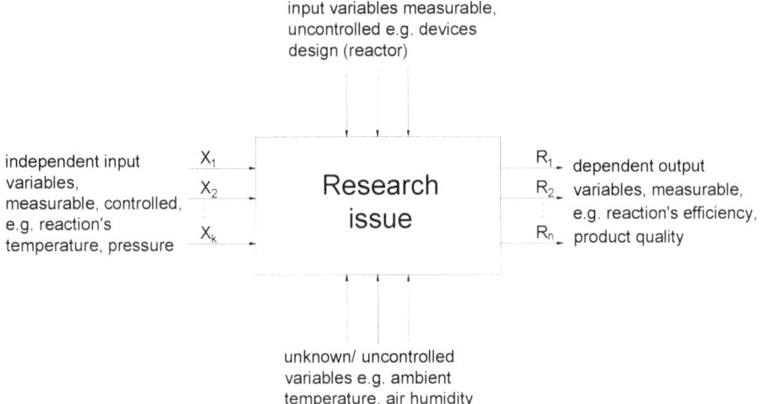

Figure 1. "Black–box" model of the research issue in design of experiments methodology.

The choice of experiment plan depends mainly on the issue which is the subject matter of investigations as well as on objectives which are set. The most commonly applied experiments plans include: full or fractional factorial, Plackett-Burman, central composite, Box-Behnken and Taguchi designs.

As a result of modelling of the data obtained, empirical equations with statistically significant importance are received which describe the effect of process variables (independent variables) on the process result (response variable).

Desirability function (*DF*) can be applied in search for optimal operational parameters. The method proposed by Derringer and Suich [18] involves the construction response surface model and then finding the values of independent variables which ensure the most desirable value. The objective of the presented studies was the analysis of the impact of reactor's operational parameters on ammonia oxidation reaction. To the best of our knowledge, the approach presented here to describe ammonia oxidation process is published for the first time.

2. Results and Discussion

2.1. Design of Experiments

Ammonia oxidation reaction depends on a few process variables. In this study, the effect of the reactor's load (X_1), the temperature of nitrous gas specifying the temperature of reaction (X_2) and the number of catalytic gauzes (X_3) on ammonia oxidation reaction was investigated. The oxidation efficiency of NH_3 to NO (R_1) and N_2O concentration in nitrous gases (R_2) were selected as measures for ammonia oxidation reaction. The matrix of 15 experiments including particular levels of coded variables and achieved values of response variables R_1 and R_2 are presented in Table 1. In the regarded experimental area of independent variables, ammonia oxidation efficiency ranged from 91.4% to 96.4%, whereas N_2O concentration in nitrous gases ranged from 1011 to 1762 ppm.

Table 1. The Box-Behnken design matrix and experimental data.

Standard Order	Run Order	X_1 Reactor's Load, kg $NH_3/(m^2h)$	X_2 Temperature, °C	X_3 No. of Gauzes, pcs	R_1 Ammonia Oxidation Efficiency, %	R_2 N_2O Concentration in Nitrous Gases, ppm
3	1	−1	1	0	96.1	1011
15	2	0	0	0	96.1	1279
7	3	−1	0	1	96.2	1238
6	4	1	0	−1	91.4	1620
9	5	0	−1	−1	92.0	1762
5	6	−1	0	−1	93.7	1348
14	7	0	0	0	96.2	1265
12	8	0	1	1	96.4	1074
2	9	1	−1	0	95.6	1457
8	10	1	0	1	96.2	1312
1	11	−1	−1	0	96.3	1423
4	12	1	1	0	95.8	1114
13	13	0	0	0	96.2	1271
11	14	0	−1	1	96.0	1506
10	15	0	1	−1	92.7	1207

2.2. Model Fitting

The first task was to find out which equation would allow to obtain the best correlation between independent variables and responses. Analysis of Variance (ANOVA) was carried out for most frequently applied equations: linear, two-factor interaction (2FI), quadratic and cubic. Table 2 includes the summary statistics of both responses for different mathematical equations.

Table 2. Model summary statistics for response variables R_1 and R_2.

Response Variable: R_1—Ammonia Oxidation Efficiency						
Source	Std. Dev.	R^2	Adjusted R^2	Predicted R^2	PRESS	
Linear	1.08	0.6969	0.6142	0.4268	24.38	
2FI	1.20	0.7294	0.5265	−0.1215	47.70	
Quadratic	0.2790	0.9908	0.9744	0.8557	6.14	Suggested
Cubic	0.0577	0.9998	0.9989		*	Aliased
Response variable: R_2—N_2O concentration in nitrous gases						
Source	Std. Dev.	R^2	Adjusted R^2	Predicted R^2	PRESS	
Linear	87.82	0.8524	0.8121	0.6993	1.728×10^5	
2FI	93.58	0.8781	0.7867	0.4155	3.359×10^5	
Quadratic	49.78	0.9784	0.9396	0.6574	1.969×10^5	Suggested
Cubic	7.02	0.9998	0.9988		*	Aliased

* - case(s) with leverage of 1.0000; PRESS statistic not defined.

Based on the achieved results, it was found that the experimental data is described best with quadratic and cubic equations. For both responses, high values of R^2 and adjusted R^2 were achieved. The number of conducted experiments caused that the cubic model was aliased. It means that the experimental matrix contains an insufficient number of experimental points for independent estimation of all effects for these models. Therefore, quadratic equation was selected for further analysis.

The statistical significance of these equations and their particular terms was specified based on Analysis of Variance (ANOVA). Results of this analysis are presented in Tables 3 and 4, for response variables R_1 and R_2 respectively. Large F-value indicates that most changes of independent variable can be explained with the developed regression equation. The correlated probability p-value is used to estimate whether F-value is large enough to show statistical significance.

Table 3. ANOVA results for response variable R_1.

Source	Sum of Squares	df	Mean Square	F-Value	p-Value	Significance
Model	42.14	9	4.68	60.16	0.0001	highly significant
X_1	1.36	1	1.36	17.49	0.0086	significant
X_2	0.1512	1	0.1512	1.94	0.2221	not significant
X_3	28.13	1	28.13	361.35	<0.0001	highly significant
X_1X_2	0.0400	1	0.0400	0.5139	0.5055	not significant
X_1X_3	1.32	1	1.32	16.99	0.0092	significant
X_2X_3	0.0225	1	0.0225	0.2891	0.6139	not significant
${X_1}^2$	0.0126	1	0.0126	0.1614	0.7044	not significant
${X_2}^2$	0.0926	1	0.0926	1.19	0.3252	not significant
${X_3}^2$	11.09	1	11.09	142.53	< 0.0001	highly significant
Residual	0.3892	5	0.0778			
Lack of Fit	0.3825	3	0.1275	38.25	0.0256	significant
Pure Error	0.0067	2	0.0033			
Corrected total SS	42.53	14				

$p < 0.0001$—highly significant, $0.0001 < p < 0.05$—significant, $p > 0.05$—not significant

The probability p-value for the achieved model of variable R_1 is 0.0001. It means that the model is statistically significant but some terms of equation are statistically not significant. Coefficients: R^2, adjusted R^2 and predicted R^2 are very high: 0.9908, 0.9744 and 0.8577, respectively. There is also high compliance between coefficients: predicted R^2 and adjusted R^2 (difference <0.2). The achievement of statistically significant value lack of fit (0.0256) is the incompliance of this model as this parameter should be statistically not significant.

Table 4. ANOVA results for response variable R_2.

Source	Sum of Squares	df	Mean Square	F-Value	p-Value	Significance
Model	5.623×10^5	9	62,480.28	25.21	0.0012	highly significant
X_1	29,161.12	1	29,161.12	11.77	0.0186	significant
X_2	3.793×10^5	1	3.793×10^5	153.05	<0.0001	highly significant
X_3	81,406.13	1	81,406.13	32.85	0.0023	significant
X_1X_2	1190.25	1	1190.25	0.4803	0.5192	not significant
X_1X_3	9801.00	1	9801.00	3.95	0.1034	not significant
X_2X_3	3782.25	1	3782.25	1.53	0.2716	not significant
${X_1}^2$	732.33	1	732.33	0.2955	0.6101	not significant
${X_2}^2$	148.10	1	148.10	0.0598	0.8166	not significant
${X_3}^2$	54,881.26	1	54,881.26	22.14	0.0053	significant
Residual	12,391.92	5	2478.38			
Lack of Fit	12,293.25	3	4097.75	83.06	0.0119	significant
Pure Error	98.67	2	49.33			
Corrected total SS	5.74710^5	14				

$p < 0.0001$—highly significant, $0.0001 < p < 0.05$—significant, $p > 0.05$—not significant

In case of response variable R_2, the probability p-value (0.0012) indicates that the assumed quadratic equation is statistically significant but some of its terms are statistically not significant. High coefficients R^2, adjusted R^2 and predicted R^2 are also achieved for the second response variable and they are: 0.9784, 0.9396 and 0.6574, respectively. However, the difference between predicted R^2 and adjusted R^2 is larger than the recommended one (>0.2). This may demonstrate a large block effect or problems with model or data. This model is also characteristic of statistically significant parameter lack of fit ($p = 0.0119$).

At a further stage of analysis, statistically not significant terms of initial equation were eliminated from the analysis. The reduction was made using step-by-step method (from the most insignificant term). For both these response variables, only statistically significant terms were left and higher R^2, adjusted R^2 and predicted R^2 coefficients were achieved. Results of Analysis of Variance (ANOVA) are presented in Tables 5 and 6.

Table 5. ANOVA results for reduced model of the response variable R_1.

Source	Sum of Squares	df	Mean Square	*F*-Value	*p*-Value	Significance
Model	41.83	4	10.46	148.66	<0.0001	highly significant
X_1	1.36	1	1.36	19.35	0.0013	significant
X_3	28.13	1	28.13	399.85	<0.0001	highly significant
X_1X_3	1.32	1	1.32	18.80	0.0015	significant
$X_3{}^2$	11.02	1	11.02	156.63	<0.0001	highly significant
Residual	0.7034	10	0.0703			
Lack of Fit	0.6967	8	0.0871	26.13	0.0374	significant
Pure Error	0.0067	2	0.0033			
Corrected total SS	42.53	14				
R^2	0.9835					
Adjusted R^2	0.9768					
Predicted R^2	0.9550					

The obtained mathematical model for response R_1 is highly significant (p-value < 0.0001). The dependence on linear terms X_1, X_3, interaction X_1X_3 and quadratic term $X_3{}^2$ are significant. High determination coefficients are obtained l (R^2 = 0.9835, adjusted R^2 = 0.9768, predicted R^2 = 0.9550).

The final model is presented in Equation (1).

$$R = 96.04 - 0.4125X_1 + 1.88X_3 + 0.575X_1X_3 - 1.72X_3^2 \quad (1)$$

Table 6. ANOVA results for reduced model of the response variable R_2.

Source	Sum of Squares	df	Mean Square	*F*-Value	*p*-Value	Significance
Model	5.467×10^5	4	1.367×10^5	48.81	<0.0001	highly significant
X_1	29,161.12	1	29,161.12	10.41	0.0091	significant
X_2	3.793×10^5	1	3.793×10^5	135.47	<0.0001	highly significant
X_3	81,406.13	1	81,406.13	29.07	0.0003	significant
$X_3{}^2$	56,826.52	1	56,826.52	20.30	0.0011	significant
Residual	28,000.13	10	2800.01			
Lack of Fit	27,901.46	8	3487.68	70.70	0.0140	significant
Pure Error	98.67	2	49.33			
Corrected total SS	5.747×10^5	14				
R^2	0.9513					
Adjusted R^2	0.9318					
Predicted R^2	0.8743					

The obtained mathematical model for response R_2 is highly significant (p-value < 0.0001). The dependence on linear terms X_1, X_2, X_3 and quadratic term $X_3{}^2$ are significant. High determination coefficients (R^2 = 0.9513, adjusted R^2 = 0.9318, predicted R^2 = 0.8743) were achieved for the model. The final model is presented in Equation (2).

$$R = 1260 + 60.37X_1 - 217.75X_2 - 100.88X_3 + 123.37X_3^2 \quad (2)$$

2.3. Model Diagnostics

Before the process optimization, the model diagnostics for both equations was performed because of occurrence of statistically significant *Lack of fit* parameter. Results of model diagnostics: a normal probability of the residuals, residuals analysis and actual data versus predicted values plots were analysed.

Figure 2 presents model diagnostics for response variable R_1, whereas Figure 3 presents model diagnostics for response variable R_2. Normal plot of studentised residuals should be approximately a straight line, whereas studentised residuals versus predicted response values and versus run should be a random scatter. Points in plots of real response values with reference to predicted response values line up accurately along the axis at the angle of 45°.

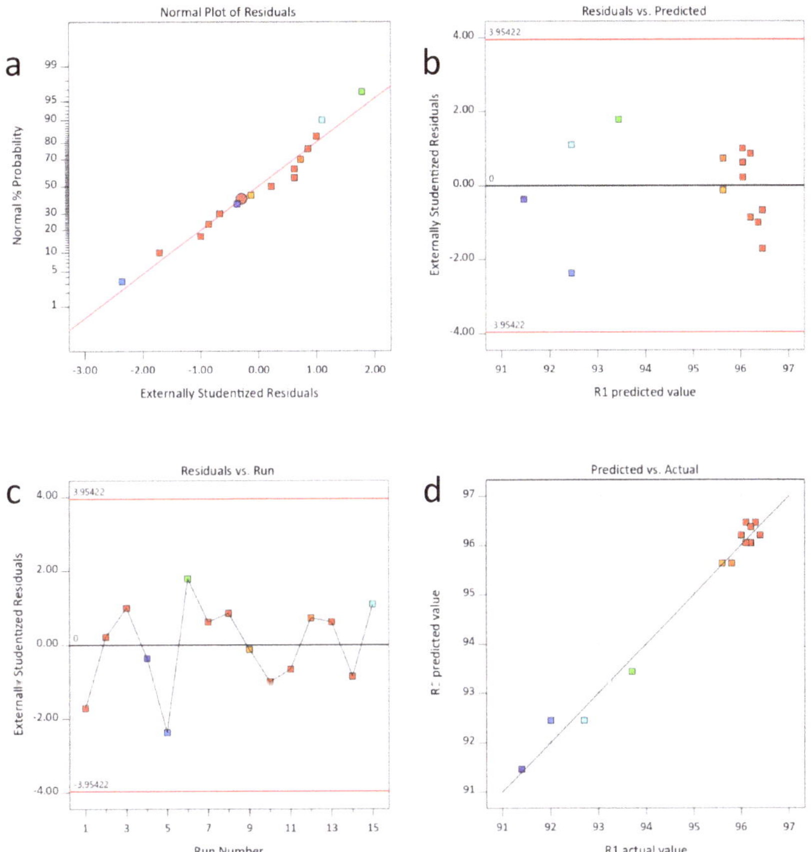

Figure 2. Model diagnostics for response variable R_1 (**a**) A normal probability plot of the residuals; (**b**) Residuals versus predicted value of R_1; (**c**) Residuals versus run number; (**d**) Predicted versus actual value of response variable R_1. Points on graphs correspond to results of particular experiments and colour points correspond to the value in accordance with the scale.

These diagnostics show that despite the fact that *Lack of fit* parameter is statistically significant, experimental and predicted points for both equations correlate well with each other.

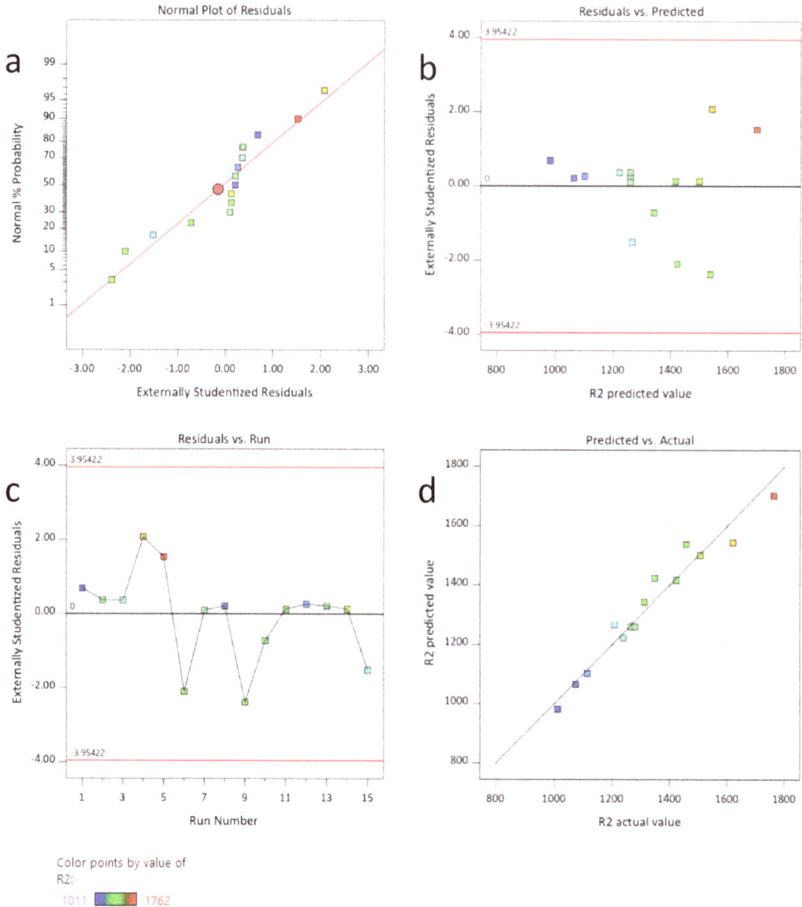

Figure 3. Model diagnostics for response variable R_2 (**a**) A normal probability plot of the residuals; (**b**) Residuals versus predicted value of R_2; (**c**) Residuals versus run number; (**d**) Predicted versus actual value of response variable R_2. Points on graphs correspond to results of particular experiments and colour points correspond to the value in accordance with the scale.

2.4. The Effect of Independent Variables

In case of response variable R_1 (ammonia oxidation efficiency), the mathematical model shows a strong linear effect of the reactor's load (X_1) and the number of catalytic gauzes (X_3) and interaction between these two variables (X_1X_3) and the quadratic term number of catalytic gauzes (X_3^2) on the achieved response variable. The temperature of reaction in the studied range does not affect the ammonia oxidation efficiency. The effect of X_1 and X_3 on response R_1 were shown as contour plot (Figure 4). According to the presented plot of variable of R_1, a small number of catalytic gauzes causes lower ammonia oxidation efficiency for the entire range of studies reactor's load. The increase in the number of catalytic gauzes to $X_3 = 0$ causes increase in oxidation efficiency within the entire range of studied reactor's load.

Studies related to dependency of N_2O concentration in nitrous gases on operating parameters are relatively new research issue. Therefore, there is a lack of scientific reports dedicated to systematic studies in this field. In case of the N_2O concentration in nitrous gases, the achieved mathematical model demonstrates a significant effect of the selected process variables (X_1, X_2, X_3) and the quadratic term number of catalytic gauzes (X_3^2) on the achieved response variable. This effect is illustrated in Figures 5–7. The analysis of Equation (2) and Figures 5–7 indicates that the temperature of reaction has the biggest quantitative effect on N_2O concentration in nitrous gases. From the comparison of plots (Figure 5a–c) it can be concluded that despite the presence of statistically significant terms of equations derived from variable X_3, plots of contour line corresponding to levels 0 and 1 are similar. Only for $X_3 = -1$, higher values of R_2 are achieved. Profiles of response variable R_2 presented in Figures 6a–c and 7a–c confirm the effect of the number of catalytic gauzes. Both these figures show that the number of catalytic gauzes has little effect on the amount of N_2O being formed. For the level of X_3 = 0–0.4 (10–12 gauzes), the local optimum is observed. For this number of gauzes, increasing the reactor's load (X_1 = 1) at the fixed reaction temperature (Figure 6a–c) and decreasing the reaction temperature at the fixed reactor's load (Figure 7a–c) does not cause a significant decrease in N_2O concentration in nitrous gases.

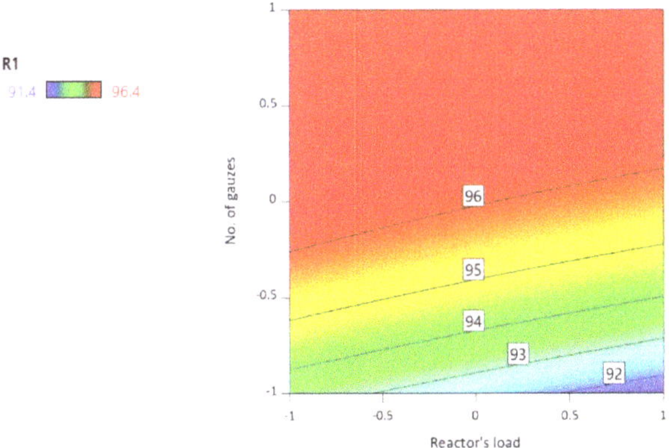

Figure 4. Interaction effect between the reactor's load (X_1) and the number of catalytic gauzes (X_3) on ammonia oxidation efficiency (R_1) by contour plot.

2.5. Multi-Response Desirability Optimization

The major optimization task is to find the number of catalytic gauzes and the permissible reactor's load ensuring the maximization of ammonia oxidation efficiency (R_1) and the minimization of N_2O concentration in nitrous gases. Results of experiments discussed in Section 2.4. indicate that statistically, the temperature has no significant effect on ammonia oxidation efficiency but on the other hand, the amount of N_2O formed is reversely proportional to the temperature of reaction. For desirability function, it was assumed that independent variables are in the variability range. Assumptions for the optimization are presented in Table 7.

Table 7. Assumptions for the optimization of the ammonia oxidation process using desirability function. Variables symbol identification according to the Table 1.

Name	Goal	Lower Limit	Upper Limit	Lower Weight *	Upper Weight *	Importance **
X_1	in range	−1	1	1	1	3
X_2	in range	−1	1	1	1	3
X_3	in range	−1	1	1	1	3
R_1	maximize	91.4	96.4	1	1	3
R_2	minimize	1011	1762	1	1	3

* Weight: 1—linear change of values in the range from 0 to 1; ** Importance: 5—high, 3—medium, 1—low.

For such optimization assumptions, the area of detailed set of parameters was achieved. It confirms that optimization assumptions are met. Desirability functions for three temperature levels are presented as contour plot in Figure 8.

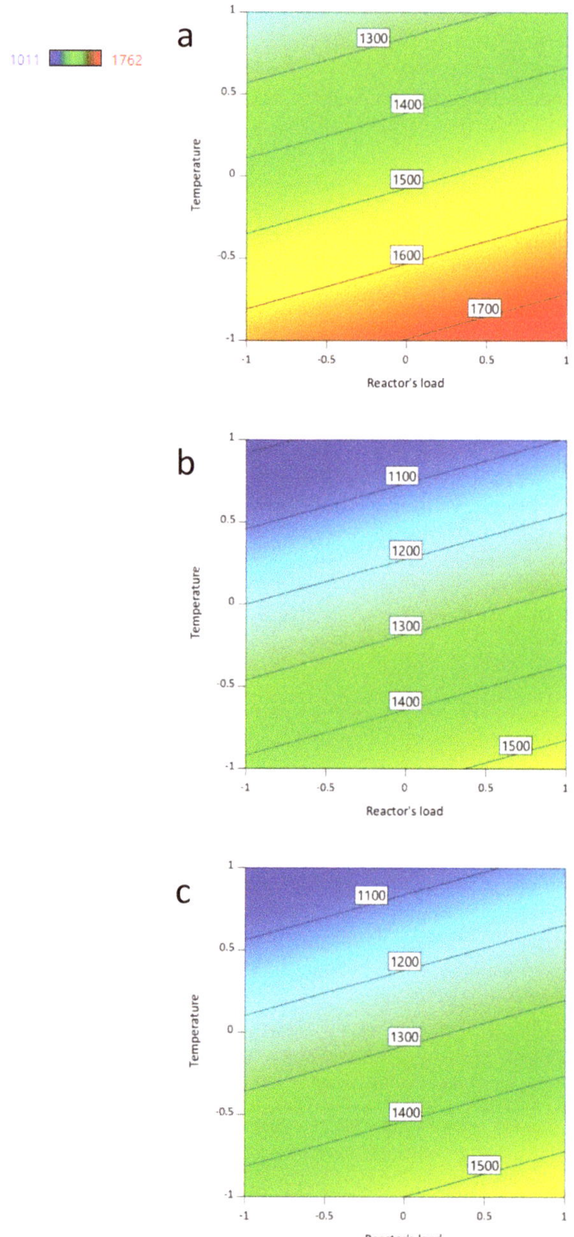

Figure 5. Interaction effect between the reactor's load (X_1) and the temperature (X_2) at fixed number of catalytic gauzes (X_3) on N_2O concentration in nitrous gases (R_2) by contour plot. (**a**) $X_3 = -1$; (**b**) $X_3 = 0$; (**c**) $X_3 = 1$.

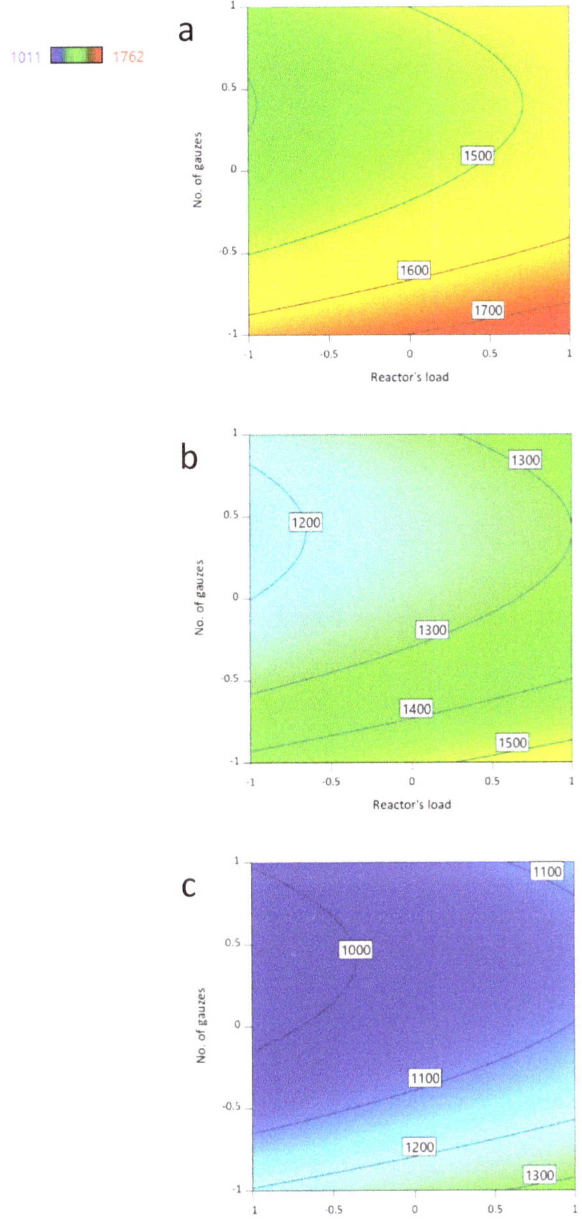

Figure 6. Interaction effect between the reactor's load (X_1) and the number of catalytic gauzes (X_3) at fixed temperature (X_2) on N_2O concentration in nitrous gases (R_2) by contour plot. (**a**) $X_2 = -1$; (**b**) $X_2 = 0$; (**c**) $X_2 = 1$.

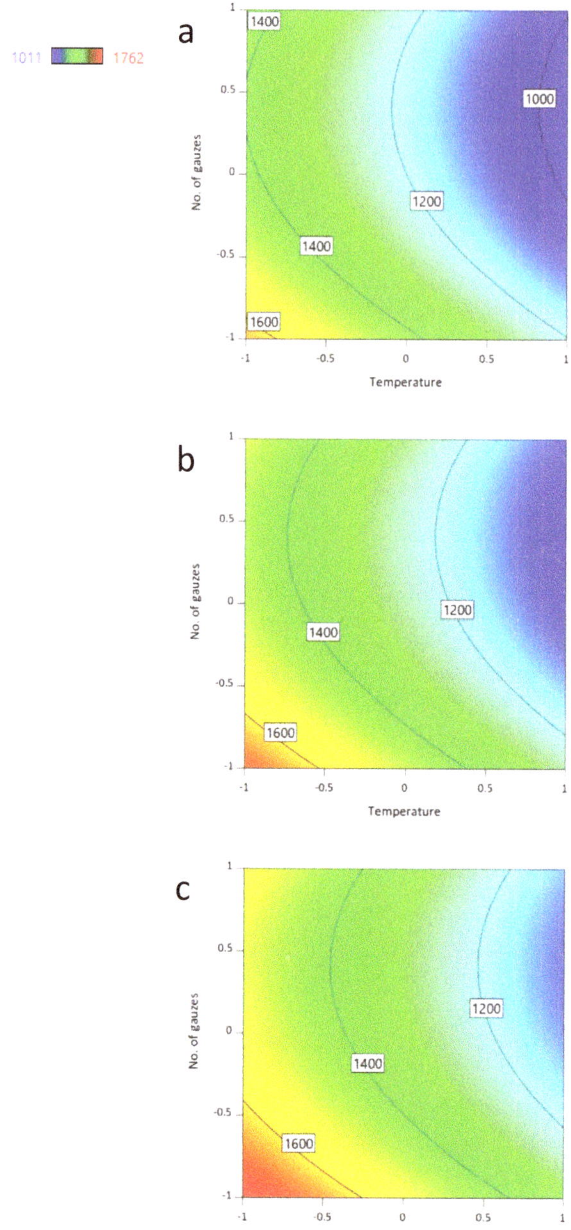

Figure 7. Interaction effect between the temperature (X_2) and the number of catalytic gauzes (X_3) at fixed reactor's load (X_1) on N_2O concentration in nitrous gases (R_2) by contour plot. (**a**) $X_1 = -1$; (**b**) $X_1 = 0$; (**c**) $X_1 = 1$.

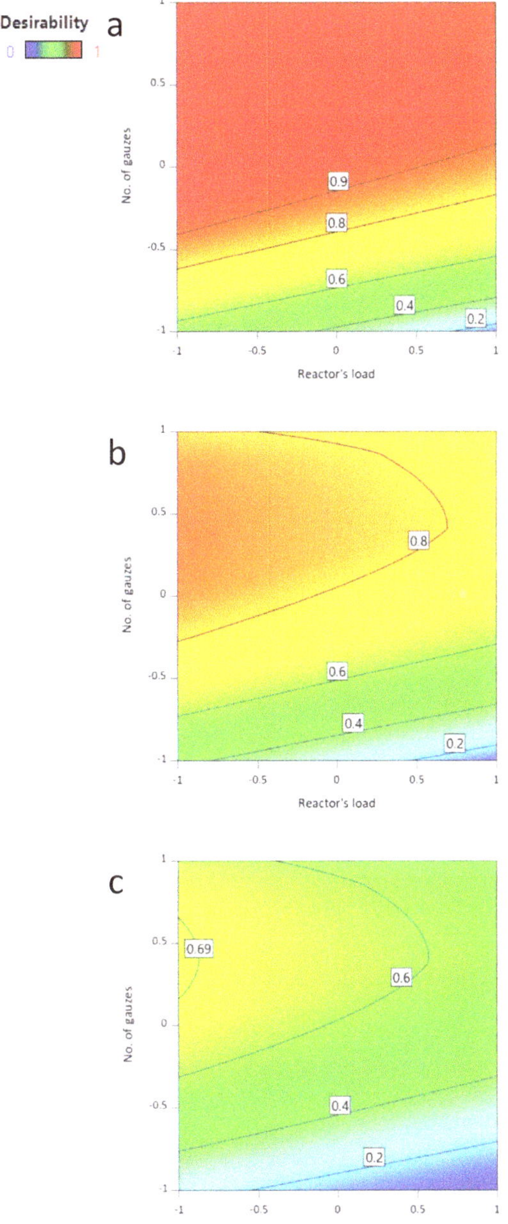

Figure 8. Desirability function plots. Effect of the reactor's load (X_1) and the number of catalytic gauzes (X_3) at three levels of temperature (**a**) $X_2 = -1$; (**b**) $X_2 = 0$; (**c**) $X_2 = 1$.

High values of desirability function ($DF > 0.9$) at 910 °C are described with dependency according to which for the load of 456 kg $NH_3/(m^2h)$, the sufficient number of catalytic gauzes is 8. However, for the maximum load studied, 10 catalytic gauzes should be applied. At the temperature of 910 °C and when all optimization criteria are met, the expected value of N_2O concentration ranges from

1000 ppm to 1100 ppm (Figure 8a). Lowering the reaction temperature to 890 °C means that desirability function *DF* > 0.8 is within the region where the minimum catalyst gauzes is 9 for the loading not higher than 480 kg $NH_3/(m^2h)$ and 12 gauzes for load of 645 kg $NH_3/(m^2h)$ (Figure 8b). At this temperature, the expected concentration of N_2O in nitrous gases ranges from 1180 to 1200 ppm. At the lowest temperature (within the studied range) of 870 °C, the highest value of desirability function is 0.69. For 12 gauzes and the load of 456 kg/(m^2h), the expected concentration of N_2O in nitrous gases is 1400 ppm.

Taking into account the amount of the primary emissions of N_2O (the environmental aspect), it is favourable to conduct the reaction at the temperature of 910 °C. However, this leads to the increased platinum losses. Platinum losses at 910 °C are higher by approx. 25% as compared to losses at 890 °C and by 45% as compared to losses at 870 °C [1,2]. Lowering the reaction temperature to 890 °C with maintaining the optimal range of other parameters causes the increase of N_2O concentration in nitrous gases by 100–200 ppm.

The assumption of other values of 'weight' and 'importance' for particular variables leads to obtain other profiles of desirability function. Under industrial conditions, the assumed value of 'weight' and 'importance' should take into account the process economics with regard to platinum losses.

2.6. Validation

Validation of the developed optimization model should be carried out under conditions specified as optimal. Optimization results indicate a wide set of parameters for which desirability function achieved high values. Therefore, in order to carry out additional measurements, the point with the independent variables value of: 1, 1, 1 was selected. This point is in the range of high desirability function value. In Table 8 levels of independent variables, results of validation experiment and predicted mean values of response variables with standard deviation are presented.

Table 8. The assumed levels of independent variables in validation studies and predicted responses values.

Independent Variable	**Reactor's Load**	**Temperature**	**Number of Gauzes**
Level	1	1	1
	Two-sided Confidence = 95%; Population = 99%		
Response variable	Experimental Data	Predicted Mean Value	Std Dev
R_1	96.2	96.3625 ± 0.4672	0.265216
R_2	1120	1125.12 ± 83.37	52.9151

For the assumed independent variables, the values of predicted mean with 95% two-sided confidence intervals met by 99% of population were estimated based on the achieved mathematical model. High conformity of results expected according to mathematical models with the obtained measurements results was achieved.

3. Materials and Methods

3.1. Materials

Standard knitted gauzes made of platinum alloy with the addition of 10% wt. Rh made of 0.076 mm wire and specific weight of 600 g/m^2 were used for ammonia oxidation studies. The catchment gauzes which are most commonly used in industrial process were not used in studies. The prefiltered compressed air and gaseous ammonia were used as raw materials for ammonia oxidation reaction.

3.2. Experimental Procedure

Ammonia oxidation studies were performed in a pilot plant equipped with a flow reactor (inner diameter: 100 mm). The Pt-Rh catalytic gauzes were installed inside the basket. After initiating the reaction, the stable ammonia and air ratio was maintained in reaction mixture amounting to approx. 10.9% vol. The air–ammonia mixture temperature was controlled in such a manner as to obtain the temperature of nitrous gas as assumed in the experiment plan. Air–ammonia mixture temperature was variable in the range of 135–195 °C. The flow of the air-ammonia mixture was also controlled in order to obtain the assumed reactor's load. All experiments were conducted under the pressure of 0.5 MPa. The range of temperature of reaction at which studies were carried out is similar to that applied in industrial practice. The reactor's load was selected in such a manner as to ensure that the gas flow through the catalytic package in the range applied for medium-pressure industrial reactor namely 1–3 m/s. The scheme of the pilot plant is presented in Figure 9. For measuring ammonia oxidation efficiency, samples of ammonia-air mixture were taken at the inlet to the reactor and samples of nitrous gases were taken at the outlet of the reactor. For determination of N_2O concentration in nitrous gases, only samples of nitrous gases were taken.

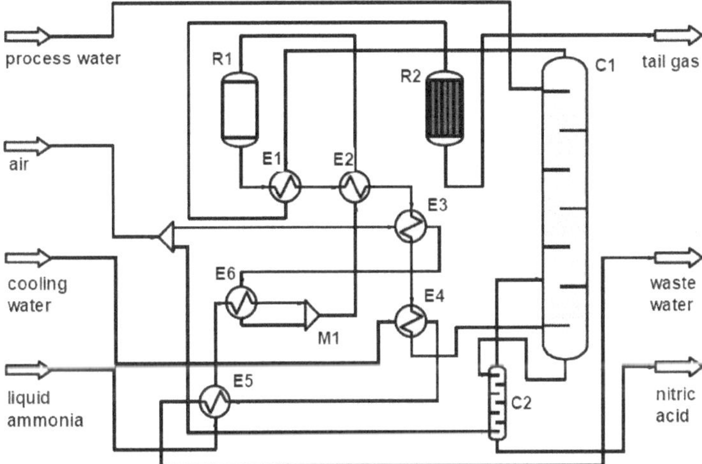

Figure 9. Scheme of the pilot plant. Symbols: C1—absorption column, C2—bleaching column, E1–E6—heat exchangers, M1—air–ammonia mixer, R1—ammonia oxidation reactor, R2—selective catalytic reduction reactor.

3.3. Analytical Methods

The ammonia oxidation efficiency was calculated based on concentrations of ammonia in the air-ammonia mixture at the inlet and concentrations of NO in nitrous gases at the outlet of the reactor. Both analysis were determined according to a titration method. Ammonia from air-ammonia mixture samples was absorbed in water with the formation of ammonia-water solution which was then titrated with sulphuric acid. The nitrous gases samples were absorbed in 3% water solution of hydrogen peroxide. After ensuring the sufficient period of time, NO oxidized completely to NO_2 and then, it reacted with water to HNO_3. The formed HNO_3 was titrated with the sodium hydroxide solution in the presence of an indicator.

Ammonia oxidation efficiency (R_1) was calculated according to the following formula:

$$R_1 = \left(\frac{C_2}{C_1}\right) \cdot 100\% \quad (3)$$

where: C_1—ammonia concentration in ammonia-air mixture, % w/w; C_2—concentration of oxidized ammonia, % w/w.

The result of each measurement is an average value, calculated from 7 independent samplings. The difference in the extreme individual values were not greater than ±0.3% in comparison to the average one.

N_2O concentration in nitrous gases (R_2) was determined by gas chromatography using a Unicam 610 system with a discharge ionization detector. Gaseous samples were collected in the vacuum flasks containing 3% water solution of hydrogen peroxide. After the absorption of nitrous gases and water vapor condensation, exhaust gas from the flasks was injected to the gas chromatograph through 1 mL sample loop. The result of each measurement was an average value calculated from 3 independent samplings. The difference in the extreme individual values was not greater than ±35 ppm in comparison to the average one.

3.4. Statistical Methods

The experimental procedure was carried out according to Box–Behnken design matrix. The reactor's load (X_1), the temperature of nitrous gas specifying the temperature of reaction (X_2) and the number of catalytic gauzes (X_2) were selected as independent variables. Ammonia oxidation efficiency (R_1) and N_2O concentration in nitrous gas (R_2) were specified as response variables. Each level of independent variables were coded according to the Equation (4).

$$X_i = \frac{x_i - x_0}{\Delta x_i} \tag{4}$$

where, X_i is the dimensionless, coded level of independent variable (−1, 0 or 1), x_i is the actual value of the independent variable, x_0 is the value of the independent variable at the centre point, Δx_i is the step change in x_i.

Ranges and levels of independent variables are presented in Table 9.

Table 9. Coded and uncoded levels of independent variables used in experiments.

Independent Variable:	X_1	X_2	X_3
	Reactor's Load, kg $NH_3/(m^2h)$	**Temperature of Reaction, °C**	**Number of Catalytic Gauzes, pcs**
Coded variable level:			
Low level (−1)	456	870	5
Mid-level (0)	582	890	10
High level (+1)	708	910	15

The total number of the experiments (N) was calculated using the Equation (5).

$$N = 2k(k-1) + c_0 \tag{5}$$

where, k is the number of independent variables, c_0—number of the replicates run of the centre point (in our research c_0 = 3). For three independent variables, the total number of experiments assumed in the plan was 15.

The experiments were conducted in a randomized order to avoid the influence of uncontrolled variables on the dependent responses.

A mathematical relationship between the independent variables and response variables was determined by fitting the experimental data with second-order polynomial Equation (6).

$$R_i = b_0 + \sum_{i=1}^{3} b_i X_i + \sum_{i=1}^{3} \sum_{j=1,\, i<j}^{3} b_{ij} X_i X_j + \sum_{i=1}^{3} b_{ii} X_i^2 \tag{6}$$

where R_i is the estimate response variable, b_0, b_i, b_{ii}, b_{ij} are regression coefficients fitted from the experimental data, X_i, X_j are coded independent variables, listed in Table 9.

The significance of the model equation, individual parameters, were evaluated through ANOVA with the confidence interval (CI) of 95%. A simultaneous optimization of several dependent variables requires the application of multi-criteria methodology. In this case, the desirability function (*DF*) was used. The particular desirability functions are combined using the geometric mean which allows to achieve overall desirability function [16], according to Equation (7).

$$DF = \left((d_1)^{w_1} \times (d_2)^{w_2} \times \ldots x\ (d_n)^{w_n})\right)^{1/\sum w_i} \tag{7}$$

where n is the number of responses, d_i is an individual response desirability, w_i is a response 'weight'.

The adjustment of the shape of particular desirability function can be performed by assigning the specified 'weight.' Setting a different 'importance' for each objective with respect to the remaining objectives is also possible. For these studies, identical 'weight' for all the independent variables and response variables was assumed. Desirability function assigns values from 0 to 1 where 1 means meeting all the optimization criteria. It is not always necessary to search for the solution aiming at achievement of the highest value of desirability function but it is vital to search for the set of parameters which would meet the optimization objectives to the particular extent (e.g., $DF > 0.75$). The statistical software used to experimental design and analysis was Design Expert 11.0.6.0 version (Stat-Ease, Inc., Minneapolis, MN, USA).

4. Conclusions

The conducted studies allowed us to develop statistically significant mathematical models describing the course of variables of ammonia oxidation efficiency and N_2O concentration in nitrous gases depending on three selected independent variables.

The design of the experiment allowed the reduction of the costs of studies and to achieve a number of results accurate for modelling. It was found that, within the studied range of variability, the temperature of reaction has no significant effect statistically on the achieved ammonia oxidation efficiency, whereas it has the effect on the amount of N_2O formed in the side reaction (primary emission of N_2O).

The developed models were used to optimize the process. As a result of this optimization, the set of the independent variables was developed for which optimization assumptions are met, which are expressed as a high value of desirability functions. It is possible to specify the optimum number of gauzes with the determined reactor's load for the studied package of catalytic gauzes.

In validation experiments, the developed model of desirability function achieved the high conformity of experimental values with the expected ones.

The presented methodology can be used to minimize the primary N_2O emission at high ammonia oxidation efficiency. It can be applied for optimization of operating parameters of ammonia oxidation reactor with two types of catalysts: catalytic gauzes and catalyst for high temperature of N_2O decomposition. As a result, it is possible to obtain the set of independent variables ensuring low N_2O emission and to meet the binding environmental regulations.

Author Contributions: Conceptualization, M.I.; Formal analysis, M.I. and A.D.-I.; Investigation, M.I. and J.R.; Methodology, M.I.; Supervision, M.W.; Visualization, A.D.-I.; Writing—original draft, M.I.

Funding: This research received no external funding.

Conflicts of Interest: The authors declare no conflict of interest.

References

1. Thiemann, M.; Scheibler, E.; Wiegand, K.W. Nitric Acid, Nitrous Acid, and Nitrogen Oxides. In *Ullmann's Encyclopedia of Industrial Chemistry*; Wiley-VCH Verlag GmbH & Co. KGaA: Weinheim, Germany, 2012.

2. Sadykow, V.A.; Isupova, L.A.; Zolotarskii, I.A.; Bobrova, L.N.; Noskov, A.S.; Parmon, V.N.; Brushtein, E.A.; Telyatnikova, T.V.; Chernyshev, V.I.; Lunin, V.V. Oxide catalysts for ammonia oxidation: Properties and perspectives. *Appl. Catal. A Gen.* **2000**, *204*, 59–87. [CrossRef]
3. Biausque, G.; Schuurman, Y. The reaction mechanism of the high temperature ammonia oxidation to nitrous oxide over $LaCoO_3$. *J. Catal.* **2010**, *276*, 306–313. [CrossRef]
4. Petryk, J.; Kołakowska, E. Cobalt oxide catalysts for ammonia oxidation activated with cerium and lanthanum. *Appl. Catal. A Environ.* **2000**, *24*, 121–128. [CrossRef]
5. Isupova, L.A.; Sutormina, E.F.; Kulikovskaya, N.A.; Plyasova, L.M.; Rudina, N.A.; Ovsyannikova, I.A.; Zolotarskii, I.A.; Sadykov, V.A. Honeycomb supported perovskite catalysts for ammonia oxidation processes. *Catal. Today* **2005**, *105*, 429–435. [CrossRef]
6. Isupova, L.A.; Sutormina, E.F.; Zakharov, V.P.; Rudina, N.A.; Kulikovskaya, N.A.; Plyasova, L.M. Cardierite-like mixed oxide monolith for ammonia oxidation process. *Catal. Today* **2009**, *1475*, 5319–5323. [CrossRef]
7. Marret, S.; du Chatelier, L. Recent advances in catalyst technology used in nitric acid production. *Proc. Fertil. Soc.* **1998**, *413*, 1–44.
8. Kay, O.; Buennagel, T. Targeting improving performance and conversion efficiency in nitric acid plants. *Proc. Fertil. Soc.* **2016**, *787*, 1–31.
9. Anonymous. New advances in platinum gauze systems. *Nitrogn+Syngas* **2016**, *344*, 34–41.
10. Anonymous. Catalysts for ammonia oxidation. *Nitrogen+Syngas* **2017**, *349*, 48–50.
11. Kyoto Protocol. Available online: https://unfccc.int/process/the-kyoto-protocol (accessed on 26 November 2017).
12. Brown, S.M.; Sneyd, J.R. Nitrous oxide in modern anaesthetic practice. *BJA Educ.* **2016**, *16*, 87–91. [CrossRef]
13. Van Amsterdam, J.; Nabben, T.; van den Brink, W. Recreational nitrous oxide use: Prevalence and risks. *Regul. Toxicol. Pharmacol* **2015**, *73*, 790–796. [CrossRef]
14. Pérez-Ramirez, F.; Kapteijn, K.; Schöffel, J.A. Moulijn, Formation and control of N_2O in nitric acid production: Where do we stand today? *Appl. Catal. B Environ.* **2003**, *44*, 117–151. [CrossRef]
15. Marzo, L.M. Nitric acid production and abatement technology including azeotropic acid. *Proc. Fertil. Soc.* **2004**, *540*, 1–28.
16. Reference Document on Best Available Techniques for the Manufacture of Large Volume Inorganic Chemicals—Ammonia, Acids and Fertilisers, Chapter 3 Nitric Acid, European Commission Document 2007. Available online: http://eippcb.jrc.es (accessed on 16 March 2012).
17. Anonymous. Emission monitoring in nitric acid plants. *Nitrogn+Syngas* **2014**, *328*, 48–53.
18. Myers, R.H.; Montgomery, D.C.; Anderson-Cook, C.M. *Response Surface Methodology: Process and Product Optimization Using Designed Experiments*, 3rd ed.; John Wiley & Sons: Hoboken, NJ, USA, 2009.
19. Granato, D.; Ares, G. *Mathematical and Statistical Methods in Food Science and Technology*; John Wiley & Sons, Ltd.: Chicago, IL, USA, 2014.
20. Capaci, F.; Bergquist, B.; Kulahci, M.; Vanhatalo, E. Exploring the use of design of experiments in industrial processes operating under closed-loop control. *Qual. Reliab. Eng. Int.* **2017**, *33*, 1601–1614. [CrossRef]
21. Zhang, L.; Mao, S. Application of quality by design in the current drug development. *Asian J. Pharm. Sci.* **2017**, *12*, 1–8. [CrossRef]
22. Huang, S.-M.; Wu, P.-Y.; Chen, J.-H.; Kou, C.-H.; Shieh, C.-J. Develpoing a High-Temperature Solvent-free System for Efficient Biocatalysis of Octyl Ferulate. *Catalysts* **2018**, *8*, 338. [CrossRef]
23. Lin, Y.-P.; Mehrvar, M. Photocatalytic Treatment of An Actual Confectionery Wastewater Using $Ag/TiO_2/Fe_2O_3$: Optimization of Photocatalytic Reactions Using Surface Response Methodology. *Catalysts* **2018**, *8*, 409. [CrossRef]
24. Tafreshi, N.; Sharifnia, S.; Dehaghi, S.M. Box–Behnken experimental design for optimization of ammonia photocatalytic degradation by ZnO/Oak charcoal composite. *Proc. Saf. Environ. Prot.* **2017**, *106*, 203–210. [CrossRef]
25. Nobandegani, M.S.; Birjandi, M.R.S.; Darbandi, T.; Khalilipour, M.M.; Shahraki, F.; Mohebbi-Kalhori, D. An industrial Steam Methane Reformer optimization using response surface methodology. *J. Nat. Gas Sci. Eng.* **2016**, *36*, 540–549. [CrossRef]

26. Hafizi, A.; Rahimpour, M.R.; Hassanajili, S. Hydrogen production by chemical looping steam reforming of methane over Mg promoted iron oxygen carrier: Optimization using design of experiments. *J. Taiwan Inst. Chem. Eng.* **2016**, *62*, 140–149. [CrossRef]
27. Hoseiny, S.; Zare, Z.; Mirvakili, A.; Setoodeh, P.; Rahimpour, M.R. Simulation–based optimization of operating parameters for methanol synthesis process: Application of response surface methodology for statistical analysis. *J. Nat. Gas Sci. Eng.* **2016**, *34*, 439–448. [CrossRef]
28. Duraković, B. Design of experiments application, concepts, examples: State of the art. *Period. Eng. Nat. Sci.* **2017**, *5*, 421–439. [CrossRef]
29. NIST/SEMATECH. e-Handbook of Statistical Methods. 2012. Available online: http://www.itl.nist.gov/div898/handbook/ (accessed on 29 November 2018).

Article

Tuning the Catalytic Properties of Copper-Promoted Nanoceria via a Hydrothermal Method

Konstantinos Kappis [1], Christos Papadopoulos [1], Joan Papavasiliou [1,2], John Vakros [3], Yiannis Georgiou [4], Yiannis Deligiannakis [4] and George Avgouropoulos [1,*]

1 Department of Materials Science, University of Patras, GR-26504 Patras, Greece; kostaskapphs@gmail.com (K.K.); c_papadopoulos@upatras.gr (C.P.); jpapav@iceht.forth.gr (J.P.)
2 Foundation for Research and Technology-Hellas (FORTH), Institute of Chemical Engineering Sciences (ICE-HT), P.O. Box 1414, GR-26504 Patras, Greece
3 Department of Chemistry, University of Patras, GR-26504 Patras, Greece; vakros@chemistry.upatras.gr
4 Department of Physics, University of Ioannina, GR-45110 Ioannina, Greece; yiannisgeorgiou@hotmail.com (Y.G.); ideligia@cc.uoi.gr (Y.D.)
* Correspondence: geoavg@upatras.gr; Tel.: +30-2610-969-811

Received: 8 January 2019; Accepted: 21 January 2019; Published: 1 February 2019

Abstract: Copper-cerium mixed oxide catalysts have gained ground over the years in the field of heterogeneous catalysis and especially in CO oxidation reaction due to their remarkable performance. In this study, a series of highly active, atomically dispersed copper-ceria nanocatalysts were synthesized via appropriate tuning of a novel hydrothermal method. Various physicochemical techniques including electron paramagnetic resonance (EPR) spectroscopy, X-ray diffraction (XRD), N_2 adsorption, scanning electron microscopy (SEM), Raman spectroscopy, and ultraviolet-visible diffuse reflectance spectroscopy (UV-Vis DRS) were employed in the characterization of the synthesized materials, while all the catalysts were evaluated in the CO oxidation reaction. Moreover, discussion of the employed mechanism during hydrothermal route was provided. The observed catalytic activity in CO oxidation reaction was strongly dependent on the nanostructured morphology, oxygen vacancy concentration, and nature of atomically dispersed Cu^{2+} clusters.

Keywords: copper-ceria catalysts; hydrothermal method; CO oxidation; copper clusters; nanoceria

1. Introduction

Carbon monoxide (CO) is a harmful, toxic gas that is present in many industrial processes. Due to its negative impact for both humans and the environment, the catalytic oxidation of CO into CO_2 has always been a research topic of great interest [1–3]. Moreover, the catalytic oxidation of CO is an important reaction in the technological fields of fuel cells [4–6], gas sensors [7], and CO_2 lasers [8].

Cerium oxide or ceria (CeO_2) has been thoroughly studied as a catalyst or support in CO oxidation reaction due to its defective structure enriched with oxygen vacancies and high oxygen storage capacity (OSC) resulting from the interaction between Ce^{3+} and Ce^{4+} [9–12]. In the case of ceria synthesized in a nanosized form, more remarkable functions can be obtained due to the nanosize effects. For this reason, research has focused on the understanding of the properties of nanoceria as well as improving its OSC, surface to volume ratio, and redox properties [13–16]. Computational studies have shown that the catalytic activity of nanoceria is strongly associated with the exposed surface plane. Sayle et al. [17] predicted that the (110) and (100) surfaces are catalytically more active for CO oxidation than the (111) surface, due to more oxygen vacancies located in the former. According to Conesa [18], the formation of oxygen vacancies on the (110) and (100) surfaces requires less energy than the (111) surface. It is noteworthy that the exposition of the reactive surface plane is dependent on the morphology of the material at the nanoscale. Zhou et al. [19] have shown that CeO_2 nanorods, which exposed the (110) and (100) planes achieved higher catalytic activity for CO oxidation than nanoparticles exposing the (111) planes. Wu et al. [20] studied the morphology dependence of CO oxidation over ceria nanocrystals. They discovered that the activity for CO oxidation of those CeO_2 nanostructures follows the order: Rods > cubes > octahedra, whereas the activity of different planes follows the order: (110) > (100) > (111). These results were also confirmed by Tana et al. [21]. In order to prepare various shapes of nanoceria, a number of methods have been applied, such as sol-gel [22], precipitation [23], hydrothermal or solvothermal methods [24–27], and electrochemical deposition [28]. Among these methods, the hydrothermal method has attracted great interest because a desired morphology can be obtained via appropriate control of the hydrothermal parameters such as reaction time, temperature, and concentration [29–31].

Despite the attractive physicochemical properties of ceria, poor catalytic activity of pure ceria [32] can by highly promoted via doping with a series of metal ions, in order to change its surface chemistry and promote the active oxygen content [33]. It is well known that the reduction behavior of ceria can be rapidly altered by the addition of a minimal amount of Au, Pd, and Pt precious metals and/or transition metals [34–38]. While the activity of the catalysts is improved by the addition of precious metals, their high cost prohibits their application. Numerous reports have indicated that the activity of ceria in oxidation reactions is enhanced by transition metals like copper. The copper–ceria system presents a cost-effective material with unique catalytic properties, comparable to noble metal catalysts, in many catalytic reactions and especially in the CO oxidation and in the preferential oxidation of CO in excess of hydrogen [39–44]. The superiority of $CuCeO_x$ catalytic system has been attributed to a synergistic effect. Reports on the mechanism of CO oxidation reaction over these catalysts have demonstrated the significance of both copper and ceria species in the adsorption of CO and CO_2 production, as the former takes place in the copper-ceria interface [45]. Particularly, the main reasons that trigger the highly catalytic performance of these Cu-Ce catalysts are the large amount of well-dispersed copper species in the ceria support, the creation of oxygen vacancies due to incorporation of Cu^{2+} ions into the ceria structure and the presence of high concentration of active lattice oxygen [45–47]. Several Cu^{2+} entities (e.g., amorphous clusters, isolated ions, dimers, and discrete crystallites) have been detected, which can take part in the catalytic mechanism, displaying high levels of activity [5,39,48–50]. In order to form these entities, conventional preparation methods have been used such as the deposition of Cu^{2-} species onto pre-synthesized ceria support or the coprecipitation of Cu and Ce precursors. The obtained materials are calcined at high temperatures, which enable the dispersion of copper species and the chemical bonding between the Cu and Ce components. For example, Harrison et al. [39] prepared CuO/CeO_2 catalysts via coprecipitation and impregnation routes, and tested in CO oxidation. After thermal pretreatment of materials at 400 °C, the copper content on the surface of ceria consisted of amorphous clusters of Cu^{2+} ions, which presented high catalytic activity for CO oxidation. In several cases, the high temperatures required for the activation of copper component (>600 °C for 4 h) may lead to particle sintering and phase segregation, which facilitates the formation of tenorite particles of CuO, which are inactive for CO oxidation [51].

In the present work, copper clusters were atomically dispersed in ceria nanostructures via a novel hydrothermal route, yielding highly active catalysts in CO oxidation reaction. Tuning of the physicochemical and catalytic properties was studied by varying the hydrothermal parameters (temperature and concentration). A set of analytical techniques such as electron paramagnetic resonance (EPR) spectroscopy, X-ray diffraction (XRD), N_2 adsorption-desorption, scanning electron microscopy (SEM), Raman spectroscopy, and ultraviolet-visible diffuse reflectance spectroscopy (UV-Vis DRS) was used to assess the physicochemical characteristics of the materials and correlate with the catalytic performance.

2. Results

2.1. EPR Measurements

Figure 1 illustrates the EPR spectra of the hydrothermally prepared ceria-based catalysts, using various concentrations of NaOH. The concentrations of the detected species are shown in Table 1. In general, the EPR spectra present three characteristic signals; A signal at ca. 1580 Gauss (g = 4) is attributed to Ce^{3+} ions in high spin (S = 5/2) state, corresponding to reduced Ce^{3+} ceria centers located inside the lattice of ceria particles [15,52]. The sharp signal at around 3400 Gauss (visualized better in the right column spectra of Figure 1) is ascribed to [Ce^{3+}-O^--Ce^{4+}] (S = 1/2) units localized on the surface of the ceria particles [5,53,54]. According to the geometrical disposition of the two types of Ce^{3+} centers—lattice and surface—it is considered that surface Ce^{3+} ions can be correlated with high catalytic activity, i.e., since they may result to the formation of oxygen vacancies [55–57]. Regarding the Cu^{2+} signals, the EPR signals of the sample treated at 180 °C show a strong signal at 2600–3300 Gauss, which is attributed to Cu^{2+} (S = 1/2, I = 3/2) ions. The line shape of this EPR spectrum indicates that the copper atoms show dipolar Cu-Cu interactions, i.e., the Cu^{2+} is clustered within 10 ± 3 Angstroms from each other [58]. The g-and A values of Cu^{2+}, ($g_\perp$ = 2.035, $g_{//}$ = 2.305, $A_\perp$ = 15, $A_{//}$=150) suggest that the Cu^{2+} ions are located in the octahedral sites in ceria with a tetragonal distortion [39,49,55,56,59–61].

Table 1. Concentrations of the species as detected by EPR spectroscopy.

Samples	Bulk Ce^{3+} (μM/gr)	Surface Ce^{3+} (μM/gr)	Cu^{2+} Species (wt %)
Ce-120-0	0.43	0.1	N.D.
CuCe-120-0.05	0.50	1.3	0.20 (isolated ions)
Ce-120-0.1	0.40	0.4	N.D.
CuCe-120-0.5	0.50	4.0	0.002 (isolated ions)
CuCe-120-1	0.29	4.0	0.02 (isolated ions)
Ce-120-5	0.50	4.0	N.D.
CuCe-150-0	1.00	1.3	0.15 (isolated ions)
CuCe-150-0.05	0.20	1.1	0.3 (>95% clusters)
CuCe-150-0.1	0.40	0.8	3.9 (>95% clusters)
Ce-150-0.5	0.50	3.0	N.D.
CuCe-150-1	0.27	4.0	0.002 (isolated ions)
Ce-150-5	0.50	4.0	N.D.
CuCe-180-0	0.02	<0.05	2.1 (>95% clusters)
CuCe-180-0.05	N.D. [1]	<0.05	2.8 (>95% clusters)
CuCe-180-0.1	0.40	0.1	1.4 (>95% clusters)
CuCe-180-0.5	0.50	3.0	0.02 (isolated ions)
CuCe-180-1	0.25	3.0	0.01 (isolated ions)
Ce-180-5	0.40	4.0	N.D.

[1] Not-detected.

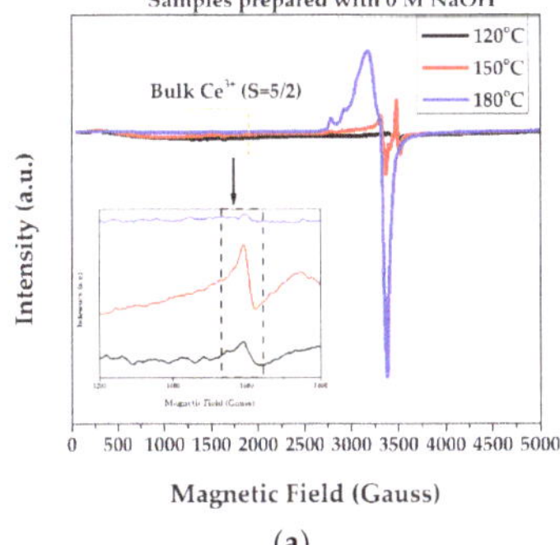

(**a**)

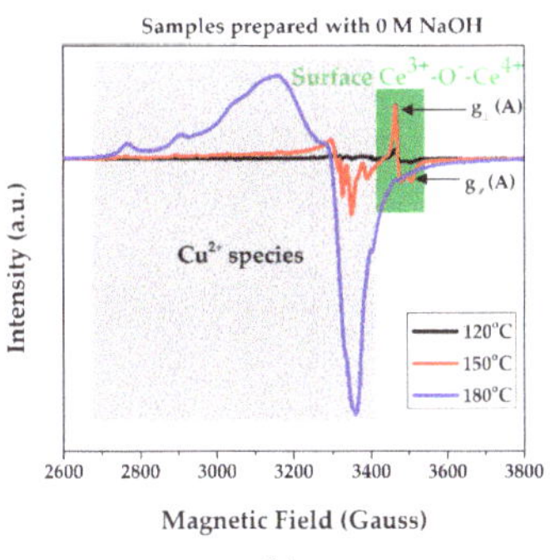

(**b**)

Figure 1. *Cont.*

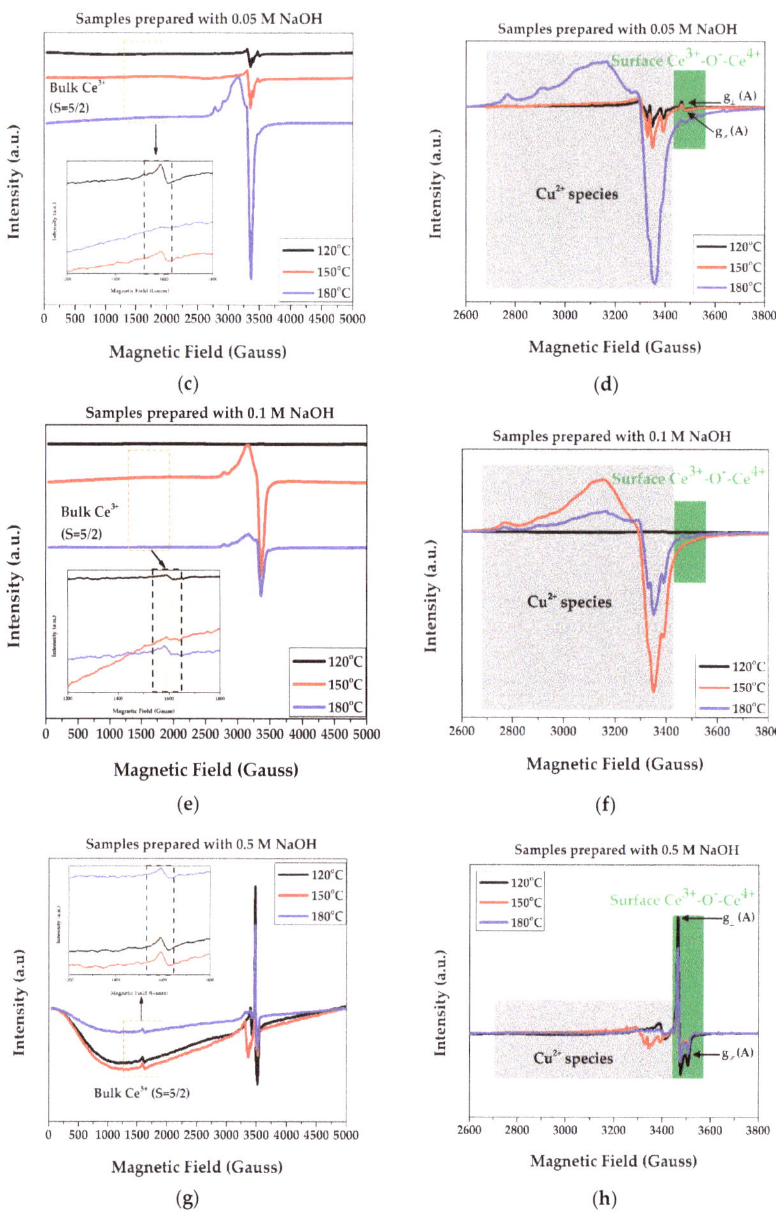

Figure 1. *Cont.*

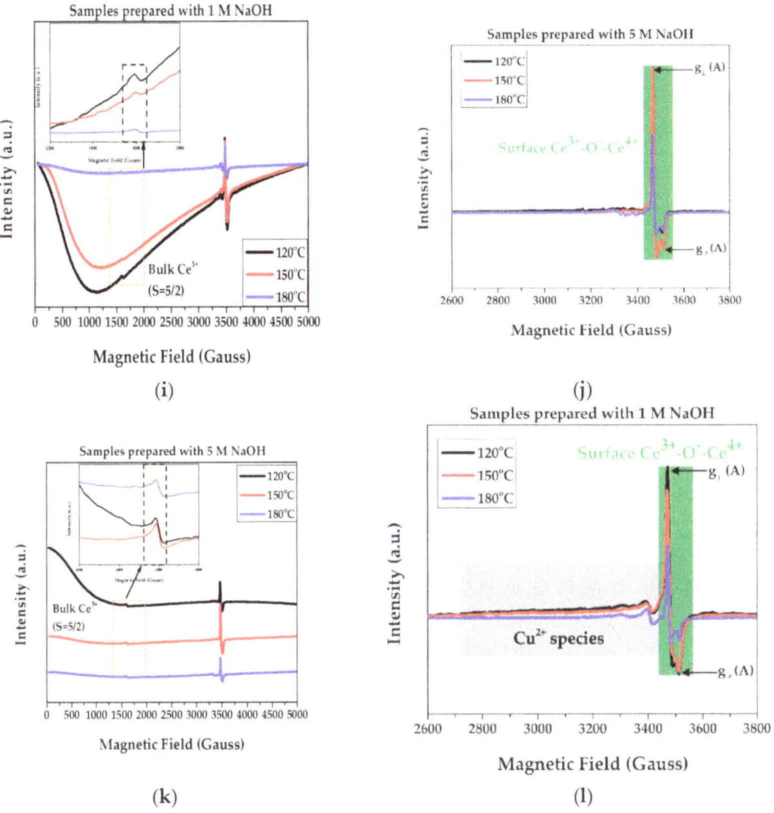

Figure 1. EPR spectra of the catalysts prepared with 0–5 M NaOH. Magnification of the region in which the bulk Ce^{3+} ions were detected, is shown in the inset figure; the figures on the right side present an enlarged region where the Cu^{2+} species and the surface Ce^{3+} ions were detected.

Quantitate data on the Ce^{3+} and Cu^{2+}, estimated from the EPR spectra, are listed in Table 1. Copper loading was also confirmed via XRF measurements (within an experimental error of ±10%; not shown here), in line with EPR results. Concerning the surface Ce^{3+} ions, there is a clear trend vs. the NaOH concentration: Their concentrations increases when concentrated solutions of sodium hydroxide were employed in the hydrothermal route. With regard to the copper entities, the combination of low concentrations of NaOH and high hydrothermal temperatures caused the formation of a high amount of copper species (especially Cu^{2+} clusters). On the other hand, the high concentrations of NaOH halted the dispersion of copper species in the ceria phase, due to the high basicity of NaOH resulting in $Cu(OH)_n$ clusters' formation in the CeO_2 phase.

2.2. XRD Measurements

The XRD diffractograms of all the samples are shown in Figure S1. Noticeably, all peaks can be indexed to (111), (200), (220), (331), (222), (400), (331), (420), and (422) planes corresponding to the pure cubic phase [space group: Fm-3m, JCPDS: 00-043-1002, α = 0.54113 nm] of CeO_2 [30]. No diffraction peaks of crystalline copper species can be observed, due to the presence of highly dispersed amorphous copper species or/and the low copper loading, which can be hardly detected at XRD [62,63]. However, as it was discussed in EPR section, the copper species might be atomically dispersed in the ceria matrix. The calculated average crystallite size and the lattice parameter of the ceria-based materials are presented in Table 2. It has to be noted that there is no clear relationship between Cu content (see Table 1) and crystallite size or lattice parameter of the catalysts.

Table 2. Average crystallite size and lattice parameter of the catalysts based on the crystal plane of (111) of CeO_2.

Samples	d_{111} (nm)	α (nm)
Ce-120-0	8.6	0.54078
CuCe-120-0.05	6.9	0.54228
Ce-120-0.1	10.0	0.54136
CuCe-120-0.5	13.2	0.54118
CuCe-120-1	13.8	0.54123
Ce-120-5	8.2	0.54185
CuCe-150-0	7.2	0.53966
CuCe-150-0.05	7.5	0.54147
CuCe-150-0.1	19.0	0.54211
Ce-150-0.5	9.5	0.54164
CuCe-150-1	15.2	0.54127
Ce-150-5	25.3	0.54133
CuCe-180-0	37.7	0.54168
CuCe-180-0.05	29.2	0.54168
CuCe-180-0.1	18.2	0.54142
CuCe-180-0.5	11.3	0.54134
CuCe-180-1	13.9	0.54138
Ce-180-5	23.5	0.54159

The dependence of the crystallite size from the hydrothermal parameters is depicted in Figure 2. It can be seen in Figure 2a that, for concentrations of NaOH ≤0.05 M, small-size crystallites were formed at temperatures of 120–150 °C, while at 180 °C, a dramatic increase of their size can be observed. A more rapid increase of crystallites size was obtained in the case of 0.1 M NaOH. For instance, the sample Ce-120-0.1 presents an average crystallite size at 10 nm, while at 150 °C, the average size is ca. 20 nm. In the range of 0.5–1 M NaOH, the crystallites size varies from ~10 to 15 nm following an increase in the hydrothermal temperature. Finally, for 5 M NaOH, the combination of high concentration and high temperatures caused the formation of large crystallites. In Figure 2b, at 120 and 150 °C, a fluctuation of the crystallites size is indicated as the concentration of NaOH increases. At 180 °C, a decrease of the crystallite size is observed for concentrations ≤0.5 M. Further increase of the concentration caused the rise of the crystallites size.

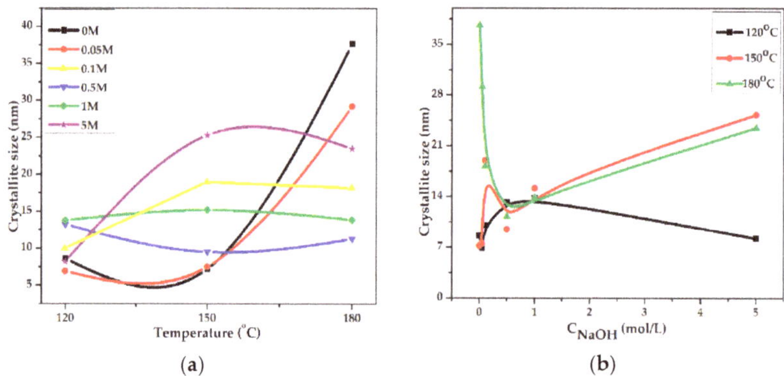

Figure 2. Dependence of the crystallites size of the catalysts with respect to the hydrothermal parameters: (**a**) Based on the temperature of the hydrothermal treatment; and (**b**) based on the concentration of NaOH.

Concerning the lattice parameter, the vast majority of the samples present higher values of lattice parameter than the pure ceria, suggesting the lattice expansion for the obtained materials. According to several studies, the lattice expansion is closely correlated with the presence of Ce^{3+} ions in the crystal lattice because the radius of Ce^{3+} ions (0.114 nm) is higher than the radius of Ce^{4+} ions (0.097 nm), inducing the lattice expansion [27,64–66]. On the other hand, only two samples (Ce-120-0 and CuCe-150-0) presented a smaller value of lattice parameter than the pure ceria. According to Pan et al. [30], the hydroxyl groups may stabilize the smaller nanoparticles resulting in the smaller value of lattice parameter.

2.3. N_2 Adsorption Measurements

The N_2 adsorption/desorption isotherms and the pore size distribution for the ceria-based materials are illustrated in Figure 3. Additionally, the specific surface areas (SSA), the pore volume and the pore size of all samples are shown in Table 3. For hydrothermal solutions of NaOH $\leq$0.1 M, the materials present type II isotherms with type H2 hysteresis loops, independently of the hydrothermal temperature. Materials that present type H2 hysteresis loops are often disordered without well-defined pore distribution. However, for NaOH concentrations $\geq$0.5 M, it can observed that the type of isotherm becomes type IV, which is characteristic of mesoporous materials [67] with type H2 hysteresis loops. It is worth mentioning that a different type of hysteresis loops is revealed for Ce-150-5 and Ce-180-5 samples. Specifically, the type of hysteresis loop is type H1, which is associated with well-defined cylindrical pores.

Table 3. Specific Surface Area, Pore Volume, and Pore Size of the catalysts.

Samples	SSA (m^2/g)	Pore Volume (cm^3/g)	Pore Size (nm)
Ce-120-0	36.3	0.0860	10.34
CuCe-120-0.05	67.0	0.0920	5.33
Ce-120-0.1	22.2	0.0420	10.02
CuCe-120-0.5	70.6	0.1286	5.17
CuCe-120-1	78.8	0.1447	5.63
Ce-120-5	137.1	0.2294	5.55
CuCe-150-0	63.7	0.1070	5.51
CuCe-150-0.05	73.3	0.1050	4.80
CuCe-150-0.1	40.4	0.0320	5.23
Ce-150-0.5	86.2	0.1462	4.97
CuCe-150-1	58.6	0.1390	5.88
Ce-150-5	37.3	0.2114	18.06
CuCe-180-0	9.2	0.0450	15.30
CuCe-180-0.05	14.8	0.0530	12.30
CuCe-180-0.1	27.8	0.0730	10.00
CuCe-180-0.5	76.5	0.1437	5.50
CuCe-180-1	59.4	0.1343	6.50
Ce-180-5	35.0	0.1952	16.80

The pore size distribution diagrams (Figure 3) denote that these ceria-based catalysts have not got a well-defined pore size. Indeed, the combination of hydrothermal parameters affected this distribution. Pores ranging in the mesoporous region were formed in the case of NaOH concentration $\geq$0.5 M, while different distributions can be observed with NaOH concentrations $\leq$0.1 M. For example, the CuCe-150-0 catalyst mainly presents mesopores, whereas the CuCe-180-0 sample mainly consists of macroporous.

The effect of the hydrothermal parameters on the specific surface area (SSA) of the obtained materials is illustrated in Figure 4. It can be seen in Figure 4a that an increase of the hydrothermal temperature from 120 to 150 °C resulted in higher SSA when the NaOH concentration was $\leq$0.5 M. Further increase of the temperature lowered the surface area of the catalysts. On the other hand, for higher NaOH concentration ($\geq$1 M), the highest surface area was obtained at 120 °C. The highlight of this trend was the SSA of the Ce-120-5 catalyst (137.1 $m^2 g^{-1}$). A general trend depicted in Figure 4b, suggests that for elevated concentration of NaOH, higher surface areas can be obtained for the catalysts prepared hydrothermally at 120 °C. A similar trend can be seen at 150 and 180 °C, however a maximum of SSA corresponds to NaOH concentration of 0.5 M. Therefore, it can be concluded that the combination of high concentrations and high hydrothermal temperatures favors the formation of catalysts with poor surface area.

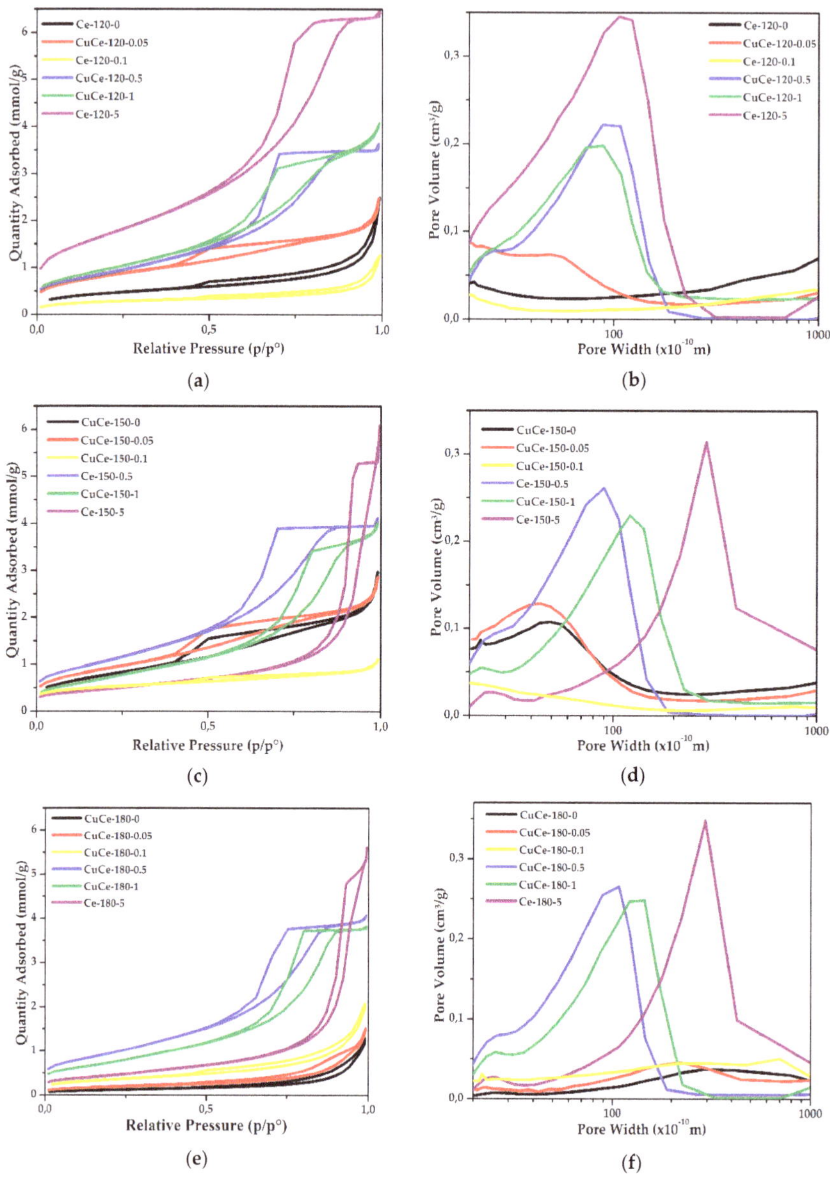

Figure 3. N_2 adsorption/desorption isotherms and pore size distribution curves of CeO_2 and Cu/CeO_2 catalysts: (**a**,**c**,**e**) N_2 adsorption/desorption isotherms; and (**b**,**d**,**f**) pore size distribution curves.

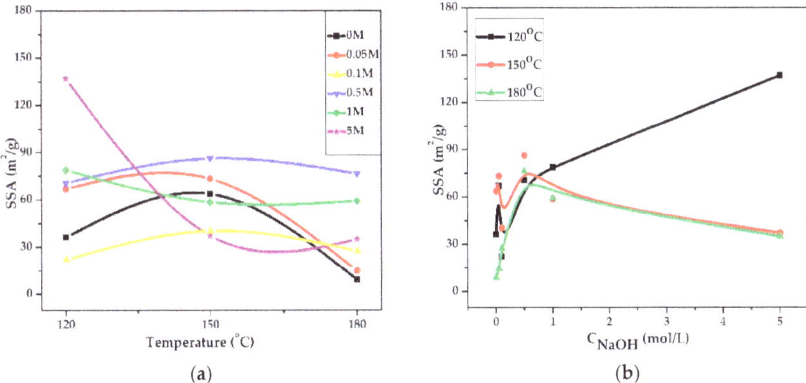

Figure 4. Dependence of the specific surface area of the catalysts with respect to the hydrothermal parameters: (**a**) Based on the temperature of the hydrothermal treatment; and (**b**) based on the concentration of NaOH.

2.4. SEM Measurements

Figure 5 illustrates representatives SEM images of the materials which were hydrothermally synthesized at 120 °C. A morphology of bulk rods with various aggregates onto their surface was obtained for the Ce-120-0 sample. In the presence of NaOH, the morphology changed and spheres with a diameter of 4–8 µm were formed for the CuCe-120-0.05 sample. Spherical aggregates with a size of few hundred nanometers were formed onto this material surface. Further increases of the NaOH concentration resulted in rods with various lengths (2–10 µm) for the Ce-120-0.1 sample. No particles or aggregates can be observed on the surface of these rods, a fact that confirms the high crystallinity of the rods. These rods disappeared at elevated concentrations of NaOH, and very big aggregates with non-defined morphology were formed. In the case of hydrothermal synthesis at 150 °C, a spherical morphology dominates (Figure 6), while at high NaOH concentrations ($\geq$ 0.5 M), bulky aggregates with particles without well-defined geometry, were formed. Figure 7 illustrates representative SEM images of the materials, which were treated hydrothermally at 180 °C. The CuCe-180-0 sample maintained the spherical morphology. The spheres composed of particles without well-defined geometry with size of ca. 70 nm. Adding low amounts of NaOH (0.05 M) resulted in a mixed morphology with rods of few micrometers, spheres and to a smaller extent sheets. All these structures contained irregular particles with an average size of a few dozen nanometers. The CuCe-180-0.1 sample appears to possess a spherical morphology, while similar structures with the lower hydrothermal temperatures were obtained at elevated NaOH concentrations ($\geq$0.5 M).

2.5. Raman Measurements

Raman spectra of all the catalysts are shown in Figure 8. Ceria presents a fluorite structure with only one allowed Raman mode, which has an F_{2g} symmetry and can be viewed to the symmetrical stretching mode of oxygen ions around Ce^{4+} ions [68–70]. For bulk ceria, this band appears at 465 cm^{-1}. However, a shift to lower frequencies (Table 4), together with a non-linear linewidth, can be viewed for all the samples. According to Spanier et al. [71], a large number of factors can contribute to the changes in the Raman peak position and linewidth of the 465 cm^{-1} peak, including phonon confinement, broadening associated with size distribution, defects, strain, and variations in phonon relaxation as a function of particle size. Apart from this main band, several other bands can be clearly distinguished in the corresponding spectra. The band at ca. 265 cm^{-1} is attributed to the tetrahedral displacement of oxygen from the ideal fluorite lattice [25,69,72]. The band at ca. 600 cm^{-1} is assigned to the defect-induced mode (D), related to the presence of lattice defects, mostly oxygen vacancies [5,42,68,73]. Noticeably, the samples synthesized at 150 °C with low concentration of NaOH (CuCe-150-0, CuCe-150-0.05, and CuCe-150-0.1) exhibit one additional band at ca. 830 cm^{-1}. According to Choi et al. [72], this band is associated with peroxo-like oxygen species adsorbed on the oxygen vacancies, in close relation with reduced ceria species. No separated copper phase (CuO) can be confirmed via the appearance of the corresponding Raman peaks at ca. 295 cm^{-1} and ca. 350 cm^{-1}. Therefore, both the XRD and Raman results clearly indicate that Cu species onto ceria are highly dispersed.

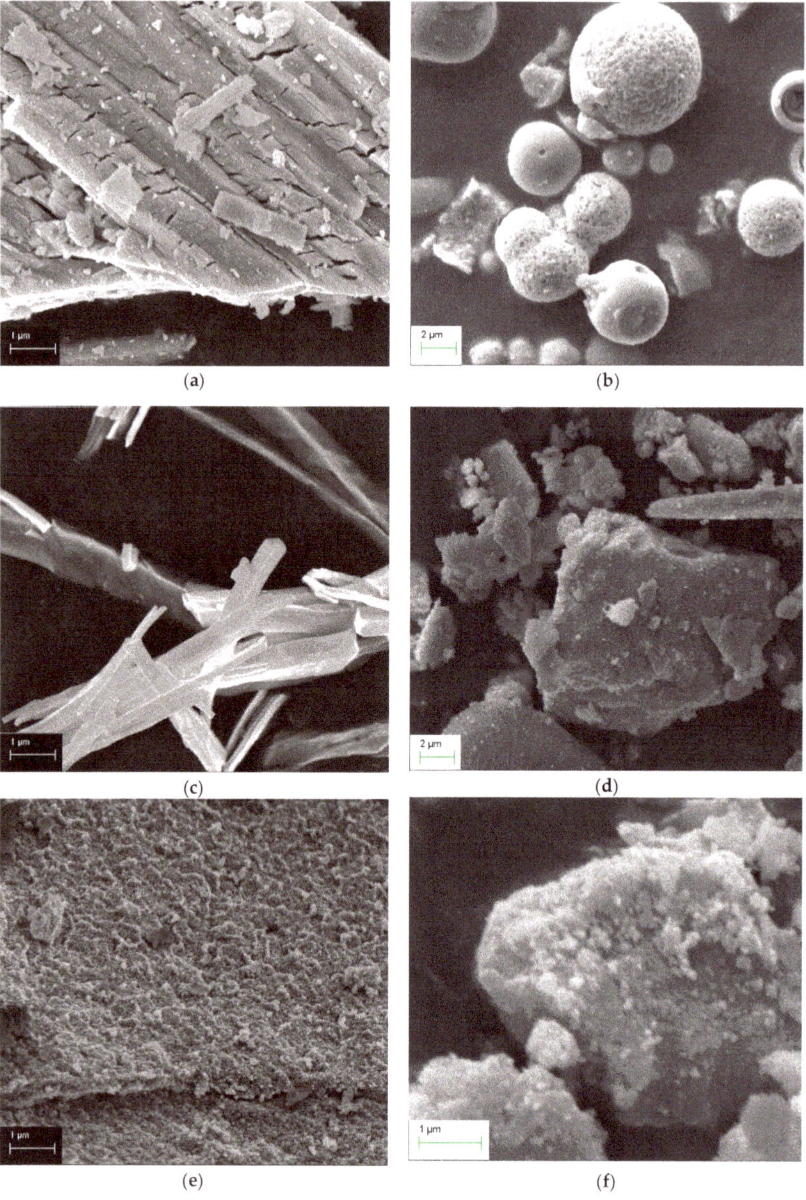

(a) (b) (c) (d) (e) (f)

Figure 5. Catalysts prepared hydrothermally at 120 °C: (**a**) Ce-120-0, (**b**) CuCe-120-0.05, (**c**) Ce-120-0.1, (**d**) CuCe-120-0.5, (**e**) CuCe-120-1, and (**f**) Ce-120-5.

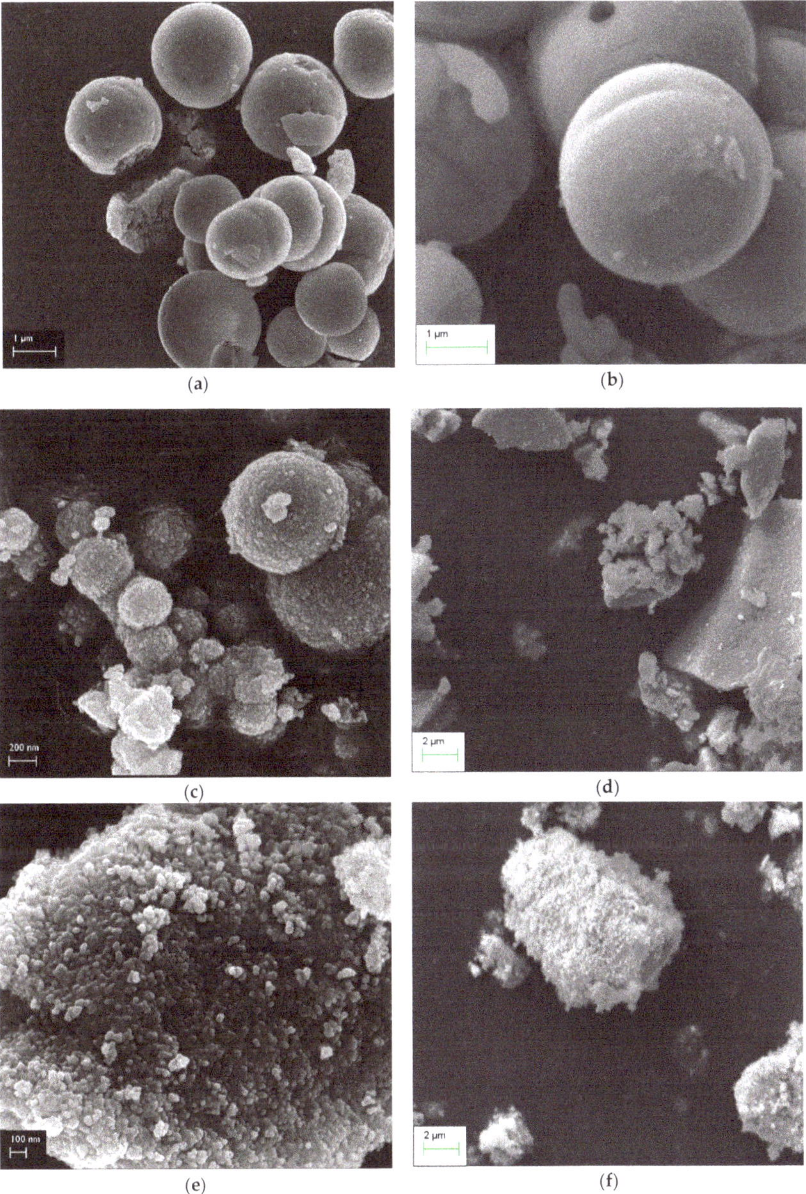

(a) (b) (c) (d) (e) (f)

Figure 6. Catalysts prepared hydrothermally at 150 °C: (**a**) CuCe-150-0, (**b**) CuCe-150-0.05, (**c**) CuCe-150-0.1, (**d**) Ce-150-0.5, (**e**) CuCe-150-1, and (**f**) Ce-150-5.

The position and the FWHM for the F_{2g} band, and the relative intensity ratio of I_D/I_{F2g} for the ceria-based catalysts are shown in Table 4. In general, the ratio of I_D/I_{F2g}, where I_{F2g} and I_D correspond to the maximum intensity of F_{2g} and D bands, respectively, can roughly reflect the amount of lattice defects (oxygen vacancies) in the obtained materials [72,74,75]. With appropriate tuning of the employed hydrothermal parameters, materials with high concentration of oxygen vacancies are obtained. For instance, the Ce-120-5 sample presents the highest value of this ratio, i.e., 0.106, suggesting a material with high perspectives in catalytic CO oxidation reaction. According to previous studies, the FWHM of the F_{2g} band is influenced to a great extent by the crystallite size of ceria and the concentration of oxygen vacancies [75–77]. However, there is no specific trend in this work and mainly the high FWHM may be closely correlated with the high concentration of oxygen vacancies, since more factors might also play a role in these features.

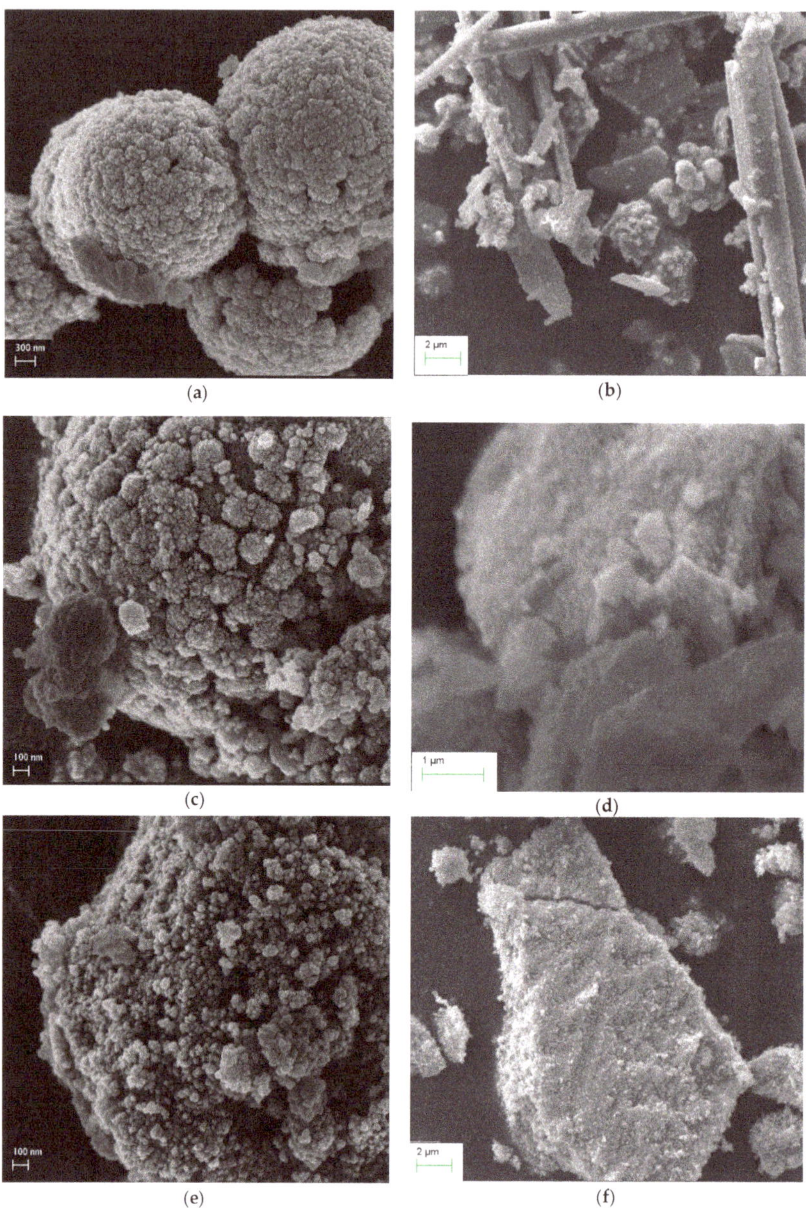

Figure 7. Catalysts prepared hydrothermally at 180 °C: (**a**) CuCe-180-0, (**b**) CuCe-180-0.05, (**c**) CuCe-180-0.1, (**d**) CuCe-180-0.5, (**e**) CuCe-180-1, and (**f**) Ce-180-5.

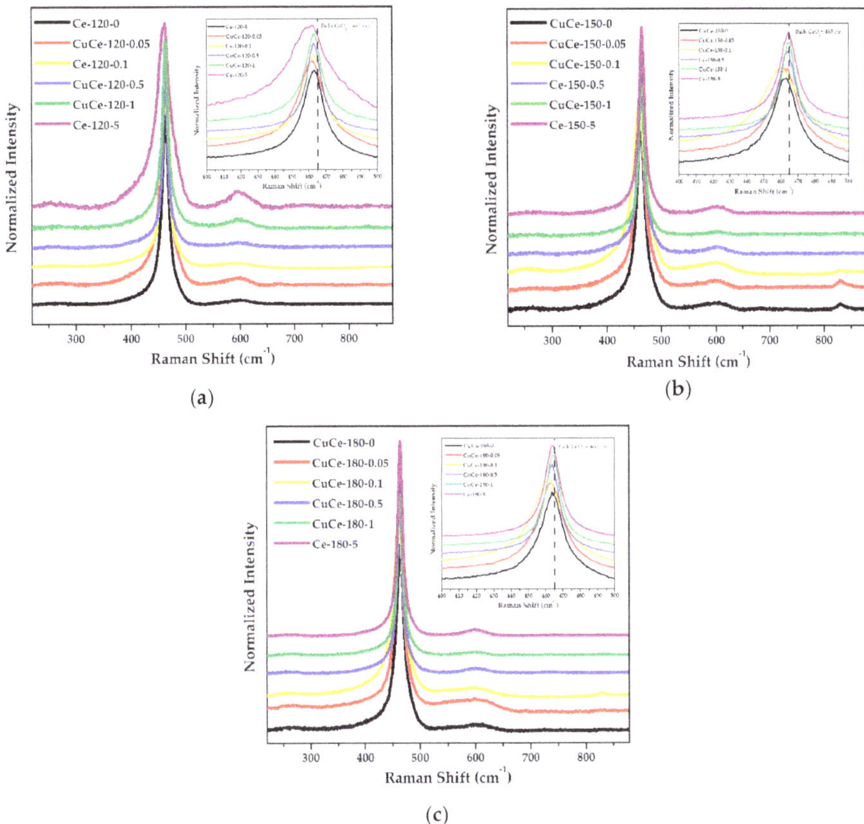

Figure 8. Raman spectra of the catalysts synthesized hydrothermally (**a**) 120, (**b**) 150, and (**c**) 180 °C.

Table 4. F_{2g} peak position, full width at half maximum (FWHM) for the F_{2g} band, relative intensity ratio of I_D/I_{F2g}, and energy band gap (E_g) of the catalysts.

Samples	F_{2g} (cm^{-1})	FWHM (cm^{-1})	I_D/I_{F2g}	E_g (eV)
Ce-120-0	463.0	14.8	0.027	3.05
CuCe-120-0.05	462.3	22.8	0.046	3.03
Ce-120-0.1	463.1	17.7	0.023	3.13
CuCe-120-0.5	463.0	13.3	0.035	3.18
CuCe-120-1	463.0	12.5	0.060	3.24
Ce-120-5	462.3	36.5	0.106	3.12
CuCe-150-0	463.0	24.2	0.031	3.12
CuCe-150-0.05	462.3	26.7	0.050	3.06
CuCe-150-0.1	460.7	36.0	0.046	3.07
Ce-150-0.5	463.8	17.6	0.032	3.20
CuCe-150-1	464.6	12.7	0.014	3.28
Ce-150-5	464.6	17.0	0.046	3.29
CuCe-180-0	463.7	22.8	0.047	3.11
CuCe-180-0.05	463.0	23.2	0.080	3.00
CuCe-180-0.1	461.4	28.1	0.035	3.21
CuCe-180-0.5	463.7	15.8	0.038	3.20
CuCe-180-1	464.5	13.1	0.020	3.25
Ce-180-5	463.8	14.8	0.034	3.29

2.6. UV-Vis DRS Measurements

Figure S2 presents the UV-Vis DRS spectra of the materials. Each spectrum indicates three different bands, which according to the literature correspond to different electronic transitions between cerium and oxygen ions. The band at ca. 360 nm is ascribed to the $O^{2-} \rightarrow Ce^{4+}$ interband-transfer transition. The bands at ca. 270 and ca. 230 nm are attributed to the $O^{2-} \rightarrow Ce^{4+}$ and $O^{2-} \rightarrow Ce^{3+}$ charge-transition, respectively [25,78]. Interestingly, the CuCe-180-0, CuCe-180-0.05, CuCe-180-0.1, and CuCe-150-0.1 samples present one extra wide band at 670 nm. It is important to mention that there is a lot of controversy about this band. According to Rakai et al. [79], this band is correlated with a surface redox couple of cerium ions (Ce^{3+}/Ce^{4+}). A similar band was found from Bensalem et al. [80] and Binet et al. [81]. However, there is a number of studies which ascribe this band to d-d transitions of Cu^{2+} in an octahedral environment [2,82,83]. It also has to be noted that, for the CuCe-150-0.1, CuCe-180-0, CuCe-180-0.05, and CuCe-180-0.1, catalysts a shoulder appears in the spectra at ca. 400–500 nm. This feature represents the charge transfer between "Support ← Oxygen—Active Phase", suggesting strong interactions between the copper species and ceria [43,44].

The energy band gap (E_g) of the catalysts, calculated via the Tauc plots method, is summarized in Table 4. Taking into account the reference value for the bulk CeO_2 as a direct band gap semiconductor (E_g = 3.19 eV), the obtained values of the ceria-based materials, especially the ones synthesized with low concentrations of NaOH ($\leq$0.1 M), are smaller than the energy band gap of bulk ceria. On the other hand, the catalysts prepared with higher concentrations of NaOH ($\geq$0.5 M) resulted in a higher value of E_g. The former values indicate the presence of defects and especially oxygen vacancies [26,84], while the latter values of E_g are closely related to the quantum confinement effect [77,85].

2.7. Formation Mechanism

Over the years, the utilization of organic additives such as polyvinylpyrrolidone (PVP) [86], cetyltrimethylammonium bromide (CTAB) [87] and oleic acid [88] has been established in the hydrothermal method in order to obtain well-defined particles and morphologies. However, due to adsorption effects on the surface of particles these additives lead to quenching because of their high-energy vibration [89]. Compared with the above surfactants, citric acid (CA) presents weaker morphology control ability. Therefore, when the citric acid is used as a chelating agent, products with various morphologies and sizes can be obtained [90,91].

According to Levien [92], citric acid separates into different ionic species:

$$H_3cit \text{ (citric acid)} \rightarrow H^+ + H_2cit^-, \ (pK_\alpha = 3.2), \tag{1}$$

$$H_2cit^- \rightarrow H^+ + Hcit^{2-}, \ (pK_\alpha = 4.9), \tag{2}$$

$$Hcit^{2-} \rightarrow H^+ + cit^{3-}, \ (pK_\alpha = 6.4), \tag{3}$$

The concentrations of these ionic species are closely depended on the pH value. In an acid solution (ca. pH = 2) the citric acid is not effectively dissociated to citric ions. Further increase of pH at values of 4, 6, and 8, the main citric species are the H_2cit^-, $Hcit^{2-}$ and cit^{3-}, respectively. The citric ions can form complexes with the cerium ions, which are depended on the pH of solution [93].

The pH values from each step of synthesis route are illustrated in Table 5. It can be observed that the low concentrations of NaOH ($\leq$0.1 M) did not increase the pH of the final solution (acid solution). As a result, the citric acid was not effectively dissociated, as there were H_3cit and H_2cit^- species in the final solution. Concerning the partial dissociation of citric acid, it is not possible to form several and stable complexes with the cerium ions under ambient conditions. On the other hand, the high concentrations of NaOH ($\geq$0.5 M) caused the formation of an alkaline solution. Under these conditions, the citric acid has been completely dissociated and can form several and stable complexes with the cerium ions, even under ambient conditions.

Table 5. pH values from the different part of the catalyst's preparation.

Samples	pH (Ce^{3+}) [2]	pH (CA) [3]	pH (Ce^{3+} + CA) [4]	pH (NaOH) [5]	pH Total [6]
0 M [1]	3.75	1.56	0.95	0	1.68
0.05 M	3.75	1.56	0.95	12.76	2.10
0.1 M	3.75	1.56	0.95	13.03	2.52
0.5 M	3.75	1.56	0.95	13.35	13.11
1 M	3.75	1.56	0.95	13.51	13.46
5 M	3.75	1.56	0.95	13.91	13.86

[1] The sample 0M is referred to the sample that no addition of NaOH is occurred and so on; [2] pH of the solution that contains Ce^{3+} ions; [3] pH of the solution that contains citric ions; [4] pH during the mixing of the previous solutions; [5] pH of the NaOH solution; and [6] pH of the final solution before the hydrothermal treatment.

It is believed that the increase of the temperature (from room temperature to the desired hydrothermal temperature) ensures the complete complexation of citric ions with the cerium ions [93–95], while the formed complex is polymerized and becomes stable in the solution. As the hydrothermal treatment proceeds the elevated pressure and temperature triggers the appearance of two events:

- The weakening of the complex stability and precipitation of the latter as hydroxide into the solution.
- Depending on the hydrothermal parameters, the leaching and migration of copper species from the copper ring to the solution where their deposition in the ceria phase is taking place.

The final catalyst is obtained after the steps of filtration, drying, and calcination.

Concerning the various morphologies which are illustrated in Figures 5–7, it is proposed that the successful combination of hydrothermal parameters (temperature and concentration of NaOH) is the main reason behind these morphologies. It has been mentioned in Section 2.4 that the combination of low concentrations (≤0.1 M) and hydrothermal temperatures caused the formation spheres and/or rods. Given that the pH of the solution at these specific concentration was acidic, Ostwald Ripening seems to be the most dominant particle's formation mechanism [96–98]. Once the particles have been formed, a process of self-organization starts to happen, which results in a spherical and/or rod-like morphology depending on the hydrothermal parameters. On the other hand, the high concentrations of NaOH (≥0.5 M) and the hydrothermal temperatures resulted in the formation of bulky aggregates consisting of particles with a spherical and/or irregular geometry. At these conditions, the pH of the solution was alkaline and so the proposed formation mechanism of particles appears to be the oriented attachment [87,99]. Once the formation of particles has been completed, self-organization process initiates, but simultaneously the excess of OH^- groups and the hydrothermal temperatures cause the formation of bulky aggregates.

2.8. CO Oxidation Catalytic Studies

The conversion of CO to CO_2 as a function of the temperature of the reaction over the ceria-based catalysts is shown in Figure 9. As a general trend, it can be commented that the catalysts prepared with low concentrations of NaOH (≤0.1 M) and contained Cu^{2+} species onto their surface, presented better catalytic behavior than the catalysts which were hydrothermally treated with high concentrations of NaOH (≥0.5 M). Noticeably, the latter catalysts illustrated a significant amount of surface Ce^{3+} ions (see Table 1), which is usually correlated with high catalytic activity. However, this is not the case in this work and other factors control the catalytic activity, as will be discussed in the next section.

An indicator of the catalytic activity behavior are the temperatures where 50% and 90% CO conversion is achieved (T_{50} and T_{90}, respectively) (Figure 9). For the samples treated hydrothermally at 120 °C, it can be seen that the most active catalysts are the CuCe-120-0.05 and the Ce-120-0, presenting T_{50} equal to 194 °C and 224 °C, respectively, and T_{90} equal to 272 °C and 263 °C, respectively. Additionally, the above samples achieved full removal of CO. The less active sample was the Ce-120-0.1, which revealed a T_{50} at 306 °C and T_{90} at 351 °C. One possible explanation for the poor catalytic activity of this sample is its morphology, which presented rods with high crystallinity. Increasing the concentration of NaOH (≥0.5 M), the catalytic activity of the samples was improved, but it cannot be comparable with the activities of the CuCe-120-0.05 and Ce-120-0. The morphology of the samples prepared with high concentrations of NaOH (see Figure 5) seems to influence the catalytic activity in a negative way. Moreover, although the sample Ce-120-5 illustrated the highest SSA among all catalysts (see Table 3), this is not related to the achievement of high catalytic activity.

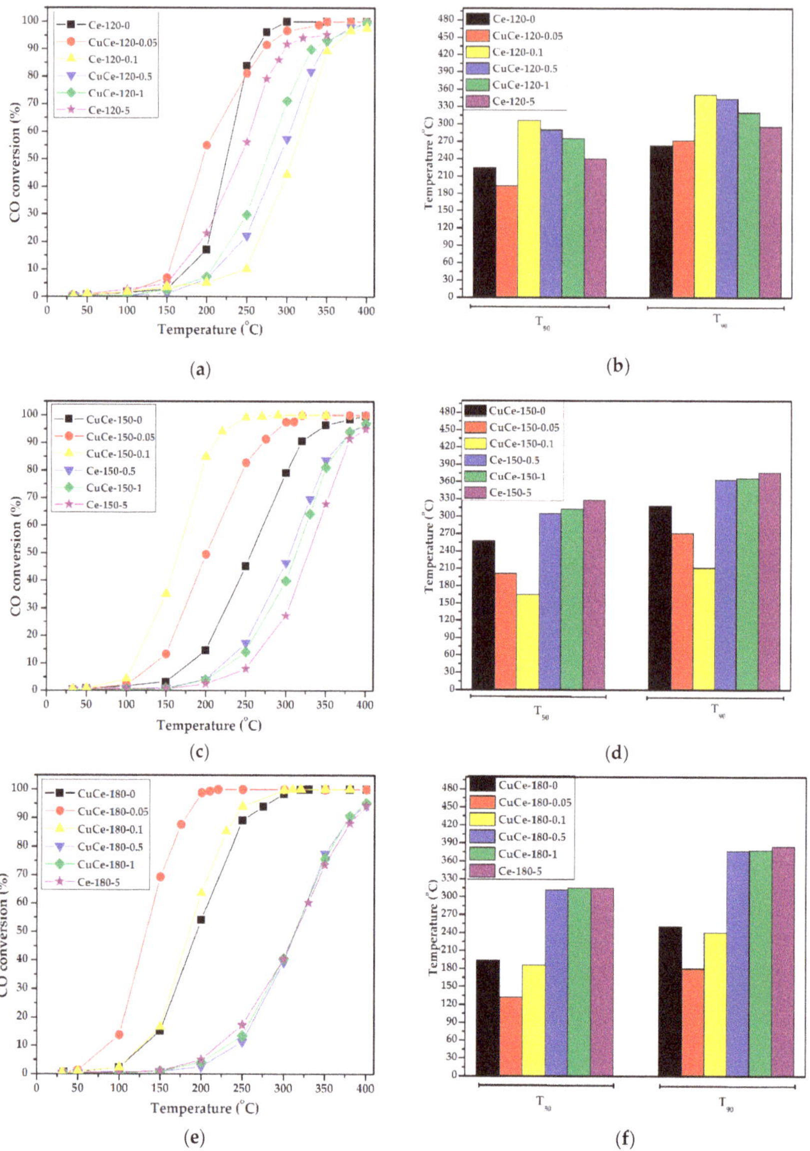

Figure 9. (**a**,**c**,**e**) CO conversion diagrams of the catalysts which were treated hydrothermally at 120, 150 and 180 °C, respectively; and (**b**,**d**,**f**) Bar graphs which are shown the temperatures where 50% and 90% conversion of CO to CO_2 is achieved.

The catalytic activity of the samples synthesized at 150 °C, was dramatically improved, especially for NaOH concentrations ≤0.1 M, reaching in some cases 100% of CO conversion. For instance, the CuCe-150-0 sample presents a T_{50} at 257 °C and the sample CuCe-150-0.1 shows T_{50} at 165 °C. It should be mentioned that the former sample contained Cu^{2+} isolated ions dispersed in ceria, as confirmed via EPR spectroscopy, while the latter sample presented Cu^{2+} clusters onto ceria surface. Higher reactivity of Cu^{2+} clusters than Cu^{2+} isolated ions, was also reported by Harrison et al. [39]. Higher concentrations of NaOH (≥0.5 M) resulted in catalytic materials with poor activity.

Similar trends can be also depicted in the catalytic behavior of the materials synthesized with a hydrothermal temperature of 180 °C. Overall, the highest catalytic activity was achieved over the CuCe-180-0.05 sample which present in Figure 9, T_{50} and T_{90} at 132 °C and 180 °C, respectively. The rod-like morphology, the presence of high-content Cu^{2+} clusters (see Table 1) and high concentration of oxygen vacancies (see Table 4) seems to be the crucial reasons behind this extraordinary catalytic activity. It is noteworthy that the CuCe-180-0.05 as both CuCe-180-0 and CuCe-180-0.1 illustrated 100% conversion of CO. On the other hand, the samples that were prepared with high concentrations (≥0.5 M) presented similar catalytic behavior with the samples hydrothermally treated at 120 °C and 150 °C.

3. Discussion

Correlation of the Physicochemical Characteristics with the Catalytic Activity

Taking into consideration both the physicochemical characterization and catalytic evaluation, it can be proposed that the high catalytic activity is controlled by the successful combination of specific materials characteristics such as the morphology, oxygen vacancies, and type of copper entities. The specific surface area seems to be the feature with less impact in order to achieve high activity. More specifically:

- The most active samples presented spherical and/or rod-like morphology (samples with concentrations of NaOH ≤0.1 M), which was attributed to the effective combination of hydrothermal parameters. Exception to this trend was the sample Ce-120-0.1, which illustrated rods with high crystallinity. The bulky aggregates which, were revealed by the samples with high concentrations (≥0.5 M), seemed to negatively affect the catalytic activity.
- While the samples with high NaOH concentrations (≥0.5 M) showed high SSA, they could not compete the activities of the samples prepared with low concentrations of NaOH (≤0.1 M). In general, the high SSA is a desirable parameter for enhanced catalytic activity, but in our work, it is not a determining factor, as the observed catalytic performance is a combination of the copper content, copper species nature, morphology, oxygen vacancies and Cu-Ce interactions.
- Two different copper species were detected in ceria-based materials, which promoted the catalytic activity. This effect was more pronounced in the case of Cu^{2+} clusters as compared with the effect caused by the presence of Cu^{2+} isolated ions. The former species were favored under high hydrothermal temperatures and low NaOH concentrations.
- The presence of lattice defects and especially oxygen vacancies, which were certified by the spectroscopic techniques, affected the catalytic behavior, but not following a conventional trend. As shown via EPR spectroscopy, the catalysts, which were prepared with high concentrations of NaOH (≥0.5 M), presented a high concentration of Ce^{3+} ions (see Table 1), they did not result in a high catalytic activity. On the other hand, the samples with low concentrations of NaOH (≤0.1 M), although possessing small concentrations of Ce^{3+} ions, they exhibited better catalytic behavior. Strong copper-ceria concentrations were confirmed via DRS analysis for several samples prepared at 150 and 180 °C with low concentrations of NaOH. Conclcuding the above remarks, it can be commented that the formation of a high-defective structure is not a crucial factor for high catalytic activity, but the formation of a structure that will have a suitable concentration of oxygen vacancies. Similar results have been illustrated by Lin et al. [99]. Pulse calorimetry

measurements showed that the high concentration of oxygen vacancies promoted the buildup of adsorbed carbonates that prohibit the adsorption and activation of CO, but not O_2. According to several studies, oxygen vacancies tend to form oxygen vacancy clusters [100,101]. Wang et al. [102] showed that oxygen vacancy clusters with suitable size and distribution are responsible for high activity. Thus, it is believed that the various morphologies, which were obtained in this study, can present oxygen vacancies with different size and distribution.

4. Materials and Methods

4.1. Catalysts Preparation

All the chemicals used in this work were of analytical reagent grade. Cerium(III) nitrate hexahydrate $Ce(NO_3)_3 \cdot 6H_2O$ (purity 99.99%, Sigma-Aldrich) and a pure metallic copper ring were used as precursors for the preparation of ceria and copper-promoted ceria nanomaterials. Moreover, citric acid monohydrate $C_6H_8O_7 \cdot H_2O$ (purity 99.5–101.0%, Ing. Petr Švec PENTA, Prague, Czech Republic) and sodium hydroxide NaOH (purity $\geq$98.0%, Ing. Petr Švec PENTA, Prague, Czech Republic) were also employed in the synthesis procedure as a chelating agent and precipitating agent, respectively.

At first, 3.7834 gr of $Ce(NO_3)_3 \cdot 6H_2O$ and 1.8314 gr of $C_6H_8O_7$ H_2O were dissolved under continuous stirring into 15 mL of triple distilled (3D) water, respectively. The molar ratio of citric acid to metal nitrate was adjusted according to the stoichiometry of the reaction (citric acid/Ce = 1/1). When the dissolution of the compounds was completed, the two aqueous solutions were mixed under continuous stirring, while at the same time, a 150 mL aqueous solution of NaOH (C_{NaOH} = 0–5 M) was prepared. Both solutions were mixed in a Teflon beaker, and this mixture was kept under stirring for 20 min. The Teflon beaker was placed in a lab-made stainless-steel autoclave (chamber volume of 200 mL), where the copper ring was placed over the beaker (not immersed into the solution). The autoclave was sealed tightly and heated at various temperatures (120, 150, and 180 °C) for 24 h. After the hydrothermal treatment, the autoclave was opened, excess water was decanted, and the precipitates were filtered, washed several times with triple distilled water (until pH = 7), and dried under vacuum at 70 °C overnight. Finally, the obtained powders were calcined at 400 °C for 2 h (heating ramp = 2 °C min^{-1}).

In order to facilitate the presentation of results, the following encoding of catalysts is used: Ce-T-M and CuCe-T-M, where T represents the temperature (°C) of the hydrothermal reaction and M the molarity of NaOH (M). For example, the Ce-120-5 catalyst was synthesized hydrothermally at 120 °C, using 5 M NaOH.

4.2. Catalysts Characterization

An X-ray powder diffractometer (Bruker D8 Advance, Bruker, Birmingham, UK) employing Cu K_a radiation (λ = 0.15418 nm) at 40 kV and 40 mA was used to analyze the crystalline structure of the catalysts.

The specific surface area (SSA), the pore volume and the pore size distribution of the materials were determined from the adsorption and desorption isotherms of nitrogen at -196 °C using a TriStar 3000 Micromeritics instrument (Norcross, GA, U.S.A). Prior to the measurements, the samples were outgassed at 150 °C for 1 h, under N_2 flow.

The morphology of the obtained materials was observed with scanning electron microscopy (SEM, Leo Supra 35VP (Carl Zeiss SMT AG Company, Oberkochen, Germany).

Raman spectra were accumulated with the 441.6 nm laser line as the excitation source emerging from a He–Cd laser (Kimon). The scattered light was analyzed by the Lab-Ram HR800 (Jobin-Yvon, Horiba, Montpellier, France) micro Raman spectometer at a spectral resolution of about 2.0 cm^{-1}. A microscope objective with magnification 50× was used to focus the light onto a spot of ~3 μm in diameter. Low laser intensities were used (~0.37 mW on the sample) to avoid spectral changes due to heat-induced effects. The Raman shift was calibrated using the 520 cm^{-1} Raman band of crystalline Si.

The diffuse reflectance spectra of the obtained materials were recorded in the range 200–800 nm at room temperature using a UV-vis spectrophotometer (Varian Cary 3; Varian Inc. Palo Alto, CA, USA) equipped with an integration sphere. The DR spectra were collected on calcined samples with PTFE disks. The powder samples were mounted in a quartz cell, which provided a sample thickness >3 mm to guarantee the "infinite" sample thickness.

Ceria-based nanomaterials were also characterized by Electron Paramagnetic Resonance (EPR) spectroscopy employing a Bruker ER200D spectrometer (Billerica, MA, USA) equipped with an Agilent 5310A frequency counter (Agilent Technologies, Santa Clara, CA, USA). EPR spectra were recorded at 77 K in suprasil-quarz tubes (3 mm inner diameter; Willmad Glass). 10 mg of nano-powders were introduced into the sample EPR tube followed by outgassing at 300 K for 10 min under 10-4 bar vacuum. The spectrometer was running under home-made software based on LabView. Numerical simulation of experimental EPR spectra was performed with EasySpin 5.2.21 software (The MathWorks Inc., Natick, MA, USA) [103]. Quantification of Ce3+-O–Ce4+ (S = 1/2) was performed using 2,2-diphenyl-1-picrylhydrazyl (DPPH) [104] as a spin standard (Sigma Aldrich). Quantification of Ce^{3+} (S = 5/2) was performed using FeIII (S = 5/2)-EDTA complex [105,106], while the quantification of the Cu^{2+} centers was done using a $Cu(NO_3)_2$ standard. Copper content was also determined via WDXRF analysis (Wavelength Dispersive X-Ray Fluorescence; ZSX PRIMUS II, RIGAKU, Austin, TX, USA).

4.3. Catalytic Studies

Activity measurements for the catalytic oxidation of CO were conducted in a conventional fixed-bed reactor (described in detail elsewhere [107]) at atmospheric pressure, in the temperature range of 30–400 °C. The catalyst was in the form of powder with a mass of 120 mg and the total flow rate of the reaction mixture was 25 cm^3 min^{-1}, yielding a contact time of W/F = 0.288 g s cm^{-3}, where W is the weight of catalyst and F the total flow rate of the reactant gas. The reaction feed stream contained 1 vol.% CO, 20 vol.% O_2 and He as balance. Product and reactant analysis was carried out by a gas chromatograph (Shimadzu GC-14B) equipped with a thermal conductivity detector. The CO conversion calculation was based on the CO_2 formation or CO consumption:

$$\text{CO conversion } (\%) = \frac{[CO_2]_{OUT}}{[CO]_{IN}} \times 100\%, \tag{4}$$

5. Conclusions

In this study, highly active ceria and copper-promoted ceria catalysts were synthesized via a novel hydrothermal method and evaluated in CO oxidation reaction. The physicochemical characteristics, and as a consequence, the catalytic properties were controlled via appropriate combination of hydrothermal parameters (temperature and concentration of the precipitating agent). EPR spectroscopy demonstrated the presence of different copper species (isolated ions and amorphous clusters), dispersed in ceria nanostructrures (mainly nanorods and nanospheres, depending on the pH of the hydrothermal solution). Elevated hydrothermal temperatures and NaOH concentrations favored the formation of isolated copper ions, which resulted to be less active, as compared with copper clusters. Moreover, the catalysts prepared with high concentrations of NaOH (≥0.5 M) presented a significant amount of surface Ce^{3+} ions, while Raman spectroscopy and UV-Vis DRS measurements revealed the presence of lattice defects and especially oxygen vacancies. Overall, ceria-based catalysts prepared at elevated temperatures with low concentrations of NaOH (≤0.1 M) were more active in CO oxidation, and this behavior can be mainly related with the obtained morphology and the nature of oxygen vacancies and dispersed copper species, and to a lesser extent, with the specific surface area and the concentration of defects.

Supplementary Materials: The following are available online at http://www.mdpi.com/2073-4344/9/2/138/s1, Figure S1. XRD diffractograms of CeO_2 and Cu/CeO_2 catalysts: (**a**) Catalysts prepared hydrothermally at 120 °C; (**b**) Catalysts prepared hydrothermally at 150 °C; (**c**) Catalysts prepared hydrothermally at 180 °C, Figure S2. UV-Vis DRS spectra of CeO_2 and Cu/CeO_2 catalysts: (**a**) Catalysts prepared hydrothermally at 120 °C; (**b**) Catalysts prepared hydrothermally at 150 °C; (**c**) Catalysts prepared hydrothermally at 180 °C.

Author Contributions: Conceptualization, G.A.; investigation and formal analysis, K.K., C.P., Y.G., J.V., J.P. and Y.D.; writing—original draft preparation, K.K.; writing—review and editing, G.A.

Funding: This research was funded by Research Committee of the University of Patras via "K. Karatheodori" program, grant number (E.610). This research was also co-financed by Greece and the European Union (European Social Fund- ESF) through the Operational Programme "Human Resources Development, Education and Lifelong Learning" in the context of the project "Strengthening Human Resources Research Potential via Doctorate Research" (MIS-5000432), implemented by the State Scholarships Foundation (IKY).

Conflicts of Interest: The authors declare no conflict of interest.

References

1. Liu, Y.; Wen, C.; Guo, Y.; Lu, G.; Wang, Y. Effects of surface area and oxygen vacancies on ceria in CO oxidation: Differences and relationships. *J. Mol. Catal. Chem.* **2010**, *316*, 59–64. [CrossRef]
2. Rao, K.N.; Bharali, P.; Thrimurthulu, G.; Reddy, B.M. Supported copper–ceria catalysts for low temperature CO oxidation. *Catal. Commun.* **2010**, *11*, 863–866. [CrossRef]
3. He, H.; Yang, P.; Li, J.; Shi, R.; Chen, L.; Zhang, A.; Zhu, Y. Controllable synthesis, characterization, and CO oxidation activity of CeO_2 nanostructures with various morphologies. *Ceram. Int.* **2016**, *42*, 7810–7818. [CrossRef]
4. Avgouropoulos, G.; Ioannides, T.; Matralis, H.K.; Batista, J.; Hocevar, S. CuO-CeO_2 mixed oxide catalysts for the selective oxidation of carbon monoxide in excess hydrogen. *Catal. Lett.* **2001**, *73*, 33–40. [CrossRef]
5. Gamarra, D.; Cámara, A.L.; Monte, M.; Rasmussen, S.B.; Chinchilla, L.E.; Hungría, A.B.; Munuera, G.; Gyorffy, N.; Schay, Z.; Corberán, V.C.; et al. Preferential oxidation of CO in excess H_2 over CuO/CeO_2 catalysts: Characterization and performance as a function of the exposed face present in the CeO_2 support. *Appl. Catal. B Environ.* **2013**, *130–131*, 224–238. [CrossRef]
6. Guo, X.; Zhou, R. A new insight into the morphology effect of ceria on CuO/CeO_2 catalysts for CO selective oxidation in hydrogen-rich gas. *Catal. Sci. Technol.* **2016**, *6*, 3862–3871. [CrossRef]
7. Paliwal, A.; Sharma, A.; Tomar, M.; Gupta, V. Carbon monoxide (CO) optical gas sensor based on ZnO thin films. *Sens. Actuators B Chem.* **2017**, *250*, 679–685. [CrossRef]
8. Lin, Q.; Han, C.; Su, H.; Sun, L.; Ishida, T.; Honma, T.; Sun, X.; Zheng, Y.; Qi, C. Remarkable enhancement of Fe–V–O_x composite metal oxide to gold catalyst for CO oxidation in the simulated atmosphere of CO_2 laser. *RSC Adv.* **2017**, *7*, 38780–38783. [CrossRef]
9. Kašpar, J.; Fornasiero, P.; Graziani, M. Use of CeO_2-based oxides in the three-way catalysis. *Catal. Today* **1999**, *50*, 285–298. [CrossRef]
10. Mai, H.-X.; Sun, L.-D.; Zhang, Y.-W.; Si, R.; Feng, W.; Zhang, H.-P.; Liu, H.-C.; Yan, C.-H. Shape-Selective Synthesis and Oxygen Storage Behavior of Ceria Nanopolyhedra, Nanorods, and Nanocubes. *J. Phys. Chem. B* **2005**, *109*, 24380–24385. [CrossRef]
11. Qin, J.; Lu, J.; Cao, M.; Hu, C. Synthesis of porous CuO–CeO_2 nanospheres with an enhanced low-temperature CO oxidation activity. *Nanoscale* **2010**, *2*, 2739–2743. [CrossRef] [PubMed]
12. Lykaki, M.; Pachatouridou, E.; Iliopoulou, E.; Carabineiro, S.A.; Konsolakis, M. Impact of the synthesis parameters on the solid state properties and the CO oxidation performance of ceria nanoparticles. *RSC Adv.* **2017**, *7*, 6160–6169. [CrossRef]
13. Campbell, C.T.; Peden, C.H.F. Oxygen Vacancies and Catalysis on Ceria Surfaces. *Science* **2005**, *309*, 713–714. [CrossRef] [PubMed]
14. Yang, Z.; Zhou, K.; Liu, X.; Tian, Q.; Lu, D.; Yang, S. Single-crystalline ceria nanocubes: Size-controlled synthesis, characterization and redox property. *Nanotechnology* **2007**, *18*, 185606–185609. [CrossRef]
15. Xu, J.; Harmer, J.; Li, G.; Chapman, T.; Collier, P.; Longworth, S.; Chi Tsang, S. Size dependent oxygen buffering capacity of ceria nanocrystals. *Chem. Commun.* **2010**, *46*, 1887–1889. [CrossRef] [PubMed]

16. Ishikawa, Y.; Takeda, M.; Tsukimoto, S.; Nakayama, K.S.; Asao, N. Cerium Oxide Nanorods with Unprecedented Low-Temperature Oxygen Storage Capacity. *Adv. Mater.* **2016**, *28*, 1467–1471. [CrossRef] [PubMed]
17. Sayle, T.X.T.; Parker, S.C.; Catlow, C.R.A. The role of oxygen vacancies on ceria surfaces in the oxidation of carbon monoxide. *Surf. Sci.* **1994**, *316*, 329–336. [CrossRef]
18. Conesa, J. Computer modeling of surfaces and defects on cerium dioxide. *Surf. Sci.* **1995**, *339*, 337–352. [CrossRef]
19. Zhou, K.; Wang, X.; Sun, X.; Peng, Q.; Li, Y. Enhanced catalytic activity of ceria nanorods from well-defined reactive crystal planes. *J. Catal.* **2005**, *229*, 206–212. [CrossRef]
20. Wu, Z.; Li, M.; Overbury, S.H. On the structure dependence of CO oxidation over CeO_2 nanocrystals with well-defined surface planes. *J. Catal.* **2012**, *285*, 61–73. [CrossRef]
21. Tana; Zhang, M.; Li, J.; Li, H.; Li, Y.; Shen, W. Morphology-dependent redox and catalytic properties of CeO_2 nanostructures: Nanowires, nanorods and nanoparticles. *Catal. Today* **2009**, *148*, 179–183. [CrossRef]
22. Periyat, P.; Laffir, F.; Tofail, S.A.M.; Magner, E. A facile aqueous sol–gel method for high surface area nanocrystalline CeO_2. *RSC Adv.* **2011**, *1*, 1794–1798. [CrossRef]
23. Suresh, R.; Ponnuswamy, V.; Mariappan, R. Effect of annealing temperature on the microstructural, optical and electrical properties of CeO_2 nanoparticles by chemical precipitation method. *Appl. Surf. Sci.* **2013**, *273*, 457–464. [CrossRef]
24. Kepenekçi, O.; Emirdag-Eanes, M.; Demir, M.M. Effect of alkali metal hydroxides on the morphological development and optical properties of ceria nanocubes under hydrothermal conditions. *J. Nanosci. Nanotechnol.* **2011**, *11*, 3565–3577. [CrossRef] [PubMed]
25. Liu, L.; Cao, Y.; Sun, W.; Yao, Z.; Liu, B.; Gao, F.; Dong, L. Morphology and nanosize effects of ceria from different precursors on the activity for NO reduction. *Catal. Today* **2011**, *175*, 48–54. [CrossRef]
26. Zdravković, J.; Simović, B.; Golubović, A.; Poleti, D.; Veljković, I.; Šćepanović, M.; Branković, G. Comparative study of CeO_2 nanopowders obtained by the hydrothermal method from various precursors. *Ceram. Int.* **2015**, *41*, 1970–1979. [CrossRef]
27. Zhang, W.; Niu, X.; Chen, L.; Yuan, F.; Zhu, Y. Soot Combustion over Nanostructured Ceria with Different Morphologies. *Sci. Rep.* **2016**, *6*, 29062. [CrossRef]
28. Lu, X.; Zhai, T.; Cui, H.; Shi, J.; Xie, S.; Huang, Y.; Liang, C.; Tong, Y. Redox cycles promoting photocatalytic hydrogen evolution of CeO_2 nanorods. *J. Mater. Chem.* **2011**, *21*, 5569–5572. [CrossRef]
29. Terribile, D.; Trovarelli, A.; Llorca, J.; de Leitenburg, C.; Dolcetti, G. The Synthesis and Characterization of Mesoporous High-Surface Area Ceria Prepared Using a Hybrid Organic/Inorganic Route. *J. Catal.* **1998**, *178*, 299–308. [CrossRef]
30. Pan, C.; Zhang, D.; Shi, L.; Fang, J. Template-Free Synthesis, Controlled Conversion, and CO Oxidation Properties of CeO_2 Nanorods, Nanotubes, Nanowires, and Nanocubes. *Eur. J. Inorg. Chem.* **2008**, *2008*, 2429–2436. [CrossRef]
31. Shan, W.; Guo, H.; Liu, C.; Wang, X. Controllable preparation of CeO_2 nanostructure materials and their catalytic activity. *J. Rare Earths* **2012**, *30*, 665–669. [CrossRef]
32. Montini, T.; Melchionna, M.; Monai, M.; Fornasiero, P. Fundamentals and Catalytic Applications of CeO_2-Based Materials. *Chem. Rev.* **2016**, *116*, 5987–6041. [CrossRef] [PubMed]
33. Kehoe, A.B.; Scanlon, D.O.; Watson, G.W. Role of Lattice Distortions in the Oxygen Storage Capacity of Divalently Doped CeO_2. *Chem. Mater.* **2011**, *23*, 4464–4468. [CrossRef]
34. Mahammadunnisa, S.; Reddy, P.M.K.; Karuppiah, J.; Subrahmanyam, C. Facile Synthesis of Au/CeO_2 Catalyst for Low Temperature CO Oxidation. *Adv. Chem. Lett.* **2013**, *1*, 264–271. [CrossRef]
35. Hu, Z.; Liu, X.; Meng, D.; Guo, Y.; Guo, Y.; Lu, G. Effect of Ceria Crystal Plane on the Physicochemical and Catalytic Properties of Pd/Ceria for CO and Propane Oxidation. *ACS Catal.* **2016**, *6*, 2265–2279. [CrossRef]
36. Gatla, S.; Aubert, D.; Agostini, G.; Mathon, O.; Pascarelli, S.; Lunkenbein, T.; Willinger, M.G.; Kaper, H. Room-Temperature CO Oxidation Catalyst: Low-Temperature Metal–Support Interaction between Platinum Nanoparticles and Nanosized Ceria. *ACS Catal.* **2016**, *6*, 6151–6155. [CrossRef]
37. Elias, J.S.; Stoerzinger, K.A.; Hong, W.T.; Risch, M.; Giordano, L.; Mansour, A.N.; Shao-Horn, Y. In Situ Spectroscopy and Mechanistic Insights into CO Oxidation on Transition-Metal-Substituted Ceria Nanoparticles. *ACS Catal.* **2017**, *7*, 6843–6857. [CrossRef]

38. Zhou, L.; Li, X.; Yao, Z.; Chen, Z.; Hong, M.; Zhu, R.; Liang, Y.; Zhao, J. Transition-Metal Doped Ceria Microspheres with Nanoporous Structures for CO Oxidation. *Sci. Rep.* **2016**, *6*, 23900. [CrossRef] [PubMed]
39. Harrison, P.G.; Ball, I.K.; Azelee, W.; Daniell, W.; Goldfarb, D. Nature and Surface Redox Properties of Copper(II)-Promoted Cerium(IV) Oxide CO-Oxidation Catalysts. *Chem. Mater.* **2000**, *12*, 3715–3725. [CrossRef]
40. Avgouropoulos, G.; Ioannides, T. Effect of synthesis parameters on catalytic properties of CuO-CeO_2. *Appl. Catal. B Environ.* **2006**, *67*, 1–11. [CrossRef]
41. Elias, J.S.; Artrith, N.; Bugnet, M.; Giordano, L.; Botton, G.A.; Kolpak, A.M.; Shao-Horn, Y. Elucidating the Nature of the Active Phase in Copper/Ceria Catalysts for CO Oxidation. *ACS Catal.* **2016**, *6*, 1675–1679. [CrossRef]
42. Wang, W.-W.; Yu, W.-Z.; Du, P.-P.; Xu, H.; Jin, Z.; Si, R.; Ma, C.; Shi, S.; Jia, C.-J.; Yan, C.-H. Crystal Plane Effect of Ceria on Supported Copper Oxide Cluster Catalyst for CO Oxidation: Importance of Metal–Support Interaction. *ACS Catal.* **2017**, *7*, 1313–1329. [CrossRef]
43. Papavasiliou, J.; Rawski, M.; Vakros, J.; Avgouropoulos, G. A Novel Post-Synthesis Modification of CuO-CeO_2 Catalysts: Effect on Their Activity for Selective CO Oxidation. *ChemCatChem* **2018**, *10*, 2096–2106. [CrossRef]
44. Papavasiliou, J.; Vakros, J.; Avgouropoulos, G. Impact of acid treatment of CuO-CeO_2 catalysts on the preferential oxidation of CO reaction. *Catal. Commun.* **2018**, *115*, 68–72. [CrossRef]
45. Landi, G.; Barbato, P.S.; Di Benedetto, A.; Lisi, L. Optimization of the preparation method of CuO/CeO_2 structured catalytic monolith for CO preferential oxidation in H_2-rich streams. *Appl. Catal. B Environ.* **2016**, *181*, 727–737. [CrossRef]
46. Konsolakis, M. The role of Copper-Ceria interactions in catalysis science: Recent theoretical and experimental advances. *Appl. Catal. B Environ.* **2016**, *198*, 49–66. [CrossRef]
47. Gong, X.; Liu, B.; Kang, B.; Xu, G.; Wang, Q.; Jia, C.; Zhang, J. Boosting Cu-Ce interaction in Cu_xO/CeO_2 nanocube catalysts for enhanced catalytic performance of preferential oxidation of CO in H_2-rich gases. *Mol. Catal.* **2017**, *436*, 90–99. [CrossRef]
48. Adamski, A.; Zając, W.; Zasada, F.; Sojka, Z. Copper ionic pairs as possible active sites in N_2O decomposition on CuO_x/CeO_2 catalysts. *Catal. Today* **2012**, *191*, 129–133. [CrossRef]
49. Kydd, R.; Teoh, W.Y.; Wong, K.; Wang, Y.; Scott, J.; Zeng, Q.-H.; Yu, A.-B.; Zou, J.; Amal, R. Flame-Synthesized Ceria-Supported Copper Dimers for Preferential Oxidation of CO. *Adv. Funct. Mater.* **2009**, *19*, 369–377. [CrossRef]
50. Wang, F.; Büchel, R.; Savitsky, A.; Zalibera, M.; Widmann, D.; Pratsinis, S.E.; Lubitz, W.; Schüth, F. In Situ EPR Study of the Redox Properties of CuO–CeO_2 Catalysts for Preferential CO Oxidation (PROX). *ACS Catal.* **2016**, *6*, 3520–3530. [CrossRef]
51. Knauth, P.; Tuller, H.L. Solute segregation, electrical properties and defect thermodynamics of nanocrystalline TiO_2 and CeO_2. *Solid State Ion.* **2000**, *136–137*, 1215–1224. [CrossRef]
52. Abi-aad, E.; Bechara, R.; Grimblot, J.; Aboukais, A. Preparation and characterization of ceria under an oxidizing atmosphere. Thermal analysis, XPS, and EPR study. *Chem. Mater.* **1993**, *5*, 793–797. [CrossRef]
53. Mendelovich, L.; Tzehoval, H.; Steinberg, M. The adsorption of oxygen and nitrous oxide on platinum ceria catalyst. *Appl. Surf. Sci.* **1983**, *17*, 175–188. [CrossRef]
54. Fierro, J.L.G.; Soria, J.; Sanz, J.; Rojo, J.M. Induced changes in ceria by thermal treatments under vacuum or hydrogen. *J. Solid State Chem.* **1987**, *66*, 154–162. [CrossRef]
55. Aboukaïs, A.; Bennani, A.; Aïssi, C.F.; Wrobel, G.; Guelton, M.; Vedrine, J.C. Highly resolved electron paramagnetic resonance spectrum of copper(II) ion pairs in CuCe oxide. *J. Chem. Soc. Faraday Trans.* **1992**, *88*, 615–620. [CrossRef]
56. Lamonier, C.; Bennani, A.; D'Huysser, A.; Aboukaïs, A.; Wrobel, G. Evidence for different copper species in precursors of copper–cerium oxide catalysts for hydrogenation reactions. An X-ray diffraction, EPR and X-ray photoelectron spectroscopy study. *J. Chem. Soc. Faraday Trans.* **1996**, *92*, 131–136. [CrossRef]
57. Li, Y.; Cai, Y.; Xing, X.; Chen, N.; Deng, D.; Wang, Y. Catalytic activity for CO oxidation of Cu–CeO_2 composite nanoparticles synthesized by a hydrothermal method. *Anal. Methods* **2015**, *7*, 3238–3245. [CrossRef]
58. Grigoropoulou, G.; Christoforidis, K.C.; Louloudi, M.; Deligiannakis, Y. Structure-catalytic function relationship of SiO_2-immobilized mononuclear Cu complexes: An EPR study. *Langmuir* **2007**, *23*, 10407–10418. [CrossRef]

59. Courcot, D.; Abi-Aad, E.; Capelle, S.; Aboukaïs, A. Investigation of copper-cerium oxide catalysts in the combustion of diesel soot. In *Studies in Surface Science and Catalysis*; Kruse, N., Frennet, A., Bastin, J.-M., Eds.; Catalysis and Automotive Pollution Control IV.; Elsevier: Amsterdam, The Netherlands, 1998; Volume 116, pp. 625–634.
60. Aboukaïs, A.; Bennani, A.; Lamonier-Dulongpont, C.; Abi-Aad, E.; Wrobel, G. Redox behaviour of copper(II) species on CuCe oxide catalysts: Electron paramagnetic resonance (EPR) study. *Colloids Surf. Physicochem. Eng. Asp.* **1996**, *115*, 171–177. [CrossRef]
61. Soria, J.; Conesa, J.C.; Martínez-Arias, A.; Coronado, J.M. ESR study of the clustering of Cu ions on the ceria surface in impregnated CuO/CeO_2. *Solid State Ion.* **1993**, *63–65*, 755–761. [CrossRef]
62. Wang, J.; Liu, Q.; Liu, Q. Ceria- and Cu-Doped Ceria Nanocrystals Synthesized by the Hydrothermal Methods. *J. Am. Ceram. Soc.* **2008**, *91*, 2706–2708. [CrossRef]
63. Sakai, M.; Nagai, Y.; Aoki, Y.; Takahashi, N. Investigation into the catalytic reduction of NOx at copper–ceria interface active sites. *Appl. Catal. Gen.* **2016**, *510*, 57–63. [CrossRef]
64. Xu, B.; Zhang, Q.; Yuan, S.; Zhang, M.; Ohno, T. Morphology control and characterization of broom-like porous CeO_2. *Chem. Eng. J.* **2015**, *260*, 126–132. [CrossRef]
65. Zhou, X.-D.; Huebner, W. Size-induced lattice relaxation in CeO_2 nanoparticles. *Appl. Phys. Lett.* **2001**, *79*, 3512–3514. [CrossRef]
66. Tsunekawa, S.; Ishikawa, K.; Li, Z.-Q.; Kawazoe, Y.; Kasuya, A. Origin of Anomalous Lattice Expansion in Oxide Nanoparticles. *Phys. Rev. Lett.* **2000**, *85*, 3440–3443. [CrossRef]
67. Lu, B.; Li, Z.; Kawamoto, K. Synthesis of mesoporous ceria without template. *Mater. Res. Bull.* **2013**, *48*, 2504–2510. [CrossRef]
68. Guo, T.; Du, J.; Li, J. The effects of ceria morphology on the properties of Pd/ceria catalyst for catalytic oxidation of low-concentration methane. *J. Mater. Sci.* **2016**, *51*, 10917–10925. [CrossRef]
69. Araiza, D.G.; Gómez-Cortés, A.; Díaz, G. Partial oxidation of methanol over copper supported on nanoshaped ceria for hydrogen production. *Catal. Today* **2017**, *282 Pt 2*, 185–194. [CrossRef]
70. McBride, J.R.; Hass, K.C.; Poindexter, B.D.; Weber, W.H. Raman and X-ray studies of $Ce_{1-x}RE_xO_{2-y}$, where RE = La, Pr, Nd, Eu, Gd, and Tb. *J. Appl. Phys.* **1994**, *76*, 2435–2441. [CrossRef]
71. Spanier, J.E.; Robinson, R.D.; Zhang, F.; Chan, S.-W.; Herman, I.P. Size-dependent properties of CeO_{2-y} nanoparticles as studied by Raman scattering. *Phys. Rev. B* **2001**, *64*, 245407. [CrossRef]
72. Wu, Z.; Li, M.; Howe, J.; Meyer, H.M.; Overbury, S.H. Probing Defect Sites on CeO_2 Nanocrystals with Well-Defined Surface Planes by Raman Spectroscopy and O_2 Adsorption. *Langmuir* **2010**, *26*, 16595–16606. [CrossRef]
73. Choi, Y.M.; Abernathy, H.; Chen, H.-T.; Lin, M.C.; Liu, M. Characterization of O_2-CeO_2 Interactions Using In Situ Raman Spectroscopy and First-Principle Calculations. *ChemPhysChem* **2006**, *7*, 1957–1963. [CrossRef]
74. Matei-Rutkovska, F.; Postole, G.; Rotaru, C.G.; Florea, M.; Pârvulescu, V.I.; Gelin, P. Synthesis of ceria nanopowders by microwave-assisted hydrothermal method for dry reforming of methane. *Int. J. Hydrogen Energy* **2016**, *41*, 2512–2525. [CrossRef]
75. López, J.M.; Gilbank, A.L.; García, T.; Solsona, B.; Agouram, S.; Torrente-Murciano, L. The prevalence of surface oxygen vacancies over the mobility of bulk oxygen in nanostructured ceria for the total toluene oxidation. *Appl. Catal. B Environ.* **2015**, *174–175*, 403–412. [CrossRef]
76. Tok, A.I.Y.; Du, S.W.; Boey, F.Y.C.; Chong, W.K. Hydrothermal synthesis and characterization of rare earth doped ceria nanoparticles. *Mater. Sci. Eng. A* **2007**, *466*, 223–229. [CrossRef]
77. Naganuma, T.; Traversa, E. Stability of the Ce^{3+} valence state in cerium oxide nanoparticle layers. *Nanoscale* **2012**, *4*, 4950–4953. [CrossRef]
78. Ho, C.; Yu, J.C.; Kwong, T.; Mak, A.C.; Lai, S. Morphology-Controllable Synthesis of Mesoporous CeO_2 Nano- and Microstructures. *Chem. Mater.* **2005**, *17*, 4514–4522. [CrossRef]
79. Rakai, A.; Bensalem, A.; Muller, J.C.; Tessier, D.; Bozon-Verduraz, F. Revisiting Diffuse Reflectance Spectroscopy for the Characterization of Metal and Semiconducting Oxide Catalysts. In *Studies in Surface Science and Catalysis*; Guczi, L., Solymosi, F., Tétényi, P., Eds.; New Frontiers in Catalysis—Proceedings of the 10th International Congress on Catalysis, Budapest, Hungary, 19–24 July 1992; Elsevier: Amsterdam, The Netherlands, 1993; Volume 75, pp. 1875–1878.
80. Bensalem, A.; Muller, J.C.; Bozon-Verduraz, F. From bulk CeO_2 to supported cerium-oxygen clusters: A diffuse reflectance approach. *J. Chem. Soc. Faraday Trans.* **1992**, *88*, 153–154. [CrossRef]

81. Binet, C.; Badri, A.; Lavalley, J.-C. A Spectroscopic Characterization of the Reduction of Ceria from Electronic Transitions of Intrinsic Point Defects. *J. Phys. Chem.* **1994**, *98*, 6392–6398. [CrossRef]
82. Aguila, G.; Guerrero, S.; Araya, P. Effect of the preparation method and calcination temperature on the oxidation activity of CO at low temperature on CuO–CeO_2/SiO_2 catalysts. *Appl. Catal. Gen.* **2013**, *462–463*, 56–63. [CrossRef]
83. Pintar, A.; Batista, J.; Hočevar, S. TPR, TPO, and TPD examinations of $Cu_{0.15}Ce_{0.85}O_{2-y}$ mixed oxides prepared by co-precipitation, by the sol–gel peroxide route, and by citric acid-assisted synthesis. *J. Colloid Interface Sci.* **2005**, *285*, 218–231. [CrossRef]
84. Mena, E.; Rey, A.; Rodríguez, E.M.; Beltrán, F.J. Nanostructured CeO_2 as catalysts for different AOPs based in the application of ozone and simulated solar radiation. *Catal. Today* **2017**, *280 Pt 1*, 74–79. [CrossRef]
85. Tsunekawa, S.; Fukuda, T.; Kasuya, A. Blue shift in ultraviolet absorption spectra of monodisperse CeO_{2-x} nanoparticles. *J. Appl. Phys.* **2000**, *87*, 1318–1321. [CrossRef]
86. Itoh, T.; Uchida, T.; Izu, N.; Matsubara, I.; Shin, W. Effect of Core–Shell Ceria/Poly(vinylpyrrolidone) (PVP) Nanoparticles Incorporated in Polymer Films and Their Optical Properties. *Materials* **2013**, *6*, 2119–2129. [CrossRef]
87. Pan, C.; Zhang, D.; Shi, L. CTAB assisted hydrothermal synthesis, controlled conversion and CO oxidation properties of CeO_2 nanoplates, nanotubes, and nanorods. *J. Solid State Chem.* **2008**, *181*, 1298–1306. [CrossRef]
88. Taniguchi, T.; Katsumata, K.; Omata, S.; Okada, K.; Matsushita, N. Tuning Growth Modes of Ceria-Based Nanocubes by a Hydrothermal Method. *Cryst. Growth Des.* **2011**, *11*, 3754–3760. [CrossRef]
89. Wang, C.; Cheng, X. Hydrothermal Synthesis and Upconversion Properties of α-$NaYF_4$:Yb^{3+}, Er^{3+} Nanocrystals Using Citric Acid as Chelating Ligand and $NaNO_3$ as Mineralizer. *J. Nanosci. Nanotechnol.* **2015**, *15*, 9656–9664. [CrossRef]
90. Liu, H.; Liu, H. Preparing micro/nano core-shell sphere CeO_2 via a low temperature route for improved lithium storage performance. *Mater. Lett.* **2016**, *168*, 80–82. [CrossRef]
91. Jiang, T.; Qin, W.; Di, W.; Yang, R.; Liu, D.; Zhai, X.; Qin, G. Citric acid-assisted hydrothermal synthesis of α-$NaYF_4$: Yb^{3+}, Tm^{3+} nanocrystals and their enhanced ultraviolet upconversion emissions. *CrystEngComm* **2012**, *14*, 2302–2307. [CrossRef]
92. Levien, B.J. A Physicochemical Study of Aqueous Citric Acid Solutions. *J. Phys. Chem.* **1955**, *59*, 640–644. [CrossRef]
93. Masui, T.; Hirai, H.; Imanaka, N.; Adachi, G.; Sakata, T.; Mori, H. Synthesis of cerium oxide nanoparticles by hydrothermal crystallization with citric acid. *J. Mater. Sci. Lett.* **2002**, *21*, 489–491. [CrossRef]
94. Ma, J.; Qian, K.; Huang, W.; Zhu, Y.; Yang, Q. Facile One-Step Synthesis of Double-Shelled CeO_2 Hollow Spheres and Their Optical and Catalytic Properties. *Bull. Chem. Soc. Jpn.* **2010**, *83*, 1455–1461. [CrossRef]
95. Zhang, Y.; Lin, Y.; Jing, C.; Qin, Y. Formation and Thermal Decomposition of Cerium-Organic Precursor for Nanocrystalline Cerium Oxide Powder Synthesis. *J. Dispers. Sci. Technol.* **2007**, *28*, 1053–1058. [CrossRef]
96. Wu, N.-C.; Shi, E.-W.; Zheng, Y.-Q.; Li, W.-J. Effect of pH of Medium on Hydrothermal Synthesis of Nanocrystalline Cerium(IV) Oxide Powders. *J. Am. Ceram. Soc.* **2002**, *85*, 2462–2468. [CrossRef]
97. Lin, M.; Fu, Z.Y.; Tan, H.R.; Tan, J.P.Y.; Ng, S.C.; Teo, E. Hydrothermal Synthesis of CeO_2 Nanocrystals: Ostwald Ripening or Oriented Attachment? *Cryst. Growth Des.* **2012**, *12*, 3296–3303. [CrossRef]
98. Bezkrovnyi, O.S.; Lisiecki, R.; Kepinski, L. Relationship between morphology and structure of shape-controlled CeO_2 nanocrystals synthesized by microwave-assisted hydrothermal method. *Cryst. Res. Technol.* **2016**, *51*, 554–560. [CrossRef]
99. Lin, J.; Huang, Y.; Li, L.; Wang, A.; Zhang, W.; Wang, X.; Zhang, T. Activation of an Ir-in-CeO_2 catalyst by pulses of CO: The role of oxygen vacancy and carbonates in CO oxidation. *Catal. Today* **2012**, *180*, 155–160. [CrossRef]
100. Nörenberg, H.; Briggs, G.A.D. Defect Structure of Nonstoichiometric CeO_2 (111) Surfaces Studied by Scanning Tunneling Microscopy. *Phys. Rev. Lett.* **1997**, *79*, 4222–4225. [CrossRef]
101. Esch, F.; Fabris, S.; Zhou, L.; Montini, T.; Africh, C.; Fornasiero, P.; Comelli, G.; Rosei, R. Electron Localization Determines Defect Formation on Ceria Substrates. *Science* **2005**, *309*, 752–755. [CrossRef]
102. Wang, L.; Yu, Y.; He, H.; Zhang, Y.; Qin, X.; Wang, B. Oxygen vacancy clusters essential for the catalytic activity of CeO_2 nanocubes for o-xylene oxidation. *Sci. Rep.* **2017**, *7*, 12845. [CrossRef]
103. Stoll, S.; Schweiger, A. EasySpin, a comprehensive software package for spectral simulation and analysis in EPR. *J. Magn. Reson.* **2006**, *178*, 42–55. [CrossRef] [PubMed]

104. Stathi, P.; Gournis, D.; Deligiannakis, Y.; Rudolf, P. Stabilization of phenolic radicals on graphene oxide: An XPS and EPR study. *Langmuir* **2015**, *31*, 10508–10516. [CrossRef] [PubMed]
105. Scullane, M.; White, L.; Chasteen, N. An efficient approach to computer simulation of EPR spectra of high-spin Fe(III) in rhombic ligand fields. *J. Magn. Reson.* **1969**, *47*, 383–397. [CrossRef]
106. Knijnenburg, J.T.N.; Seristatidou, E.; Hilty, F.M.; Krumeich, F.; Deligiannakis, Y. Proton-promoted iron dissolution from nanoparticles and the influence by the local iron environment. *J. Phys. Chem. C* **2014**, *118*, 24072–24080. [CrossRef]
107. Avgouropoulos, G.; Ioannides, T.; Papadopoulou, C.; Batista, J.; Hocevar, S.; Matralis, H.K. A comparative study of Pt/γ-Al_2O_3, Au/α-Fe_2O_3 and CuO–CeO_2 catalysts for the selective oxidation of carbon monoxide in excess hydrogen. *Catal. Today* **2002**, *75*, 157–167. [CrossRef]

Article

Catalytic Properties of Double Substituted Lanthanum Cobaltite Nanostructured Coatings Prepared by Reactive Magnetron Sputtering

Mohammad Arab Pour Yazdi [1], Leonardo Lizarraga [2,3], Philippe Vernoux [2], Alain Billard [1] and Pascal Briois [1,*]

1 FEMTO-ST Institute (UMR CNRS 6174), Université Bourgogne Franche-Comté, UTBM, 2 Place Lucien Tharradin, F-25200 Montbéliard CEDEX, France; mohammad.arab-pour-yazdi@utbm.fr (M.A.P.Y.); alain.billard@utbm.fr (A.B.)

2 Université Lyon, Université Claude Bernard Lyon 1, CNRS—IRCELYON—UMR 5256, 2 avenue A. Einstein, 69626 Villeurbanne, France; leonardo.lizarraga@cibion.conicet.gov.ar (L.L.); philippe.vernoux@ircelyon.univ-lyon1.fr (P.V.)

3 CIBION, Centro de Investigaciones en Bionanociencias, CONICET, Consejo Nacional de Investigaciones Científicas y Técnicas CONICET, Godoy Cruz 2390, Buenos Aires C1425FQD, Argentina

* Correspondence: pascal.briois@utbm.fr; Tel.: +33-384-583-701

Received: 29 March 2019; Accepted: 17 April 2019; Published: 23 April 2019

Abstract: Lanthanum perovskites are promising candidates to replace platinum group metal (PGM), especially regarding catalytic oxidation reactions. We have prepared thin catalytic coatings of Sr and Ag doped lanthanum perovskite by using the cathodic co-sputtering magnetron method in reactive condition. Such development of catalytic films may optimize the surface/bulk ratio to save raw materials, since a porous coating can combine a large exchange surface with the gas phase with an extremely low loading. The sputtering deposition process was optimized to generate crystallized and thin perovskites films on alumina substrates. We found that high Ag contents has a strong impact on the morphology of the coatings. High Ag loadings favor the growth of covering films with a porous wire-like morphology showing a good catalytic activity for CO oxidation. The most active composition displays similar catalytic performances than those of a Pt film. In addition, this porous coating is also efficient for CO and NO oxidation in a simulated Diesel exhaust gas mixture, demonstrating the promising catalytic properties of such nanostructured thin sputtered perovskite films.

Keywords: perovskite; catalytic coating; CO oxidation; cathodic sputtering method

1. Introduction

The family of perovskite oxides is known for its catalytic properties, hydrothermal stability, high recyclability and low cost compared with Platinum Group Metal (PGM) [1–3]. The general formula of perovskites is ABO_3 where the larger size A-cation presents a 12-coordination number and the B-cation coordinates with 6 neighboring atoms [4]. Partial substitution of A and/or B atoms with other elements showing redox properties may enhance the catalytic activity due to the generation of structural defects such as anionic or cationic vacancies and/or modification of the oxidation state of B cations to maintain the electro-neutrality [5]. Lanthanum perovskites are promising candidates to replace noble metals (Pt, Pd, etc.) [6,7], especially regarding catalytic oxidation reactions. Lanthanum cobaltite ($LaCoO_3$) is one of the most promising catalysts for the oxidation of gaseous pollutants such as carbon monoxide, unburnt hydrocarbons and nitrogen oxide [7,8]. Lanthanum cobaltites are used in many others fields due to their magnetic properties as well as their mixed ionic and electronic conductivity [9–12]. Most of these applications require the implementation of thin films that act as electrodes for fuel cells [13], thermoelectric processes [14], sensors [15], and magnetoresistance devices [16]. These electric, magnetic and electrocatalytic properties depend on the composition but also on the microstructure and morphology of coatings, which strongly depend on the preparation method [17,18]. Several techniques are reported for the synthesis of perovskite thin films such as the ground-frost [7], casting in band [18], chemical vapor deposition (CVD) [17], spray painting [19], sol-gel [7], atomic layer deposition (ALD) [20], laser pulsed deposition (PLD) [21], and physical vapor deposition (PVD) [22,23]. Regarding catalytic applications, the development of submicrometric catalytic films may optimize the surface/bulk ratio to save raw materials. Indeed, a thin catalytic coating can combine a large exchange surface with the gas phase with an extremely low loading. This is particularly suitable for air cleaning since gaseous pollutants only lick the surface of the catalyst. In addition, thin catalytic coatings could be deposited on hot substrates such as collector walls in thermal engine exhausts for removing pollutants or of radiators for improving indoor air quality. Few studies on the development of thin perovskite coatings for catalytic applications are reported in the literature [24–26]. The challenge lies in preparing adherent, thermally stable and pure crystallized perovskite films without any secondary parasite phase [18], showing appropriate compositions and nanostructures for catalysis [7,8,16,18]. The porosity and the specific surface areas have to be optimized to counter-balance the low quantity of materials involved in submicrometric catalytic coatings. In the present study, we used the cathodic co-sputtering magnetron method in reactive condition to prepare submicrometric nanostructured coatings of perovskites. This technique allows the deposition of pure and adherent coatings of complex oxides with a controlled and reproducible manner, while respecting the environment [27,28]. We have chosen to prepare Sr and Ag-doped lanthanum cobaltite as this perovskite composition is one of the most active for CO oxidation [8]. The partial substitution of La^{3+} cations by Sr^{2+} ones improves the thermal stability and then the specific surface area of pure $LaCoO_3$ [29] and also enhances the surface concentration of oxygen vacancies involved in the oxidation catalytic mechanism [30]. The partial substitution of La^{3+} by Ag^{+} can also increase the catalytic properties of lanthanum cobaltites for oxidation reactions due to the formation of oxygen vacancies [31] and the stabilisation of Ag nanoparticles on the oxide surface [6,32]. The different parameters (intensities of current applied on the metallic targets, total pressure in the chamber, oxygen partial pressure, etc.) of the reactive magnetron sputtering preparation method were tuned to achieve pure and adherent coatings of $LaCoO_3$ on alumina dense membranes. After a calcination at 500 °C, crystallized and dense cubic perovskite films of around 1.5 µm thick were achieved with the targeted La/Co stoichiometry. Different compositions of $La_{1-x-y}Sr_xAg_yCoO_{3-\alpha}$ (x = 0.13–0.28, y = 0.14–0.48) doped perovskites were sputtered on alumina disks. We found that the incorporation of high contents of Ag can strongly modify the morphology of the coatings, increasing their porosity. The catalytic performance of the perovskite catalytic coatings for CO oxidation was found to be improved by the double substitution of Sr and Ag. The most active composition, $La_{0.40}Sr_{0.1}Ag_{0.48}Co_{0.93}O_3$, displays similar catalytic performances than those of a Pt film, for CO oxidation. In addition, this porous coating is also active for CO and NO oxidation in a simulated Diesel exhaust gas mixture, demonstrating the promising catalytic properties of such nanostructured thin sputtered perovskite films.

2. Results

2.1. Preparation and Characterization of $LaCoO_3$ Catalytic Coatings

Layers of $LaCoO_3$ were synthetized on alumina substrates by using the reactive magnetron sputtering preparation method. Depositions have been performed in a reactive mode, mixing oxygen and argon in the chamber. To achieve the suitable La/Co ratio, the discharge current applied to the Co target was adjusted while maintaining a constant current of 1 A on the La target (Table 1). As expected, the La/Co ratio, estimated by Energy Dispersive Spectroscopy (EDS), decreases with the current dissipated on the Co target (Figure 1). We determined that a current intensity of 0.3 A is required to reach the targeted La/Co ratio. The diffractogram recorded on this as-deposited coating

(Figure 2) shows that the layer is amorphous as no peak is detected. The difference between the radii of La^{3+} (136 pm) and Co^{3+} cations (72 pm) suggests an amorphous as-deposited coating due to steric effects as predicted by the confusion principle [33–35]. Then, a calcination step was performed at different temperatures (from 100 °C to 500 °C) for 2 h in air. X-Ray Diffraction (XRD) patterns (Figure 2), obtained under a Bragg-Brentano configuration (θ/2θ), evidence that the oxide coating crystallizes from 500 °C as a cubic perovskite phase (JCPDS 01-075-0279). This temperature is approximately 100 °C lower than those reported by H. Seim et al. [20] and H.J. Hwang et al. [36] on $LaCoO_3$ films prepared by PLD and sol-gel methods, respectively. This demonstrates that the reactive magnetron sputtering technique is efficient for preparing crystallized lanthanum cobaltite films at rather low temperatures.

Table 1. Sputtering parameters used for $LaCoO_3$ coating.

Target	Intensity (A)	Pulse Frequency (kHz)	Dead Time (Toff µs)	Ar Flow Rate (sccm) *	O_2 Flow Rate (sccm) *	Total Pressure (Pa)	Draw Distance (mm)	Sputtering Time (h)
La Co	1 0.2 to 0.4	50	5	50	20	1.5	45	3

* sccm = Standard Cubic Centimeter per Minutes.

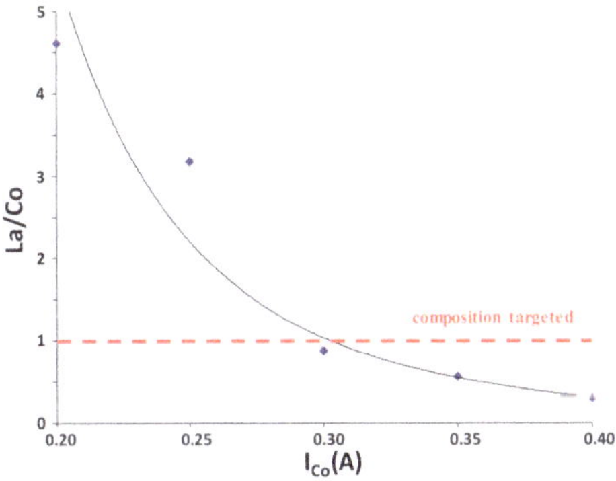

Figure 1. Evolution of the atomic La/Co ratio measured by Energy Dispersive Spectroscopy (EDS) as a function of the current dissipated on the Co target (I_{La} = 1 A). The total pressure is 1.5 Pa.

This calcination step at 500 °C was performed for all coatings (Figure 3). The thin film with the highest La/Co ratio is not crystallized and only presents small peaks of La_2O_3 (JCDPS 00-005-0602), probably coming from a high excess of La. On the opposite, for larger atomic ratios, coatings are crystallised with various space groups of the perovskite-type structure. Indeed, the film is cubic (Pm3m space group) for a ratio of 0.87, exhibits a rhomboedric symmetry (R-3c space group) for a ratio of 0.56 and again becomes cubic for a ratio of 0.3.

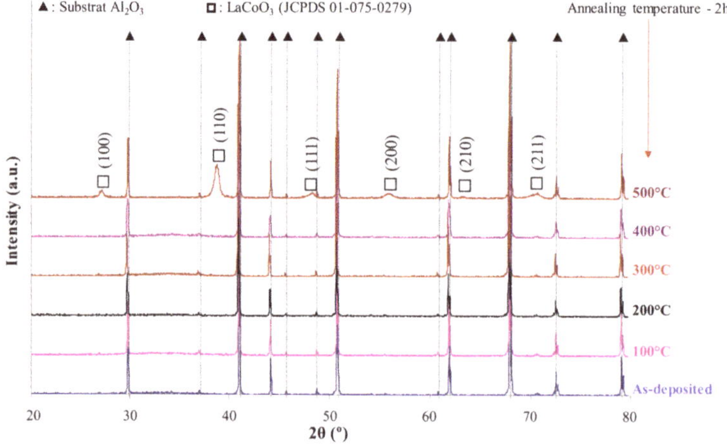

Figure 2. X-Ray diffractograms of $LaCoO_3$ sputtered films prepared with a current dissipated on the Co target of 0.3 A (La/Co = 0.87) after different post-calcination treatments during 2 h in air from 100 °C to 500 °C.

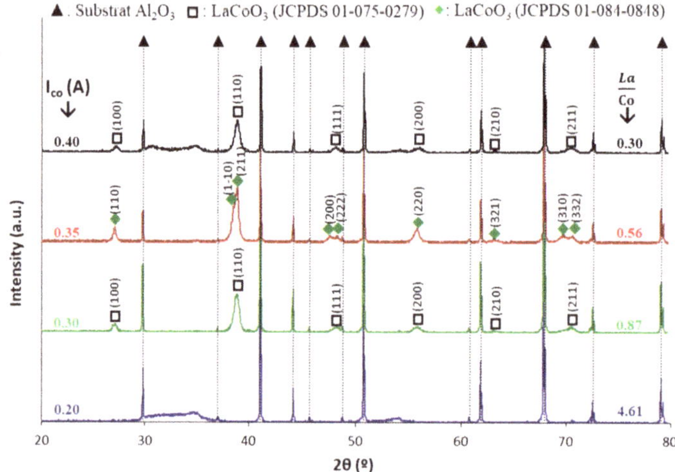

Figure 3. XRD patterns as a function of the atomic composition ratio after calcination treatment at 500 °C for 2 h under air.

The morphology of the $LaCoO_3$ coating with the suitable ratio (La/Co = 0.87) was observed by SEM. Figure 4 shows SEM images of the top view and the brittle cross section of the sample before and after the calcination step at 500 °C. The as-deposited coating (Figure 4a,c) covers the surface of the alumina substrate and follows its morphology. This film is quite dense, as shown in the cross-section image (Figure 4a) and adherent with a vitreous appearance characteristic of an amorphous material, in agreement with XRD (Figure 2) [37]. After calcination (Figure 4b), some cracks can be observed especially in the cross section (Figure 4d). The crystallized film is not sticking well with the alumina support and remains quite dense. The cracks were probably formed under stresses during the crystallization or the thermal treatment. In this latter case, the mismatch of thermal expansion coefficients between the perovskite film and the alumina substrate could be at the origin of the delamination ($\alpha_{LaCoO3} \approx 20 \times 10^{-6}$ °C^{-1} [38] and $\alpha_{Al2O3} \approx 7 \times 10^{-6}$ °C^{-1} [39]) of the coating. The thickness of the annealed coating is around 1.5 μm (Figure 4c,d), resulting in a 500 nm.h^{-1} deposition rate. This result is in agreement with the thickness measured by tactile profilometry.

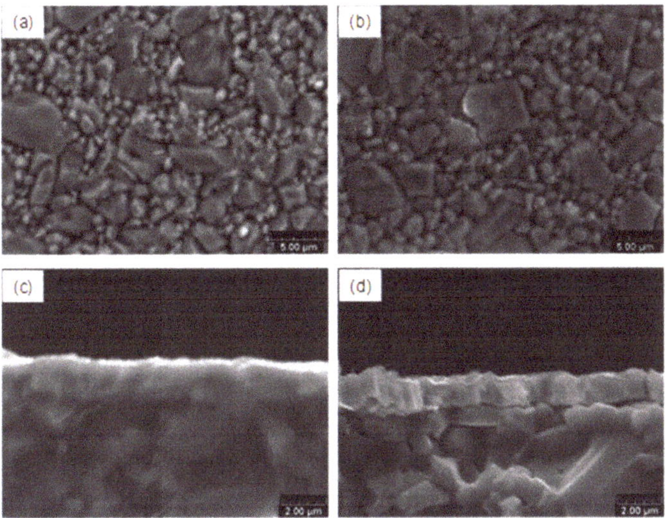

Figure 4. SEM images of the surface and the brittle cross section of the $LaCoO_3$ sample synthesized with I_{Co} = 0.3 A and I_{La} = 1 A: (**a**,**c**) as deposited coating and (**b**,**d**) after calcination treatment at 500 °C for 2 h under air.

2.2. Preparation and Characterization of $La_{1-x-y}Sr_xAg_yCoO_{3-\alpha}$ (LSACO) Catalytic Coatings

The experimental parameters used for the synthesis of double substituted cobaltite (denoted as LSACO) coatings on alumina supports are reported in Table 2. The protocol is similar to that developed for the synthesis of pure lanthanum cobaltite (Ar flow rate = 50 sccm, O_2 flow rate = 20 sccm, total pressure = 1.5 Pa, draw distance = 45 mm) except the sputtering time which was extended to 4 h to achieve thicker films. The control of the substitution rates of Sr and Ag was achieved by tuning the intensity of the currents dissipated on the metallic targets (Table 2). We assumed that Sr^{2+} and also Ag^+ cations will partially substitute La^{3+} cations in A sites of the perovskite. This hypothesis is based on the similarity of the ionic radius of La^{3+} (136 pm) and Ag^+ (128 pm) [40]. The Ag content in the perovskites was gradually enhanced by increasing the current applied to the Ag target (LSACO-1 to 4). This was counter-balanced by a progressive decay of the current dissipated to the La target. The Co content has been slightly increased in LSACO-5 compared with LSACO-4, while maintaining constant the Ag, La, and Sr contents (Table 2)

Table 2. Sputtering parameters used for *LSACO* coatings.

Coating	I_{La} (A) Pulse (kHz)/Toff (μs)	I_{Sr} (A) Pulse (kHz)/Toff (μs)	I_{Ag} (A) Pulse (kHz)/Toff (μs)	I_{Co} (A) Pulse (kHz)/Toff (μs)
LSACO-1	1.2 50/5	1.2 350/1.4	0.008 0	0.35 50/5
LSACO-2	0.8 50/5	1.1 350/1.4	0.009 0	0.35 50/5
LSACO-3	0.75 50/5	1.1 350/1.4	0.01 0	0.35 50/5
LSACO-4	0.6 50/5	1.1 350/1.4	0.013 0	0.35 50/5
LSACO-5	0.6 50/5	1.1 350/1.4	0.013 0	0.4 50/5

The chemical composition of the as-prepared LSACO coatings was estimated by EDS analysis. Table 3 displays the variations of the atomic percent of each element, without considering the oxygen level which is tricky to quantity by using EDS. The Sr and Ag contents inversely vary in LSACO-1, 2 and 3 while La and Co loadings are fairly stable. Let us note that the chemical composition of LSACO-2 and LSACO-3 is fairly similar with 11 at% of Ag and around 10 and 30 at% of Sr and La, respectively. The high current applied to the Ag target during the preparation of LSACO-4 and LSACO-5 leads to a significant increase in the Ag concentration and a concomitant drop of the La content. In LSACO-4 and LSACO-5, the Ag atomic percent reaches around 25% while the La level is approximately 20%. Surprisingly, the Co concentration decreases in LSACO-5 despite the largest applied current to the Co target. Except LSACO-5, the atomic ratio between cations located in A sites (theoretically La^{3+}, Sr^{2+} and Ag^{+}) and Co^{3+} ones located in B sites is close to the target of 1. The morphology of the as-deposited coatings (Figure 5) was found to drastically change with the chemical composition. The LSACO-1 coating only contains few perovskite clusters (Figure 5a, white spots at the top right) dispersed on the surface of the micrometric alumina grains (Figure 5, dark grey), despite of the 4 h deposition time. For LSACO-2 and LSACO-3, the number of perovskite clusters significantly increases on the surface of the alumina grains (white spots in Figure 5b,c). Interestingly, larger starfish shaped perovskites islands are growing on the substrate defects (cavity of alumina substrate or grain boundaries) showing arms like filaments. The morphology of the surface of LSACO-4 and LSACO-5 coatings (Figure 5d,e) are quite different. The alumina substrate is now fairly fully covered by a porous film with a wire-like morphology (Figure 5d,e). These wires have grown parallel to each other, leading void between each other, reaching a length of the order of 1 μm and a diameter of around 100 nm. Therefore, high Ag contents seem to strongly enhance the porosity and then the coverage of the films.

Table 3. Chemical composition determined by EDS of the as-prepared $La_{1-x-y}Sr_xAg_yCoO_{3-\alpha}$ (LSCAO) perovskite coatings.

Coatings	La (at%)	Sr (at%)	Ag (at%)	Co (at%)	Chemical Formula
LSACO-1	29 ± 0.29	14 ± 0.14	7 ± 0.07	50 ± 0.5	$La_{0.58}Sr_{0.28}Ag_{0.14}Co_1O_3$
LSACO-2	29 ± 0.29	11 ± 0.11	11 ± 0.11	49 ± 0.5	$La_{0.56}Sr_{0.23}Ag_{0.21}Co_{0.94}O_3$
LSACO-3	32 ± 0.32	9 ± 0.1	11 ± 0.11	48 ± 0.5	$La_{0.61}Sr_{0.18}Ag_{0.21}Co_{0.92}O_3$
LSACO-4	20 ± 0.20	7 ± 0.1	25 ± 0.25	48 ± 0.5	$La_{0.39}Sr_{0.13}Ag_{0.49}Co_{0.93}O_3$
LSACO-5	21 ± 0.21	9 ± 0.1	28 ± 0.28	42 ± 0.4	$La_{0.40}Sr_{0.20}Ag_{0.48}Co_{0.70}O_3$

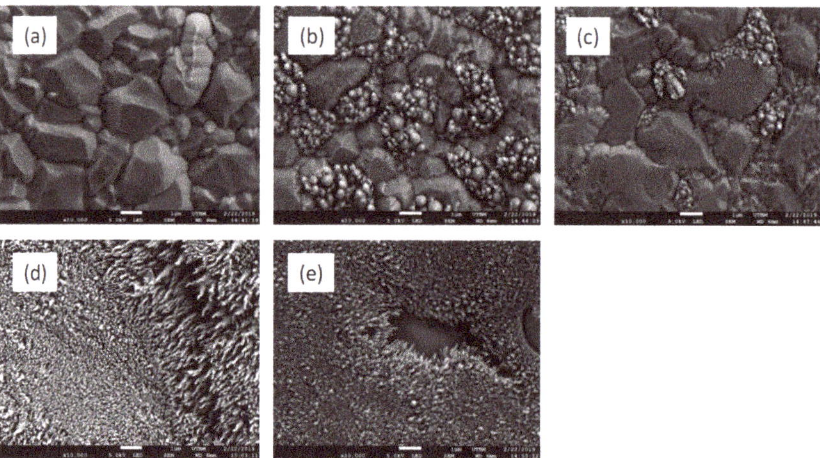

Figure 5. SEM images of the surface morphology of as-deposited LSACO coatings on alumina substrate: (**a**) LSACO-1, (**b**) LSACO-2, (**c**) LSACO-3, (**d**) LSACO-4, and (**e**) LSACO-5.

These series of LSACO coatings were annealed at 500 °C for 2 h in air. This calcination step was sufficient to crystallize all the films as a cubic perovskite phase (JCPDS 01-075-0279) whatever the composition. On the other hand, an additional XRD pattern at 44.3° on the diffractogramm of LSACO-4 and LSACO-5, corresponding to (2 0 0) planes of fcc metallic silver, proves the presence of metallic Ag, out of the perovskite structure. These results demonstrate the limited solubility of Ag in the perovskite in agreement with previous studies [32]. For high contents of Ag, reaching 25 at. %, i.e., around 10 at% in the overall oxide including the oxygen atoms, part of Ag is not incorporated into the perovskite structure. No XRD peaks corresponding to Ag° were observed for LSAC0-1, LSACO-2, and LSACO-3 (Figure 6), but the low content of Ag, below 5 at%, makes difficult their detection. Therefore, we cannot exclude the presence of metallic Ag on the surface of these samples.

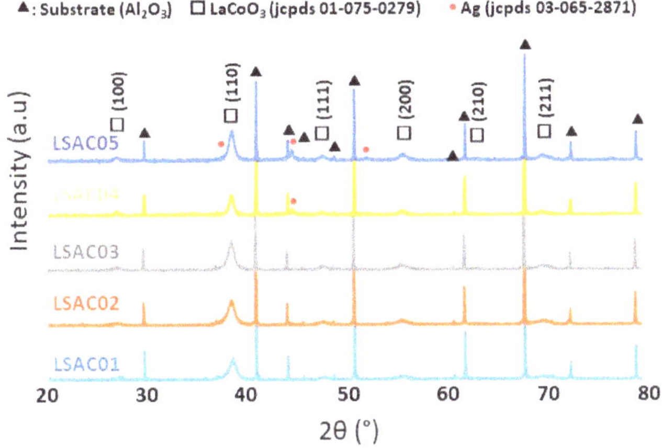

Figure 6. XRD patterns after the calcination treatment at 500 °C for 2 h under air of the different LSACO coatings.

Figure 7 displays the surface morphology of the calcinated films. The alumina substrates are cracked probably due to the thermal treatment. The LSACO-1 coating only contains isolated perovskites clusters on the surface of alumina grains. As before calcination, the morphology of LSACO-2 and LSACO-3 coatings are similar with perovskite islands mainly located in the cavity and interstices of the alumina substrate. Filaments that were present around of the perovskite clusters have disappeared. Coatings containing high Ag loadings (LSACO-4 and LSACO-5) are the only ones able to fully cover the alumina substrate with a porous wire-like morphology. These nanowires are less ordered than before calcination and interlock, then decreasing the porosity of the film. Ag particles (Figure 7e) can be observed on the surface of LSACO-5 (white particles), confirming that a part of Ag was not incorporated into the perovkiste structure.

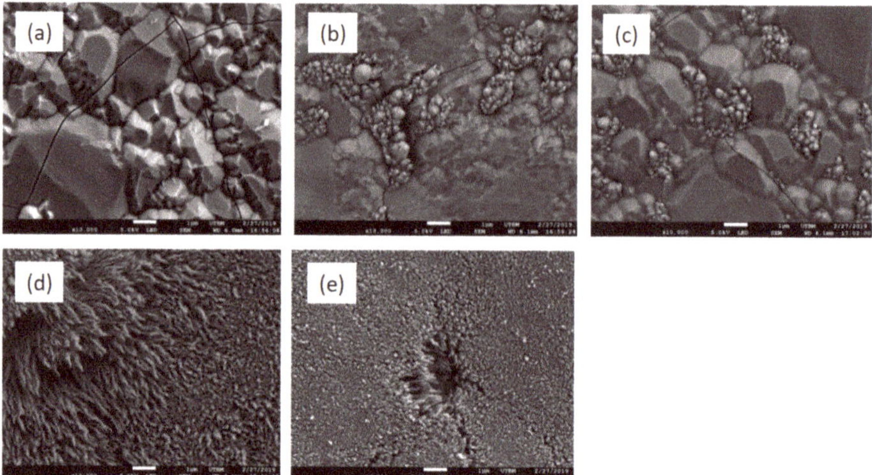

Figure 7. SEM images of the surface morphology of calcined LSACO coatings on alumina substrate: (**a**) LSACO-1, (**b**) LSACO-2, (**c**) LSACO-3, (**d**) LSACO-4, and (**e**) LSACO-5.

2.3. *Catalytic Performances of the Perovskite Coatings*

2.3.1. Catalytic Performances for CO Oxidation

The catalytic activity of LSACO-2, LSACO-4, and LSACO-5 coatings, containing respectively 11, 25 and 28 at% of Ag, was measured for CO oxidation during heating ramps up to 400 °C. We have selected LSACO-4 and LSACO-5 as these films exhibit a porous and wire-like morphology. We have also tested the catalytic performances for CO oxidation of pure lanthanum cobaltite coatings. These samples have shown no activity below 500 °C in good agreement with their dense morphology. The catalyst LSACO-2 was also tested for comparison as a non-porous and non-covering representative layer. In addition, we have also tested, for comparison, a Pt coating (3 μm thick, 5 mg Pt/cm^2) prepared by spray-painting and annealed at 500 °C. Figure 8 displays the Light-off (LO) curves of the different catalytic coatings for CO oxidation. The results show that the catalytic performances increase with the silver content in the film. Values of T20, temperature at 20% conversion, significantly decrease from 280 to 215 °C when increasing the Ag content. Furthermore, the catalytic performances of LSACO-5 are rather close to those of the Pt coating up to 220 °C. This underlines the promising catalytic activity of the sputtered Ag-doped perovskite coating.

Two successive LO up to 400 °C have been performed on LSACO-4 (Figure 9a). The value of T20 increases from 215 to 242 °C during the second LO. This indicates that a modification of the coating morphology during the first LO, most probably due to the Ag particles sintering. Nevertheless, the onset temperature slightly decreases from 175 to 150 °C (Figure 9a). Similar catalytic experiments have been carried out on LSACO-5 (Figure 9b) without any significant modification of the catalytic performances between the two successive LO. This indicates a better stability of the morphology of LSACO-5 in the presence of the reactive mixture up to 400°C compared with LSACO-4. XPS measurements performed after catalytic tests evidence a La surface seggregation with a concommitant drop of the Co concentration. The two catalytic coatings show a similar Ag surface atomic concentration, i.e., around 10 at%, which is far lower from the loading estimated by EDS before the catalytic tests (Table 3). The Ag3d XPS peaks (Figure 10) show a binding energy of Ag3d5/2 at 367.7 eV, attributed to metallic silver. The catalytic properties during the second LO of the two catalytic perovskite coatings are fairly similar, in good agreement with an equivalent Ag surface concentration (Table 4). This confirms the direct link between the surface Ag concentration and the catalytic activity for CO oxidation.

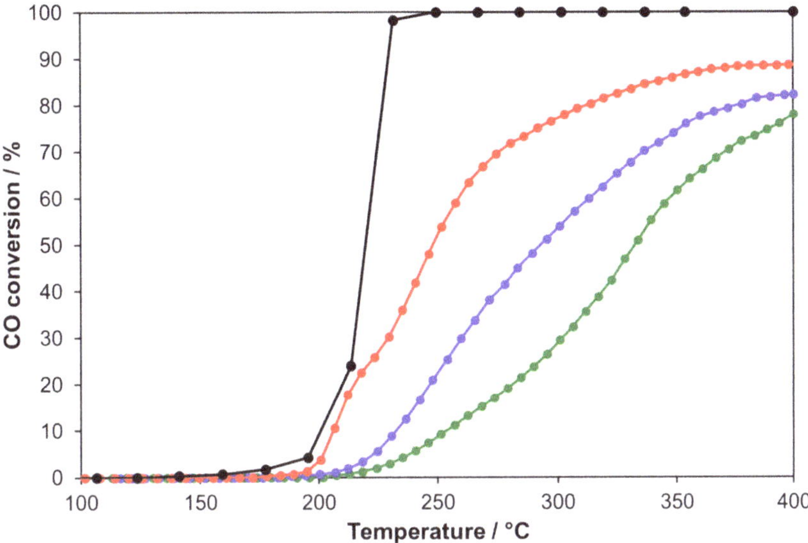

Figure 8. Light-off curves for CO oxidation recorded on the sputtered perovskite coatings and on Pt film. Reactive mixture: CO/O_2 = 3000 ppm/3%. Overall flow = 3.6 L/h.

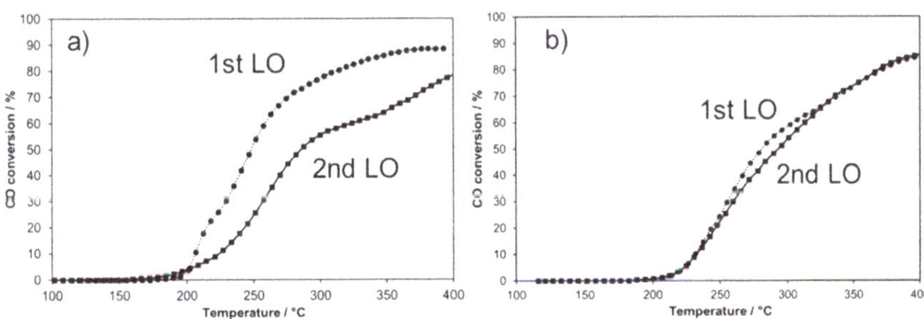

Figure 9. Two successive LO recorded on (**a**) LSACO-4 and (**b**) LSACO-5. Reactive mixture: CO/O_2 = 3000 ppm/3%. Overall flow = 3.6 L/h.

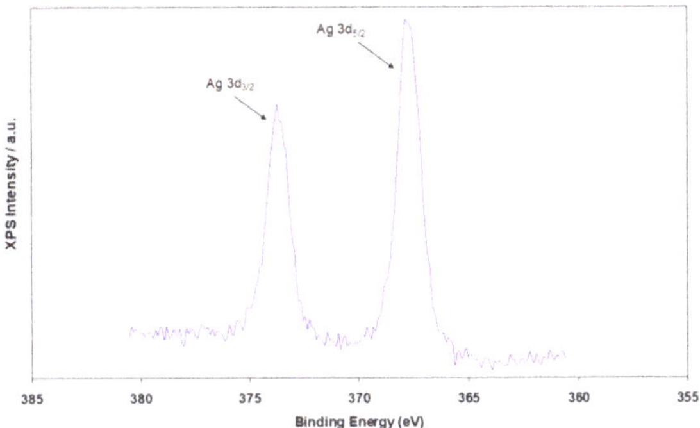

Figure 10. Ag3d XPS spectrum of LSAC-5.

Table 4. Surface composition determined by XPS of LSACO-4 and LSACO-5 after catalytic tests.

Sample	La at%	Sr at%	Ag at%	Co at%
LSACO-4	47	11	12	30
LSACO-5	51	13	10	26

2.3.2. Catalytic Performances in a Model Lean Diesel Exhaust Gas

The catalytic performances of LSACO-4 were explored in a model lean diesel exhaust gas mixture containing 8% O_2, 950 ppm CO, 270 ppm NO, 1000 ppm C_3H_8 and 10% H_2O. The LO was recorded up to 375 °C to investigate the low temperature activity. Figure 11 shows the ability of the catalytic coating to oxidize CO and NO into CO_2 and NO_2, respectively. Below 350 °C, we did not observed any propane oxidation. However, the catalytic coating shows a remarkable activity for NO oxidation from around 225 °C. The NO conversion achieves 30% at 375 °C. CO conversion starts from 250 °C but seems to reach a maximum at only 18% from 350 °C. These results show that, despite a very low mass of catalyst and a complex mixture including the inhibiting effect of H_2O, the perovkite catalytic coating can be active at low temperature for low temperature NO and CO oxidation.

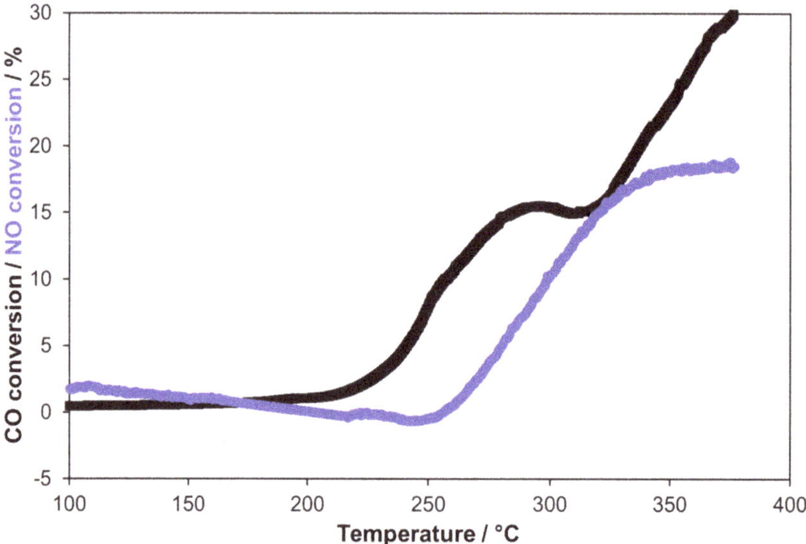

Figure 11. Variation of CO and NO conversion as a function of temperature on LSACO-4. Reactive mixture: $CO/O_2/NO/C_3H_8/H_2O$: 950 ppm/8%/270 ppm/1000 ppm/10%. Overall flow: 3.6 L/h.

3. Materials and Methods

3.1. Preparation of the Catalytic Coatings

$LaCoO_3$ perovskite coatings were synthesized by magnetron sputtering from two metallic targets of La (kurt J. Lekser, purity 99.9%, Hastings, England) and Co (kurt J. Lekser, purity 99.9%). The Alcatel SCM 604 reactor, described elsewhere [33], was a 90-l sputtering chamber where a pressure of 10^{-4} Pa was maintained by a primary pump assisted by a molecular turbo pump. The La metallic target (thickness = 3 mm; diameter = 50 mm) and the Co one (thickness = 1 mm; diameter = 50 mm) were attached to two magnetrons distant from 120 mm and connected to a pulsed direct current generator (PINACLE+ purchased by advance energy). The distance between the two targets and the substrate holder (DT-S) can be independently modified. The draw distance between the targets and the alumina substrate was adjusted to 45 mm (Table 1). The targets are powered with pulsed currents to avoid any electric micro-arcs that can damage the quality of the films [41]. To avoid any accumulation of positive charge (Ar+) on the targets, the current is periodically stopped for very short dead time of 5 μS. Gases were introduced with mass-flow controllers (Brooks, 5850 SLA, Hatfield, PA, USA) and the total working pressure was measured with a MKS Baratron gauge. Films were deposited on dense alumina pellets (Keral 99, diameter = 16 mm, thickness = 0.60 mm purchased by Kerafol Gmbh, (Koppe Platz 1, Eschenbach i. d. Opf). The synthesis of doped perovskites with Sr and Ag was performed with the addition of two magnetrons into the sputtering chamber and metallic targets of Sr and Ag (thickness = 3 mm diameter = 50 mm, kurt J. Lekser, purity 99.9%). The Sr and Ag targets were powered by a two pulsed direct current generator (PINACLE+ and MDX500 both purchased by advance energy, Metzingen, Germany).

3.2. Characterizations of the Catalytic Coatings

The crystal structure of the coatings was determined thanks to a BRUKER D8 focus X-Ray diffractometer (Co Kα1+α2 radiations, in Bragg Brentano configuration and equipped with a LynxEye linear detector (Bruker, Billerica, MA, USA). XRD patterns were collected under air during 10 min in the (20–80°) scattering angle range by steps of 0.019°. The surface and the morphology of films were observed with a scanning electron microscope (JEOL JSM 7800F, Akishima, Japan) equipped with an EDS detector allowing the estimation of the chemical composition of samples. The observation of brittle cross section by SEM led also to the determination of the thickness of films. Coating thickness was also determined by the step method with an Altysurf profilometer produced by Altimet society (Marin, France) equipped with tungsten micro force probe inductive allowing an accuracy of about 20 nm. Before each measurement, the calibration of the experimental device was realized with a reference sample number 787569 accredited by CETIM organization.

XPS spectra were recorded for each catalysts on an AXIS Ultra DLD from Kratos Analytical (Manchester, UK) using a monochromatized Al X-ray source ($h\upsilon$ = 1486.6 eV) between 0 and 1200 eV with a pass energy of 40 eV. Sample were pretreated at 200 °C in He before measurements to clean the surface. Peaks were referenced using C1S peak of carbon (BE = 284.6 eV).

3.3. Catalytic Activity Measurements

A quartz reactor was operated under continuous flowing conditions at atmospheric pressure. The samples were placed on a fritted quartz, 18 mm in diameter, with the catalytic coating side facing the fritted quartz [42]. Gases were mixed by using mass-flow controllers (Brooks) to generate the different reactive mixtures. H_2O vapor was introduced using an atmospheric pressure Pyrex saturator heated at 46 °C. Reactants and products were analyzed using a gas micro-chromatograph (SRA 3000 equipped with two TCD detectors, a molecular sieve and a Porapak Q column for O_2, CO, C_3H_6, and CO_2 analysis) and a CO_2 Infra-Red analyzer (Horiba VA 3000, Horiba Europe Gmbh, Leichlingen, Germany). NO and NO_2 concentrations were measured with IR and UV online analyzers (EMERSON NGA2000).

Reactants were Air Liquide certified standards of NO in He (8000 ppm), C_3H_8 in He (8005 ppm), CO in He (1%), O_2 (99.999%), which could be further diluted in (99.999%). The carbon and nitrogen balance closure was found to be within 2%.

4. Conclusions

Different compositions of $La_{1-x-y}Sr_xAg_yCoO_{3-\alpha}$ (x = 0.13–0.28, y = 0.14–0.48) doped perovskites were synthetized as thin coatings deposited on alumina disks with the cathodic co-sputtering magnetron method in reactive conditions. The control of different parameters during the sputtering process can tune the morphology of the catalytic films. In particular, we found that the incorporation of high Ag loadings can generate covering films with a porous wire-like morphology showing good catalytic activity for CO oxidation. The most active composition, $La_{0.40}Sr_{0.1}Ag_{0.48}Co_{0.93}O_3$, displays similar catalytic performances than those of a Pt film. In addition, this porous coating is also efficient for CO and NO oxidation in a simulated Diesel exhaust gas mixture, demonstrating the promising catalytic properties of such nanostructured thin sputtered perovskite films.

Author Contributions: Investigation, M.A.P.Y. and L.L.; data curation, P.B. and P.V.; writing—original draft preparation, M.A.P.Y, P.B., and P.V.; writing—review and editing, P.B. and P.V.; formal analysis: A.B., P.V. supervision, A.B., P.B. and P.V.; project administration, P.B.; funding acquisition, P.B., P.V. and A.B.

Funding: The research was funded by "ADEME" in the frame of the ADIABACAT project and "Pays de Montbéliard Agglomération".

Conflicts of Interest: The authors declare no conflict of interest. The founding sponsors had no role in the design of the study; in the collection, analyses, or interpretation of data; in the writing of the manuscript, or in the decision to publish the results.

References

1. Fino, D. Diesel emission control: Catalytic filters for particulate removal. *Sci. Technol. Adv. Mater.* **2007**, *8*, 93–100. [CrossRef]
2. Wang, X.; Zhang, Y.X.; Li, Q.; Wang, Z.P.; Zhang, Z.L. Identification of active oxygen species for soot combustion on LaMnO3 perovskite. *Catal. Sci. Technol.* **2012**, *2*, 1822–1824. [CrossRef]
3. Hernández, W.Y.; Tsampas, M.N.; Zhao, C.; Boreave, A.; Bosselet, F.; Vernoux, P. La/Sr-based perovskites as soot oxidation catalysts for Gasoline Particulate Filters. *Catal. Today* **2015**, *258*, 525–534. [CrossRef]
4. Pena, M.A.; Fierro, J.L.G. Chemical Structures and Performance of Perovskite Oxides. *Chem. Rev.* **2001**, *101*, 1981–2017. [CrossRef] [PubMed]
5. Ciambelli, P.; Cimino, S.; Lisis, L.; Faticanti, M.; Minelli, G.; Pettiti, I.; Porta, P. La, Ca and Fe oxide perovskites: Preparation, characterization and catalytic properties for methane combustion. *Appl. Catal. B* **2001**, *33*, 193–203. [CrossRef]
6. Song, K.S.; Cui, H.X.; Kim, S.D.; Kang, S.K. Catalytic combustion of CH_4 and CO on $La_{1-x}M_xMnO_3$ perovskites. *Catal. Today* **1999**, *47*, 155–160. [CrossRef]
7. Hwang, H.J.; Awano, M. Preparation of $LaCoO_3$ catalytic thin film by the sol–gel process and its NO decomposition characteristics. *J. Eur. Ceram. Soc.* **2001**, *21*, 2103–2107. [CrossRef]
8. Teng, F.; Liang, S.; Gaugeu, B.; Zong, R.; Yao, W.; Zhu, Y. Carbon nanotubes-templated assembly of $LaCoO_3$ nanowires at low temperatures and its excellent catalytic properties for CO oxidation. *Catal. Commun.* **2007**, *8*, 1748–1754. [CrossRef]
9. Hattori, T.; Matsui, T.; Tsuda, H.; Mabuchi, H.; Morii, K. Fabrication and electric properties of $LaCoO_3$ thin films by ion-beam sputtering. *Thin Solid Films* **2001**, *388*, 183–188. [CrossRef]
10. López, M.L.; Arillo, M.A.; Álvarez-Serrano, I.; Martín, P.; Rodríguez, E.; Pico, C.; Veiga, M.L. Random spin configurations of Co cations in $LaCo_{1-x}Mg_xO_3$ ($0 < x \leq 0.20$) perovskite oxides: Magnetic and transport properties. *Mater. Chem. Phys.* **2010**, *120*, 387–392. [CrossRef]
11. Uhlenbruck, S.; Tietz, F. High-temperature thermal expansion and conductivity of cobaltites: Potentials for adaptation of the thermal expansion to the demands for solid oxide fuel cells. *Mater. Sci. Eng. B* **2004**, *107*, 277–282. [CrossRef]
12. Ullmann, H.; Trofimenko, N.; Tietz, F.; Stöver, D.; Ahmad-Khanlou, A. Correlation between thermal expansion and oxide ion transport in mixed conducting perovskite-type oxides for SOFC cathodes. *Solid State Ion.* **2000**, *138*, 79–90. [CrossRef]
13. Ringuedé, A.; Fouletier, J. Oxygen reaction on strontium-doped lanthanum cobaltite dense electrodes at intermediate. *Solid State Ion.* **2001**, *139*, 167–177. [CrossRef]
14. Weidenkaff, A.; Robert, R.; Aguirre, M.; Bocher, L.; Lippert, T.; Canulescu, S. Development of thermoelectric oxides for renewable energy conversion technologies. *Renew. Energy* **2008**, *33*, 342–347. [CrossRef]
15. Anh, D.T.V.; Olthuis, W.; Bergveld, P. Sensing properties of perovskite oxide La0.5Sr0.5CoO3−δ obtained by using pulsed laser deposition. *Sens. Actuators B* **2004**, *103*, 165–168. [CrossRef]
16. Popa, M.; Hong, L.V.; Kakihana, M. Particle morphology characterization and magnetic properties of $LaMnO_{3+d}$ perovskites. *Phys. B Condens. Matter* **2003**, *327*, 237–240. [CrossRef]
17. Ngamou, P.H.T.; Bahlawane, N. Chemical vapor deposition and electric characterization of perovskite oxides LaMO3 (M=Co, Fe, Cr and Mn) thin films. *J. Solid State Chem.* **2009**, *182*, 849–854. [CrossRef]

18. Popa, M.; Calderón-Moreno, J.M. Lanthanum cobaltite thin films on stainless steel. *Thin Solid Films* **2009**, *517*, 1530–1533. [CrossRef]
19. Roche, V.; Siebert, E.; Steil, M.C.; Deloume, J.P.; Roux, C.; Pagnier, T.; Revel, R.; Vernoux, P. Electrochemical promotion of propane deep oxidation on doped lanthanum manganites. *Ionics* **2008**, *14*, 235–241. [CrossRef]
20. Seim, H.; Nieminen, M.; Niinisto, L.; Fjellvag, H.; Johansson, L.S. Growth of $LaCoO_3$ thin films from β-diketonate precursors. *Appl. Surf. Sci.* **1997**, *112*, 243–250. [CrossRef]
21. Robert, R.; Aguirre, M.H.; Bocher, L.; Trottmann, M.; Heiroth, S.; Lippert, T.; Döbeli, M.; Weidenkaff, A. Thermoelectric properties of $LaCo_{1-x}Ni_xO_3$ polycrystalline samples and epitaxial thin films. *Solid State Sci.* **2008**, *10*, 502–507. [CrossRef]
22. Karoum, R.; Roche, V.; Pirovano, C.; Vannier, R.; Billard, A.; Vernoux, P. CGO-based electrochemical catalysts for low temperature combustion of propene. *J. Appl. Electrochem.* **2010**, *40*, 1867–1873, ISSN: 0021-891X (Print) 1572-8838 (Online). [CrossRef]
23. Cui, W.Y.; Li, P.; Bai, H.L. Spin-state configuration induced faster spin dynamics in epitaxial $La_{1-x}Sr_xCoO_3$ thin films. *Solid State Commun.* **2015**, *209–210*, 49–54. [CrossRef]
24. Tsiakaras, P.; Athanasiou, C.; Marnellos, G.; Stoukides, M.; Elshof, J.E.T.; Bouwmeester, H.J.M. Methane activation on a $La_{0.6}Sr_{0.4}Co_{0.8}Fe_{0.2}O_3$ perovskite: Catalytic and electrocatalytic results. *Appl. Catal. A Gen.* **1998**, *169*, 249–261. [CrossRef]
25. Gaillard, F.; Li, X.; Uray, M.; Vernoux, P. Electrochemical Promotion of Propene Combustion in Air Excess on Perovskite Catalyst. *Catal. Lett.* **2004**, *96*, 177–183, ISSN: 1011-372X (Print) 1572-879X (Online). [CrossRef]
26. Vernoux, P.; Lizarraga, L.; Tsampas, M.N.; Sapountzi, F.M.; de Lucas-Consuegra, A.; Valverde, J.L.; Souentie, S.; Vayenas, C.G.; Tsiplakides, D.; Balomenou, S.; et al. Ionically Conducting Ceramics as Active Catalyst Supports. *Chem. Rev.* **2013**, *113*, 8192–8260. [CrossRef]
27. Monaghan, D.; Arnell, R.D. Novel PVD films by unbalanced magnetron sputtering. *Vacuum* **1992**, *43*, 77–81. [CrossRef]
28. Kelly, P.J.; O'Brien, J.; Arnell, R.D. The production of porous and chemically reactive coatings by magnetron sputtering. *Vacuum* **2004**, *74*, 1–10. [CrossRef]
29. Rousseau, S.; Loridant, S.; Delichere, P.; Boreave, A.; Deloume, J.P.; Vernoux, P. $La_{(1-x)}Sr_xCo_{1-y}Fe_yO_3$ perovskites prepared by sol–gel method: Characterization and relationships with catalytic properties for total oxidation of toluene. *Appl. Catal. B* **2009**, *88*, 438–447. [CrossRef]
30. O'Connell, M.; Norman, A.K.; Huttermann, C.F.; Morris, M.A. Catalytic oxidation over lanthanum-transition metal perovskite materials. *Catal. Today* **1999**, *47*, 123–132. [CrossRef]
31. Pecchi, G.; Campos, C.M.; Jiliberto, M.G.; Delgado, E.J.; Fierro, J.L.G. Effect of additive Ag on the physicochemical and catalytic properties of $LaMn_{0.9}Co_{0.1}O_{3.5}$ perovskite. *Appl. Catal. A* **2009**, *371*, 78–84. [CrossRef]
32. Hernández, W.Y.; Lopez-Gonzalez, D.; Ntais, S.; Zhao, C.; Boréave, A.; Vernoux, P. Silver-modified manganite and ferrite perovskites for catalyzed gasoline particulate filters. *Appl. Catal. B* **2018**, *226*, 202–212. [CrossRef]
33. Yazdi, M.A.P.; Briois, P.; Georges, S.; Lapostolle, F.; Billard, A. Structural and electrical characterisation of strontium zirconate proton conductor co-sputter deposited coatings. *Ionics* **2008**, *14*, 285–291, ISSN: 0947-7047 (Print) 1862-0760 (Online). [CrossRef]
34. Yazdi, M.A.P.; Briois, P.; Billard, A. Influence of the annealing conditions on the structure of $BaCe_{1-x}YxO_{3-\alpha}$ coatings elaborated by DC magnetron sputtering at room temperature. *Mater. Chem. Phys.* **2009**, *117*, 178–182. [CrossRef]
35. Creus, J.; Berziou, C.; Cohendoz, S.; Perez, A.; Rébéré, C.; Reffass, M.; Touzain, S.; Allely, C.; Gachon, Y.; Héau, C.; et al. Reactivity classification in saline solution of magnetron sputtered or EBPVD pure metallic, nitride and Al-based alloy coatings. *Corros. Sci.* **2012**, *57*, 162–173. [CrossRef]
36. Hwang, H.J.; Moon, J.; Awano, M.; Maeda, K. Sol–Gel Route to Porous Lanthanum Cobaltite ($LaCoO_3$) Thin Films. *J. Am. Ceram. Soc.* **2004**, *83*, 2852–2854. [CrossRef]
37. Sanchette, F.; Billard, A.; Frantz, C. Mechanically reinforced and corrosion-resistant sputtered amorphous aluminium alloy coatings. *Surf. Coat. Technol.* **1998**, *98*, 1162–1168. [CrossRef]
38. Tietz, F. Thermal expansion of SOFC materials. *Ionics* **1999**, *5*, 129–139. [CrossRef]
39. Abe, O.; Taketa, Y.; Haradome, M. The Effect of Various Factors on the Resistance and TCR of RuO_2 Thick Film Resistors—Relation Between the Electrical Properties and Particle Size of Constituents, the Physical Properties of Glass and Firing Temperature. *Act. Passiv. Electron. Compon.* **1988**, *13*, 67–83. [CrossRef]

40. Shannon, R.D. Revised effective ionic radii and systematic studies of interatomic distances in halides and chalcogenides. *Acta Crystallogr.* **1976**, *A32*, 751–767. [CrossRef]
41. Billard, A.; Perry, F. Techniques de l'ingenieur, M1654. Available online: https://www.techniques-ingenieur.fr/base-documentaire/materiaux-th11/traitements-de-surface-des-metaux-par-voie-seche-et-en-milieu-fondu-42360210/pulverisation-cathodique-magnetron-m1654/ (accessed on 22 April 2019).
42. Vernoux, P.; Gaillard, F.; Bultel, L.; Siebert, E.; Primet, M. Electrochemical Promotion of Propane and Propene Oxidation on Pt/YSZ. *J. Catal.* **2002**, *208*, 412–421. [CrossRef]

Article

Effect of Preparation Method of Co-Ce Catalysts on CH_4 Combustion

Sofia Darda [1,2], Eleni Pachatouridou [1], Angelos Lappas [1] and Eleni Iliopoulou [1,*]

[1] Chemical Process & Energy Resources Institute (CPERI), Centre for Research & Technology Hellas (CERTH), 6th km. Charilaou—Thermi Rd., Thermi, GR-57001 Thessaloniki, Greece; dardasofia@gmail.com (S.D.); e_pahat@cperi.certh.gr (E.P.); angel@cperi.certh.gr (A.L.)

[2] Department of Chemical Engineering, Aristotle University of Thessaloniki, P.O. Box 1517, 54006 Thessaloniki, Greece

* Correspondence: eh@cperi.certh.gr; Tel.: +30-2310-498312

Received: 21 January 2019; Accepted: 22 February 2019; Published: 27 February 2019

Abstract: Transition metal oxides have recently attracted considerable attention as candidate catalysts for the complete oxidation of methane, the main component of the natural gas, used in various industrial processes or as a fuel in turbines and vehicles. A series of novel Co-Ce mixed oxide catalysts were synthesized as an effort to enhance synergistic effects that could improve their redox behavior, oxygen storage ability and, thus, their activity in methane oxidation. The effect of synthesis method (hydrothermal or precipitation) and Co loading (0, 2, 5, and 15 wt.%) on the catalytic efficiency and stability of the derived materials was investigated. Use of hydrothermal synthesis results in the most efficient Co/CeO_2 catalysts, a fact related with their improved physicochemical properties, as compared with the materials prepared via precipitation. In particular, a CeO_2 support of smaller crystallite size and larger surface area seems to enhance the reducibility of the Co_3O_4/CeO_2 materials, as evidenced by the blue shift of the corresponding reduction peaks (H_2-TPR, H_2-Temperature Programmed Reduction). The limited methane oxidation activity over pure CeO_2 samples is significantly enhanced by Co incorporation and further improved by higher Co loadings. The optimum performance was observed over a 15 wt% Co/CeO_2 catalyst, which also presented sufficient tolerance to water presence.

Keywords: Co_3O_4; CeO_2; complete CH_4 oxidation; hydrothermal synthesis; precipitation

---

1. Introduction

Today, CH_4, N_2O, and CO_2 emissions represent approximately 98% of the total greenhouse gas (GHG) inventory worldwide, a percentage expected to increase even further in the 21st century. Methane is the main compound of natural gas (NG) and is today used in various industrial processes and also as energy source in gas turbines or natural gas fuelled vehicles. Heavy-duty (HD) natural gas–fuelled fleet represents today less than 2% of the total fleet, however this percentage is expected to grow to represent as much as 50% in the near future. Moreover, methane is considered a substantially more potent greenhouse gas (GHG) than CO_2 and, hence, leakage of only a small percentage of methane from the supply chain could alter the net climate benefits of NG, which is lately suggested as an alternative fuel for heavy-duty (HD) transportation. Methane can be also a result of incomplete combustion processes, similarly to CO, while both of them are well-known environmentally detrimental pollutants and, thus, proper catalytic oxidation converters are needed for their abatement [1,2]. Noble metals and transition metals are the most commonly used components of such catalysts, investigated for methane and/or CO complete oxidation. Both these groups of catalysts have been extensively studied in order to develop efficient catalytic systems for combustion of methane, recognized as the lowest reactivity molecule among alkanes. All up to date research efforts have disclosed the superiority of the noble metals (especially Pt, Pd, and their mixtures) on oxide supports [3–5]. However, there are many factors defining the effectiveness of these supported catalysts, including the nature and properties of the support, the metal loading, the size, shape, and electronic state of metal nanoparticles and their interaction/synergy with the support, their preparation method and, finally, any pretreatment conditions possibly applied for their improvement [6]. However, in spite of their excellent activity, their applications at a larger scale have been limited because of their prohibitively high cost, their shortage in worldwide reserves, and their prompt deactivation during methane oxidation, mainly due to sintering of supported noble metal nanoparticles [7]. On the other hand, transition metal oxides have recently attracted considerable attention as alternative catalytic materials, among which cobalt oxide is considered as the most promising catalyst for methane combustion. Most crucial property of the early transition metal oxides is the fact that they can generate oxygen vacancies, suggested as the necessary sites that can activate molecular oxygen for oxidation applications [5]. Co_3O_4 nanosheets are reported to show better catalytic activity than Co_3O_4 nanobelts and nanocubes for CH_4 oxidation, in spite of their low specific surface area. Co_3O_4 catalysts with various structures (nanoparticles, two-dimensional and three-dimensional structures) were also investigated for methane catalytic combustion, exhibiting that enhanced catalytic performance of methane of the 2D-Co_3O_4 and 3D-Co_3O_4 catalysts is related with their noticeable reducibility and existence of abundant active Co^{3+} species [8,9]. Highly porous Co_3O_4 nanorods, with narrow pore-size distribution and a high surface area were synthesized by simple hydrothermal methods and similarly exhibited an enhanced catalytic performance for CH_4 oxidation, especially at high GHSV (Gas Hourly Space Velocity) conditions [10]. Co_3O_4 nanomaterials were also synthesized by hydrothermal method and tested for low temperature CO oxidation and methane combustion in another report. In that case, addition of ethylene glycol and its varying concentration significantly influenced the size and shape of the Co_3O_4 oxides (resulting in nanosheets or nanospheres). During catalyst evaluation, it was evidenced that the Co_3O_4 nanosheets exhibited more active oxygen species than the Co_3O_4 nanospheres, a fact accounting for their higher activity in the oxidative reactions [11].

In addition to cobalt, several ceria-based, low-cost catalysts have recently attracted much attention as a reliever of costly, unstable noble metal catalysts, since they exhibit good performance in oxidation reactions of carbon monoxide and lower hydrocarbons. Once more, the synthesis method is very crucial; for example nanocrystalline CeO_2, as small as 5 nm, were prepared by the precipitation method, using hydrogen peroxide as an oxidizer and is reported to present even 100 °C lower methane oxidation temperature, as compared with microcrystalline CeO_2. Nabih et al. has also reported the synthesis of mesoporous ceria nanoparticles (NPs) by combining the sol-gel (SG) process with the inverse mini-emulsion technique. These materials were of high specific surface area and exhibited an even better catalytic performance than the reference samples, synthesized by precipitation [12,13]. The addition of MO_x (M: di- or tri-valent transition metal ion) in CeO_2 is interestingly reported to promote the catalytic activity for the oxidation reaction, a fact possibly related with improved redox properties and high oxygen storage capacity, induced by the synergistic effect between CeO_2 and MO_x [14–17]. A nanocomposite catalyst Co_3O_4/CeO_2 is very recently reported to exhibit high activity in complete oxidation of CH_4, suggesting a synergistic effect of CeO_2 nanorods and the supported Co_3O_4 nanoparticles [7].

In addition to oxidation performance, catalyst tolerance to deactivation, caused by poisoning and thermal aging, is also a problem in all natural gas applications. Poisoning is either due to adsorption of impurities, e.g., S, P, Zn, Ca, and Mg, present in the exhaust gases, on the catalytic active sites or their reaction with the active sites causing the formation of non-active compounds. Sulfur in the NG vehicles exhaust gas actually originates from odorants and lubricating oils and less from gas itself. Sulfur is considered as the most noxious component readily deactivating Pd catalysts, recognized as the most efficient for methane oxidation. Only small amounts of SO_2 can significantly inhibit catalyst activity by blocking the active noble metal sites by sulfur compounds [18–21]. Moreover, water is always present in the process of the catalytic combustion of methane, either as gas humidity or as a combustion product. Water presence may affect the reaction kinetics, resulting in a serious inhibition of the oxidation reactions [14]. Thus, activity of Pd/Al_2O_3 is reported as severely inhibited by water at temperatures below 450 °C, a fact attributed to the formation of surface hydroxyl groups, which blocks active catalyst sites, while suppressed oxygen exchange between support and PdO via the formed hydroxyl groups is another recent explanation. Based on experiments and kinetic modelling two are the suggested routes of the water inhibition effect; rapid adsorption of water species on the active sites and slow build-up of hydroxyl species. Formation of these surface hydroxyls is enhanced by high concentration of oxygen and high concentration of water vapor [22]. In general, while various mechanisms have been proposed in the literature to explain the deactivation observed due to water presence, all studies agree on a strong effect of the support. Thus, tuning opportunely the metal-support interaction and the oxygen exchange capability of the support is a suggested way to deal with water inhibition problem over Pd/CeO_2 catalysts, otherwise very highly active for methane oxidation in dry conditions [23]. In a recent study, Pt addition was suggested to improve the performance of Pd-modified Mn-hexaaluminate catalyst in the high-temperature oxidation of methane, especially in SO_2 and water presence, correlating the improved water and sulfur resistance with the presence of particles of PtPd alloy [24]. In a similar study, presence of water vapor is reported to significantly inhibit the catalytic performance of CeO_2-MO_x (M = Cu, Mn, Fe, Co, and Ni) mixed oxide catalysts for CH_4 combustion, while CeO_2-NiO samples showed the best durability for CH_4 wet combustion (T50: 528 °C, when 20% water vapor was added in the reaction stream) among all materials tested [14]. Interestingly, water presence in the reaction feed is even reported to promote methane conversion over a NiO catalyst, prepared from thermal decomposition of the corresponding metal nitrate. As activation of oxygen plays a crucial role in methane oxidation, in the surface kinetics controlled region, the observed high catalytic efficiency of NiO is attributed to its larger capability for oxygen adsorption, even at 500 °C. The promoting role of water could be due to the fact that H_2O molecules could modify the NiO surface and, thus, promote further activation of O_2 and/or CH_4 on the surface [25].

In a series of recent studies, we have investigated the impact of synthesis parameters on the solid state properties of CeO_2 materials. In detail, we followed four different, time- and cost-effective, preparation methods, i.e., thermal decomposition, precipitation, and hydrothermal method of low and high NaOH concentration, employing in all cases $Ce(NO_3)_3{\cdot}6H_2O$ as the cerium precursor. A thorough characterization of all samples was carried out to gain insight into the impact of synthesis route on the textural, structural, morphological, and redox properties of the derived materials. The results revealed the superiority of the hydrothermal method towards the development of ceria nanoparticles of high specific surface area (>90 m^2 g^{-1}), well-defined geometry (nanorods) and improved redox properties. Employing CO oxidation as a probe reaction evidenced a direct quantitative correlation between the catalytic activity and the abundance of easily reduced, loosely bound oxygen species. More specifically, the rod-like morphology of the hydrothermally synthesized CeO_2 nanoparticles, with well-defined (100) and (110) reactive planes, favored the enhanced reducibility and lattice oxygen mobility, rendering this material appropriate as catalyst or supporting carrier [26–28]. The current study aims in the preparation of novel Co-Ce catalytic materials targeting in the complete combustion of CH_4. Co incorporation on an optimum selected CeO_2 support aims to promote the catalytic activity for the oxidation reactions, via an attempted improvement of the redox property and high oxygen storage capacity, deliberately induced by the synergistic effect between CeO_2 and CoO_x. We purposely explored the effect of both the synthesis method of the CeO_2 support (following either the hydrothermal or precipitation technique) and the metal loading (0, 2, 5, and 15 wt% Co) on the catalytic performance and durability for complete methane oxidation. Deactivation studies were also performed examining tolerance of the optimum selected catalysts to thermal aging and presence of water vapor in the reaction feed (wet CH_4 combustion).

2. Results and Discussion

2.1. Catalyst Characterization

2.1.1. Textural and Structural Characterization

The physicochemical properties of catalytic materials are presented in Table 1. As expected [26], the synthesis method of CeO_2 greatly affects the specific surface area, the pore volume and the average pore diameter of the carrier. In addition, the mean particle size of CeO_2 presents important differences between the two synthesis methods. Applying the hydrothermal method (CeO_2-H), the formed ceria presents higher surface area (110.9 m^2/g) and smaller CeO_2 crystallites (11.8 nm) than the precipitated ceria (CeO_2-P), where the growing up of the ceria crystallites (18.6 nm) resulted in decreased surface area (49.5 m^2/g). The effect of synthesis method on the crystallite size is probably related with slower rates of crystal migration and crystallites grow up, when following the hydrothermal synthesis method. Incorporation of cobalt, on both CeO_2 carriers, as well as the increase of its metal loading, affects the texture of the CeO_2 support, decreasing the specific surface area of the derived, CeO_2-supported, Co catalysts. On the contrary, in the case of pure Co_3O_4, the different synthesis method does not appear to significantly affect the specific surface area. However, applying the hydrothermal method (Co_3O_4-H), the formed Co oxide presents lower pore volume and average pore diameter and similarly smaller Co_3O_4 crystallites than Co_3O_4-P (33.3 nm as compared to 42.2 nm, respectively). Lower crystallite sizes (16.1–16.8 nm) of Co_3O_4 nanomaterials, synthesized by hydrothermal method are reported in the recent literature, probably related with lower calcination temperature, thus limiting aggregation of Co_3O_4 crystallites in the sample [11]. It is noteworthy that much smaller Co_3O_4 crystallites are also formed on the CeO_2-H rather than the CeO_2-P support (16.0 as compared to 76.6 nm, respectively), when incorporating 15 wt% Co on either support via wet impregnation, a fact indicating synergetic effects between the CeO_2-H support and the incorporated Co oxide.

Table 1. Physicochemical properties of catalytic materials.

Catalyst	Co (wt%)	S_{BET} (m^2/g)	Pore Volume (cm^3/g)	Average Pore Diameter (nm)	Mean Particle Size (nm) [1]	
					CeO_2	Co_3O_4
CeO_2-P	-	**49.5**	**0.06**	5.2	18.6	-
CeO_2-H	-	110.9	0.55	19.9	11.8	-
2Co/CeO_2-P	2.1	43.3	0.06	5.8	16.3	-
2Co/CeO_2-H	2.3	100.3	0.49	19.5	8.1	-
5Co/CeO_2-P	5.1	39.6	0.05	5.3	20.5	-
5Co/CeO_2-H	5.1	92.1	0.23	9.5	9.4	17.8
15Co/CeO_2-P	14.6	38.8	0.09	9.3	16.2	76.6
15Co/CeO_2-H	15.4	78.9	0.14	7.3	6.4	16.0
Co_3O_4-P	-	15.5	0.14	47.6	-	42.2
Co_3O_4-H	-	12.8	0.04	11.4	-	33.3

[1] Measured by XRD diffractograms using the Scherrer equation, at (2θ) 28.6°, 47.5° for CeO_2 and 31.3°, 36.9° for Co_3O_4.

Representative XRD (X-Ray Diffraction) patterns of bare CeO_2, Co_3O_4 and xCo/CeO_2 prepared with either the precipitation or the hydrothermal method, are shown in Figure 1a,b respectively. As expected, both samples of CeO_2 supports (CeO_2-P and CeO_2-H) exhibit characteristic peaks predominately located at 2θ: 28.6°, 33.1°, 47.5°, 56.36°, 59.1°, 69.4°, 76.9°, and 79.2° and have face centered cubic structure with the lattice parameters $a = b = c = 5.411$ Å and $\alpha = \beta = \gamma = 90°$ [29], which are also preserved in the Co-supported catalysts.

Bare Co oxides (Co_3O_4-P and Co_3O_4-H) show diffraction peaks at 2θ: 19.0°, 31.3°, 36.9°, 38.5°, 44.8°, 55.9°, 59.4°, and 65.2°, which are attributed to the cobalt oxide Co_3O_4 and suggest that the cobalt precursor led to the formation of a face centered cubic unit cell of Co_3O_4 (space group *Fd3m*) with a spinel type structure after calcination [9,11]. At low Co loadings (2 wt.%), no diffraction peak related to Co_3O_4 was observed, while increasing Co loading (to 5 wt.% and 15 wt.%) leads to the development of low intensity Co_3O_4 peaks (2θ: 31.3°, 36.9°, and 44.8°) for Co/CeO_2-P (Figure 1a) and Co/CeO_2-H (Figure 1b) catalytic materials, respectively.

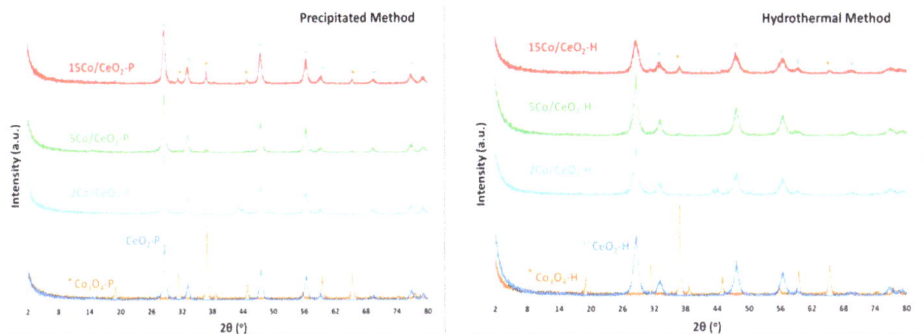

Figure 1. XRD patterns of CeO_2, Co_3O_4 and xCo/CeO_2 prepared following the (**a**) precipitation method; and (**b**) the hydrothermal method. (x: 2, 5 or 15 wt% Co loading).

2.1.2. Morphological Characterization

The HR-TEM (High Resolution-Transmission Electron Microscopy) images further confirm the exposed planes of the two 15 wt.%. Co catalysts, supported on differently synthesized CeO_2 carriers. In both cases, two different face-centered cubic phases are identified, on taken electron diffraction patterns: a face-centered cubic phase of CeO_2 with a = 0.541 nm, and a second face-centered cubic phase of Co_3O_4, with a = 0.808 nm. However, the morphology of the Co_3O_4/CeO_2 catalysts is further explored with HR-TEM. In detail, low magnification images in Figure 2 (Figure 2a,b correspond to CeO_2-P and CeO_2-H, respectively) show an overview of the two catalytic samples. A very well defined, crystalline hexagonal formation (nanocrystalline structures of CeO_2 are forming plates) is observed in the first case, while in the second case the CeO_2 nanocrystals show a nanorod structure, an expected morphology related with the hydrothermal synthesis applied during ceria preparation [26]. Moreover, in both HR-TEM micrographs showing a very well defined hexagonal formation of CeO_2-P (Figure 2c), and a nanorod structure of CeO_2-H (Figure 2d) respectively, the characteristic d-spacings of the (111) planes are evidenced for both CeO_2 supports, in agreement with the XRD patterns (characteristic peak, predominately located at 2θ: 28.6°) [7].

As shown in Figure 3, both CeO_2 nanocrystals and Co_3O_4 nanoparticles are well crystallized, as evidenced by the clear lattice fringes in these HR-TEM micrographs. The coexistence of the Co_3O_4 and CeO_2 cubic phases is evidenced by fast Fourier transformation analysis (FFT). However, Co_3O_4 nanoparticles are supported on bigger CeO_2 nanocrystals with a size of about 25–30 nm (slightly larger than estimated via XRD patterns) over the Co/CeO_2-P sample (Figure 3a). The experimental lattice spacings of 0.285 nm and 0.243 nm, as well as the lattice spacings of 0.312 nm, can be assigned to (220) planes and (311) planes of the Co_3O_4 cubic phase, and to the (111) planes of the CeO_2 respectively. On the contrary, HR-TEM image of the Co/CeO_2-H sample (Figure 3b) shows a much smaller CeO_2 nanorod (<20 nm), a trend in agreement with the XRD characterization. A smaller Co_3O_4 nanoparticle with a size of about 15 nm, seems to be attached to these nanorod support. In this case, the experimental lattice spacings of 0.285 nm and 0.202 nm, as well as the lattice spacings of 0.312 nm, can be defined to (220) planes and (400) planes of the Co_3O_4 cubic phase, and to the (111) planes of the CeO_2, respectively [7,9].

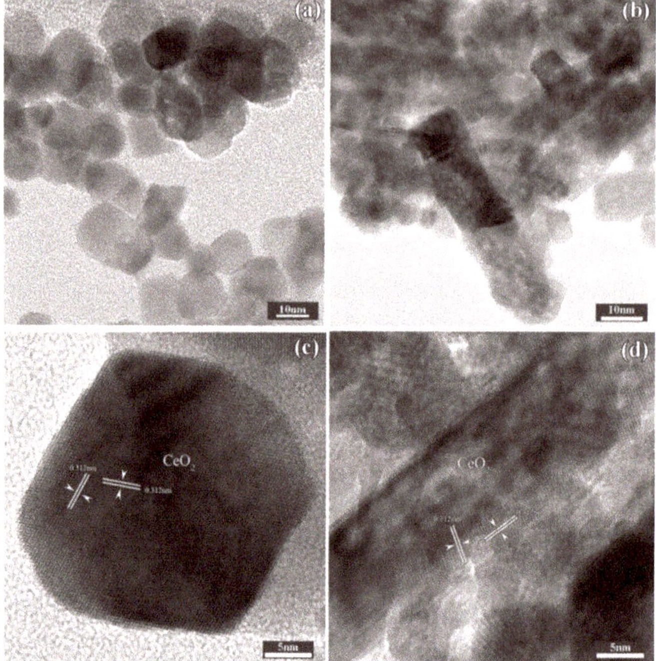

Figure 2. Low magnification images (overview) of the CeO_2 specimens, synthesized using the precipitation, CeO_2-P (**a**) or the hydrothermal CeO_2-H (**b**) method. The corresponding HR-TEM micrographs of the CeO_2-P sample (**c**), and the CeO_2-H sample (**d**), respectively.

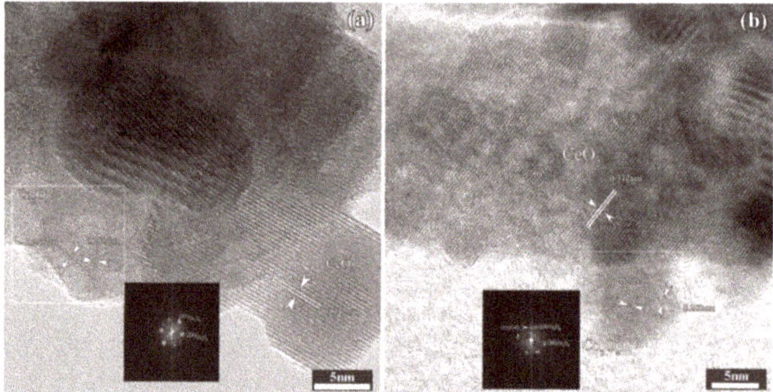

Figure 3. HR-TEM micrographs of the CeO_2-P (**a**) and CeO_2-H (**b**) specimens. As insets, FFT images taken at the white marked area are present for the two different cases, CeO_2-P (**a**) and CeO_2-H (**b**) samples, respectively.

2.1.3. H_2-TPR and O_2-TPD Studies

The reducibility of the catalysts was also investigated with H_2-TPR experiments and the corresponding reduction profiles are presented in Figure 4. Bare CeO_2-P and CeO_2-H (Figure 4a,b insets, respectively) exhibit two peaks; one at 500–515 °C and one at 780–790 °C. The former is attributed to the reduction of surface oxygen of ceria, while the second one to the reduction of the bulk oxygen of CeO_2 [29,30]. Regarding pure Co oxides synthesized by different methods (Co_3O_4-P and Co_3O_4-H in Figure 4a,b insets, respectively), a broad double peak is observed in both cases, which is associated with the reduction of Co_3O_4 to CoO (T: 330–340 °C) and CoO to Co^0 (T: 400–460 °C) [30,31]. It is worth to mention that for both materials prepared with hydrothermal method (Co_3O_4-H and CeO_2-H) a blue shift is observed; the reduction peaks are shifted to lower temperatures, which indicates enhanced reducibility of both oxide materials under study.

Concerning the H_2-TPR profiles of xCo/CeO_2 catalysts, all samples present three reduction peaks, which are also shifted to lower temperatures, in comparison with the corresponding bare CeO_2 and Co_3O_4 oxides (Figure 4a,b), an observation assigned to the interaction of Co with ceria. The first two peaks are related to the reduction of Co_3O_4, while the third one is attributed to the reduction of bulk oxygen of ceria support. In addition, the corresponding H_2 consumption tends to increase as expected, as Co loading increases from 2 to 5, and finally to 15 wt.%.

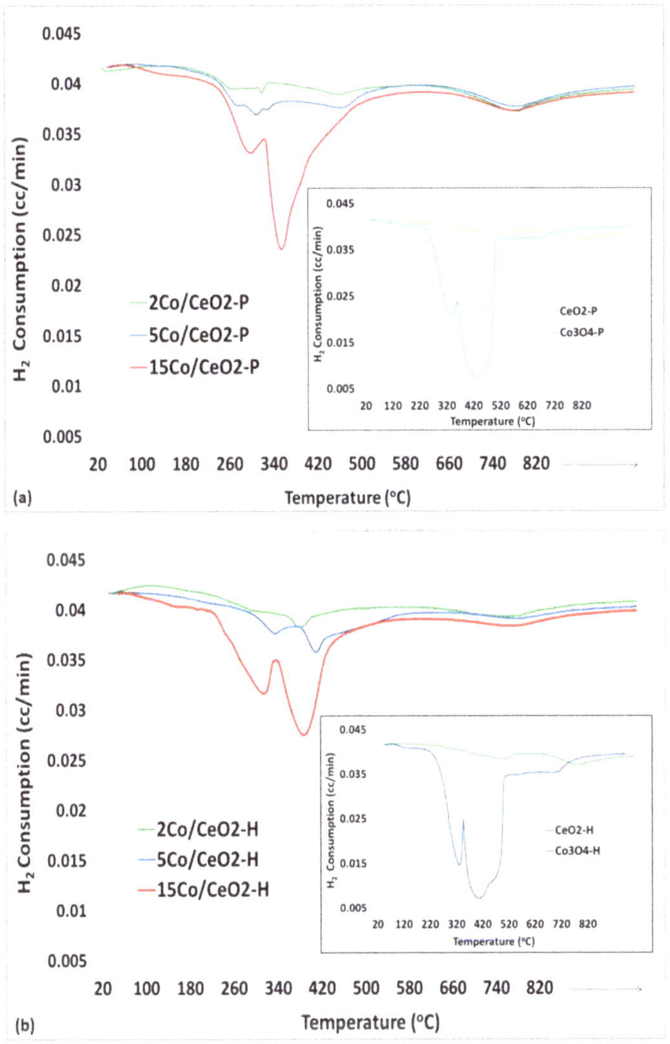

Figure 4. TPR profiles of bare CeO_2, Co_3O_4 (insets) and xCo/CeO_2 catalysts prepared with: (**a**) Precipitation and (**b**) Hydrothermal Method. (x: 2, 5m or 15 wt% Co loading).

As suggested in several studies, during methane combustion the lattice oxygen in the oxide surface contributes to the oxidation of CH_x, thus leading to the formation of an oxygen vacancy, available for oxygen adsorption from the gas-phase. Both the adsorbed oxygen, as well as the surface lattice oxygen, are considered crucial for oxidation reactions, as they contribute to the adsorption/activation of oxygen and, thus, the desired oxidation of intermediates on the catalyst surface [11,32,33]. Oxygen-temperature programmed desorption (O_2-TPD) experiments have evidenced that oxygen molecules were released from Co_3O_4 nanomaterials, suggesting that these oxygen molecules were adsorbed on the surface oxygen vacancies of the oxides [32]. These oxygen vacancies facilitate oxygen mobility of the oxide, forming the active oxygen species and, thus, enhancing the reaction activity with reductive molecules like H_2 and CH_4 [8,34]. Tuning the shape of the oxide catalyst, e.g. nanorods, nanocubes and nanosheets, is suggested to improve availability and abundance of these active oxygen species [26,35]. Especially, synthesis of Co_3O_4 catalysts of specific shape in order to provide more active oxygen species in the catalyst surface is reported in the recent literature [11].

Figure 5 shows the profiles of O_2 temperature-programmed desorption (O_2-TPD) of bare CeO_2 (prepared either following the precipitation or the hydrothermal method) and the corresponding 15 wt.% Co/CeO_2 supported catalysts. The bare ceria supports exhibit one peak at 418 °C, related with the adsorbed atomic oxygen evolved from the bulk of CeO_2 support [36]. The incorporation of 15 wt.% Co to CeO_2 carriers shifted the peak to lower temperatures, suggesting that O_2 desorption is facilitated, especially for the 15Co/CeO_2-H catalyst (observed shift from 418 to 376 °C) [30]. In addition, both Co-based catalysts exhibit two extra peaks at a higher temperature area (500–650 °C), which are related with the oxygen desorption from the surface of the Co oxides, while the peak observed at T > 700 °C corresponds to O_2 desorption from the bulk of Co_3O_4 [11].

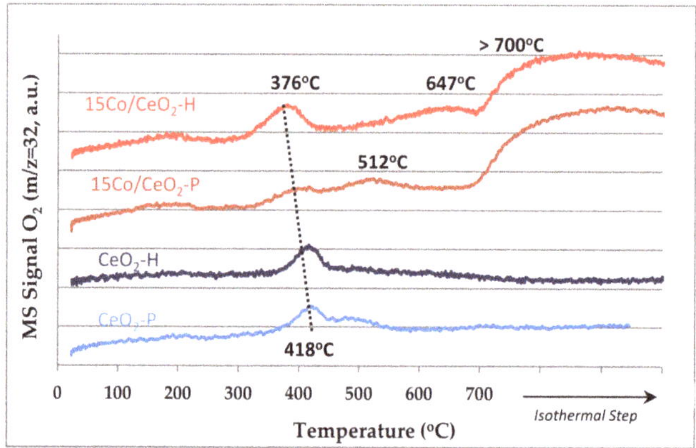

Figure 5. O_2-TPD profiles of the two bare CeO_2 supports (prepared with either the precipitation or the hydrothermal method) and the corresponding 15 wt.% Co/CeO_2 supported catalysts.

2.2. *Catalytic Activity of Pure CeO_2, Pure Co_3O_4 and Co_3O_4/CeO_2 Oxides for Methane Oxidation*

The catalytic performances of all materials (pure CeO_2, pure Co_3O_4 and Co_3O_4/CeO_2 composite catalysts) in complete oxidation of methane were investigated using a fixed-bed reactor and are presented in Figure 6a. Below 500°C, CH_4 conversion was <10% for pure CeO_2 supports, while the CeO_2 nanorods (CeO_2-H) exhibit a slightly improved performance, finally reaching 41% conversion at 600 °C (as compared to 31% conversion achieved with the CeO_2-P sample, respectively). Co incorporation via deposition of Co_3O_4 nanoparticles on both CeO_2 supports largely increased their catalytic activity for CH_4 combustion, which is further enhanced as Co loading increases from 2 to 5 and, finally, 15 wt.%. However, an improved catalytic performance is always observed on the CeO_2-H (ceria nanorods) supported materials. Co incorporation on CeO_2 nanorods achieves 10% CH_4 conversion from 420 °C and finally reaches up to ~90% CH_4 conversion at 600 °C (15Co/CeO_2-H sample). It is worth to mention that on higher Co loadings (5 and 15 wt.%) a significant improvement (>15% higher conversion) is always observed, when using the CeO_2 nanorods support. This is probably related with the optimized physicochemical properties of this CeO_2-H support: smaller CeO_2 cystallites, combined with higher surface area and possibly some synergetic effects between the deposited Co_3O_4 and the CeO_2 carrier. These catalysts exhibit similar performance ($T_{50\%}$ = 520 °C) with materials reported in the recent literature [7], evaluated however at milder conditions (lower GHSV:18,000 h^{-1} and different feedstock ~6.67% CH_4 and ~33.3% O_2). Efficiency at high GHSV feeds is very important and still remains a significant challenge, as reported in the literature [10].

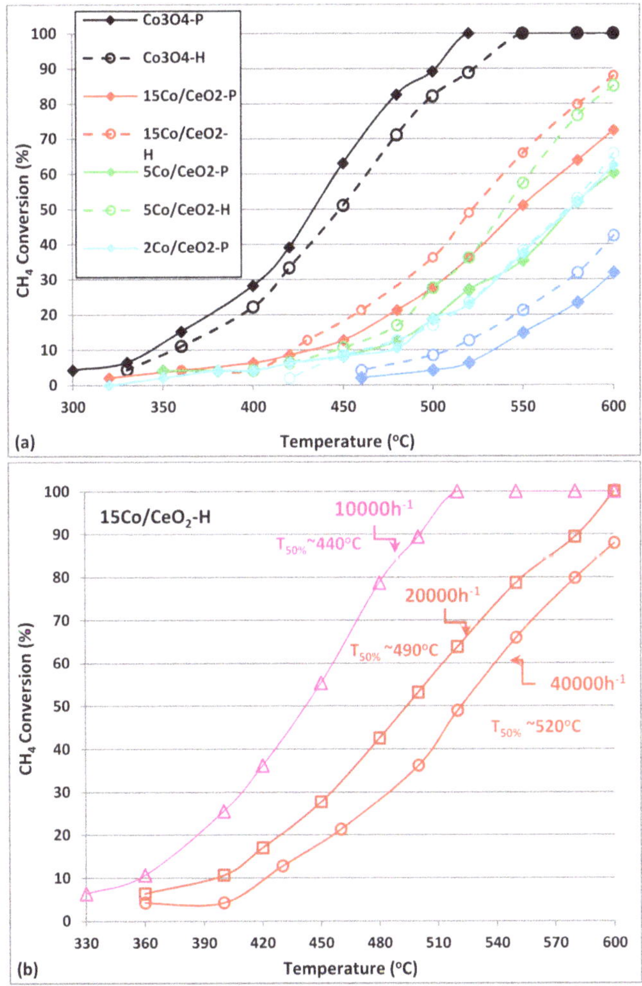

Figure 6. Catalytic performance in complete CH_4 oxidation on pure CeO_2, Co_3O_4 and nanocomposite Co/CeO_2 catalysts. (**a**) CH_4 conversion vs. temperature Flow rate: 900 cm^3/min; GHSV~40,000 h^{-1}; (**b**) Effect of WHSV on CH_4 conversion for the optimum selected 15Co/CeO_2-H catalyst. Feed: 0.5% vol. CH_4 and 10% vol. O_2, balanced with He.

In order to further exhibit superiority of our optimum material, $15Co/CeO_2$-H was further evaluated differentiating the reaction conditions used; varying GHSV between 10,000, 20,000, and 40,000 h^{-1} (Figure 6b). As expected, $15Co/CeO_2$-H catalyst exhibits a significantly improved performance at lower GHSV reaching $T_{50\%}$ = 440 °C, which is obviously much better when compared with similar reported catalysts (Co_3O_4/CeO_2 exhibiting $T_{50\%}$ = 475 °C at WHSV of 9000 mL g^{-1} h^{-1}) [7]. Moreover, the much lower temperature for 50% conversion of CH_4 ($T_{50\%}$) on Co_3O_4/CeO_2-H, as compared with the bare CeO_2-H support ($T_{50\%}$ = 440 °C as compared with $T_{50\%}$ = 575 °C, respectively, at low GHSV: 10,000 h^{-1} conditions), once more supports the significant role of the loaded Co_3O_4 nanoparticles for complete oxidation of methane. Finally, the superior performance of pure Co_3O_4 catalysts is very interesting, as these materials reached complete (100%) oxidation of CH_4 from 520 °C independently of their synthesis method and thus deserve further investigation.

2.3. Catalyst Stability Tests

Thermal stability of a catalyst used in the CH_4 catalytic combustion is very important because of its expected operation at high temperatures. In general, thermal aging may cause crystal growth and loss of catalytic surface. Thus, catalyst stability over time of the optimum selected material ($15Co/CeO_2$-H) was tested under the flow of mixture of 0.5 vol% CH_4 and 10 vol% O_2, balanced with He. As shown in Figure 7, the conversion at 600 °C is rather stable (only a small decrease from 88 to 81% is observed) after 10 hours on stream, under a GHSV of 40,000 h^{-1}. This is probably attributed to the thermal stability of the particular oxides (CeO_2 and Co_3O_4), which are expected to deteriorate at higher operating windows (700–750 °C). Preservation of finely dispersed Co_3O_4/CeO_2 species after aging at 750 °C is reported to retain high methane oxidation activity and tolerance to thermal ageing, suggesting once more the importance of the preparation method to achieve intimate contact between the cobalt–ceria species leading to good oxidation activity [17].

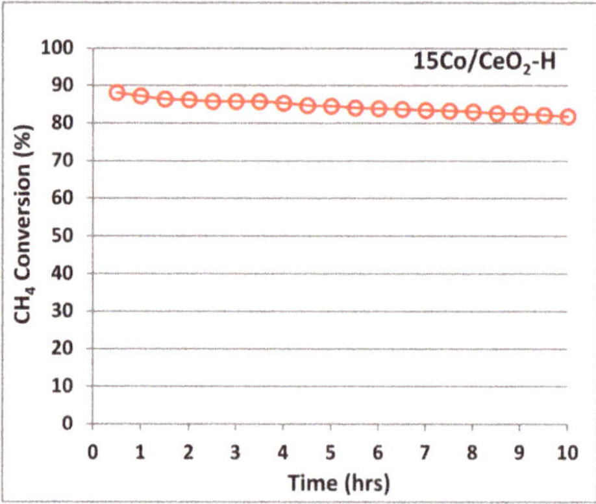

Figure 7. Response of CH_4 combustion to long-term stability tests over the optimum selected $15Co/CeO_2$-H catalyst; Flow rate: 900 cm^3/min; GHSV~40,000 h^{-1}.; Reaction temperature 600 °C; Feed:. 0.5% vol. CH_4 and 10% vol. O_2, balanced with He.

The presence of ~10 vol.% of water in the NG exhausts still remains one of the main challenges for the required catalytic system to control the emissions of NG vehicles. Thus, in order to further test the water vapor effect on the catalytic performance of our optimum catalyst ($15Co/CeO_2$-H) for methane oxidation, 5% and 10% water vapor was added in the process of CH_4 combustion and the corresponding catalytic performance is shown in Figure 8 in comparison with the case using the dry reaction gas. The amount of water added significantly exceeds the stoichiometric amount produced during combustion, under the conditions used in this work (0.5% CH_4, 10% O_2). Increasing the water vapor content in the reaction stream, slightly descends its catalytic performance, as $T_{50\%}$ is increased from 520 to 530 °C, when water vapor is added, while increase of water content (from 5 to 10 wt.%) does not induce any further inhibition effect. Thus, water presence does not lead to a significant activity fall (observed increase in $T_{50\%}$ is less than 20 °C), while water effect is even more limited at higher temperatures of methane combustion. On the contrary, effect of water presence is most evident on the initiation temperature of the CH_4 combustion, increasing T10 value from 420 to 450 °C [14]. Removing water from the reaction feed completely restored the initial activity of the catalyst, without the need to subject the catalyst to any regeneration/thermal treatment. The restored catalyst efficiency of the used catalyst suggests that the limited water inhibition effect is probably related with adsorption of water species on the active sites at the low temperature range. There is no evidence of hydroxyl species' build-up or sintering of Co species, which could lead to a more severe/permanent deactivation, in spite the high concentration of oxygen (10 wt%) and high concentration of water vapor (10 wt%) used, expected to facilitate formation of surface hydroxyls [22]. The steam reforming reaction of methane might be involved, during the CH_4 catalytic combustion in the presence of water vapor, in which CH_4 is partially oxidized to CO and H_2. However, analysis of the effluent gas hardly detected any CO, confirming that CH_4 is totally oxidized to CO_2 and H_2O over 15Co/CeO_2-H, even in the presence of water vapor (wet CH_4 catalytic combustion).

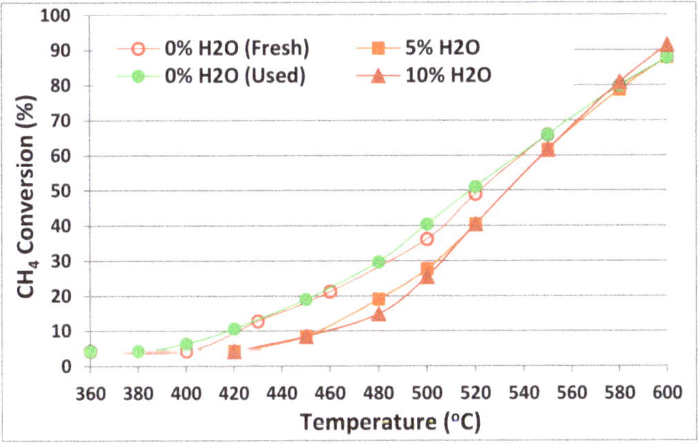

Figure 8. CH_4 catalytic combustion over 15Co/CeO_2-H (fresh and used) with different contents of water vapor addition (0, 5 and 10 wt%). Flow rate: 900 cm^3/min; GHSV~40,000 h^{-1}.

3. Materials and Methods

The CeO_2 carrier was prepared with two different methods; the precipitation method (CeO_2-P) and the hydrothermal method (CeO_2-H). Cerium nitrate hexahydrate ($Ce(NO_3)_3{\cdot}6H_2O$, 99% purity, supplied by Fluka, Steinheim, Germany) was used as a precursor salt for both synthesis methods. For the final production of 25 g of CeO_2-P we initially prepared an aqueous solution (0.5 M) of cerium nitrate dissolving ~63 g of $Ce(NO_3)_3{\cdot}6H_2O$ in 290 mL of double distilled water. Then a precipitating agent (25 vol% NH_3 solution) was added at room temperature to the continuously stirred solution of the cerium nitrate, until the pH of the solution reached the value of 10. Then the pH remained stable at this value for additional 3 h. Following precipitation, the resulting precipitate was filtered, dried overnight at 110 °C and then calcined at 500 °C for 5 h under air flow. For the preparation of CeO_2-H, 0.023 mol of cerium nitrate was dissolved in 46 mL of double distilled water. Then, 125 mL of a 1 M NaOH solution was added rapidly under vigorous stirring. The mixed solution was placed in a 1 L Teflon bottle, which was sealed and placed in an oven at 110 °C for 5 h. The as-obtained material was washed several times with double distilled water and ethanol, and then dried and calcined, following the same thermal treatment with the precipitated one.

The incorporation of cobalt over the CeO_2 supports was conducted by applying the incipient wetness impregnation method. A series of Co catalysts supported on CeO_2-P and CeO_2-H, were prepared in three different metal loadings (2, 5 and 15 wt%). Incorporation of the Co active phase was accomplished by using $Co(NO_3)_2{\cdot}6H_2O$ (99% purity, supplied by Merck, Darmstadt, Germany). Aqueous solution of the metal salt was impregnated on CeO_2 supports, in successive stages, with intermediate drying in an oven at 100 °C for 30 min. The derived samples were dried at 110 °C overnight and finally calcined under air flow at 500 °C for 5 h. The as prepared composites are herein labeled as xCo/CeO_2-P or H, where x is the cobalt loading.

For comparison reasons, bare Co_3O_4 was prepared by precipitation (Co_3O_4-P) and hydrothermal (Co_3O_4-H) method. $Co(NO_3)_2{\cdot}6H_2O$ was used as a precursor salt and the same precipitation method was followed as for the synthesis of CeO_2-P. For the preparation of Co_3O_4-H, 0.0515 mol of cobalt nitrate was dissolved in 51 mL of double distilled water and 125 mL of a 1 M NaOH solution was added rapidly under vigorous stirring. Then, the next steps (aging, washing, and thermal treatment) were the same with the hydrothermal synthesis of ceria.

The physicochemical characteristics of as-synthesized materials were evaluated by various complimentary techniques. The total Co loading (wt%) of the final catalysts was determined by the Inductively Coupled Plasma Atomic Emission Spectroscopy (ICP-AES -on a Perkin-Elmer Optima 4300DV apparatus, Waltham, MA, USA). The textural characteristics of the as prepared catalysts were determined by the N_2 adsorption–desorption isotherms at −196 °C (Nova 2200e Quantachrome flow apparatus, Boynton Beach, FL, USA). Specific surface areas (m^2g^{-1}) were obtained according to the Brunauer-Emmett-Teller (BET) method at relative pressures in the 0.05–0.30 range. The specific pore volume (cm^3g^{-1}) was calculated based on the highest relative pressure, whereas the average pore size diameter (dp, nm) was determined by the Barrett-Joyner-Halenda (BJH) method. Prior to measurements the samples were degassed at 250 °C under vacuum. The crystalline structure of the catalysts was determined by powder X-ray diffraction (XRD) on a Siemens D 500 diffractometer (Bruker, Karlsruhe, Germany) operated at 40 kV and 30 mA with Cu Kα radiation (λ = 0.154 nm). Diffractograms were recorded in the 5–80° 2θ range and at a scanning rate of 0.02° s^{-1}. The Scherrer equation was employed to determine the primary particle size of a given crystal phase based on the most intense diffraction peaks of CeO_2 (28.6°, 47.5°) and Co_3O_4 (31.3°, 36.9°). The size, morphology and structure of the two 15 wt% Co-supported catalysts (either on CeO_2-H or CeO_2-P supports) were also investigated by transmission electron microscopy using a (JEOL JEM 2011 TEM, Zaventem, Belgium) operating at 200 kV with an atomic resolution of 0.194nm. The HR-TEM micrographs were analyzed using Digital Micrograph v2 software (Gatan, Inc., München, Germany).

Hydrogen temperature-programmed reduction (H_2-TPR) experiments were performed in a different bench-scale unit, also available in our facilities for TPX studies, loading a fixed bed tubular reactor with 0.1 g of sample, while the reactor exit was coupled with a mass spectrometer (MS). A flow (50 cm^3/min) of 5 vol% H_2/He gas mixtures was used, while the temperature of the solid catalyst was increased to 800°C at the rate of 10 °C/min. The effluent gas from the reactor was analysed using the MS detector (m/z = 2 was used for H_2). Prior to reduction, the as prepared sample was loaded to the reactor and pre-treated at 300°C for 1h under He flow. The catalyst sample was then cooled in He gas flow to room temperature, before recording the H_2-TPR trace. Oxygen temperature-programmed desorption was performed (O_2-TPD) in the same unit. The temperature was increased to 400 °C (heating rate 5 °C/min) under 20 vol% O_2/N_2 (50 cm^3/min), for 30 min and were cooled to room temperature to adsorb oxygen for 30 min. After that, the samples were purged under He flow for 1 h, in order to purge physically adsorbed oxygen. The desorption step was conducted under He flow (50 cm^3/min) up to 750 °C at a rate of 10 °C/min and the signal of O_2 (m/z = 32) were detected by the mass spectrometer.

The catalytic combustion of CH_4 was performed in a fixed-bed reactor loaded with 0.6 g of catalyst. The total gas flow rate was 900cm^3/min, corresponding to a gas hourly space velocity (GHSV) of ~40,000 h^{-1}. The composition of the feed was 0.5 vol% CH_4 and 10 vol% O_2, balanced with He.

The effect of GHSV was also investigated at 40,000, 20,000, and 10,000 h^{-1}. The CH_4 conversion was monitored in the 300–600 °C range (in a decreasing temperature mode) and was held constant at each temperature for ~20 min prior taking measurements. The composition of the effluent gas from the reactor was analysed using gas chromatography (GC-FID, Agilent 7890, Walborn, Germany). Stability of optimum catalyst's performance (time-on-stream) was tested at 600 °C for 10 h (under 0.5 vol% CH_4, 10 vol% O_2 in He flow). In addition, the effect of H_2O content (0, 5, and 10% vol. H_2O in He) was studied on CH_4 oxidation efficiency.

4. Conclusions

Various novel Co-Ce catalysts were prepared following different preparation techniques for the synthesis of ceria support (precipitation or hydrothermal method) and incorporating different loadings of Co on either support. The synthesis route results in ceria carriers of different physicochemical properties, further modified after Co incorporation. Hydrothermal synthesis leads to an improved CeO_2 support, with smaller crystallite size, larger surface area and enhanced reducibility. The ceria support and, thus, the synthetic procedure is important in terms of dispersing the active Co component (smaller Co_3O_4 crystallites formed on the CeO_2-H support) and exhibiting higher oxygen mobility (O_2 desorption peak of larger area, additionally appearing at lower temperature), as deduced by comparing the 15Co/CeO_2-H and 15Co/CeO_2-P catalytic samples. In summary, a nanocomposite catalyst Co_3O_4/CeO_2-H exhibits higher activity in complete oxidation of CH_4, than Co_3O_4/CeO_2-P and pure CeO_2 materials. The higher dispersion of the deposited Co species and the enhanced reducibility of Co-Ce catalysts advocate synergistic effects of CeO_2 nanorods and the supported Co_3O_4 nanoparticles. Moreover, the addition of water vapor in the reaction stream only slightly inhibited the catalytic activity of the optimum selected 15Co/CeO_2-H catalyst, a fact mainly observed at lower temperature ranges. Thus, the corresponding material is very promising for methane oxidation purposes, as it additionally presented good water resistance properties in wet CH_4 combustion.

Author Contributions: The experimental work was conceptualized and designed by E.P. and E.I.; S.D. and E.P. synthesized all catalytic materials, while S.D. helped in performing the evaluation tests and E.P. helped in characterization studies; A.L. provided all resources, while E.P. and E.I. analyzed all experimental results and wrote the original draft. The manuscript was amended and supplemented by all authors. All authors have given approval for the final version of the manuscript.

Funding: This research received no external funding.

Acknowledgments: The authors would like to thank Ioannis Tsiaousis for performing the HR-TEM characterization.

Conflicts of Interest: The authors declare no conflict of interest.

References

1. Clark, N.N.; Johnson, D.R.; McKain, D.L.; Wayne, W.S.; Li, H.; Rudek, J.; Mongold, R.A.; Sandoval, C.; Covington, A.N.; Hailer, J.T. Future methane emissions from the heavy-duty natural gas transportation sector for stasis, high, medium, and low scenarios in 2035. *J. Air Waste Manag.* **2017**, *67*, 1328–1341. [CrossRef] [PubMed]
2. Farrauto, R.J. Chemistry. Low-temperature oxidation of methane. *Science* **2012**, *337*, 659–660. [CrossRef] [PubMed]
3. Gélin, P.; Primet, M. Complete oxidation of methane at low temperature over noble metal based catalysts: A review. *Appl. Catal. B* **2002**, *39*, 1–37. [CrossRef]
4. Filip, M.; Todorova, S.; Shopska, M.; Ciobanu, M.; Papa, F.; Somacescu, S.; Munteanu, C.; Parvulescu, V. Effects of Ti loading on activity and redox behavior of metals in PtCeTi/KIT-6 catalysts for CH_4 and CO oxidation. *Catal. Today* **2018**, *306*, 138–144. [CrossRef]
5. Chen, J.; Arandiyan, H.; Gao, X.; Li, J. Recent Advances in Catalysts for Methane Combustion. *Catal. Surv. Asia* **2015**, *19*, 140–171. [CrossRef]

6. Liu, Y.; Deng, J.; Xie, S.; Wang, Z.; Dai, H. Catalytic removal of volatile organic compounds using ordered porous transition metal oxide and supported noble metal catalysts. *Chin. J. Catal.* **2016**, *37*, 1193–1205. [CrossRef]
7. Dou, J.; Tang, Y.; Nie, L.; Andolina, C.M.; Zhang, X.; House, S.; Li, Y.; Yang, J.; Tao, F. Complete oxidation of methane on Co_3O_4/CeO_2 nanocomposite: A synergic effect. *Catal. Today* **2018**, *311*, 48–55. [CrossRef]
8. Hu, L.; Peng, Q.; Li, Y. Selective Synthesis of Co_3O_4 Nanocrystal with Different Shape and Crystal Plane Effect on Catalytic Property for Methane Combustion. *J. Am. Chem. Soc.* **2008**, *130*, 16136–16137. [CrossRef] [PubMed]
9. Jia, Y.; Wang, S.; Lu, J.; Luo, M. Effect of structural properties of mesoporous Co_3O_4 catalysts on methane combustion. *Chem. Res. Chin. Univ.* **2016**, *32*, 808–811. [CrossRef]
10. Teng, F.; Chen, M.D.; Li, G.Q.; Teng, Y.; Xu, T.G.; Hang, Y.C.; Yao, W.Q.; Santhanagopalan, S.; Meng, D.D.; Zhu, Y.F. High combustion activity of CH_4 and catalluminescence properties of CO oxidation over porous Co_3O_4 nanorods. *Appl. Catal. B* **2011**, *110*, 133–140. [CrossRef]
11. Wang, F.; Zhang, L.; Xu, L.; Deng, Z.; Shi, W. Low temperature CO oxidation and CH_4 combustion over Co_3O_4 nanosheets. *Fuel* **2017**, *203*, 419–429. [CrossRef]
12. Choi, H.J.; Moon, J.; Shim, H.B.; Han, K.S.; Lee, E.G.; Jung, K.D. Preparation of Nanocrystalline CeO_2 by the Precipitation Method and Its Improved Methane Oxidation Activity. *J. Am. Ceram. Soc.* **2006**, *89*, 343–345. [CrossRef]
13. Nabih, N.; Schiller, R.; Lieberwirth, I.; Kockrick, E.; Frind, R.; Kaskel, S.; Weiss, C.K.; Landfester, K. Mesoporous CeO_2 nanoparticles synthesized by an inverse miniemulsion technique and their catalytic properties in methane oxidation. *Nanotechnology* **2011**, *22*, 135606. [CrossRef] [PubMed]
14. Qiao, D.; Lu, G.; Guo, Y.; Wang, Y.; Guo, Y. Effect of water vapor on the CO and CH_4 catalytic oxidation over CeO_2-MOx (M = Cu, Mn, Fe, Co, and Ni) mixed oxide. *J. Rare Earths* **2010**, *28*, 742. [CrossRef]
15. Shi, L.M.; Chu, W.; Qu, F.F.; Hu, J.Y.; Li, M.M. Catalytic performance for methane combustion of supported Mn-Ce mixed oxides. *J. Rare Earths* **2008**, *26*, 836. [CrossRef]
16. Li, Y.; Guo, Y.; Xue, B. Catalytic combustion of methane over M (Ni, Co, Cu) supported on ceria–magnesia. *Fuel Process. Technol.* **2009**, *90*, 652. [CrossRef]
17. Liotta, L.F.; Carlo, G.D.; Pantaleo, G.; Deganello, G. Co_3O_4/CeO_2 and Co_3O_4/CeO_2–ZrO_2 composite catalysts for methane combustion: Correlation between morphology reduction properties and catalytic activity. *Catal. Commun.* **2005**, *6*, 329. [CrossRef]
18. Honkanen, M.; Kärkkäinen, M.; Kolli, T.; Heikkinen, O.; Viitanen, V.; Zeng, L.; Jiang, H.; Kallinen, K.; Huuhtanen, M.; Keiski, R.L.; et al. Accelerated deactivation studies of the natural-gas oxidation catalyst—Verifying the role of sulfur and elevated temperature in catalyst aging. *Appl. Catal. B Environ.* **2016**, *182*, 439–448. [CrossRef]
19. Colussi, S.; Arosio, F.; Montanari, T.; Busca, G.; Groppi, G.; Trovarelli, A. Study of sulfur poisoning on Pd/Al_2O_3 and $Pd/CeO_2/Al_2O_3$ methane combustion catalysts. *Catal. Today* **2010**, *155*, 59–65. [CrossRef]
20. Monai, M.; Montini, T.; Melchionna, M.; Duchoň, T.; Kús, P.; Chen, C.; Tsud, N.; Nasi, L.; Prince, K.C.; Veltruská, K.; et al. The effect of sulfur dioxide on the activity of hierarchical Pd-based catalysts in methane combustion. *Appl. Catal. B Environ.* **2017**, *202*, 72–83. [CrossRef]
21. Wilburn, M.S.; Epling, W.S. Sulfur deactivation and regeneration of mono- and bimetallic Pd-Pt methane oxidation catalysts. *Appl. Catal. B Environ.* **2017**, *206*, 589–598. [CrossRef]
22. Friberg, I.; Sadokhina, N.; Olsson, L. Complete methane oxidation over Ba modified Pd/Al_2O_3: The effect of water vapor. *Appl. Catal. B Environ.* **2018**, *231*, 242–250. [CrossRef]
23. Toso, A.; Colussi, S.; Padigapaty, S.; de Leitenburg, C.; Trovarelli, A. High stability and activity of solution combustion synthesized Pd-based catalysts for methane combustion in presence of water. *Appl. Catal. B Environ.* **2018**, *230*, 237–245. [CrossRef]
24. Yashnik, S.A.; Chesalov, Y.A.; Ishchenko, A.V.; Kaichev, V.V.; Ismagilov, Z.R. Effect of Pt addition on sulfur dioxide and water vapor tolerance of Pd-Mn-hexaaluminate catalysts for high-temperature oxidation of methane. *Appl. Catal. B Environ.* **2017**, *204*, 89–106. [CrossRef]
25. Liu, F.; Sang, Y.; Ma, H.; Li, Z.; Gao, Z. Nickel oxide as an effective catalyst for catalytic combustion of methane. *J. Nat. Gas Sci. Eng.* **2017**, *41*, 1–6. [CrossRef]

26. Lykaki, M.; Pachatouridou, E.; Iliopoulou, E.F.; Carabineiro, S.A.C.; Konsolakis, M. Impact of the synthesis parameters on the solid state properties and the CO oxidation performance of ceria nanoparticles. *RSC Adv.* **2017**, *7*, 6160–6169. [CrossRef]
27. Lykaki, M.; Pachatouridou, E.; Carabineiro, S.A.C.; Iliopoulou, E.; Andriopoulou, C.; Kallithrakas-Kontos, N.; Boghosian, S.; Konsolakis, M. Ceria nanoparticles shape effects on the structural defects and surface chemistry: Implications in CO oxidation by Cu/CeO_2 catalysts. *Appl. Catal. B Environ.* **2018**, *230*, 18–28. [CrossRef]
28. Konsolakis, M.; Carabineiro, S.A.C.; Marnellos, G.E.; Asad, M.F.; Soares, O.S.G.P.; Pereira, M.F.R.; Órfão, J.J.M.; Figueiredo, J.L. Effect of cobalt loading on the solid state properties and ethyl acetate oxidation performance of cobalt-cerium mixed oxides. *J. Colloid Interf. Sci.* **2017**, *496*, 141–149. [CrossRef] [PubMed]
29. Wang, X.; Zhao, S.; Zhang, Y.; Wang, Z.; Feng, J.; Song, S.; Zhang, H. CeO_2 nanowires self-inserted into porous Co_3O_4 frameworks as high-performance "noble metal free" hetero-catalysts. *Chem. Sci.* **2016**, *7*, 1109–1114. [CrossRef] [PubMed]
30. Xue, L.; Zhang, C.; He, H.; Teraoka, Y. Catalytic decomposition of N_2O over CeO_2 promoted Co_3O_4 spinel catalyst. *Appl. Catal. B Environ.* **2007**, *75*, 167–174. [CrossRef]
31. Bai, B.; Arandiyan, H.; Li, J. Comparison of the performance for oxidation of formaldehyde on nano-Co_3O_4, 2D-Co_3O_4, and 3D-Co_3O_4 catalysts. *Appl. Catal. B Environ.* **2013**, *142–143*, 677–683. [CrossRef]
32. Lou, Y.; Ma, J.; Cao, X.; Wang, L.; Dai, Q.; Zhao, Z. Promoting Effects of In_2O_3 on Co_3O_4 for CO Oxidation: Tuning O_2 Activation and CO Adsorption Strength Simultaneously. *ACS Catal.* **2014**, *4*, 4143–4152. [CrossRef]
33. Sun, Y.; Liu, J.; Song, J.; Huang, S.; Yang, N.; Zhang, J. Exploring the Effect of Co_3O_4 Nanocatalysts with Different Dimensional Architectures on Methane Combustion. *ChemCatChem* **2016**, *8*, 540. [CrossRef]
34. Hu, L.H.; Peng, Q.; Li, Y.D. Low-temperature CH_4 Catalytic Combustion over Pd Catalyst Supported on Co_3O_4 Nanocrystals with Well-Defined Crystal Planes. *ChemCatChem* **2011**, *3*, 868–874. [CrossRef]
35. Li, Y.; Shen, W. Morphology-dependent nanocatalysts: Rod-shaped oxides. *Chem. Soc. Rev.* **2014**, *43*, 1543–74. [CrossRef] [PubMed]
36. Manto, M.J.; Xie, P.; Wang, C. Catalytic Dephosphorylation Using Ceria Nanocrystals. *ACS Catal.* **2017**, *7*, 1931–1938. [CrossRef]

Article

Electrochemical Promotion of Nanostructured Palladium Catalyst for Complete Methane Oxidation

Yasmine M. Hajar, Balaji Venkatesh and Elena A. Baranova *

Department of Chemical and Biological Engineering, Centre for Catalysis Research and Innovation (CCRI), University of Ottawa, 161 Louis-Pasteur, Ottawa, ON K1N 6N5, Canada; yhaja059@uottawa.ca (Y.M.H.); bvenk044@uottawa.ca (B.V.)

* Correspondence: elena.baranova@uottawa.ca; Tel.: +1-6135625800 (ext. 6302); Fax: +1-6135625172

Received: 4 December 2018; Accepted: 4 January 2019; Published: 6 January 2019

Abstract: Electrochemical promotion of catalysis (EPOC) was investigated for methane complete oxidation over palladium nano-structured catalysts deposited on yttria-stabilized zirconia (YSZ) solid electrolyte. The catalytic rate was evaluated at different temperatures (400, 425 and 450 °C), reactant ratios and polarization values. The electrophobic behavior of the catalyst, i.e., reaction rate increase upon anodic polarization was observed for all temperatures and gas compositions with an apparent Faradaic efficiency as high as 3000 (a current application as low as 1 μA) and maximum rate enhancement ratio up to 2.7. Temperature increase resulted in higher enhancement ratios under closed-circuit conditions. Electrochemical promotion experiments showed persistent behavior, where the catalyst remained in the promoted state upon current or potential interruption for a long period of time. An increase in the polarization time resulted in a longer-lasting persistent promotion (p-EPOC) and required more time for the reaction rate to reach its initial open-circuit value. This was attributed to continuous promotion by the stored oxygen in palladium oxide, which was formed during the anodic polarization in agreement with p-EPOC mechanism reported earlier.

Keywords: electrochemical promotion; NEMCA; palladium; ionic promoter; nanoparticles; yttria-stabilized zirconia

---

1. Introduction

Natural Gas Vehicles (NGVs) have gained considerable attention in the last decade due to much lower greenhouse gas emissions and lower price of methane compared to diesel or gasoline. Not only CH_4 is abundant in natural gas form, but methane can also be produced using anaerobic digestion technologies of bio-derived sources [1–4]. Despite lower emissions of NGVs they often suffer from incomplete methane combustion. Because methane is also 23 times more potent in warming the atmosphere than carbon dioxide its complete conversion to CO_2 is paramount. Therefore, the development of efficient low temperature catalysts for deep oxidation of methane (CH_4) has recently attracted significant attention [5–10].

Palladium-based catalysts are considered the most efficient for methane activation in excess of oxygen and their activity depends on temperature, methane/oxygen ratio, and catalyst surface oxidation state and composition [5]. The nature of the active surface sites PdO_x vs. Pd was a subject of several studies [6,7]. It was shown that chemisorbed oxygen on Pd metal is poorly active, whereas Pd oxidation with an optimum of 3–4 monolayers forms an active PdO catalyst. Chemisorption of a first layer of oxygen is fast; however, partial bulk oxidation is relatively slow [7].

The innovative field of catalysis that could boost complete methane oxidation reaction over Pd is electrochemical promotion of catalysis (EPOC) or non-Faradaic electrochemical modification of catalytic activity (NEMCA). This general, well-established phenomenon in catalysis aims at controlling in-situ both the activity and the selectivity of a catalyst through application of electric stimuli [8–11].

EPOC is observed with solid electrolyte materials that serve as catalyst support. Ions contained in these electrolytes (O^{2-}, H^+, Na^+ OH^-, etc.) are electrochemically pumped to the catalyst surface, where they act as promoting species leading to modification of catalyst electronic properties and as a result its catalytic activity and selectivity. More precisely, applying an anodic polarization results in the strengthening of electron-donor adsorbates, e.g., chemisorbed methane, and weakening of the binding strength of electron-accepting adsorbates, e.g., dissociatively chemisorbed oxygen [12]. The resulting electrochemical activation magnitude is much higher than that predicted by Faraday's law. [13,14]. The increase in the catalytic rate, Δr (mol O/s) divided by the electrochemical rate, I/nF (I is current, F is Faraday's constant and n is number of electrons, 2 for O^{2-}) is denoted as the apparent Faradaic efficiency, Λ, and the process is considered non-Faradaic when $|\Lambda|$ is greater than 1. Another parameter used to quantify EPOC is the rate enhancement ratio, ρ, which is the ratio between the promoted closed-circuit catalytic rate, r and the unpromoted open-circuit catalytic rate, r_o.

The electrochemical promotion of complete methane oxidation was investigated on palladium catalysts prepared using various methods as summarized in Table 1 [15–23]. Electrochemical promotion of Pd thick film catalyst electrode prepared using wet impregnation was investigated in the temperature range of 470–600 °C [18]. The rate enhancement of 40% and an apparent Faradaic efficiency of 1.85 were found in this work. The addition of CeO_2 layer between the YSZ solid-electrolyte and Pd film catalyst increased the open-circuit catalytic rate but decreased the apparent Faradaic efficiency due to the higher electric resistance [18]. Another study on a Pd film catalyst prepared using commercial organometallic paste showed higher enhancement ratio (ρ = 5.6) and Faradaic efficiency (Λ = 579) at 560 °C; however, instability of the catalyst over time and rapid deactivation within 900 min of experiment was also noted [20]. Furthermore, the effect of CeO_2 layer was studied in [24]. It was shown that presence of ceria increased catalytic activity of Pd due to the formation of an active PdO phase, which was stabilized by CeO_2 acting as a continuous source of oxygen, similarly to the oxygen migration from the YSZ electrolyte under EPOC. In another study, an addition of porous YSZ layer between Pd film-catalyst and the dense YSZ solid-electrolyte resulted in a high open-circuit catalytic rate. The authors reported an enhancement ratio of 1.2 and apparent Faradaic efficiency, Λ, of 17 under fuel-rich conditions [23]. EPOC of sputtered Pd catalyst-electrode was compared to impregnated Pd for methane complete oxidation. The sputtered catalysts showed slightly higher Λ of 12 but similar enhancement ratio (ρ = 1.6) [21]. The effect of metal loading and catalyst thickness on EPOC of Pd was studies on Pd catalyst prepared by physical vapor deposition (PVD). It was found that metal loading and catalyst thickness have significant effect on the open-circuit rate and electrochemical promotion, where the thinner films resulted in the highest reaction rates per gram of catalyst at 500 °C [17]. A scaling-up of the system was attempted by electroless deposition of Pd in the channels of a YSZ monolith honeycomb; however, decrease in the conversion of methane occurred under positive and negative polarization [19].

Therefore, from previous EPOC studies on Pd for complete CH_4 oxidation, it is clear that the catalyst preparation method has a strong influence on Pd morphology, structure, oxidation state and as a result, on its catalytic activity, degree of promotion and stability under open and closed circuit conditions. Furthermore, from the practical point of view it is essential to work with low loadings of noble-metal catalysts that exhibit high dispersion and large active surface area In the present work, we report electrochemical promotion of nano-structured, highly dispersed Pd catalyst prepared by polyol reduction method for CH_4 complete oxidation in the temperature range of 400 and 450 °C and various gas compositions.

Table 1. Summary of EPOC tests performed for methane oxidation on Pd-YSZ support.

Catalyst Synthesis Method	Loading	I or U Applied	T/°C	P_{CH4}/kPa	P_{O2}/kPa	Total Flow/ccm	Rate Change/mol O.s^{-1} 10^{-8}	ρ	Λ	Authors, Year & Ref.
Paste coating	n/a	+300 µA	400	2.75	1.55	n/a	4.52 to 20.5	4.5	103	[a] Giannikos et al. (1998) [15]
Paste coating	n/a	+1 V	400	2.6	1.9	n/a	0.295 to 20	68	153	[a] Frantzis et al. (2000) [16]
PVD	24 µg	+100 µA	500	2	10	166	7.3 to 20.4	2.8	258	Roche et al. (2008) [17]
Thermal decomposition	1.1 mg/cm^2	+10 mA	600	0.4	1	150	0.47 to 0.68	2.6	<1	[a,b] Jimenez-Borja et al. (2009) [18]
Electroless deposition	5 mg total	+100 µA	400	2	10	166	136.4 to 135.2	0.99	−23	Roche et al. (2010) [19]
Paste coating	7 mg/cm^2	+25 µA	560	0.4	1.2	150	1.6 to 9.1	5.6	579	[a] Jimenez-Borja et al. (2011) [20]
Sputtered	0.4 mg/cm^2	+1 mA	350	1.3	4.5	200	11.4 to 18.2	1.6	12	Matei et al. (2012) [21]
Impregnation	0.4 mg/cm^2	+300 µA	350	1.3	4.5	200	22 to 26	1.18	25	Jimenez-Borja et al. (2012) [22]
Impregnation	0.4 mg/cm^2	+5 mA	400	1.4	2.8	200	135 to 158	1.2	17	Matei et al. (2013) [23]
Polyol method	0.3 mg/cm^2	+0.5 V	425	2	4	100	6 to 16	2.66	383	[a] This work

[a] Continuous increase in closed-circuit rate post EPOC; [b] Electrolysis effect.

2. Results and Discussion

Transmission electron microscopy (TEM) was used to determine the palladium morphology and particle size (Figure 1a,b). The resulting Pd particles are spherical in shape with a diameter of approximately 5 nm, that are coalesced together in larger aggregates of roughly 50 nm in size. Figure 1c,d shows SEM images of as-prepared Pd/YSZ catalysts and the same catalyst after catalytic measurements under open circuit and EPOC conditions. It can be seen that as-prepared catalyst-electrode forms a highly dispersed, non-continuous layer on YSZ surface that consists of fine grains and pores. After the reaction, the "spent" catalyst shows much larger catalyst islands indicating a change in the morphology due to the catalyst agglomeration that takes place during the reaction. The resulting energy dispersive X-ray spectroscopy (EDS) spectrum (Figure 2a) of as-prepared Pd shows that the only element present is Pd. The x-ray diffraction pattern (XRD) contains several diffraction peaks of Pd (111), (200) and (220) corresponding to face-centered cubic (fcc) structure typical for bulk palladium metal. The crystallite size of Pd found from Pd (111) was 8 nm.

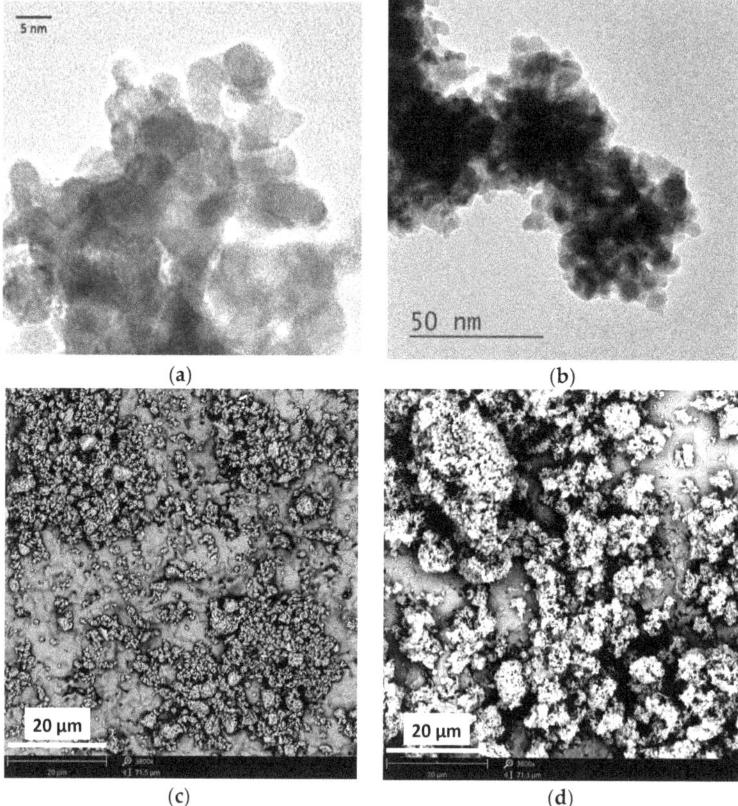

Figure 1. (**a**,**b**) TEM images of stand-alone Pd NPs at 5 nm and 50 nm scale; (**c**,**d**) SEM images of as-prepared (**c**) and post-experiment (**d**) Pd catalyst deposited on YSZ solid electrolyte.

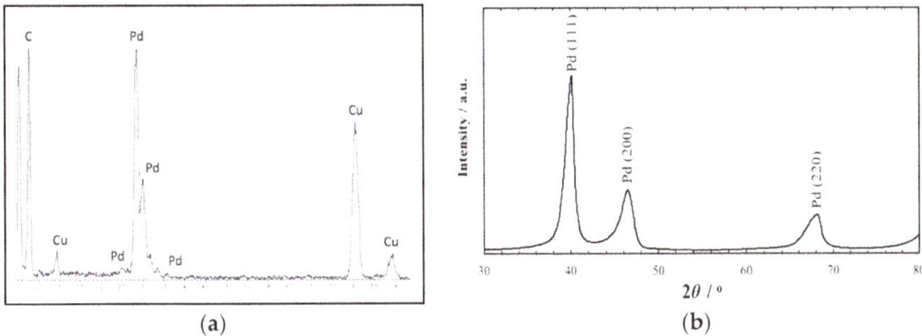

Figure 2. (**a**) EDS spectrum and (**b**) XRD pattern of Pd catalyst.

The open-circuit catalytic rate of methane oxidation was tested at two different temperatures and at various gas compositions (Figure 3). It can be seen that for both temperatures the rate increased as a function of oxygen partial pressure. The increase was more significant at 4 kPa of methane where oxygen-to-methane partial pressures ratios were lower. The rate increase in fuel-rich condition is indicative of a Langmuir-Hinshelwood mechanism [15] where methane is able to competitively adsorb on palladium as seen at 450 °C under 4 kPa of CH_4, while the quasi-stable value at higher ratio of oxygen (at 2 kPa of CH_4) can be explained by an Eley-Rideal mechanism as CH_4 reacts on the oxygen covered surface [25].

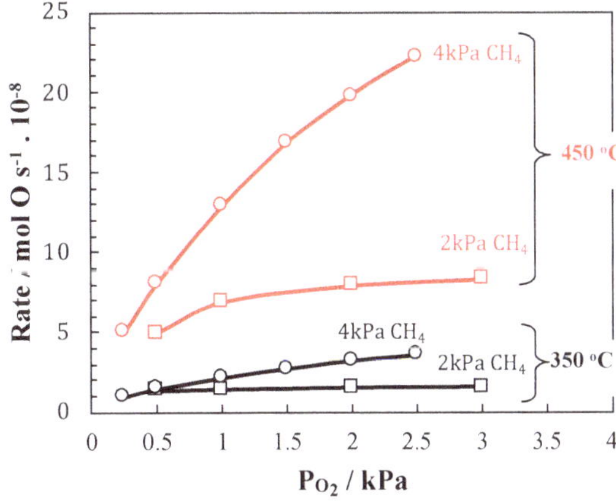

Figure 3. Effect of oxygen-to-methane ratio on catalytic rate at 350 and 450 °C.

Figure 4 shows the transient rate response to an application of positive current (20 μA) between the Pd working electrode and the counter electrode at 450 °C. Under open-circuit conditions (t < 0.5 h), the catalytic rate was at 8.35×10^{-8} mol/s. When a constant current was imposed, the reaction rate gradually increased due to pumping of O^{2-} promoters to the catalyst surface. After 2 h of polarization, the closed-circuit rate increase was 180% higher than its corresponding open-circuit value. In addition, the non-Faradaic behavior resulted in an apparent Faradaic efficiency, Λ, of 610 denoting that the back-spillover of O^{2-} at a I/2F rate gave a 610 times increase in the overall oxidation rate [26]. It should be noted that slight increase in the catalyst-working electrode potential (U_{WR}) was observed upon positive polarization (from 0.39 to 0.42 V). Furthermore, the slow reaction rate increase did not reach a steady-state value even after 2 h of applied polarization and took over 1 h to reach the open circuit value observed before polarization. According to the mechanism of EPOC the reaction rate is due to the supply of oxygen ionic species from YSZ and the formation of an effective double layer at the surface of the catalyst that changes the work function of Pd leading to weakening of the chemisorbed oxygen bond strength thus facilitating the $-O_2C$ desorption from the catalyst surface.

A similar behavior was observed under potentiostatic conditions (U_{WR} = 0.25 V) and a stoichiometric flow of reactants at 425 °C (Figure 5). The catalytic rate slowly increased in value until it reached 7.0×10^{-8} mol s^{-1} after 2 h. As in the galvanostatic conditions (Figure 4) the closed-circuit reaction rate was continuously increasing with time without reaching a steady-state. After 2 h, the rate enhancement ratio of 1.31 and, an apparent Faradaic efficiency of 1107 were obtained. The high Λ value is due to the low current that passed through the cell, which was sufficient to promote Pd catalyst.

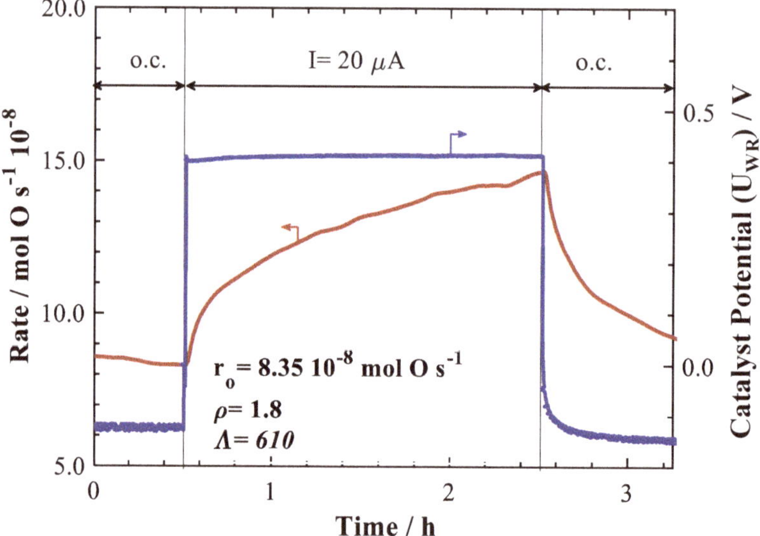

Figure 4. Transient rate response of Pd nanoparticles to a current step change. o.c.: open-circuit. Conditions: T = 450 °C, 2 kPa of CH_4 and 4 kPa of O_2. Flow rate: 100 ccm.

The continuous increase of the catalytic rate in Figures 4 and 5 can be explained by the continuous oxidation of palladium catalyst to PdO_x, which makes the catalyst more active for methane oxidation. In Figure 5, this continuous increase in catalytic rate occurred simultaneously with a decrease in current at a constant applied potential value indicating Pd oxide formation. Upon current or potential interruption the rate slowly returned to its initial state, because oxygen species stored in PdO_x continued acting as sacrificial promoters for methane oxidation reaction.

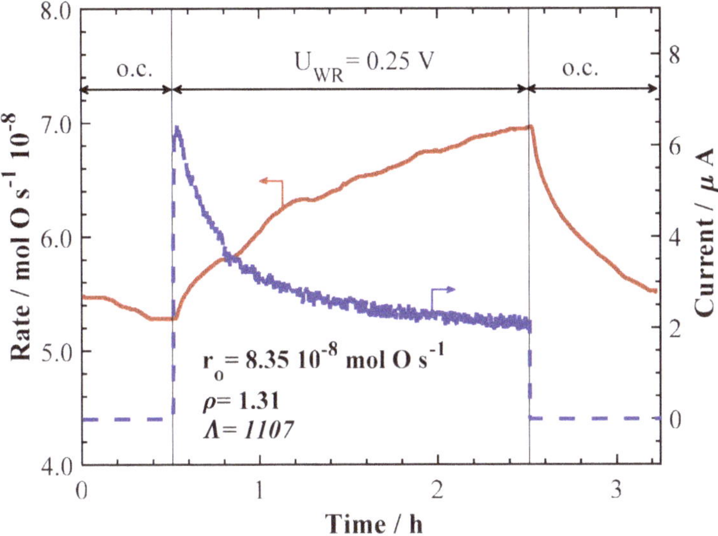

Figure 5. Transient rate response of Pd nanoparticles to potential step changes. o.c.: open-circuit. Conditions: T = 425 °C, 2 kPa of CH_4 and 4 kPa of O_2. Flow rate: 100 ccm.

To confirm PdO_x formation and oxygen storage effect on p-EPOC, the catalyst was polarized for a different duration 3, 6 and 10 h. As seen in Figure 6, the closed-circuit catalytic rate was continuously increasing with time under constant current (40 μA for 3 h) and potential (0.5 V for 6 and 10 h) application. It can be seen that even after ten hours of polarization, the closed-circuit rate kept on rising without reaching a steady-state. At the same time, the longer polarization time resulted in longer decrease of the open-circuit rate after polarization was stopped. A proportional relationship (as shown in the inset Figure 6) was found between the duration of EPOC and the time to reach the open-circuit rate $r_{o.c.}$. The slope of this relationship was 0.5. This persistent electrochemical promotion (p-EPOC) is due to the stored oxygen ions in PdO_x that act as sacrificial promoters when the electrical circuit is open [27].

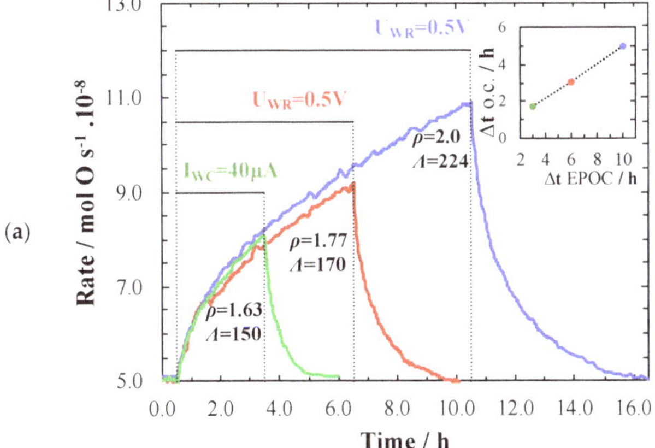

Figure 6. *Cont.*

(b)

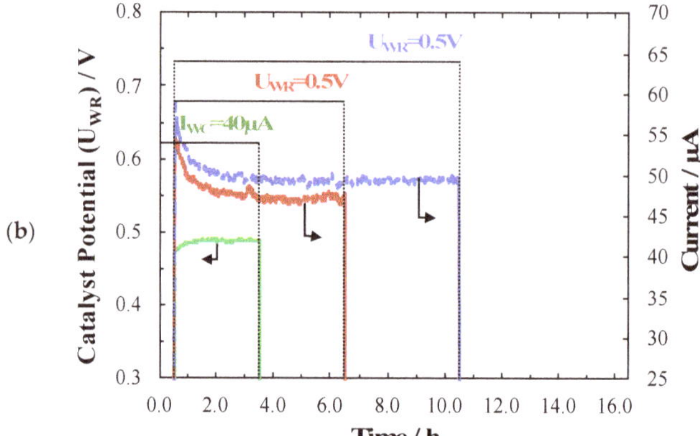

Figure 6. (**a**) Transient rate response of Pd at different duration of EPOC and (**b**) the potential/current read at potentiostatic or galvanostatic application. Conditions: T = 425 °C, 2 kPa of CH_4 and 4 kPa of O_2. Flow rate: 100 ccm.

This indicates that during the positive polarization two parallel processes take place: i. Migration of $O^{\delta-}$ promoters to the gas exposed catalyst surface and ii PdO_x formation at the three-phase boundary (tpb) according to the electrochemical reaction:

$$Pd + xO^{2-} \rightarrow PdO_x + xe^- \quad (1)$$

Current decrease and potential increase upon positive potentiostatic (Figures 5 and 6) and glavanostatic polarization (Figure 4), respectively, confirms the formation of PdO_x. Palladium oxide has lower conductivity than Pd metal, therefore current that flows through the solid-state cell or potential difference of the working catalyst-electrode (U_{WR}) are the clear indication of an electrochemical oxide formation [28].

Figure 7 shows a transient rate response at a constant applied potential 0.5 V for 24 h. The reaction rate continuously increased for up to 20 h followed by 2 h of a steady-state rate and then a slight rate decrease. The continuous rate increase indicates constant catalyst activation due to the growth of PdO_x, whereas somewhat rate decrease after 22 h of polarization at 425 °C may be linked with morphology change observed for the "spent" catalysts (Figure 1d). The open-circuit rate took more than 6 h to return to its initial value showing a persistent promotional effect.

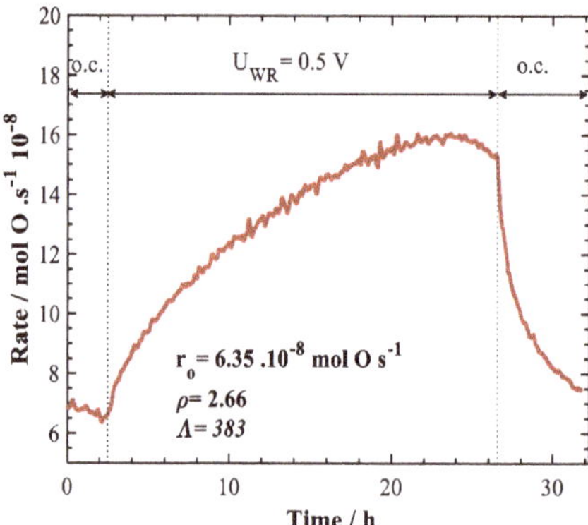

Figure 7. Long period transient rate response of Pd to a potentiostatic step change. o.c.: open-circuit. Conditions: T = 425 °C, 2 kPa of CH_4 and 4 kPa of O_2. Flow rate: 100 ccm.

Figure 8 shows the catalytic rate and catalyst potential change upon application of constant current as low as 1 μA. This resulted in the continuous catalytic rate increase for 1, 2 and 4 kPa of O_2, accompanied by catalyst potential (U_{WR}) increase, confirming palladium oxide formation. The corresponding Λ values are summarized in Figure 9.

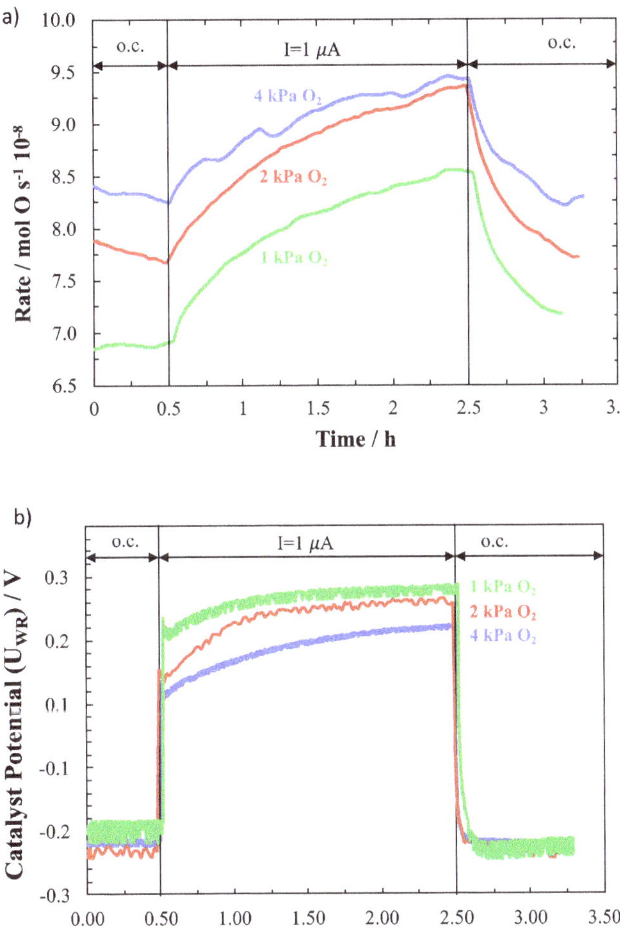

Figure 8. (**a**)Transient rate response of Pd catalyst at different O_2 partial pressure under 1 μA galvanostatic application and (**b**) the corresponding catalyst potential response. Conditions: T = 450 °C, 2 kPa of CH_4. Flow rate: 100 ccm.

In Figure 9, the effect of partial pressure on closed-circuit reaction rate was tested at different galvanostatic conditions. The highest increase in catalytic rate was found at slightly fuel-rich conditions resulting in a ρ value of 1.3; the higher rate can be explained by the advantaged adsorption of gaseous methane over oxygen. At this condition, gaseous methane can be expected to directly adsorb onto Pd, resulting in a competition between oxygen and methane adsorption following a Langmuir-Hinshelwood mechanism. In addition, the desorption of oxygen from the surface becomes facilitated at lower oxygen partial pressure in the atmosphere as the overall chemical potential of oxygen is reduced [29]. At a partial pressure ratio higher than the stoichiometric ratio, it is perceived that the catalytic rate increase is slightly lower. The slight decrease in the enhancement is due to the competing adsorption of oxygen on the surface of Pd, putting slight mass-transfer limitations on the chemisorption of CH_4.

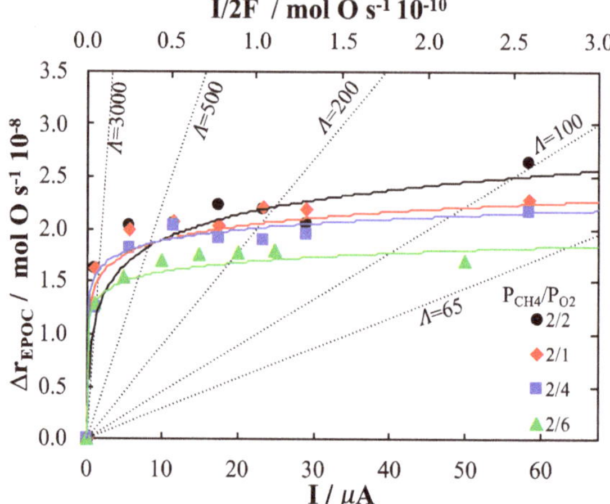

Figure 9. Current effect on catalytic rate in function of methane/oxygen ratio. T = 450 °C. Flow rate: 100 ccm.

Similarly, the effect of temperature on closed-circuit reaction rate was tested at different galvanostatic conditions. Figure 10 shows that there is an increase in catalytic rate as a function of temperature at all applied positive current values for 400, 425 and 450 °C. It can be noticed that upon application of small current of 1 μA, a significant rate increase was detected, resulting in a logarithmic-shape relationship of rate increase versus applied current. In addition, a highest value of ~2400 was found for the apparent Faradaic efficiency, constructing that a very minimal current was able to result in a change of the Pd surface oxidation state and hence the adsorption strength of methane reactant [15,16,18,20].

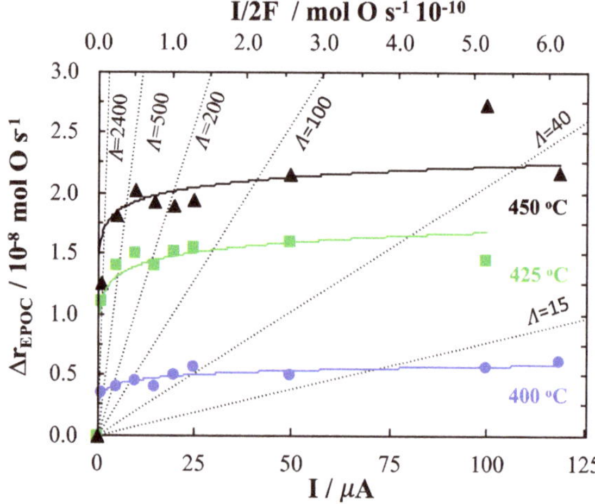

Figure 10. Current effect on catalytic rate in function of temperature: 400, 425 and 450 °C. P_{CH4} = 2 kPa and P_{O2} = 4 kPa. Flow rate: 100 ccm.

Table 1 compares EPOC of methane oxidation on Pd/YSZ found in this work to previous studies carried out on Pd catalyst-electrode deposited on YSZ solid-electrolyte. The table depicts where our results fall in comparison with previous experiments. It should be noted that the metal loading used in this work is the lowest in the temperature range of interest (T $\leq$ 450 °C), which is important for cold-start emission application. In agreement with previous work, Pd nanostructured catalyst synthesized by polyol method shows electrophobic type of EPOC, where only positive polarization promotes the reaction. Applied polarization led to the supply of $O^{\delta-}$ promoters from YSZ electrolyte to the catalyst surface, resulting in the formation of a more active phase of PdO, on the surface first and in the bulk gradually.

Ionic oxygen migration to the surface altered the adsorption properties of the catalyst surface, resulting in the weakening of gaseous oxygen adsorption and strengthening that of electron-donor methane. The alteration of the catalytic oxidation state was similar under both potentiostatic and galvanostatic application, which have resulted in a continuous increase in catalytic rate and a very slow open circuit rate decrease when the circuit was interrupted. This was explained by the formation of PdO during polarization and oxygen storage in its bulk, which was continuously providing the promoting oxygen species to the surface post-polarization according to p-EPOC mechanism [27].

3. Experimental

3.1. Synthesis of Pd Nanoparticles

Mono-metallic Pd NPs were synthesized using 0.133 g of palladium chloride (Fisher Scientific®, Canada) precursor salts dissolved in 25 mL of ethylene glycol and 0.8 M NaOH. The mixture was heated up to 160 °C and kept under stirring conditions for 3 h. The final colloidal solution was cooled down and washed repeatedly with ethanol.

3.2. Catalyst Characterization

X-ray diffraction (XRD) was performed on the fresh Pd sample using Rigaku Ultima IV multipurpose diffractometer (Rigaku, The Woodlands, TX, USA). The diffractometer was equipped with an X'Celerator detector with monochromatic CuKα radiation (λ = 1.5418 Å) at 40 kV and 44 mA with a divergence slit of 2/3 degree, a scan speed of 0.03 ^{o}/s and a scan step of 0.02 degrees between 30 and 80° 2θ.

The transmission electron microscopy (TEM) micrographs were obtained using JEOL JEM 2100F FETEM (JEOL, Peabody, MA, USA) operating with a field emission gun at an acceleration voltage of 200 kV. SEM micrographs were recorded using Phenom^{TM} SEM (Nanoscience Instruments, Virginia, USA). Additional elemental analysis was performed using the energy dispersive X-ray spectroscopy (EDS) attachment.

3.3. Electrochemical Cell and Reactor

The solid electrolyte is a 19 mm diameter and 1 mm thickness disk of 8 mol % Y_2O_3-stabilized ZrO_2 (YSZ) (TOSOH®, Grove city, OH, USA) fabricated following the procedure reported earlier [30]. Inert gold reference and counter electrodes were deposited on one side of the disk by applying thin gold paste coating (Gwent Group, Pontypool, UK) of 0.2 and 1 cm^2 surface areas, respectively. This was followed by annealing in air at 500 °C. The catalyst-working electrode was deposited on the other side of the solid electrolyte disk (1 cm^2 surface area) opposing to the counter electrode. To this end, mono-metallic Pd were dispersed in isopropanol and 10 µL of a suspension were deposited at a time with intermediate drying at room temperature. The resulting total metal loading was 0.5 mg of Pd on YSZ. Catalytic measurements were carried out at atmospheric pressure in the single-chamber capsule reactor reported earlier [31,32]. The working electrode side of the electrolyte was pressed against a gold mesh (1 cm^2) that served as a current collector, while the counter and reference electrodes were pressed directly against gold wires [33]. Two type K thermocouples (Omega®, Quebec, Canada) were placed in vicinity of the electrochemical cell, one for temperature control and one for data acquisition.

3.4. Catalytic and Electrochemical Measurements

The reaction gases were CH_4 (Linde, 99.99%), O_2 (Linde, 99.99%), and He (Linde, 99.997%) as a carrier gas. The total flow rate was constant at 100 mL min^{-1}. Gas composition was varied using MKS, 1259 C and 1261-C series flow meters and detected using non-dispersive infrared (NDIR) CO_2 gas analyzer (Horiba, VA-3000, Burlington, Canada). Constant electric current or potential were applied using a potentiostat-galvanostat (Arbin Instruments®, MSTAT, College Station, TX, USA) connected to the electrodes of solid-electrolyte electrochemical cell.

4. Conclusions

Electrochemical promotion of Pd nanostructured catalyst was investigated for the methane oxidation reaction in the 400–450 °C temperature range. The promotion of the catalytic rate of Pd NPs was achieved under anodic polarization. Upon various potentiostatic and galvanostatic tests, non-Faradaic enhancement was achieved, most notably at 450 °C, under the application of 1 μA, where Λ was equal to ~3000, higher than any previous EPOC study on Pd. Continuous increase in the reaction rate was found under EPOC conditions, due to the Pd oxide formation in the vicinity of the tpb. A proportional relationship was found between the duration of polarization and the post-polarization time required to reach the initial open-circuit rate value. Post polarization, a persistent promotion (p-EPOC) was observed due to the promotion of the reaction by the stored oxygen, which was accumulated during positive polarization. Overall, our work revealed interesting behavior of Pd synthesized by polyol method, providing further insight into the application of electrochemical promotion for complete methane oxidation with highly dispersed Pd.

Author Contributions: Conceptualization, E.A.B.; Data curation, Y.M.H and B.V.; Formal analysis, B.V.; Funding acquisition, E.A.B.; Methodology, Y.M.H.; Project administration, E.A.B.; Supervision, E.A.B.; Validation, Y.M.H.; Writing—original draft, Y.M.H.; Writing—review & editing, E.A.B.

Funding: Financial support from Natural Science and Engineering Research Council (NSERC) Canada is greatly acknowledged.

Conflicts of Interest: The authors declare no conflict of interest.

References

1. Demirbaş, A. Biomass resource facilities and biomass conversion processing for fuels and chemicals. *Energy Convers. Manag.* **2001**, *42*, 1357–1378. [CrossRef]
2. Cheng, H.; Hu, Y. Municipal solid waste (MSW) as a renewable source of energy: Current and future practices in China. *Bioresour. Technol.* **2010**, *101*, 3816–3824. [CrossRef] [PubMed]
3. Tripathi, M.; Sahu, J.N.; Ganesan, P. Effect of process parameters on production of biochar from biomass waste through pyrolysis: A review. *Renew. Sustain. Energy Rev.* **2016**, *55*, 467–481. [CrossRef]
4. Xiao, Y.; He, P.; Cheng, W.; Liu, J.; Shan, W.; Song, H. Converting solid wastes into liquid fuel using a novel methanolysis process. *Waste Manag.* **2016**, *49*, 304–310. [CrossRef] [PubMed]
5. Farrauto, R.J.; Hobson, M.C.; Kennelly, T.; Waterman, E.M. Catalytic chemistry of supported palladium for combustion of methane. *Appl. Catal. A Gen.* **1992**, *81*, 227–237. [CrossRef]
6. Ciuparu, D.; Lyubovsky, M.R.; Altman, E.; Pfefferle, L.D.; Datye, A. Catalytic combustion of methane over palladium-based catalysts. *Catal. Rev. Sci. Eng.* **2002**, *44*, 593–649. [CrossRef]
7. Burch, R.; Urbano, F.J. Investigation of the active state of supported palladium catalysts in the combustion of methane. *Appl. Catal. A Gen.* **1995**, *124*, 121–138. [CrossRef]
8. Boudart, M.; Djega-Mariadassou, G. *Kinetics of Heterogeneous Catalytic Reactions*; Princeton University Press: Princeton, NJ, USA, 1984.
9. Campbell, I.M. *Catalysis at Surfaces*; Springer Science & Business Media: New York, NY, USA, 1988.

10. Vayenas, C.G.; Bebelis, S.; Neophytides, S. Non-Faradaic Electrochemical Modification of Catalytic Activity. *J. Phys. Chem.* **1988**, *92*, 5083–5085. [CrossRef]
11. Ladas, S.; Kennou, S.; Bebelis, S.; Vayenas, C.G. Origin of non-faradaic electrochemical modification of catalytic activity. *J. Phys. Chem.* **1993**, *97*, 8845–8848. [CrossRef]
12. Nicole, J.; Tsiplakides, D.; Pliangos, C.; Verykios, X.E.E.; Comninellis, C.; Vayenas, C.G.G. Electrochemical Promotion and Metal–Support Interactions. *J. Catal.* **2001**, *204*, 23–34. [CrossRef]
13. Vayenas, C.G.; Bebelis, S.; Ladas, S. Dependence of catalytic rates on catalyst work function. *Nature* **1990**, *343*, 625–627. [CrossRef]
14. Vayenas, C.G.; Bebelis, S.; Pliangos, C.; Brosda, S.; Tsiplakides, D. *Electrochemical Activation of Catalysis: Promotion, Electrochemical Promotion, and Metal-Support Interactions*; Springer: New York, NY, USA, 2001. [CrossRef]
15. Giannikos, A.; Frantzis, A.D.; Pliangos, C.; Bebelis, S.; Vayenas, C.G. Electrochemical promotion of CH_4 oxidation on Pd. *Ionics (Kiel)* **1998**, *4*, 53–60. [CrossRef]
16. Frantzis, A.D.; Bebelis, S.; Vayenas, C.G. Electrochemical promotion (NEMCA) of CH_4 and C_2H_4 oxidation on Pd/YSZ and investigation of the origin of NEMCA via AC impedance spectroscopy. *Solid State Ionics* **2000**, *136–137*, 863–872. [CrossRef]
17. Roche, V.; Karoum, R.; Billard, A.; Revel, R.; Vernoux, P. Electrochemical promotion of deep oxidation of methane on Pd/YSZ. *J. Appl. Electrochem.* **2008**, *38*, 1111–1119. [CrossRef]
18. Jiménez-Borja, C.; Dorado, F.; de Lucas-Consuegra, A.; García-Vargas, J.M.; Valverde, J.L.J.L. Complete oxidation of methane on Pd/YSZ and Pd/CeO_2/YSZ by electrochemical promotion. *Catal. Today* **2009**, *146*, 326–329. [CrossRef]
19. Roche, V.; Revel, R.; Vernoux, P. Electrochemical promotion of YSZ monolith honeycomb for deep oxidation of methane. *Catal. Commun.* **2010**, *11*, 1076–1080. [CrossRef]
20. Jiménez-Borja, C.; Dorado, F.; Consuegra, A.D.; Vargas, J.M.G.; Valverde, J.L. Electrochemical Promotion of CH_4 Combustion over a Pd/CeO_2-YSZ Catalyst. *Fuel Cells* **2011**, *11*, 131–139. [CrossRef]
21. Matei, F.; Ciuparu, D.; Jiménez-Borja, C.; Dorado, F.; Valverde, J.L.; Brosda, S. Electrochemical promotion of methane oxidation on impregnated and sputtered Pd catalyst-electrodes deposited on YSZ. *Appl. Catal. B Environ.* **2012**, *127*, 18–27. [CrossRef]
22. Jimenez-Borja, C.; Brosda, S.; Matei, F.; Makri, M.; Delgado, B.; Sapountzi, F.; Ciuparu, D.; Dorado, F.; Valverde, J.L.; Vayenas, C.G. Electrochemical promotion of methane oxidation on Pd catalyst-electrodes deposited on Y_2O_3-stabilized-ZrO_2. *Appl. Catal. B Environ.* **2012**, *128*, 48–54. [CrossRef]
23. Matei, F.; Jimenez-Borja, C.; Canales-Vazquez, J.; Brosda, S.; Dorado, F.; Valverde, J.L.; Ciuparu, D. Enhanced electropromotion of methane combustion on palladium catalysts deposited on highly porous supports. *Appl. Catal. B Environ.* **2013**, *132–133*, 80–89. [CrossRef]
24. Jimenez-Borja, C.; Matei, F.; Dorado, F.; Valverde, J.L. Characterization of Pd catalyst-electrodes deposited on YSZ: Influence of the preparation technique and the presence of a ceria interlayer. *Appl. Surf. Sci.* **2012**, *261*, 671–678. [CrossRef]
25. Garbowski, E.; Feumi-Jantou, C.; Mouaddib, N.; Primet, M. Catalytic combustion of methane over palladium supported on alumina catalysts: Evidence for reconstruction of particles. *Appl. Catal. A Gen.* **1994**, *109*, 277–292. [CrossRef]
26. Vayenas, C.G.; Bebelis, S.; Yentekakis, I.V.; Lintz, H.-G. Non-faradaic electrochemical modification of catalytic activity: A status report. *Catal. Today* **1992**, *11*, 303–442. [CrossRef]
27. Nicole, J.; Comninellis, C.H. Electrochemical promotion of IrO_2 catalyst activity for the gas phase combustion of ethylene. *J. Appl. Electrochem.* **1998**, *28*, 223–226. [CrossRef]
28. Vayenas, C.; Brosda, S.; Pliangos, C. The double-layer approach to promotion, electrocatalysis, electrochemical promotion, and metal–support interactions. *J. Catal.* **2003**, *216*, 487–504. [CrossRef]
29. Reuter, K.; Scheffler, M. Composition, structure, and stability of RuO_2 (110) as a function of oxygen pressure. *Phys. Rev. B* **2001**, *65*, 035406. [CrossRef]
30. Gibson, I.R.; Dransfield, G.P.; Irvine, J.T.S. Sinterability of commercial 8 mol % yttria-stabilized zirconia powders and the effect of sintered density on the ionic conductivity. *J. Mater. Sci.* **1998**, *33*, 4297–4305. [CrossRef]

31. Hajar, Y.M.; Dole, H.A.; Couillard, M.; Baranova, E.A. Investigation of heterogeneous catalysts by electrochemical method: Ceria and titania supported iridium for ethylene oxidation. *ECS Trans.* **2016**, *72*, 161–172. [CrossRef]
32. Hajar, Y.M.; Patel, K.D.; Tariq, U.; Baranova, E.A. Functional equivalence of electrochemical promotion and metal support interaction for Pt and RuO_2 nanoparticles. *J. Catal.* **2017**, *352*, 42–51. [CrossRef]
33. Hajar, Y.M.; Houache, M.S.; Tariq, U.; Vernoux, P.; Baranova, E.A. Nanoscopic Ni interfaced with oxygen conductive supports: Link between electrochemical and catalytic studies. *Electrochem. Soc.* **2017**, *77*, 51–66. [CrossRef]

Article

Synthesis of Sulfur-Resistant TiO_2-CeO_2 Composite and Its Catalytic Performance in the Oxidation of a Soluble Organic Fraction from Diesel Exhaust

Na Zhang [1], Zhengzheng Yang [1,*], Zhi Chen [1], Yunxiang Li [1], Yunwen Liao [1,2], Youping Li [1], Maochu Gong [3] and Yaoqiang Chen [3,*]

1 College of Environmental Science and Engineering, China West Normal University, Nanchong 637009, China; nzhang@cwnu.edu.cn (N.Z.); ZCHEN0211@gmail.com (Z.C.); yx_li@cwnu.edu.cn (Y.L.); liao-yw@163.com (Y.L.); lyp920@163.com (Y.L.)

2 Chemical Synthesis and Pollution Control Key Laboratory of Sichuan Province, College of Chemistry and Chemical Engineering, China West Normal University, Nanchong 637009, China

3 Key Laboratory of Green Chemistry & Technology of the Ministry of Education, College of Chemistry, Sichuan University, Chengdu 610064, China; gongmaochu@scu.edu.cn

* Correspondence: zyang@cwnu.edu.cn (Z.Y.); nic7501@scu.edu.cn (Y.C.); Tel./Fax: +86-81-7256-8646 (Z.Y.); +86-28-8541-8451 (Y.C.)

Received: 13 May 2018; Accepted: 11 June 2018; Published: 14 June 2018

Abstract: Sulfur poisoning is one of the most important factors deteriorating the purification efficiency of diesel exhaust after-treatment system, thus improving the sulfur resistibility of catalysts is imperative. Herein, ceria oxygen storage material was introduced into a sulfur-resistant titania by a co-precipitation method, and the sulfur resistibility and catalytic activity of prepared TiO_2-CeO_2 composite in the oxidation of diesel soluble organic fraction (SOF) were studied. Catalytic performance testing results show that the CeO_2 modification significantly improves the catalytic SOF purification efficiency of TiO_2-CeO_2 catalyst. SO_2 uptake and energy-dispersive X-ray (EDX) results suggest that the ceria doping does not debase the excellent sulfur resistibility of bare TiO_2, the prepared TiO_2-CeO_2 catalyst exhibits obviously better sulfur resistibility than the CeO_2 and commercial CeO_2-ZrO_2-Al_2O_3. X-ray powder diffraction (XRD) and Raman spectra indicate that cerium ions can enter into the TiO_2 lattice and not form complete CeO_2 crystals. X-ray photoelectron spectroscopy (XPS), H_2-temperature programmed reduction (H_2-TPR) and oxygen storage capacity (OSC) testing results imply that the addition of CeO_2 in TiO_2-CeO_2 catalyst can significantly enhance the surface oxygen concentration and oxygen storage capacity of TiO_2-CeO_2.

Keywords: cerium-doped titania; sulfur-tolerant materials; organic compounds purification; diesel oxidation catalyst; vehicle exhaust

1. Introduction

Diesel engines have been widely used in passenger cars and vans, due to excellent fuel efficiency and durability. However, diesel exhaust gases, such as carbon monoxide (CO), unburned hydrocarbons (HC_x), nitrogen oxides (NO_x), particulate matter (PM) and soluble organic fraction (SOF), are considered major sources of air pollutants [1–3]. Among these hazardous pollutants, SOF are the heavy liquid hydrocarbons (C > 16 [4,5], aromatics and oxygenated compounds [6]) adsorbed on soot, which mainly come from unburned fuel and lube oils [7,8]. The content of SOF is known to vary with engine operating conditions and can reach about 5–60% of the whole mass of the particulate matter [6,9–11]. Due to the fact that diesel SOF contains types of polycyclic aromatic hydrocarbons (PAHs) [12,13] which are recognized as strong carcinogens [14,15], purifying the SOF from diesel exhaust is an important and essential work.

Diesel oxidation catalyst (DOC) was employed to accelerate the oxidation and purification of diesel exhaust gases of CO, HC_x and SOF. In recent decades, CeO_2-ZrO_2, Al_2O_3 and their mixed oxide-based catalysts were widely used as commercial DOC and displayed excellent catalytic CO and HC_x oxidation activity. Focusing on the purification of SOF, CeO_2-based oxygen storage materials (OSM) have been greatly impressed by considerable researchers due to the superior catalytic activity on hydrocarbons and SOF oxidation [16–18]. Meanwhile, controlled synthesized of nanostructured CeO_2-based catalysts and their catalytic performance in diesel soot oxidation are lucubrated [19,20]. However, CeO_2-based catalysts are easily poisoned by SO_2 [21–24]. And SO_2 is a subsistent in the diesel exhaust, since sulfur is present in almost all commercial diesel fuels [25–27], sulfur in fuels would be oxidized to SO_2 and then emitted from diesel engines [28,29]. Furthermore, sulfur poisoning resulting from sulfur species accumulation is more destructive, since even using ultra-low sulfur diesel (ULSD), cumulative exposure of a catalyst over its lifetime in a heavy-duty diesel may amount to kilograms of sulfur [30]. A large amount of sulfur species accumulation inevitably results in the blocking of channels of monolithic catalyst, and hence the strike of diesel exhaust after-treatment system. Thus, it is important and realistic for a diesel oxidation catalyst to improve the sulfur resistibility.

TiO_2 is known as an effective sulfur-resistant material [31–33], and our previous works [34–36] also prove that TiO_2-based diesel oxidation catalysts display excellent sulfur resistibility. However, TiO_2 is not active enough for catalytic diesel SOF combustion. Reports about the sulfur resistance catalyst for diesel SOF oxidation are still scarce. Considering all of this, in this work, CeO_2 was introduced in TiO_2 by a co-precipitation method, and its effects on sulfur resistibility and catalytic activity for diesel SOF combustion were investigated.

2. Results and Discussion

2.1. Sulfur Resistibility

The sulfur resistibility values of catalysts were measured by sulfur uptake testing. As shown in Figure 1, under SO_2 exposure, the weight of all samples increased with time; final weight increments tended to be flat after 1–3 h SO_2 exposure, except for CeO_2. After 4 h, chosen as representative exposure time of simulative 160,000 km vehicle aged catalyst, the final weight increments of TiO_2, TiO_2-CeO_2 and CeO_2 catalysts are 1.63 wt. %, 2.01 wt. % and 4.72 wt. %, respectively. The normalized sulfur uptake results are calculated by supposing 1 g of sample as standard, and the results are listed in Table 1. The normalized sulfur uptake values of TiO_2, TiO_2-CeO_2 and CeO_2 catalysts are 166 μg/m^2, 170 μg/m^2, and 891 μg/m^2, respectively. From the results it can be seen that the sulfur species accumulation is severe on CeO_2 catalyst but is slight on both TiO_2 and TiO_2-CeO_2 catalysts, which implies that the TiO_2-based catalysts (TiO_2 and TiO_2-CeO_2) present obviously better sulfur resistibility than the CeO_2 catalyst. Since the non-sulfating material of TiO_2 displays low SO_2 adsorption and hence relieves sulfate generation and exhibits superior sulfur resistibility [32,37]. Furthermore, the introduction of moderate amounts of ceria in TiO_2 has essentially no effect on the naturally excellent sulfur resistibility of TiO_2.

Table 1. Sulfur accumulation and normalized sulfur uptake of the TiO_2, TiO_2-CeO_2 and CeO_2 samples.

Samples	Surface Area [a] (m^2/g)	Sulfur Content (wt. %)		Normalized Sulfur Uptake [c] (μg/m^2)
		Sulfur Uptake	EDX [b]	
TiO_2	98	1.63	1.02	166
TiO_2-CeO_2	118	2.01	1.04	170
CeO_2	53	4.72	4.26	891

[a] Surface area was calculated by BET method from the N_2 adsorption-desorption results; [b] EDX results were obtained by detecting the simulative 160,000 km vehicle aged samples; [c] The normalized sulfur uptake = sulfur uptake/(100 × surface area) [37,38].

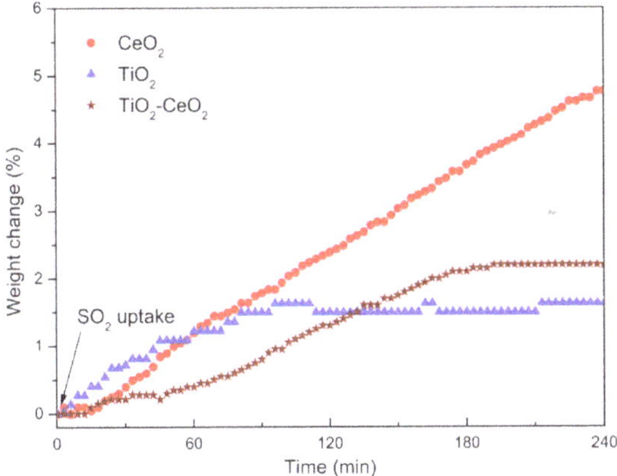

Figure 1. Sulfur uptake of the TiO_2, TiO_2-CeO_2 and CeO_2 catalysts.

Additionally, the accumulation amounts of sulfur species on simulative 160,000 km vehicle aged TiO_2, TiO_2-CeO_2 and CeO_2 catalysts tested by EDX (Table 1) are 1.02 wt. %, 1.04 wt. % and 4.26 wt. %, respectively, which shows the same trend with the sulfur uptake results. It is clear that the prepared TiO_2 and TiO_2-CeO_2 present significantly less sulfur species accumulation and better sulfur resistibility than the CeO_2 catalyst under the long-term exposure of diesel exhaust ambiences.

2.2. Catalytic Performance

Figure 2 shows the thermogravimetry-differential thermal analysis-differential thermogravimetry (TG-DTA-DTG) curves of the bulk lube without catalysts and the lube impregnated on catalysts; all of the DTA curves are the positive peaks indicating exothermic peaks, and all of the originally DTG negative peaks are inverted into positive peaks for a better readability of the graph. The combustion of lube without catalyst under air flow is shown in Figure 2a; the weight loss of bulk lube is tersely distinguished into two stages; about 90% of lube is deflagrated at 220–350 °C, and then the rest of 10% lubricating oil is consumed tardily after 350 °C till 500 °C, which implies that the commercial lube contains a fraction of hydrocarbons hard to pyrolyze (may come from the lubricant additive). The onset combustion temperature of $T_{10\%}$ (the temperature at which 10% of the initial lube is converted) is about 264 °C, the lube combustion fastest temperature of T_m (the temperature of weightlessness fastest point in DTG curves) is about 324 °C, and the final reaction temperature of T_f (the temperature of lube is completely converted) is about 507 °C. As shown in DTA curves of Figure 2a, a sharp and large exothermic peak is seen at about 325 °C which result from the rapid pyrolysis of bulk lube. Due to the decomposition of lube being an exothermic reaction, once ignition occurs, the heat continually increases and accumulates, and hence, most of lube is removed rapidly. The multiple peaks at 400–500 °C imply that the commercial lube contains multifarious hydrocarbons (lubricant additives) that are hard to decompose. Figure 2b plots the lubricant oxidation on CeO_2 catalyst. About 93% of the lube is rapidly oxidized between 140 °C and 280 °C, and the rest of 7% of lube is fully burnt at 280–340 °C. The onset combustion temperature of $T_{10\%}$ is about 162 °C, the fastest weightlessness temperature of T_m is about 186 °C and the final reaction temperature of T_f is about 322 °C. Figure 2c shows the decomposition of lubricant with TiO_2, the TG-DTG curves can be divided into four stages with different decomposition rates. About 16% of lube is decomposed at 205–266 °C, another 28% of the lube is burnt at 266–324 °C, about 49% of the lubricant is consumed at 324–396 °C and the rest of 7% of uninflammable lube is ignited between 396 °C and 420 °C. The onset combustion temperature of $T_{10\%}$ is about 249 °C, the final reaction temperature of T_f is about 420 °C, and three obviously fast weightlessness peaks of lube combustion are observed at about 252, 289 and 363 °C. For the TiO_2-CeO_2 catalyzed lube combustion (Figure 2d), about 97% of the lubricating oil is rapidly combusted between 180 °C and 330 °C, and the rest of about 3% of lube is burnt out at 334–362 °C. The $T_{10\%}$, T_m and T_f is about 212, 261 and 362 °C, respectively.

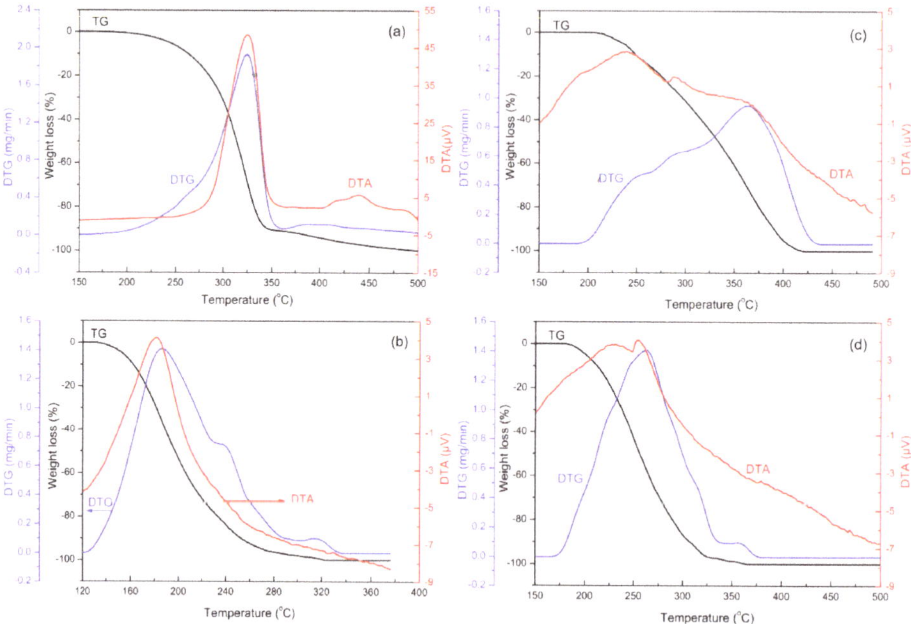

Figure 2. Simultaneous TG-DTA-DTG curves for simulating the catalytic performance for the combustion of SOF on (**a**) without catalyst; (**b**) CeO_2; (**c**) TiO_2 and (**d**) TiO_2-CeO_2 catalysts.

Due to the fact that the weight loss of lube can be ascribed either to evaporation or to combustion, the activity of prepared catalyst is also identified by the integrated area of the exotherm. The normalized DTA peak areas are described in units of μV·°C/(mg lube)/(mg sample), and then different catalysts can be directly compared on a common basis [39,40]. The larger the value of normalized DTA exotherm peak area, the greater the fraction of lubricant combusted verses evaporated, and the better the catalytic performance [39–41]. The oxidation activity data of catalysts for lube combustion are listed in Table 2. CeO_2 catalyst exhibits an outstanding catalytic lube oxidation activity, which is consistent with our previous reports [40,41]. Although the introduction of TiO_2 slightly lowers the combustion temperature of lube, bare TiO_2 is not active enough for catalytic SOF oxidation. About 60% of lube is burnt at 350 °C in the lube/TiO_2 sample, while the value of lube without catalyst sample is about 90%. This is because the lube oxidation is an exothermic reaction; once ignition occurs, the heat continually increases and accumulates. Thus, the burn of lube without catalyst (containing more lube oil) is more violently than the lube/TiO_2. The TiO_2-CeO_2 catalyst obviously lowers the onset temperature of lube combustion and considerably promote the removing of lube resulting from combustion. Compared to TiO_2, the prepared TiO_2-CeO_2 catalyst presents obviously lower SOF removal temperature and larger exothermal peak area, which indicates that the TiO_2-CeO_2 catalyst presents better catalytic lube combustion activity.

Table 2. Catalytic performances for the combustion of SOF over the prepared catalysts.

Samples	$T_{10\%}$ (°C)	T_m (°C)	T_f (°C)	Exothermal Peak Area (μV·°C/mg Lube)	Exothermal Peak Area (μV·°C/(mg Lube)/(mg Sample))
lube	264	324	507	283.9	-
lube/CeO_2	162	186	322	1068.1	119.5
lube/TiO_2	249	- [a]	420	901.6	102.6
lube/TiO_2-CeO_2	212	261	362	1029.5	115.5

[a] Three obviously fast weightlessness peaks are observed over the lube/TiO_2 sample.

2.3. Catalyst Characterization

2.3.1. XRD and Raman Spectra

The XRD patterns of prepared catalysts are shown in Figure 3; both TiO_2 and TiO_2-CeO_2 display only characteristic peaks which refer to the typical anatase structure of TiO_2. The peaks of TiO_2 are sharper than those of TiO_2-CeO_2, which indicates that the addition of CeO_2 impedes the crystal growth and sintering and lower crystallinity of the TiO_2-CeO_2 composite materials. In the case of TiO_2-CeO_2, the typical reflections of CeO_2 crystals at 28.7°, 33.2°, 47.7°, 56.6° and 77.1° are not observed, and the positions of typical anatase structure of TiO_2 shift obviously to smaller angles, which suggest that a complete CeO_2 crystal is not formed and Ce ions (Ce^{4+} radius: 0.087 nm) possibly enter into the TiO_2 (Ti^{4+} radius: 0.06 nm) lattice and resulting in an expansion of TiO_2 unit cell, the unit cell volume of anatase tetragonal cell of TiO_2 is 134.95 $Å^3$, for TiO_2-CeO_2, the value enlarges to 135.15 $Å^3$. Thus, it can be inferred that Ce ions entered into the TiO_2 unit cell, and this is a possible reason why the addition of ceria into TiO_2 has no effect on the naturally excellent sulfur resistibility of TiO_2.

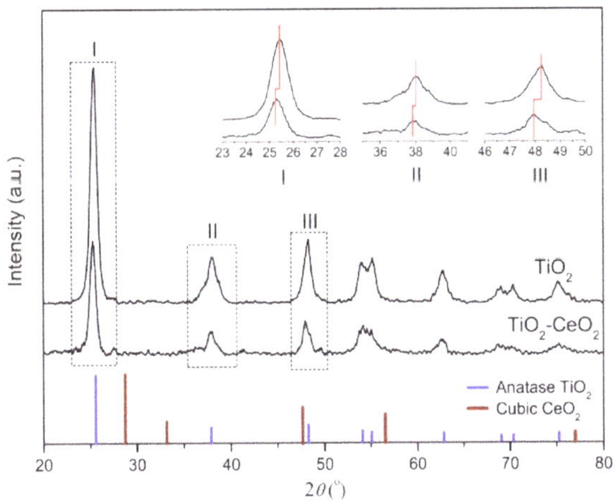

Figure 3. XRD patterns of the TiO_2 and TiO_2-CeO_2 catalysts.

To further confirm the above conjecture, Raman spectra were employed. As can be seen in Figure 4, CeO_2 presents a strong peak at about 464 cm^{-1}, which can be associated with the cubic CeO_2 [42]. TiO_2 and TiO_2-CeO_2 catalysts show five visible peaks at 145, 196, 397, 517 and 639 cm^{-1}, which are the A1g + 2B1g + 3Eg Raman-active modes of TiO_2 anatase phase (the peak at 517 cm^{-1} is complex of A_{1g} and B_{1g}) [43], for the TiO_2-CeO_2 catalyst, the characteristic Raman peak of CeO_2 at 464 cm^{-1} is not observed. This result further confirmed that a CeO_2 crystal is not formed in the TiO_2-CeO_2 catalyst, which is consistent with the XRD result.

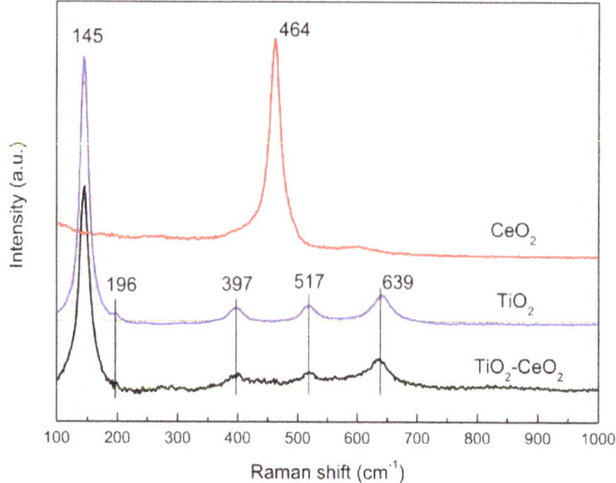

Figure 4. Raman spectra of the CeO_2, TiO_2 and TiO_2-CeO_2 catalysts.

2.3.2. Nitrogen Adsorption-Desorption

The nitrogen adsorption–desorption isotherms and pore size dispersion of TiO_2 and TiO_2-CeO_2 are shown in Figure 5; both TiO_2 and TiO_2-CeO_2 are mesoporous materials and show distinct H3 and H4 complex hysteresis loops indicating slit pore features; compared to TiO_2, the prepared TiO_2-CeO_2 displays obviously larger pore size; the textual features are listed in Table 3.

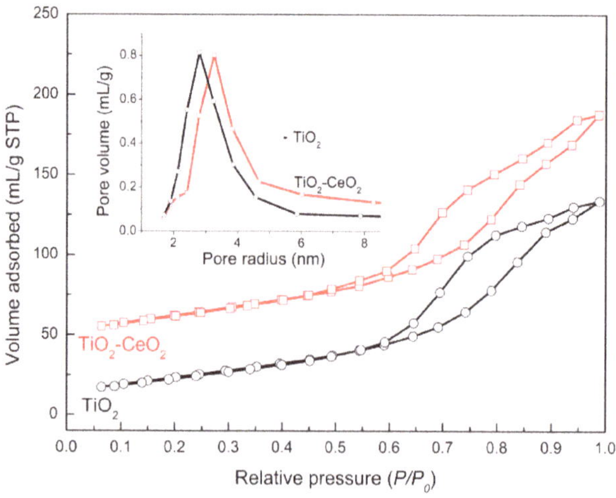

Figure 5. Nitrogen adsorption-desorption isotherms and pore size dispersion (inset) of the TiO_2 and TiO_2-CeO_2 catalysts.

Table 3. The texture properties of TiO_2 and TiO_2-CeO_2 catalysts.

Samples	Surface Area (m^2/g)	Pore Volume (cm^3/g)	Average Pore Diameter (nm)
TiO_2	98	0.22	7.2
TiO_2-CeO_2	118	0.26	8.6

The surface areas of TiO_2 and TiO_2-CeO_2 catalysts are about 98 m^2/g and 118 m^2/g, respectively, which indicates that the introduction of CeO_2 slightly increases the surface area of TiO_2. The pore volume and average pore size of TiO_2 are about 0.22 cm^3/g and 7.2 nm, respectively; the pore volume and average pore diameter of TiO_2-CeO_2 catalyst are 0.26 cm^3/g and 8.6 nm, respectively. It can be seen that the addition of CeO_2 to TiO_2 increases its surface area, pore volume and average pore size; this is possibly due to the addition of CeO_2 which impedes crystal growth and sintering and lowers crystallinity of the TiO_2-CeO_2 composite materials, and this speculation is consistent with the XRD results (Figure 3). The larger surface area, pore volume and pore size can be advantages to the contacting, transmitting and diffusion of the lube molecules on the catalyst, and hence, be beneficial to the catalytic SOF oxidation activity.

2.3.3. XPS

Figure 6 shows the XPS spectra of O 1s region of the CeO_2, TiO_2 and TiO_2-CeO_2 catalysts, all samples show two XPS peaks, the peak at about 530.1 eV can be assigned to lattice oxygen, and the peak with a binding energy of 532.1 eV is characteristic of surface adsorbed oxygen [37,44]. Due to the superior oxygen storage capacity, the peak of surface adsorbed oxygen of CeO_2 is obviously stronger than the lattice oxygen; the surface adsorbed oxygen ratio is about 55%. For the TiO_2 catalyst, the surface adsorbed oxygen ratio is about 38%, and the addition of CeO_2 obviously increases the surface adsorbed oxygen ratio, where the value of TiO_2-CeO_2 is about 42%. Usually, surface adsorbed oxygen is more reactive than lattice oxygen due to its higher mobility [45,46], and our previous work also confirms that the adsorbed oxygen is more active than the lattice oxygen in the catalytic SOF oxidation reaction [40]. Thus, the addition of CeO_2 in TiO_2 enhances the amounts of surface adsorbed oxygen of TiO_2-CeO_2, which can be responsible for the improved catalytic activity of SOF oxidation.

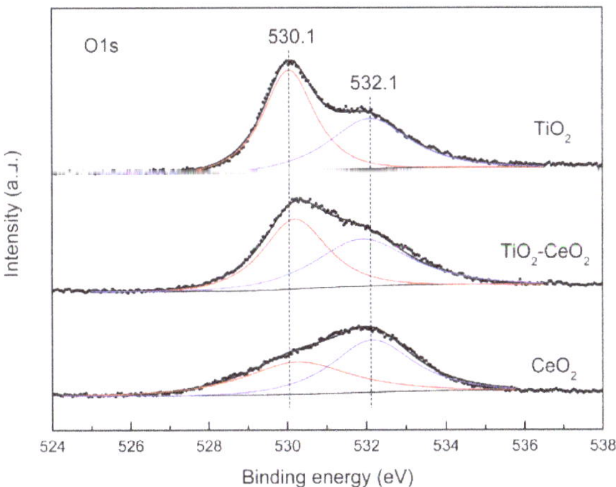

Figure 6. XPS (O 1s) spectra of the CeO_2, TiO_2 and TiO_2-CeO_2 catalysts.

Additionally, the XPS (Ti2p) spectra of TiO_2 and TiO_2-CeO_2 catalysts are both located at 458.5 eV ($2p_{3/2}$) and 464.2 eV ($2p_{1/2}$), which are characteristic of TiO_2 species. Compared to bare CeO_2, the cerium peaks of TiO_2-CeO_2 are very weak and indiscernible, which indicates that the surface concentration of Ce in TiO_2-CeO_2 catalyst is very low; this phenomenon further confirms that Ce dopants are not gathering on the surface and are possibly entering into TiO_2 lattice, which is consistent with the XRD results (Figure 3).

2.3.4. H_2-TPR

The redox property of a catalyst is closely related to the catalytic performance. The redox behavior of the prepared catalysts is described by hydrogen temperature-programmed reduction (H_2-TPR) and shown in Figure 7.

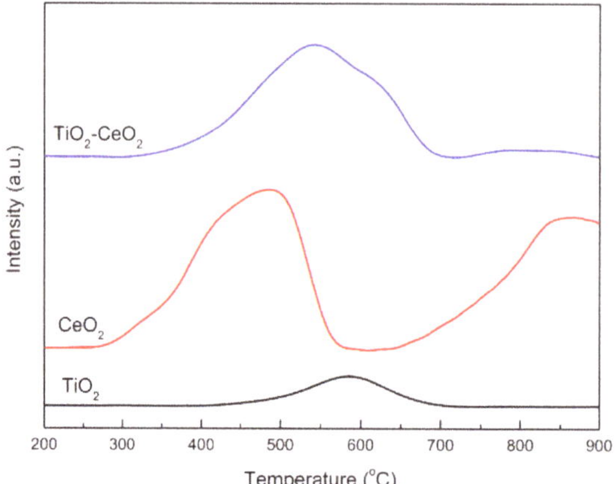

Figure 7. H_2-TPR profiles of the CeO_2, TiO_2 and TiO_2-CeO_2 catalysts.

The TPR peak of individual TiO_2 shows a weak peak at about 500–700 °C which may be ascribed to the reduction of TiO_2, and this phenomenon has been observed by other researchers [47,48]. The CeO_2 obviously shows a two-step reduction, the multiple low-temperature reduction peak at about 300–600 °C can be assigned to the reduction of surface oxygen, and the peak after 700 °C is ascribed to the reduction of CeO_2 bulk oxygen [47]. For the TiO_2-CeO_2, the reduction peak at 300–700 °C can be ascribed to the reduction of TiO_2 and surface oxygen of CeO_2; interestingly, the reduction peak of bulk oxygen of CeO_2 after 700 °C is poorly identified, which indicates that the TiO_2-CeO_2 catalyst exhibits lots of surface oxygen, and this result is consistent with XPS result. Consequently, the prepared TiO_2-CeO_2 catalyst with lots of surface oxygen species presents superior catalytic activity, due to the surface oxygen is highly reactive [45,46].

2.3.5. OSC

The CeO_2-based catalyst presents excellent oxygen storage capacity (OSC) and exhibits superior catalytic SOF oxidation activity [16,41]. The OSC of prepared catalysts were tested by an oxygen pulse injection technique and the results are listed in Table 4.

Table 4. The oxygen storage capacity (OSC) of TiO_2, TiO_2-CeO_2 and CeO_2-based catalysts.

Samples	OSC (μmol O_2/g Sample)	Normalized OSC (μmol O_2/g CeO_2)
TiO_2	2.9	-
TiO_2-CeO_2 [a]	101	524
CeO_2	73	73
$Ce_{0.35}Zr_{0.60}Nd_{0.05}O_2$	269 [49]	638

[a] The prepared TiO_2-CeO_2 catalyst in this work is $Ce_{0.1}Ti_{0.9}O_2$.

It can be seen that the OSC of TiO_2 is about 2.9 μmol/g, which is very slight and may be within the measurement uncertainties. While the addition of CeO_2 significantly increases the OSC of TiO_2-CeO_2 catalyst, the OSC is about 101 (μmol O_2)/(g sample), and the normalized OSC of TiO_2-CeO_2 catalyst is about 524 (μmol O_2)/(g CeO_2). For the pure CeO_2 sample, the OSC is about 73 (μmol O_2)/(g sample). It should be mentioned that the static oxygen storage capacity testing for OSC of pure CeO_2 is an undervalued result, due to the oxygen mobility of bare CeO_2 being very low. Thus, a CeO_2-based oxygen storage materials in our previous work [49] is used as reference, the normalized OSC of $Ce_{0.35}Zr_{0.60}Nd_{0.05}O_2$ is about 638 (μmol O_2)/(g CeO_2). It can be seen that the normalized OSC of prepared TiO_2-CeO_2 catalyst is almost as good as the CeO_2-based oxygen storage materials. Based on the results, it can be suggested that although Ce ions enter into the TiO_2 lattice, the OSC of CeO_2 in TiO_2 is not degraded, the TiO_2-CeO_2 still exhibits good OSC. The good OSC of TiO_2-CeO_2 is one of the reasons that TiO_2-CeO_2 catalyst presents excellent catalytic SOF oxidation activity, this is consistent with the catalytic performance results (Figure 2 and Table 2) and the reports [16,41].

3. Materials and Methods

3.1. Catalyst Preparation

The TiO_2-CeO_2 catalyst was prepared by a co-precipitation method. Desired $TiOSO_4{\cdot}2H_2O$ and $Ce(NO_3)_3{\cdot}9H_2O$ mixture solutions with the molar ratio of Ti:Ce = 9:1, which was the optimal ratio to expose the single TiO_2 crystal structure in our previous studies [35,36], were slowly added to $NH_3{\cdot}H_2O$ solutions under vigorous stirring. And then the precipitate was filtered and washed many times, after dried at 120 °C overnight and calcined for 3 h at 500 °C under airflow, the TiO_2-CeO_2 catalyst powder was obtained. The TiO_2 and CeO_2 catalysts were prepared by the same method.

The simulative 160,000 km vehicle aged sample was obtained by following reference [50] and our previous works [35,37]. Fresh catalyst was placed in a reactor and aged at 670 °C for 15 h and then at 250 °C for 15 h in the gases mixture at flow rate of 800 mL/min: 600 ppm C_3H_6, 1500 ppm CO, 200 ppm NO, 50 ppm SO_2, 5% O_2, 4% CO_2, 8% vapor and N_2 balance.

3.2. Catalyst Evaluation

The catalytic activity for SOF combustion of prepared catalysts were tested using TG-DTA method [16,39]. Due to the fact that the diesel SOF is comprised primarily of lube with a small amount of unburned fuel [51], commercial lubricating oil was often used to simulate the diesel SOF catalytic combustion [16,39].

For this test, the prepared powder catalyst was dried overnight at 120 °C to remove the effects of surface adsorbed water, and then impregnated with 5.0 wt. % commercial lube (Shell Helix HX7 5W-40, Shell Petrochemicals Company Limited, Jiaxing, China), the slurry of lube/catalyst mixture was stirred and mixed till a homogeneous state was obtained. About 10 mg of the lube/catalyst mixture powder was placed in the sample pan of TG-DTA unit (HCT-2, Beijing Henven Instruments, Beijing, China) and dried at 120 °C for 1 h to eliminate the effects of adsorbed substances (water, volatile matters etc.), and then heated to 550 °C with a temperature rate of 5 °C/min under airflow at 30 mL/min. The TG-DTA curves were recorded to determine the catalytic performance for SOF combustion of the prepared catalysts.

The test on lube without catalyst was carried out as a reference. About 10 mg of the lube was placed in the sample pan of TG-DTA unit and dried at 120 °C for 1 h, and then heated to 550 °C with a temperature rate of 5 °C/min under airflow at 30 mL/min.

3.3. Catalyst Characterization

Sulfur uptake was tested on a thermogravimetric analyzer (TGA) HCT-2 (Henven Instruments, Beijing, China). Consulting the references [38,52] and our previous works [37,53], about 15 mg of catalyst was placed in the sample crucible and pretreated under a 35 mL/min of N_2 flow for 5 h at 300 °C, and then the gas mixture of 43 mL/min SO_2(0.05 vol. %)-N_2 and 31 mL/min O_2(15.0 vol. %)-N_2 was introduced at 300 °C for 4 h, the weight increase as a function of time was recorded by the TGA.

The sulfur cumulant of catalysts after mimicking 160,000 km vehicle aging were analyzed by an energy-dispersive X-ray (EDX) spectroscopy (IE-250, Oxford Instruments, Oxford, UK).

X-ray powder diffraction (XRD) patterns were collected on a powder X-ray diffractometer (DX-1000, Dandong Fangyuan Instrument Ltd., Dandong, China) employing Cu Kα radiation (λ = 0.1542 nm).

Raman spectra were recorded by a LabRAM HR Laser Raman spectrometer (HORIBA Jobin Yvon Inc., Paris, France) with an excitation wavelength of 532 nm.

N_2 adsorption-description isotherms were measured on a QUADRASORB SI automated surface area and pore size analyzer (Quantachrome Instruments Ltd., Boynton Beach, FL, USA). The specific surface area and pore size were calculated by the Brunauer-Emmett-Teller (BET) method and Barret-Joyner-Halenda (BJH) method, respectively. Before adsorption measurements, the samples were degassed at 300 °C for 3 h under vacuum.

X-ray photoelectron spectroscopy (XPS) data were collected on a Kratos XSAM 800 spectrometer (Kratos Analytic Inc., Manchester, UK) with Al Kα radiation. The binding energy shifts of the samples were calibrated by fixing the C1s binding energy (BE 284.8 eV).

H_2-temperature programmed reduction (H_2-TPR) experiments were performed in a quartz tubular reactor. Samples were pretreated at 450 °C for 1 h under the N_2 flow (35 mL/min) and then cooled to room temperature; after that, the samples were heated from room temperature to 800 °C with a heating rate of 10 °C/min under the flow of H_2 (5.0 vol. %)-N_2 mixture. The hydrogen consumption as a function of reduction temperature was recorded by a thermal conductivity detector (TCD) cell.

The oxygen storage capacity (OSC) of the samples was measured by a pulse injection technique [54]. The sample was firstly reduced in a H_2 flow (30 mL/min) at 550 °C for 1 h; after cooling to 200 °C, an oxygen pulse was injected every 5 min to obtain a breakthrough curve, from which the OSC was calculated.

4. Conclusions

From the aforementioned results, it can be concluded that moderate amounts of ceria dopants in titania can obviously enhance the catalytic SOF oxidation activity of TiO_2-CeO_2 catalyst. Meanwhile, the prepared TiO_2-CeO_2 catalyst can maintain the naturally excellent sulfur resistibility of titania; the sulfur resistibility of TiO_2-CeO_2 is as well as the bare TiO_2. The prepared TiO_2-CeO_2 catalyst significantly enhances the sulfur tolerance of conventional CeO_2-based SOF oxidation catalysts and displays a good catalytic SOF oxidation activity. The TiO_2-CeO_2 catalyst exhibits a typical phase of anatase, and the cerium ions can enter into the TiO_2 unit cell, impede the crystal growth and sintering and lower crystallinity of the TiO_2-CeO_2 composite materials and, hence, improve the surface area, pore volume and pore size of TiO_2-CeO_2 catalyst. Moreover, the addition of CeO_2 in TiO_2 can significantly enhance the surface oxygen concentration and oxygen storage capacity of TiO_2-CeO_2; the normalized oxygen storage capacity of TiO_2-CeO_2 is almost as good as the CeO_2-based oxygen storage materials. The improvement of textual features and surface oxygen concentration of TiO_2-CeO_2 catalyst are the main reasons for the enhancement of catalytic SOF oxidation activity.

Author Contributions: Z.Y. and Y.C. conceived the project; N.Z., Z.Y. and Z.C. performed the experiments; Y.L. (Yunxiang Li), Y.L. (Yunwen Liao), Y.L. (Youping Li) and M.G. carried out the data analysis; N.Z. and Z.Y. wrote the paper.

Acknowledgments: This work was supported by the National Natural Science Foundation of China (21703174), Meritocracy Research Funds of China West Normal University (17YC146) and Doctor Startup Foundation of China West Normal University (15E012). Zhi Chen is particularly grateful to the Sichuan Provincial Students' Innovation and Entrepreneurship Training Program (201610638094).

Conflicts of Interest: The authors declare no conflict of interest.

References

1. Johnson, T.; Joshi, A. *Review of Vehicle Efficiency and Emissions*; SAE Technical Paper, 2017-01-0907; SAE: Warrendale, PA, USA, 2017.
2. Wang, X.; Westermann, A.; Shi, Y.; Cai, N.; Rieu, M.; Viricelle, J.-P.; Vernoux, P. Electrochemical Removal of NO_x on Ceria-Based Catalyst-Electrodes. *Catalysts* **2017**, *7*, 61. [CrossRef]
3. Godoy, M.; Banús, E.; Sanz, O.; Montes, M.; Miró, E.; Milt, V. Stacked Wire Mesh Monoliths for the Simultaneous Abatement of VOCs and Diesel Soot. *Catalysts* **2018**, *8*, 16. [CrossRef]
4. Williams, P.T.; Abbass, M.K.; Tarn, L.P.; Andrews, G.E.; Ng, K.L.; Bartle, K.D. *A Comparison of Exhaust Pipe, Dilution Tunnel and Roadside Diesel Particulate SOF and Gaseous Hydrocarbon Emissions*; SAE Technical Paper, 880351; SAE: Warrendale, PA, USA, 1988.
5. Osada, H.; Aoyagi, Y.; Shimada, K.; Akiyama, K.; Goto, Y.; Suzuki, H. *SOF Component of Lubricant Oil on Diesel PM in a High Boosted and Cooled EGR Engine*; SAE Technical Paper, 2007-01-0123; SAE: Warrendale, PA, USA, 2007.
6. Collura, S.; Chaoui, N.; Azambre, B.; Finqueneisel, G.; Heintz, O.; Krzton, A.; Koch, A.; Weber, J. Influence of the soluble organic fraction on the thermal behaviour, texture and surface chemistry of diesel exhaust soot. *Carbon* **2005**, *43*, 605–613. [CrossRef]
7. Atribak, I.; Bueno-López, A.; García-García, A. Uncatalysed and catalysed soot combustion under NO_x + O_2: Real diesel versus model soots. *Combust. Flame* **2010**, *157*, 2086–2094. [CrossRef]
8. Farrauto, R.J.; Mooney, J.J. *Effects of Sulfur on Performance of Catalytic Aftertreatment Devices*; SAE Technical Paper, 920557; SAE: Warrendale, PA, USA, 1992.
9. Spezzano, P.; Picini, P.; Cataldi, D. Gas-and particle-phase distribution of polycyclic aromatic hydrocarbons in two-stroke, 50-cm^3 moped emissions. *Atmos. Environ.* **2009**, *43*, 539–545. [CrossRef]
10. Stanmore, B.R.; Brilhac, J.F.; Gilot, P. The oxidation of soot: A review of experiments, mechanisms and models. *Carbon* **2001**, *39*, 2247–2268. [CrossRef]
11. Clague, A.D.H.; Donnet, J.B.; Wang, T.K.; Peng, J.C.M. A comparison of diesel engine soot with carbon black1. *Carbon* **1999**, *37*, 1553–1565. [CrossRef]
12. Westerholm, R.; Christensen, A.; de Serves, C.; Almén, J. *Determination of Polycyclic Aromatic Hydrocarbons (PAH) in Size Fractionated Diesel Particles from a Light Duty Vehicle*; SAE Technical Paper, 1999-01-3533; SAE: Warrendale, PA, USA, 1999.
13. Hansen, K.F.; Jensen, M.G. *Chemical and Biological Characteristics of Exhaust Emissions from a DI Diesel Engine Fuelled with Rapeseed Oil Methyl Ester (RME)*; SAE Technical Paper, 971689; SAE: Warrendale, PA, USA, 1997.
14. Callén, M.; Iturmendi, A.; López, J.; Mastral, A. Source apportionment of the carcinogenic potential of polycyclic aromatic hydrocarbons (PAH) associated to airborne PM_{10} by a PMF model. *Environ. Sci. Pollut. Res.* **2014**, *21*, 2064–2076. [CrossRef] [PubMed]
15. Benbrahim-Tallaa, L.; Baan, R.A.; Grosse, Y.; Lauby-Secretan, B.; el Ghissassi, F.; Bouvard, V.; Guha, N.; Loomis, D.; Straif, K. Carcinogenicity of diesel-engine and gasoline-engine exhausts and some nitroarenes. *Lancet Oncol.* **2012**, *13*, 663–664. [CrossRef]
16. Farrauto, R.J.; Voss, K.E. Monolithic diesel oxidation catalysts. *Appl. Catal. B Environ.* **1996**, *10*, 29–51. [CrossRef]
17. Cao, Y.; Lan, L.; Feng, X.; Yang, Z.; Zou, S.; Xu, H.; Li, Z.; Gong, M.; Chen, Y. Cerium promotion on the hydrocarbon resistance of a Cu-SAPO-34 NH_3-SCR monolith catalyst. *Catal. Sci. Technol.* **2015**, *5*, 4511–4521. [CrossRef]
18. Huang, Y.-F.; Zhang, H.-L.; Yang, Z.-Z.; Zhao, M.; Huang, M.-L.; Liang, Y.-L.; Wang, J.-L.; Chen, Y.-Q. Effects of CeO_2 Addition on Improved NO Oxidation Activities of Pt/SiO_2-Al_2O_3 Diesel Oxidation Catalysts. *Acta Phys.-Chim. Sin.* **2017**, *33*, 1242–1252.
19. Sudarsanam, P.; Hillary, B.; Amin, M.H.; Rockstroh, N.; Bentrup, U.; Bruckner, A.; Bhargava, S.K. Heterostructured Copper-Ceria and Iron-Ceria Nanorods: Role of Morphology, Redox, and Acid Properties in Catalytic Diesel Soot Combustion. *Langmuir* **2018**, *34*, 2663–2673. [CrossRef] [PubMed]
20. Putla, S.; Amin, M.H.; Reddy, B.M.; Nafady, A.; Al Farhan, K.A.; Bhargava, S.K. MnO_x Nanoparticle-Dispersed CeO_2 Nanocubes: A Remarkable Heteronanostructured System with Unusual Structural Characteristics and Superior Catalytic Performance. *ACS Appl. Mater. Interfaces* **2015**, *7*, 16525–16535. [CrossRef] [PubMed]

21. Yang, Z.-Z.; Yang, Y.; Zhao, M.; Gong, M.-C.; Chen, Y.-Q. Enhanced Sulfur Resistance of Pt-Pd/CeO_2-ZrO_2-Al_2O_3 Commercial Diesel Oxidation Catalyst by SiO_2 Surface Cladding. *Acta Phys.-Chim. Sin.* **2014**, *30*, 1187–1193.
22. Kolli, T.; Kanerva, T.; Lappalainen, P.; Huuhtanen, M.; Vippola, M.; Kinnunen, T.; Kallinen, K.; Lepisto, T.; Lahtinen, J.; Keiski, R.L. The Effect of SO_2 and H_2O on the Activity of Pd/CeO_2 and Pd/Zr-CeO_2 Diesel Oxidation Catalysts. *Top. Catal.* **2009**, *52*, 2025–2028. [CrossRef]
23. Luo, T.; Gorte, R.J. Characterization of SO_2-poisoned ceria-zirconia mixed oxides. *Appl. Catal. B Environ.* **2004**, *53*, 77–85. [CrossRef]
24. Deshmukh, S.S.; Zhang, M.; Kovalchuk, V.I.; d'Itri, J.L. Effect of SO_2 on CO and C_3H_6 oxidation over CeO_2 and $Ce_{0.75}Zr_{0.25}O_2$. *Appl. Catal. B Environ.* **2003**, *45*, 135–145. [CrossRef]
25. Xia, Z.; Fu, J.; Duan, A.; Han, L.; Wu, H.; Zhao, Z.; Xu, C.; Wang, D.; Wang, B.; Meng, Q. Post Synthesis of Aluminum Modified Mesoporous TUD-1 Materials and Their Application for FCC Diesel Hydrodesulfurization Catalysts. *Catalysts* **2017**, *7*, 141. [CrossRef]
26. Zhang, M.H.; Fan, J.Y.; Chi, K.; Duan, A.J.; Zhao, Z.; Meng, X.L.; Zhang, H.L. Synthesis, characterization, and catalytic performance of NiMo catalysts supported on different crystal alumina materials in the hydrodesulfurization of diesel. *Fuel Process. Technol.* **2017**, *156*, 446–453. [CrossRef]
27. Zhang, K.; Hu, J.; Gao, S.; Liu, Y.; Huang, X.; Bao, X. Sulfur content of gasoline and diesel fuels in northern China. *Energy Policy* **2010**, *38*, 2934–2940. [CrossRef]
28. Corro, G. Sulfur impact on diesel emission control—A review. *React. Kinet. Catal Lett.* **2002**, *75*, 89–106. [CrossRef]
29. Zelenka, P.; Ostgathe, K.; Lox, E. *Reduction of Diesel Exhaust Emissions by Using Oxidation Catalysts*; SAE Technical Paper, 902111; SAE: Warrendale, PA, USA, 1990.
30. Li, J.; Kumar, A.; Chen, X.; Currier, N.; Yezerets, A. *Impact of Different Forms of Sulfur Poisoning on Diesel Oxidation Catalyst Performance*; SAE Technical Paper, 2013-01-0514; SAE: Warrendale, PA, USA, 2013.
31. Paulson, T.; Moss, B.; Todd, B.; Eckstein, C.; Wise, B.; Singleton, D.; Zemskova, S.; Silver, R. *New Developments in Diesel Oxidation Catalysts*; SAE Technical Paper, 2008-01-2638; SAE: Warrendale, PA, USA, 2008.
32. Kanno, Y.; Hihara, T.; Watanabe, T.; Katoh, K. *Low Sulfate Generation Diesel Oxidation Catalyst*; SAE Technical Paper, 2004-01-1427; SAE: Warrendale, PA, USA, 2004.
33. Ueno, H.; Furutani, T.; Nagami, T.; Aono, N.; Goshima, H.; Kasahara, K. *Development of Catalyst for Diesel Engine*; SAE Technical Paper, 980195; SAE: Warrendale, PA, USA, 1998.
34. Yang, Z.; Li, J.; Zhang, H.; Yang, Y.; Gong, M.; Chen, Y. Size-dependent CO and propylene oxidation activities of platinum nanoparticles on the monolithic Pt/TiO_2-YO_x diesel oxidation catalyst under simulative diesel exhaust conditions. *Catal. Sci. Technol.* **2015**, *5*, 2358–2365. [CrossRef]
35. Yang, Z.; Zhang, N.; Cao, Y.; Gong, M.; Zhao, M.; Chen, Y. Effect of yttria in Pt/TiO_2 on sulfur resistance diesel oxidation catalysts: Enhancement of low-temperature activity and stability. *Catal. Sci. Technol.* **2014**, *4*, 3032–3043. [CrossRef]
36. Yang, Z.; Chen, Y.; Zhao, M.; Zhou, J.; Gong, M.; Chen, Y. Preparation and Properties of Pt/$Zr_xTi_{1-x}O_2$ Catalysts with Low-Level SO_2 Oxidation Activity for Diesel Oxidation, Cuihua Xuebao. *Chin. J. Catal.* **2012**, *33*, 819–826. (In Chinese)
37. Yang, Z.; Zhang, N.; Cao, Y.; Li, Y.; Liao, Y.; Li, Y.; Gong, M.; Chen, Y. Promotional effect of lanthana on the high-temperature thermal stability of Pt/TiO_2 sulfur-resistant diesel oxidation catalysts. *RSC Adv.* **2017**, *7*, 19318–19329. [CrossRef]
38. Koranne, M.M.; Pryor, J.N. Sulfur Tolerant Alumina Catalyst Support. U.S. Patent 8,158,257, 17 April 2012.
39. Zhang, Z.; Zhang, Y.; Mu, Z.; Yu, P.; Ni, X.; Wang, S.; Zheng, L. Synthesis and catalytic properties of $Ce_{0.6}Zr_{0.4}O_2$ solid solutions in the oxidation of soluble organic fraction from diesel engines. *Appl. Catal. B Environ.* **2007**, *76*, 335–347. [CrossRef]
40. Yang, Y.; Yang, Z.Z.; Xu, H.D.; Xu, B.Q.; Zhang, Y.H.; Gong, M.C.; Chen, Y.Q. Influence of La on CeO_2-ZrO_2 Catalyst for Oxidation of Soluble Organic Fraction from Diesel Exhaust. *Acta Phys.-Chim. Sin.* **2015**, *31*, 2358–2365.
41. Chen, Y.-D.; Wang, L.; Guan, X.-X.; Liu, Y.-B.; Gong, M.-C.; Chen, Y.-Q. Catalytic Oxidation of Soluble Organic Fraction in Diesel Exhausts Using Composite Oxides $(CeO_2)_x(La\text{-}Al_2O_3)_{1-x}$. *Acta Phys.-Chim. Sin.* **2013**, *29*, 1048–1054.

42. Fornasiero, P.; Balducci, G.; di Monte, R.; Kašpar, J.; Sergo, V.; Gubitosa, G.; Ferrero, A.; Graziani, M. Modification of the Redox Behaviour of CeO_2 Induced by Structural Doping with ZrO_2. *J. Catal.* **1996**, *164*, 173–183. [CrossRef]
43. Balachandran, U.; Eror, N.G. Raman spectra of titanium dioxide. *J. Solid State Chem.* **1982**, *42*, 276–282. [CrossRef]
44. Sellick, D.R.; Aranda, A.; García, T.; López, J.M.; Solsona, B.; Mastral, A.M.; Morgan, D.J.; Carley, A.F.; Taylor, S.H. Influence of the preparation method on the activity of ceria zirconia mixed oxides for naphthalene total oxidation. *Appl. Catal. B Environ.* **2013**, *132–133*, 98–106. [CrossRef]
45. Xu, H.; Cao, Y.; Wang, Y.; Fang, Z.; Lin, T.; Gong, M.; Chen, Y. The influence of molar ratios of Ce/Zr on the selective catalytic reduction of NO_x with NH_3 over Fe_2O_3-WO_3/$Ce_xZr_{1-x}O_2$ ($0 \leq x \leq 1$) monolith catalyst. *Chin. Sci. Bull.* **2014**, *59*, 3956–3965. [CrossRef]
46. Wu, Z.; Jin, R.; Liu, Y.; Wang, H. Ceria modified MnO_x/TiO_2 as a superior catalyst for NO reduction with NH_3 at low-temperature. *Catal. Commun.* **2008**, *9*, 2217–2220. [CrossRef]
47. Zhu, H.; Qin, Z.; Shan, W.; Shen, W.; Wang, J. Pd/CeO_2–TiO_2 catalyst for CO oxidation at low temperature: A TPR study with H_2 and CO as reducing agents. *J. Catal.* **2004**, *225*, 267–277. [CrossRef]
48. Watanabe, S.; Ma, X.; Song, C. Characterization of Structural and Surface Properties of Nanocrystalline TiO_2-CeO_2 Mixed Oxides by XRD, XPS, TPR, and TPD. *J. Phys. Chem. C* **2009**, *113*, 14249–14257. [CrossRef]
49. Guo, J.; Shi, Z.; Wu, D.; Yin, H.; Gong, M.; Chen, Y. Effects of Nd on the properties of CeO_2–ZrO_2 and catalytic activities of three-way catalysts with low Pt and Rh. *J. Alloy. Compd.* **2015**, *621*, 104–115. [CrossRef]
50. Andersson, J.; Antonsson, M.; Eurenius, L.; Olsson, E.; Skoglundh, M. Deactivation of diesel oxidation catalysts: Vehicle- and synthetic aging correlations. *Appl. Catal. B Environ.* **2007**, *72*, 71–81. [CrossRef]
51. Voss, K.E.; Lampert, J.K.; Farrauto, R.J.; Rice, G.W.; Punke, A.; Krohn, R. Catalytic oxidation of diesel particulates with base metal oxides. *Stud. Surf. Sci. Catal.* **1995**, *96*, 499–515.
52. Koranne, M.M.; Pryor, J.N.; Chapman, D.M.; Brezny, R. Sulfur Tolerant Alumina Catalyst Support. U.S. Patent 8,076,263, 13 December 2011.
53. Yang, Z.; Li, Y.; Liao, Y.; Li, Y.; Zhang, N. Preparation and properties of the Pt/SiO_2-Al_2O_3 sulfur resistance diesel oxidation catalyst. *Environ. Chem.* **2016**, *35*, 1682–1689. (In Chinese)
54. Lan, L.; Chen, S.; Cao, Y.; Gong, M.; Chen, Y. New insights into the structure of a CeO_2-ZrO_2-Al_2O_3 composite and its influence on the performance of the supported Pd-only three-way catalyst. *Catal. Sci. Technol.* **2015**, *5*, 4488–4500. [CrossRef]

Article

Low-Temperature Electrocatalytic Conversion of CO_2 to Liquid Fuels: Effect of the Cu Particle Size

Antonio de Lucas-Consuegra [1,*], Juan Carlos Serrano-Ruiz [2], Nuria Gutiérrez-Guerra [1] and José Luis Valverde [1]

1 Department of Chemical Engineering, School of Chemical Sciences and Technologies, University of Castilla-La Mancha, Avda. Camilo José Cela 12, 13005 Ciudad Real, Spain; nuriagutierrezguerra@gmail.com (N.G.-G.); joseluis.valverde@uclm.es (J.L.V.)
2 Department of Engineering, Universidad Loyola Andalucía, Energía Solar, 1. Edifs. E, F and G, 41014 Seville, Spain; jcserrano@uloyola.es
* Correspondence: antonio.lconsuegra@uclm.es; Tel.: +34-926-295-300

Received: 30 July 2018; Accepted: 14 August 2018; Published: 20 August 2018

Abstract: A novel gas-phase electrocatalytic system based on a low-temperature proton exchange membrane (Sterion) was developed for the gas-phase electrocatalytic conversion of CO_2 to liquid fuels. This system achieved gas-phase electrocatalytic reduction of CO_2 at low temperatures (below 90 °C) over a Cu cathode by using water electrolysis-derived protons generated in-situ on an IrO_2 anode. Three Cu-based cathodes with varying metal particle sizes were prepared by supporting this metal on an activated carbon at three loadings (50, 20, and 10 wt %; 50% Cu-AC, 20% Cu-AC, and 10% Cu-AC, respectively). The cathodes were characterized by N_2 adsorption–desorption, temperature-programmed reduction (TPR), and X-ray diffraction (XRD) and their performance towards the electrocatalytic conversion of CO_2 was subsequently studied. The membrane electrode assembly (MEA) containing the cathode with the largest Cu particle size (50% Cu-AC, 40 nm) showed the highest CO_2 electrocatalytic activity per mole of Cu, with methyl formate being the main product. This higher electrocatalytic activity was attributed to the lower Cu–CO bonding strength over large Cu particles. Different product distributions were obtained over 20% Cu-AC and 10% Cu-AC, with acetaldehyde and methanol being the main reaction products, respectively. The CO_2 consumption rate increased with the applied current and reaction temperature.

Keywords: CO_2 electroreduction; CO_2 valorization; Cu catalyst; particle size; PEM; acetaldehyde production; methanol production

---

1. Introduction

Fossil fuels and biomass are the most common feedstocks for the production of liquid fuels. Since the burning of these feedstocks results in CO_2 emissions into the atmosphere, it is necessary to develop strategies for the upgrading of this gas into useful products. One of these approaches involves the recycling of CO_2 into sustainable hydrocarbon fuels [1] by different methods such as catalytic processes (e.g., hydrogenation to alkanes, alkenes or other oxygenated, or reforming with hydrocarbons) [2], biological processes [3], microwave and plasma systems [4–6], and photocatalytic and electrocatalytic routes [7]. Among these methods, the electrochemical reduction of CO_2 is highly interesting since it allows to directly transform CO_2 to syngas and light hydrocarbons with electricity, which may be obtained from renewable energy sources [1,8,9]. This approach is advantageous in that electrochemical cell reactors are typically compact, modular, and easy to scale-up. While electrolysis of CO_2 and/or H_2O can be carried out in solid oxide cells (SOCs) [10], the high temperatures required for these systems to operate (above 600 °C) usually result in catalyst sintering and stability losses. In addition, high-temperature CO_2-H_2O co-electrolysis produces syngas as the only product, and further conversion steps (e.g., Fischer–Tropsch synthesis) are required to produce hydrocarbon fuels. Alternatively, low-temperature electrolyzers containing protonic exchange membranes (PEM) have been proposed to directly transform CO_2 into hydrocarbons and oxygenates [11–13]. Despite the overall single-pass conversions being typically low in these reactors, the unreacted CO_2 can be easily separated from the liquid fuels and recycled again to the reactor. The mild working conditions of these systems (typically below 90 °C and atmospheric pressure) facilitate the utilization of renewable energies such as solar heating and electrical energy for driving the electrochemical process. Moreover, unlike conventional catalytic hydrogenation of CO_2, PEM-based electrolyzers do not require external hydrogen since CO_2 directly reacts with the protons produced in-situ by water electrolysis [7]. These advantages have motivated researchers to investigate on gas-phase low-temperature CO_2 electroreduction [9,13–16], opening the way for incorporating renewable energies into the value chain of chemical industries. Professor Centi has developed most of these works with Pt, Fe, and Cu catalyst supported on a variety of carbonaceous materials such as carbon black, carbon nanofibers, and graphene, among others.

In this work, we carried out the electroreduction of CO_2 in the gas phase at low temperature over Cu cathodes. We performed a systematic study with three different Cu-based cathodic catalysts supported on a high-surface-area activated carbon. The main objective of the work was to study the influence of the Cu particle size on the electrocatalytic activity of the system. With this aim, three different membrane electrode assemblies (MEAs) were fabricated with three Cu cathodic catalysts having different metal loadings and particle sizes. These MEAs were characterized and subsequently used for the electrocatalytic conversion of CO_2 into synthetic fuels.

2. Results and Discussion

2.1. Characterization of The Cu Powder Catalysts and The Cu Electrodes

The different Cu powders and Cu cathodic-catalysts were characterized by N_2 adsorption–desorption, atomic absorption spectroscopy (AAS), temperature-programmed reduction (TPR), and X-ray diffraction (XRD). Table 1 shows the physicochemical properties of the activated carbon support (AC) and the three Cu cathodic-catalysts.

Table 1. Physicochemical properties of the support material, the catalysts, and the fresh electrodes.

Sample	Powder Metal Loading/wt %	Electrode Metal Weight/mg·cm^{-2}	Surface Area/m^2·g^{-1}	Total Pore Volume/cm^3·g^{-1}	TPR-T$_{max}$/°C	Mean Particle Size from XRD/nm
AC	-	-	866	0.293	-	-
50% Cu-AC	55	0.22	773	0.186	171	40
20% Cu-AC	19	0.18	797	0.260	185	14
10% Cu-AC	12	0.16	817	0.274	193	12

The main difference between the three Cu cathodic catalysts is the metal loading. Thus, atomic absorption spectroscopy revealed metal loadings of 55, 19, and 12 wt % corresponding to electrode metal weights of 0.22, 0.18, and 0.16 mg·cm^{-2} for the 50% Cu-AC, 20% Cu-AC, and 10% Cu-AC cathodic-catalysts, respectively. The electrocatalytic rates discussed below were normalized to the corresponding metal weight of the electrode.

The surface area and total pore volume of the different Cu cathodic-catalysts were determined by N_2 adsorption–desorption (Table 1). The activated carbon support showed high Langmuir areas and total pore volumes (866 m^2·g^{-1} and 0.293 cm^3·g^{-1}, respectively), as previously reported in the literature [17,18]. As shown in Table 1, metal addition resulted in an important decrease of both the surface area and the total pore volume. As expected, the surface area and the total pore volume decreased with the metal loading, probably as a result of partial pore blockage by the metal particles [18,19]. Combined IUPAC (International Union of Pure and Applied Chemistry) types I and

IV N_2 adsorption–desorption isotherms (not shown) were obtained in all cases, revealing the presence of a microporous structure.

The TPR profiles of 50% Cu-AC, 20% Cu-AC, and 10% Cu-AC and the activated carbon support are shown in Figure 1. These TPR profiles result from the following sequential Cu reduction $Cu^{2+} \rightarrow Cu^{+} \rightarrow Cu^{0}$ [20]. The first two reduction peaks at 171–193 °C and 223–276 °C (depending on the catalyst) can be attributed to the reduction of the more dispersed Cu particles and the reduction of CuO (II), respectively. The third reduction peak that appeared at 316–380 °C can be assigned to the reduction of Cu_2O (I). Finally, the peaks at higher temperatures are typically associated with the gasification of activated carbon and the reduction of surface oxygenated groups on the activated carbon support [21,22]. The temperature of the most intense consumption peak (T_{max}) is given in Table 1. The TPR profiles revealed that the interaction between the metal phase and the support varied depending on the metal particle size. Thus, stronger interactions were obtained for those catalysts having smaller Cu particles since T_{max} decreased with the Cu particle size (Table 1) [20,23,24]. Thus, the reducibility of the catalysts followed the sequence: 50% Cu-AC > 20% Cu-AC > 10% Cu-AC. Furthermore, given the TPR profiles, 400 °C was selected as a suitable reduction temperature for ensuring complete metal activation while maintaining the surface properties of the support.

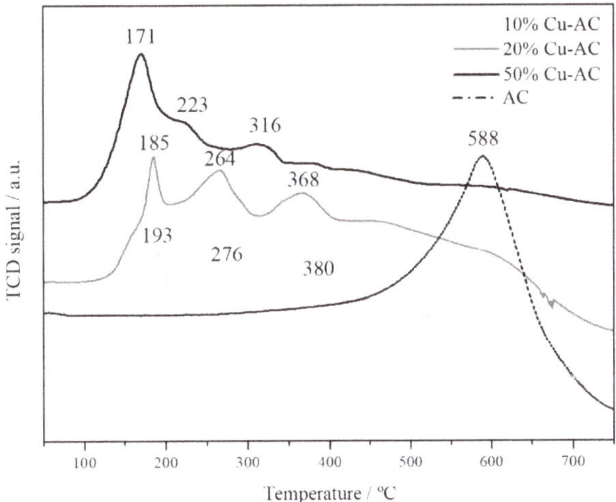

Figure 1. Temperature-programmed reduction (TPR) profiles of the fresh catalysts and the activated carbon support.

Figure 2 shows the XRD patterns of the powder catalyst 50% Cu-AC before and after the reduction treatment at 400 °C. No significant differences were appreciated between the three Cu-AC powder catalysts. As shown in Figure 2a, diffraction peaks corresponding to metallic copper (Cu) and copper oxide (Cu_2O and CuO) were observed before the reduction treatment. The main diffraction Cu peaks were (111), (200), (220), and (331) observed at 43.3, 50.4, 74.1, and 89.8°, respectively. These peaks are associated with a metallic Cu phase with face-centered cubic (FCC) crystalline structure (JCPDS, 85–1326). Diffraction peaks corresponding to Cu_2O (JCPDS, 78–2076) and CuO (JCPDS, 80–1917) were also observed, in line with the TPR results (Figure 1). However, these peaks were lower in intensity as compared to those of metallic copper. Figure 2b shows the XRD patterns of the catalysts after the reduction treatment. As shown by the XRD patterns, only peaks corresponding to metallic Cu were observed at $2\theta = 43.3$, 50.4, 74.1, and 90° for the (111), (200), (220), and (331) planes. These results revealed that Cu was completely reduced at 400 °C, as anticipated by the TPR profiles. Additionally, the metal precursor was completely calcined, since no peaks corresponding to the metal precursor were observed in the XRD patterns. The mean Cu particle sizes of the different catalysts were estimated using the Scherrer equation, and the results are summarized in Table 1. As expected, the mean Cu particle size increased significantly with the metal loading. The Cu particle sizes obtained herein (10–40 nm) were similar to those measured for similar electrocatalytic systems prepared by direct impregnation with Cu precursor solutions [16,17,25].

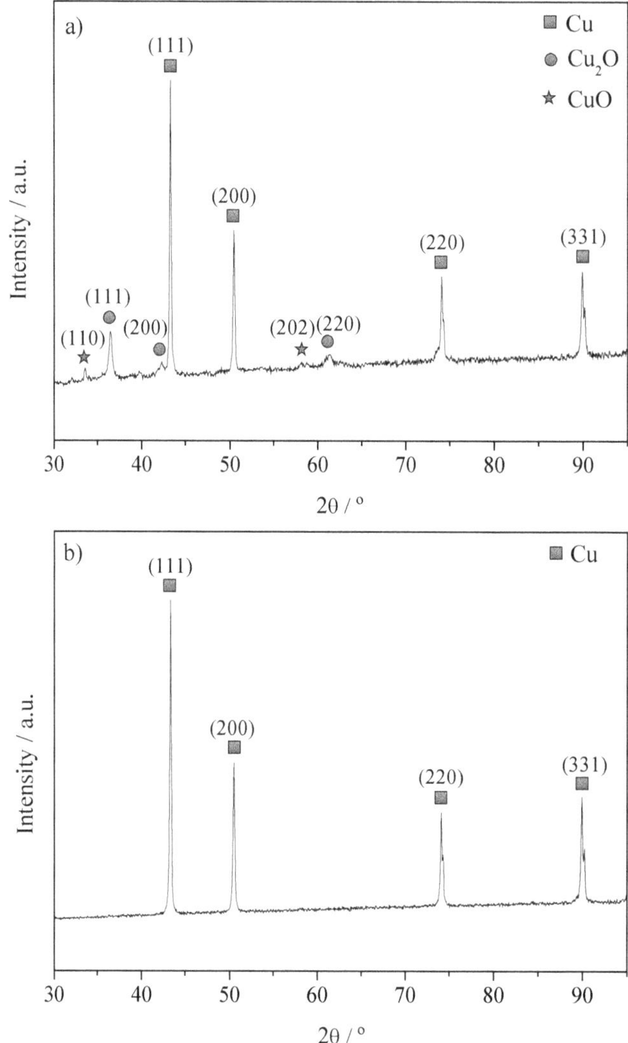

Figure 2. X-ray diffraction (XRD) patterns of 50% Cu-AC supported on carbon paper substrates: (**a**) before reduction at 400 °C and (**b**) after reduction at 400 °C.

2.2. CO_2 *Conversion Electrocatalytic Experiments*

Figure 3 shows the production rates of different compounds as a function of the time on stream during the electroreduction of CO_2 at a constant applied current of −20 mA and 90 °C. Note that no products were obtained under open circuit conditions (OCC, no current application). A constant current of −20 mA was subsequently applied at 90 °C for approximately 350 min under the same reaction atmosphere. This polarization was maintained until products were obtained at a steady state rate. During this current application step, hydrogen (not shown in Figure 3, reaction (1)) and different products such as methanol, acetaldehyde, methyl formate, acetone, and n-propanol were obtained by reactions (2)–(6), respectively:

$$2H^{+}+2e^{-} \rightarrow H_2 \tag{1}$$

$$CO_2+6H^{+}+6e^{-} \rightarrow CH_3OH + H_2O \tag{2}$$

$$2CO_2+10H^{+}+10e^{-} \rightarrow C_2H_4O + 3H_2O \tag{3}$$

$$2CO_2+8H^{+}+8e^{-} \rightarrow C_2H_4O_2+2H_2O \tag{4}$$

$$3CO_2+16H^{+}+16e^{-} \rightarrow C_3H_6O + 5H_2O \tag{5}$$

$$3CO_2+18H^{+}+18e^{-} \rightarrow C_3H_8O + 5H_2O \tag{6}$$

Most of these products have been previously obtained during the electrocatalytic conversion of CO_2 on Fe, Co, Pt, and Cu supported on carbon nanotubes [15,16,26–28] and Cu supported on different carbonaceous supports (activated carbon, carbon nanofibers, and graphite) [17] at similar temperatures. These products reached maximum production rates after ca. 300 min on stream and decreased upon OCC. A slow dynamic behavior was observed (the steady state was reached after 4–5 h of reaction), and this can be attributed to the high residence times used herein. At this point it should be mentioned that low Faradaic efficiencies to hydrocarbon products (below 10%) were obtained in all the experiments. This is due to the high kinetics of the hydrogen evolution reaction (reaction (1)) vs. hydrocarbon formation reactions (reactions (2)–(6)) due to the low surface concentration of CO_2.

The configuration used herein is advantageous in that it allows the direct supply of H^+ (more reactive than H_2) to the cathodic side of the cell. This configuration allows lower temperatures (around 90 °C) as compared to catalytic CO_2 hydrogenation processes (above 250 °C) [21,29–31].

Finally, in all experiments, the cathodic side of the cell was purged with N_2 (30 NmL·min^{-1}) and returned to OCC to remove all the products for subsequent reaction experiments. Sample 50% Cu-AC showed the highest intrinsic electrocatalytic activity among the three cathodic-catalyst studied herein. This higher electrocatalytic activity can be associated with the higher Cu particle size of 50% Cu-AC. Thus, large Cu particles have been previously reported to favor the formation of reaction products by reducing the strength of the metal–CO interaction [32]. In the electroreduction of CO_2, CO_2 is adsorbed on active centers and subsequently converted into CO and O_2. Since this adsorbed CO is more reactive than CO_2, it reacts with protons to generate the different products observed. This reaction is favored on large Cu particles. Since small Cu particles adsorb CO more strongly than large particles, the subsequent reaction of CO is hindered on small Cu particles, leading to CO and H_2 as the main products [32].

With regard to the composition of the reaction products, methyl formate was the main reaction product over 50% Cu-AC, followed by acetaldehyde and methanol. Acetaldehyde and methanol were the main reaction products obtained over samples 20% Cu-AC and 10% Cu-AC, respectively. In line with our results, acetaldehyde and methanol were previously obtained as main products over Cu-AC catalysts [17]. In the case of catalyst 50% Cu-AC, the high Cu particle size favored the production of methyl formate by dehydrogenation of methanol Reaction (7), as previously reported [33].

$$2CH_3OH \rightarrow HCOOCH_3+2H_2 \tag{7}$$

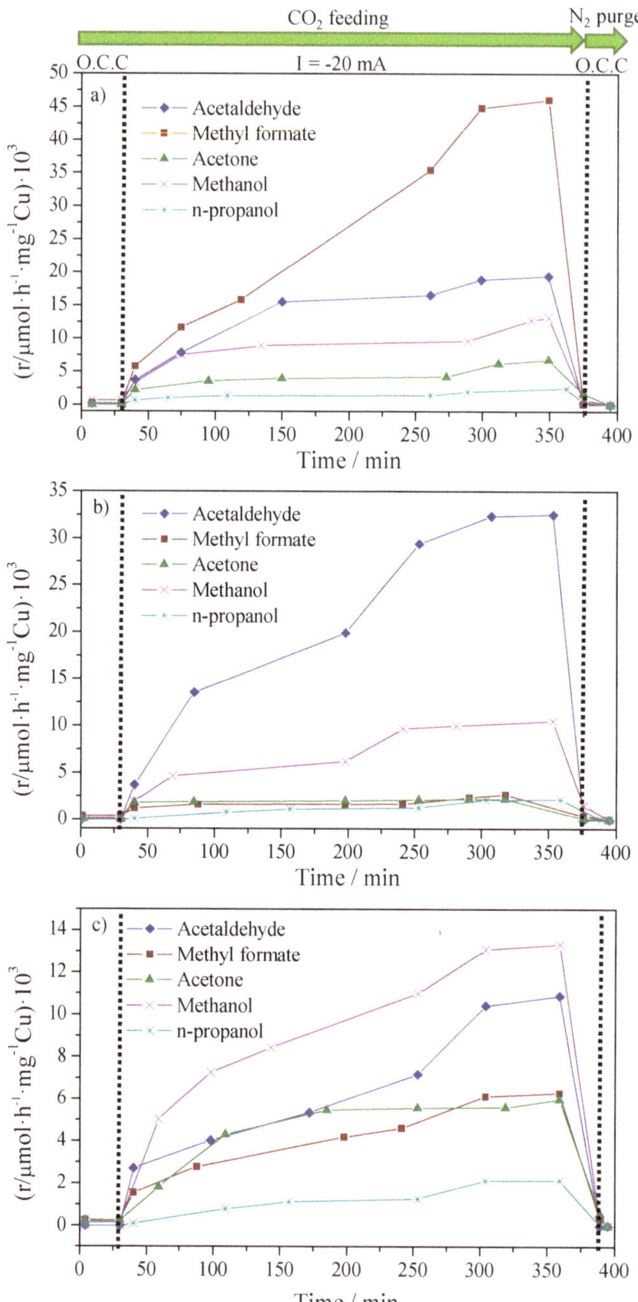

Figure 3. Time-on-stream variation of the rate of production for the different products at a constant current of −20 mA over the cathodic-catalysts on carbon paper substrates: (**a**) 50% Cu-AC, (**b**) 20% Cu-AC, and (**c**) 10% Cu-AC. Conditions: temperature = 90 °C, $F_{(CO_2)}$, cathode = 1.65 NmL·min^{-1} y $F_{(H_2O)}$, anode = 6 NmL·min^{-1}.

Figure 4a,b summarize the effect of the applied current and the reaction temperature on the steady state CO_2 consumption rate (after 350 min of polarization), respectively. The reaction rates were normalized by the amount of Cu deposited on each cathodic-catalyst. In line with the previous experiments, sample 50% Cu-AC showed larger electrocatalytic activities than samples 20% Cu-AC and 10% Cu-AC for all the reaction conditions studied. As expected, the consumption rate of CO_2 increased with the applied current, most likely because of an increase in the electrochemical supply of H^+. In line with previous studies [15], an increase in the reaction temperature also resulted in higher electrocatalytic activities for all the catalysts tested. While the kinetics of the electrochemical reactions can be improved by increasing the temperature [34], the protonic membrane prevented us from testing the system above 90 °C since its stability and conductivity under appropriate humidity conditions is not ensured at these conditions.

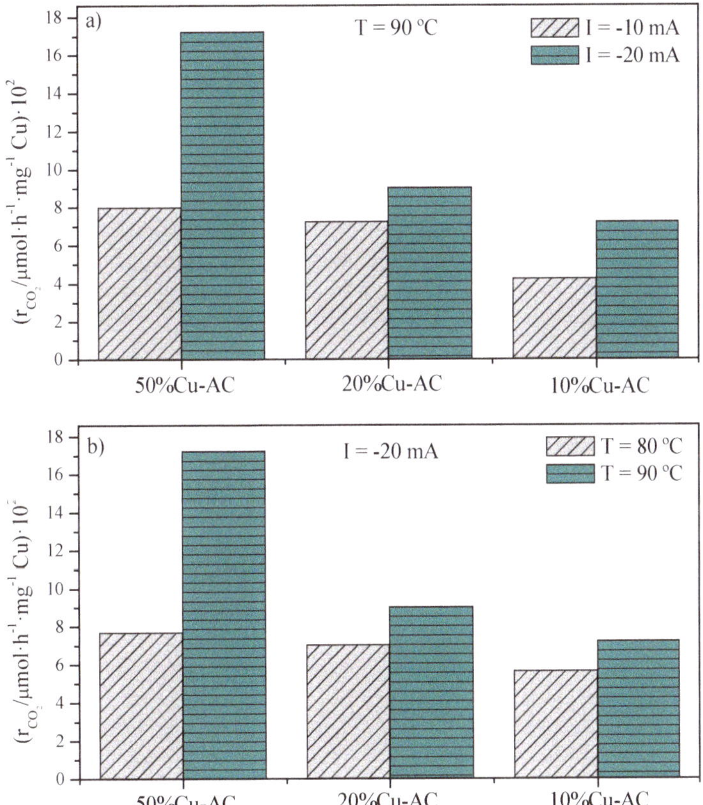

Figure 4. (**a**) Effect of the current at 90 °C and (**b**) temperature at I = −20 mA on the steady state CO_2 consumption rate for the cathodic-catalysts 50% Cu-AC, 20% Cu-AC, and 10% Cu-AC. Conditions: $F_{\_(CO\_2)}$, cathode = 1.65 NmL·min^{-1} y $F_{\_(H\_2O)}$, anode = 6 NmL·min^{-1}.

Finally, we calculated the energy consumption for the three different cathodic-catalysts evaluated. The three different MEAs were compared at 90 °C and −20 mA. The overall energy consumption for CO_2 conversion (kW·h·mol^{-1} CO_2) and the energy consumption for the production of methanol (kW·h·mol^{-1} CH_3OH), acetaldehyde (kW·h·mol^{-1} CH_3CHO), and methyl formate (kW·h·mol^{-1} HCO_2CH_3) were calculated. As indicated above, the cathodic-catalyst with the highest metal loading, 50% Cu-AC, showed the highest electrocatalytic activity. This catalyst showed the lowest energy consumption for the conversion of CO_2 (119.01 kW·h·mol^{-1}) among the catalysts tested. In addition, the MEA containing 50% Cu-AC consumed less energy per kg of methanol, acetaldehyde, and methyl formate (1549, 1054, and 444 kW·h·mol^{-1}, respectively) than the other two.

3. Materials and Methods

Cu catalysts supported on activated carbon were used as cathodes for the electrochemical reduction of CO_2, while Ir (IV) oxide (IrO_2) supported on carbon was used as a cell anode.

A commercial high surface area activated carbon (Sigma Aldrich, St. Louis, MO, USA) was used as a support. Metal particles were impregnated on activated carbon under vacuum at room conditions using an ethanolic precursor solution of $Cu(NO_3)_3 \cdot 3H_2O$ (Panreac, Barcelona, Spain) in a rotary evaporator. The catalysts were dried at 120 °C overnight and then calcined for 2 h at 350 °C in an N_2 atmosphere. A final reduction step under H_2 at 400 °C for 2 h was performed with a heating rate of 5 °C·min^{-1}. Three different catalysts were prepared with total Cu loadings of 50, 20, and 10 wt % (50% Cu-AC, 20% Cu-AC, and 10% Cu-AC, respectively).

The catalyst inks were prepared by mixing appropriate amounts of the different catalysts; IrO_2 commercial catalyst powders (Alfa Aesar, Haverhill, MA, USA, 99% for the anode, and Cu-activated-carbon powder for the cathode) with a Nafion solution (5 wt %, Aldrich Chemistry, St. Louis, MO, USA, Nafion® 117 solution) in isopropanol (Sigma Aldrich) containing a Nafion/solvent volume ratio of 0.04. IrO_2 was selected as an anode as in conventional PEM water electrolyzers [17]. The different inks were deposited on carbon paper (Fuel Cell Earth, Woburn, MA, USA) substrates with a geometric surface area for both electrodes of 12.56 cm^2 (circular electrode) at 65 °C. A commercial proton conducting Sterion® membrane (Hydrogen Works) was used as the electrolyte. Prior to use, the Sterion® membrane was successively pre-treated at 100 °C for 2 h in H_2O_2 and then into H_2SO_4 (to promote activation), and finally in deionized water to wash it. Then, the membrane electrode assembly (MEA) was prepared by sandwiching the membrane between the electrodes. Finally, the whole system was hot-pressed under heating conditions at 120 °C, and a final pressure of 1 metric ton was applied for 3 min.

The Cu metal loading on the powder catalysts was measured by Atomic Absorption Spectroscopy (AAS) using a SPECTRA 220FS analyzer. The sample (ca. 0.5 g) was dissolved in a mixture containing 2 mL of HCl, 3 mL of HF, and 2 mL of H_2O_2 followed by microwave digestion at 250 °C. The surface area and volume porosity of the different materials were measured using a Micromeritics ASAP 2010. N_2 was used as the sorbate at −193 °C, and the microporosity was measured by the Howath–Kawazoe (HK) method. Prior to the analyses, the samples were outgassed at 180 °C under vacuum (5 × 10^{-3} Torr) for 12 h. Temperature Programmed Reduction (TPR) experiments were conducted on a commercial Micromeritics AutoChem 2950 HP unit provided with a thermal conductivity (TCD) detector. The samples (ca. 0.15 g) were loaded into a U-shaped reactor and ramped from room temperature up to 900 °C (10 °C·min^{-1}) under a reducing H_2/Ar gas mixture of 17.5% v/v (60 cm^3·min^{-1}). X-Ray diffraction analysis (XRD) were conducted on the Cu-AC powder catalysts prior and after reduction with a Philips PW-1710 instrument, using Ni-filtered Cu Kα radiation (λ = 1.5404 Å). The samples were analyzed at a rate of 0.02°·step^{-1} over a 2θ range of 20–80° (scan time 2 s·step^{-1}) and the obtained spectra were compared with the corresponding JCPDS-ICDD references.

CO_2 electro-reduction experiments were performed in an electrochemical cell reactor working at room pressure conditions [17]. Water was introduced into the anode side using a saturator. The water content in the anodic feed stream (25% H_2O/N_2) was controlled by the heating the saturator at 65 °C. All the pipelines downstream from the saturator were heated to avoid condensation. On the anode side, water electro-oxidation was carried out on IrO_2 to generate protons across the Sterion® membrane. The water stream was also used to hydrate the Sterion® membrane and to keep its ionic conductivity under working conditions [16]. The cathodic side of the cell operates under a gas flow of pure CO_2 (Praxair Inc. certified standards 99.999% purity, Danbury, CT, USA). Both gas flow rates (N_2 for the anode and CO_2 for the cathode) were controlled by mass flowmeters (Brooks, Seattle, WA, USA).

The electrocatalytic experiments were carried out at atmospheric pressure with an overall gas flow rate of 0.5 $NmL{\cdot}min^{-1}$ of CO_2 for the cathodic stream and 6 $NmL{\cdot}min^{-1}$ for the anodic stream working at 80 and 90 °C. The reactant and products outer stream from the cathodic chamber were analyzed by using a double channel gas chromatograph (Bruker 450-GC) equipped with Hayesep Q-Molsieve 13X consecutive columns and flame ionization detectors (FIDs). Hydrogen, methanol, acetaldehyde, methyl formate, acetone, and n-propanol were the reaction products detected. The carbon atom balance closed within a 10% error. A potentiostat/galvanostat (Voltalab 21, Radiometer Analytical, Lyon, France) was used for the different applied polarizations, and the two electrodes were connected to the potentiostat by using gold wires.

4. Conclusions

Three different Cu-based cathodic-catalysts with Cu loadings of 50, 20, and 10 wt % were synthesized by impregnation, characterized, and tested in the electrocatalytic conversion of CO_2.

50% Cu-AC showed the highest CO_2 electrocatalytic activity among the catalysts tested under all the explored reaction conditions. These results could be explained by the higher Cu particle size of this material. Considering that CO is an important intermediate in the process, large Cu particles are believed to favor electroreduction by weakening the metal–CO interaction.

Methyl formate was the main reaction product for 50% Cu-AC, while acetaldehyde and methanol were the main products for 20% Cu-AC and 10% Cu-AC, respectively. This fact can be attributed to the higher particle size of Cu that favored the production of methyl formate via dehydrogenation of methanol.

The CO_2 consumption rate increased with the applied current and the reaction temperature due to an enhancement of the kinetic of the electrocatalytic reactions.

Finally, 50% Cu-AC showed the highest electrocatalytic activity and the lowest energy consumption values for the conversion of CO_2 (119 $kW{\cdot}h{\cdot}mol^{-1}$). In addition, the MEA containing 50% of Cu consumed less energy per kg of methanol, acetaldehyde, and methyl formate than the other two MEAs containing less amount of Cu.

Author Contributions: J.L.V. conceived the project, N.G.-G. performed the experiments and J.C.S.-R. and A.d.L.-C. wrote the manuscript.

Funding: Financial Support received from the Spanish "Ministerio de Ciencia e Innovación" (Project CTQ2016-75491-R) and from Abengoa Research is gratefully acknowledged. JCSR would like to acknowledge the Spanish Ministry of Science, Innovation and Universities for financial support under the Ramón y Cajal Program (grant RYC-2015-19230).

Conflicts of Interest: The authors declare no conflict of interest. The founding sponsors had no role in the design of the study; in the collection, analyses, or interpretation of data; in the writing of the manuscript, and in the decision to publish the results.

References

1. Umeyama, T.; Imahori, H. Carbon nanotube-modified electrodes for solar energy conversion. *Energy Environ. Sci.* **2008**, *1*, 120–133. [CrossRef]
2. Bansode, A.; Urakawa, A. Towards full one-pass conversion of carbon dioxide to methanol and methanol-derived products. *J. Catal.* **2014**, *309*, 66–70. [CrossRef]
3. Sekar, N.; Ramasamy, R.P. Recent advances in photosynthetic energy conversion. *J. Photochem. Photobiol. C-Photochem. Rev.* **2015**, *22*, 19–33. [CrossRef]
4. Khiabani, N.H.; Yaghmaee, M.S.; Sarani, A.; Shokri, B. Synthesis-gas production from CH_4-CO_2-Ar via microwave plasma torch. *Adv. Stud. Theor. Phys.* **2012**, *6*, 1273–1287.
5. Yu, Q.; Kong, M.; Liu, T.; Fei, J.; Zheng, X. Characteristics of the decomposition of CO_2 in a dielectric packed-bed plasma reactor. *Plasma Chem. Plasma Process.* **2012**, *32*, 153–163. [CrossRef]
6. Rahemi, N.; Haghighi, M.; Babaluo, A.A.; Jafari, M.F.; Allahyari, S. CO_2 reforming of methane over Ni-Cu/Al_2O_3-ZrO_2 nanocatalyst: The influence of plasma treatment and process conditions on catalytic properties and performance. *Korean J. Chem. Eng.* **2014**, *31*, 1553–1563. [CrossRef]

7. Centi, G.; Perathoner, S. Opportunities and prospects in the chemical recycling of carbon dioxide to fuels. *Catal. Today* **2009**, *148*, 191–205. [CrossRef]
8. Albo, J.; Alvarez-Guerra, M.; Castaño, P.; Irabien, A. Towards the electrochemical conversion of carbon dioxide into methanol. *Green Chem.* **2015**, *17*, 2304–2324. [CrossRef]
9. Centi, G.; Quadrelli, E.A.; Perathoner, S. Catalysis for CO_2 conversion: A key technology for rapid introduction of renewable energy in the value chain of chemical industries. *Energy Environ. Sci.* **2013**, *6*, 1711–1731. [CrossRef]
10. Graves, C.; Ebbesen, S.D.; Mogensen, M. Co-electrolysis of CO_2 and H_2O in solid oxide cells: Performance and durability. *Solid State Ionics* **2011**, *192*, 398–403. [CrossRef]
11. Blomen, E.; Hendriks, C.; Neele, F. Capture technologies: improvements and promising developments. *Energy Procedia* **2009**, *1*, 1505–1512. [CrossRef]
12. Yamanaka, I.; Tabata, K.; Mino, W.; Furusawa, T. Electroreduction of Carbon Dioxide to Carbon Monoxide by Co-pthalocyanine Electrocatalyst under Ambient Conditions. *ISIJ Int.* **2015**, *55*, 399–403. [CrossRef]
13. Ampelli, C.; Genovese, C.; Perathoner, S.; Centi, G.; Errahali, M.; Gatti, G.; Marchese, L. An electrochemical reactor for the CO_2 reduction in gas phase by using conductive polymer based electrocatalysts. *Chem. Eng. Trans.* **2014**, *41*, 13–18.
14. Lee, K.; Zhang, J.; Wang, H.; Wilkinson, D.P. Progress in the synthesis of carbon nanotube-and nanofiber-supported Pt electrocatalysts for PEM fuel cell catalysis. *J. Appl. Electrochem.* **2006**, *36*, 507–522. [CrossRef]
15. Genovese, C.; Ampelli, C.; Perathoner, S.; Centi, G. Electrocatalytic conversion of CO_2 to liquid fuels using nanocarbon-based electrodes. *J. Energy Chem.* **2013**, *22*, 202–213. [CrossRef]
16. Genovese, C.; Ampelli, C.; Perathoner, S.; Centi, G. Electrocatalytic conversion of CO_2 on carbon nanotube-based electrodes for producing solar fuels. *J. Catal.* **2013**, *308*, 237–249. [CrossRef]
17. Gutiérrez-Guerra, N.; Moreno-López, L.; Serrano-Ruiz, J.C.; Valverde, J.L.; de Lucas-Consuegra, A. Gas phase electrocatalytic conversion of CO_2 to syn-fuels on Cu based catalysts-electrodes. *Appl. Catal. B-Environ.* **2016**, *188*, 272–282.
18. Rodriguez-Reinoso, F. The role of carbon materials in heterogeneous catalysis. *Carbon* **1998**, *36*, 159–175. [CrossRef]
19. Molina-Sabio, M.; Perez, V.; Rodriguez-Reinoso, F. Impregnation of activated carbon with chromium and copper salts: Effect of porosity and metal content. *Carbon* **1994**, *32*, 1259–1265. [CrossRef]
20. Zhang, G.; Li, Z.; Zheng, H.; Fu, T.; Ju, Y.; Wang, Y. Influence of the surface oxygenated groups of activated carbon on preparation of a nano Cu/AC catalyst and heterogeneous catalysis in the oxidative carbonylation of methanol. *Appl. Catal. B-Environ.* **2015**, *179*, 95–105. [CrossRef]
21. Díez-Ramírez, J.; Sánchez, P.; Rodríguez-Gómez, A.; Valverde, J.L.; Dorado, F. Carbon nanofiber-based palladium/zinc catalysts for the hydrogenation of carbon dioxide to methanol at atmospheric pressure. *Ind. Eng. Chem. Res.* **2016**, *55*, 3556–3567. [CrossRef]
22. Román-Martínez, M.C.; Cazorla-Amorós, D.; Linares-Solano, A.; De Lecea, C.S.M. TPD and TPR characterization of carbonaceous supports and Pt/C catalysts. *Carbon* **1993**, *31*, 895–902. [CrossRef]
23. Gil, S.; Muñoz, L.; Sánchez-Silva, L.; Romero, A.; Valverde, J.L. Synthesis and characterization of Au supported on carbonaceous material-based catalysts for the selective oxidation of glycerol. *Chem. Eng. J.* **2011**, *172*, 418–429. [CrossRef]
24. Trépanier, M.; Dalai, A.K.; Abatzoglou, N. Synthesis of CNT-supported cobalt nanoparticle catalysts using a microemulsion technique: Role of nanoparticle size on reducibility, activity and selectivity in Fischer–Tropsch reactions. *Appl. Catal. A-Gen.* **2010**, *374*, 79–86. [CrossRef]
25. Karelovic, A.; Ruiz, P. The role of copper particle size in low pressure methanol synthesis via CO_2 hydrogenation over Cu/ZnO catalysts. *Catal. Sci. Technol.* **2015**, *5*, 869–881. [CrossRef]
26. Genovese, C.; Ampelli, C.; Perathoner, S.; Centi, G. A gas-phase electrochemical reactor for carbon dioxide reduction back to liquid fuels. *Chem. Eng. Trans.* **2013**, *32*, 289–294.
27. Gangeri, M.; Perathoner, S.; Caudo, S.; Centi, G.; Amadou, J.; Begin, D.; Schlögl, R. Fe and Pt carbon nanotubes for the electrocatalytic conversion of carbon dioxide to oxygenates. *Catal. Today* **2009**, *143*, 57–63. [CrossRef]
28. Centi, G.; Perathoner, S.; Winè, G.; Gangeri, M. Electrocatalytic conversion of CO_2 to long carbon-chain hydrocarbons. *Green Chem.* **2007**, *9*, 671–678. [CrossRef]

29. Ahouari, H.; Soualah, A.; Le Valant, A.; Pinard, L.; Magnoux, P.; Pouilloux, Y. Methanol synthesis from CO_2 hydrogenation over copper based catalysts. *React. Kinet. Mech. Catal.* **2013**, *110*, 131–145. [CrossRef]
30. Díez-Ramírez, J.; Valverde, J.L.; Sánchez, P.; Dorado, F. CO_2 hydrogenation to methanol at atmospheric pressure: influence of the preparation method of Pd/ZnO catalysts. *Catal. Lett.* **2016**, *146*, 373–382. [CrossRef]
31. Díez-Ramírez, J.; Dorado, F.; de la Osa, A.R.; Valverde, J.L.; Sánchez, P. Hydrogenation of CO_2 to methanol at atmospheric pressure over Cu/ZnO catalysts: influence of the calcination, reduction, and metal loading. *Ind. Eng. Chem. Res.* **2017**, *56*, 1979–1987. [CrossRef]
32. Reske, R.; Mistry, H.; Behafarid, F.; Roldan Cuenya, B.; Strasser, P. Particle size effects in the catalytic electroreduction of CO_2 on Cu nanoparticles. *J. Am. Chem. Soc.* **2014**, *136*, 6978–6986. [CrossRef] [PubMed]
33. Tonner, S.P.; Trimm, D.L.; Wainwright, M.S.; Cant, N.W. Dehydrogenation of methanol to methyl formate over copper catalysts. *Ind. Eng. Chem. Prod. Res. Dev.* **1984**, *23*, 384–388. [CrossRef]
34. Caravaca, A.; Sapountzi, F.M.; de Lucas-Consuegra, A.; Molina-Mora, C.; Dorado, F.; Valverde, J.L. Electrochemical reforming of ethanol–water solutions for pure H_2 production in a PEM electrolysis cell. *Int. J. Hydrogen Energy* **2012**, *37*, 9504–9513. [CrossRef]

Article

CuO Nanoparticles Supported on TiO_2 with High Efficiency for CO_2 Electrochemical Reduction to Ethanol

Jing Yuan, Jing-Jie Zhang, Man-Ping Yang, Wang-Jun Meng, Huan Wang * and Jia-Xing Lu *

Shanghai Key Laboratory of Green Chemistry and Chemical Processes, School of Chemistry and Molecular Engineering, East China Normal University, Shanghai 200062, China; yuanjing158yj@126.com (J.Y.); zjj18221019582@163.com (J.-J.Z.); YMP2018@126.com (M.-P.Y.); mwjynl@outlook.com (W.-J.M.)

* Correspondence: hwang@chem.ecnu.edu.cn (H.W.); jxlu@chem.ecnu.edu.cn (J.-X.L.); Tel.: +86-21-5213-4935 (H.W.); +86-21-6223-3491 (J.-X.L.)

Received: 7 March 2018; Accepted: 18 April 2018; Published: 21 April 2018

Abstract: Non-noble metal oxides consisting of CuO and TiO_2 (CuO/TiO_2 catalyst) for CO_2 reduction were fabricated using a simple hydrothermal method. The designed catalysts of CuO could be in situ reduced to a metallic Cu-forming Cu/TiO_2 catalyst, which could efficiently catalyze CO_2 reduction to multi-carbon oxygenates (ethanol, acetone, and n-propanol) with a maximum overall faradaic efficiency of 47.4% at a potential of −0.85 V vs. reversible hydrogen electrode (RHE) in 0.5 M $KHCO_3$ solution. The catalytic activity for CO_2 electroreduction strongly depends on the CuO contents of the catalysts as-prepared, resulting in different electrochemistry surface areas. The significantly improved CO_2 catalytic activity of CuO/TiO_2 might be due to the strong CO_2 adsorption ability.

Keywords: electrochemical reduction; CO_2; CuO; TiO_2; ethanol

1. Introduction

Electrochemical reduction of CO_2 (CO_2ER) is a promising and feasible method with which to sustainably transform this waste stream into value-added low-carbon fuels [1–5], which has the following advantages: (1) An electrochemical system can be operated under moderate reaction conditions [5,6]; (2) The reaction process involves highly complex multiple protons and electrons transfer steps, which lead to wide product distribution, such as methanol, ethanol, acetone, and so on [7,8]; (3) The electricity could be supplied by clean and sustainable energy, such as solar, wind, and hydropower [9,10]; and (4) This reaction requires minimal chemical intake and is convenient for large-scale applications [5,8]. However, CO_2ER is facing severe challenges, including poor faradaic efficiency (FE), high overpotential, and low selectivity [11,12], which urges us to design new catalysts to address the above efficiency and selectivity issues.

Over the past few decades, numerous trials have been made to explore catalysts with distinguished performance for CO_2ER. Until now, multifarious catalysts including metals [5,11–13], metal oxides [14–16], and metal complexes [17,18] have been reported. Among these catalysts, Cu, as a relatively low-cost and earth abundant metal, has a unique capacity to produce hydrocarbons through a multiple protons and electrons transfer pathway; however, the traditional Cu catalysts show high overpotential [16] and low selectivity for diversiform products [5,12,19–21]. Substantial efforts have been pursued to enhance energetic efficiency for CO_2ER through altering surface structures, morphologies, and the nature compositions. Recently, Cu-based catalysts have been reported to possess enhanced FEs for CO_2ER; for instance, Takanabe et al. [22] used Cu-Sn alloy for the efficient and selective reduction of CO_2 to CO over a wide potential range. Yu and her team [23] reported that Cu nanoparticle (NP) interspersed MoS_2 nanoflowers facilitate CO_2ER to hydrocarbon, such as

CH_4 and C_2H_4 with high FE at low overpotentials. Sun et al. [24] introduced a Core/Shell Cu/SnO_2 structure that shows high selectivity to generate CO with FE reaching 93% at −0.7 V (vs. the reversible hydrogen electrode (RHE)).

Normally, TiO_2, as a semiconductor material, is one of the most widely used photocatalysts and electrocatalysts for CO_2 reduction because of its nontoxicity, low cost, and high chemical stability [25,26]. Furthermore, TiO_2 has been reported to act as a redox electron carrier to facilitate a variety of reduction reactions, including CO_2 conversion [27,28], and to assist in CO_2 adsorption [29,30]; thus, it may stabilize the CO_2ER intermediate and reduce overpotential. Some early works could verify this point in the CO_2ER, such as Ag/TiO_2 [31], Cu/TiO_2 [32], and Cu/TiO_2/N-graphene [33].

CuO/TiO_2 has been previously prepared using different synthetic methods and is used for various applications, such as photodriven reduction of CO_2 [34] and hydrogen production reaction [35]. Nevertheless, to the author's knowledge, CuO/TiO_2 as electrocatalysts for CO_2ER have been rarely investigated. In previous works, we have demonstrated that CuO with various morphologies can be highly efficient electrocatalysts for CO_2 reduction to ethanol in simple aqueous medium, but at high overpotential (−1.7 V vs. the saturated calomel electrode (SCE)) [36]. Inspired by previous studies, in this work, we systematically and carefully synthesized a series of nano-sized CuO/TiO_2 catalysts for CO_2ER. The prepared CuO/TiO_2 catalysts were optimized by varying the amount of loaded CuO NPs. CuO/TiO_2 with an intended CuO content of 60%, as an efficient electrocatalyst, exhibits the most outstanding activity (achieved total FE of 47.4% at the potential of −0.85 V vs. RHE) for CO_2ER in 0.5 M $KHCO_3$ solution at room temperature among all as-prepared CuO/TiO_2 catalysts.

2. Results and Discussion

2.1. Catalyst Characterizations

A simple and mild hydrothermal synthesis method was used to prepare six different weight ratios of CuO/TiO_2 catalysts, which are defined as CuO/TiO_2-1, CuO/TiO_2-2, CuO/TiO_2-3, CuO/TiO_2-4, CuO/TiO_2-5, and CuO/TiO_2-6, corresponding to the intended CuO content of 5 wt %, 10 wt %, 20 wt %, 40 wt %, 60 wt %, and 80 wt %, respectively. X-ray diffractometer (XRD) patterns shown in Figure 1 indicated that these catalysts consist of both CuO and TiO_2, which were represented by solid lines and dashed lines, respectively. It can be clearly seen that all as-prepared catalysts with different amounts of CuO loadings present similar XRD patterns. The diffraction peaks ascribed to CuO were significantly shown with the increasing of CuO contents in the XRD patterns. Meanwhile, the actual CuO contents were analyzed by inductively coupled plasma atomic emission spectroscopy (ICP-AES) and summarized in Table S1. These values are in good agreement with the intended values.

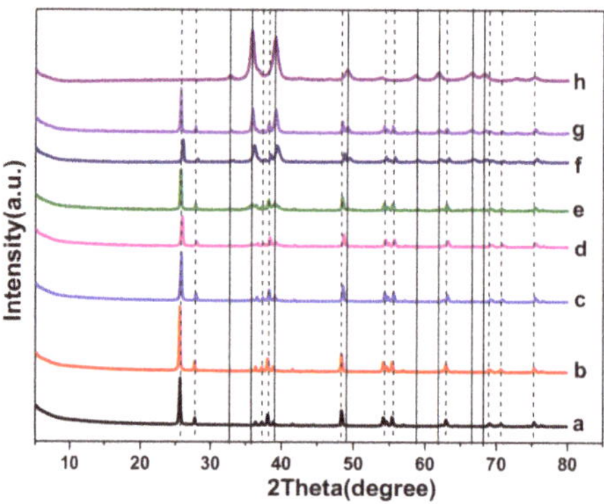

Figure 1. XRD patterns of (**a**) pure TiO_2, (**b**) CuO/TiO_2-1, (**c**) CuO/TiO_2-2, (**d**) CuO/TiO_2-3, (**e**) CuO/TiO_2-4, (**f**) CuO/TiO_2-5, (**g**) CuO/TiO_2-6, and (**h**) pure CuO.

To thoroughly examine the element proportion and chemical state of CuO/TiO_2 catalysts, X-ray photoelectron spectroscopy (XPS) of CuO/TiO_2-5 catalyst was investigated. As expected, Cu, Ti, and O elements from CuO/TiO_2 catalyst were observed in the full spectrum from Figure 2a. Furthermore, Figure 2b displays the high resolution spectrum of Cu 2p, separated into Cu $2p^{3/2}$ and Cu $2p^{1/2}$ at 933.8 eV and 953.8 eV, respectively. The distance between these Cu 2p main peaks positions is 20.0 eV, which agrees well with previous reports about CuO spectrum [37]. Moreover, additional confirmation of CuO state was seen with the broad satellite peaks at a higher binding energy than the main peaks. The main peak of Cu $2p^{3/2}$ at 933.8 eV was accompanied by two satellite peaks on the higher binding energy side at about 943.8 eV and 941.5 eV, which suggests the existence of CuO [38–41]. From this figure, we can clearly see that the main peak of Cu $2p^{1/2}$ at 953.8 eV and its satellite peak at 962.5 eV were separated by about 9.0 eV, which also confirms the presence of CuO [42]. Both Ti $2p^{3/2}$ and Ti $2p^{1/2}$ peaks at 458.7 eV and 464.4 eV, respectively, were observed (Figure 2c) with a separation of 5.7 eV, indicating that TiO_2 existed in this catalyst [43]. The obtained XPS spectrum of O 1 s was presented in Figure 2d. An obvious peak appeared at 529.7 eV, which can be indexed to O^{2-} in the CuO and TiO_2. Notably, there are other three weak O 1 s peaks. One located at 530.7 eV is identified with surface hydroxyls, which is likely the by-product from the synthesis process of CuO/TiO_2. Additionally, the remaining two peaks observed at 531.5 eV and 532.8 eV are confirmed to C=O and C-O [44,45]. A slight peak of C 1 s was detected in Figure 2a, which is always observed in XPS spectra of real-world solids [46,47]. The high resolution spectrum of C 1 s was shown in Figure S1, consistent with the results of O 1 s.

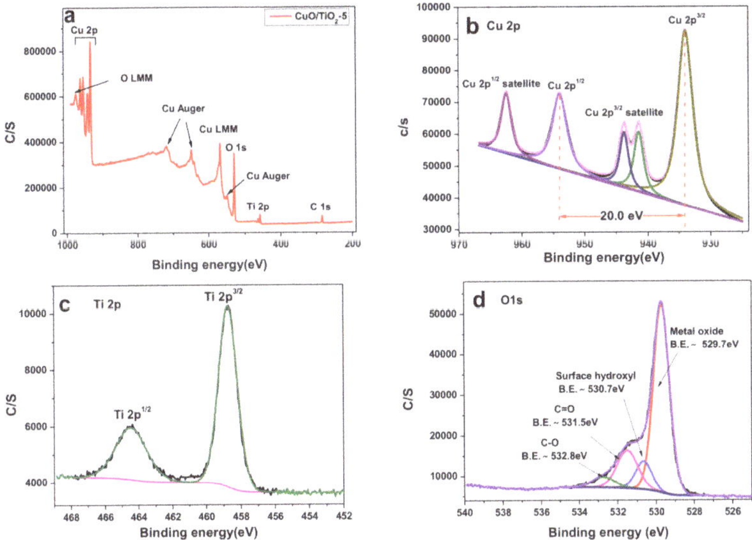

Figure 2. (**a**) The full XPS spectrum of CuO/TiO_2-5 catalyst and high-resolution XPS spectra of (**b**) Cu 2p; (**c**) Ti 2p and (**d**) O 1 s.

To obtain the morphological information regarding the CuO/TiO_2 catalysts, scanning electron microscope (SEM) images were firstly shown in Figure S2. One can see that the catalysts were composed of 3-dimension NPs with a certain amount of CuO nano-floc supported on the surface of TiO_2. Further from the SEM images, for low CuO content of CuO/TiO_2 (5 wt %, 10 wt %, and 20 wt %), CuO NPs were dotted sporadically on the surface of TiO_2. Subsequently, for high CuO content of CuO/TiO_2 (40 wt %, 60 wt %, and 80 wt %), it is evident that the CuO NPs completely covered on the TiO_2 surface forming a massive and compact layer. Pure CuO NPs presented very porous, sponge-like structures. Moreover, transmission electron microscopy (TEM) investigation was used to gain deeper insight into the structural feature of the catalysts. TEM of CuO/TiO_2-5 was shown in Figure 3a. Numerous CuO NPs were irregularly interspersed on the TiO_2 surface, in accordance with the SEM results. High-resolution TEM images of CuO/TiO_2-5 were displayed in Figure 3b to acquire more detailed information about the structure of the catalyst. The fast Fourier transform (FFT, inset of Figure 3b) pattern of CuO and TiO_2 shows concentric rings and bright discrete diffraction spots, which are indicative of high crystallinity. From Figure 3b, distinctive lattice fringes were found in both TiO_2 and CuO NPs. The lattice fringes with interplanar spacing of 0.188 nm were assigned to the (200) plane of the TiO_2, expressed by orange lines (inserted in Figure 3b). Additionally, the lattice fringes with d-spacings of 0.256 nm corresponded to the plane (111) of CuO phase, represented by the yellow lines (inserted in Figure 3b).

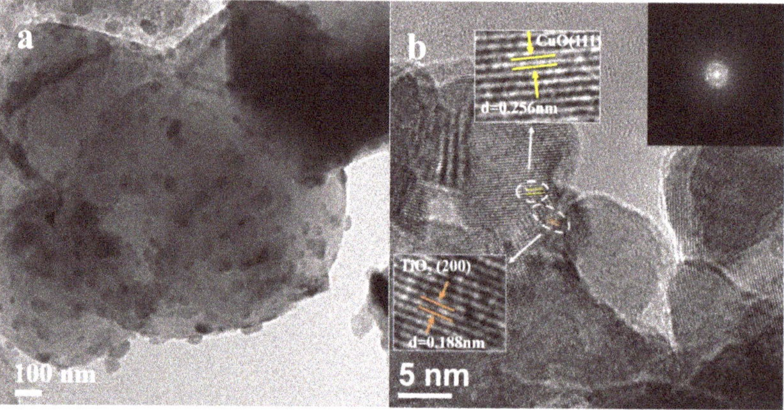

Figure 3. TEM images of CuO/TiO_2-5 catalyst in (**a**) low magnification and (**b**) high magnification (the insets are FFT pattern and the detailed images of white dashed circle).

2.2. Catalyst Properties

The electrocatalytic activity of CuO/TiO_2-5 catalyst was carried out in the typical three-electrode system through cyclic voltammetry (CV) measurement, which was firstly tested at a scan rate of 50 mV/s in N_2-saturated and CO_2-saturated 0.5 M $KHCO_3$ solution, respectively, as shown in Figure 4a. In both N_2 and CO_2 atmosphere, two obvious reduction peaks were observed, corresponding to the in situ reduction of CuO to Cu_2O and sequentially Cu_2O to Cu [48]. As expected, XRD characterization could verify the formation of Cu metal from CuO/TiO_2-5 catalyst after 2 h electrolysis, depicted in Figure S3. Three significant diffraction peaks for metallic Cu (denoted by solid diamond) appeared, suggesting that CuO NPs in CuO/TiO_2-5 catalyst could be in situ electroreduced to metallic Cu-forming Cu/TiO_2 in the electrolysis process and subsequently serves as an effective catalyst for CO_2 reduction, verified by previous reports [16,36,47]. In the more negative potential region, an obvious increase of current density (J) after −0.65 V vs. RHE was shown, relating to the hydrogen evolution reaction (HER) in the N_2-saturated 0.5 M $KHCO_3$ solution (pH = 8.63, black line), while more dramatic increase of J was observed in the CO_2-saturated 0.5 M $KHCO_3$ solution (pH = 7.21, red line). To avoid the pH effect, a CV curve was recorded in the N_2-saturated solution with HCl solution (pH = 7.21, same as that of the CO_2-saturated 0.5 M $KHCO_3$ solution, blue line), which shows lower J than that in CO_2-saturated 0.5 M $KHCO_3$ solution after −0.65 V vs. RHE. It demonstrates that CO_2 reduction is more favorable than HER on CuO/TiO_2-5.

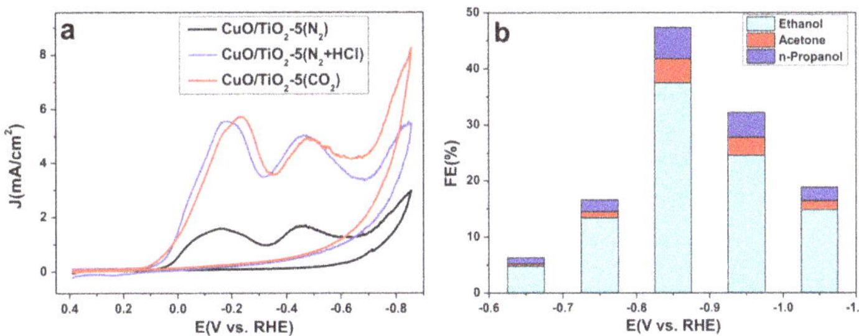

Figure 4. (**a**) CV curves of CuO/TiO$_2$-5 catalyst over glassy carbon electrode (GCE) in N$_2$-saturated without (black line) and with (blue line) HCl solution, and CO$_2$-saturated (red line) 0.5 M KHCO$_3$ solution; (**b**) FEs for different products over CuO/TiO$_2$-5 catalyst at various potentials.

The CO$_2$ER electrocatalytic activity on CuO/TiO$_2$-5 catalyst was also evaluated by controlled potential electrolysis (from −0.65 V to −1.05 V vs. RHE) in CO$_2$-saturated 0.5 M KHCO$_3$ solution. The main liquid products of the electrolysis are ethanol, acetone, and n-propanol, which were detected by ^{1}H NMR. The overall FEs (Figure 4b) presented a sharply incremental tendency at an applied potential range from −0.65 V to −0.85 V vs. RHE and achieved a maximum of 47.4% (37.5% for ethanol, 4.3% for acetone, and 5.6% for n-propanol) at the potential of −0.85 V vs. RHE. As the potentials shift more negatively, the overall FEs significantly decreased to 18.8% at the potential of −1.05 V vs. RHE because of the competition from HER. Furthermore, a similar variation trend for $FE_{ethanol}$, $FE_{acetone}$, and $FE_{n\text{-}propanol}$ was observed and depicted in detail in Figure 4b. Moreover, Figure S4 shows the total current vs. time curve for the CuO/TiO$_2$-5 electrode at −0.85 V vs. RHE, which exhibited an initial current of 85 mA as the CuO was reduced and, subsequently, a sharp decline current in a short time. Finally, a stable current of 16 mA in the long test appeared. Notably, the FE for ethanol was maintained at approximately 35% throughout the electrolysis. This finding suggests not only efficient but also stable activity for CO$_2$ reduction on this electrode.

We also studied the effect of CuO contents in various CuO/TiO$_2$ catalysts on CO$_2$ER activities, as shown in Figure 5a. All CuO/TiO$_2$ catalysts showed two obvious reduction peaks in the CO$_2$-saturated 0.5 M KHCO$_3$ solution; however, the position and size of the reduction peaks varied greatly among the catalysts, which, assigned to in situ, reduce CuO to Cu. Noteworthily, in the more negative potential region (below −0.65 V vs. RHE), all CuO/TiO$_2$ catalysts show different catalytic abilities for CO$_2$ reduction. The J of all CuO/TiO$_2$ catalysts at a potential of −0.85 V vs. RHE were summarized in Table S2. CuO/TiO$_2$-5 showed the largest J. The dramatic difference of electrocatalytic performance on various CuO/TiO$_2$ catalysts might be closely related to the composition of these catalysts. Figure 5a shows that increasing the CuO content (from 5 wt % to 60 wt %) in the CuO/TiO$_2$ catalysts can facilitate the reduction of CO$_2$, leading to enhanced activity for CO$_2$ reduction. However, on the CuO/TiO$_2$-6 (with the intended CuO content of 80 wt %) catalyst for sequentially increasing the content of CuO, CO$_2$ reduction was subsequently suppressed. The results clearly indicate that an optimum content of CuO in the catalysts is required to achieve the maximum activity for CO$_2$ reduction.

Potentiostatic electrolysis (at −0.85 V vs. RHE) of CO$_2$ on various CuO/TiO$_2$ catalysts was performed in CO$_2$-saturated 0.5 M KHCO$_3$ solution to further investigate the catalytic activities of CuO/TiO$_2$ catalysts with different CuO contents. Figure 5b summarizes the FEs of ethanol, acetone, and n-propanol achieved over all CuO/TiO$_2$ catalysts. As expected, the CuO/TiO$_2$-5 with the intended CuO content of 60 wt % gives the highest total FE, which reached 47.4%, when compared with other CuO/TiO$_2$ catalysts, which is in a well agreement with CV results as shown in Figure 5a. However, the total FEs decreased along with the higher CuO content (80 wt % and even 100 wt %) of CuO/TiO$_2$, which is ascribed to the large number of CuO NPs that accumulated on the surface of TiO$_2$, as shown in Figure S2f,g, which probably generated the decrease of active areas for CO$_2$ER. Yet, low CuO content (5 wt %–40 wt %) of CuO/TiO_2 also exhibited slightly poor electrocatalytic performance for CO_2ER due to the low content of CuO NPs. It demonstrates further that the catalytic activity for CO_2ER strongly depends on the CuO contents of the catalysts as-prepared, which results in different electrochemistry surface areas (ECSA). To confirm this finding, ECSA of a variety of CuO/TiO_2 electrodes were examined by CV in a potential range in which the faradaic process did not occur with the double layer capacitance in N_2 atmosphere 0.1 M $HClO_4$ solution (Figure S5 and Table S3). As expected, CuO/TiO_2-5 catalyst showed a noticeable performance improvement for the double layer capacitance (C_{dl}), which gives the positive correlation with ECSA. The C_{dl} in Table S3 indicated the ECSA increased with increasing CuO content up to 60 wt %; however, a slight decrease was observed at 80 wt %, which reconfirms the CuO content of CuO/TiO_2 catalysts plays a vital role in ECSA that can affect CO_2ER, which is in accordance with the results of Brunauer– Emmett–Teller (BET) specific surface area (Table S4).

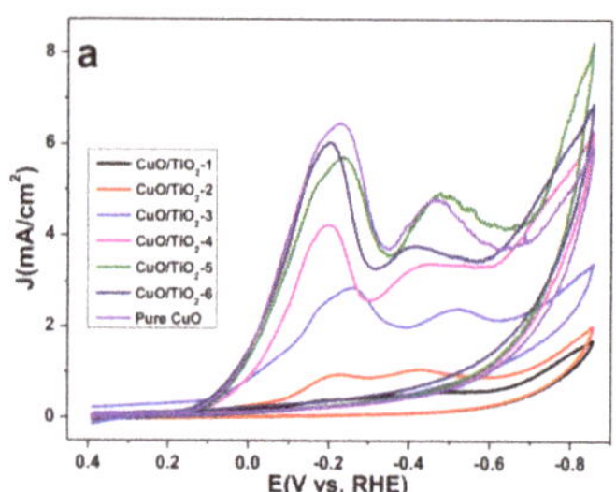

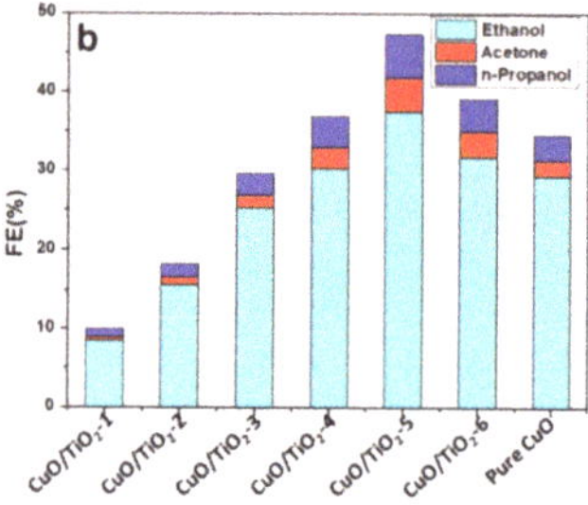

Figure 5. (**a**) CV curves of various CuO/TiO_2 catalysts over GCE in CO_2-saturated 0.5 M $KHCO_3$ solution. Scan rate, 50 mV/s; (**b**) FEs for different products over various CuO/TiO_2 catalysts at −0.85 V vs. RHE in CO_2-saturated 0.5 M $KHCO_3$ aqueous solution.

To further study the improved performance of CuO/TiO_2-5 catalyst, we compared the electrocatalytic abilities of CO_2 reduction on CuO/TiO_2-5 and CuO/C catalysts. The CuO/C was successfully synthesized and characterized by XRD (Figure S6), SEM, and TEM (Figure S7). CuO/C catalyst represents the onset potential for CO_2 reduction at −0.75 V vs. RHE, corresponding to a 0.10 V more negative onset potential than CuO/TiO_2-5, as shown in Figure S8. This finding suggests that TiO_2 as the supporter is more beneficial to facilitating CO_2ER. The J in CO_2-saturated solution of CuO/TiO_2-5 at the potential of −0.85 V vs. RHE was 8.307 mA/cm^2, which is larger than that of CuO/C (1.402 mA/cm^2). Additionally, potentiostatic electrolysis in CO_2-saturated 0.5 M $KHCO_3$ solution for the CuO/C was performed at the potential of −0.85 V vs. RHE, which shows poor active towards CO_2 reduction. Only 16.3% for overall FE (12.9% for ethanol, 1.1% for acetone, and 2.3% for n-propanol) was analyzed by ^{1}H NMR. In contrast, CuO/TiO_2-5 catalyst presents excellent activity for CO_2 reduction, which achieved 47.4% for overall FE. Besides, as seen from Figure S5 and Table S3, the ECSA of CuO/TiO_2-5 reveals a larger value than the one of CuO/C. As is well known, CO_2 adsorption on the active sites is the prerequisite for subsequent CO_2 reduction reaction. The relatively larger amount of CO_2 adsorption on active sites may offer more original reactants such as CO_2 [49]. The CO_2 adsorption abilities of CuO/TiO_2-5 and CuO/C were evaluated by CO_2 adsorption, which were displayed in Figure S9. CuO/TiO_2-5 catalyst possessed a more remarkably improved CO_2 adsorption capacity than CuO/C, which was 5.80 mg/g and 1.34 mg/g, respectively. It is reasonable to assume that the enhanced CO_2 adsorption capacity can make a significant contribution to the superior performance of CO_2 reduction. By comparing the electrochemical and material characterizations of CuO/TiO_2 with those of CuO/C, we gained insight that TiO_2 plays an important role in promoting the electrocatalytic performance of CuO/TiO_2. Hence, on the basis of the above analysis, a brief illustration of CO_2ER over CuO/TiO_2 catalysts is stated. Large amount of CO_2 is first adsorbed on TiO_2. CuO/TiO_2 catalysts could be in situ electroreduced to Cu/TiO_2 at less negative cathode potentials than the ones of CO_2ER. Then, the adsorbed CO_2 gets one electron from the achieved Cu/TiO_2 electrode and is converted to CO_2^-, which can be dimerized to $^*C_2O_2^-$. Notably, the C-C bond-making step on the Cu surface is a key step for CO_2 reduction to ethanol and even the C3 product (acetone and n-propanol) [50]. Interestingly, the catalytic ability of CuO/TiO_2-5 at −0.85 V vs. RHE expresses more obvious performance than the one of 40 wt % Cu/TiO_2 (the optimal catalyst in X wt % Cu/TiO_2 system) at −1.45 V vs. RHE [34], which indicates that in situ electroreduced Cu NPs have greater activity to catalyze CO_2 reduction. The low-cost CuO/TiO_2 catalyst is an efficient alternative to expensive materials for the application of CO_2ER in industry.

3. Materials and Methods

3.1. Materials and Instruments

Cupric acetate monohydrate [$Cu(Ac)_2 \cdot H_2O$, Analytical Reagent (AR) grade], ammonium carbonate [$(NH_4)_2CO_3$, AR grade], potassium bicarbonate [$KHCO_3$, AR grade], and titanium dioxide [TiO_2, AR grade] were purchased from Sinopharm Chemical Reagent Co. (SCR, Shanghai, China) with 99% purity and used as received. Nafion® 117 solution (5%) and Nafion® 117 membrane were obtained from Dupont (Wilmington, DE, USA). Carbon paper (CP, HCP010) and conductive carbon black (C, VXC72R) were purchased from Shanghai Hesen Electrical CO. (Shanghai, China).

Crystal-phase X-ray diffraction (XRD) patterns of CuO/TiO_2 catalysts were recorded using an Ultima IV X-ray powder diffractometer (Kurary, Tokyo, Japan) equipped with Cu Kα radiation (k = 1.5406 Å). The values for actual CuO loadings of the synthesized catalysts were determined on an inductively coupled plasma atomic emission spectroscopy (ICP-AES, IRIS Intrepid II XPS, Waltham, MA, USA). The scanning electron microscope (SEM) images of the catalysts were obtained using a Hitachi S-4800 field-emission scanning electron microscope (Tokyo, Japan). Transmission electron microscopy (TEM) patterns were recorded by TECNAI G_2F30 transmission electron microscope (Tokyo, Japan). The X-ray photoelectron spectroscopy (XPS, Thermo Fisher Scientific, Waltham, MA, USA) analysis was performed on the Thermo Scientific ESCA Lab 250Xi using 200 W monochromatic Al Kα radiation. The 500 µm X-ray spot was used. The base pressure in the analysis chamber was about 3×10^{-10} mbar. Typically, the hydrocarbon C1s line at 284.8 eV from adventitious carbon was used for energy referencing. Brunauer−Emmett−Teller (BET) specific surface area was characterized by nitrogen adsorption in a BELSORP-MAX instrument (MicrotracBEL, Tokyo, Japan) after outgassing the samples for 10 h under vacuum at 573 K. CO_2 adsorption was obtained in a BELSORP-MAX instrument (MicrotracBEL, Japan) under CO_2 atmosphere at 298 K. Potentiostatic electrolysis and cyclic voltammetry (CV) were performed using a CHI 660 C electrochemical station (Shanghai Chenhua Instrument Co. Ltd., Shanghai, China).

3.2. Materials Synthesis

A simple and mild hydrothermal synthesis method was used to prepare six different weight ratios of CuO/TiO_2 catalysts, which are defined as CuO/TiO_2-1, CuO/TiO_2-2, CuO/TiO_2-3, CuO/TiO_2-4, CuO/TiO_2-5, and CuO/TiO_2-6, corresponding to the CuO content of 5 wt %, 10 wt %, 20 wt %, 40 wt %, 60 wt %, and 80 wt %, respectively. Typically, due to similar fabrication procedure, for simplicity, CuO/TiO_2-5 acts as the specimen to illustrate the following experiment. The detailed synthesis procedure for CuO/TiO_2-5 is depicted as follows: 0.05 M of $Cu(Ac)_2$ aqueous solution (60 mL) was mixed with 127 mg of TiO_2. After stirring for several minutes, an appropriate amount of 0.05 M of $(NH_4)_2CO_3$ aqueous solution was added drop-wise with stirring to control the synthetic rate. Then, the mixture was stirred gently for 3 h at the ambient temperature. Subsequently, the above mixture was centrifuged, washed with distilled water, and stored in a vacuum oven with 60 °C for 4 h to dry the precipitate. Finally, the collected powder was transferred to a crucible, further maintaining at 220 °C for 3 h to obtain the final CuO/TiO_2 powder. Pure CuO catalyst was acquired by the same method, without adding TiO_2. Additionally, CuO/C catalyst with the CuO content of 60 wt % got through the similar synthetic process, but with C replacing TiO_2.

3.3. Electrode Preparation and Electrochemical Test

Potentiostatic electrolysis was carried out in an H-type three-electrode cell with a piece of Nafion® 117 cation exchange membrane (H^+ form) as a separator. The working electrode was manufactured using the following route: 10 mg of CuO/TiO_2 catalyst was suspended in a mixture solution with 20 μL of 5 wt % Nafion® solution and 40 μL of distilled water. After sonification for 20 min, the mixture was spread on a porous CP (2 × 2 cm) by micropipette and then dried in air. The counter and reference electrodes were Pt sheet and SCE, respectively. A 0.5 M of $KHCO_3$ aqueous solution serves as electrolyte, which was saturated with CO_2 by bubbling for 30 min before the electrolysis experiment. CO_2 was bubbled continuously throughout overall experiment time. CV tests were measured in a single cell system using a standard three-electrode setup at 0.5 M $KHCO_3$ aqueous solution under CO_2 and N_2 atmosphere with a scan rate of 50 mV/s, respectively. In this system, 2 μL of the above prepared catalyst suspension was coated on glassy carbon electrode (GCE, diameter = 2 mm) playing as the working electrode, and a Pt mesh and an SCE acted as the counter electrode and the reference electrode, respectively. All potentials were based on RHE as reference potentials, and converted by the following equation:

$$E\ (\text{vs. RHE}) = E\ (\text{vs. SCE}) + 0.059 \times \text{pH} + 0.241.$$

3.4. Product Analysis

Liquid phase products were analyzed by ^{1}H-nuclear magnetic resonance (NMR) spectra recorded on an Ascend 400 (400 MHz, Bruker, Germany) spectrometer in D_2O with Me_4Si as an internal standard.

4. Conclusions

In this study, a simple synthesis strategy was employed to fabricate CuO/TiO_2 material. A series of characterizations has demonstrated that CuO NPs were uniformly distributed on the surface of TiO_2. Different CuO/TiO_2 catalysts displayed dramatically different electrocatalytic performance for CO_2ER, depending strongly on the CuO contents of the catalysts as-prepared. High FEs of CO_2 reduction products (ethanol, acetone, and n-propanol) reach up to 47.4% on CuO/TiO_2-5 at the potential of −0.85 V vs. RHE among all CuO/TiO_2 and CuO/C catalysts. The CO_2 adsorption ability test suggests that CuO/TiO_2-5 exhibits more superior CO_2 adsorption performance than CuO/C, indicating that CuO/TiO_2-5 is more beneficial to CO_2 reduction. The fact that the as-synthesized CuO/TiO_2 catalysts exhibit high activity for CO_2ER is attributed to the special roles of TiO_2 that show excellent CO_2 adsorption ability. Our work may provide some concepts for designing cheap and effective catalysts for highly efficient CO_2ER.

Supplementary Materials: The following are available online at http://www.mdpi.com/2073-4344/8/4/171/, Table S1: CuO composition of the synthesized CuO/TiO_2 catalysts. Figure S1: The high-resolution XPS spectrum of C1s for CuO/TiO_2-5 catalyst. Figure S2: SEM images of (a) pure TiO_2, (b) CuO/TiO_2-1, (c) CuO/TiO_2-2, (d) CuO/TiO_2-3, (e) CuO/TiO_2-4, (f) CuO/TiO_2-5, (g) CuO/TiO_2-6, and (h) pure CuO. Figure S3: XRD patterns of CuO/TiO_2-5 catalyst after 2 h electrolysis. The solid diamond represents the in situ generated Cu metal. Figure S4: CO_2 reduction electrolysis data at −0.85 V vs. RHE for CuO-TiO_2-5 catalyst. Table S2: The current density (J) for all the CuO/TiO_2 catalysts in CO_2-saturated environment at the potential of −0.85 V vs. RHE. Figure S5: Plot of I vs. potential of (A) CuO/TiO_2-1, (B) CuO/TiO_2-2, (C) CuO/TiO_2-3, (D) CuO/TiO_2-4, (E) CuO/TiO_2-5, (F) CuO/TiO_2-6, (G) pure CuO, and (H) CuO/C in N_2-saturated 0.1 M $HClO_4$ solution cycled between 0.35 and 0.55 V vs. SCE at scan rates in the range of 5–60 mV/s. Insets show plot of I_c vs. ν in which the linear regressions give capacitance information. Table S3: The analysis of C_{dl} of various CuO/TiO_2 and CuO/C catalysts. Table S4: The analysis of specific surface area of CuO/TiO_2 catalysts. Figure S6: XRD patterns of CuO/C and C catalysts. Figure S7: SEM (a) and TEM (b) images of CuO/C catalyst. Figure S8: CV curve of CuO/C catalyst over glassy carbon electrode in CO_2-saturated (red line) 0.5 M $KHCO_3$ solution. Figure S9: CO_2 adsorption-desorption isotherms at 297 K of CuO/TiO_2-5 (red square) and CuO/C (blue triangle). Filled and empty symbols represent adsorption and desorption, respectively.

Acknowledgments: We gratefully acknowledge the financial support from the National Natural Science Foundation of China (21673078, 21473060, 21773071).

Author Contributions: Jing Yuan and Jing-Jie Zhang conceived and designed the experiments; Jing Yuan and Wang-Jun Meng performed the experiments; Jing Yuan, Man-Ping Yang, and Huan Wang analyzed the data; Jing Yuan, Huan Wang, and Jia-Xing Lu wrote the paper.

Conflicts of Interest: The authors declare no conflict of interest.

References

1. Kumar, B.; Brian, J.P.; Atla, V.; Kumari, S.; Bertram, K.A.; White, R.T.; Spurgeon, J.M. New trends in the development of heterogeneous catalysts for electrochemical CO_2 reduction. *Catal. Today* **2016**, *270*, 19–30. [CrossRef]
2. Qiao, J.L.; Liu, Y.Y.; Hong, F.; Zhang, J.J. A review of catalysts for the electroreduction of carbon dioxide to produce low-carbon fuels. *Chem. Soc. Rev.* **2014**, *43*, 631–675. [CrossRef] [PubMed]
3. Costentin, C.; Robert, M.; Saveant, J.M. Catalysis of the electrochemical reduction of carbon dioxide. *Chem. Soc. Rev.* **2013**, *42*, 2423–2436. [CrossRef] [PubMed]
4. Jitaru, M.; Lowy, D.A.; Toma, M.; Toma, B.C.; Oniciu, L. Electrochemical reduction of carbon dioxide on flat metallic cathodes. *J. Appl. Electrochem.* **1997**, *27*, 875–889. [CrossRef]
5. Kumar, B.; Brian, J.P.; Atla, V.; Kumari, S.; Bertram, K.A.; White, R.T.; Spurgeon, J.M. Controlling the Product Syngas H_2:CO Ratio through Pulsed-Bias Electrochemical Reduction of CO_2 on Copper. *ACS Catal.* **2016**, *6*, 4739–4745. [CrossRef]
6. Kondratenko, E.V.; Mul, G.; Baltrusaitis, J.; Larrazabal, G.O.; Perez-Ramírez, J. Status and perspectives of CO_2 conversion into fuels and chemicals by catalytic, photocatalytic and electrocatalytic processes. *Energy Environ. Sci.* **2013**, *6*, 3112–3135. [CrossRef]
7. Reske, R.; Duca, M.; Oezaslan, M.; Schouten, K.J.P.; Koper, M.T.M.; Strasser, P. Controlling Catalytic Selectivities during CO_2 Electroreduction on Thin Cu Metal Overlayers. *J. Phys. Chem. Lett.* **2013**, *4*, 2410–2413. [CrossRef]
8. Zhao, K.; Liu, Y.M.; Quan, X.; Chen, S.; Yu, H.T. CO_2 Electroreduction at Low Overpotential on Oxide-Derived Cu/Carbons Fabricated from Metal Organic Framework. *ACS Appl. Mater. Interfaces* **2017**, *9*, 5302–5311. [CrossRef] [PubMed]
9. Ohya, S.; Kaneco, S.; Katsumata, H.; Suzuki, T.; Ohta, K. Electrochemical reduction of CO_2 in methanol with aid of CuO and Cu_2O. *Catal. Today* **2009**, *148*, 329–334. [CrossRef]
10. Kaneco, S.; Iiba, K.; Katsumata, H.; Suzuki, T.; Ohta, K. Electrochemical reduction of high pressure carbon dioxide at a Cu electrode in cold methanol with CsOH supporting salt. *Chem. Eng. J.* **2007**, *128*, 47–50. [CrossRef]
11. Whipple, D.T.; Kenis, P.J.A. Prospects of CO_2 Utilization via Direct Heterogeneous Electrochemical Reduction. *J. Phys. Chem. Lett.* **2010**, *1*, 3451–3458. [CrossRef]
12. Kaneco, S.; Iiba, K.; Katsumata, H.; Suzuki, T.; Ohta, K. Electrochemical reduction of high pressure CO_2 at a Cu electrode in cold methanol. *Electrochim. Acta.* **2006**, *51*, 4880–4885. [CrossRef]
13. Kuhl, K.P.; Cave, E.R.; Abram, D.N.; Jaramillo, T.F. New insights into the electrochemical reduction of carbon dioxide on metallic copper surfaces. *Energy Environ. Sci.* **2012**, *5*, 7050–7059. [CrossRef]
14. Oh, Y.; Vrubel, H.; Guidoux, S.; Hu, X. Electrochemical reduction of CO_2 in organic solvents catalyzed by MoO_2. *Chem. Commun.* **2014**, *50*, 3878–3881. [CrossRef] [PubMed]
15. Chen, Y.; Kanan, M.W. Tin Oxide Dependence of the CO_2 Reduction Efficiency on Tin Electrodes and Enhanced Activity for Tin/Tin Oxide Thin-Film Catalysts. *J. Am. Chem. Soc.* **2012**, *134*, 1986–1989. [CrossRef] [PubMed]
16. Li, C.W.; Kanan, M.W. CO_2 Reduction at Low Overpotential on Cu Electrodes Resulting from the Reduction of Thick Cu_2O Films. *J. Am. Chem. Soc.* **2012**, *134*, 7231–7234. [CrossRef] [PubMed]
17. Dubois, M.R.; Dubois, D.L. Development of Molecular Electrocatalysts for CO_2 Reduction and H_2 Production/Oxidation. *Acc. Chem. Res.* **2009**, *42*, 1974–1982. [CrossRef] [PubMed]
18. Lin, S.; Diercks, C.S.; Zhang, Y.B.; Kornienko, N.; Nichols, E.M.; Zhao, Y.; Paris, A.R.; Kim, D.; Yang, P.; Yaghi, O.M. Covalent organic frameworks comprising cobalt porphyrins for catalytic CO_2 reduction in water. *Science* **2015**, *349*, 1208–1213. [CrossRef] [PubMed]

19. Qiu, Y.L.; Zhong, H.X.; Zhang, T.T.; Xu, W.B.; Li, X.F.; Zhang, H.M. Copper Electrode Fabricated via Pulse Electrodeposition: Toward High Methane Selectivity and Activity for CO_2 Electroreduction. *ACS Catal.* **2017**, *7*, 6302–6310. [CrossRef]
20. Hori, Y.; Takahashi, I.; Koga, O.; Hoshi, N. Electrochemical reduction of carbon dioxide at various series of copper single crystal electrodes. *J. Mol. Catal. A Chem.* **2003**, *199*, 39–47. [CrossRef]
21. Nie, X.; Esopi, M.R.; Janik, M.J.; Asthagiri, A. Selectivity of CO_2 Reduction on Copper Electrodes: The Role of the Kinetics of Elementary Steps. *Angew. Chem. Int. Ed.* **2013**, *52*, 2459–2462. [CrossRef] [PubMed]
22. Sarfraz, S.; Esparza, A.T.G.; Jedidi, A.; Cavallo, L.; Takanabe, K. Cu–Sn Bimetallic Catalyst for Selective Aqueous Electroreduction of CO_2 to CO. *ACS Catal.* **2016**, *6*, 2842–2851. [CrossRef]
23. Shi, G.D.; Yu, L.; Ba, X.; Zhang, X.S.; Zhou, J.Q.; Yu, Y. Copper nanoparticle interspersed MoS_2 nanoflowers with enhanced efficiency for CO_2 electrochemical reduction to fuel. *Dalton Trans.* **2017**, *46*, 10569–10577. [CrossRef] [PubMed]
24. Li, Q.; Fu, J.J.; Zhu, W.L.; Chen, Z.Z.; Shen, B.; Wu, L.H.; Xi, Z.; Wang, T.Y.; Lu, G.; Zhu, J.J.; et al. Tuning Sn-Catalysis for Electrochemical Reduction of CO_2 to CO via the Core/Shell Cu/SnO_2 Structure. *J. Am. Chem. Soc.* **2017**, *139*, 4290–4293. [CrossRef] [PubMed]
25. Ni, M.; Leung, M.K.H.; Leung, D.Y.C.; Sumathy, K. A review and recent developments in photocatalytic water-splitting using TiO_2 for hydrogen production. *Renew. Sust. Energ. Rev.* **2007**, *11*, 401–425. [CrossRef]
26. Liu, G.; Wang, L.; Yang, H.G.; Cheng, H.M.; Lu, G.Q. Titania-based photocatalysts—Crystal growth, doping and heterostructuring. *J. Mater. Chem.* **2010**, *20*, 831–843. [CrossRef]
27. Ravichandran, C.; Kennady, C.J.; Chellammal, S.; Thangavelu, S.; Anantharaman, P.N. Indirect electroreduction of o-nitrophenol to o-aminophenol on titanium dioxide coated titanium electrodes. *J. Appl. Electrochem.* **1991**, *21*, 60–63. [CrossRef]
28. Chu, D.; Qin, G.; Yuan, X.; Xu, M.; Zheng, P.; Lu, J. Fixation of CO_2 by Electrocatalytic Reduction and Electropolymerization in Ionic Liquid-H_2O Solution. *ChemSusChem* **2008**, *1*, 205–209. [CrossRef] [PubMed]
29. Thompson, T.L.; Diwald, O.; Yates, J.T. CO_2 as a Probe for Monitoring the Surface Defects on TiO_2 (110) Temperature-Programmed Desorption. *J. Phys. Chem. B* **2003**, *107*, 11700–11704. [CrossRef]
30. Cueto, L.F.; Hirata, G.A.; Sanchez, E.M. Thin-film TiO_2 electrode surface characterization upon CO_2 reduction processes. *J. Sol-Gel Sci. Technol.* **2006**, *37*, 105–109. [CrossRef]
31. Ma, S.C.; Lan, Y.C.; Perez, G.M.J.; Moniri, S.; Kenis, P.J.A. Silver Supported on Titania as an Active Catalyst for Electrochemical Carbon Dioxide Reduction. *ChemSusChem* **2014**, *7*, 866–874. [CrossRef] [PubMed]
32. Yuan, J.; Liu, L.; Guo, R.R.; Zeng, S.; Wang, H.; Lu, J.X. Electroreduction of CO_2 into Ethanol over an Active Catalyst: Copper Supported on Titania. *Catalysts* **2017**, *7*, 220. [CrossRef]
33. Yuan, J.; Yang, M.P.; Hu, Q.L.; Li, S.M.; Wang, H.; Lu, J.X. Cu/TiO_2 nanoparticles modified nitrogen-doped graphene as a highly efficient catalyst for the selective electroreduction of CO_2 to different alcohols. *J. CO_2 Util.* **2018**, *24*, 334–340. [CrossRef]
34. Fang, B.Z.; Xing, Y.L.; Bonakdarpour, A.; Zhang, S.C.; Wilkinson, D.P. Hierarchical CuO-TiO_2 Hollow Microspheres for Highly Efficient Photodriven Reduction of CO_2 to CH_4. *ACS Sustain. Chem. Eng.* **2015**, *3*, 2381–2388. [CrossRef]
35. Choi, H.J.; Kang, M. Hydrogen production from methanol/water decomposition in a liquid photosystem using the anatase structure of Cu loaded TiO_2. *Int. J. Hydrog. Energy* **2007**, *32*, 3841–3848. [CrossRef]
36. Chi, D.H.; Yang, H.P.; Du, Y.F.; Lv, T.; Sui, G.J.; Wang, H.; Lu, J.X. Morphology-controlled CuO nanoparticles for electroreduction of CO_2 to ethanol. *RSC Adv.* **2014**, *4*, 37329–37332. [CrossRef]
37. Kulkarni, P.; Mahamuni, S.; Chandrachood, M.; Mulla, I.S.; Sinha, A.P.B.; Nigavekar, A.S.; Kulkarni, S.K. Photoelectron spectroscopic studies on a silicon interface with $Bi_2Sr_2CaCu_2BO_{8+\delta}$ high T_c superconductor. *J. Appl. Phys.* **1990**, *67*, 3438–3442. [CrossRef]
38. Ethiraj, A.S.; Kang, D.J. Synthesis and characterization of CuO nanowires by a simple wet chemical method. *Nanoscale Res. Lett.* **2012**, *7*, 70. [CrossRef] [PubMed]
39. Chang, T.Y.; Liang, R.M.; Wu, P.W.; Chen, J.Y.; Hsieh, Y.C. Electrochemical reduction of CO_2 by Cu_2O-catalyzed carbon clothes. *Mater. Lett.* **2009**, *63*, 1001–1003. [CrossRef]
40. Gupta, K.; Bersani, M.; Darr, J.A. Highly efficient electro-reduction of CO_2 to formic acid by nano-copper. *J. Mater. Chem. A* **2016**, *4*, 13786–13794. [CrossRef]

41. Qiao, J.L.; Fan, M.Y.; Fu, Y.S.; Bai, Z.Y.; Ma, C.Y.; Liu, Y.Y.; Zhou, X.D. Highly-active copper oxide/copper electrocatalysts induced from hierarchical copper oxide nanospheres for carbon dioxide reduction reaction. *Electrochim. Acta.* **2015**, *153*, 559–565. [CrossRef]
42. Ghijsen, J.; Tjeng, L.H.; Van-Elp, J.; Esks, H.; Westerink, J.; Sawatzky, C.A.; Czyzyk, M.T. Electronic structure of Cu_2O and CuO. *Phys. Rev. B* **1988**, *38*, 11322–11330. [CrossRef]
43. Mohamed, A.M.; Aljaber, A.S.; AlQaradawi, S.Y.; Allam, N.K. TiO_2 nanotubes with ultrathin walls for enhanced water splitting. *Chem. Commun.* **2015**, *51*, 12617–12620. [CrossRef] [PubMed]
44. Bai, L.; Qiao, S.; Fang, Y.; Tian, J.; Mcleod, J.; Song, Y.; Huang, H.; Liu, Y.; Kang, Z. Third-order nonlinear optical properties of carboxyl group dominant carbon nanodots. *J. Mater. Chem. C* **2016**, *4*, 8490–8495. [CrossRef]
45. Yu, B.Y.; Kwak, S.Y. Carbon quantum dots embedded with mesoporous hematite nanospheres as efficient visible light-active photocatalysts. *J. Mater. Chem.* **2012**, *22*, 8345–8353. [CrossRef]
46. Li, H.; Zhang, X.; MacFarlane, D.R. Carbon Quantum Dots/Cu_2O Heterostructures for Solar-Light-Driven Conversion of CO_2 to Methanol. *Adv. Energy Mater.* **2015**, *5*, 1401077. [CrossRef]
47. Li, C.W.; Ciston, J.; Kanan, M.W. Electroreduction of carbon monoxide to liquid fuel on oxide-derived nanocrystalline copper. *Nature* **2014**, *508*, 504–507. [CrossRef] [PubMed]
48. Wan, Y.; Zhang, Y.; Wang, X.; Wang, Q. Electrochemical formation and reduction of copper oxide nanostructures in alkaline media. *Electrochem. Commun.* **2013**, *36*, 99–102. [CrossRef]
49. Gao, S.; Lin, Y.; Jiao, X.C.; Sun, Y.F.; Luo, Q.Q.; Zhang, W.H.; Li, D.Q.; Yang, J.L.; Xie, Y. Partially oxidized atomic cobalt layers for carbon dioxide electroreduction to liquid fuel. *Nature* **2016**, *529*, 68–82. [CrossRef] [PubMed]
50. Kortlever, R.; Shen, J.; Schouten, K.J.P.; Calle-Vallejo, F.; Koper, M.T.M. Catalysts and Reaction Pathways for the Electrochemical Reduction of Carbon Dioxide. *J. Phys. Chem. Lett.* **2015**, *6*, 4073–4082. [CrossRef] [PubMed]

Article

Hydrogen Production from Chemical Looping Reforming of Ethanol Using Ni/CeO_2 Nanorod Oxygen Carrier

Lin Li, Bo Jiang, Dawei Tang *, Zhouwei Zheng and Cong Zhao

School of Energy and Power Engineering, Key Laboratory of Ocean Energy Utilization and Energy Conservation of Ministry of Education, Dalian University of Technology, Dalian 116024, China; lilinnd@dlut.edu.cn (L.L.); ericchiang@mail.dlut.edu.cn (B.J.); 953596340zzw@mail.dlut.edu.cn (Z.Z.); zc287643861@mail.dlut.edu.cn (C.Z.)

* Correspondence: dwtang@dlut.edu.cn; Tel.: +86-411-8470-8460

Received: 8 June 2018; Accepted: 22 June 2018; Published: 25 June 2018

Abstract: Chemical looping reforming (CLR) technique is a prospective option for hydrogen production. Improving oxygen mobility and sintering resistance are still the main challenges of the development of high-performance oxygen carriers (OCs) in the CLR process. This paper explores the performance of Ni/CeO_2 nanorod (NR) as an OC in CLR of ethanol. Various characterization methods such as N_2 adsorption-desorption, X-ray diffraction (XRD), Raman spectra, transmission electron microscopy (TEM), X-ray photoelectron spectroscopy (XPS), H_2 temperature-programmed reduction (TPR), and H_2 chemisorption were utilized to study the properties of fresh OCs. The characterization results show the Ni/CeO_2-NR possesses high Ni dispersion, abundant oxygen vacancies, and strong metal-support interaction. The performance of prepared OCs was tested in a packed-bed reactor. H_2 selectivity of 80% was achieved by Ni/CeO_2-NR in 10-cycle stability test. The small particle size and abundant oxygen vacancies contributed to the water gas shift reaction, improving the catalytic activity. The covered interfacial Ni atoms closely anchored on the underlying surface oxygen vacancies on the (111) facets of CeO_2-NR, enhancing the anti-sintering capability. Moreover, the strong oxygen mobility of CeO_2-NR also effectively eliminated surface coke on the Ni particle surface.

Keywords: chemical looping reforming; hydrogen; oxygen carrier; CeO_2; nanorod

1. Introduction

Hydrogen is considered an efficient energy carrier that is environmentally benign [1]. Chemical looping reforming (CLR) is a prospective alternative for hydrogen production due to its energy efficiency and inherent CO_2 capture [2,3]. The fixed-bed reactor configuration CLR process (Figure 1) is performed by alternatively switching the feed gases; the oxygen carriers (OCs) are stationary and periodically exposed to redox atmosphere [4]. The key to developing a CLR process is to screen high-performance OCs. Ni-based OCs have been widely investigated because of their ability for carbon–carbon and carbon–hydrogen bonds cleavage [5–7]. Zafar et al. [8] prepared a series of OCs including Fe, Cu, Mn, and Ni supported by SiO_2 and $MgAl_2O_4$ and concluded that NiO supported on SiO_2 exhibited high H_2 selectivity in CLR. Löfberg et al. [9] demonstrated that Ni plays two essential roles in CLR process, i.e., the anticipated activation of reactants as well as the regulation of the oxygen supply rate from solids. Dharanipragada et al. [10] reported that the Ni-ferrites OC suffers from the loss of oxygen storage capacity due to Ni sintering in chemical looping with alcohols. Improving oxygen mobility and sintering resistance remain the major challenges of the development of Ni-based OCs in the CLR process due to the high activation energy of oxygen anion diffusion in NiO (2.23 eV in CLR) and the low Tammann temperature of Ni (691 °C) [11–14].

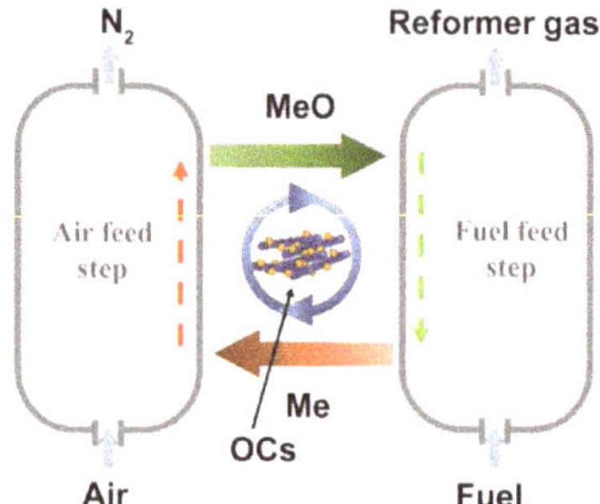

Figure 1. Schematic description of CLR of ethanol.

Recent research has revealed that CeO_2 exhibits excellent redox property due to abundant oxygen storage capacity and the strong capability to stabilize Ni nanoparticles because of strong metal support interaction (MSI) [15,16]. CeO_2 can readily release lattice oxygen under reducing conditions, thus creating oxygen vacancies which are associated with oxygen mobility [17]. The MSI between the Ni and CeO_2 could tune the physiochemical properties of Ni, contributing to high catalytic reactivity and stability [18]. Jiang et al. [19] proposed that the oxygen vacancies of CeO_2 could effectively eliminate surface coke deposition, activate steam, and shorten the "dead time" in the CLR process. Dou et al. [20] revealed that the surface oxygen originating from the CeO_2 lattice can oxidize coke precursors, keeping the OC surface free of coke deposition. It has been reported that MSI and the mobility of lattice oxygen show strong dependence on the morphology of CeO_2 [21]. Lykaki et al. [22] have demonstrated CeO_2 morphology dominates reducibility and oxygen mobility, which follow the sequence: nanorod (NR) > nanopolyhedra (NP) > nanocube (NC). Moreover, the apparent activation energy of these three CeO_2 shapes for the CO oxidation in a hydrogen-rich gas shows the opposite trend, implying the potential highest water gas shift (WGS) activity of NR [23].

Therefore, in this work, a Ni/CeO_2-NR OC for CLR of ethanol for hydrogen production is prepared by a hydrothermal method. The physicochemical properties are investigated by N_2 adsorption-desorption, X-ray diffraction (XRD), inductively coupled plasma optical emission spectroscopy (ICP-OES), Raman spectra, high resolution transmission electron microscopy (HRTEM), X-ray photoelectron spectroscopy (XPS), and H_2 temperature-programmed reduction (H_2-TPR). The performances of the Ni/CeO_2-NR OC are tested in a packed-bed reactor and compared with the other reference bulk OC, i.e., Ni/CeO_2.

2. Results and discussion

2.1. Characterization of OCs

Predominant physicochemical properties of fresh Ni/CeO_2-NR and CeO_2-NR are tabulated in Table 1. The introduction of Ni species has a pronounced influence on the texture of CeO_2-NR. The CeO_2-NR support showed higher Brunauer-Emmett-Teller (BET) surface area than that of Ni/CeO_2-NR. The average pore diameter and pore volume also exhibited the same trends. It has been reported that the BET surface of CeO_2-NR decreased by 13–18% after the incorporation of 7.5 wt % CuO [22]. The actual Ni content of Ni/CeO_2-NR determined by ICP-OES was 9.7 wt %.

Table 1. Physicochemical properties of fresh Ni/CeO_2-NR and CeO_2-NR.

Sample	Surface Area (m^2/g)	Average Pore Diameter (nm)	Pore Volume (cm^3/g)	Ni Content (wt %)	Ni Dispersion (m^2/g_{Ni})	Ni Crystal Size/Ni Average Particle Size (nm)	CeO_2 Crystal Size (nm)
Ni/CeO_2-NR	75.1	18.2	0.37	9.7 [1]	6.5 [2]	8.9 [3]/8.1 [4]	16.1 [3]
CeO_2-NR	90.3	26.5	0.80	N/A	N/A	N/A	14.8

[1] Determined by ICP-OES; [2] Determined by H_2 chemisorption; [3] Determined by XRD from Ni (111) and CeO_2 (111) plane; [4] Calculated from the HRTEM images.

The XRD profiles of fresh Ni/CeO_2-NR and CeO_2-NR OCs are shown in Figure 2. The reflection peaks at 28.5°, 33.1°, 47.5°, 56.3°, 59.1°, 69.4°, 76.7°, and 79.1° could be indexed to (111), (200), (220), (311), (222), (400), (331), (420), and (422) planes of a fluorite-structure CeO_2 (Space group: *Fm3m*), respectively. There were no other impurity peaks of the hexagonal structure of $Ce(OH)_3$ or $Ce(OH)CO_3$ detected, indicating the excellent crystalline purity of CeO_2. Two peaks at 44.5° and 51.8°, which could be respectively attributed to (111) and (200) facets of Ni, were observed for Ni/CeO_2-NR. The mean crystal sizes of Ni and CeO_2 were obtained from the Scherrer Equation (listed in Table 1). The crystal sizes of CeO_2 for CeO_2-NR and Ni/CeO_2-NR were 14.8 and 16.1 nm, respectively, indicating that the incorporation of Ni would not significantly change the structure characteristics of CeO_2 support. Similar crystallite sizes of CeO_2-NR prepared by the hydrothermal method have been reported in other work [22,24].

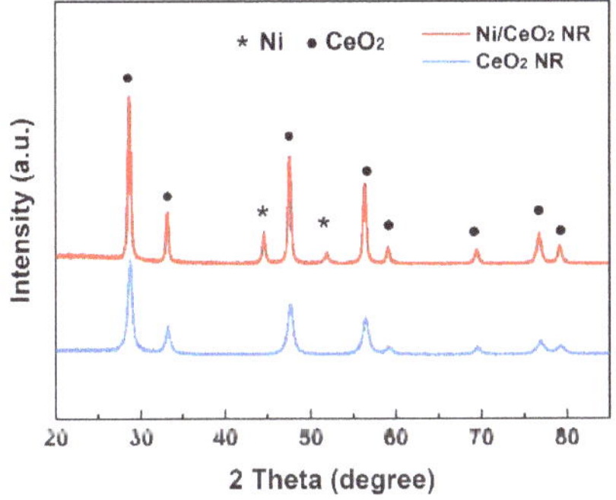

Figure 2. XRD profiles of Ni/CeO_2-NR and CeO_2-NR.

Raman spectroscopy was employed to characterize surface and bulk defects. As shown in Figure 3, both samples present four characteristic peaks. The appreciable peak at 462.8 cm^{-1} resulted from the first order F_{2g} mode of the fluorite cubic structure. The Raman shift at 596.1 cm^{-1} could be assigned to the defect-induced band (D band). The relocation of O atom from the interior of tetrahedral cationic sub-lattice to the interior of ideally empty octahedral cationic sites (Frenkel interstitial sites) would lead to the deformation of the anionic lattice of CeO_2 [25]. The intensity of the D band is a gauge of the distortion of ionic lattice, which results in punctual defects and oxygen vacancies [22]. Therefore, the value of $I_D/I_{F_{2g}}$ is commensurate with the number of defect sites in ceria. The relative intensity of $I_D/I_{F_{2g}}$ decreased after the incorporation of Ni, suggesting NiO species suppress the surface oxygen vacancy of CeO_2-NR. Li et al. [26] reported that Vanadium atom would bond to the surface of CeO_2-NR by generating V–O–Ce species, thus covering oxygen defects and stabilizing adjacent Ce atoms. The inconspicuous peak at 258.1 cm^{-1} results from the second order transverse acoustic mode (2TA) or doubly degenerate transverse optical mode (TO), which are Raman inactive in a perfect crystal [27]. The peak at 1179.9 cm^{-1} may be related to the stretching mode of the short terminal Ce=O. Moreover, the Raman shift of Ni/CeO_2-NR moved towards a low wavenumber in comparison with that of CeO_2-NR, indicating the incorporation of Ni affects the symmetry of Ce–O bonds [21].

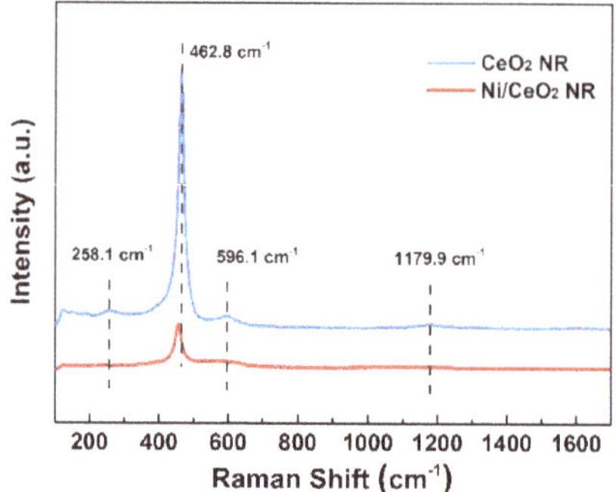

Figure 3. Visual Raman spectroscopy of Ni/CeO_2-NR and CeO_2-NR.

The TEM of CeO_2-NR and Ni/CeO_2-NR are illustrated in Figure 4. The Ni/CeO_2-NR maintained the original nanorod shape after Ni incorporation. The CeO_2-NR was approximately 14 nm in diameter and several hundreds nm in length. As shown in Figure 4a, the calculated interference fringe spacings (d) are 0.27 and 0.31 nm, revealing the CeO_2-NR predominantly exposes the (100) and (111) facets. The preferable exposure crystal facets of CeO_2-NR are consistent with those of samples derived from the $CeCl_3$ precursor in other studies [28,29]. Theory calculation has demonstrated that the (111) is the least reactive facet, followed by (100) and (110). In addition to the preferable crystal facets, there are other aspects degerming CeO_2 activity. Several 'dark pits' are shown in the box of Figure 4. Liu et al. [29] have reported that these dark pits are related to the surface reconstruction and defects; their study also concluded that these defects play a more important role in determining the CeO_2 activity than the exposure planes. Sayle et al. [30] have performed a molecular dynamic modeling to simulate the synthesis of CeO_2-NR and discovered that the atomistic sphere model exhibits many steps on the (111) planes of CeO_2-NR. The migration of oxygen in CeO_2 occurs by a vacancy hopping mechanism; therefore, the clusters of defects are conducive to oxygen transfer. If the diffusion rate of oxygen anions becomes adequately high, a consecutive oxygen flow is generated, resulting in high reducibility. The Ni particle size distribution, which is calculated from 52 particles, was inserted in Figure 4, and the average particle size was 8.1 nm (shown in Table 1). Shen et al. [31] have elucidated that the strong interfacial anchoring effect, which exists between the surface oxygen vacancies on (111) planes of CeO_2-NR and the gold particles, only allows the gold particles to locally rotate or vibrate but not to migrate to form aggregates. Therefore, the strong MSI would significantly improve the Ni dispersion on the CeO_2-NR surface, resulting in a small particle size of Ni.

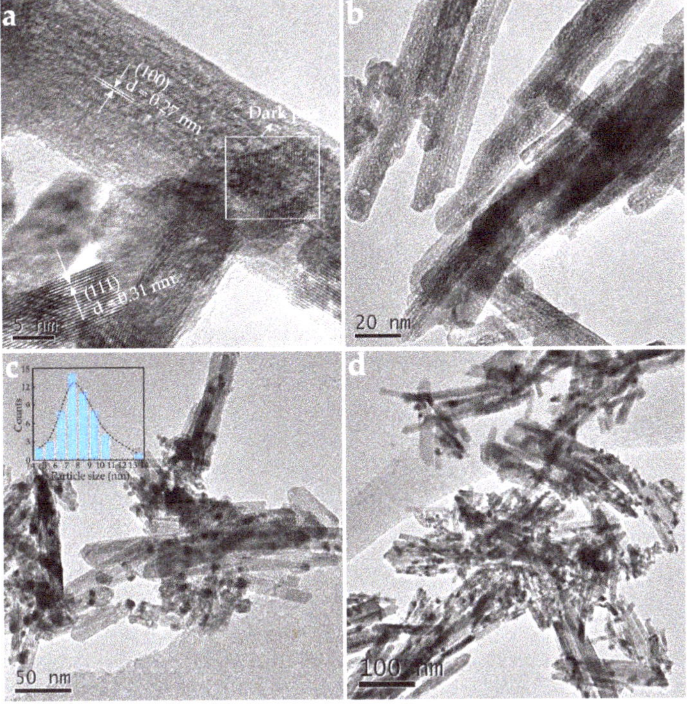

Figure 4. (**a**) HRTEM image of CeO_2-NR, (**b**) TEM image of CeO_2-NR, (**c**) and (**d**) TEM images of Ni/CeO_2-NR.

XPS was employed to investigate the valences of Ce and Ni cations. The XPS spectra of Ce 3d of Ni/CeO_2-NR (shown in Figure 5a) could be deconvoluted into two spin-orbit series, i.e., $3d_{5/2}$ (u) and $3d_{3/2}$ (v). The multiplet splitting components labeled u^0, u^1, u^2, v, v^1 and v^2 corresponds to the $3d^{10}4f^0$ state of Ce^{4+}, while the u_0 and v_0 are related to $3d^{10}4f^1$ state of Ce^{3+}. These two Ce species in Ni/CeO_2-NR indicate the OC surface was partly reduced because of oxygen desorption and the formation of oxygen vacancies. It is widely accepted that oxygen vacancies are produced to maintain electrostatic balance once Ce^{3+} exists in fluorite Ce (Equation 1). The percentage of Ce^{3+} cations to the total Ce cations is determined by the area ratio of different Ce species in XPS spectra. The Ce^{3+} ratio of Ni/CeO_2-NR (16.9%) was lower than that of pure CeO_2-NR (24.3%) reported in Lykaki et al. [22], suggesting NiO species could inhibit the formation of surface-unsaturated Ce^{3+}. The decrease in surface Ce^{3+} species of Ni/CeO_2-NR was consistent in the trend of $I_D/I_{F_{2g}}$ in the Raman result. The Ni 2p XPS spectra (illustrated in Figure 5b) were characterized by two spin-orbit groups, i.e., $2p_{3/2}$ (855.2 and 856.4 eV) and $2p_{1/2}$ (873.3 eV), as well as a shake-up peak at 861.8 eV. The photoelectron peak of $2p_{3/2}$ over Ni/CeO_2-NR shifted towards high-binding energy in contrast to those over pure NiO (844.4 eV, reported in Lemonidou et al. [32]) and Ni/CeO_2 (854.5 eV, reported in Tang et al. [33]). This result implies there is an enhanced MSI between Ni and CeO_2-NR. The Ni/Ce atom ratio of the outer surface of CeO_2-NR (0.49) is higher than the nominal one (0.32), suggesting a Ni species enrichment from bulk to surface.

$$4Ce^{4+} + O^{2-} \rightarrow 4Ce^{4+} + \frac{2e^-}{\delta} + 0.5O_2 \rightarrow 2Ce^{4+} + 2Ce^{3+} + \delta + 0.5O_2 \quad (1)$$

where δ represents an empty position derived from the removal of O^{2-} from an oxygen tetrahedral site (Ce_4O).

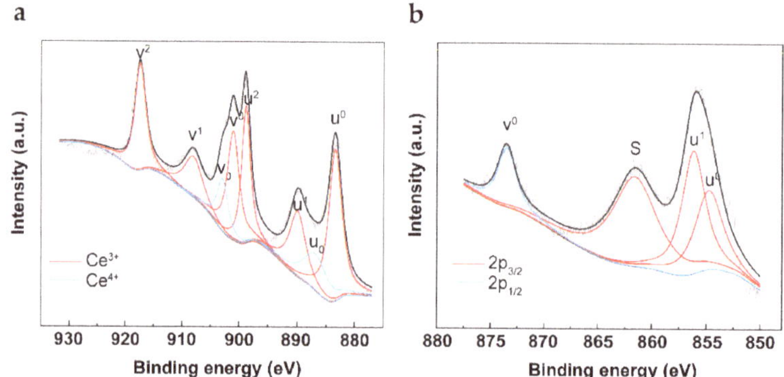

Figure 5. (**a**) XPS spectra of Ce 3d; (**b**) XPS spectra of Ni 2p.

H_2-TPR was performed to investigate MSI. As illustrated in Figure 6, the H_2-TPR patterns of CeO_2-NR are comprised of two reduction peaks. An inconspicuous peak at 262 °C, which can be assigned to the reduction of surface-adsorbed oxygen, was observed, while a broad peak ranging from 400 °C to 550 °C could be ascribed to the reduction of Ce^{4+} to Ce^{3+} [22,34]. With regard to Ni/CeO_2-NR, the TPR profiles consisted of three peaks. The small peak at 220 °C may have been attributable to the reduction of the NiO species, which slightly interacted with CeO_2-NR supports. Zhang et al. [21] have reported that this NiO species is characterized by small radius and could incorporate into CeO_2-NR surface. The second peak at 356 °C could be ascribed to the reduction of NiO species, with a strong interaction with CeO_2-NR supports. The H_2 consumption of the second peak was highest among the three peaks, indicating most of the NiO strongly interacted with the support. The third peak at 494 °C was also attributed to the reduction of Ce^{4+} to Ce^{3+}. However, the broad reduction peak of Ni/CeO_2-NR shifted to a lower temperature than that of pure CeO_2-NR, suggesting the introduction of Ni improves the reducibility of CeO_2-NR. It has demonstrated that the promoted reduction behavior is caused by the generated Ni-Ce-O species which improves the deformation of CeO_2 [35].

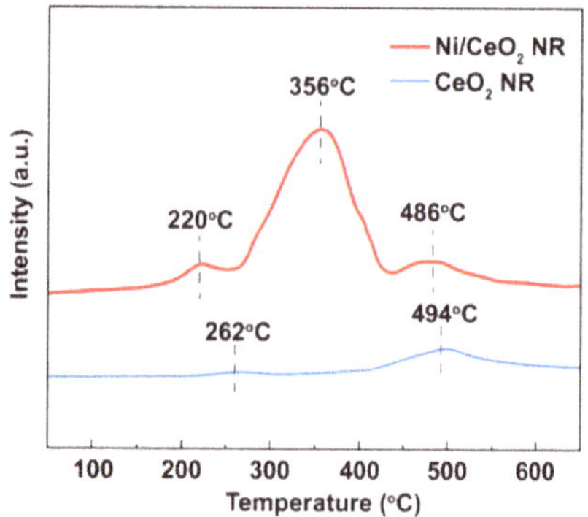

Figure 6. H_2-TPR patterns of Ni/CeO_2-NR and CeO_2-NR.

2.2. Activity Tests of OCs

The activity of Ni/CeO_2-NR was tested in CLR of ethanol and compared with Ni/CeO_2. Figure 7 displays the performance of both OCs in CLR: H_2 selectivity and ethanol conversion at the fuel feed step as well as the concentration of CO, CO_2 and O_2 at the air feed step. The H_2 selectivity and ethanol conversion of both OCs remained relatively stable during the process. With regard to Ni/CeO_2-NR, the average ethanol conversion remained at 88.0%, and the average H_2 selectivity was 78.9%. However, with the average ethanol conversion and H_2 selectivity of conventional Ni/CeO_2 at 72.9% and 60.5% respectively, the activity of Ni/CeO_2-NR was superior to the Ni/CeO_2. As the Ni content and other experimental conditions were the same for both OCs, the improved activity therefore resulted from the support.

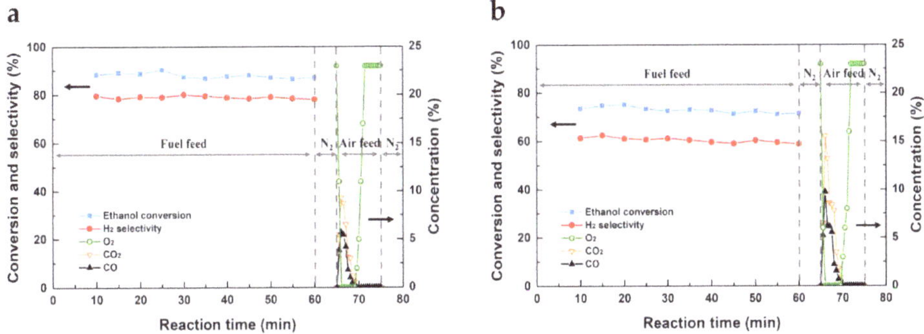

Figure 7. Activity tests of (**a**) Ni/CeO_2-NR and (**b**) Ni/CeO_2.

Various factors contributed to the enhanced catalytic activity. Notably, the CeO_2-NR with high specific-to-volume ratio guarantees excellent Ni dispersion, as evidenced by TEM and TPR results. Such a one-dimensional nanostructure enables uniform and small Ni nanoparticles to be finely dispersed on supports, thus generating many accessible catalytic active sites. Additionally, the high concentration oxygen vacancies of CeO_2-NR, which was proven via Raman and XPS analysis, can activate and produce OH groups from steam. H_2 and CO_2 can be generated via the reaction between the formed OH groups as well as intermediate species, thus improving H_2 selectivity. It has been proposed that the interface of the metal/CeO_2 is the main site for steam reforming reaction [36].

The air feed process is highly oxygen consuming and is accompanied by the generation of CO_2 and CO due to coke oxidation and partial oxidation. As illustrated in Figure 7, all the samples exhibited CO_2 and CO evolution at the air feed stage. The integration areas of C-containing gas concentrations corresponded to the quantity of carbon deposition. Ni/CeO_2-NR exhibited small peak areas compared with the Ni/CeO_2 OC, indicating that the carbon deposition on Ni/CeO_2-NR was effectively suppressed. This is a result of the abundant oxygen vacancies and its strong oxygen storage capacity, which enhanced the oxygen mobility and thereby facilitated the removal of carbon deposition. In addition, the depleted CeO_2-NR could be partially replenished by the air feed step. The O_2 concentration increased from 0%to 23% at the end of the air feed stage, indicating the end of coke elimination and replenishment of lattice oxygen.

2.3. Stability Tests

The durability of Ni/CeO_2-NR and Ni/CeO_2 was tested in a 10-cycle CLR process. As shown in Figure 8, the H_2 selectivity of Ni/CeO_2 slowly declined with the increase of the running cycles, while the Ni/CeO_2 NR maintained its activity throughout the test. During the tests, the H_2 selectivity of Ni/CeO_2-NR decreased from 81.7% to 79.2%, while Ni/CeO_2 exhibited a significant decrease in H_2

selectivity from 61.8% to 51.4%. The deactivation of OCs predominantly caused Ni sintering and carbon deposition. Notably, the Ni/CeO_2-NR OC exhibited a more durable performance than the Ni/CeO_2, revealing that the CeO_2-NR supported metal was more resistant to deactivation. The superior durability of the Ni/CeO_2-NR sample resulted from the strong MSI between Ni and CeO_2-NR support, which improved the metal-sintering resistance. The abundant oxygen vacancies of CeO_2-NR are essential units for the anchoring of metal particles. The essential role of CeO_2 is to disperse and stabilize Ni particles over its surface oxygen vacancies, which depend on the morphology of CeO_2. As evidenced by XPS, the (111) plane of CeO_2-NR plays an important role in anchoring the Ni particles. It has been proposed that the interfacial Au atoms which are located away from the particle perimeter (covered interfacial atoms) would closely interact with the underlying surface oxygen vacancies on the (111) facets of CeO_2-NR [31]. Because this interfacial region was not involved in the reforming reaction, this strong interaction would effectively stabilize the Ni particles on CeO_2-NR. Additionally, small size Ni particles can lower the driving force for coke diffusion and thereby help to reduce the carbon deposition. Furthermore, the strong oxygen mobility of CeO_2-NR is indispensable to removing the carbon deposition at the metal surface. It has been demonstrated that the oxygen deposited in CeO_2 lattice can react with the carbon species left over from the steam reforming reaction [37]. Overall, CeO_2-NR support not only promotes the anti-sintering capability of OCs but also reduces the carbon deposition, thus improving the durability of OCs. A comparison of several investigations regarding CLR over different Ni-based OCs is tabulated in Table 2. The Ni/CeO_2-NR in this work showed higher average fuel conversion in long-term tests despite the low Ni loading, indicating the strong sintering resistance and high reforming activity of this OC.

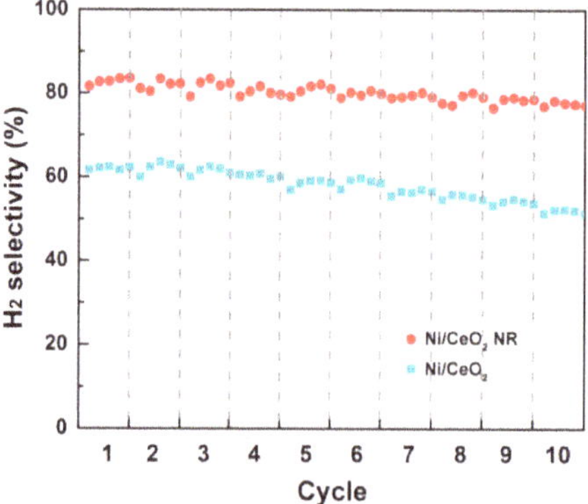

Figure 8. Stability tests of OCs.

Table 2. Comparison of CLR over different OCs.

OCs	Ni Loading (wt %)	T (°C)	S/C	Tested Cycles	Reactant	Average Conversion (%)	Average H_2 Yield (%)	References
CeNi/MCM-41	6	650	1.5	10	Glycerol	~90	6.2	[37]
$LaNiO_3$/MMT	13	650	3	10	Ethanol	~90	-	[3]
Ni/MMT	19.9	650	4	20	Ethanol	~78	-	[13]
CeNi/PSNT	24.7	650	3	10	Glycerol	~100	12.5	[19]
CeNi/SBA-15	12	650	3	14	Ethanol	~84	-	[20]
Ni/CeO_2-NR	9.7	650	4	10	Ethanol	~90	-	This work

3. Materials and Methods

3.1. Preparation of OCs

CeO_2-NR was synthesized on the basis of our previous work with small modifications [38]. Firstly, 1 g of $CeCl_3$ (99.9%, Aladdin, Shanghai, China) and 19.3 g of NaOH (97%, Aladdin) were dissolved in 50 mL deionized water and continuously stirred for 25 min. The resulting mixture was then put into a Teflon-lined autoclave and kept at 100 °C for 24 h. Subsequently, the sample was separated, washed, then dried at 85 °C for 10 h and finally calcined at 400 °C for another five hours. The Ni/CeO_2-NR oxygen carrier was synthesized by a wet-impregnation method. $Ni(NO_3)_2 \cdot 6H_2O$ (98%, Aladdin) of 0.2 g, equivalent to 10 wt % Ni loading, was first dispersed in 5 mL deionized water, and the prepared CeO_2-NR of 0.5 g was added into the solution. The mixture was stirred under sonication for three hours at 60 °C and then dried at 85 °C for 10 h. The as-synthesized sample was then calcined at 700 °C for two hours.

The other reference bulk Ni/CeO_2 was also synthesized by a wet-impregnation method, and the procedure has been reported in Xu et al. [39]. The Ni content of this OC was also set to 10 wt %.

3.2. Characterization of OCs

The texuture of OCs were analyzed by a Micrometric Acusorb 2100E apparatus (Ottawa, ON, Canada) at 77 K. The OCs were degassed at 573 K for three hours before tests.

ICP-OES (Nijmegen, The Netherlands) was applied to measure the elemental composition. Before tests, the sample was solved by hydrofluoric acid solution.

XRD was performed, using a Shimadzu XRD-600 instrument (Kyoto, Japan), to identify the phase composition. The scanning 2θ degrees ranged from 10° to 80°, and the scanning rate was set to 4°/min. A graphite-filtered Cu Kα radiation (λ = 1.5406 Å) was applied as a radiation source.

Raman spectra were employed by a Renishaw Spectroscopy (Gloucestershire, UK) with a visible 514 nm Ar-ion laser under ambient condtions. During the mesurement, the flowing gas was He, and the test temperature was maintained at 300 °C.

TEM was conducted, using FEI Tecnai F30 (Cleveland, OH, USA), to investigate the morphology of the Ocs. The samples were first solved in ethanol under sonication, followed by dispersal on a copper grid-supported carbon foil and dried in air.

XPS was carried out by a ThermoFisher K-Alpha system with a 150 W Al Kα source (Waltham, MA, USA). The samples were placed on the holder, and the scanning step was set to 0.15 eV.

H_2-TPR was employed by a Micrometrics AutoChem 2920 instrument (Ottawa, ON, Canada). In a typical procedure, a sample of 90 mg was preheated at 450 °C for one hour in an Ar flow of 35 mL/min and subsequently cooled to 80 °C. A mixture of 10 vol % H_2 in Ar flow (35 mL/min) was then inserted, and the temperature was increased from 80 °C to 100 °C at a rate of 5 °C/min simultaneously. H_2 chemisorption was also performed by the same apparatus to analyze the Ni dispersion. The sample was first reduced at 700 °C in a H_2/Ar flow (30 mL/min) for one hour, followed by cooling to 100 °C under Ar flow. Subsequently, H_2 pulses were introduced until the eluted peaks of successive pulses became steady. The Ni active surface area was obtained from the H_2 adsorbption volume considering the stoichiometric ratio $H_{adsorbed}/Ni_{surface} = 1$ and a surface area of 6.5×10^{-20} m^2 per Ni atom [40].

3.3. Activity and Stability Tests

The performance of CLR of ethanol by Ni/CeO_2-NR and reference conventional bulk OCs, i.e., Ni/CeO_2, were conducted in a packed-bed reactor, whose schematic was described in our previous work [41,42]. The tested OCs of 0.5 g were loaded at the center of the quartz reactor. During a fuel feed step, an ethanol solution (4 mL/h) with a steam to carbon ratio (S/C) of four was preheated to 150 °C and then inserted into the reactor in a N_2 flow of 180 mL/min. The reacter temperature of the fuel feed step was 650 °C. An Agilent 7890A (Santa Clara, CA, USA) chromatography with two detectors, thermal conductivity detector (TCD) and flame ionization detector (FID), was applied to verify the effluents. TCD with a TDX-01 column was applied to detect N_2, H_2, CO, CO_2, and CH_4; the FID with a Porapak-Q column was applied to measure the concentration of $C_3H_8O_3$, H_2O, and CH_3CHO. In an air feed step, the air flow was 100 mL/min, and the reaction temperature was also 650 °C. The exhausted gases were detected by another GC (Agilent 7890A) with a TCD detector. A 5A molecular sieve column was used to detect the O_2, and the TDX-01 column was applied to detect CO_2, CO, and N_2. The durations of the fuel feed step and the air feed step were 60 and 10 min, respectively. A N_2 purge process of five minutes was performed between the fuel feed step and the air feed step to eliminate the residue gas in the reactor.

The conversion and H_2 selectivity were calculated as follows:

$$X = \frac{F_{in} - F_{out}}{F_{in}} \times 100\% \tag{2}$$

$$S_{H_2} = \frac{1}{6} \times \frac{moles\ H_2\ produced}{moles\ ethanol\ feed \times X} \times 100\% \tag{3}$$

4. Conclusions

A Ni/CeO_2-NR OC was synthesized by hydrothermal method and tested in CLR of ethanol process in this work. H_2 selectivity of 80% was achieved by Ni/CeO_2-NR in a 10-cycle stability test. The characterization results show that the Ni/CeO_2-NR possesses high Ni dispersion, abundant oxygen vacancies, and strong MSI. The small particle size and abundant oxygen vacancies contributed to the WGS reaction, thus improving the catalytic activity. The buried interfacial Ni atoms strongly anchored on the underlying surface oxygen vacancies on the (111) facets of CeO_2-NR, therefore enhancing the anti-sintering capability. Moreover, the strong oxygen mobility of CeO_2-NR also effectively eliminated surface coke on the Ni particle surface.

Author Contributions: Experiment, L.L., Z.Z. and C.Z.; Data Curation, Z.Z. and C.Z.; Writing-Original Draft Preparation, L.L. and B.J.; Writing-Review & Editing, B.J.; Supervision, D.T. and L.L.; Project Administration, D.T. and L.L.; Funding Acquisition, L.L. and D.T.

Funding: This research was funded by (National Natural Science Foundation of China) grant number (51706030), (Fundamental Research Funds for Central Universities) grant number (DUT18JC11) and (China Postdoctoral Science Foundation) grant number (2017M611219).

Acknowledgments: We deeply appreciate the kind assistance from the Key Laboratory of Ocean Energy Utilization and Energy Conservation of Ministry of Education (China).

Conflicts of Interest: The authors declare no conflict of interest. The founding sponsors had no role in the design of the study; in the collection, analyses, or interpretation of data; in the writing of the manuscript, and in the decision to publish the results.

References

1. Wang, C.; Chen, Y.; Cheng, Z.; Luo, X.; Jia, L.; Song, M.; Jiang, B.; Dou, B. Sorption-Enhanced Steam Reforming of Glycerol for Hydrogen Production over a NiO/$NiAl_2O_4$ Catalyst and Li_2ZrO_3-Based sorbent. *Energy Fuels* **2015**, *29*, 7408–7418. [CrossRef]
2. Adanez, J.; Abad, A.; Garcia-Labiano, F.; Gayan, P.; de Diego, L.F. Progress in chemical-looping combustion and reforming technologies. *Prog. Energy Combust. Sci.* **2012**, *38*, 215–282. [CrossRef]
3. Li, L.; Song, Y.; Jiang, B.; Wang, K.; Zhang, Q. A novel oxygen carrier for chemical looping reforming: $LaNiO_3$ perovskite supported on montmorillonite. *Energy* **2017**, *131*, 58–66. [CrossRef]
4. Dou, B.; Zhang, H.; Cui, G.; Wang, Z.; Jiang, B.; Wang, K. Hydrogen production by sorption-enhanced chemical looping steam reforming of ethanol in an alternating fixed-bed reactor: Sorbent to catalyst ratio dependencies. *Energy Convers. Manage.* **2018**, *155*, 243–252. [CrossRef]

5. Dou, B.; Zhang, H.; Cui, G.; Wang, Z.; Jiang, B.; Wang, K.; Chen, H.; Xu, Y. Hydrogen production and reduction of Ni-based oxygen carriers during chemical looping steam reforming of ethanol in a fixed-bed reactor. *Int. J. Hydrogen Energy* **2017**, *42*, 26217–26230. [CrossRef]
6. de Diego, L.F.; Ortiz, M.; García-Labiano, F.; Adánez, J.; Abad, A.; Gayán, P. Hydrogen production by chemical-looping reforming in a circulating fluidized bed reactor using Ni-based oxygen carriers. *J. Power Sources* **2009**, *192*, 27–34. [CrossRef]
7. Johansson, M.; Mattisson, T.; Lyngfelt, A.; Abad, A. Using continuous and pulse experiments to compare two promising nickel-based oxygen carriers for use in chemical-looping technologies. *Fuel* **2008**, *87*, 988–1001. [CrossRef]
8. Zafar, Q.; Mattisson, T.; Gevert, B. Redox investigation of some oxides of transition-state metals Ni, Cu, Fe, and Mn supported on SiO_2 and $MgAl_2O_4$. *Energy Fuels* **2006**, *20*, 34–44. [CrossRef]
9. Löfberg, A.; Guerrero-Caballero, J.; Kane, T.; Rubbens, A.; Jalowiecki-Duhamel, L. Ni/CeO_2 based catalysts as oxygen vectors for the chemical looping dry reforming of methane for syngas production. *Appl. Catal. B-Environ.* **2017**, *212*, 159–174. [CrossRef]
10. Dharanipragada, N.V.R.A.; Galvita, V.V.; Poelman, H.; Buelens, L.C.; Detavernier, C.; Marin, G.B. Bifunctional Co- and Ni- ferrites for catalyst-assisted chemical looping with alcohols. *Appl. Catal. B-Environ.* **2018**, *222*, 59–72. [CrossRef]
11. García-Labiano, F.; de Diego, L.F.; Adánez, J.; Abad, A.; Gayán, P. Temperature variations in the oxygen carrier particles during their reduction and oxidation in a chemical-looping combustion system. *Chem. Eng. Sci.* **2005**, *60*, 851–862. [CrossRef]
12. Tang, M.; Xu, L.; Fan, M. Progress in oxygen carrier development of methane-based chemical-looping reforming: A review. *Appl. Energy* **2015**, *151*, 143–156. [CrossRef]
13. Jiang, B.; Dou, B.; Wang, K.Q.; Zhang, C.; Song, Y.C.; Chen, H.; Xu, Y. Hydrogen production by chemical looping steam reforming of ethanol using NiO/montmorillonite oxygen carriers in a fixed-bed reactor. *Chem. Eng. J.* **2016**, *298*, 96–106. [CrossRef]
14. Feng, Y.; Guo, X. Study of reaction mechanism of methane conversion over Ni-based oxygen carrier in chemical looping reforming. *Fuel* **2017**, *210*, 866–872. [CrossRef]
15. Li, L.; Tang, D.; Song, Y.C.; Jiang, B.; Zhang, Q. Hydrogen production from ethanol steam reforming on Ni-Ce/MMT catalysts. *Energy* **2018**, *149*, 937–943. [CrossRef]
16. Li, D.; Zeng, L.; Li, X.; Wang, X.; Ma, H.; Assabumrungrat, S. Ceria-promoted Ni/SBA-15 catalysts for ethanol steam reforming with enhanced activity and resistance to deactivation. *Appl. Catal. B-Environ.* **2015**, *176–177*, 532–541. [CrossRef]
17. Campbell, C.T.; Peden, C.H.F. Oxygen vacancies and catalysis on ceria surfaces. *Science* **2005**, *309*, 713–714. [CrossRef] [PubMed]
18. Senanayake, S.D.; Rodriguez, J.A.; Stacchiola, D. Electronic metal–support interactions and the production of hydrogen through the water-gas shift reaction and ethanol steam reforming: Fundamental studies with well-defined model catalysts. *Top. Catal.* **2013**, *56*, 1488–1498. [CrossRef]
19. Jiang, B.; Li, L.; Bian, Z.F.; Li, Z.; Othman, M.; Sun, Z.; Tang, D.; Kawi, S.; Dou, B. Hydrogen generation from chemical looping reforming of glycerol by Ce-doped nickel phyllosilicate nanotube oxygen carriers. *Fuel* **2018**, *222*, 185–192. [CrossRef]
20. Wang, K.; Dou, B.; Jiang, B.; Song, Y.C.; Zhang, C.; Zhang, Q.; Chen, H.; Xu, H. Renewable hydrogen production from chemical looping steam reforming of ethanol using *x*CeNi/SBA-15 oxygen carriers in a fixed-bed reactor. *Int. J. Hydrogen Energy* **2016**, *41*, 12899–12909. [CrossRef]
21. Du, X.; Zhang, D.; Shi, L.; Gao, R.; Zhang, J. Morphology dependence of catalytic properties of Ni/CeO_2 nanostructures for carbon dioxide reforming of methane. *J. Phys. Chem. C* **2012**, *116*, 10009–10016. [CrossRef]
22. Lykaki, M.; Pachatouridou, E.; Carabineiro, S.A.C.; Iliopoulou, E.; Andriopoulou, C.; Kallithrakas-Kontos, N.; Boghosian, S.; Konsolakis, M. Ceria nanoparticles shape effects on the structural defects and surface chemistry: Implications in CO oxidation by Cu/CeO_2 catalysts. *Appl. Catal. B-Environ.* **2018**, *230*, 18–28. [CrossRef]
23. Yi, G.; Xu, Z.; Guo, G.; Tanaka, K.I.; Yuan, Y. Morphology effects of nanocrystalline CeO_2 on the preferential CO oxidation in H_2-rich gas over Au/CeO_2 catalyst. *Chem. Phys. Lett.* **2009**, *479*, 128–132. [CrossRef]

24. Pan, C.; Zhang, D.; Shi, L. CTAB assisted hydrothermal synthesis, controlled conversion and CO oxidation properties of CeO_2 nanoplates, nanotubes, and nanorods. *J. Solid State Chem.* **2008**, *181*, 1298–1306. [CrossRef]
25. Mamontov, E.; Egami, T.; Brezny, R.; Koranne, M.; Tyagi, S. Lattice defects and oxygen storage capacity of nanocrystalline ceria and ceria-zirconia. *J. Phys. Chem. B* **2000**, *104*, 11110–11116. [CrossRef]
26. Peng, Y.; Wang, C.; Li, J. Structure–activity relationship of VO_x/CeO_2 nanorod for NO removal with ammonia. *Appl. Catal. B-Environ.* **2014**, *144*, 538–546. [CrossRef]
27. Lee, Y.; He, G.; Akey, A.J.; Si, R.; Flytzani-Stephanopoulos, M.; Herman, I.P. Raman analysis of mode softening in nanoparticle $CeO_{2-\delta}$ and Au-$CeO_{2-\delta}$ during CO oxidation. *J. Am. Chem. Soc.* **2011**, *133*, 12952–12955. [CrossRef] [PubMed]
28. Huang, P.; Wu, F.; Zhu, B.; Gao, X.; Zhu, H.; Yan, T.; Huang, W.; Wu, S.; Song, D. CeO_2 nanorods and gold nanocrystals supported on CeO_2 nanorods as catalyst. *J. Phys. Chem. B* **2005**, *109*, 19169–19174. [CrossRef] [PubMed]
29. Liu, X.; Zhou, K.; Wang, L.; Wang, B.; Li, Y. Oxygen vacancy clusters promoting reducibility and activity of ceria nanorods. *J. Am. Chem. Soc.* **2009**, *131*, 3140–3141. [CrossRef] [PubMed]
30. Sayle, D.C.; Feng, X.; Ding, Y.; Wang, Z.L.; Sayle, T.X. "Simulating synthesis": Ceria nanosphere self-assembly into nanorods and framework architectures. *J. Am. Chem. Soc.* **2007**, *129*, 7924–7935. [CrossRef] [PubMed]
31. Ta, N.; Liu, J.; Chenna, S.; Crozier, P.A.; Li, Y.; Chen, A.; Shen, W. Stabilized gold nanoparticles on ceria nanorods by strong interfacial anchoring. *J. Am. Chem. Soc.* **2012**, *134*, 20585–20588. [CrossRef] [PubMed]
32. Lemonidou, A.; Goula, M.; Vasalos, I. Carbon dioxide reforming of methane over 5 wt % nickel calcium aluminate catalysts–Effect of preparation method. *Catal. Today* **1998**, *46*, 175–183. [CrossRef]
33. Tang, S.; Lin, J.; Tan, K.L. Partial oxidation of methane to syngas over Ni/MgO, Ni/CaO and Ni/CeO2. *Catal. Lett.* **1998**, *51*, 169–175. [CrossRef]
34. Maitarad, P.; Han, J.; Zhang, D.; Shi, L.; Namuangruk, S.; Rungrotmongkol, T. Structure–activity relationships of NiO on CeO_2 nanorods for the selective catalytic reduction of NO with NH_3: Experimental and DFT studies. *J. Phys. Chem. C* **2014**, *118*, 9612–9620. [CrossRef]
35. Zhang, S.; Muratsugu, S.; Ishiguro, N.; Tada, M. Ceria-doped Ni/SBA-16 catalysts for dry reforming of methane. *ACS Catal.* **2013**, *3*, 1855–1864. [CrossRef]
36. Cai, W.; Wang, F.; Van Veen, A.; Provendier, H.; Mirodatos, C.; Shen, W. Autothermal reforming of ethanol for hydrogen production over a Rh/CeO_2 catalyst. *Catal. Today* **2008**, *138*, 152–156. [CrossRef]
37. Jiang, B.; Dou, B.; Wang, K.Q.; Song, Y.C.; Chen, H.; Zhang, C.; Xu, Y.; Li, M. Hydrogen production from chemical looping steam reforming of glycerol by Ni based Al-MCM-41 oxygen carriers in a fixed-bed reactor. *Fuel* **2016**, *183*, 170–176. [CrossRef]
38. Li, L.; Tang, D.; Song, Y.C.; Jiang, B. Dual-film optofluidic microreactor with enhanced light-harvesting for photocatalytic applications. *Chem. Eng. J.* **2018**, *339*, 71–77. [CrossRef]
39. Xu, W.; Liu, Z.; Johnston-Peck, A.C.; Senanayake, S.D.; Zhou, G.; Stacchiola, D.; Zhou, G.; Stacchiola, D.; Stach, E.; Rodriguez, J. Steam reforming of ethanol on Ni/CeO_2: Reaction Pathway and interaction between Ni and the CeO_2 support. *ACS Catal.* **2013**, *3*, 975–984. [CrossRef]
40. Jiang, B.; Zhang, C.; Wang, K.Q.; Dou, B.; Song, Y.C.; Chen, H.; Xu, Y. Highly dispersed Ni/montmorillonite catalyst for glycerol steam reforming: Effect of Ni loading and calcination temperature. *Appl. Therm. Eng.* **2016**, *109*, 99–108. [CrossRef]
41. Jiang, B.; Dou, B.; Song, Y.C.; Zhang, C.; Du, B.; Chen, H.; Xu, Y. Hydrogen production from chemical looping steam reforming of glycerol by Ni-based oxygen carrier in a fixed-bed reactor. *Chem. Eng. J.* **2015**, *280*, 459–467. [CrossRef]
42. Jiang, B.; Dou, B.; Wang, K.Q.; Zhang, C.; Li, M.; Chen, H.; Xu, Y. Sorption enhanced steam reforming of biodiesel by-product glycerol on Ni-CaO-MMT multifunctional catalysts. *Chem. Eng. J.* **2017**, *313*, 207–216. [CrossRef]

Article

Experimental Clarification of the RWGS Reaction Effect in H_2O/CO_2 SOEC Co-Electrolysis Conditions

Evangelia Ioannidou [1,2], Stylianos Neophytides [1] and Dimitrios K. Niakolas [1,*]

1 Foundation for Research and Technology, Institute of Chemical Engineering Sciences (FORTH/ICE-HT), Patras, GR-26504, Greece; eioannidou@iceht.forth.gr (E.I.); neoph@iceht.forth.gr (S.N.)
2 Department of Chemical Engineering, University of Patras, GR-56504, Greece
* Correspondence: niakolas@iceht.forth.gr

Received: 31 December 2018; Accepted: 29 January 2019; Published: 2 February 2019

Abstract: In the present investigation, modified X-Ni/GDC electrodes (where X = Au, Mo, and Fe) are studied, in the form of half-electrolyte supported cells, for their performance in the RWGS through catalytic-kinetic measurements. The samples were tested at open circuit potential conditions in order to elucidate their catalytic activity towards the production of CO (r_{co}), which is one of the products of the H_2O/CO_2 co-electrolysis reaction. Physicochemical characterization is also presented, in which the samples were examined in the form of powders and as half cells with BET, H_2-TPR, Air-TPO and TGA re-oxidation measurements in the presence of H_2O. In brief, it was found that the rate of the produced CO (r_{co}) increases by increasing the operating temperature and the partial pressure of H_2 in the reaction mixture. In addition, the first results revealed that Fe and Mo modification enhances the catalytic production of CO, since the 2wt% Fe-Ni/GDC and 3wt% Mo-Ni/GDC electrodes were proven to perform better compared to the other samples, in the whole studied temperature range (800–900 °C), reaching thermodynamic equilibrium. Furthermore, carbon formation was not detected.

Keywords: SOECs; RWGS reaction kinetics; Au–Mo–Fe-Ni/GDC electrodes; high temperature H_2O/CO_2 co-electrolysis

---

1. Introduction

Solid oxide electrolysis is a contemporary process for CO_2 capture/recycling, which is proven as an attractive method to provide CO_2 neutral synthetic hydrocarbon fuels. In particular, co-electrolysis of H_2O and CO_2 in a solid oxide electrolysis cell (SOEC) yields synthesis gas (CO + H_2), which in turn can be used towards the formation of various types of synthetic fuels [1–3] by applying the Fischer-Tropsch process. According to thermodynamics, SOECs offer benefits for endothermic reactions, such as H_2O and/or CO_2 electrolysis at temperatures >700 °C, because a larger part of the required electrical energy can be substituted by thermal energy [4,5]. In addition, high temperature can lead to a significant decrease in the internal resistance of the cell and acceleration of the electrode reaction processes due to fast reaction kinetics [4–7].

H_2O/CO_2 Co-electrolysis is a much more complex process compared to pure steam or CO_2 electrolysis. This is because three reactions take place simultaneously, namely H_2O electrolysis, CO_2 electrolysis, and the catalytic Reverse Water Gas Shift reaction (RWGS). More specifically, at the cathode of SOECs, two electrochemical reactions take place in parallel at the triple phase boundaries, i.e., H_2O and CO_2 electrochemical reductions. The oxygen ions (O^{2-}), produced by these reactions, are moved to the anode, through an oxygen ion-conducting electrolyte (Yttria-Stabilized Zirconia – YSZ), where oxygen (O_2) gas is formed [8,9]:

$$\text{Cathode:} \quad H_2O + 2e^- \rightarrow H_2 + O^{2-} \quad \Delta H_{800\,^\circ C} = 151.8\ \text{kJ/mol} \tag{1}$$

$$CO_2 + 2e^- \rightarrow CO + O^{2-} \quad \Delta H_{800\,^\circ C} = 185.9\ kJ/mol \tag{2}$$

$$\text{Anode:} \quad O^{2-} \rightarrow \frac{1}{2} O_2 + 2e^- \tag{3}$$

Besides the electrolysis reactions mentioned above, the reversible water gas shift (RWGS) catalytic reaction also occurs at the cathode with the co-existence of CO_2 and H_2O:

$$\text{Cathode:} \quad H_2 + CO_2 \rightarrow H_2O + CO \quad \Delta H_{800\,^\circ C} = 36.8\ kJ/mol \tag{4}$$

It thus appears that CO can be produced either electrocatalytically by the reduction of CO_2 or catalytically via the RWGS reaction. Questionable conclusions have been proposed in the literature about the role of the RWGS reaction for CO production. Some studies mention that CO is mainly produced electrocatalytically and RWGS has a small participation [10–15], while other studies mention that CO is absolutely produced catalytically via the RWGS reaction [16,17]. However, it has not been shown conclusively to which degree the RWGS is responsible for CO production in an SOEC.

Specifically, Mogensen and co-researchers found that the performance of co-electrolysis on an Ni/YSZ cathode varied between the highest pure H_2O electrolysis and the lowest CO_2 electrolysis, being much closer to the prior process [13,14]. This implies the significant contribution of the RWGS reaction in the H_2O/CO_2 SOEC co-electrolysis process and also confirms the co-existence of CO_2 electrolysis. The above conclusion was supported by recently published results on computational modelling of the direct and indirect (with electro-generated H_2) reduction of CO_2 [18,19]. The direction of the Water Gas Shift (WGS) reaction could be forward or backward, depending on the operating conditions, indicating that at high temperatures (>838 °C), CO was produced electrochemically and the WGS was shifted in reverse towards H_2 and CO_2.

In contrast, Yue and Irvine [20] suggested that the RWGS reaction did not contribute critically in catalysing H_2O/CO_2 co-electrolysis on s (La,Sr)(Cr,Mn)O_3 (LSCM) perovskite based cathode. The LSCM cathode displayed higher catalytic activity for pure CO_2 electrolysis than pure H_2O electrolysis, and the performance for H_2O/CO_2 co-electrolysis was much closer to pure CO_2 electrolysis.

On the other hand, Bae et al., Hartvigsen et al., Kee et al., and Zhao et al. considered that CO production on Ni-based cathodes was mainly from the RWGS reaction, whereas the electrochemical processes are dominated by H_2O electrolysis with identical performances for H_2O electrolysis and H_2O/CO_2 co-electrolysis [16,17,21].

According to the above mentioned reports, until now, no agreement has been reached regarding the role of the RWGS reaction in the production of CO. As a result, it is crucial to quantify the degree of CO formation for each reaction. The discrepancies regarding the CO production route can be affected both by the structural characteristics of the electrode/catalyst (specific surface area, reducibility and re-oxidation behaviour, porosity, particle size, ionic and electronic conductivity), but also by the operating conditions (gas composition, temperature, etc.) [9,22]. As reported by Li's study [10], the heterogeneous thermochemical reactions occur at the external surface of the cathode and they are 20–100 times faster than the electrochemical reactions, which occur close to the electrolyte at the three phase boundaries. Furthermore, structural modifications of the cathode could enhance mass transport and promote CO production through the catalytic RWGS reaction, resulting in H_2O/CO_2 co-electrolysis performance close to that of H_2O electrolysis.

The majority of studies on high temperature H_2O/CO_2 co-electrolysis utilize Ni-containing ceramic cathodes with YSZ and Gadolinia-Doped Ceria (GDC), similarly to the case of H_2O electrolysis [10–14]. Ni-based materials indeed are cheap and exhibit porous structure, high electronic conductivity, appropriate catalytic activity, and a similar Thermal Expansion Coefficient (TEC) with the electrolyte. Consequently, they could act as excellent SOECs cathodes for H_2O/CO_2 co-electrolysis [7,22–26].

However, SOECs comprising Ni-based cathodes face some critical degradation issues, which are more pronounced with an increasing current density [3,12,14,27–33]. The main reasons for degradation have been reported to be microstructural changes that take place after prolonged co-electrolysis, resulting in passivation and blocking of the Three-Phase Boundaries (TPB) area [12,13,28,30,31,33,34]. Post-mortem microscopy investigations in the Ni/YSZ electrode have shown irreversible damages of the electrode's microstructure, such as loss of Ni-YSZ contact, decomposition of YSZ, Ni grain growth, loss of Ni percolation (loss of Ni–Ni contact), and even migration of Ni from the fuel electrode [28,30,31,33]. Taking into account the above, it is critical to develop alternative cathode materials for H_2O/CO_2 co-electrolysis with improved structural properties.

Previous studies reported that the electrocatalytic efficiency of SOECs can be improved by alloying transition metals with the state of the art (SoA) Ni catalyst [35–39]. Modification by means of alloying may be a promising strategy to promote the catalytic activity due to ligand and strain effects that change the electronic structure [40] of the active element. Recent studies focused on the Ni–Co alloy with Sm-doped ceria (Ni–Co/SDC) as an alternative material, with enhanced performance for H_2O electrolysis in SOECs [36]. Specifically, the addition of Co increased the intrinsic catalytic activity of pure Ni and simultaneously expanded the active reaction region [36,41]. Other reports have shown that an Ni–Fe bimetallic cathode, mixed with $Ba_{0.6}La_{0.4}CoO_3$ on $La_{0.9}Sr_{0.1}Ga_{0.8}Mg_{0.2}O_3$ (LSGM) electrolyte, improved the electrochemical performance of the electrode for H_2O electrolysis [38] and CO_2 electrolysis [39], due to prevention of Ni aggregation. Furthermore, the enhanced performance of the Ni–Fe formulations could be attributed to the increased stability and improved catalytic activity, due to the addition of Fe, as this was concluded by combining experimental results with theoretical Density Functional Theory (DFT) calculations [41].

Our research group have studied the effect of Au and/or Mo doping on the physicochemical and catalytic properties of Ni/GDC for the reactions of catalytic CH_4 dissociation and methane steam reforming [23,42–45] in the presence of H_2 and H_2S impurities. These modifications resulted in SOFC fuel electrodes with high sulfur and carbon tolerance as well as with improved electrocatalytic activity. Regarding the doping level of Au and Mo, the amount of 3wt% was indicated as an appropriate loading for achieving the most promising results under the examined SOFC conditions. Recently, the 3wt% Au-Ni/GDC electrode was tested under SOEC conditions for the H_2O electrolysis process and exhibited improved electrocatalytic performance compared to Ni/GDC [46]. The enhanced performance of the Au-doped cathode was attributed to the creation of a surface Ni–Au solid solution, which causes weaker interplay of absorbed H_2O_{ads} and O_{ads} species with the modified cathode. The result is a durable and resistant electrode to surface oxidation with an improved "three phase boundaries" area, especially at temperatures higher than 800 °C [46].

The presented investigation deals with modified X-Ni/GDC electrodes (where X = Au, Mo, Fe), in the form of half-Electrolyte Supported Cells (ESCs), for their performance in the RWGS through catalytic-kinetic measurements. Ni/GDC, 3wt% Au-Ni/GDC, 3wt% Mo-Ni/GDC, 3wt% Au–3wt% Mo-Ni/GDC, and 2wt% Fe-Ni/GDC modified cathodes were tested at Open Circuit Potential (OCP) conditions to elucidate their catalytic activity towards the production of CO (r_{co}), which is one of the products from the H_2O/CO_2 co-electrolysis reaction. The latter approach is considered as an attempt to create a reference profile for the catalytic performance of the candidate electrodes, by applying the same $H_2O/CO_2/H_2$ feed conditions as those under the co-electrolysis operational mode. The samples, both in the form of powders and as half cells, were physicochemically characterized, including specific redox stability measurements in the presence of H_2O.

2. Results and Discussion

2.1. Physicochemical Characterization

2.1.1. Specific Surface Area Values of Au–Mo–Fe-Modified Powders

The specific surface area values (SSA) for the modified-NiO/GDC samples, calcined in air at 600 and 1100 °C, as well as after reduction with H_2 at 850 °C, are presented in Table 1.

The first remark is that all samples exhibit quite similar and low SSA values, which decrease further by increasing the calcination temperature from 600 °C to 1100 °C and after H_2 reduction at

850 °C. Furthermore, Mo and Au–Mo modification slightly decrease the SSA of Ni/GDC, while the addition of Fe causes a significant increase in the SSA values up to approximately 70%. Taking into account the generally low values and the ±0.2 $m^2 g^{-1}$ accuracy limits of the measurements, it can be concluded that the reduced Mo and Au–Mo-Ni/GDC powders have similar SSAs with Ni/GDC. On the other hand, the 70% increase in the SSA of the Fe modified sample can be primarily ascribed to the formation of FeO_x species, in the oxidized form of the sample, which inhibit the decrease of SSA during reduction.

Table 1. BET Specific Surface Area (SSA_{BET}) of commercial NiO/GDC, and X-modified NiO/GDC samples (where X = Au, Mo, and Fe). The measurements were performed on powders, calcined at 600 °C, 1100 °C, and after H_2-reduction at 850 °C. Error/accuracy of SSA_{BET} = ±0.2 $m^2 g^{-1}$.

Sample Powder	SSA_{BET} ($m^2 g^{-1}$)		
	T = 600 °C, (oxidized)	T = 1100 °C, (oxidized)	After H_2-reduction at T = 850 °C
NiO/GDC	5.3	2.9	2.0
3Au-NiO/GDC	5.4	2.7	2.1
3Mo-NiO/GDC	5.0	2.2	1.7
3Au–3Mo-NiO/GDC	5.1	2.4	1.8
2Fe-NiO/GDC	8.8	4.0	3.4

2.1.2. H_2 Reducibility and air Re-Oxidation Behaviour of Au–Mo–Fe-Modified Powders

Temperature Programmed Reduction followed by Oxidation (H_2-TPR, Air-TPO) measurements were performed, in order to investigate the reducibility and re-oxidation behavior of the prepared composite powders, as well as the existence of possible bulk effects due to the modification with Au, Mo, and Fe. The corresponding H_2-TPR and Air-TPO TGA profiles of modified-NiO/GDC samples, calcined at 1100 °C, are presented in Figure 1.

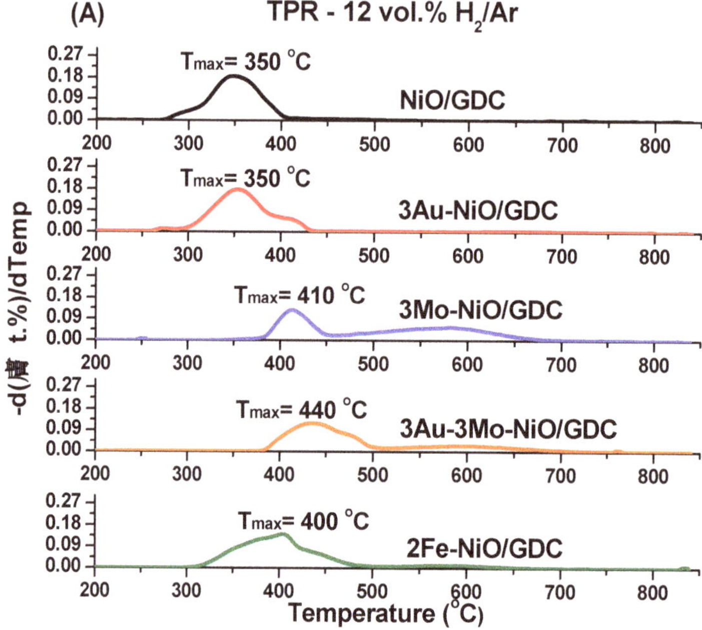

Figure 1. *Cont.*

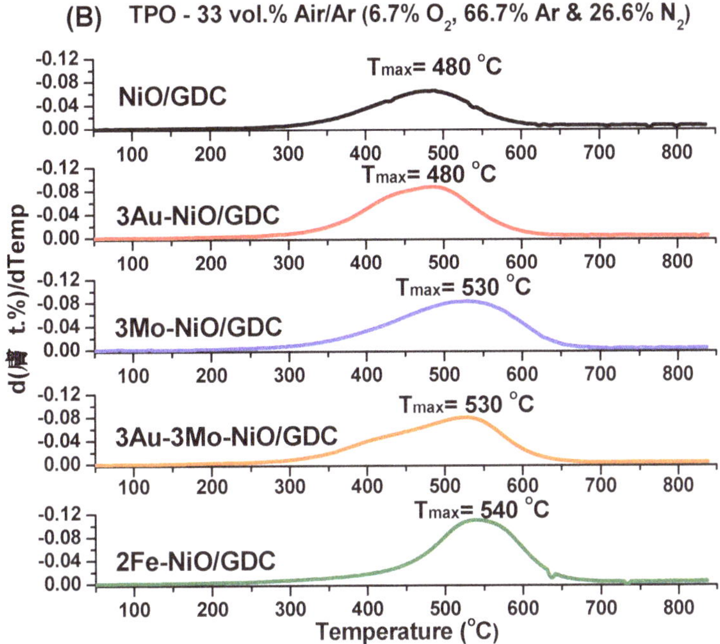

Figure 1. TGA (**A**) H_2-TPR (12 vol.% H_2/Ar), and (**B**) O_2-TPO (33 vol.% air) profiles of X-modified NiO/GDC samples (where X= Au, Mo, and Fe). The measurements were performed on powders, calcined at 1100 °C. F_{total} = 60 cm^3/min (STP conditions: 0 °C, 1 atm).

H_2-TPR and air-TPO measurements showed different reduction and oxidation behavior in each of the examined modified samples. In particular, two main reduction peaks (Figure 1A) and one oxidation peak (Figure 1B) are mainly observed, which can be attributed to the reduction of NiO species to Ni and the reverse [42,47]. Moreover, the presence of Au does not seem to affect significantly the reduction of NiO. On the other hand, MoO_3 and, to a certain extent, Fe inhibit the NiO reduction, implying a stronger Ni–O bond. The inhibition effect of MoO_3 is also observed in the case of the ternary 3Au–3Mo-NiO/GDC sample and indicates a bulk interaction between Ni–Au–Mo, which has been thoroughly investigated in previous studies [23,42,43]. In particular, the second broad peak ("shoulder") at 580 °C, which is observed for the 3Mo-NiO/GDC and 3Au–3Mo-NiO/GDC samples, may be associated with the reduction of MoO_3 species, which are reduced at higher temperatures in the range of 600–770 °C [42,47]. The presence of Fe seems to affect the main reduction peak of NiO, in a similar way as MoO_3, suggesting the possible formation of a solid solution between Ni and Fe. The latter observation is currently under further clarification.

Upon re-oxidation of the samples, the unmodified Ni/GDC as well as the Au modified sample are oxidized practically at the same temperature with a similar TPO profile. The reduced state of the Fe, Mo-, and Au–Mo- modified samples proved to be more resistant to re-oxidation in 33 vol.% air/Ar (6.7 vol.% O_2/Ar), since they had to reach higher temperature for complete re-oxidation.

2.1.3. H_2O Re-Oxidation Profiles of Au-Mo-Fe–Modified Powders

Isothermal-TGA measurements, under 15.5 vol.% H_2O/Ar conditions, were carried out at 650 °C, 700 °C, 750 °C, and 800 °C and the results are depicted in Figures 2 and 3. These measurements investigate the activity of the powders for the H_2O dissociation reaction and their concomitant re-oxidation rate. H_2O acts as an oxidative agent and interacts with the Ni atoms on the surface of each sample towards H_2 and NiO [reaction (5)] [46].

$$H_2O + Ni \rightarrow H_2 + NiO \tag{5}$$

Thereafter, there is a progressive diffusion of the absorbed oxygen species (O_{ads}) from the surface in the bulk phase of the sample, which is oxidized further [46,48–50]. Figure 2 depicts the H_2O re-oxidation profiles, as an increase of the weight (Δwt%), of the pre-reduced Ni/GDC, 3wt% Au-Ni/GDC, 3wt% Mo-Ni/GDC, 3wt% Au–3wt% Mo-Ni/GDC, and 2wt% Fe-Ni/GDC samples in the temperature range between 650–800 °C.

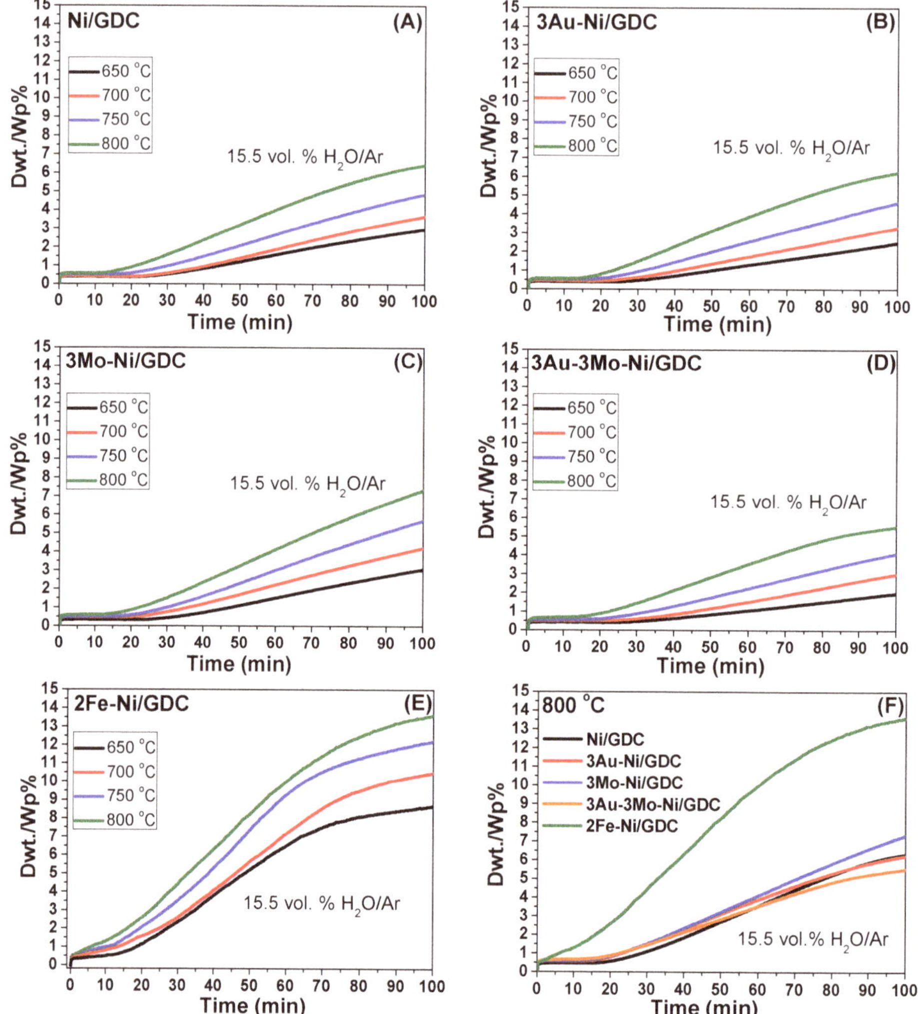

Figure 2. Isothermal TG analysis of the pre-reduced, (**A**) Ni/GDC, (**B**) 3wt% Au-Ni/GDC, (**C**) 3wt% Mo-Ni/GDC, (**D**) 3wt% Au–3wt% Mo-Ni/GDC, and (**E**) 2wt% Fe-Ni/GDC samples in the temperature range between 650–800 °C. (**F**) Comparative TG profiles of the samples at 800 °C. Feed Conditions: 15.5 vol.% H_2O/Ar, F_{total} = 100 cm^3/min (STP conditions: 0 °C, 1 atm).

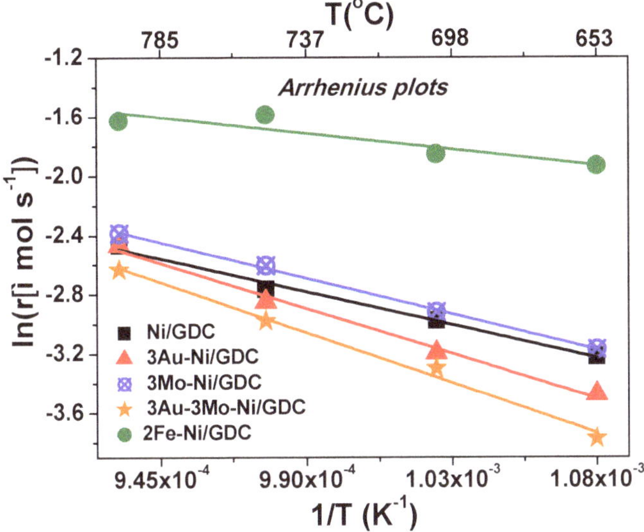

Figure 3. Arrhenius plots of the H_2O re-oxidation rate (resulting from the slope of the isothermal TG profiles in Figure 2) for: Ni/GDC, 3wt% Au-Ni/GDC, 3wt% Mo-Ni/GDC, 3wt% Au–3wt% Mo-Ni/GDC, and 2wt% Fe-Ni/GDC Feed Conditions: 15.5 vol.% H_2O/Ar, F_{total} = 110 cm^3/min (STP conditions: 0 °C, 1 atm).

The re-oxidation profiles of the samples in Figure 2 can be separated in three sections [46]. The first section corresponds to the initial sharp increase in weight and is ascribed [46] to the complete, bulk, re-oxidation of the partially reduced CeO_2 in GDC and specifically of Ce^{3+} to Ce^{4+}. The second section occurs during the following 12–25 min of the reaction, where the samples, except for 2Fe-Ni/GDC, keep their reduced state without any changes in their weight. This step is associated [46] to the dissociation of bulk NiH species, which are formed during the prior reduction period, and maintains the samples reduction. NiH dissociation is considered as an activated process, because by decreasing the operating temperature from 800 °C to 650 °C, the specific "delay" period increases from 12 min to 25 min. The first two sections are quite similar for the pre-reduced powders, with identical low SSA values (Table 1). On the other hand, the initial re-oxidation behavior, in the first section, of the pre-reduced 2Fe-Ni/GDC is more intense, indicating the strong effect of H_2O.

In the third section of the TG profiles, further re-oxidation takes place [reaction (5)] and the main discrepancies are detected among the samples. In the range of 650–800 °C, the ternary 3Au–3Mo-Ni/GDC sample is the most tolerant in bulk re-oxidation by H_2O, while the binary 2Fe-Ni/GDC sample is the least tolerant. The decrease in temperature (800 → 650 °C) leads to inhibition of the bulk re-oxidation, as observed by the reduced slopes in the TG profiles, whereas the trend among the samples does not change. The high tolerance of the 3Au–3Mo-Ni/GDC sample against bulk re-oxidation by H_2O is attributed to the synergistic interaction of nickel with gold and molybdenum [23,42].

The slope on the linear part of the third section in the TG curves represents the intrinsic dissociation rate of H_2O and correspondingly the re-oxidation rate of Ni atoms on each sample [reaction (5)] [46]. Figure 3 depicts the Arrhenius plots of the oxidation rates of the samples, which were calculated from the slopes on the linear parts of the TG profiles (Figure 2). The Arrhenius plots (Figure 3) show that the binary 2Fe-Ni/GDC sample has the highest H_2O re-oxidation rates, while the ternary 3Au–3Mo-Ni/GDC has the lowest. This is further confirmed by the calculated apparent activation energies ($E_{a,ap}$), which are defined from the Arrhenius plots and are reported in Table 2.

Table 2. Apparent activation energies (kJ mol^{-1}) for the dissociation of H_2O on commercial Ni/GDC [46], and modified Ni/GDC samples. The values correspond to the slopes of the Arrhenius plots in Figure 3.

Sample Powder	$E_{a,\ app}$ (kJ mol^{-1})
Ni/GDC [46]	41
3Au-Ni/GDC [46]	56
3Mo-Ni/GDC	45
3Au–3Mo-Ni/GDC	63
2Fe-Ni/GDC	20

The calculated values verify that the binary 2Fe-Ni/GDC sample exhibits the lowest $E_{a,ap}$ (20 kJ/mol) for the catalytic dissociation of H_2O, whereas the ternary 3Au–3Mo-Ni/GDC sample exhibits the highest $E_{a,ap}$ (63 kJ/mol). The above behavior can be interpreted by the fact that Ni-Fe interaction reinforces the bond of H_2O_{ads}, resulting in a higher re-oxidation rate (lower $E_{a,ap}$). On the other hand, following the same interpretation, Ni–Au–Mo interaction weakens the bond of H_2O_{ads}, resulting in a lower re-oxidation rate (higher $E_{a,ap}$) [46].

The calculated $E_{a,ap}$ values are in good agreement with our previous study [46] and other studies [51–53] in the literature, which focus on theoretical investigations. The binary 2Fe-Ni/GDC sample is the most active for the thermochemical dissociation of H_2O and is less resistant against bulk re-oxidation. On the contrary, the ternary 3Au–3Mo-Ni/GDC sample exhibits the lowest activity for the thermochemical dissociation of H_2O and the highest tolerance against bulk re-oxidation.

Taking into account the above, H_2O, apart from being the main reactant of the co-electrolysis process, is also considered as a potential poisoning agent of the electrode. This means that the bonding strength of the adsorbed oxygen species, which result from H_2O decomposition, may induce the re-oxidation of the electrode and finally the deactivation of the sample [46,53]. The H_2O poisoning effect in electrolysis and co-electrolysis processes can also be correlated to the CH_4 poisoning effect in SOFC applications, where degradation is enhanced by the strong adsorption bond of CH_4 on the Ni surface [23,42–45,54,55]. Thus, the stronger binding of the adsorbed oxygen species, which result from the catalytic dissociation of H_2O_{ads}, can similarly cause a poisoning effect on the Ni surface, leading to faster re-oxidation and finally deactivation of the sample [46,53]. According to Besenbacher et al. [56], the surface solid solution between Au and Ni induces significant modifications in the electronic properties of Ni (Fermi level, work function, and d-band center) and affects the bonding strength of the adsorbed species on the surface of the sample.

In this respect, the Au- and Au–Mo- doping of Ni should shift the d-band center to lower energies, with respect to the Fermi level of nickel, thus inhibiting the interaction and the dissociation of H_2O. Therefore, the 3Au-Ni/GDC and 3Au–3Mo-Ni/GDC samples are suggested to be the least active samples for the dissociation of H_2O, having the highest $E_{a,app}$ and consequently the lowest binding energies for both H_2O and O_{ads}. The binary 2Fe-Ni/GDC is suggested to be the most active sample for the dissociation of H_2O, with the lowest $E_{a,app}$, resulting in O_{ads} species with the highest binding energy and the potential to cause faster re-oxidation/poisoning of the electrode [46,53] during SOEC operation.

2.2. *Catalytic-Kinetic Measurements of the RWGS Reaction*

All the prepared samples, in the form of half-electrolyte supported cells, were tested at Open Circuit Potential (OCP) conditions to elucidate their catalytic activity towards the production of CO, which is one of the products from the H_2O/CO_2 co-electrolysis reaction. Furthermore, the effect of the modification (type of dopant: Au, Mo, Fe) on the catalytic activity for the production of CO, through the RWGS reaction, was also investigated. The latter approach is considered as an attempt to create a reference profile for the catalytic performance of the candidate electrodes, under the same experimental conditions as in co-electrolysis (including the presence of current collector). This is an important step, because it will provide detailed experimental feedback on the possible contribution of the RWGS reaction to the production of CO and the extent of this contribution to the electrochemical reduction of CO_2, during the co-electrolysis mode.

In regards to the "homogenous" catalytic production of CO, the rate was found to be negligible, in the range of $r_{co, homogenous}$ = 0.05 μmol/s, for a fuel feed comprising 30 vol.% He − 24.5 vol.% H_2O − 24.5 vol.% CO_2 − 21 vol.% H_2. Concerning the current collector, Ni-mesh shows some catalytic activity for the production of CO (r_{CO}), which increases by increasing the operating temperature and the partial pressure of H_2. However, comparative r_{CO} measurements (see Figure S1 in Supplementary Material) of an Ni/GDC electrode with and without the presence of Ni mesh as the current collector, as well as of the bare Ni mesh, suggest that there is no direct catalytic correlation/contribution of the Ni mesh to the activity of the electrocatalysts. This is mainly explained by the fact that Ni/GDC is a porous electrocatalyst with SSA_{BET} and thus with more active sites, compared to the metallic nickel mesh, which does not possess similar properties. Consequently, from the point where they co-exist as the electrode and current collector, the catalytic activity is mainly attributed to the electrocatalyst.

Figure 4 presents the comparison of the produced r_{CO} for the Ni/GDC and for modified-Ni/GDC electrodes, under three different H_2O-CO_2-H_2 mixtures. The corresponding $\%CO_2$ conversions are depicted in Figure 5. The values are low enough to be considered in the differential region, apart from the case of 2Fe-Ni/GDC in the reaction mixture (A), which is relatively high. The % conversion of CO_2 was calculated by the following formula:

$$CO_2 \text{ conversion } (\%) = \frac{F_{CO}^{out}}{F_{CO_2}^{in}} \cdot 100 \tag{6}$$

where: F_{CO}^{out} and $F_{CO_2}^{in}$ correspond to the rate (μmol/s) of the produced CO and of the introduced CO_2 in the reactants feed, respectively.

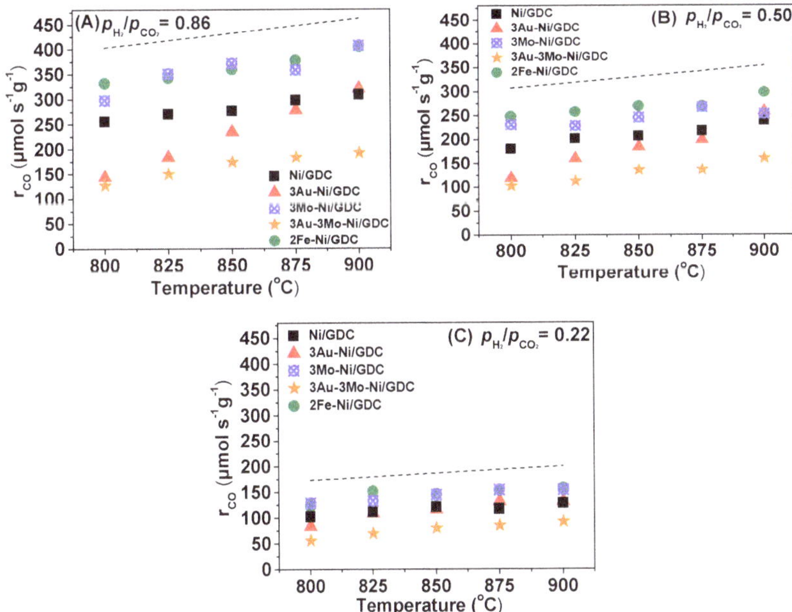

Figure 4. CO production rates (μmol s^{-1} g^{-1}) on ESCs comprising: Ni/GDC, 3Au-Ni/GDC, 3Mo-Ni/GDC, 3Au–3Mo-Ni/GDC, and 2Fe-Ni/GDC, as fuel electrodes, in the range of 800–900 °C under three different mixtures: (**A**) 24.5% H_2O − 24.5% CO_2 − 21% H_2 (P_{H2}/P_{CO_2} = 0.86), (**B**) 28% H_2O − 28% CO_2 − 14% H_2 (P_{H2}/P_{CO_2} = 0.50), and (**C**) 31.5% H_2O − 31.5% CO_2 − 7% H_2 (P_{H2}/P_{CO_2} = 0.22). Dilution of He: 30 vol.% and F_{total} = 140 cm^3/min (STP conditions: 0 °C, 1 atm) in all cases. The dash line (– –) corresponds to the r_{CO} values in thermodynamic equilibrium. All studied electrodes have similar loading in the range of 10–12 mg/cm^2.

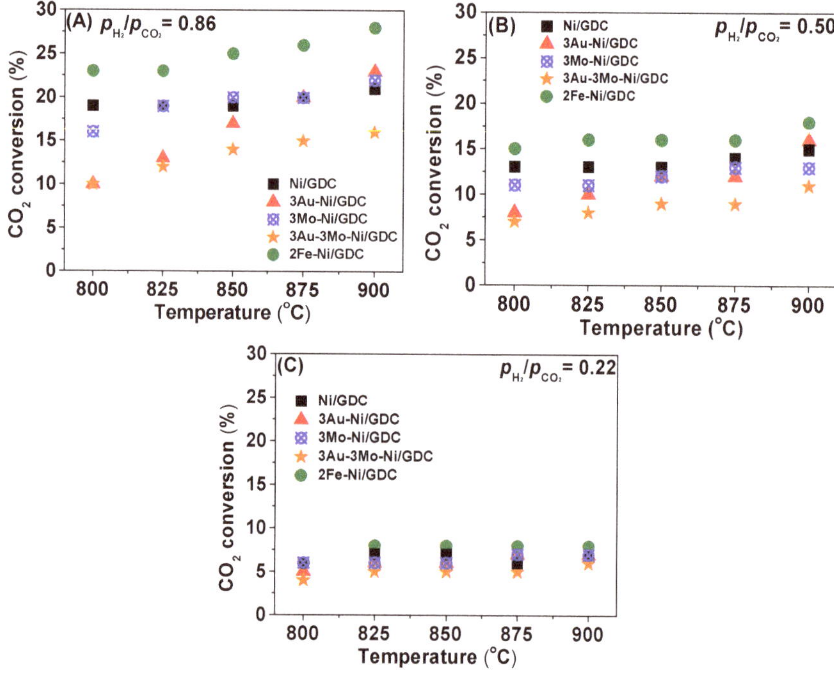

Figure 5. The corresponding $\%CO_2$ conversions of ESCs comprising: Ni/GDC, 3Au-Ni/GDC, 3Mo-Ni/GDC, 3Au–3Mo-Ni/GDC, and 2Fe-Ni/GDC, as fuel electrodes, in the range of 800–900 °C under three different mixtures: (**A**) 24.5% H_2O – 24.5% CO_2 – 21% H_2 (P_{H2}/P_{CO_2} = 0.86), (**B**) 28% H_2O – 28% CO_2 – 14% H_2 (P_{H2}/P_{CO_2} = 0.50), and (**C**) 31.5% H_2O – 31.5% CO_2 – 7% H_2 (P_{H2}/P_{CO_2} = 0.22). Dilution of He: 30 vol.% and F_{total} = 140 cm^3/min (at STP conditions: 0 °C, 1 atm) in all cases. All studied electrodes have similar loading in the range of 10–12 mg/cm^2.

It is shown (Figures 4 and 5) that 2Fe-Ni/GDC and 3Mo-Ni/GDC have the highest catalytic activity of the examined electrodes for the RWGS reaction. In fact, 2Fe-Ni/GDC is the most active in terms of the produced CO. The above performance is observed for all applied fuel feeds (P_{H2}/P_{CO_2} = 0.86, 0.50 and 0.22), whereas it is enhanced by increasing (i) the operating temperature and (ii) the partial pressure of H_2 in the fuel feed. The enhanced catalytic performance of the Fe-modified sample, can be primarily ascribed to the possible stronger adsorption and consequent catalytic dissociation of CO_2 on the active sites of the catalyst. This first conclusion is going to be further examined with specific CO_2 Temperature Programmed Desorption (TPD) measurements.

Another noteworthy remark is that the produced CO rates were compared to the thermodynamic equilibrium rates (dash line in Figure 4) for the three different H_2O-CO_2-H_2 reaction mixtures. The equilibrium rates were calculated by using the equilibrium constant (K_{eq}) formula that is reported in [57,58]. It was found (Figure 4) that Ni/GDC, the binary Au-, and the ternary Au–Mo- modified samples exhibit CO production rates, which are lower than the thermodynamic equilibrium for all the examined reaction conditions. The performances of 2Fe-Ni/GDC and 3Mo-Ni/GDC are closer to the equilibrium, but cannot be considered as thermodynamically limited.

The results from the measurements in Figure 4 can also be presented as Arrhenius plots, depicted in Figure 6, where the derived apparent activation energies ($E_{a,\ app}$) for the production of CO and consequently for the RWGS reaction are listed in Table 3.

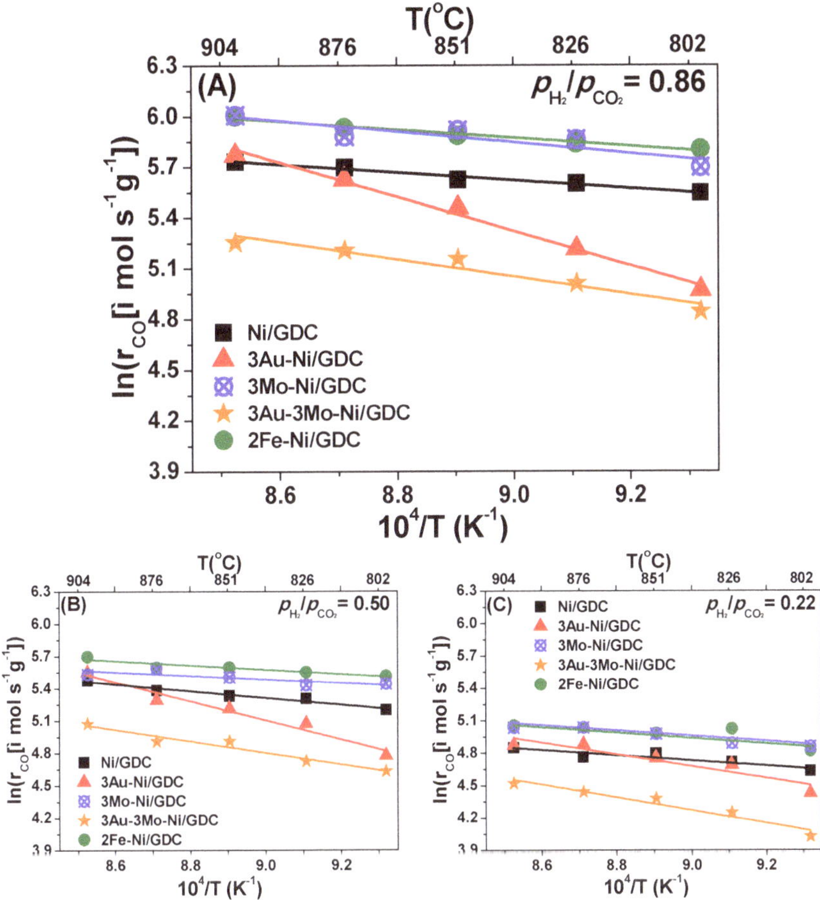

Figure 6. Arrhenius plots of the CO production rates (μmol s^{-1} g^{-1}) on ESCs comprising: Ni/GDC, 3Au-Ni/GDC, 3Mo-Ni/GDC, 3Au–3Mo-Ni/GDC, and 2Fe-Ni/GDC, in the range of 800–900 °C, under three different mixtures: (**A**) 24.5% H_2O – 24.5% CO_2 – 21% H_2 (P_{H2}/P_{CO_2} = 0.86), (**B**) 28% H_2O – 28% CO_2 – 14% H_2 (P_{H2}/P_{CO_2} = 0.50), and (**C**) 31.5% H_2O – 31.5% CO_2 – 7% H_2 (P_{H2}/P_{CO_2} = 0.22). Dilution of He: 30 vol.% and F_{total} = 140 cm^3/min (at STP conditions: 0 °C, 1 atm) in all cases.

Table 3. Apparent activation energies ($E_{a,app}$, kJ/mol) for the RWGS reaction on ESCs for three different mixtures: (**A**) 24.5% H_2O – 24.5% CO_2 – 21% H_2 (P_{H2}/P_{CO_2} = 0.86), (**B**) 28% H_2O – 28% CO_2 – 14% H_2 (P_{H2}/P_{CO_2} = 0.50), and (**C**) 31.5% H_2O – 31.5% CO_2 – 7% H_2 (P_{H2}/P_{CO_2} = 0.22).

Sample	**$E_{a,\ apparent}$ (kJ mol^{-1}) and A^* ($\mu mol\ s^{-1}\ g^{-1}$), per reaction mixture**					
	P_{H2}/P_{CO_2} = 0.86		**P_{H2}/P_{CO_2} = 0.50**		**P_{H2}/P_{CO_2} = 0.22**	
Ni/GDC	**20**	*2.3×10^3*	**22**	*3.6×10^3*	**20**	*1.0×10^3*
3Au-Ni/GDC	**85**	*1.9×10^6*	**74**	*5.0×10^5*	**46**	*1.5×10^4*
3Mo-Ni/GDC	**27**	*6.5×10^3*	**14**	*1.1×10^3*	**21**	*1.4×10^3*
3Au–3Mo-Ni/GDC	**43**	*1.7×10^4*	**45**	*1.5×10^4*	**50**	*1.6×10^4*
2Fe-Ni/GDC	**21**	*3.3×10^3*	**19**	*1.6×10^3*	**21**	*1.3×10^3*

* *A*: The pre-exponential factor in the Arrhenius equation, $r = A \cdot \exp(-\frac{E_{a,app}}{R \cdot T})$.

The Arrhenius plots and the calculated $E_{a, app}$ show that 2Fe-Ni/GDC, 3Mo-Ni/GDC, and Ni/GDC have practically the same and the lowest $E_{a, app}$. However, the Fe- and Mo- modified samples exhibit higher pre-exponential factors, compared to Ni/GDC, which explains the higher production rates of CO. 3Au-Ni/GDC shows overall the highest $E_{a, app}$, which is an additional indication for its worst catalytic activity. Finally, the ternary 3Au–3Mo-Ni/GDC sample shows an apparent activation energy, which lies between that for 2Fe-Ni/GDC and 3Au-Ni/GDC. However, the catalytic activity of the ternary sample is the lowest, due to the lower pre-exponential factor compared to the binary Au-modified sample. According to the knowledge of the authors, there are no literature data available for experimentally measured $E_{a, app}$, for similar samples and reaction conditions. The so far available data come from theoretical investigations and there is a recent study from Cho et al. [41], who performed DFT calculations to evaluate the ability of various transition metals to increase the activity of Ni for the H_2O/CO_2 co-electrolysis. In particular, they computed the activation energies of specific elementary reaction steps and in the case of the RWGS on Ni(111), the E_a was found to be approximately 46 kJ/mole, which is very close to the values that were experimentally calculated in the present study.

The effect of H_2 partial pressure on the catalytic rate of CO production is further verified in Figure 7 for all samples at 900 °C and 800 °C. The 2Fe-Ni/GDC sample is the most active and 3Au–3Mo-Ni/GDC the least one. The 3Au-Ni/GDC catalyst at 900 °C shows similar performance with that of Ni/GDC. In addition, by decreasing the temperature at 800 °C, the catalytic activity of 3Au-Ni/GDC exhibits the highest reduction. This is apparent from the significant decrease in the slope of the corresponding curve (Figure 7) from 800 °C to 900 °C and can be further explained from the calculated $E_{a, app}$ (Table 3), which is the highest from all samples. Finally, the kinetic behavior of all samples suggests that the production rate of CO exhibits a positive order dependence on the partial pressure of H_2.

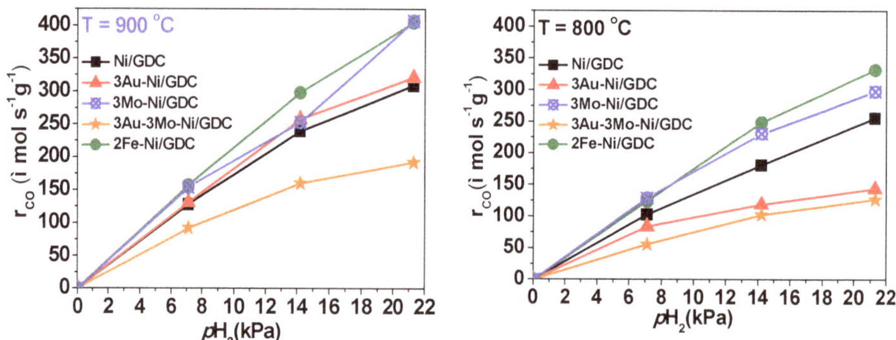

Figure 7. Steady state effect of the H_2 partial pressure (pH_2, kPa) on the CO production rates (μmol s^{-1} g^{-1}) on ESCs comprising: Ni/GDC, 3Au-Ni/GDC, 3Mo-Ni/GDC, 3Au–3Mo-Ni/GDC, and 2Fe-Ni/GDC, for 900 °C and 800 °C. The pH_2O/pCO_2 ratios are fixed and their values are presented in Table 3.

The fact that in the majority of the samples there is no trend in the $E_{a, app}$, by varying the pH_2/pCO_2 ratios, primarily shows that all samples, apart from the binary Au-modified, have similar intrinsic catalytic activity for the RWGS reaction. This is also concluded from the remark that the effect of the H_2 partial pressure on the CO production rate is almost linear. On the other hand, in the case of 3Au-Ni/GDC, there is an increase in $E_{a, app}$, by increasing the pH_2/pCO_2 ratio. This is corroborated from the results in Figure 7, where r_{CO} seems to reach a plateau at high pH_2. The latter remarks indicate that the majority of the samples have actives sites that follow a similar reaction mechanism, in regards to the dissociative adsorption of H_2 and CO_2 towards the RWGS. On the other hand, the reaction mechanism seems to be different in the case of 3Au-Ni/GDC. In particular, it is implied that the high coverage of adsorbed H_2 may inhibit the CO_2 dissociative adsorption and thus decreases the intrinsic catalytic activity of the specific sample. This suggestion is going to be further clarified through the currently performed H_2O/CO_2 co-electrolysis measurements.

Carbon formation, through the Boudouard reaction (6), was also investigated and the results are presented in Figure 8. The slight scattering of the experimentally measured rates in combination with the theoretical values, where $r[CO_2]_{inlet}$ is equal to $r[CO + CO_2]_{outlet}$, suggest that there is no carbon deposition under the examined reaction conditions. This result is also in accordance with the fact that the Boudouard reaction is not thermodynamically favored above 750 °C [59].

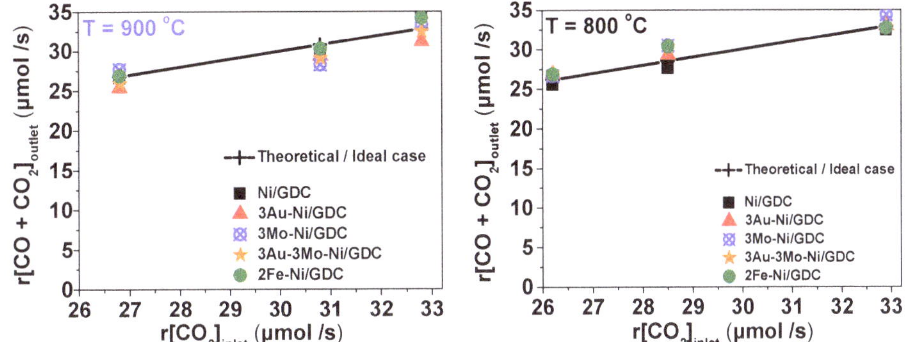

Figure 8. Investigation of the Boudouard reaction at 900 °C and 800 °C under OCP conditions on ESCs comprising: Ni/GDC, 3Au-Ni/GDC, 3Mo-Ni/GDC, 3Au–3Mo-Ni/GDC, and 2Fe-Ni/GDC. Theoretical case where $[CO_2]_{inlet}$ is equal to $[CO+CO_2]_{outlet}$ is also depicted with a solid line.

Previous studies [1,59,60] reported that at realistic CO_2/CO concentrations, during CO_2 electrolysis in the range of 650–750 °C, the equilibrium of (6) is shifted towards CO production (reaction (7)) and therefore carbon will not be formed, catalytically, during CO_2 electrolysis. In the case where the cell is operated at OCP conditions, as in our case, any coke deposited within the porous, modified or not, Ni/GDC electrode would be oxidized to CO according to the reverse Boudouard reaction (7) and thereby removed. Furthermore, the addition of steam in the feed is reported to remove carbon depositions according to reaction (8) [1,8]:

$$\text{Boudouard reaction:} \quad 2CO \rightarrow CO_2 + C \tag{7}$$

$$\text{Reverse Boudouard reaction:} \quad CO_2 + C \rightarrow 2CO \tag{8}$$

$$\text{Reaction of coke with } H_2O\text{:} \quad C + H_2O \rightarrow CO + H_2 \tag{9}$$

3. Materials and Methods

3.1. Preparation of Electrocatalysts

The modified cathode powders were prepared via the Deposition-Precipitation (D.P.) and Deposition-Co Precipitation (D.CP.) methods by using the commercial NiO/GDC cermet (65wt% NiO-35wt% GDC, Marion Technologies, Verniolle, France) as the support. The precursors for the 3wt% Au-NiO/GDC, 3wt% Mo-NiO/GDC, 3wt% Au−3wt% Mo-NiO/GDC, and 2wt% Fe-NiO/GDC samples were the $HAuCl_4$ (Sigma-Aldrich) or/and $(NH_4)_6Mo_7O_{24}$ (Sigma-Aldrich, St. Louis, MO, USA) and $Fe(NO_3)_3x9H_2O$ (Sigma-Aldrich, St. Louis, MO, USA) solutions, respectively. Full details about the synthesis can be found elsewhere [42,43]. After filtering, the precipitate was dried at 110 °C for 24 h. All dried powders were calcined at 600 °C/90 min and a part of them at 1100 °C/75 min. The first batch was used for the paste preparation, which is described in the following paragraph. The batch at 1100 °C was used for the physicochemical characterization. In this way, the prepared catalysts were studied at similar thermal stress as the calcined electrode-electrolyte assemblies.

3.2. Preparation of Electrolyte-Supported Half Cells

The electrolyte-supported half cells consisted of circular shaped planar 8YSZ electrolyte (by Kerafol) with a 25 mm diameter and 300 µm thickness. The fuel electrode was deposited by using the screen printing technique as reported in previous studies [23,46]. In particular, a paste was prepared by using an appropriate amount of the electrocatalyst (modified NiO/GDC powder), terpineol (Sigma-Aldrich, St. Louis, MO, USA) as the dispersant, and PVB (polyvinylbutyral, Sigma-Aldrich, St. Louis, MO, USA) as binder. After the deposition of the paste, the cell was sintered at 1150 °C with a heating/cooling ramp rate of 2 °C/min. The last temperature is the lowest possible in order to obtain proper adherence of the electrode on the electrolyte, whereas it is equivalent with the calcination temperature of the characterized powders (1100 °C). The examined electrodes were approximately 20 µm thick and their loading varied in the region of 10–12 mg/cm^2 with a 1.8 cm^2 geometric surface area (Figure 9). The prepared half cells were adjusted on a ceramic YSZ tube and sealed airtight with a glass sealing material manufactured by Kerafol.

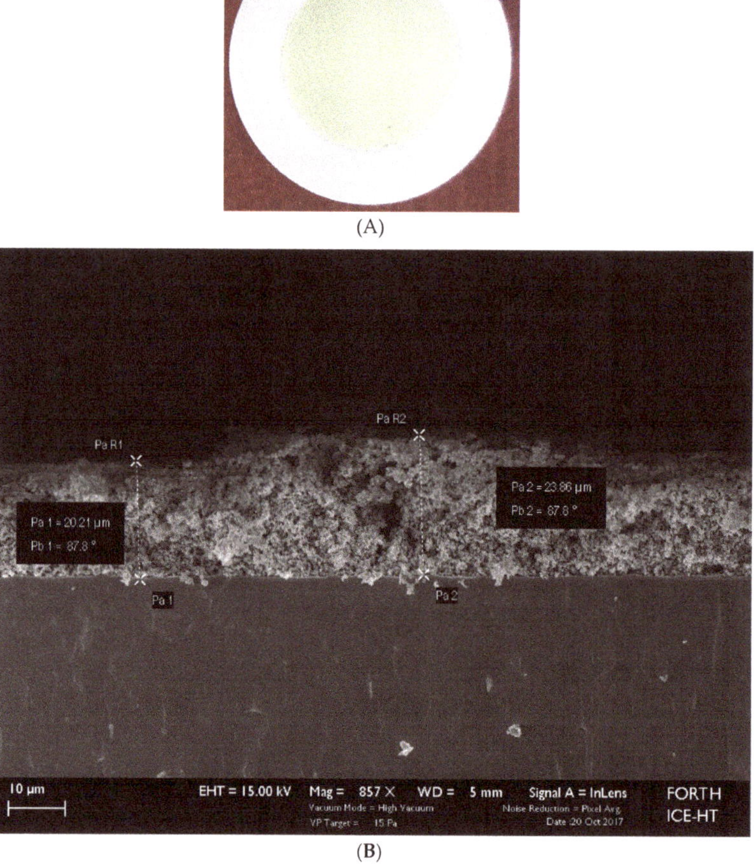

Figure 9. Images of a (half) electrolyte supported cell (ESC) prepared by screen printing: (**A**) NiO/GDC electrode and (**B**) SEM cross section perpendicular to the NiO/GDC//YSZ (half) electrolyte-supported cell.

3.3. Physicochemical Characterization

The samples, in the form of powders, were characterized with BET, H_2-TPR, Air-TPO, and TGA re-oxidation measurements in the presence of H_2O.

The BET specific surface area values of the samples were determined from the adsorption isotherms of nitrogen at −196 °C, recorded with a Micromeritics TriStar 3000 apparatus (Micromeritics, Norcross, GA, USA).

The re-oxidation measurements in the presence of H_2O, as well as H_2-TPR and Air-TPO measurements were carried out with a TA Q50 instrument. The H_2O re-oxidation properties of the powders were studied with TGA measurements, at a constant temperature in the range of 650–800 °C. Before the measurement, the samples were pre-reduced in-situ with 80 vol.% H_2/Ar at 800 °C for 100 min and then the feed was changed to 15.5 vol.% H_2O/Ar. The total flow rate (at STP conditions: 0 °C, 1 atm) was 100 cm^3/min and the loading of the measured samples was approximately 50 mg. Steam was introduced in the reactor by passing Ar through a saturator, which was maintained at a fixed temperature (65 °C).

3.4. Catalytic-Kinetic Measurements

The prepared half cells were catalytically investigated at Open Circuit Potential (OCP) conditions for the RWGS reaction, in the presence of Ni mesh. The catalytic experiments were accomplished at temperatures between 800–900 °C under various $H_2O/CO_2/H_2$ mixtures by keeping in all cases the P_{H_2O}/P_{CO_2} ratio constant. In regards to the experimental part of these measurements, H_2O was added and handled in the feed in the form of steam, as in the SOEC measurements. Before its evaporation, liquid H_2O was pressurized in a container and circulated in the system by means of a liquid water mass flow controller. Then, liquid H_2O was evaporated through lines and valves, heated at 160 °C, to prevent water condensation. The flow rate was fixed at 140 cm^3/min (at STP conditions: 0 °C, 1 atm), avoiding any mass transfer limitation effects in the reactor. Reactants and products were determined by using an on-line gas chromatograph (Varian CP-3800) with a thermal conductivity detector. Further details regarding the experimental parameters are indicated in the corresponding Figures.

4. Conclusions

The presented study deals with the kinetic investigation of Ni-based (modified or not) electrodes towards their performance for the RWGS reaction. The samples were examined in the form of electrolyte-supported (half) cells and the measured kinetic parameter was the production rate of CO. The main objective was to clarify the effect of the modification on the catalytic activity for the RWGS, which is considered as a key reaction for the CO production under H_2O/CO_2 co-electrolysis operation. The reaction conditions were similar to those that are applied under co-electrolysis mode.

Redox stability measurements in the presence of H_2O showed that the ternary 3Au–3Mo-Ni/GDC electrode is the least active sample for the dissociation of H_2O, having the highest $E_{a,app}$ and consequently the lowest binding energy for the H_2O_{ads}. On the other hand, the binary 2Fe-Ni/GDC is the most active sample for the dissociation of H_2O, thus having the potential to experience faster re-oxidation. Complementary characterization suggests that the interaction of Ni and Fe (FeO_x species in the oxidized form of the sample) during the H_2-reduction process increases the SSA_{BET} and affects the bulk properties of the binary Ni-Fe/GDC. The interaction can be realized through the possible formation of an Ni-Fe solid solution, which is currently under further clarification and may be responsible for enhancing the catalytic dissociation of H_2O_{ads}.

The kinetic study of the candidate electrocatalysts showed that Au modification inhibits the catalytic production of CO, through the RWGS reaction, while modification with Fe or Mo induces an enhancement of r_{CO}. In fact, the 2wt% Fe-Ni/GDC sample is the most active both in terms of $\%CO_2$ conversion and of the produced CO. In addition, a negative synergy was observed for the ternary Au–Mo-Ni modified sample. Specifically, the 2wt% Fe-Ni/GDC and 3wt% Mo-Ni/GDC samples showed similar apparent activation energy for the RWGS reaction as that of Ni/GDC, while the 3wt% Au-Ni/GDC and 3wt% Au–3wt% Mo-Ni/GDC samples showed lower $E_{a,app}$. The above performance is observed for all applied fuel feeds (P_{H2}/P_{CO_2} = 0.86, 0.50, and 0.22), whereas it is enhanced by increasing (i) the operating temperature and (ii) the partial pressure of H_2.

Consequently, the kinetic behavior of all samples suggests that the production rate of CO exhibits a positive order dependence on the partial pressure of H_2, whereas carbon formation was not detected. In the case of the most active 2Fe-Ni/GDC sample, it is proposed that Fe-modification may enhance the catalytic dissociative adsorption of CO_2 towards the production of CO and this is further investigated. Finally, it is worth mentioning that the RWGS catalytic performance of both Fe- and Mo- modified samples is close to the equilibrium, but cannot be considered as thermodynamically limited.

Overall, the presented results correspond to a reference catalytic profile of the examined modified Ni/GDC samples for the RWGS reaction. These candidate electrocatalysts are currently being examined, as full electrolyte supported cells, in SOEC measurements for the H_2O/CO_2 co-electrolysis reaction to elucidate any additional effects by the applied current. Moreover, further investigation of the Fe-modification on NiO/GDC for the H_2O/CO_2 electrolysis processes will occur in the future.

Supplementary Materials: The following are available online at http://www.mdpi.com/2073-4344/9/2/151/s1, **Figure S1.** Kinetic study of the R.W.G.S. reaction on the nickel mesh (■), Ni/GDC with Ni mesh (●) and Ni/GDC without Ni mesh (△), in the range of 800–900 °C, under two different mixtures: (**A**) 24.5% H_2O – 24.5% CO_2 – 21% H_2 (P_{H2}/P_{CO_2} = 0.86) and (**B**) 28% H_2O – 28% CO_2 – 14% H_2 (P_{H2}/P_{CO_2} = 0.50). Dilution of He: 30 vol.% and F_{total} = 140 cm^3/min (at STP conditions: 0 °C, 1 atm) in all cases. The loading of Ni/GDC is 11 mg/cm^2.

Author Contributions: Conceptualization, S.N. and D.N.; Data curation, E.I.; Formal analysis, E.I.; Funding acquisition, S.N. and D.N.; Investigation, E.I.; Methodology, S.N. and D.N.; Supervision, S.N. and D.N.; Validation, S.N. and D.N.; Writing—original draft, D.N.; Writing—review & editing, S.N., D.N. and E.I.

Funding: The research leading to these results has received funding from the Fuel Cells and Hydrogen 2 Joint Undertaking under the project SElySOs with Grant Agreement No: 671481. This Joint Undertaking receives support from the European Union's Horizon 2020 Research and Innovation Programme and Greece, Germany, Czech Republic, France, and Norway.

Conflicts of Interest: The authors declare no conflict of interest.

References

1. Ebbesen, S.D.; Mogensen, M. Electrolysis of carbon dioxide in Solid Oxide Electrolysis Cells. *J. Power Sources* **2009**, *193*, 349–358. [CrossRef]
2. Graves, C.; Ebbesen, S.D.; Mogensen, M.; Lackner, K.S. Sustainable hydrocarbon fuels by recycling CO_2 and H_2O with renewable or nuclear energy. *Renew. Sustain. Energy Rev.* **2011**, *15*, 1–23. [CrossRef]
3. Tao, Y.; Ebbesen, S.D.; Mogensen, M.B. Carbon Deposition in Solid Oxide Cells during Co-Electrolysis of H_2O and CO_2. *J. Electrochem. Soc.* **2014**, *161*, F337–F343. [CrossRef]
4. Moçoteguy, P.; Brisse, A. A review and comprehensive analysis of degradation mechanisms of solid oxide electrolysis cells. *Int. J. Hydrogen Energy* **2013**, *38*, 15887–15902. [CrossRef]
5. Laguna-Bercero, M.A. Recent advances in high temperature electrolysis using solid oxide fuel cells: A review. *J. Power Sources* **2012**, *203*, 4–16. [CrossRef]
6. Petipas, F.; Brisse, A.; Bouallou, C. Benefits of external heat sources for high temperature electrolyser systems. *Int. J. Hydrogen Energy* **2014**, *39*, 5505–5513. [CrossRef]
7. Zhang, X.; Song, Y.; Wang, G.; Bao, X. Co-electrolysis of CO_2 and H_2O in high-temperature solid oxide electrolysis cells: Recent advance in cathodes. *J. Energy Chem.* **2017**, *26*, 839–853. [CrossRef]
8. Sun, X.; Chen, M.; Jensen, S.H.; Ebbesen, S.D.; Graves, C.; Mogensen, M. Thermodynamic analysis of synthetic hydrocarbon fuel production in pressurized solid oxide electrolysis cells. *Int. J. Hydrogen Energy* **2012**, *37*, 17101–17110. [CrossRef]
9. Kim, S.-W.; Kim, H.; Yoon, K.J.; Lee, J.-H.; Kim, B.-K.; Choi, W.; Lee, J.-H.; Hong, J. Reactions and mass transport in high temperature co-electrolysis of steam/CO_2 mixtures for syngas production. *J. Power Sources* **2015**, *280*, 630–639. [CrossRef]
10. Li, W.; Wang, H.; Shi, Y.; Cai, N. Performance and methane production characteristics of H_2O–CO_2 co-electrolysis in solid oxide electrolysis cells. *Int. J. Hydrogen Energy* **2013**, *38*, 11104–11109. [CrossRef]

11. Mahmood, A.; Bano, S.; Yu, J.H.; Lee, K.-H. Effect of operating conditions on the performance of solid electrolyte membrane reactor for steam and CO_2 electrolysis. *J. Membr. Sci.* **2015**, *473*, 8–15. [CrossRef]
12. Sun, X.; Chen, M.; Liu, Y.-L.; Hjalmarsson, P.; Ebbesen, S.D.; Jensen, S.H.; Mogensen, M.B.; Hendriksen, P.V. Durability of Solid Oxide Electrolysis Cells for Syngas Production. *J. Electrochem. Soc.* **2013**, *160*, F1074–F1080. [CrossRef]
13. Ebbesen, S.D.; Graves, C.; Mogensen, M. Production of Synthetic Fuels by Co-Electrolysis of Steam and Carbon Dioxide. *Int. J. Green Energy* **2009**, *6*, 646–660. [CrossRef]
14. Graves, C.; Ebbesen, S.D.; Mogensen, M. Co-electrolysis of CO_2 and H_2O in solid oxide cells: Performance and durability. *Solid State Ion.* **2011**, *192*, 398–403. [CrossRef]
15. Diethelm, S.; Herle, J.V.; Montinaro, D.; Bucheli, O. Electrolysis and Co-Electrolysis Performance of SOE Short Stacks. *Fuel Cells* **2013**, *13*, 631–637. [CrossRef]
16. Kim-Lohsoontorn, P.; Bae, J. Electrochemical performance of solid oxide electrolysis cell electrodes under high-temperature coelectrolysis of steam and carbon dioxide. *J. Power Sources* **2011**, *196*, 7161–7168. [CrossRef]
17. Stoots, C.; O'Brien, J.; Hartvigsen, J. Results of recent high temperature coelectrolysis studies at the Idaho National Laboratory. *Int. J. Hydrogen Energy* **2009**, *34*, 4208–4215. [CrossRef]
18. Ni, M. 2D thermal modeling of a solid oxide electrolyzer cell (SOEC) for syngas production by H_2O/CO_2 co-electrolysis. *Int. J. Hydrogen Energy* **2012**, *37*, 6389–6399. [CrossRef]
19. Ni, M. An electrochemical model for syngas production by co-electrolysis of H_2O and CO_2. *J. Power Sources* **2012**, *202*, 209–216. [CrossRef]
20. Yue, X.; Irvine, J.T.S. (La,Sr)(Cr,Mn)O_3/GDC cathode for high temperature steam electrolysis and steam-carbon dioxide co-electrolysis. *Solid State Ion.* **2012**, *225*, 131–135. [CrossRef]
21. Zhan, Z.; Zhao, L. Electrochemical reduction of CO_2 in solid oxide electrolysis cells. *J. Power Sources* **2010**, *195*, 7250–7254. [CrossRef]
22. Ebbesen, S.D.; Jensen, S.H.; Hauch, A.; Mogensen, M.B. High Temperature Electrolysis in Alkaline Cells, Solid Proton Conducting Cells, and Solid Oxide Cells. *Chem. Rev.* **2014**, *114*, 10697–10734. [CrossRef]
23. Neofytidis, C.; Dracopoulos, V.; Neophytides, S.G.; Niakolas, D.K. Electrocatalytic performance and carbon tolerance of ternary Au-Mo-Ni/GDC SOFC anodes under CH_4-rich Internal Steam Reforming conditions. *Catal. Today* **2018**, *310*, 157–165. [CrossRef]
24. Liang, M.; Yu, B.; Wen, M.; Chen, J.; Xu, J.; Zhai, Y. Preparation of NiO–YSZ composite powder by a combustion method and its application for cathode of SOEC. *Int. J. Hydrogen Energy* **2010**, *35*, 2852–2857. [CrossRef]
25. Kleiminger, L.; Li, T.; Li, K.; Kelsall, G.H. CO_2 splitting into CO and O_2 in micro-tubular solid oxide electrolysers. *RSC Adv.* **2014**, *4*, 50003–50016. [CrossRef]
26. Marina, O.A.; Pederson, L.R.; Williams, M.C.; Coffey, G.W.; Meinhardt, K.D.; Nguyen, C.D.; Thomsen, E.C. Electrode Performance in Reversible Solid Oxide Fuel Cells. *J. Electrochem. Soc.* **2007**, *154*, B452–B459. [CrossRef]
27. Hjalmarsson, P.; Sun, X.; Liu, Y.-L.; Chen, M. Durability of high performance Ni–yttria stabilized zirconia supported solid oxide electrolysis cells at high current density. *J. Power Sources* **2014**, *262*, 316–322. [CrossRef]
28. Tao, Y.; Ebbesen, S.D.; Mogensen, M.B. Degradation of solid oxide cells during co-electrolysis of steam and carbon dioxide at high current densities. *J. Power Sources* **2016**, *328*, 452–462. [CrossRef]
29. Hauch, A.; Ebbesen, S.D.; Jensen, S.H.; Mogensen, M. Solid Oxide Electrolysis Cells: Microstructure and Degradation of the Ni/Yttria-Stabilized Zirconia Electrode. *J. Electrochem. Soc.* **2008**, *155*, B1184–B1193. [CrossRef]
30. Chen, M.; Liu, Y.-L.; Bentzen, J.J.; Zhang, W.; Sun, X.; Hauch, A.; Tao, Y.; Bowen, J.R.; Hendriksen, P.V. Microstructural Degradation of Ni/YSZ Electrodes in Solid Oxide Electrolysis Cells under High Current. *J. Electrochem. Soc.* **2013**, *160*, F883–F891. [CrossRef]
31. Tietz, F.; Sebold, D.; Brisse, A.; Schefold, J. Degradation phenomena in a solid oxide electrolysis cell after 9000 h of operation. *J. Power Sources* **2013**, *223*, 129–135. [CrossRef]
32. Knibbe, R.; Traulsen, M.L.; Hauch, A.; Ebbesen, S.D.; Mogensen, M. Solid Oxide Electrolysis Cells: Degradation at High Current Densities. *J. Electrochem. Soc.* **2010**, *157*, B1209–B1217. [CrossRef]
33. Hauch, A.; Brodersen, K.; Chen, M.; Mogensen, M.B. Ni/YSZ electrodes structures optimized for increased electrolysis performance and durability. *Solid State Ion.* **2016**, *293*, 27–36. [CrossRef]

34. Papaefthimiou, V.; Niakolas, D.K.; Paloukis, F.; Teschner, D.; Knop-Gericke, A.; Haevecker, M.; Zafeiratos, S. Operando observation of nickel/ceria electrode surfaces during intermediate temperature steam electrolysis. *J. Catal.* **2017**, *352*, 305–313. [CrossRef]
35. Liu, S.; Chuang, K.T.; Luo, J.-L. Double-Layered Perovskite Anode with in Situ Exsolution of a Co–Fe Alloy To Cogenerate Ethylene and Electricity in a Proton-Conducting Ethane Fuel Cell. *ACS Catal.* **2016**, *6*, 760–768. [CrossRef]
36. Nishida, R.; Puengjinda, P.; Nishino, H.; Kakinuma, K.; Brito, M.E.; Watanabe, M.; Uchida, H. High-performance electrodes for reversible solid oxide fuel cell/solid oxide electrolysis cell: Ni-Co dispersed ceria hydrogen electrodes. *RSC Adv.* **2014**, *4*, 16260–16266. [CrossRef]
37. Kim, S.-W.; Park, M.; Kim, H.; Yoon, K.J.; Son, J.-W.; Lee, J.-H.; Kim, B.-K.; Lee, J.-H.; Hong, J. Catalytic Effect of Pd-Ni Bimetallic Catalysts on High-Temperature Co-Electrolysis of Steam/CO_2 Mixtures. *J. Electrochem. Soc.* **2016**, *163*, F3171–F3178. [CrossRef]
38. Ishihara, T.; Jirathiwathanakul, N.; Zhong, H. Intermediate temperature solid oxide electrolysis cell using $LaGaO_3$ based perovskite electrolyte. *Energy Environ. Sci.* **2010**, *3*, 665–672. [CrossRef]
39. Wang, S.; Inoishi, A.; Hong, J.-E.; Ju, Y.-W.; Hagiwara, H.; Ida, S.; Ishihara, T. Ni–Fe bimetallic cathodes for intermediate temperature CO_2 electrolyzers using a La0.9Sr0.1Ga0.8Mg0.2O_3 electrolyte. *J. Mater. Chem. A* **2013**, *1*, 12455–12461. [CrossRef]
40. Back, S.; Jung, Y. Importance of Ligand Effects Breaking the Scaling Relation for Core–Shell Oxygen Reduction Catalysts. *ChemCatChem* **2017**, *9*, 3173–3179. [CrossRef]
41. Cho, A.; Ko, J.; Kim, B.-K.; Han, J.W. Electrocatalysts with Increased Activity for Coelectrolysis of Steam and Carbon Dioxide in Solid Oxide Electrolyzer Cells. *ACS Catal.* **2018**, 967–976. [CrossRef]
42. Niakolas, D.K.; Athanasiou, M.; Dracopoulos, V.; Tsiaoussis, I.; Bebelis, S.; Neophytides, S.G. Study of the synergistic interaction between nickel, gold and molybdenum in novel modified NiO/GDC cermets, possible anode materials for CH_4 fueled SOFCs. *Appl. Catal. A Gen.* **2013**, *456*, 223–232. [CrossRef]
43. Niakolas, D.K.; Ouweltjes, J.P.; Rietveld, G.; Dracopoulos, V.; Neophytides, S.G. Au-doped Ni/GDC as a new anode for SOFCs operating under rich CH_4 internal steam reforming. *Int. J. Hydrogen Energy* **2010**, *35*, 7898–7904. [CrossRef]
44. Neofytidis, C.; Athanasiou, M.; Neophytides, S.G.; Niakolas, D.K. Sulfur Tolerance of Au–Mo–Ni/GDC SOFC Anodes Under Various CH_4 Internal Steam Reforming Conditions. *Top. Catal.* **2015**, *58*, 1276–1289. [CrossRef]
45. Niakolas, D.K.; Neofytidis, C.S.; Neophytides, S.G. Effect of Au and/or Mo Doping on the Development of Carbon and Sulfur Tolerant Anodes for SOFCs—A Short Review. *Front. Environ. Sci.* **2017**, *5*. [CrossRef]
46. Ioannidou, E.; Neofytidis, C.; Sygellou, L.; Niakolas, D.K. Au-doped Ni/GDC as an Improved Cathode Electrocatalyst for H_2O Electrolysis in SOECs. *Appl. Catal. B Environ.* **2018**, *236*, 253–264. [CrossRef]
47. Wandekar, R.V.; Ali, M.; Wani, B.N.; Bharadwaj, S.R. Physicochemical studies of NiO–GDC composites. *Mater. Chem. Phys.* **2006**, *99*, 289–294. [CrossRef]
48. Sarantaridis, D.; Atkinson, A. Redox Cycling of Ni-Based Solid Oxide Fuel Cell Anodes: A Review. *Fuel Cells* **2007**, *7*, 246–258. [CrossRef]
49. Ettler, M.; Timmermann, H.; Malzbender, J.; Weber, A.; Menzler, N.H. Durability of Ni anodes during reoxidation cycles. *J. Power Sources* **2010**, *195*, 5452–5467. [CrossRef]
50. Atkinson, A.; Smart, D.W. Transport of Nickel and Oxygen during the Oxidation of Nickel and Dilute Nickel/Chromium Alloy. *J. Electrochem. Soc.* **1988**, *135*, 2886–2893. [CrossRef]
51. Ghosh, S.; Hariharan, S.; Tiwari, A.K. Water Adsorption and Dissociation on Copper/Nickel Bimetallic Surface Alloys: Effect of Surface Temperature on Reactivity. *J. Phys. Chem. C* **2017**, *121*, 16351–16365. [CrossRef]
52. Gan, L.-Y.; Tian, R.-Y.; Yang, X.-B.; Lu, H.-D.; Zhao, Y.-J. Catalytic Reactivity of CuNi Alloys toward H_2O and CO Dissociation for an Efficient Water–Gas Shift: A DFT Study. *J. Phys. Chem. C* **2012**, *116*, 745–752. [CrossRef]
53. Gu, X.-K.; Nikolla, E. Fundamental Insights into High-Temperature Water Electrolysis Using Ni-Based Electrocatalysts. *J. Phys. Chem. C* **2015**, *119*, 26980–26988. [CrossRef]
54. Niakolas, D.K. Sulfur poisoning of Ni-based anodes for Solid Oxide Fuel Cells in H/C-based fuels. *Appl. Catal. A Gen.* **2014**, *486*, 123–142. [CrossRef]

55. Sapountzi, F.M.; Zhao, C.; Boreave, A.; Retailleau-Mevel, L.; Niakolas, D.; Neofytidis, C.; Vernoux, P. Sulphur tolerance of Au-modified Ni/GDC during catalytic methane steam reforming. *Catal. Sci. Technol.* **2018**, *8*, 1578–1588. [CrossRef]
56. Besenbacher, F.; Chorkendorff, I.; Clausen, B.S.; Hammer, B.; Molenbroek, A.M.; Norskov, J.K.; Stensgaard, I. Design of a surface alloy catalyst for steam reforming. *Science* **1998**, *279*, 1913–1915. [CrossRef]
57. Water-Gas Shift Reaction. Available online: https://en.wikipedia.org/wiki/Water-gas_shift_reaction (accessed on 10 December 2018).
58. Caitlin, C. *Kinetics and Catalysis of the Water-Gas-Shift Reaction: A Microkinetic and Graph Theoretic Approach*; Worcester Polytechnic Institute: Worcester, MA, USA, 2006.
59. Stoots, C.M.; O'Brien, J.E.; Herring, J.S.; Hartvigsen, J.J. Syngas Production via High-Temperature Coelectrolysis of Steam and Carbon Dioxide. *J. Fuel Cell Sci. Technol.* **2008**, *6*. [CrossRef]
60. Sasaki, K.; Teraoka, Y. Equilibria in Fuel Cell Gases: II. The C-H-O Ternary Diagrams. *J. Electrochem. Soc.* **2003**, *150*, A885–A888. [CrossRef]

MDPI
St. Alban-Anlage 66
4052 Basel
Switzerland
Tel. +41 61 683 77 34
Fax +41 61 302 89 18
www.mdpi.com

Catalysts Editorial Office
E-mail: catalysts@mdpi.com
www.mdpi.com/journal/catalysts

www.ingramcontent.com/pod-product-compliance
Lightning Source LLC
LaVergne TN
LVHW071620170726
843515LV00009B/2440

* 9 7 8 3 0 3 9 3 6 0 3 6 9 *